Technician Mathematics 1

M. G. PAGE

B.Sc. (Hons.) (Eng.), C.Eng., F.I.Mech.E.,
F.I.Prod.E., F.S.S.

Head of the Department of
Production Engineering at
The Polytechnic,
Wolverhampton

Member of Programme Committee A1 of the Technician
Education Council.
Chief Examiner in Mathematics and Science for Course 255
for Mechanical Engineering Technicians
of the City and Guilds of London Institute

Cassell · London

CASSELL & COMPANY LIMITED

35 Red Lion Square, London WC1R 4SG
and at Sydney, Auckland, Toronto, Johannesburg

An affiliate of the Macmillan Publishing Co. Inc., New York

First published 1977

I.S.B.N. 0 304 29862 x

FILMSET BY COMPOSITION HOUSE LTD.,
SALISBURY, WILTSHIRE

PRINTED IN GREAT BRITAIN BY
THE CAMELOT PRESS LIMITED, SOUTHAMPTON

Preface

The task facing authors writing textbooks for programmes of study leading to the awards of the Technician Education Council is formidable to say the least. The syllabus content of the majority of Technician Courses of the past was relatively well defined, particularly those leading to the awards of the City and Guilds of London Institute with the associated external examinations. The majority of the remainder were associated with guide syllabuses and internal examinations, externally assessed, and in many cases the guide syllabuses were used with little or no modification.

Circumstances have now changed. T.E.C. programmes of study are now virtually all unit based, and theoretically there is no set content for a T.E.C unit which bears a particular title. The content of a unit is primarily a matter for agreement between an educational institution and its local industry, although before a programme can commence, approval is required from the T.E.C., who, when considering whether or not approval is given, takes note of the views made by validating sub-committees of appropriate T.E.C. Programme Committees.

When the T.E.C. issued its *Policy Statement*, educational institutions became aware that an immense amount of work would be necessary when constructing units, particularly when it became evident that the T.E.C. favoured units being written in terms of objectives. As could be expected, the task of compiling units was attacked in a variety of ways. Some individuals and institutions accepted the freedom with avidity. Others saw some advantage in making better use of resources by some form of collective activity, while the balance, in the main, felt it advisable to wait for developments from the individual Programme Committees of the T.E.C.

As time progressed, the Programme Committees proceeded with their deliberations, which were assisted by the provision of suggested units which emanated from a working party drawn from the various curriculum development groups which had appeared. These were referred to as Standard Units, but perhaps it would have been better had they have been referred to as proposals for standard units. The Programme Committees received these proposals with various degrees of satisfaction. Some were accepted without modification, others with reservations, and in certain cases, were found to be totally unacceptable. In the latter case, certain Programme Committees decided to construct their own units.

iii

In the case of mathematics, the author supports strongly the point of view expressed in many quarters that there is a body of study in mathematics at Level 1 which is relevant to the needs of all interested parties, and there should be relatively little difficulty in constructing a unit entitled Mathematics I which could be studied by all technician students, irrespective of the programme being undertaken. At the time of compiling the manuscript for this book, the T.E.C. had commenced some deliberations on common units, in which the author was participating, but no firm proposal had appeared. With a significant number of T.E.C. Courses starting in September 1976, and a substantially increased amount commencing in September 1977, the need for textbooks written specifically for T.E.C. courses became a matter of considerable urgency. At the time that the manuscript of this book was compiled, all that any author could do would be to survey the immediate position and make a considered judgement of an appropriate content.

Consequently, this book has been written with the aim of satisfying completely the objectives of the proposed Standard Unit T.E.C. U75/005, in the state it was in January 1977. In addition, some topics have been added which reflect the author's personal viewpoint of items which may well appear should a Common Unit entitled Mathematics 1 eventually be published. The opinion is far from a guess, as the author has had opportunities of studying proposals for units entitled Mathematics 1 which have emanated from a variety of sources, and he wishes to acknowledge, with considerable gratitude, the assistance of his colleagues in providing the information.

The style follows that of the author's previous work, *Mathematics for Mechanical Technicians* Book 1, which, judging from its sales, has been found to be welcome. However, the problems and examples have been the subject of considerable revision, to ensure that as far as is reasonably possible, the problems are relevant to technician students in general rather than to a specific group.

The text closely follows the sequence of the presentation of the topics in unit T.E.C. U75/005, the only significant variation being statistics, which has been included in the first chapter, which deals with computation. The remaining chapters deal with algebra, diagrammatic representations, geometry and trigonometry. Particular attention has been given to the fact that the pocket-size electronic calculator is now a recognised working tool of the typical student.

Students enter programmes from varying patterns of previous education and with various degrees of competency. The book has been written on the assumption that it is necessary to commence from basic fundamentals, and for many readers certain topics will be merely a consolidation of previous knowledge. A further advantage will accrue for colleges who adopt a policy of placing a considerable proportion of a new entry into a diagnostic stream. Because the book commences from basic fundamentals, and is intended for students who progress to technician courses, it will also be found suitable for a considerable proportion of craft courses. It will be particularly valuable for those students who have used the book on craft courses, and who change to technician courses at some later stage.

It is the intention of the author that the present book will be the first of a series of textbooks in mathematics for technicians, and fortunately there is some evidence to suggest that the content of units at Levels 2 and beyond will be in a more settled position than was that of Mathematics Level 1 when the manuscript of the present book was written.

It will be interesting to compare the content of this book with that of a common unit in Mathematics Level 1, if and when it appears. In the opinion of the author this is a virtual certainty. One of the delights he has enjoyed as a result of his writings is the kindly help which has been given by his colleagues which has resulted in improvements, and he sincerely trusts this assistance will continue. In the first instance, correspondence should be directed via the publisher.

Finally, the author wishes to acknowledge the help of Mrs Janet Phelps who typed portions of the manuscript with her usual competence and skill, and above all, patience. Also of the help of Mr Francis Fletcher, B.Sc., A.F.I.M.A., who read the original manuscript and made extremely useful comments, all of which have resulted in improvements to the original draft.

GEORGE PAGE

Wolverhampton 1977

Technician
Mathematics
1

This book has been written to provide background material for a student who is engaged upon a programme of studies leading to an award of the Technician Education Council (TEC), and whose programme includes a unit bearing a title of, or similar to, Mathematics Level 1. In particular, it covers all the topics of Standard Unit 005, which has been recommended for adoption in the individual guidelines issued by several Programme Committees of the TEC. The learning objective structure of Unit 005 has been closely followed. There are numerous worked examples and more than sufficient problems (with answers) to satisfy the coursework demands of a typical course of study. The book is the first of a series for Mathematics Units at all levels, and will be immediately followed by Books 2 and 3, which will cover the objectives in Mathematics up to Certificate and Diploma Level.

M. G. Page is Head of the Department of Production Engineering at Wolverhampton Polytechnic. He is a member of the Programme Committee A1 of the T.E.C. dealing with Certificate and Diploma Programmes in General Engineering and common units at Certificate and Diploma level in other engineering disciplines. Mr. Page is also Chief Examiner in Mathematics and Science for the C.G.L.I. Mechanical Engineering Technicians' course (255).

Contents

Introduction

It would not come amiss, before commencing a consideration of the topics of Technician Mathematics at Level 1, to discuss the use of pocket-size electronic calculators. The remarkable reductions in the costs of such calculators since their introduction now means that every student can be expected to possess this type of calculator, and this book has been written with that factor kept firmly in view.

The author, in his capacity as one of the Chief Examiners of the City and Guilds of London Institute, has observed, with considerable pleasure, two results of the use of calculators in examinations. The first is the dramatic reduction in the number of arithmetical errors. The second is that due to rapidity with which a calculator provides a result, when compared with traditional arithmetic, more time can be devoted to aspects of mathematics other than numerical computation.

The author will state quite categorically that if the student for whom this book is intended has a calculator available which has a capacity for providing the solution to a particular computational problem, that calculator should be used for that purpose. There is no doubt that this viewpoint may be the subject of criticism by skilled mathematicians, many of whom pass derogatory comments about students who use calculators for the most elementary of computational processes, such as adding together two simple whole numbers. This book is not intended for skilled mathematicians. The use of a calculator for simple processes, which admittedly could be performed more rapidly mentally, has one important advantage. The confirmation from a calculator of a result achieved mentally gives more confidence that the calculator is being used correctly. Furthermore, there are very few computational operations which can be accomplished mentally quicker than by using a calculator, and there is ample evidence to suggest that computations undertaken with a calculator tend to produce less errors than those undertaken mentally, or those made with the aid of what may be termed traditional arithmetic.

What has been said should not be interpreted as an invitation to give scant regard to traditional arithmetic, and especially to mental arithmetic. A person may be placed in a situation where a computational problem has to be solved and no calculator is available. Or, as often proves an embarrassment, to have a calculator available which lacks the necessary motive power. Consequently, it is just as necessary now as it was in the past to be competent with traditional arithmetic, with mathematic tables and with a slide rule.

A further point in favour of calculators, since in general errors occur less frequently when traditional arithmetic is used, is that it will tend to give less support to the viewpoint that the major consideration in solving a problem is to get the method right, any resulting computational work being of lesser importance. It must be agreed that this policy may help a student to satisfy assessments, where quite often a portion of the maximum allocation of marks to a problem is devoted to the method, but students who carry this attitude into a working life quickly realise that accuracy is just as important as methodology. In the scholastic situation a student should be encouraged to consider that the accuracy of a solution is just as important as the method by which that solution was obtained.

The price of a calculator which is capable of providing a particular range of computations has decreased remarkably over recent years, although at the time the manuscript of this book was written the trend of decreasing prices was much less noticeable. The rate of decrease in price was apparently fairly close to that of inflation, and the general level of prices seems now to have 'bottomed-out'. One calculator the author uses was purchased some five years ago at a cost of well over £50. A calculator with virtually the same capabilities could be purchased at the time the manuscript of this book was written, for well under £20, but there seems to be very little possibility of a significant reduction in price in the immediate future. Calculators are very reliable pieces of equipment, and as distinct from purchasing cars, no particular make can be considered superior to others as regards reliability. Once our reader has decided on a specification which satisfies his or her requirements, and any one of a particular group will satisfy that specification, the cheapest of that group should be purchased.

In all probability, the first decision that will have to be made is in respect of the motive power, which can be either a disposable dry battery, a rechargeable battery, mains only, or a facility for either or both of the latter. This is reflected in the price, and if it can be afforded, since the calculator may be used in a variety of situations, a calculator having a rechargeable battery which is supplied with an adaptor for recharging while it is being used from the mains is preferable. If the cheapest possible calculator has to be purchased the motive power will invariably be a disposable dry battery. In which case our reader is urged to have a replacement battery readily available, rather than being placed in the embarrassing position of having a calculator which is devoid of motive power.

The next decision will very likely be in respect of the number of digits available in the display. For the present and immediate future needs of our reader a display of more than eight digits is something of a luxury, as is a facility for computation in vulgar fractions. The display should be capable of indicating values in both ordinary form and in standard form, the latter giving a display indicating a coefficient and the power of 10 which it multiplies.

The programme of study on which our reader is engaged is constructed from a series of objectives to be attained. In particular, the course is designed to lead to further studies. These factors provide some guidance in the selection of a suitable calculator. It should be chosen to complement the objectives of the immediate course, and, if possible, some attention should be given to future requirements, so that an investment in a calculator does not prove to be unsound. As applied to a calculator, the word function is used to describe a computational process of which the calculator is capable. Most manufacturers will indicate how many functions a particular calculator can provide. A typical statement is that a particular model is a '28 function' calculator. In general terms, the greater the number of functions, the greater will be the price.

The computational requirements of Technician Mathematics at Certificate and/or Diploma level suggest the following functions of a calculator would be to advantage, and are given in some general, but not exact, order of priority:

1. The four basic functions of addition, subtraction, multiplication and division.
2. A square rooting function.
3. A squaring function.
4. A reciprocal function.
5. Logarithmic functions; natural and common.
6. Trigonometrical functions; sine, cosine and tangent.
7. An inverse function; determining antilogs, arcsin, arccos, and arctan.
8. Degree or radian notation for angles.
9. Change sign function, $+$ to $-$, and vice versa.
10. Power function; determining y^x and $\sqrt[x]{y}$.
11. At least one store for retaining values for subsequent use, usually called a memory.

Of recent times, it has been noticed that calculators which have a capability for the functions previously listed gradually tended to become known as 'scientific' models. To cater for a large number of functions, most of the 'scientific' calculators are provided with a changeover arrangement, so that some keys are allocated to more than one function.

Bearing in mind the preceding information, at the time at which the manuscript of this book was written, the Commodore Scientific Notation Calculator SR 7919D and the Texas Instruments Scientific Model TI 30 would most certainly have evoked the interest of a prospective purchaser, particularly when the price was compared with the functions available. The author prefers and now uses the TI30.

However, the design of calculators is a rapidly changing technology. Other models introduced since the manuscript for this book was written may well prove a 'better buy'. In any case, once our reader has decided which model to purchase, or to change to, it would be most advisable to 'shop around'. There may be remarkable differences between retailers of the selling price of a particular model.

1 Computation

1.1
The manipulation of numbers

1.1.1
Directed numbers

Mathematics is that branch of science which has evolved from the discovery that it was possible to represent concepts such as 'how many' and 'what shape' by means of symbols. A typical example is that portion of mathematics which deals with counting, the symbols that are used being called *numbers*. In early days small stones were used as an aid to counting. The Latin word for pebble is calculus, from which is derived the word calculation. Over a period of time the word *calculation* has become associated with the obtaining of a result from data by any mathematical procedure, numerical or otherwise, which does not involve the use of measurement or graphical methods. In more recent times the word *computation* has been adopted to mean calculation with numbers. We shall commence our studies of mathematics with computation.

The word number can have many interpretations. The interpretation most familiar to our reader is associated with the use of numbers as a counting system. This enables us to answer problems such as how many bricks there are in a particular pile of bricks. The answer is given by selecting one of the *natural numbers* which we represent nowadays by the symbols 1, 2, 3, 4, 5 and so on. It is convenient in life to refer to a collection of things as a *set*. The set of natural numbers has no limit, there is an infinite amount of natural numbers. If we use numbers as an aid to counting, then we use *zero* (symbolic representation 0) to indicate that there are none of the particular things which have to be counted.

Algebra is a general term used in mathematics to describe the methods of reasoning about numbers by employing letters to represent those numbers and arithmetical signs to represent their relationships. Let us proceed to using numbers as an aid to measurement. Let us imagine we are measuring the vertical heights of features with respect to some basic feature such as ground level. A feature such as the top of a building will be above ground level, the foot of a mine shaft will lie below ground level. Our measurements will be facilitated if we erect a vertical *line of numbers* where distances are proportional to the magnitude of numbers.

1

We can ease our task of making measurements if we accept the convention that proceeding in a vertically upward direction is known as proceeding in a positive direction, and to use the symbol +, known as the positive sign, or plus, to indicate this upward direction. Conversely, we adopt the convention that proceeding in a vertically downward direc-

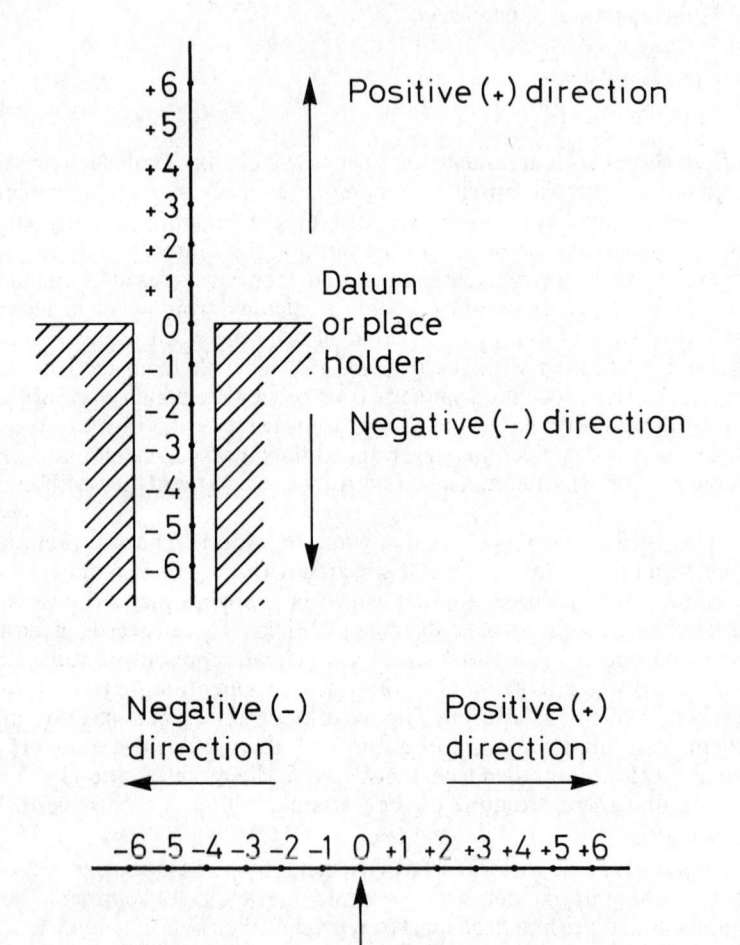

Fig. 1.1

tion is known as proceeding in a negative direction, and to use symbol −, known as the negative sign, or minus, to indicate this downward direction. We can now use the symbol for zero, 0, as a datum, or *place holder*, on a line of numbers to indicate the boundary between positive and negative numbers. We use the expression *directed numbers* to indicate those numbers prefixed by either a positive or a negative sign and which can be represented by points on a *directed line of numbers*.

A directed line of numbers need not be vertical, it can lie in any direction. We commenced our discussion with a vertical line of numbers simply for convenience in measuring distances above or below a reference point. The most commonly used directions for directed lines of numbers are a vertical line or a horizontal line. Using zero as the place holder on a horizontally directed line of numbers, we adopt the convention that moving in a rightward direction is moving in the positive direction while moving in the leftward direction is moving in the negative direction. Illustrations of directed lines of numbers are shown in fig. 1.1. Only portions of the lines are shown. Since there is no limit to the magnitude of a number the lines are theoretically of infinite length. A *whole number* is any one of the natural numbers 1, 2, 3, 4, 5, 6, etc. An *integer* is any one of the numbers +1, +2, +3, +4, etc., which are called positive integers, and −1, −2, −3, −4, etc., which are called negative integers. The value of zero, which is neither negative nor positive, can be considered to be a place holder which separates positive integers from negative integers. It is conventional practice to assume that if a number is not preceded by either a positive or a negative sign, a positive value is inferred. For example, the value 7 is presumed to be +7 when dealing with directed numbers.

A directed number is comprised of two items. The first item is an arithmetical sign, stated or inferred, which will be either positive or negative. The second part is a numerical quantity, which mathematically we call the *modulus* of the directed number. For example, the modulus of +15 is 15 and the modulus of −6 is 6. The mathematical way of indicating the modulus of a directed number is to use two vertical lines, one on each side of the directed number. As an illustration $|-18|$ is 18. The plural of modulus is moduli.

1.1.2
The addition of directed numbers

In the previous article we referred to the *set* of natural numbers. Each of the natural numbers is called an *element* of that set. In mathematics, a *binary operation* is a process which combines two elements of a set into a third element of the same set. *Addition* is a general term for the binary operation of combining two equal or unequal groups into a whole. The result of that combination is usually described as either a *sum*, a *total* or an *aggregation*.

There are many occasions in mathematics where we require to indicate clearly that an aggregation is to be treated as one complete thing, or, using the terminology of mathematics, as an *entity*. The conventional way of doing this is to put the items which form the entity within *aggregation signs*. The most common form of an indication of an entity are strictly called *parentheses*, and are indicated ( ). In a somewhat loose manner, parentheses are often called *brackets*, and for the time being there is no point in our reader changing a nomenclature which has perhaps been used in the past. We shall use the expression brackets for the time being, until it becomes necessary to introduce the fact that brackets are only one of many ways of indicating an entity.

Algebraic addition is the collecting together of two equal or unequal directed numbers according to certain particular rules. The arithmetical sign for moving in a positive direction is $+$, read as plus, while that for moving in the negative direction is $-$, read as minus. The problem symbolised by

$$8 + 5$$

can be interpreted as

'starting at $+8$ on a directed line of numbers, move five numbers in the positive direction'.

The answer is that we arrive at the point $+13$ on our directed line of numbers. If numerically, the difference between two values is zero, we say that a state of equality exists. The usual mathematical form of an indication of equality is called an *equation*, and we use the equals sign ($=$) between the two features which are in a state of equality. Hence a typical symbolic representation of an addition operation is

$$8 + 5 = 13.$$

Before we can list the rules of algebraic addition it will be necessary for us to appreciate two more mathematical symbols. The statement that 7 is greater than 4 is written

$$7 > 4$$

while the statement that 6 is less than 22 is written

$$6 < 22$$

It will be noted that the point, in both cases, is in the direction of the smaller quantity. Or, if we like, we can say that the symbol 'opens up' towards the larger quantity.

Let us set down six examples of the algebraic addition of integers.

(1) $(+8) + (+23) = +(8 + 23) = (+31)$
(2) $(+16) + (-4) = +(16 - 4) = (+12)$
(3) $(+3) + (-18) = -(18 - 3) = (-15)$
(4) $(-7) + (-19) = -(7 + 19) = (-26)$
(5) $(-25) + (+11) = -(25 - 11) = (-14)$
(6) $(-2) + (+15) = +(15 - 2) = (+13)$

The six examples cover all the variations of algebraic addition, and we can erect six rules to cover general rather than specific cases, for instance, if we let a be a number, and b be some other number, then

(1) $(+a) + (+b) = +(a + b)$
(2) $(+a) + (-b) = +(a - b)$,   if $a > b$
(3) $(+a) + (-b) = -(b - a)$,   if $a < b$
(4) $(-a) + (-b) = -(a + b)$
(5) $(-a) + (+b) = -(a - b)$,   if $a > b$
(6) $(-a) + (+b) = +(b - a)$,   if $a < b$

We will ask our reader not to be unduly perturbed about remembering these rules, as shortly we will introduce a considerable simplification. It will be quite evident to our reader that from the rules that

$$a + b = b + a,$$

or, by using numbers, that

$$8 + 23 = 23 + 8, \qquad \text{a total of 31.}$$

This equation is one example of what we call the *commutative law* of mathematics.

The general form of the commutative law is

$$a \odot b = b \odot a$$

in which a and b are elements of a set and $\odot$ represents a binary operation. Algebraic addition is one of the mathematical operations which conforms to the commutative law.

If we now extend addition to more than two numbers, we shall find that

$$(a + b) + c = a + (b + c) = (a + b + c).$$

or, by using numbers, that

$$(6 + 7) + 3 = 6 + (7 + 3) = (6 + 7 + 3), \qquad \text{a total of 16.}$$

This continued equation is one example of what we call the *associative* law of mathematics.

The general form of the associative law is

$$(a \odot b) \odot c = a \odot (b \odot c) = (a \odot b \odot c)$$

in which a, b and c are elements of a set and $\odot$ represents a binary operation. Algebraic addition therefore not only conforms to the commutative law but also to the associative law.

Our reader may feel that we are using somewhat unwarranted difficult mathematical terminology for something which is apparently self-evident, and that all that has really been stated is that if a set of numbers are added together the operation can be done in any sequence. This has a certain ring of truth, but we are hoping to lay down a firm basis for subsequent studies in mathematics and our reader is encouraged not to dismiss, in a cursory manner, the whole basis upon which computation is founded.

1.1.3
The subtraction of directed numbers

Subtraction is a general term for the operation of determining the remainder when the part of a whole is removed. It can also be interpreted as the operation required to determine the numerical difference between two groups. It can also be considered to be the reverse operation to addition. If a and b are two different numbers, the subtraction of b from a is represented by the expression $(a - b)$. The arithmetical sign is minus, which is the Latin word for less. The number a that we start with is called the *minuend*, and the number we subtract from the minuend is called the *subtrahend*. The end result is called a *remainder*.

Let us set down six examples of the algebraic subtraction of integers.

$$
\begin{aligned}
&(1) \quad (+13) - (+7) &&= +(13 - 7) &&= (+6) \\
&(2) \quad (+2) - (+7) &&= -(7 - 2) &&= (-5) \\
&(3) \quad (+16) - (-4) &&= +(16 + 4) &&= (+20) \\
&(4) \quad (-5) - (-2) &&= -(5 - 2) &&= (-3) \\
&(5) \quad (-14) - (-31) &&= +(31 - 14) &&= (+17) \\
&(6) \quad (-7) - (+4) &&= -(7 + 4) &&= (-11)
\end{aligned}
$$

Just as we did with algebraic addition, let us now use the examples to formulate the general rules for algebraic subtraction, which is the application of the subtraction process to directed numbers, and pays particular attention to arithmetical signs. If we let a be a number and b be some other number.

(1) $(+a) - (+b) = +(a - b)$,   if $a > b$
(2) $(+a) - (+b) = -(b - a)$,   if $a < b$
(3) $(+a) - (-b) = +(a + b)$
(4) $(-a) - (-b) = -(a - b)$,   if $a > b$
(5) $(-a) - (-b) = +(b - a)$,   if $a < b$
(6) $(-a) - (+b) = -(a + b)$

Once again we shall ask our reader not to be unduly perturbed about remembering these rules, as shortly we will introduce a considerable simplification.

It will be observed that a subtraction such as $7 - 5$ does not produce the same result as $5 - 7$. Just as we represent a state of equality by the equals sign $(=)$, we represent a state of inequality by using a symbol where the equals sign is struck through $(\neq)$.

The symbolic representation

$$7 - 5 \neq 5 - 7$$

is read as

'seven minus five is not equal to five minus seven',

and hence algebraic subtraction does not conform to the commutative law. *Algebraic subtraction has to be undertaken in the particular sequence indicated.*

1.1.4
Algorithms

In general, a technician uses mathematics as a tool to be used when solving a variety of problems. The usual method is to present the problem in a mathematical form called a *mathematical model*. A typical form of a mathematical model is the statement of equality called an equation. It could prove to be extremely time-consuming to go into the detail of every mathematical rule when determining a solution of the mathematical model of a problem. What is more useful is a step-by-step approach which will lead us to a correct answer rapidly and accurately. An *algorithm* is any standardised procedure for the solution of a particular type of problem. Let us illustrate this approach by stating the algorithm for addition and subtraction, so that our reader does not have to memorise in detail the twelve rules previously stated.

If two integers are both positive or both negative they are said to have *like signs*. If one sign is positive and the other sign is negative, irrespective of the order, they are said to have *unlike signs*.

The algorithm for the addition or subtraction of two directed numbers is as follows:

Step 1. To add two directed numbers which have like signs, add their moduli and prefix the total with their common sign.

Step 2. To add two directed numbers of unlike sign, subtract the smaller modulus from the larger modulus and prefix with the sign of the directed number which has the larger modulus.

Step 3. To subtract a second directed number from a first directed number, change the sign of the second directed number and add.

Our reader is reminded that the modulus of a directed number is that obtained by discarding the sign of that directed number, for instance

$$|8| \text{ is } 8, \qquad |+15| \text{ is } 15, \qquad |-26| = 26.$$

Let us illustrate the algorithm for the addition and subtraction of directed numbers with examples.

$(+8) + (+7)$ The integers have like signs. Add their moduli to produce the value 15. The common sign is positive. Prefix the modulus 15 with the positive sign to obtain $(+15)$.

$$(+8) + (+7) = (+15)$$

$(-27) + (-16)$ The integers have like signs. Add their moduli to produce the value 43. The common sign is negative. Prefix the modulus 43 with the negative sign to obtain (-43).

$$(-27) + (-16) = (-43)$$

$(+16) + (-7)$ The integers have unlike signs. Subtract the smaller modulus (7) from the larger modulus (16) to produce the value 9. Prefix with the sign of the directed number with the larger modulus. 16 is greater than 7, the sign of $(+16)$ is positive, so we prefix the modulus 9 with the positive sign to obtain $(+9)$.

$$(+16) + (-7) = +9$$

$(-6) + (+25)$ The integers have unlike signs. Subtract the smaller modulus (6) from the larger modulus (25) to obtain the value 19. Prefix with the sign of the directed number which has the larger modulus. 25 is larger than 6, the sign of $(+25)$ is positive, so we prefix the modulus 19 with the positive sign to obtain $(+19)$.

$$(-6) + (+25) = (+19)$$

$(+16) - (+7)$ We have to subtract $(+7)$ from $(+16)$, so we change the sign of $(+7)$ to obtain (-7) and then we add it to $(+16)$. The result is obtained from

$$(+16) + (-7)$$

and using the methodology of a previous example, the answer is $(+9)$

$$(+16) - (+7) = (+9)$$

The immediately preceding example could be thought to be a complete waste of time, with an accompanying remark of the type 'anybody knows that if you subtract 7 from 16 you get 9'. Fair comment! However, it should be noted that its purpose was to demonstrate one aspect of the general algorithm for addition and subtraction. It certainly leads us to have some confidence in an algorithm which has the advantage of consisting of but three simple rules instead of twelve far more difficult rules. Now let us go on to another subtraction which is less obvious.

$(-7) - (-15)$ We have to subtract (-15) from (-7) so we change the sign of (-15) to $(+15)$, and then we add it to (-7). The result is obtained from $(-7) + (+15)$. This is an addition of integers of unlike sign. We subtract the smaller modulus from the larger modulus to obtain the value of 8, and then prefix the modulus 8 by the sign of the directed number which has the greater modulus. $(+15)$ has a positive sign so we prefix the modulus of 8 with a positive sign.

$$(-7) - (-15) = (+8).$$

It is not suggested that our reader has to write out the steps in detail. As confidence is gained, continued practice will lead to the steps being accomplished mentally, and results written down directly.

1.1.5
The continued addition and subtraction of directed numbers

In the previous article we restricted ourselves to the addition or subtraction of two directed numbers. Let us proceed to develop the topic to more than two directed numbers, such as would occur in the evaluation of

$$(+7) - (+15) + (-3) + (+18) - (-22) - (+2).$$

We could proceed steadily from left to right using the algorithm of the previous article, to obtain

$$(+7) - (+15) = (-8)$$

then
$$(-8) + (-3) = (-11)$$

then
$$(-11) + (+18) = (+7)$$

and
$$(+7) - (-22) = (+29)$$

and
$$(+29) - (+2) = (+27),$$

the final answer could be written as 27, which implies a preceding positive sign.

We have, heretofore, used brackets to assist the allocation of arithmetical signs of the directed numbers. In actual practice, the original problem which introduced this article would normally be written as

$$7 - 15 - 3 + 18 + 22 - 2$$

and if we compare this statement with the one which introduced this article, we note:

(a) the brackets have been removed.
(b) if a minus sign preceded a directed number in brackets the minus sign is discarded and the sign of the directed number within the brackets is changed.
(c) if a positive sign preceded a directed number in brackets the positive sign is discarded, and the sign of the directed number within brackets is unchanged.

For example

$$-(+15) \quad \text{became} \quad -15$$
$$+(-3) \quad \text{became} \quad -3$$
$$+(+18) \quad \text{became} \quad +18$$
$$-(-22) \quad \text{became} \quad +22$$

and
$$-(+2) \quad \text{became} \quad -2.$$

Our reader is asked to accept that what has been done is quite proper, and the proof will be demonstrated when we give a little more attention to quantities within aggregation signs.

To repeat the problem thus far, we have

$$7 - 15 - 3 + 18 + 22 - 2$$

and grouping positive directed numbers with negative directed numbers, we can write the problem as

$$7 + 18 + 22 - 15 - 3 - 2.$$

Now, noting how a negative sign operates when placed in front of an aggregation, we can write

$$(7 + 18 + 22) - (15 + 3 + 2).$$

We now simplify the aggregations to obtain

$$(47) - (20)$$

and obtain the result of 27, which agrees with that already obtained. Consequently, the algorithm for continued addition or subtraction of directed numbers is:

Step 1. Add the moduli of the positive directed numbers.
Step 2. Add the moduli of the negative directed numbers.
Step 3. Subtract the modulus obtained in step 2 from the modulus obtained in step 1, using the rules of algebraic subtraction.

For example, with

$$-18 + 7 + 15 - 3 - 29 + 8 - 41$$

Step 1. $$7 + 15 + 8 = 30$$

Step 2. $$18 + 3 + 29 + 41 = 91$$

Step 3. $$30 - 91 = -61$$

which is far less time consuming than proceeding through the problem with a sequence of individual operations.

1.1.6
The multiplication and the division of directed numbers

Multiplication is a general term for the operation of combining a quantity of equal groups into a whole. The mathematical symbol for multiplication is $\times$, which is read as 'multiplied by'. The general representation of a multiplication operation is

$$a \times b = c.$$

The value a is called the *multiplicand*, the value b is called the *multiplier*, and the value c the *product*. The values a and b can also be called *factors* of the product. Multiplication can be considered as successive addition. The product of 4×7 can be obtained by adding either four sevens together, or seven fours together.

Algebraic multiplication is the multiplication of directed numbers according to certain specific rules, which are

$$(+a) \times (+b) = +(a \times b)$$
$$(+a) \times (-b) = -(a \times b)$$
$$(-a) \times (+b) = -(a \times b)$$
$$(-a) \times (-b) = +(a \times b)$$

Earlier in this article it was stated that $7 \times 4 = 4 \times 7$, which we can express as $a \times b = b \times a$. Our reader may recall that we defined the commutative law as

$$a \odot b = b \odot a,$$

where $\odot$ represents a binary operation.

Hence multiplication conforms to the commutative law. Furthermore, since

$$(a \times b) \times c = a \times (b \times c) = (a \times b \times c),$$

multiplication conforms also to the associative law. What has been said, thus far, in simple terms, is that when numbers are multiplied together, and the process is just multiplication and no other, they can be multiplied together in any sequence. For instance,

$$(+5) \times (-7) \times (-2) = (-7) \times (+5) \times (-2)$$
$$= (-2) \times (-7) \times (+5)$$
$$= 70$$

The algorithm for the multiplication of two directed numbers is as follows:

Step 1.   Multiply the moduli of the directed numbers.

Step 2.   Prefix the value obtained in step 1 by an arithmetical sign. The sign will be positive if two directed numbers of like sign are multiplied together. The sign will be negative if two directed numbers of unlike sign are multiplied together.

Let us illustrate the algorithm for multiplication with some examples.

$(+8) \times (+5)$ Multiply the moduli of the directed numbers to obtain $8 \times 5 = 40$. The integers both have positive signs, that is, they have like signs. Prefix the modulus 40 with a positive sign.

$$(+8) \times (+5) = (+40)$$

$(+16) \times (-2)$ Multiply the moduli of the directed numbers to obtain $16 \times 2 = 32$. $(+16)$ and (-2) have unlike signs. Prefix the modulus 32 with a negative sign to obtain (-32).

$$(+16) \times (-2) = (-32)$$

$(-4) \times (-17)$ Multiply the moduli of the directed numbers to obtain $4 \times 17 = 68$. (-4) and (-17) have like signs. Prefix the modulus 68 with a positive sign to obtain $(+68)$.

$$(-4) \times (-17) = (+68)$$

$(-3) \times (+9)$ Multiply the moduli of the directed numbers to obtain $3 \times 9 = 27$. (-3) and $(+9)$ have unlike signs. Prefix the modulus 27 with a negative sign to obtain (-27).

$$(-3) \times (+9) = -27$$

Division is the general term for the binary operation of breaking up a whole into equal parts. The mathematical symbol for division is $\div$, read as divided by. If two integers are represented by a and b, then the operation of division can be symbolised by

$$a \div b = c$$

where a is called the *dividend*, b is called the *divisor* and c is called the *quotient*. *Algebraic division* is the division of directed numbers according to certain rules, which are

$$(+a) \div (+b) = +(a \div b)$$
$$(+a) \div (-b) = -(a \div b)$$
$$(-a) \div (+b) = -(a \div b)$$
$$(-a) \div (-b) = +(a \div b).$$

Algebraic division does not conform to the commutative law

$$a \odot b = b \odot a$$

since $$25 \div 5 \neq 5 \div 25$$

The algorithm for a division operation closely follows that for multiplication.

Step 1.   Divide the modulus of the dividend by the modulus of the divisor to obtain the modulus of the quotient.

Step 2.   Prefix the value obtained in step 1 with an arithmetical sign. If the directed numbers have like signs the sign of the quotient is positive. If the directed numbers have unlike signs, the sign of the quotient is negative.

As examples

$(-30) \div (-6)$ Dividing the modulus of the dividend, (30), by the modulus of the divisor, (6), we obtain the modulus of the quotient, (5). The division is between directed numbers of like signs. The modulus 5 is to be prefixed by a positive sign to obtain $(+5)$.

$$(-30) \div (-6) = (+5)$$

$(-34) \div (+17)$ Dividing the modulus of the dividend, (34), by the modulus of the divisor, (17), we obtain the modulus of the quotient, (2). The division is between directed numbers of unlike signs. The modulus 2 is to be prefixed by a negative sign to obtain (-2).

$$-(34) \div (+17) = (-2).$$

We shall leave the topic of continued multiplication and/or division until we have introduced the subject of factors. A simple and elegant method of easing the solving of problems involving multiplication and/or division will then be provided, known as cancellation.

1.1.7
The distributive law and combined operations with directed numbers

We now come to a most important law of mathematics called the *distributive law*, which refers to the distribution of multiplication operations over addition operations. It can be represented symbolically in many ways. For an initial consideration it is probably best represented by

$$a \times b + c = (a \times b) + c.$$

In particular,

$$a \times b + c \neq a \times (b + c).$$

Our reader will recall that it is very useful to put certain items within aggregation signs, of which brackets are typical, to show that the collection within the aggregation signs is to be treated as an entity. An investigation of the equation and the inequality which have been used to illustrate the distributive law shows that *multiplication takes precedence over addition*. For instance, to evaluate $a \times b + c$, it being noted that there is no indication of an entity, it is first necessary to multiply a by b to obtain a sub-total, and then to add c to the sub-total to determine the final total. The sequence of signs is not necessarily that of the sequence in which operations are conducted. For instance, with

$a + b \times c$ it is incorrect to perform the addition first simply because it is the operational sign which comes first as we read from left to right.

$a + b \times c$, again noting that there is no indication of an entity, is evaluated as $a + (b \times c)$.

We multiply b by c to obtain a sub-total, and then add a to the sub-total to find the final total.

Since division is the converse of multiplication and subtraction is the converse of addition, we can extend our original interpretation of the distributive law to state

> *multiplication and/or division processes take precedence over addition and/or subtraction processes.*

Hence we evaluate $(+5) \times (-6) + (-4) \div (+2) - (-3) + (-8)$ by appreciating that multiplication and division operations must be undertaken first. Let us go through the problem from left to right. The first two integers have to be multiplied together and treated as a complete whole. We have already used parentheses, when setting out the problem, so we now use the next of the generally accepted order of sequence of aggregation signs. These are called *braces*, and are indicated $\{\ \ \}$.

Thus far, we have

$$\{(+5) \times (-6)\} + (-4) \div (+2) - (-3) + (-8)$$

After the braces we find an addition sign but later we have a division sign. This takes precedence over the addition sign, and the division must be accomplished before the addition.

We can indicate the division operation by

$$\{(-4) \div (+2)\}$$

and, since there are no more multiplications and divisions, we can indicate the application of the distributive law by restating the problem as

$$\{(+5) \times (-6)\} + \{(-4) \div (+2)\} - (-3) + (-8)$$

The first aggregate produces the product (-30). The second aggregate produces the result (-2). Hence we can restate the problem as

$$(-30) + (-2) - (-3) + (-8)$$

We now remove the parentheses, and write

$$-30 - 2 + 3 - 8$$

We rewrite the sequence as

$$3 - 30 - 2 - 8$$

We now group positive and negative integers together, with parentheses, as

$$3 - (30 + 2 + 8)$$

and simplify to

$$3 - 40$$

which gives the final result of -37.

The evaluation of $3 \times 2 + 5$, using the rule that multiplication takes precedence over addition, is undertaken by multiplying 3 by 2 to obtain 6, and then adding 5 to obtain the result of 11. If, however, for some reason, the figures have to be manipulated so as not to conform to the rule, this can be indicated by using brackets. For instance, if for some reason we wish to add 5 to 2 *before* multiplying by 3 we can adopt a symbolic representation such as $3 \times (2 + 5)$ to indicate that whatever appears within brackets must be treated as an entity, and in this case $3 \times (2 + 5) = 3 \times 7 = 21$.

Problems 1.1

1. With the following array of positive and negative integers, and the occasional zero, add up the columns and add up the rows, finally checking that the total of the rows is equal to the total of the columns

4	-7	22	-15	7	0	8	(h)
2	-1	-32	6	0	15	-8	(i)
5	18	0	-29	16	-7	14	(j)
7	6	-6	16	-22	0	-8	(k)
3	-12	18	-6	-9	14	13	(l)
5	5	-12	3	-7	-9	4	(m)
(a)	(b)	(c)	(d)	(e)	(f)	(g)	

2. Perform the subtractions indicated:
   (a) $17 - 15$ (b) $15 - 17$ (c) $-15 - 17$
   (d) $-17 - (-15)$

3. Evaluate:
   (a) 18×5 (b) $6 \times (-5)$ (c) -3×2
   (d) $(-6) \times (-8)$

4. Evaluate
   (a) $3 \times 2 + 4$ (b) $3 \times (2 + 4)$ (c) $3 \times 2 + 4 \div 2$
   (d) $3 \times (2 + 4) \div 2$ (e) $15 \times (-7) + 6 + (-16 \div 2)$
   (f) $18 + 27 \div 3 - 16 + 2 \times 6$
   (g) $-21 \times 7 \div (-3) + 16 - 5 \times (-2) + 6$
   (h) $3 \times (7 + 2) - 5 + 22 \div (-2) + 6(-2 + 3)$

Answers to Problems 1.1

1. (a) 26　(b) 9　　(c) −10　(d) −25　(e) −15　(f) 13　(g) 23
   (h) 19　(i) −18　(j) 17　　(k) −7　(l) 21　　(m) −11
   The total of the rows = the total of the columns = 21

2. (a) 2　(b) −2　(c) −32　(d) −2

3. (a) 90　(b) −30　(c) −6　(d) 48

4. (a) 10　　(b) 18　(c) 8　(d) 9
   (e) −153　(f) 23　(g) 81　(h) 17

1.2
The factorisation of positive integers

1.2.1
Factorisation

We have previously stated that if a number is not preceded by an arithmetical sign it is an agreed convention to regard that number as a positive number. For the purposes of this article, in which we shall deal only with positive integers, we shall dispense with the positive sign. Furthermore, no advantages will be gained by placing integers in parentheses, we will also dispense with their use. In previous articles we may have represented a positive integer by symbols such as (+210). In this article we shall adopt the convention of representing that number by 210.

When two or more numbers are multiplied together the result is called the *product* of those numbers. The reverse process, that is, of finding the numbers, which, when multiplied together, give a particular product, is called *factorisation*. The numbers which are found are called *factors*.

Hence, since

$$2 \times 3 \times 5 \times 7 = 210$$

then 2, 3, 5 and 7 are the factors of 210.

However, since $6 \times 35 = 210$, then 6 and 35 could be considered to be the factors of 210, as could 10 and 21, or 3, 7 and 10.

Theoretically, a *prime number* is any positive integer which has no factors other than itself and unity. If, however, we adopt the convention that unity is not considered to be a factor, then a prime number can be considered to be a number which has no factors. The set of prime numbers is infinite. The first few are

2   3   5   7   11   13   17   19   23   29   31   37   41   43   and   47.

The difficulty of a given number having several groups of factors, such as was illustrated previously with the value of 210, is overcome by adopting a convention that the factors of a positive integer are expressed as positive prime numbers, which we call *prime factors*. They are the set of prime factors which, when multiplied together, produce the original number. A particular prime factor may therefore occur more than once in a set of factors. It is the usual practice to list factors in ascending order of magnitude.

Factorisation of a positive integer is accomplished by taking the first prime number 2 and attempting to divide it into the positive integer. If the number is even, i.e., the last figure is 0, 2, 4, 6 or 8, that number can be divided by 2 without leaving a remainder. If the quotient is itself even, we divide by 2 again. Eventually a time will come when the quotient is odd, i.e., the last figure is 1, 3, 5, 7 or 9. Dividing by 2 would not give an integer, there would be a remainder of unity. We therefore cease dividing by 2.

We then proceed by attempting to divide the quotient, which is now an odd integer, by the next prime number after 2, which is 3. A little earlier we referred to 'the last figure of a number being even'. The mathematical term for figure, in the sense of numbers, is *digit*. A number such as 83 consists of two digits. The number of digits in the number 465 is three. An integer will divide by 3 without leaving a remainder if its digits add up to an integer which is itself divisible by 3. For instance,

81 will divide by 3 because $8 + 1 = 9$, which divides by 3

and

489 will divide by 3 because $4 + 8 + 9 = 21$, which divides by 3.

Returning to our factorisation procedure, we continue to divide by 3, until a quotient is obtained which will not divide by 3 without leaving a remainder. We then go on to the next prime number after 3, which is 5. At this stage, if 5 is a factor of the original number, the last digit will be a 5. If it is, we divide by 5, and look at the quotient. If the last digit is again a 5, we divide by 5 again. A time will come when the last quotient is not divisible by 5. We then proceed to the next prime number after 5 which is 7. The whole process is continued until eventually the quotient is unity. The factors of the original integer are the various divisors, in sequence.

In the development which follows our reader will observe that thousands are separated from hundreds by what a printer would refer to as a 'half-space'. In particular, a comma is quite specifically *not* used to separate thousands and hundreds because of a possible misinterpretation of what the comma indicates. A more detailed explanation of the reason will be given later in a more appropriate article.

Example:

Factorise 11 466

11 466 is an even number, and will divide by 2 without leaving a remainder.

$$11\,466 = 2 \times 5\,733.$$

5 733 is an odd number and will not divide by 2 without leaving a remainder. Proceed to try the next prime number, which is 3. 5 733 will divide by 3 without leaving a remainder because $5 + 7 + 3 + 3 = 18$ will divide by 3.

$$5\,733 = 3 \times 1\,911$$

and so $11\,466 = 2 \times 3 \times 1\,911$. 1 911 will divide by 3 without leaving a remainder because $1 + 9 + 1 + 1 = 12$ will divide by 3.

$$1\,911 = 3 \times 637$$

and so $11\,466 = 2 \times 3 \times 3 \times 637$. 637 will not divide by 3 without leaving a remainder because $6 + 3 + 7 = 16$ will not divide by 3.

Try the next prime number, which is 5. 637 will not divide by 5 without leaving a remainder because the last digit is neither 0 nor 5.

Try the next prime number which is 7.

$$637 = 7 \times 91$$

and so $\quad 11\,466 = 2 \times 3 \times 3 \times 7 \times 91$

$$91 = 7 \times 13, \text{ which are both prime numbers.}$$

Hence, finally, $11\,466 = 2 \times 3 \times 3 \times 7 \times 7 \times 13$.

Answer:

The factors of 11 466 are 2, 3, 3, 7, 7 and 13.

The factors are all prime numbers, and it is quite possible that on occasions a particular prime number may occur more than once. The process of factorisation of a positive integer consists of determining

ALL the prime numbers which, when multiplied together, give the original number. It is not sufficient to list only those prime numbers which can divide into a given number without leaving a remainder.

In the preceding worked example the steps were given in considerable detail to fully indicate the method. Once experience has been gained, it is quite probable that a tabular layout will be found to provide a neater and quicker solution, thus:

$$
\begin{array}{r|r}
2 & 11\ 466 \\ \hline
3 & 5\ 733 \\ \hline
3 & 1\ 911 \\ \hline
7 & 637 \\ \hline
7 & 91 \\ \hline
13 & 13 \\ \hline
 & 1
\end{array}
$$

The vertical row of divisors on the left are the factors, in this case

$$2, 3, 3, 7, 7 \quad \text{and} \quad 13.$$

1.2.2
The highest common factor of a set of positive integers

The *highest common factor* (usually abbreviated to H.C.F.) of a given set of positive integers is the largest integer that will divide into every one of the given set of integers without leaving a remainder. The principle can be extended into weights and measures, where it is then known as the *greatest common measure* (G.C.M.).

To determine the highest common factor of a set of integers it is first necessary to determine the factors of each of the set of integers. Let us use for example, the determination of the H.C.F. of the integers 504, 720 and 1 320. By applying the method of factorisation described in the preceding article, we should find that

$$504 = 2 \times 2 \times 2 \times 3 \times 3 \times 7$$

$$720 = 2 \times 2 \times 2 \times 2 \times 3 \times 3 \times 5$$

while			$$1\ 320 = 2 \times 2 \times 2 \times 3 \times 5 \times 11$$

We shall now find it convenient to tabulate these factors in such a manner so that we can see by inspection whether or not a particular factor, or a similar number of the same factor, appears in each set of factors. We note that the greatest number of twos in any set of factors is four, the greatest number of threes is two, the greatest number of fives is one, that of sevens is one and that of elevens is one. We have to make provision, whether we use them fully or not, for four vertical columns of twos, two vertical columns of threes, and one column each for fives, sevens and elevens. We set out the factors thus:

$$504 = 2 \times 2 \times 2 \quad\quad \times 3 \times 3 \quad\quad \times 7$$
$$720 = 2 \times 2 \times 2 \times 2 \times 3 \times 3 \times 5$$
$$1\,320 = 2 \times 2 \times 2 \quad\quad \times 3 \quad\quad \times 5 \quad\quad \times 11$$

It will be seen that the first, the second and the third 2 are all common to the three integers, but the fourth 2 is not. Similarly, the first 3 is common to all, but the second is not. The factors 5, 7 and 11 are not common to all. Thus the factors which *are* common to all are 2, 2, 2 and 3, and the highest common factor is obtained by multiplying these together to obtain $2 \times 2 \times 2 \times 3 = 24$. An alternative approach is to observe which vertical columns are complete. The highest common factor is the continued product of one each of these factors which appear in complete vertical columns.

Let us check if 24 is a factor of each of the three numbers

$$504 = 24 \times 21$$
$$720 = 24 \times 30$$
$$1\,320 = 24 \times 55$$

We conclude that our H.C.F. of 24 is correct since the associated values of 21, 30 and 55 have no factor which is common to all three. If it is necessary to determine the highest common factor of two relatively large numbers, the method previously described could prove to be quite time consuming. In some cases a different method is often quicker. The different method is based on the fact that a common factor of two integers is also a common factor of their difference. The method requires a knowledge of long division, a topic not covered thus far in this book. However, our reader may already be competent in this process and can understand the working. If not, the topic can be deferred for a while and a return made to it after a study of long division.

The algorithm is:

Step 1. Divide the smaller number into the larger and establish a remainder.

Step 2. Divide the remainder into the previous divisor and establish a new remainder. Continue this process until a remainder of zero occurs.

Step 3. The highest common factor is the last of the divisors from step 2.

Example:

Determine the H.C.F. of 312 and 741.

$$312 \; \boxed{741} \; 2 \qquad\qquad \text{Step 1}$$

624

117 ⌐ 312 ⌐2   Step 2 (first)

234

78 ⌐ 117 ⌐1   Step 2 (second)

78

39 ⌐ 78 ⌐2   Final step 2

78

0

The H.C.F. is the last of the divisors, in this case 39. Our reader may now care to check that this is so by using the method which introduced this article.

1.2.3
The least common multiple of a set of positive integers

A *common multiple* of a set of positive integers is any one of an infinite number of multiples, all of which can be divided by every one of the set of positive integers without leaving a remainder. As an example, consider the positive integers 2, 3, 4, 5 and 6. The common multiples of this particular group of positive integers are numbers such as 180, 600, 720, 60, 900 and 3 600. Of the various common multiples, there will be one in particular which has a magnitude lower than any of the others. This particular common multiple is called the least common multiple, an expression usually abbreviated to L.C.M.

An L.C.M. is determined in a manner somewhat similar to the determination of an H.C.F. If we wish to determine the L.C.M. of a set of positive integers, we first establish their factors, should they have any. Let us take for example the numbers 2, 3, 4, 5 and 6. We note that 2, 3 and 5 are prime numbers, and we can make the following statements of equality:

$$2 = 2$$
$$3 = 3$$
$$4 = 2 \times 2$$
$$5 = 5$$
$$6 = 2 \times 3$$

We now arrange the factors in vertical columns, providing sufficient vertical columns to ensure that every factor is included, thus:

$$2 = 2$$
$$3 = \qquad\qquad 3$$
$$4 = 2 \times 2$$
$$5 = \qquad\qquad\qquad\qquad 5$$
$$6 = 2 \qquad\times 3$$

The L.C.M. is the continued product of the factor represented by every column, whether that column be complete or not.

In our case, we have $2 \times 2 \times 3 \times 5 = 60$, and hence the L.C.M. of 2, 3, 4, 5 and 6 is 60.

Example:

Find the L.C.M. of 12, 15, 27, 36 and 84.
Factorise:

$$12 = 2 \times 2 \times 3$$
$$15 = 3 \times 5$$
$$27 = 3 \times 3 \times 3$$
$$36 = 2 \times 2 \times 3 \times 3$$
$$84 = 2 \times 2 \times 3 \times 7$$

Tabulate in columns. We shall need two columns of twos, three columns of threes and one column for each of the fives and sevens, a total of seven columns in all.

$$12 = 2 \times 2 \times 3$$
$$15 = \qquad\qquad 3 \qquad\qquad\qquad \times 5$$
$$27 = \qquad\qquad 3 \times 3 \times 3$$
$$36 = 2 \times 2 \times 3 \times 3$$
$$84 = 2 \times 2 \times 3 \qquad\qquad\qquad\qquad \times 7$$

The L.C.M. is $2 \times 2 \times 3 \times 3 \times 3 \times 5 \times 7 = 3\,780$.

Before we proceed to the next topic, our reader may care to note that the H.C.F. is 3, being the only factor which produces a complete vertical row. A further point which could be made is that the L.C.M. of a set of prime numbers is the product of those numbers.

Problems 1.2

1. Write down all the prime numbers between 50 and 100.

2. Determine the prime factors of each of the following numbers:
   (a) 66   (b) 140   (c) 450   (d) 725
   (e) 1 001 (f) 5 168   (g) 9 450   (h) 31 500

3. Determine the highest common factor of each of the following sets of numbers:
   - (a) 42 and 48
   - (b) 32 and 72
   - (c) 60 and 84
   - (d) 12, 21 and 57
   - (e) 42, 70 and 616
   - (f) 240, 330 and 1 995.

4. Determine, by the method of division, the highest common factor of:
   - (a) 138 and 805
   - (b) 730 and 949
   - (c) 1 147 and 1 333
   - (d) 765 and 2 890
   - (e) 450 and 1 998

5. Three consignments consisting of 23 100, 30 030 and 39 270 items are to be split into batches so that each consignment can be split into the minimum number of batches, each of the same whole number of items for all the three consignments. Determine:
   - (a) the batch size
   - (b) the number of batches in each consignment.

6. A rectangular space 4 500 units × 5 175 units has to be covered with square tiles so that no tile has to be cut. Determine:
   - (a) the maximum size of tile
   - (b) the number of tiles required.

7. Determine the greatest common measure of three tanks whose capacities are 560, 1 400 and 2 520 litres.

8. Determine the least common multiple of the following sets of numbers:
   - (a) 2 4 6 8 and 10
   - (b) 2 3 4 5 6 7 8 9 and 10
   - (c) 5 7 9 11 and 13
   - (d) 3 5 9 10 and 18
   - (e) 5 13 65 and 325
   - (f) 13 16 65 and 160

9. Find the lowest number which is exactly divisible by
   - (a) 42 and 56
   - (b) 54 and 72
   - (c) 65 and 117
   - (d) 37 and 43

10. The product of two numbers is equal to the product of their highest common factor and their least common multiple. Test the validity of this statement with the numbers:
    - (a) 6 and 15
    - (b) 32 and 72
    - (c) 63 and 147

11. Two gear wheels in mesh have 42 teeth on one gear and 120 teeth on the other gear. If two teeth are in contact at a particular instant, how many revolutions of each wheel occur until they are next in contact?

Answers to Problems 1.2

1. 53, 59, 61, 67, 71, 73, 79, 83, 89, 97

2. (a) $2 \times 3 \times 11$
   (b) $2 \times 2 \times 5 \times 7$
   (c) $2 \times 3 \times 3 \times 5 \times 5$
   (d) $5 \times 5 \times 29$
   (e) $7 \times 11 \times 13$
   (f) $2 \times 2 \times 2 \times 2 \times 17 \times 19$
   (g) $2 \times 3 \times 3 \times 3 \times 5 \times 5 \times 7$
   (h) $2 \times 2 \times 3 \times 3 \times 5 \times 5 \times 5 \times 7$

3. (a) 6   (b) 8   (c) 12
   (d) 3   (e) 14   (f) 15

4. (a) 23   (b) 73   (c) 31   (d) 85   (e) 18

5. (a) 2 310   (b) 10, 13   and 17 batches

6. (a) 225 units   (b) 460 tiles

7. 280 litres

8. (a) 120   (b) 2 520   (c) 45 045
   (d) 90   (e) 325   (f) 2 080

9. (a) 168   (b) 216   (c) 585   (d) 1 591

10. (a) H.C.F. $= 3$ L.C.M. $= 30$, $3 \times 30 = 6 \times 15 = 90$
    (b) H.C.F. $= 8$ L.C.M. $= 288$, $8 \times 288 = 32 \times 72 = 2\ 304$
    (c) H.C.F. $= 21$ L.C.M. $= 441$, $21 \times 441 = 63 \times 147 = 9\ 261$

11. 20 revolutions of 42 tooth wheel, 7 revolutions of 120 tooth wheel.

1.3
Vulgar fractions

1.3.1
Proper vulgar fractions, improper vulgar fractions and mixed numbers.

If a whole thing is divided into D equal parts, each of those parts can be symbolised by $\dfrac{1}{D}$, where D is called the *denominator*. If N of these equal parts are then combined to make a new thing, the new thing can be symbolised by $\dfrac{N}{D}$, where N is called the *numerator*.

An alternative approach is to interpret $\dfrac{N}{D}$ as the quotient obtained when a quantity of N things is divided into D equal parts. A splitting, or fracturing into parts, such as a division indicated by $\dfrac{N}{D}$ is called a *fraction*. A fraction is given different names according to the values of N and D. The horizontal line between the numerator and the denominator of a fraction is called a *vinculum*. In a little while we shall see that a vinculum can have a different interpretation, and its use is not restricted to fractions.

If N and D are both integers, the fraction is referred to as a *common fraction*, a *simple fraction* or a *vulgar fraction*. The last of these expressions is in greater use than the others, and consequently we shall use the expression *vulgar fraction* henceforward. If N is less than D we have what is known as a *proper vulgar fraction*, but it is conventional practice to delete the word proper and to use the expression vulgar fraction. We use the expression *improper vulgar fraction* for those cases where N is greater than D.

To recapitulate, $\frac{13}{37}$ is an example of a proper vulgar fraction, more usually referred to as a vulgar fraction. The numerator is 13, the denominator is 37. The numerator and denominator are separated by a vinculum, which can be interpreted as a symbol for division.

$\frac{47}{15}$ is an example of an improper vulgar fraction. The numerator is 47, the denominator is 15. The numerator and denominator are separated by a vinculum, which can be interpreted as a symbol for division.

Any improper vulgar fraction can be converted into a combination of an integer and a vulgar fraction. For example $47 \div 15 = 3$ with a remainder of 2. If we divide 47 things into groups each of 15 things we shall have 3 complete groups each of 15 things, and have a remainder of 2 things. The remainder can be considered to be a vulgar fraction of a group equal to two-fifteenths of a group. The mathematical indication is to place the integer first and follow it, without a space, by the vulgar fraction. Hence

$$\frac{47}{15} = 3 + \frac{2}{15}, \text{ which is conventially written as } 3\tfrac{2}{15}.$$

A number which is a combination of an integer and a vulgar fraction is called a *mixed number*.

A mixed number can be converted into an improper vulgar fraction by:

(a) forming the numerator by multiplying the whole number by the denominator in the mixed number, and then adding the numerator of the fraction portion of the mixed number
(b) using the denominator in the mixed number as the denominator of the equivalent improper vulgar fraction.

Example:

Convert the mixed number $3\frac{2}{15}$ to an improper vulgar fraction.

$$\text{Numerator} = 3 \times 15 + 2 = 45 + 2 = 47$$
$$\text{Denominator} = 15$$

Answer: $3\frac{2}{15} = \frac{47}{15}$.

Our reader should note the evaluation of $3 \times 15 + 2$ by the distributive law of mathematics. Multiplication operations take precedence over addition operations. Henceforward we shall give special emphasis to the distributive law by putting the multiplication operation within parentheses, e.g. $(3 \times 15) + 2$.

1.3.2
The cancellation of common factors in vulgar fractions

Three-sevenths of a set of 14 things is 6 things, which we can represent by the vulgar fraction $\frac{6}{14}$. Three sevenths of a set of 350 things is 150 things, which we can represent by the vulgar fraction $\frac{150}{350}$. This type of statement can be continued indefinitely, producing vulgar fractions such as $\frac{36}{84}$, $\frac{21}{49}$, $\frac{12}{28}$, $\frac{30}{70}$ and so on. Let us represent all these equalities by a continued equation. For convenience, we will arrange the vulgar fractions so that as we read from left to right, although the vulgar fractions are equal to each other, there is an increase in the general magnitude of the numbers being used.

$$\frac{3}{7} = \frac{6}{14} = \frac{12}{28} = \frac{21}{49} = \frac{30}{70} = \frac{150}{350}.$$

Let us express the various vulgar fractions in the continued equation with numerators and denominators which are products of factors.

$$\frac{3}{7} = \frac{2 \times 3}{2 \times 7} = \frac{2 \times 2 \times 3}{2 \times 2 \times 7} = \frac{3 \times 7}{7 \times 7} = \frac{2 \times 3 \times 5}{2 \times 5 \times 7} = \frac{2 \times 3 \times 5 \times 5}{2 \times 5 \times 5 \times 7}$$

Since the multiplication of integers can be accomplished in any sequence, let us rearrange the products so that in every case factors which are common to numerators and denominators are vertically in alignment, to produce

$$\frac{3}{7} = \frac{2 \times 3}{2 \times 7} = \frac{2 \times 2 \times 3}{2 \times 2 \times 7} = \frac{7 \times 3}{7 \times 7} = \frac{2 \times 5 \times 3}{2 \times 5 \times 7} = \frac{2 \times 5 \times 5 \times 3}{2 \times 5 \times 5 \times 7}$$

and we note that the value of $\frac{3}{7}$ will occur in every case if we eliminate factors which are common to both the numerator and the denominator.

We say that a vulgar fraction is given in its *simplest form* when either the numerator *OR* the denominator is a prime number. Of the various vulgar fractions which are equal to three-sevenths, the vulgar fraction $\frac{3}{7}$ is the simplest form. It is conventional practice to express vulgar fractions in their simplest form. If a vulgar fraction is not in its simplest form, it can be converted to its simplest form by a process called, in full, the cancellation of the common factors of vulgar fractions, which we usually shorten to *cancellation*.

Theoretically, in cancelling, we divide the numerator and the denominator, by the prime numbers in sequence, omitting any prime numbers which will not divide into the numerator and the denominator. Consider, for example, reducing the vulgar fraction $\frac{150}{350}$ to its simplest form. We note first of all that both the numerator and the denominator are even, hence 2 is a common factor and

$$\frac{150}{350} = \frac{150 \div 2}{350 \div 2} = \frac{75}{175}.$$

The usual indication omits an actual division by 2. We simply strike out the former numerator and the former denominator and replace them with the new quotients, thus:

$$\frac{75}{\underset{175}{\cancel{350}}} \\ \overline{\underset{175}{\cancel{350}}}$$

75 and 175 are both odd, so we cannot divide both the new numerator and new denominator by 2. We next try to divide both 75 and 175 by the next prime number after 2, which is 3, and find that 3 is not a common factor. We therefore proceed to the next prime number after 3, which is 5, and find that 5 is a common factor.

$$\frac{75}{175} = \frac{75 \div 5}{175 \div 5} = \frac{15}{35}$$

and we continue with the striking-out and replacing with a new numerator and new denominator to obtain

$$\frac{15}{7\cancel{5}}$$
$$\frac{1\cancel{5}\cancel{0}}{3\cancel{5}\cancel{0}}$$
$$\frac{1\cancel{7}\cancel{5}}{35}$$

15
7̸5̸
1̸5̸0̸
3̸5̸0̸
1̸7̸5̸
35

Considering $\frac{15}{35}$ we note that both the numerator and the denominator will again both divide by 5, so once again 5 is a common factor

$$\frac{15}{35} = \frac{15 \div 5}{35 \div 5} = \frac{3}{7},$$ and we 'divide and strike-out' as before, to proceed to

3
1̸5̸
7̸5̸
$$\frac{1̸5̸0̸}{3̸5̸0̸} = \frac{3}{7}$$
1̸7̸5̸
3̸5̸
7

We now note that at least one of the integers in the numerator and the denominator is a prime number, and so the vulgar fraction is in its simplest form.

We commenced explaining cancellation by stating that theoretically we divide the numerator and the denominator by prime numbers in sequence. Multiplication and division are computational processes which conform to the commutative law, hence division can be performed in any sequence. Cancellation to the simplest form can therefore be accomplished in one step just by dividing both the numerator and the denominator by their highest common factor. In practice, we usually do not determine the highest common factor, but first cancel by the common factor which is immediately apparent. For instance, since 150 and 350 both end with a zero, we know that 10 is a common factor. Cancelling by 10 produces the value of $\frac{15}{35}$, and now, since both numbers end in 5, then 5 is a common factor. The more agile mind would recognise that with 150 and 350, 50 is a common factor, and straight away the simplest form is obtained by a direct cancelling by 50, thus

3
$$\frac{1̸5̸0̸}{3̸5̸0̸} = \frac{3}{7}$$
7

Proceeding in the converse direction, a vulgar fraction in its simplest form can be converted into an equivalent by multiplying the numerator and the denominator by the same number, with one notable and important exception, which is when the number is zero.

For instance, $\dfrac{3}{7} = \dfrac{3 \times 7}{7 \times 7} = \dfrac{21}{49}$

or $\dfrac{3}{7} = \dfrac{3 \times 41}{7 \times 41} = \dfrac{123}{287}.$

However, $\dfrac{3}{7} \neq \dfrac{3 \times 0}{7 \times 0}.$

The right-hand side would produce the result $\dfrac{0}{0}$, which has no meaning in arithmetic, since $\dfrac{0}{0} = \dfrac{\text{any number} \times 0}{\text{any number} \times 0}$, and we cannot cancel the zeros, otherwise we would have the illogical result that

$$\frac{3}{7} = \frac{\text{any number}}{\text{any number}}.$$

Consequently our reader should note that when cancelling is used to reduce a vulgar fraction to its simplest form, *cancellation by zero is not permissible*. Furthermore, when multiplying the numerator and the denominator of a vulgar fraction to obtain an equivalent, *multiplying by zero is not permitted*.

1.3.3
The common denominator of a set of vulgar fractions

Multiplying or dividing both the numerator and the denominator of a vulgar fraction by the same quantity, with the especial exception of multiplying or dividing by zero, makes no change to the value of a vulgar fraction.

For example,

$$\frac{3}{7} = \frac{3 \times 3}{7 \times 3} = \frac{9}{21}$$

$$\frac{3}{7} = \frac{3 \times 5}{7 \times 5} = \frac{15}{35}$$

$$\frac{3}{7} = \frac{3 \times 19}{7 \times 19} = \frac{57}{153}$$

and $\dfrac{3}{7} = \dfrac{9}{21} = \dfrac{15}{35} = \dfrac{57}{153}.$

It is often very convenient when solving problems of computation involving a set of vulgar fractions to express that set of vulgar fractions as a set of equivalent vulgar fractions all having the same denominator. We call it a *common denominator*, and it is a common multiple of the various denominators. Any common multiple of the various denominators can be used, but the general magnitude of the numbers which will be involved will be reduced if the common multiple is the least common multiple. If the least common multiple is used, it is referred to as the *least common denominator*, usually abbreviated to L.C.D.

As a typical example let us suppose we had to arrange the vulgar fractions

$$\frac{2}{5} \quad \frac{7}{16} \quad \frac{1}{4} \quad \frac{4}{9} \quad \text{and} \quad \frac{1}{3}$$

in order of magnitude. A common multiple of the various denominators is the continued product of all the various denominators. However, our reader can accept that the least common multiple is 720. The least common denominator of our equivalent fractions will therefore be 720.

Let us commence with the vulgar fraction $\frac{2}{5}$ and determine the equivalent vulgar fraction which has a denominator of 720. We set up the model:

$$\frac{2}{5} = \frac{2 \times \text{a number}}{5 \times \text{the same number}} = \frac{\text{new numerator}}{720}.$$

The number we are seeking is 144, obtained from $720 \div 5$.

Hence,
$$\frac{2}{5} = \frac{2 \times 144}{5 \times 144} = \frac{288}{720}.$$

The algorithm we have used is:

Step 1. Divide the original denominator into the least common denominator to obtain a quotient.

Step 2. Multiply both the numerator and the denominator of the original fraction by the quotient obtained in step 1.

For example, to express $\frac{7}{16}$ as an equivalent vulgar fraction with a denominator of 720, we proceed:

Step 1. $720 \div 16 = 45$

Step 2. $\frac{7}{16} = \frac{7 \times 45}{16 \times 45} = \frac{315}{420}.$

Proceeding to the remaining vulgar fractions, we obtain

$$720 \div 4 = 180$$

and
$$\frac{1}{4} = \frac{1 \times 180}{4 \times 180} = \frac{180}{720}$$

$$720 \div 9 = 80$$

$$\frac{4}{9} = \frac{4 \times 80}{9 \times 80} = \frac{320}{720}$$

$$720 \div 3 = 240$$

$$\frac{1}{3} = \frac{1 \times 240}{3 \times 240} = \frac{240}{720}.$$

Our equivalent fractions are

$$\frac{288}{720} \quad \frac{315}{720} \quad \frac{180}{720} \quad \frac{320}{720} \quad \text{and} \quad \frac{240}{720}$$

which can be arranged in order of magnitude, by comparing numerators and using the mathematical symbol $<$ to indicate 'is less than', as

$$\frac{180}{720} < \frac{240}{720} < \frac{288}{720} < \frac{315}{720} < \frac{320}{720}$$

and so
$$\frac{1}{4} < \frac{1}{3} < \frac{2}{5} < \frac{7}{16} < \frac{4}{9}$$

The real value of the preceding work is to demonstrate the algorithm for converting a given vulgar fraction to its equivalent having a particular denominator. Our reader will be introduced to a far simpler method of ranging vulgar fractions in order of magnitude when we have considered another type of fraction called a decimal fraction.

1.3.4
Computations involving vulgar fractions

We will commence our work on computations involving vulgar fractions with the addition of vulgar fractions. Vulgar fractions can be added with the aid of a common denominator. It is repeated that the common denominator can be any common multiple of the various denominators, but the general magnitude of the numbers involved will be reduced if the common multiple is the least common multiple, which forms the least common denominator.

For example, consider $\frac{3}{5} + \frac{2}{7}$.

Since both denominators are prime numbers, the L.C.D. is the product of those prime numbers, which is 35.

$$\frac{3}{5} = \frac{3 \times 7}{5 \times 7} = \frac{21}{35}$$

$$\frac{2}{7} = \frac{2 \times 5}{7 \times 5} = \frac{10}{35}.$$

We now add twenty-one thirty-fifths to ten thirty-fifths to give a total of thirty-one thirty-fifths, the computation usually being set out as

$$\frac{3}{5} + \frac{2}{7} = \frac{21}{35} + \frac{10}{35} = \frac{21 + 10}{35} = \frac{31}{35}.$$

Since 31 and 35 do not have a common factor the result is in its simplest form. Our reader will note that with the statement

$$\frac{21 + 10}{35}$$

the vinculum is not only used as an indication of division, it is also used as an aggregation sign. The value of $21 + 10$ is treated as an entity.

Since addition conforms to the commutative law, the addition process can be extended to more than two vulgar fractions, and to mixed numbers, by adding the integers and adding the vulgar fractions.

Example:

Evaluate $2\frac{1}{7} + 5\frac{3}{5} + 8\frac{1}{3}$.

$$2\frac{1}{7} + 5\frac{3}{5} + 8\frac{1}{3} = (2 + 5 + 8) + (\tfrac{1}{7} + \tfrac{3}{5} + \tfrac{1}{3})$$

$$= 15 + \frac{15 + 63 + 35}{105}$$

$$= 15 + \tfrac{113}{105}$$

$$= 15 + 1\tfrac{8}{105}$$

$$= 16\tfrac{8}{105}$$

the result cannot be simplified further since 8 and 105 have no common factor.

Answer: $\qquad\qquad 16\tfrac{8}{105}$

We will now proceed to a chain of computations involving vulgar fractions and/or mixed numbers which consists of addition and subtractions only. Our reader may recall that with integers, such as

$$5 + 7 - 15 - 6 + 3 - 9$$

we grouped positive and negative operations together, as

$$5 + 7 + 3 - 15 - 6 - 9$$

then used brackets, noting carefully of the effect of a negative sign, to obtain

$$(5 + 7 + 3) - (15 + 6 + 9).$$

We follow the same general principles with vulgar fractions and/or mixed numbers.

Example:

Evaluate $5\frac{1}{3} - 2\frac{3}{4} + 3\frac{1}{6} + 8\frac{1}{3} - 16\frac{1}{2}$

$$5\frac{1}{3} - 2\frac{3}{4} + 3\frac{1}{6} + 8\frac{1}{3} - 16\frac{1}{2} = 5\frac{1}{3} + 3\frac{1}{6} + 8\frac{1}{3} - 2\frac{3}{4} - 16\frac{1}{2}$$

$$= (5\frac{1}{3} + 3\frac{1}{6} + 8\frac{1}{3}) - (2\frac{3}{4} + 16\frac{1}{2})$$

$$5\frac{1}{3} + 3\frac{1}{6} + 8\frac{1}{3} = 5 + 3 + 8 + \frac{1}{3} + \frac{1}{6} + \frac{1}{3}$$

$$= 16 + \frac{2 + 1 + 2}{6}$$

$$= 16\frac{5}{6}$$

$$2\frac{3}{4} + 16\frac{1}{2} = 2 + 16 + \frac{3}{4} + \frac{1}{2}$$

$$= 18 + \frac{3 + 2}{4} = 18 + \frac{5}{4}$$

$$= 18 + 1\frac{1}{4} = 19\frac{1}{4}.$$

We now have to evaluate $16\frac{5}{6} - 19\frac{1}{4}$.

We proceed just as we did with integers. The modulus of $(-19\frac{1}{4})$ is greater than the modulus of $(+16\frac{5}{6})$ and so the arithmetical sign of the answer will be negative. We now subtract the smaller modulus from the larger modulus, and evaluate

$$19\frac{1}{4} - 16\frac{5}{6}$$

We first subtract the integers to obtain

$$19 - 16 = 3$$

and proceed to $\frac{1}{4} - \frac{5}{6}$ so that the modulus of our answer would be $3 + (\frac{1}{4} - \frac{5}{6})$.

Now $\frac{5}{6}$ is greater than $\frac{1}{4}$, and so $\frac{1}{4} - \frac{5}{6}$ would nominally have been a negative quantity. However,

$$3 + (\tfrac{1}{4} - \tfrac{5}{6}) \quad \text{can be written as } 3\tfrac{1}{4} - \tfrac{5}{6}$$

which itself can be written as $2 + 1\frac{1}{4} - \frac{5}{6}$ where we have 'borrowed one from the previous integer' to ease the subtraction. If we have to make an evaluation of the type

$$1\tfrac{1}{4} - \tfrac{5}{6},$$

our work is eased if the mixed number is converted to an improper vulgar fraction. Now

$$1\tfrac{1}{4} = \frac{(1 \times 4) + 1}{4} = \tfrac{5}{4}$$

and

$$1\tfrac{1}{4} - \tfrac{5}{6} = \tfrac{5}{4} - \tfrac{5}{6} = \tfrac{15}{12} - \tfrac{10}{12} = \frac{15 - 10}{12} = \tfrac{5}{12}.$$

The modulus of the answer is $2\frac{5}{12}$, and coupling the modulus with the previously determined negative sign, the final result is $-2\frac{5}{12}$.

Answer:

$$-2\tfrac{5}{12}$$

The multiplication of vulgar fractions

Any integer can be expressed in the form of an improper vulgar fraction by using that integer as the numerator together with unity as the denominator. As examples,

$$5 = \frac{5}{1} \quad \text{and} \quad -16 = \frac{-16}{1},$$

noting that if an integer is not preceded by an arithmetical sign the integer is consider to be positive.

Now let us imagine we have a set of articles. For the convenience of the development which follows we will select a number which has several factors, all of which are prime numbers. We will presume the set consists of 120 articles. Let us first of all split the set into three equal parts, each of 40 articles, and then combine two of these parts into a new assembly of 80 articles. From our definition of a fraction our new assembly is two-thirds of the original set of 120 articles. The mathematical model of what we have done is

$$\frac{2}{3} \text{ of } 120 = 80$$

or

$$\frac{2}{3} \text{ of } \frac{120}{1} = \frac{80}{1}.$$

We note that

$$\frac{2 \times 120}{3 \times 1} = \frac{240}{3} = \frac{80}{1}$$

and hence we conclude that *we can replace the word 'of' by the multiplication sign*. Furthermore, we can multiply two vulgar fractions together by multiplying their numerators together to form a new numerator and their denominators together to form a new denominator.

Let us now split the assembly of 80 articles into ten equal parts and combine three of those parts to form three-tenths of the assembly. Using the methodology of the introduction to this article we can write

$$\frac{3}{10} \text{ of } 80 = \frac{3}{10} \times 80 = \frac{3}{10} \times \frac{80}{1} = \frac{3 \times 80}{10 \times 1} = \frac{240}{10} = 24.$$

Now let us relate this final group of 24 articles to our original set of 120 articles. We have three tenths of two thirds of 120 articles, and since we can replace the word 'of' by the multiplication sign, we can write

$$\frac{3}{10} \times \frac{2}{3} \times 120 = \frac{3}{10} \times \frac{2}{3} \times \frac{120}{1}$$

and we note that

$$\frac{3 \times 2 \times 120}{10 \times 3 \times 1} = \frac{720}{30} = 24.$$

The method can be extended indefinitely for any number of multiplications. Should any mixed number occur it should first be converted to an improper vulgar fraction. Before we proceed further, when evaluating $\frac{3}{10}$ of $\frac{2}{3}$ of 120, we reached a stage represented by

$$\frac{3 \times 2 \times 120}{10 \times 3 \times 1}.$$

Although we carried out the multiplication to obtain $\frac{720}{30}$, we could have noticed that the numerator and the denominator had a common factor of three. We therefore could have cancelled by 3, to produce

$$\frac{\overset{1}{\cancel{3}} \times 2 \times 120}{10 \times \underset{1}{\cancel{3}} \times 1}.$$

Furthermore, since with both 120 and 10 the end digit is a zero, we can further cancel by 10, to produce

$$\frac{\overset{1}{\cancel{3}} \times 2 \times \overset{12}{\cancel{120}}}{\underset{1}{\cancel{10}} \times \underset{1}{\cancel{3}} \times 1}$$

and our final evaluation is

$$\frac{1 \times 2 \times 12}{1 \times 1 \times 1} = \frac{24}{1} = 24.$$

Cancellation, if possible, should always be performed because it assists in producing an answer in its simplest form.

Example:

Evaluate $3\frac{3}{7} \times 1\frac{2}{3} \times 2\frac{4}{5}$

$$3\frac{3}{7} = \frac{(3 \times 7) + 3}{7} = \frac{24}{7}$$

$$1\frac{2}{3} = \frac{(1 \times 3) + 2}{3} = \frac{5}{3}$$

$$2\frac{4}{5} = \frac{(2 \times 5) + 4}{5} = \frac{14}{5}$$

$$\frac{24}{7} \times \frac{5}{3} \times \frac{14}{5} = \frac{\overset{8}{\cancel{24}} \times \overset{1}{\cancel{5}} \times \overset{2}{\cancel{14}}}{\underset{1}{\cancel{7}} \times \underset{1}{\cancel{3}} \times \underset{1}{\cancel{5}}} = \frac{8 \times 1 \times 2}{1 \times 1 \times 1} = \frac{16}{1} = 16$$

Answer: 16

Thus far we have only considered the multiplication of positive vulgar fractions. If the problem includes one or more negative vulgar fractions, the algorithm is:

Step 1. Convert any mixed numbers, if necessary, to improper vulgar fractions.

Step 2. Multiply the positive and/or negative arithmetical signs together to obtain the arithmetical sign of the answer, noting that if no arithmetical sign precedes a fraction, that fraction is deemed to be positive. Multiplying like signs gives a positive answer. Multiplying unlike signs gives a negative answer.

Step 3.   Treat all the vulgar fractions as positive and multiply them together to obtain the modulus of the product.

Step. 4   Obtain in the final answer by combining the arithmetical sign obtained from step 2 with the modulus obtained from step 3.

Example:

Evaluate $-4\frac{2}{3} \times 1\frac{3}{5} \times -\frac{5}{16}$.

Step 1.

$$4\frac{2}{3} = \frac{(4 \times 3) + 2}{3} = \frac{12 + 2}{3} = \frac{14}{3}$$

$$1\frac{3}{5} = \frac{(1 \times 5) + 3}{5} = \frac{5 + 3}{5} = \frac{8}{5}$$

$\frac{5}{16}$, not a mixed number, can be left as it is.

Step 2.   The final arithmetical sign is obtained from

$(-) \times (+) \times (-)$, the middle fraction being positive.

The first two are unlike signs, their product is negative. Replacing $(-) \times (+)$ by $(-)$, we have to determine

$$(-) \times (-).$$

This is the product of two like signs. Their product is positive. The sign of the answer is therefore positive.

Step 3.   We now disregard the positive and negative arithmetical signs and evaluate

$$4\frac{2}{3} \times 1\frac{3}{5} \times \frac{5}{16} = \frac{14}{3} \times \frac{8}{5} \times \frac{5}{16} = \frac{\overset{7}{\cancel{14}} \times \overset{1}{\cancel{8}} \times \overset{1}{\cancel{5}}}{3 \times \underset{1}{\cancel{5}} \times \underset{2}{\cancel{16}}}$$

$$= \frac{7 \times 1 \times 1}{3 \times 1 \times 1} = \frac{7}{3} = 2\frac{1}{3}$$

Step 4.   We now combine the positive sign obtained from step 2 with the modulus $2\frac{1}{3}$ obtained from step 3 to obtain $+2\frac{1}{3}$, and conforming to agreed convention we can, as a final step, discard the positive sign.

Answer: $2\frac{1}{3}$

In a previous article it was stated that division can be considered as the reverse process to multiplication. In the multiplication of two factors, we have

first factor $\times$ second factor $=$ product.

Proceeding in the converse direction, we have

product $\div$ first factor $=$ second factor

or

product $\div$ second factor $=$ first factor.

For example, since

$$4\tfrac{2}{3} \times 1\tfrac{3}{7} = 6\tfrac{2}{3}$$

then

$$6\tfrac{2}{3} \div 4\tfrac{2}{3} = 1\tfrac{3}{7}$$

and

$$6\tfrac{2}{3} \div 1\tfrac{3}{7} = 4\tfrac{2}{3}.$$

When evaluating $4\tfrac{2}{3} \times 1\tfrac{3}{7}$ to obtain $6\tfrac{2}{3}$, we would proceed by stating

$$\frac{14}{3} \times \frac{10}{7} = \frac{140}{21}.$$

Now if we proceed in the reverse direction

$$\frac{140}{21} \div \frac{14}{3} = \frac{10}{7}$$

and that

$$\frac{140}{21} \div \frac{10}{7} = \frac{14}{3}.$$

By inspecting these figures we see that

$$\frac{140}{21} \times \frac{3}{14} = \frac{420}{304} = \frac{10}{7}$$

and that

$$\frac{140}{21} \times \frac{7}{10} = \frac{980}{210} = \frac{14}{3}.$$

If we state a division process in the general form of

dividend $\div$ divisor $=$ quotient

we note that

$$\text{dividend} \times \text{the inverted divisor} = \text{quotient.}$$

The *reciprocal* of a number is unity divided by that number. For example, the reciprocal of 5 is $1 \div 5$, and since the vinculum of a vulgar fraction is an indication of division, the reciprocal of 5 can be written as $\frac{1}{5}$.

Now let us consider the reciprocal of an improper vulgar fraction, such as $\frac{47}{15}$. From the definition of a reciprocal, the reciprocal of $\frac{47}{15}$ is $1 \div \frac{47}{15}$. In a division operation we invert the divisor and multiply. Hence $1 \div \frac{47}{15} = 1 \times \frac{15}{47} = \frac{15}{47}$. Consequently, the reciprocal of an improper vulgar fraction is the inverse of that improper vulgar fraction. For example, the reciprocal of $\frac{23}{17}$ is $\frac{17}{23}$. The same reasoning will apply to a proper vulgar fraction. The reciprocal of $\frac{2}{5}$ is $\frac{5}{2}$.

The algorithm for the division of vulgar fractions is therefore:

Step 1. Convert any mixed numbers to improper vulgar fractions.
Step 2. Determine the arithmetical sign of the answer. A division between two quantities of like sign gives a positive result. A division between two quantities of unlike sign gives a negative result.
Step 3. Obtain the modulus of the answer by multiplying the modulus of the dividend by the reciprocal of the modulus of the divisor.
Step 4. Combine the arithmetical sign obtained from step 2 with the modulus obtained from step 3.

Example:

Evaluate

$$-16\tfrac{4}{5} \div 2\tfrac{5}{8}$$

Step 1.

$$16\tfrac{4}{5} = \frac{(16 \times 5) + 4}{5} = \frac{80 + 4}{5} = \frac{84}{5}$$

$$2\tfrac{5}{8} = \frac{(2 \times 8) + 5}{8} = \frac{16 + 5}{8} = \frac{21}{8}$$

The evaluation becomes $\dfrac{-84}{5} \div \dfrac{21}{8}$

Step 2. The division operation is between quantities of unlike sign. The final answer will be negative.

Step 3.
$$\frac{84}{5} \div \frac{21}{8} = \frac{84}{5} \times \frac{8}{21}$$

$$\frac{84}{5} \times \frac{8}{21} = \frac{\overset{4}{\cancel{84}} \times 8}{5 \times \cancel{21}} = \frac{4 \times 8}{5 \times 1} = \frac{32}{5} = 6\frac{2}{5}$$

Step 4. We now combine the negative sign from step 2 with the modulus of $6\frac{2}{5}$ from step 3 to obtain $-6\frac{2}{5}$.

Answer: $\qquad\qquad -6\frac{2}{5}$

When evaluating problems dealing with any combination of the addition, subtraction, multiplication and division of vulgar fractions care must be taken to conform to the distributive law of mathematics, i.e. that multiplication or division operations take precedence over addition or subtraction operations. The word 'of' can be replaced by a multiplication sign.

For example, when evaluating

$$\tfrac{5}{16} \text{ of } 1\tfrac{2}{3} + \tfrac{3}{4}$$

which can be written

$$\tfrac{5}{16} \times 1\tfrac{2}{3} + \tfrac{3}{4}$$

the multiplication operation must be performed before the addition operation. Our reader will find it to be of considerable assistance to emphasise this with aggregation signs such as brackets, to read

$$(\tfrac{5}{16} \times 1\tfrac{2}{3}) + \tfrac{3}{4}.$$

Similarly with evaluating

$$4\tfrac{6}{7} - \tfrac{3}{5} \div 4\tfrac{1}{7}$$

The problem should be written as

$$4\tfrac{6}{7} - (\tfrac{3}{5} \div 4\tfrac{1}{7})$$

and the division operation performed before the subtraction.

The distributive law can be extended to any number of operations, in which case adjoining multiplying and dividing operations can be combined. For example, with the evaluation of

$$-3\tfrac{4}{5} + \tfrac{2}{3} \times 7\tfrac{1}{4} \div -1\tfrac{4}{5} + \tfrac{3}{7} + 2\tfrac{1}{4} \div 1\tfrac{4}{7} \times 2\tfrac{1}{8} - 4\tfrac{3}{4}$$

we use brackets to write

$$-3\tfrac{4}{5} + (\tfrac{2}{3} \times 7\tfrac{1}{4} \div -1\tfrac{4}{5}) + \tfrac{3}{7} + (2\tfrac{1}{4} \div 1\tfrac{4}{7} \times 2\tfrac{1}{8}) - 4\tfrac{3}{4}$$

We perform the operations within each set of brackets before proceeding to the addition and subtraction.

Example:

Evaluate

$$2\tfrac{1}{4} + (-\tfrac{2}{3}) \times 1\tfrac{4}{5} \div \tfrac{1}{5} - 7\tfrac{3}{4} + 4\tfrac{2}{7} \div 1\tfrac{1}{14}$$

We separate out the multiplication and/or division operations with the aid of brackets, and write

$$2\tfrac{1}{4} + (-\tfrac{2}{3} \times 1\tfrac{4}{5} \div \tfrac{1}{5}) - 7\tfrac{3}{4} + (4\tfrac{2}{7} \div 1\tfrac{1}{14})$$

We now simplify the first bracketed values.

$$1\tfrac{4}{5} = \frac{(1 \times 5) + 4}{5} = \frac{5 + 4}{5} = \tfrac{9}{5}$$

$$-\tfrac{2}{3} \times 1\tfrac{4}{5} \div \tfrac{1}{5} = -(\tfrac{2}{3} \times \tfrac{9}{5} \times \tfrac{5}{1})$$

the minus sign being obtained from $(-) \times (+) \div (+)$.

$$\frac{2 \times \overset{3}{\cancel{9}} \times \overset{1}{\cancel{5}}}{\underset{1}{\cancel{3}} \times \underset{1}{\cancel{5}} \times 1} = \frac{2 \times 3 \times 1}{1 \times 1 \times 1} = \frac{6}{1},$$

which may as well be left in that form.

To proceed to the second bracketed values.

$$4\tfrac{2}{7} = \frac{(4 \times 7) + 2}{7} = \frac{28 + 2}{7} = \frac{30}{7}$$

$$1\tfrac{1}{14} = \frac{(1 \times 14) + 1}{14} = \frac{14 + 1}{14} = \frac{15}{14}$$

$$\frac{30}{7} \div \frac{15}{14} = + \left(\frac{30}{7} \times \frac{14}{15}\right)$$

$$\frac{30}{7} \times \frac{14}{15} = \frac{2 \times 2}{1} = \frac{4}{1}$$

which again may as well be left in that form.

The problem is now reduced to:

$$2\tfrac{1}{4} - 6 - 7\tfrac{3}{4} + 4$$

which, using the methodology previously described, we evaluate as

$$(2\tfrac{1}{4} + 4) - (6 + 7\tfrac{3}{4}) = 6\tfrac{1}{4} - 13\tfrac{3}{4}$$
$$= -(13\tfrac{3}{4} - 6\tfrac{1}{4})$$
$$= -(7\tfrac{1}{2})$$

Answer: $\qquad\qquad -7\tfrac{1}{2}$

1.3.5
Computations involving aggregation signs

It has been previously stated in this textbook that in mathematics there are many ways of indicating aggregations, which, somewhat incorrectly, we group together in a set and call them *brackets*. Theoretically, the aggregation signs ( ) are *parentheses*, the aggregation signs { } are *braces*, and only the aggregation signs [ ] should be referred to as brackets. However, the usage of the word brackets for what, more correctly, should be called parentheses, of the expression 'curly brackets' for braces, and of the expression 'square brackets' instead of brackets, is of long standing. As long as the theoretical terminology is known, the mild inaccuracies of common usage can be tolerated. Generally, if only parentheses are used in an expression they are usually called brackets, but if aggregation signs other than parentheses occur, it is usual to refer to parentheses, braces and brackets.

It is often necessary to indicate the true nature of the operations that are required by using aggregation signs within aggregation signs. We try, as far as we can, to use only parentheses. However, if parentheses alone do not give a true indication of what is required, we then try to see if just parentheses and braces are sufficient. If parenthese and braces are insufficient then, and only then, do we use parentheses, braces and brackets. There will be no problems in our reader's present stage of studies on vulgar fractions where aggregation signs other than the vinculum, parentheses, braces and brackets will be required.

Certain problems require aggregation signs to be eliminated. Since we use parentheses, braces and brackets in that order, we remove them in that order. If we have numbers within 'square' brackets, then somewhere within those 'square' brackets will be numbers within braces, and somewhere within those braces will be numbers within parentheses. A typical indication is

$$2\tfrac{1}{4} + [3\tfrac{1}{4} - 4\tfrac{1}{2} + \{\tfrac{3}{7} + 2\tfrac{1}{4} - (3\tfrac{1}{8} + 4\tfrac{1}{2}) - \tfrac{1}{5}\} + 5\tfrac{1}{2}]$$

and in effecting a solution the parentheses are removed first, the braces next and the brackets last.

A collection of items within an aggregation sign has to be treated as an entity, that is, one complete thing. If that entity has been multiplied by some other thing, which we call a coefficient, the coefficient is placed immediately in front of the first part of the aggregation sign. For example, an indication such as

$$1\tfrac{4}{7}(3\tfrac{1}{4} + 5\tfrac{1}{2})$$

means that $3\tfrac{1}{4} + 5\tfrac{1}{2}$ must be treated as an entity, which is $8\tfrac{3}{4}$, and then that value of $8\tfrac{3}{4}$ is multiplied by $1\tfrac{4}{7}$.

Similarly, with

$$-\tfrac{1}{3}(2\tfrac{1}{4} + \tfrac{3}{5} \times \tfrac{2}{7})$$

it means that $2\tfrac{1}{4} + \tfrac{3}{5} \times \tfrac{2}{7}$ must be treated as an entity and then that entity is to be multiplied by $-\tfrac{1}{3}$. We shall give the use of aggregation signs a little more attention at a later stage in this book, but for the time being we will restrict our activities to the simplification of groups of mixed numbers and vulgar fractions. We will illustrate the removal of aggregation signs with some typical examples

Example:

Simplify $4\tfrac{1}{5} - 3(6 + \tfrac{3}{5} \times 2\tfrac{1}{7})$.

The items within parentheses must be treated as an entity. Let us first evaluate this entity.

$6 + \tfrac{3}{5} \times 2\tfrac{1}{7}$ indicates that the multiplication must be undertaken first,

$$2\tfrac{1}{7} = \frac{(2 \times 7) + 1}{7} = \frac{14 + 1}{7} = \frac{15}{7}$$

$$\frac{3}{5} \times \frac{15}{7} = \frac{3 \times \overset{3}{\cancel{15}}}{\underset{1}{\cancel{5}} \times 7} = \frac{9}{7} = 1\tfrac{2}{7}$$

$$6 + 1\tfrac{2}{7} = 7\tfrac{2}{7}$$

and the original problem can be restated as

$$4\tfrac{1}{5} - 3(7\tfrac{2}{7}).$$

The coefficient 3 indicates that the entity within the parentheses has to be multiplied by 3.

$$3 \times 7\tfrac{2}{7} = 21\tfrac{6}{7}$$

and we remove the parentheses to further simplify the original problem to

$$4\tfrac{1}{5} - 21\tfrac{6}{7}.$$

We now have to evaluate

$$4\tfrac{1}{5} - 21\tfrac{6}{7}$$

which, using the method previously described, we evaluate as

$$-(21\tfrac{6}{7} - 4\tfrac{1}{5})$$

The entity within parentheses can be evaluated as

$$21 + \tfrac{6}{7} - 4 - \tfrac{1}{5} = (21 - 4) + (\tfrac{6}{7} - \tfrac{1}{5})$$

$$= 17 + \left(\frac{30 - 7}{35}\right) = 17\tfrac{23}{35}$$

Associating this later value with the previously determined negative sign produces the final answer of $-17\tfrac{23}{35}$.

Answer: $\qquad\qquad -17\tfrac{23}{35}.$

It has been necessary thus far when solving problems to indicate to our reader, in detail, the steps which led to a particular answer. When our reader solves problems, it is not considered necessary to accompany a solution with notes explaining the reason *why* a calculation such as $\tfrac{4}{5} - \tfrac{2}{3}$ produces the result of $\tfrac{2}{15}$. However, it is advisable to set out all the steps neatly and logically. Our reader is urged not to take short cuts until sufficient confidence is obtained that a short cut will produce a correct result. A further suggestion is that on many occasions the introduction of aggregation signs during a solution will often be a wise policy to adopt. Let us continue with a typical solution of a problem involving vulgar fractions and mixed numbers, which includes virtually all the operations and computations involving vulgar fractions which have been introduced thus far.

Example:

Evaluate

$$-1\tfrac{3}{7}[3\tfrac{1}{4} - 2\{\tfrac{7}{8} \text{ of } 3\tfrac{1}{5} + 1\tfrac{4}{5}(2\tfrac{2}{15} + 1\tfrac{2}{5} \div 1\tfrac{1}{6})\}]$$

Simplify values in parentheses, noting that division takes precedence over addition.

$$2\tfrac{2}{15} + 1\tfrac{2}{5} \div 1\tfrac{1}{6} = 2\tfrac{2}{15} + (1\tfrac{2}{5} \div 1\tfrac{1}{6}) = 2\tfrac{2}{15} + (\tfrac{7}{5} \div \tfrac{7}{6})$$

$$= 2\tfrac{2}{15} + \left(\frac{\overset{1}{\cancel{7}}}{5} \times \frac{6}{\underset{1}{\cancel{7}}}\right) = 2\tfrac{2}{15} + \tfrac{6}{5} = 2 + (\tfrac{2}{15} + \tfrac{6}{5})$$

$$= 2 + \left(\frac{2 + 18}{15}\right) = 2 + \tfrac{20}{15} = 2 + 1\tfrac{5}{15}$$

$$= 2 + 1\tfrac{1}{3} = 3\tfrac{1}{3}$$

Expression in braces now becomes

$$\{\tfrac{7}{8} \text{ of } 3\tfrac{1}{5} + 1\tfrac{4}{5}(3\tfrac{1}{3})\}$$

Remove parentheses.

$$1\tfrac{4}{5} \times 3\tfrac{1}{3} = \frac{\overset{3}{\cancel{9}}}{\underset{1}{\cancel{5}}} \times \frac{\overset{2}{\cancel{10}}}{\underset{1}{\cancel{3}}} = \frac{6}{1} = 6$$

Simplify values in braces.

$$\tfrac{7}{8} \text{ of } 3\tfrac{1}{5} + 6 = (\tfrac{7}{8} \times 3\tfrac{1}{5}) + 6 = \left(\frac{7}{\underset{1}{\cancel{8}}} \times \frac{\overset{2}{\cancel{16}}}{5}\right) + 6$$

$$= \tfrac{14}{5} + 6 = 2\tfrac{4}{5} + 6 = 8\tfrac{4}{5}.$$

Expression in brackets now becomes

$$[3\tfrac{1}{4} - 2\{8\tfrac{4}{5}\}]$$

$$3\tfrac{1}{4} - 2(8\tfrac{4}{5}) = 3\tfrac{1}{4} - (\tfrac{2}{1} \times \tfrac{44}{5}) = 3\tfrac{1}{4} - \tfrac{88}{5}$$

$$= \frac{13}{4} - \frac{88}{5} = \frac{65 - 352}{20} = \frac{-287}{20} = -14\tfrac{7}{20}.$$

Final step is $-1\tfrac{3}{7}[-14\tfrac{7}{20}]$

$$= (+)\left[\frac{\overset{1}{\cancel{10}}}{\underset{1}{\cancel{7}}} \times \frac{\overset{41}{\cancel{287}}}{\underset{2}{\cancel{20}}}\right] = +\frac{41}{2} = 20\tfrac{1}{2}.$$

Answer: $20\tfrac{1}{2}$

Problems 1.3

1. Express the following improper vulgar fractions as mixed numbers:

   (a) $\dfrac{9}{2}$ (b) $\dfrac{4}{3}$ (c) $\dfrac{13}{4}$ (d) $\dfrac{24}{5}$

   (e) $\dfrac{22}{7}$ (f) $\dfrac{50}{9}$ (g) $\dfrac{400}{11}$ (h) $\dfrac{1\,000}{7}$

2. Express the following mixed numbers as improper vulgar fractions:

   (a) $1\frac{1}{4}$ (b) $3\frac{1}{7}$ (c) $2\frac{3}{4}$ (d) $4\frac{7}{8}$
   (e) $12\frac{1}{2}$ (f) $16\frac{2}{3}$ (g) $31\frac{1}{4}$ (h) $28\frac{4}{7}$

3. Reduce the following proper vulgar fractions to their simplest forms:

   (a) $\dfrac{6}{8}$ (b) $\dfrac{25}{45}$ (c) $\dfrac{63}{81}$ (d) $\dfrac{56}{64}$

   (e) $\dfrac{77}{132}$ (f) $\dfrac{288}{1\,008}$ (g) $\dfrac{275}{4\,400}$ (h) $\dfrac{5\,625}{10\,000}$

4. Express, as proper or improper vulgar fractions the following values with 48 as the denominator:

   (a) $\frac{2}{3}$ (b) $\frac{5}{8}$ (c) $\frac{7}{24}$ (d) $1\frac{1}{2}$
   (e) $1\frac{2}{3}$ (f) $2\frac{3}{16}$ (g) $2\frac{7}{24}$ (h) $3\frac{31}{48}$

5. Perform the operations indicated, expressing the answers as either integers, proper vulgar fractions or mixed numbers in their simplest form:

   (a) $\frac{1}{6} + \frac{1}{2}$ ✓

   (b) $\frac{2}{3} + \frac{7}{8}$

   (c) $\frac{5}{6} - \frac{2}{3}$

   (d) $\frac{9}{11} - \frac{2}{7}$

   (e) $\frac{1}{3} + \frac{1}{6} + \frac{1}{18}$

   (f) $\frac{2}{3} + \frac{3}{4} + \frac{4}{5}$

   (g) $\frac{3}{4} + \frac{5}{8} - \frac{3}{16}$

   (h) $\frac{23}{24} - \frac{1}{3} - \frac{3}{8}$

   (i) $4\frac{1}{3} + 2\frac{1}{6}$

   (j) $3\frac{1}{2} + 4\frac{5}{8} + 6\frac{5}{16}$

   (k) $8\frac{3}{4} - 2\frac{1}{6} - 5\frac{1}{3}$

   (l) $2\frac{1}{2} - 1\frac{3}{4}$

   (m) $6\frac{1}{2} - 8\frac{2}{3} + 3\frac{1}{6}$ ✓

   (n) $2\frac{1}{2} + 1\frac{3}{4} - \frac{7}{8} - 1\frac{3}{16}$ ✓

   (o) $7\frac{5}{6} - 10\frac{3}{7} - 2\frac{1}{4} + 1\frac{3}{8}$ ✓

6. Perform the operations indicated, express the answers as either integers, proper vulgar fractions, or mixed numbers in their simplest forms:

(a) $\frac{2}{5} \times \frac{3}{7}$ (b) $\frac{2}{3} \times \frac{3}{8}$ (c) $1\frac{4}{5} \times 3\frac{1}{3}$

(d) $\frac{4}{5} \times \frac{8}{9} \times \frac{15}{16}$ (e) $2 \times \frac{5}{8} \times 3\frac{1}{5}$ (f) $\frac{3}{8} \times 1\frac{2}{3} \times 3\frac{3}{5}$

(g) $\frac{14}{15} \div \frac{7}{25}$ (h) $3\frac{1}{3} \div 5$ (i) $7\frac{1}{2} \div \frac{5}{8}$

(j) $3\frac{3}{4} \div 1\frac{7}{8}$ (k) $4\frac{2}{3} \times 1\frac{1}{2} \div 1\frac{3}{4}$ (l) $5 \times 1\frac{3}{4} \div 2\frac{5}{8}$

(m) $\frac{7}{8}$ of $1\frac{3}{5} \div \frac{14}{15}$ (n) $\frac{3}{4}$ of $4\frac{4}{5} \div 6\frac{2}{3}$ (o) $\frac{1}{7}$ of $\frac{3}{4}$ of $4\frac{2}{3}$

7. Simplify, expressing the answers as either proper vulgar fractions or mixed numbers in their simplest forms:

(a) $\frac{1}{2} \times 3\frac{1}{3} + 2\frac{1}{6}$ (b) $3\frac{1}{2} \times \frac{2}{7} + 1\frac{3}{5} \times \frac{3}{16}$

(c) $6\frac{1}{4} \times \frac{1}{5} + 1\frac{1}{4} - 3\frac{1}{8} \div \frac{15}{16}$ (d) $2\frac{1}{4} \div (\frac{3}{7}$ of $1\frac{5}{9}) - 1\frac{1}{2}$

(e) $\frac{3}{4}$ of $1\frac{2}{3} \div \frac{1}{7}$ of $17\frac{1}{2}$ (f) $\dfrac{1\frac{1}{3} + 3\frac{1}{2} - 2\frac{1}{6}}{\frac{3}{5} \text{ of } 1\frac{9}{16}}$

(g) $\dfrac{\frac{7}{8} \text{ of } 1\frac{1}{14} + 2\frac{2}{5} \times 1\frac{1}{4}}{6 - \frac{13}{16} \times 2}$ (h) $\dfrac{1}{\frac{3}{5} \text{ of } 6\frac{2}{3} - 1\frac{2}{7} \times 1\frac{4}{9}}$

8. Simplify, expressing the answers as either proper vulgar fractions or mixed numbers in their simplest forms:

(a) $2(\frac{3}{4}$ of $1\frac{7}{8})$

(b) $3\frac{1}{3}(\frac{2}{3}$ of $1\frac{4}{5}) - 1\frac{2}{3}(4\frac{1}{5} - 1\frac{2}{3})$

(c) $-3\{7\frac{1}{2} - \frac{1}{2}(4\frac{2}{3} \times 1\frac{5}{7})\}$

(d) $-2\frac{2}{11}[10\frac{1}{6} - 2\{2\frac{1}{2} + 1\frac{1}{4}(\frac{3}{4}$ of $1\frac{7}{9})\}]$

(e) $-2\frac{1}{3}\{2 + 2[8\frac{1}{4} - (\frac{7}{8} + 1\frac{1}{4}) - (\frac{1}{5} \times \frac{5}{8}) - 4\frac{2}{7}] - 3[1\frac{1}{2} - \frac{7}{12}]\}$

9. The petrol tank of a car was full at the commencement of a journey and contained 60 litres. After completing three-quarters of the journey the tank contained 12 litres. The car then took on 40 litres and then completed its journey. Assuming constant petrol consumption, what fraction of a full tank remained at the end of the journey?

10. Three-quarters of a pile of bricks were used for a project. When two-thirds of the remainder had been used, 50 bricks were left. How many bricks were there in the original pile?

11. Three canisters of oil, *A*, *B* and *C*, contain 25, 48 and 60 litres respectively. Four-fifths of *A*, seven-eighths of *B* and nine-tenths of *C* were withdrawn. The remainders were then collected together and poured into another container of capacity 50 litres which was already half full. What fraction of the capacity of this last container was occupied by oil?

12. Three persons A, B and C contribute to a fund. A provides three-fifths of the total, B provides two-thirds of the remainder and C provides £4. Determine the total of the fund and the contributions made by A and by B.

13. A committee consists of two groups A and B, group A occupying five-eighths of the places on the committee. In voting for a proposal one-half of group A together with one-third of group B voted in favour and nobody abstained. The proposal was lost by six votes. How many persons formed the committee?

14. In a particular parallel grouping of resistors

$$\frac{1}{R} = \frac{1}{R_1} + \frac{1}{R_2} + \frac{1}{R_3}.$$

Find the values of:
(a) R when $R_1 = 1$, $R_2 = 2$ and $R_3 = 3$
(b) R_1 when $R = 1$, $R_2 = 7\frac{1}{2}$ and $R_3 = 5$

Answers to Problems 1.3

1. (a) $4\frac{1}{2}$ (b) $1\frac{1}{3}$ (c) $3\frac{1}{4}$ (d) $4\frac{4}{5}$
   (e) $3\frac{1}{7}$ (f) $5\frac{5}{9}$ (g) $36\frac{4}{11}$ (h) $142\frac{6}{7}$

2. (a) $\frac{5}{4}$ (b) $\frac{22}{7}$ (c) $\frac{11}{4}$ (d) $\frac{39}{8}$
   (e) $\frac{25}{2}$ (f) $\frac{50}{3}$ (g) $\frac{125}{4}$ (h) $\frac{200}{7}$

3. (a) $\frac{3}{4}$ (b) $\frac{5}{9}$ (c) $\frac{7}{9}$ (d) $\frac{7}{8}$
   (e) $\frac{7}{12}$ (f) $\frac{2}{7}$ (g) $\frac{1}{16}$ (h) $\frac{9}{16}$

4. (a) $\frac{32}{48}$ (b) $\frac{30}{48}$ (c) $\frac{14}{48}$ (d) $\frac{72}{48}$
   (e) $\frac{80}{48}$ (f) $\frac{105}{48}$ (g) $\frac{110}{48}$ (h) $\frac{175}{48}$

5. (a) $\frac{2}{3}$ (b) $1\frac{13}{24}$ (c) $\frac{1}{6}$
   (d) $\frac{41}{77}$ (e) $\frac{5}{9}$ (f) $2\frac{13}{60}$
   (g) $1\frac{3}{16}$ (h) $\frac{1}{4}$ (i) $6\frac{1}{2}$
   (j) $14\frac{7}{16}$ (k) $1\frac{1}{4}$ (l) $\frac{3}{4}$
   (m) 1 (n) $2\frac{3}{16}$ (o) $-3\frac{79}{168}$

6. (a) $\frac{6}{35}$ (b) $\frac{1}{4}$ (c) 6
   (d) $\frac{2}{3}$ (e) 4 (f) $2\frac{1}{4}$
   (g) $3\frac{1}{3}$ (h) $\frac{2}{3}$ (i) 12
   (j) 2 (k) 4 (l) $3\frac{1}{3}$
   (m) $1\frac{1}{2}$ (n) $\frac{27}{50}$ (o) $\frac{1}{2}$

7. (a) $3\frac{5}{6}$ (b) $1\frac{3}{10}$ (c) $-\frac{1}{3}$ (d) $1\frac{7}{8}$
   (e) $\frac{1}{2}$ (f) $2\frac{38}{45}$ (g) $\frac{9}{10}$ (h) $\frac{7}{15}$

8. (a) $2\frac{13}{16}$ (b) $-\frac{2}{9}$ (c) $-10\frac{1}{2}$
   (d) -4 (e) $-6\frac{1}{4}$

9. $\frac{3}{5}$

10. 600

11. $\frac{21}{25}$

12. £30, A provides £18, B provides £8.

13. 48

14. (a) $\frac{6}{11}$ (b) $1\frac{1}{2}$

1.4
Decimal fractions

1.4.1
The power of a base

Certain numbers are the continued product of identical factors. Typical examples are that

$$4 = 2 \times 2$$
$$27 = 3 \times 3 \times 3$$
$$15\,625 = 5 \times 5 \times 5 \times 5 \times 5 \times 5.$$

The mathematical way of indicating these products is 2^2, 3^3 and 5^6, respectively. The number with the larger type size is the number which is multiplied by itself two or more times, and is called the *base*. The number with the smaller type size, which is put behind and at a higher level than the base, indicates the number of identical bases that are multiplied together, and is called the *index*, or the *exponent*, the former being more common for our present level of studies. The general form is:

$$\text{base}^{\text{index}} = \text{number}.$$

so that

$$2^2 = 4, \qquad 3^3 = 27 \quad \text{and} \quad 5^6 = 15\,625.$$

Theoretically, we read 2^2 as 'two raised to the power of two', 3^3 as 'three raised to the power of three' and 5^6 as 'five raised to the power of six'. In actual practice the second power of a base is generally referred to as the square of that base, and the third power of a base as the cube of that base, so that 7^2 is read as 'seven squared' and 16^3 as 'sixteen cubed'.

Let us now proceed to the special case of the base being equal to 10, so that $10^2 = 100$, $10^3 = 1\,000$ and so on. 10^1 implies that there is just one value of the base 10 in the nominally continued product, and so $10^1 = 10$. Let us see if we can deduce the value of 10^0. We observe that

$$10^4 = 10 \times 10 \times 10 \times 10 = 10\,000$$
$$10^3 = 10 \times 10 \times 10 \qquad = 1\,000$$
$$10^2 = 10 \times 10 \qquad\qquad = 100$$

and
$$10^1 = 10 \qquad\qquad\qquad = 10.$$

In the column on the extreme left of this series of statements, the power of 10 is one less than that immediately above it. In the column on the extreme right a value is one-tenth of that immediately above it. Hence we deduce that if we continue the left-hand column downwards our next value is 10^0. If we continue the right-hand column downwards our next value is 1. Hence we conclude that $10^0 = 1$.

There are a considerable number of occasions in mathematics when we require to treat a collection of things as one complete thing, i.e. an entity. It has already been stated that a convenient way of doing this is to put the collection within aggregation signs, such as brackets. For example, (10^2) means that we wish to treat 10^2 as 100. If we put a value called a coefficient immediately proceeding an aggregation sign it is an indication that the entity within the aggregation signs is to be multiplied by that coefficient. In which case, $4(10^2)$ is evaluated as $4 \times 100 = 400$.

Now let us consider the integer we write as 2 847. It is the sum of two thousands, eight hundreds, four tens and seven units.

Setting out the previous statement in tabular form, we have

$$2\,000 = 2 \times 1\,000 = 2(10^3)$$
$$800 = 8 \times 100 = 8(10^2)$$
$$40 = 4 \times 10 = 4(10^1)$$
$$7 = 7 \times 1 = 7(10^0)$$

A total of   2 847

and we can write

$$2\,847 = 2(10^3) + 8(10^2) + 4(10^1) + 7(10^0)$$

The digits of the number 2 847 are the coefficients of powers of ten, those powers of ten being in descending order, terminating with the coefficient of 10^0. A system of numerical notation based upon powers of 10 is called a *denary system of numbers*. Our everyday means of counting therefore uses the denary system of numbers.

For the convenience of the development which follows, let us repeat the list of powers of 10.

$$10^4 = 10 \times 10 \times 10 \times 10 = 10\ 000$$
$$10^3 = 10 \times 10 \times 10 \qquad = \ 1\ 000$$
$$10^2 = 10 \times 10 \qquad\qquad = \qquad 100$$
$$10^1 = 10 \qquad\qquad\qquad = \qquad 10$$
$$10^0 = \qquad\qquad\qquad\qquad = \qquad 1.$$

Now let us continue the series further. In the extreme left hand vertical column the power of ten is one less than that immediately above it. In the extreme right-hand column a number is one-tenth of that immediately above it. Hence we conclude that 10^{-1} is equal to one-tenth, which, as a vulgar fraction, we write as $\frac{1}{10}$. Similarly, 10^{-2} is equal to one-hundredth, or $\frac{1}{100}$, 10^{-3} is one-thousandth, or $\frac{1}{1\ 000}$, and so on.

A *decimal fraction* is a fraction associated with the denary system of numbers. When we introduced directed numbers we used zero as a place holder which separated positive integers from negative integers. In a somewhat similar manner, we use a decimal marker to separate whole denary numbers from decimal fractions. The indication for a decimal marker varies in different parts of the world. The recommended practice in the U.K. is to use a point and call it the decimal point. (In many continental countries a comma is used as the decimal marker.)

The preferred position for the decimal point is midway vertically in the general alignment of figures, as for example with the number 87·42, which represents a total of eight tens, seven units, four tenths and two hundredths. Some printing machinery, particularly some typewriters, do not have a facility for printing the point midway. In which case 'the point on the line' is tolerated, such as with 87.42. However, the point on the line should be avoided if at all possible, as sometimes it is used as an indication for a multiplying operation.

It is also recommended practice that if a quantity is less than unity, a zero should prefix the decimal point to bring the decimal point into greater prominence. Hence we should use indications such as 0·437 and not ·437 or .437. Consequently, a decimal fraction can be recognised immediately by observing a zero followed by a decimal marker. It is repeated that in the United Kingdom the decimal marker is the decimal point, but other countries of the world use other indications. European practice favours the comma.

With decimal fractions the number of digits, including zeros, *after the decimal marker*, is called the *number of decimal places*. As examples,

0·4   has one decimal place
0·32   has two decimal places
0·30   has two decimal places
0·03   has two decimal places
0·436 has three decimal places.

If the number of decimal places exceeds three, it is recommended that a space be left between the third decimal place and the fourth, between the sixth and the seventh, between the ninth and the tenth, and so on, so that digits are grouped in sets of three.
For example,

$$0.457\ 006\ 21 \text{ has eight decimal places.}$$

However, a common practice is that if a particular set of calculations never requires more than four decimal places at any stage, no space is left between the third and fourth decimal place, so that

$$0.407\ 4, \text{ which may occasionally be written as } 0.4074$$
$$\text{has four decimal places.}$$

The digit in the first decimal place represents tenths, the digit in the second decimal place represents hundredths, the digit in the third decimal place represents thousandths, and so on, so that

$$0.abc \quad \text{represents} \quad a\left(\frac{1}{10}\right) + b\left(\frac{1}{100}\right) + c\left(\frac{1}{1\ 000}\right) \quad \text{and so on.}$$

Since many countries use the comma as the decimal marker, so that 4,537 would represent the value we write as 4.537, it is recommended that we discard the use of the comma to separate thousands from hundreds and millions from thousands. Just as we group digits in threes to the right of the decimal point, it is recommended that we group digits in threes to the left of the decimal point, implied, (as with 15 743), or written, (as with 5 452.64). However, just as with decimal fractions, a common practice is that if none of a set of values exceeds 9 999, no space is left between the thousands and the hundreds.

The author prefers consistency. It took him just a little while to become accustomed to following the recommendation completely, and he now always groups digits in threes, with values such as

$$4.157\ 375$$
$$18.158\ 4$$
$$472.167\ 32$$
$$537\ 146.14$$
$$28\ 467\ 853.7$$

and
$$4\ 512.12.$$

1.4.2
The conversion of a vulgar fraction to a decimal fraction

When defining a vulgar fraction it was stated that the vinculum which separated the numerator from the denominator can be interpreted as an indication for a division operation, so that

$$\tfrac{2}{5} \text{ can be interpreted as } 2 \div 5.$$

Let us perform this division on a suitable calculator, by clearing off all previous work, depressing 2 followed by ÷, followed by 5, followed by =. The input is therefore

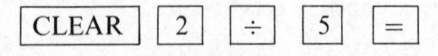

$$\boxed{\text{CLEAR}} \quad \boxed{2} \quad \boxed{÷} \quad \boxed{5} \quad \boxed{=}$$

in sequence, and the answer is revealed in the display as 0·4. Two interesting facts emerge. The calculator automatically provides the zero previous to the decimal point and the display does not use all the places of which the calculator is capable. There are no zeros after the digit 4. Let us proceed with evaluating 8 ÷ 125, which is the equivalent of $\frac{8}{125}$. After clearing the calculator of previous working, an input of

$$\boxed{\text{CLEAR}} \quad \boxed{8} \quad \boxed{÷} \quad \boxed{125} \quad \boxed{=} \,,$$

in sequence, gives a display of 0·064. Once again the calculator has not used all the places of which it is capable.

Let us determine the equivalents of $\frac{2}{5}$ and $\frac{8}{125}$ by formal arithmetic.

$$
\begin{array}{r}
5\,\underline{\smash{\big)}\,2}\ \ 0{\cdot}4 \\
0 \\
\hline
20 \\
20 \\
\hline
\end{array}
\qquad
\begin{array}{r}
125\,\underline{\smash{\big)}\,8}\ \ 0{\cdot}064 \\
0 \\
\hline
80 \\
00 \\
\hline
800 \\
750 \\
\hline
500 \\
500 \\
\hline
\end{array}
$$

We say that a value such as 0·4 is a decimal fraction which terminates after the first decimal place. In a similar manner, the decimal fraction 0·064 terminates after the third decimal place. If we evaluated $\frac{1}{64}$, we should find the answer to be 0·015 625, which terminates after the sixth decimal place. Decimal fractions obtained from vulgar fractions, and which terminate after a certain number of decimal places, are precise equivalents.

Now let us determine the decimal equivalent of the vulgar fraction $\frac{2}{7}$. Using a calculator, depending on the number of places available in the display, the answer to 2 ÷ 7 will be associated with the first few digits of

$$0{\cdot}285\ 714\ 285\ 714\ 285\ 714 \quad \text{(non-terminating)}.$$

The display will use all the places of which the calculator is capable, commencing with the decimal point. The display is restricted by the number of places of which the calculator is capable. The fact that the equivalent uses all the places in the display is a fairly good indication, particularly with a display of eight digits or more, that the decimal fraction is non-terminating.

An interesting point of the decimal fraction equivalents of the vulgar fractions for sevenths is the repeating of groups of six digits extracted from the sequence

$$14285714285714 \quad \text{etc.}$$

$$\tfrac{4}{7} = 0.571\ 428\ 571\ 428$$

$$\tfrac{6}{7} = 0.857\ 142\ 857\ 142$$

Only a proportion of vulgar fractions can be expressed precisely as decimal fractions. The others cannot be expressed precisely because the decimal fractions are non-terminating no matter how many decimal places are used. Of these latter fractions, some repeat digits at particular intervals, in a somewhat similar manner to the 'sevenths'.
For instance

$\tfrac{1}{3} = 0.333\ 333\ 333, \ldots$, the digit 3 repeating continually
$\tfrac{4}{11} = 0.363\ 636\ 363, \ldots$, the digits 36 repeating continually
$\tfrac{5}{7} = 0.714\ 285\ 714\ 285, \ldots$, the digits 714 285 repeating
continually
$\tfrac{11}{30} = 0.366\ 666, \ldots$, after the digit 3 the digit 6 repeats
continually

Decimal fractions of this type are called *recurring decimal fractions.* To avoid writing out a long list of digits a distinct dot is placed over the digits which repeat.
Hence,

$\tfrac{1}{3} = 0.\dot{3}$, read as zero point three recurring,
$\tfrac{4}{11} = 0.\dot{3}\dot{6}$, read as zero point three six, both recurring,
$\tfrac{5}{7} = 0.\dot{7}14\ 28\dot{5}$, read as zero point seven one four two eight five,
all recurring,
$\tfrac{11}{30} = 0.3\dot{6}$, read as zero point three six, six only recurring.

A decimal fraction which includes a recurring element, indicated by distinct dots over one or more digits, will have a precise vulgar fraction equivalent, as will be shown in the next article.
In a previous article dealing with vulgar fractions, we used a method involving the lowest common denominator. At that time it was stated that a more simple method would be demonstrated after a consideration of decimal fractions. The actual problem was to range the vulgar fractions.

$$\tfrac{2}{5} \quad \tfrac{7}{16} \quad \tfrac{1}{4} \quad \tfrac{4}{9} \quad \text{and} \quad \tfrac{1}{3}$$

in order of magnitude. The most rapid method of providing a result is to establish their decimal equivalents. Using a calculator,

$$\frac{2}{5} = 0.4$$
$$\frac{7}{16} = 0.4375$$
$$\frac{1}{4} = 0.25$$
$$\frac{4}{9} = 0.4444 \ldots$$
$$\frac{1}{3} = 0.3333 \ldots$$

and by direct inspection we see that

$$\frac{1}{4} < \frac{1}{3} < \frac{2}{5} < \frac{7}{16} < \frac{4}{9}.$$

1.4.3
The conversion of a decimal fraction to a vulgar fraction

Since a decimal fraction in the form

$$0.453$$

is an indication of the sum of four tenths, five hundredths, three thousandths, and so on, the algorithm for converting a *terminating* decimal fraction to its vulgar fraction equivalent is:

Step 1.   Erect a vulgar fraction where the numerator is the digit or digits of the decimal fraction, disregarding the decimal point and any zeros immediately after the decimal point.

Examples: For 0.4, the numerator is 4.
For 0.064, the numerator is 64.
For 0.312 5, the numerator is 3 125.

Step 2.   For the denominator use unity followed by a number of zeros equal to the number of decimal places.

Examples: For 0.4, the denominator is 10.
For 0.064, the denominator is 1 000.
For 0.312 5 the denominator is 10 000.

Step 3.   Reduce the vulgar fraction obtained from steps 1 and 2, if possible, to its simplest form by cancelling.

Examples:
$$0.4 = \frac{4}{10} = \frac{2}{5}.$$

$$0.064 = \frac{64}{1\ 000} = \frac{8}{125}$$

$$0.312\ 5 = \frac{3\ 125}{10\ 000} = \frac{5}{16}$$

If the decimal fraction has a recurring element, then either the whole of the figures recur, or the recurring element occurs after one or more digits after the decimal point. If all the digits recur, the algorithm is

Step 1. Erect a vulgar fraction where the numerator is the digit or combination of digits which recurs.

Examples: For $0\dot{\cdot}3$, the numerator is 3.
For $0\dot{\cdot}3\dot{6}$, the numerator is 36.
For $0\dot{\cdot}57\dot{1}\ \dot{4}2\dot{8}$, the numerator is 571 428.

Step 2. For the denominator use the number of nines equal to the number of digits in the numerator.

Examples: With a numerator of 3, the denominator is 9.
With a numerator of 36, the denominator is 99.
With a numerator of 571 428, the denominator is 999 999.

Step 3. Reduce the vulgar fraction formed from steps 1 and 2 to its simplest form by cancellation. (No matter how complicated it looks it will cancel.)

Examples:

$$0\dot{\cdot}3 = \frac{3}{9} = \frac{1}{3}$$

$$0\dot{\cdot}3\dot{6} = \frac{36}{99} = \frac{4}{11}$$

$$0\dot{\cdot}57\dot{1}\ \dot{4}2\dot{8} = \frac{571\ 428}{999\ 999} = \frac{4}{7}$$

If the number of digits is large, then since we are virtually certain that the end result will be a vulgar fraction involving simple integers, it may be advisable, before cancelling direct, to invert the vulgar fraction and divide using a calculator. In this case

$$\frac{999\ 999}{571\ 428} \quad \text{is precisely} \quad 1\cdot75 = 1\tfrac{3}{4} = \tfrac{7}{4},$$

and by inversion, the equivalent vulgar fraction is $\tfrac{4}{7}$.

If the decimal fraction has a recurring element after one or more of the digits which follow the decimal point, the fraction should first be multiplied by 10, 100, 1 000, etc. until the recurring element immediately follows the decimal point. For example, with 0·36, we first multiply by 10 to obtain 3·6. We can now proceed as previously to determine that 0·6 = $\frac{2}{3}$. Hence 3·6 = 3$\frac{2}{3}$. We now divide by the 10 we originally used as a multiplier to bring it back to its original value.

$$3\tfrac{2}{3} \div 10 = \tfrac{11}{3} \div 10 = \tfrac{11}{3} \times \tfrac{1}{10} = \frac{11 \times 1}{3 \times 10} = \tfrac{11}{30},$$

and $0\cdot3\dot{6} = \tfrac{11}{30}.$

Unless a decimal fraction is exact, or has a recurring element, there is no simple method of determining its equivalent as a vulgar fraction. This is a disadvantage of decimal fractions, but one which is heavily outweighed by their advantages. It might be mentioned, in passing, however, that there are books available for reference, dealing with gear ratios, which express the decimal equivalents of large numbers of vulgar fractions to many places of decimals.

1.4.4
Computations with denary numbers

In case a calculator is not available, it would be advisable to make reference to computations with denary numbers using the methods of what may be termed traditional arithmetic.

Addition is facilitated if the numbers to be added are written down so that decimal points, stated or implied, are aligned vertically. Individual columns of digits are then added, commencing with the extreme right hand column. If the total of a column of digits is from zero to nine inclusive, the total is put down under its particular column in the space reserved for the answer. If a total is in excess of 9, the last digit is put down under its particular column and the digit(s) to its left is/are carried forward to the next column(s). For example, if the total of a column is 27, the digit 7 is put down under its column and the digit 2 is carried forward to the next column. If the adding of a very large number of digits produced a total such as 340, then 0 would be written under the column of digits which has been added, 4 carried forward to the next column on the left and 3 to the next column on the left of the column to which 4 has been carried forward. It is a commendable practice to write digits which are carried forward under appropriate columns. Some people prefer to add a column of digits from top to bottom, others prefer to add from bottom to top. Our reader should continue the practice to which he or she has become accustomed, but a check on the total of a column of digits should always be made by totalling in the reverse direction to that originally made. A typical layout of addition is shown on page 59.

Computations with denary numbers

Addition

```
       47·172 05
        0·004 1
      200
   37 436·379 71
      847·072 93
   38 530·628 79
    1 12   211
```

Subtraction

```
   1 09    64
   2 1Ø5·47̸5 82
     348·206 9
   1 757·268 92
```

Multiplication

$152·71 \times 0·047\ 6$

$2 + 4 = 6$ decimal places in answer.

```
           15 271
              476
           91 626
        1 068 97
        6 108 4
        7 268 996
              111
```

Check

$1 + 5 + 2 + 7 + 1 = 16$ $1 + 6 = 7$
$4 + 7 + 6 = 17$ $1 + 7 = 8$
$7 \times 8 = 56$ $5 + 6 = 11$ $1 + 1 = 2$
$7 + 2 + 6 + 8 + 9 + 9 + 6 = 47$
$4 + 7 = 11$ $1 + 1 = 2$

Adjust product to give 6 decimal places.

Answer: 7·268 996

Division

$2·753 \div 0·014\ 72$, to 2 dec. pl.

$= 275·3 \div 1·472$

$=$ between $275 \div 1 = 275$

and $275 \div 2 = 137·5$

```
   1 472   2 753 | 187 024
           1 472
           12 810
           11 776
            10 340
            10 304
             3 600
             2 944
               6 560
               5 888
```

Adjust quotient to 2 decimal places.

Answer: 187·02

Square Root

$\sqrt{827·15}$ to 3 decimal places.

```
     2  | 8|27·15 | 28·760 2
        | 4
    48  | 427
     8  | 384
   567  | 4 315
     7  | 3 969
  5 746 |   34 600
     6  |   34 476
 57 520 |     12 400
     0  |         0
575 202 |    1 240 000
     2  |    1 150 404
```

Adjust answer to 3 decimal places.

$\sqrt{827·15} = 28·760$

With subtraction, our reader should recall that a subtrahend is deducted from a minuend. If the modulus of a subtrahend is larger than the modulus of the minuend the modulus of the minuend is deducted from the modulus of the subtrahend, and the modulus of the result prefixed by a negative sign. The minuend and the subtrahend are written down so that decimal points, stated or implied, are in vertical alignment. Subtraction is undertaken column by column, commencing with the column on the extreme right. If a subtraction in a particular column results in zero or a positive digit from 1 to 9 inclusive, that result is put down in the space reserved for the answer. If the result is nominally negative, 10 can be 'borrowed' from the next column on the left, and the appropriate digit in that column reduced by 1. The borrowing of 10 increases the digit in the minuend by 10 so that deducting the digit in the subtrahend produces a positive result, which is then written down under the appropriate column in the space reserved for the answer. It is a commendable practice when 10 has been borrowed to strike through the appropriate digit and reduce it by 1.

If the digit in the minuend from which 10 is borrowed is zero, it is struck through and replaced by 9, and the digit in the next column of the minuend is struck through and reduced by 1. A typical layout for subtraction is shown on page 59. A subtraction should always be checked by adding the remainder to the subtrahend to see if the total agrees with the minuend.

With multiplication it is very convenient to disregard decimal points initially and multiply an integer by an integer. The number of decimal places in the product is the sum of the decimal places of the multiplicand and the multiplier. For instance, with

$$37 \cdot 475 \times 5 \cdot 007\ 3$$

we multiply 37 475 by 50 073 and adjust the product to provide $3 + 4 = 7$ decimal places,

while with $$285 \cdot 72 \times 0 \cdot 001\ 456$$

we multiply $$28\ 572 \quad \text{by} \quad 1\ 456$$

and adjust the product to provide $2 + 6 = 8$ decimal places.

Let us use $285 \cdot 72 \times 0 \cdot 001\ 456$ referred to previously as an example of multiplication. The integers are arranged in columns, with the last integer of the multiplicand and of the multiplier in vertical alignment. In our case we have the layout

$$28\ 572$$
$$1\ 456.$$

The product is the sum of 28 572 × 6, 28 572 × 50, 28 572 × 400 and 28 572 × 1 000, which, for convenience, is set out as follows:

$$\begin{array}{r} 28\ 572 \\ 1\ 456 \\ \hline 171\ 432 \\ 1\ 428\ 60 \\ 11\ 428\ 8 \\ 28\ 572 \\ \hline 41\ 600\ 832 \\ \hline 1\ 221 \end{array}$$

With 285·72 × 0·001 456 there will be 2 + 6 = 8 decimal places in the product, and hence

$$285 \cdot 72 \times 0 \cdot 001\ 456 = 0 \cdot 416\ 008\ 32.$$

There is a rather delightful way of checking a product when two integers are multiplied together. The algorithm is as follows:

Step 1. Add the digits of the multiplicand, and if necessary, keep adding till a single digit is obtained. For example, with 28 572, 2 + 8 + 5 + 7 + 2 = 24 and 2 + 4 = 6.

Step 2. Repeat the process of step 1 with the multiplier. For example, with 1 456, 1 + 4 + 5 + 6 = 16, and 1 + 6 = 7.

Step 3. Multiply the single digit of step 1 by the single digit in step 2. If the product is greater than 9, add the digits again, and again if necessary, to produce a single digit. In our case 6 × 7 = 42 and 4 + 2 = 6.

Step 4. The result of adding the digits of the product, with repeated additions to get a single digit if necessary, should agree with that obtained from step 3.

For example, 4 + 1 + 6 + 0 + 0 + 8 + 3 + 2 = 24, and 2 + 4 = 6, and there is good reason to believe that the statement

$$28\ 572 \times 1\ 456 = 41\ 600\ 832 \text{ is correct.}$$

Our reader is now recommended to multiply 152·71 by 0·047 6 and to check that the working agrees with that on page 59.

With division, it is convenient initially to arrange that the divisor lies between zero and 10 by moving the decimal point, given or implied, a number of decimal places, either to the right or to the left. The decimal point of the dividend is then moved the same number of places in the same direction. For example, we adjust 413 672·4 ÷ 857 to 4 136·724 ÷ 8·57, and we can appreciate that the answer will be in the vicinity of 500. Similarly

$$4 \cdot 827\ 4 \div 0 \cdot 052\ 4 = 482 \cdot 75 \div 5 \cdot 24$$

and we can appreciate that the answer will be in the vicinity of 100.

The digits of the quotient can be obtained by neglecting the decimal points of both dividend and divisor, and operating on integers. For example, with

$$2\ 851 \cdot 174\ 72 \div 954 \cdot 37$$

we adjust to

$$28 \cdot 511\ 747\ 2 \div 9 \cdot 543\ 7$$

and we appreciate that the answer will be in the vicinity of 3. We now find the digits of the quotient by dividing

$$285\ 117\ 472 \quad \text{by} \quad 95\ 437.$$

We commence by setting-up the layout

$$95\ 437\ \overline{\big|\ 285\ 117\ 472\ \big|}$$

and divide the divisor of 95 437 into sufficient commencing digits of the dividend to give a quotient of 1 to 9 inclusive. In this case the quotient will be 2. We then multiply the divisor by the quotient and place it beneath the first digits of the dividend, subtract, and obtain a remainder. Thus far we have

$$
\begin{array}{r}
95\ 437\ \big|\ 285\ 117\ 472\ \big|\ 2 \\
190\ 874 \\
\hline
94\ 243
\end{array}
$$

We now bring down the next digit in the dividend, to obtain

$$
\begin{array}{r}
95\ 437\ \big|\ 285\ 117\ 472\ \big|\ 2 \\
190\ 874 \\
\hline
94\ 243\ 4
\end{array}
$$

We now divide 95 437 into the new remainder of 942 434 to obtain a quotient of 9, which is placed after the first quotient of 2. The divisor of 95 437 is multiplied by the last quotient of 9 and the product is deducted from the previous remainder. And so, thus far, we have

$$
\begin{array}{r}
95\ 437\ \big|\ 285 \cdot 117\ 472\ \big|\ 29 \\
190\ 874 \\
\hline
94\ 243\ 4 \\
85\ 893\ 3 \\
\hline
8\ 350\ 1
\end{array}
$$

We now continue by bringing down the next digit of the dividend, or a zero if all the digits of the dividend have been used, and continue until we have sufficient digits in the quotient. Let us presume that the quotient is required to three decimal places. Since the answer is about 3, we shall require 5 digits in the quotient so that we can express the answer to the nearest third decimal place. The full working is as follows:

$$95\ 437\ \overline{\smash{\big)}\ 285\ 117\ 472}\ \ 29\ 874 + \text{other digits}$$

$$
\begin{array}{r}
190\ 874 \\
\hline
94\ 243\ 4 \\
85\ 893\ 3 \\
\hline
8\ 350\ 17 \\
7\ 634\ 96 \\
\hline
715\ 212 \\
668\ 059 \\
\hline
47\ 153\ 0 \\
38\ 174\ 8 \\
\hline
8\ 978\ 2
\end{array}
$$

Hence $2\ 851\cdot147\ 2 \div 954\cdot37 = 2\cdot987$, to three decimal places.

Our reader is now recommended to establish $2\cdot753 \div 0\cdot014\ 72$ and to check the working with that shown on page 59. A slightly different check has been illustrated, by determining two values, one higher and one lower than the true value.

If a number is multiplied by itself we say that the number is squared. For example, 7 squared is $7 \times 7 = 49$. Conversely, the *square root* of a number is that number which, when multiplied by itself, gives the original number. The square root of 49 is therefore 7. To be more precise, since the product of two negative numbers gives a positive number, there are two square roots of 49, $+7$ and -7. The symbol for square root is $\sqrt{}$, hence $\sqrt{49}$ can be either $+7$ or -7, and we express this symbolically as

$$\sqrt{49} = \pm 7$$

which we read as 'root 49 equals plus or minus 7'. If we use the word root without qualification, a square root is implied. Furthermore, unless there are specific reasons to include the negative value, the positive square root is inferred. The extraction of a square root on a calculator which has a square rooting function is extremely simple, by the input of

and the answer is displayed in a fraction of a second. However, our reader should be aware of a means of determining a square root by an arithmetical method. Let us take, for example, the determination of $\sqrt{1\ 026\cdot561\ 6}$.

We first rearrange the number into pairs of digits in each direction from the decimal point, stated or implied, which could leave single digits at each end. For instance, in this case we have 10|26·56|16. With 572·167 we would have 5|72·16|7 and with 42 784 we would have 4|27|84.

Let us return to our original number, arranged in pairs of digits, viz:

$$10|26·56|16$$

We set up the pattern

$$\boxed{10|25·56|16}$$

We now decide on a digit which is the nearest perfect square below the extreme single digit or pair of digits on the left. In this case we have a pair of digits, giving the number 10. The nearest perfect square below 10 is 9, which is the square of 3, and we commence the answer by placing 3 at the top right of the pattern. We now deduct the perfect square to establish a remainder of 1 and bring down the next pair of digits. The pattern, thus far, is:

$$\begin{array}{r} \boxed{10|26·56|16}\ 3 \\ 9 \\ \hline 1\ 26 \end{array}$$

We now progress the pattern to the stage:

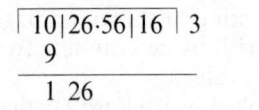

At *A* we place a number which is double our result thus far. In our case we shall have $2 \times 3 = 6$. At *B* we place a digit such that $AB \times B$ is the closest value below the remainder of 126. In our case, $B = 2$, since $62 \times 2 = 124$. The value of *B*, 2 in this case, is the next digit of the answer. The pattern has now developed to the stage

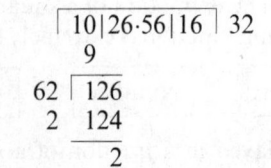

We now theoretically bring down the decimal point and at this stage we put the decimal point in the answer. The decimal point can be disregarded henceforward. We now bring down the next pair of digits and continue with the pattern:

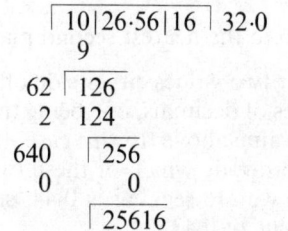

$$\overline{\ 10|26{\cdot}56|16\ }\ 32{\cdot}0$$
$$9$$
$$62\ \overline{\ 126}$$
$$2\quad 124$$
$$CD\ \overline{\ 256}$$
$$D$$

We now put at C a value of twice the digits in the answer thus far, neglecting the decimal point. In our case, $C = 32 \times 2 = 64$. At D we place a value so that $CD \times D$ is the nearest value below the remainder. In this case we have $D = 0$. The digit 0 is the next digit of the answer. We deduct $CD \times D$, in this case 0, and establish a remainder.

We now bring down the next pair of digits from the original number, and progress to:

$$\overline{\ 10|26{\cdot}56|16\ }\ 32{\cdot}0$$
$$9$$
$$62\ \overline{\ 126}$$
$$2\quad 124$$
$$640\ \overline{\ 256}$$
$$0\qquad 0$$
$$\overline{\ 25616}$$

The process of 'doubling the answer so far and finding the next digit of the answer' continues thenceforward until the remainder becomes zero, in which case the original number was a perfect square, or until sufficient digits are obtained in the answer to satisfy the accuracy required. In our case we double 320 to obtain 640. Thence 6 404 × 4 = 25 616. The digit 4 is the final digit in the answer because the remainder is zero. The number 1 026·561 6 is a perfect square, being the square of 32·04.

Our reader is now recommended to determine 827·15 to three places of decimals and to check the result with the working shown on page 59.

1.4.5
Decimal places and significant figures

It has already been stated that the decimal fraction equivalents of certain vulgar fractions are non-terminating, no matter how many decimal places are used. Fortunately, when using non-terminating decimal fractions, in most cases the use to which we put them is such that they do not have to be stated precisely, and an approximation is quite acceptable. There are two distinct methods of making such an approximation, by

    (a)   stating a number of decimal places.
    (b)   stating a number of significant figures.

Recapitulating a previous definition, the number of decimal places is the number of digits (*including* zeros should they occur) which *follow* the decimal point. As examples

        0·4 has one decimal place
        0·362 has three decimal places
        0·070 has three decimal places.

Let us presume that we are given an instruction that sufficient accuracy will be obtained if a decimal fraction is given to the nearest specific decimal place. Let us not, at present, question the validity of the instruction, but treat it as a command to be obeyed. A typical example is:

        Express 0·437 6 to the nearest second place of decimals.

We now have to put two values either side of 0·437 6 both of which are stated to two places of decimals, one being the nearest value below, the other the nearest value above. In this case, the values are 0·43 and 0·44. We now have to decide which of these two values is the nearer to 0·437 6. The value we are seeking is 0·44, since 0·437 6 is greater than the 'half-way' value of 0·435.

Answer:

        0·437 6 = 0·44 reduced to two decimal places.

There is no answer to an illogical question of the type 'express 0·435 to an accuracy of two decimal places'. Such a question should never be asked. The value should be left as it is. The value of 0·435 is just as close to 0·44 as it is to 0·43. For the purposes of this article, and work which follows, our reader is asked to dismiss completely from his mind any reference to 'if the last digit is 5, increase the previous digit by one'.

The algorithm for expressing a decimal fraction to a given number of places is:

Step 1.   Delete from the given decimal fraction, starting from the right-hand side, all digits in excess of the required number of decimal places.

        Example: If 0·508 36 has to be expressed to two decimal places, we delete the digits 6, 3, and 8 to give 0·50.

Step 2. Add one to the last digit of the answer obtained from step 1. In this case we add one to the last digit of 0·50 to obtain 0·51.

Step 3. The answer is effected by a choice between the values obtained in steps 1 and in step 2, and will be that value which is the nearer to the original value. A guide is given with a determination of the halfway value.

In the example, the half way value is 0·505 and since 0·508 36 is greater than 0·505, it is nearer to 0·51 than it is to 0·50.

Hence 0·508 36 expressed to two decimal places is 0·51. Integers remain unchanged when values are expressed to the nearest particular place of decimals. For example:

$$507·596\ 2 = 507·60, \text{ to two decimal places.}$$

Let us suppose that a coil of copper wire in a factory had to be lifted by a crane, and that the cranes available had lifting capacities of 100 kg, 500 kg and 1 000 kg. To ensure safety, the coil was weighed, and the weighbridge operator said it was 'between six and seven hundred kilograms'. This information is sufficient to decide which of the cranes to use: it would be dangerous to use either of the two cranes of lowest capacity. If someone now came along to decide the monetary value of the coil of copper wire, the difference in the value of 600 and 700 kilograms of copper would be quite appreciable. He would probably ask for the weight to be given to the nearest kilogram. Depending upon the use to which information is put, different circumstances require different degrees of accuracy, or *significance*.

Note: To the purist, the previous statements are probably disturbing. He will aver that weight is a force, and the unit of force is the newton, not the kilogram. Of course he is right. What the weighbridge operator has quoted is a mass. Nevertheless, everyday non-technical usage will still refer to 'weights' in kilograms instead of newtons, and the introduction of SI units will not immediately lead to changes. Let us exercise a little sympathetic patience.

For the purpose of mathematics, the number of significant figures is the number of figures obtained by counting rightward from the leftmost non-zero figure inclusive. The position of a decimal point is not considered when determining the number of significant figures. Thus

72 has two significant figures
72·0 has three significant figures
0·7 has one significant figure
0·005 7 has two significant figures
0·001 070 has four significant figures
1·005 has four significant figures

It should be carefully noted that only the zeros prior to the first non-zero figure are discarded. Any zeros after the first non-zero figure must be included.

The number of decimal places is no real indication of accuracy. However many decimal places are used, the answer obtained in the form of a decimal fraction for the value of $2 \div 7$ will never be as accurate as the vulgar fraction $\frac{2}{7}$. Consequently, a student should never be afraid of expressing an answer in the form of a vulgar fraction. It can often be more accurate than one using a multiplicity of decimal places. Degrees of accuracy are indicated by significant figures, not by places of decimals. Answers are sometimes requested to a particular number of decimal places for convenience, not as a general indication of accuracy.

The performing of a particular set of computations will invariably lead to a result of greater accuracy than is warranted by the original data, and it would be logical to round-off to an appropriate number of significant figures. This topic is developed further in the next article, but if rounding off has to be undertaken, it should be accomplished in a single step, and never in a series of steps. The rounding-off should be to the nearer value, up or down, which in many cases is self-evident.

Example:

Round off to three significant figures:

(a) 7·857   (b) 0·017 62
(c) 234 565   (d) 0·018 437

(a) 7·857 is nearer to 7·86 than it is to 7·85.
(b) 0·017 62 is nearer to 0·017 6 than it is to 0·017 7.
(c) 234 565 is nearer to 235 000 than it is to 234 000.
(d) 0·018 437 is nearer to 0·018 4 than it is to 0·018 5.

Answers:

(a) 7·86   (b) 0·017 6
(c) 235 000   (d) 0·018 4

In certain cases the value may fall midway, and rounding off could be in either direction. Suppose we had to round off 7·45 to two significant figures. 7·45 is just as near to 7·4 as it is to 7·5. In such cases, we round off to the even value. 7·45 rounded off to two significant figures is the even value of 7·4 and not the odd value of 7·5. By similar reasoning, 7·55 to two significant figures is 7·6. This rule (which is included in BS 1957) may seem peculiar to a student who has previously been instructed, 'If the last figure is 5, round it upwards'.

If we use the rounding off rule recommended by BS 1957, as many mid-values are rounded up as are rounded down, so that when we total rounded off values we do not introduce a significant amount of bias. Rounding by rule:

> 0·25 to one significant figure is 0·2
> 0·435 to two significant figures is 0·44
> 7·355 to three significant figures is 7·36
> 7·245 to three significant figures is 7·24

1.4.6
The feasibility and the accuracy of numerical answers

Our reader will often use mathematics to provide an answer to a scientific problem. Having obtained an answer our reader should ask two questions:

(a) is the value feasible?
(b) is the accuracy of the answer realistic?

At the time that the manuscript of this book was written the change-over from Imperial to SI units was creating a certain amount of difficulty in appreciating whether or not an answer was realistic. At that time, if a scientific problem called for the speed of car, it was easily appreciated that 30 miles per hour was the limit in a built-up area and 70 miles per hour on a motorway. A speed of 700 miles per hour for a car would be known immediately to be in error. There were many hazy impressions about realistic speeds of cars in terms of the practical metric unit of kilometres per hour, and even hazier impressions about realistic speeds in terms of the basic derived unit of the metre per second. The difficulties of appreciating realistic values in terms of metric units will gradually disappear as people 'think metric from the start' rather than 'convert from Imperial to metric'. However, the feasibility of answers is not constrained to metric units. Let us imagine that our reader was set a problem which required a determination of the time it took for an electric kettle to come to the boil and an answer was obtained in the vicinity of 150 minutes. It might well be that our reader may not be in a position to know precisely how long it would take, but everyday experience would suggest that a correct answer should not be in excess of 15 minutes. An answer of about 150 minutes is quite illogical and must be rejected as not being feasible.

In which case, it is necessary to determine where the error or errors occurred. In the first place, the author strongly recommends that a check should be made as to whether the values in the original data have been correctly introduced into the problem. It is all too easy when copying data to write down what was thought to be there rather than what was actually there. In some cases the next step might well be a check of a formula being used, and some indication of the accuracy of that formula can be obtained from a check on the units. If the answer is to be a length, then when the units are substituted in the formula, the manipulation of these units should provide a unit of length.

Presuming that the original data has been used correctly, that the formula is correct and that correct values have been substituted in the formula, the next step is to check the computation.

When a computation is made it should be accompanied by an estimate of the final result using an approximation or approximations. This is often colloquially referred to as a rough check. The word rough is somewhat unfortunate, since if the approximations are chosen with care the check can often be surprisingly accurate.

One useful method of approach is to perform two simple computations, one which will result in a value which must be lower than the true value, the other which must be higher, in which case the true value will lie between the two approximate values. For example, to provide a rough check for

$$8 \cdot 473 \times 204 \cdot 7$$

the true answer must be above

$$8 \times 200 = 1\ 600$$

and must be below

$$9 \times 210 = 1\ 890.$$

A value such as 1 734·423 1 would be quite feasible, but a value such as 134·423 1 is obviously incorrect. The error is probably a misreading of the digital display of a calculator.

If the computation involves a chain of operations, the 'too high, too low' approach would probably be far too time consuming, and a better approach could well be to make just one rough check, and chose approximations which tend to balance each other, such as approximating alternate values upwards and downwards. Let us illustrate this principle with worked examples.

Example:

A value has to be computed for $0 \cdot 785\ 4\ d^2 h$,

where $d = 9 \cdot 4$ and $h = 21.6$.

Is the answer 1 499 realistic?

Apply a rough check

$$\begin{aligned} \tfrac{3}{4} \times 10^2 \times 20 &= \tfrac{3}{4} \times 100 \times 20 \\ &= 75 \times 20 \\ &= 1\ 500. \end{aligned}$$

Answer:

$$1\ 499 \text{ is realistic.}$$

Example:

The area of a triangle is calculated from the formula:

$$\sqrt{s(s-a)(s-b)(s-c)}$$

where $$s = \frac{a+b+c}{2}.$$

A triangle has sides a, b and c of length 78·6, 117·4 and 154·8 mm. A suggested answer for the area is 45 040 mm^2.

(a) Is the answer realistic?
(b) If not, what is a more realistic value?

(a) Consider a square with a side of length equal to the longest side of the triangle, 154·8 mm. The area of the triangle must be less than the area of this square, which is about

$$150 \times 150 = 22\ 500 \text{ mm}^2.$$

The answer is quite unrealistic.
(b) Apply a rough check, with $a = 80$, $b = 120$ and $c = 150$.

$$s = \frac{a+b+c}{2} = \frac{80+120+150}{2} = \frac{350}{2} = 175.$$

$$s - a = 175 - 80 \ = 95$$
$$s - b = 175 - 80 \ = 55$$
$$s - c = 175 - 150 = 25$$

$$A = \text{about } \sqrt{175 \times 95 \times 55 \times 25}$$
$$= \text{about } \sqrt{180 \times 100 \times 50 \times 25}$$
$$= \text{about } \sqrt{225\ 000}$$
$$= \text{about } 4\ 500$$

Answers:

(a) The value of 45 040 mm^2 is unrealistic.
(b) A value of about 4 500 mm^2 is more realistic.
(Has an error occurred in the position of the decimal point?)

If an answer to a problem is obtained which is precise, that value should be quoted. If the answer involves integers and/or decimals, no more significant figures should be used than is necessary. If an answer is precisely $\frac{3}{7}$, any decimal representation is an approximation, unless a reference is made to recurrence. Even then, complexity has been introduced which is quite unnecessary.

Most computations do not result in precise values. They themselves are approximations. The value which is obtained should be expressed to a realistic number of significant figures. In everyday parlance a final result should be 'near enough', which automatically produces the question in the enquiring mind of 'how near enough is near enough?'

As an initial guide, no answer should be given to a number of significant figures greater than one more than the number of significant figures in the least accurate value in the original data. Unfortunately this latter feature may not be known. In which case it is advisable to treat all data as precise and carry out a computation as accurately as the chosen computation method allows. No intermediary computation should be rounded off, the rounding off should be a final step. If four-figure mathematical tables have been employed, the last digit in four-figure values extracted from those tables is suspect, and the final answer should be rounded off to three significant figures. At our reader's present stage of studies any answer given to three significant figures will be acceptable in more cases than not.

1.4.7
Ratio, proportion and percentage

A *ratio* is a relationship which exists between two quantities of the same kind. If, for instance, we have a first quantity x and a second quantity y then the ratio of x to y is given by $x \div y$. Conversely, the ratio of y to x is given by $y \div x$. A ratio has no unit, it has a purely numerical value.

A ratio can be expressed numerically in a number of ways. One convenient way is to use a vulgar fraction, proper or improper, depending upon the magnitudes of the actual quantities. For example, if we have a first financial amount of £2·40 and a second financial amount of £5·00, then the ratio of the first amount to the second amount is £2·40 ÷ £5·00, and

$$\frac{£2 \cdot 40}{£5 \cdot 00} = \frac{240}{500} = \frac{12}{25}.$$

The ratio could also be expressed as a decimal fraction, 0·48 in this particular case. It could also be expressed on a 'per unit' basis, such as by saying that the ratio of £2·40 to £5·00 is $\frac{12}{25}$ to 1 or 0·48 to 1. In a little while in this article we shall proceed to a 'per hundred' basis, such as by saying that this particular ratio is 48 to 100.

A *proportion* is a statement of the equality of ratios. Using the previous values we can say that

$$\frac{£2.40}{£5.00} = \frac{12}{25} = \frac{0.48}{1} = \frac{48}{100}.$$

If we consider the first two fractions, we can say that £2·40 is to £5·00 as 12 is to 25. Using standard symbols we write this mathematically as

$$£2.40 : £5.00 :: 12 : 25$$

The two inside values are known as means, the two outside values are known as extremes. With a proportionality statement of this type, the product of the means is equal to the product of the extremes. In this case

$$£5.00 \times 12 = £2.40 \times 25 \,(= £60.00).$$

Another way of looking at the proportionality statement would be to say that dividing one item on one side of the proportionality statement by the corresponding item on the other side of the statement produces a constant value. In this case

$$\frac{£2.40}{12} = \frac{£5.00}{25}.$$

A proportionality can be extended to more than two values. For instance, let us imagine three persons contributed £100, £60 and £40 to the purchase of an article, their proportionate contributions could be expressed as

$$£100 : £60 : £40.$$

Dividing each amount by a constant value, provided that value is not zero, does not change the proportionality. We can divide all through by £20, and say that the amounts contributed were in the proportion 5 : 3 : 2. If we now consider that there are ten shares, the contributions can be presented in another way by again dividing by 10, and using $\frac{1}{2} : \frac{3}{10} : \frac{1}{5}$. If the item were to be subsequently sold, for say £150, and it was agreed that the return should be proportional to the contribution, then the individual returns would be £75 : £45 : £30.

We can therefore represent the proportionalities by

$$5 : 3 : 2 :: £75 : £45 : £30$$

and we note that dividing each item by its corresponding item on the opposite side of the double colon gives a constant value.

In this case

$$\frac{£75}{5} = \frac{£45}{3} = \frac{£30}{2}$$

Example:

Electrical switches are obtained from three suppliers, A, B and C. A supplies 600, B supplies 450 and C supplies 150. The switches are mixed and placed in a bin. If 16 switches are selected at random, how many can be expected from each supplier?

Input is

$$600 : 450 : 150, \text{ and dividing by } 150, \text{ is } 4 : 3 : 1.$$

If 16 items are to be shared in the proportion of $4 : 3 : 1$, there are eight shares in total, and the proportions are

$$\tfrac{4}{8} : \tfrac{3}{8} : \tfrac{1}{8}$$

$\therefore$ A can be expected to provide $\tfrac{4}{8} \times 16 = 8$

                B can be expected to provide $\tfrac{3}{8} \times 16 = 6$

and C can be expected to provide $\tfrac{1}{8} \times 16 = 2$.

Answer:

$$A = 8, \qquad B = 6, \qquad C = 2.$$

The wording of the previous example should be noted. It is not to be interpreted that 8 *must* come from A, 6 come from B and 2 come from C. It is what is *expected* under normal circumstances, but it may not precisely occur.

A *percentage* is a fraction having 100 as the denominator. The numerator can be an integer, another vulgar fraction, a mixed number, a mixed denary number or a decimal fraction, whichever proves convenient for the purpose of expressing the particular percentage. The symbol for a percentage is %, read as 'per cent'. To convert a fraction into a percentage we multiply by 100%.

For example

$$\tfrac{2}{5} = \tfrac{2}{5} \times 100\% = 40\%.$$

Conversely, we convert a percentage into a fraction by dividing by 100%.

For example

$$12\tfrac{1}{2}\% = \frac{12\tfrac{1}{2}\%}{100\%} = \frac{25}{2} \div 100 = \frac{25 \times 1}{2 \times 100} = \frac{1}{8}.$$

A percentage is often a useful means of expressing a ratio between two quantities, such as by saying that £20 is 50% of £40. A percentage can be determined with the aid of the proportionality statement,

first quantity : second quantity :: percentage : 100

and since the product of the means is equal to the product of the extremes

$$\text{percentage} \times \text{second quantity} = \text{first quantity} \times 100$$

and

$$\text{percentage} = \frac{\text{first quantity}}{\text{second quantity}} \times 100.$$

Example:

An article costing £75 is sold for £90. Calculate the percentage profit:

(a) related to the cost price
(b) related to the selling price.

Amount of profit $= £90 - £75 = £15$.

(a) Ratio of profit to cost price $= \dfrac{£15}{£75}$

As a percentage $\dfrac{£15}{£75} = \dfrac{15 \times 100\%}{75} = 20\%$

(b) Ratio of profit to selling price $= \dfrac{£15}{£90}$

As a percentage $\dfrac{£15}{£90} = \dfrac{15 \times 100\%}{90} = 16\frac{2}{3}\%$

Answer:

(a) On a cost price basis, profit $= 20\%$
(b) One a selling price basis, profit $= 16\frac{2}{3}\%$

Problems 1.4

1. How many places of decimals has each of the following?
   (a) 6·02   (b) 10·030   (c) 0·03
   (d) 0·007 04   (e) 500·007 040   (f) 200·5

2. Convert the following proper vulgar fractions to their precise decimal equivalents:
   (a) $\frac{1}{2}$ (b) $\frac{1}{4}$ (c) $\frac{3}{4}$ (d) $\frac{5}{8}$
   (e) $\frac{9}{16}$ (f) $\frac{11}{32}$ (g) $\frac{5}{64}$ (h) $\frac{3}{10}$
   (i) $\frac{4}{25}$ (j) $\frac{37}{50}$ (k) $\frac{29}{100}$ (l) $\frac{29}{1\,000}$

3. Convert the following proper vulgar fractions to their precise decimal equivalents with the aid of the recurring indication.

(a) $\frac{1}{3}$ (b) $\frac{2}{9}$ (c) $\frac{5}{6}$ (d) $\frac{5}{12}$

(e) $\frac{7}{30}$ (f) $\frac{13}{24}$ (g) $\frac{7}{90}$ (h) $\frac{5}{7}$

4. By establishing their decimal equivalents range the following vulgar fractions in a series from the smallest to the largest:

$$\frac{3}{5} \qquad \frac{2}{3} \qquad \frac{9}{16} \qquad \frac{5}{8} \qquad \frac{8}{13} \qquad \frac{4}{7}$$

5. Convert the following decimal fractions to their precise vulgar fraction equivalents:

(a) 0·8   (b) 0·82   (c) 0·375   (d) 0·365

(e) 0·481   (f) 0·015 625   (g) 0·$\dot{6}$ (h) 0·1$\dot{6}$

(i) 0·0$\dot{3}$ (j) 0·$\dot{3}\dot{0}$ (k) 0·41$\dot{6}$ (l) 0·$\dot{8}5\dot{7}$ $\dot{1}4\dot{2}$

6. Perform the operations indicated:

(a) 125·4 + 37·72 + 5·347 + 0·05

(b) 128·92 − 72·375

(c) 4 150 − 38·62 + 1·539 − 281·7

(d) 2 868 − 47·41 − 236·2 − 0·005 8

7. Perform the operations indicated:

(a) 373 × 29   (b) 28·2 × 48

(b) 48·1 × 36·9   (d) 41·7 × 0·536

(e) 281·2 × 0·083   (f) 0·078 × 0·043

8. Perform the operations indicated, giving the answers to parts (f) to (h) inclusive to the nearest second place of decimals:

(a) 1 311 ÷ 57   (b) 1 485·06 ÷ 53   (c) 154·98 ÷ 28·7

(d) 4·692 ÷ 0·06   (e) 0·518 88 ÷ 0·564   (f) 481 ÷ 49

(g) 0·516 4 ÷ 22·4   (h) 4·565 ÷ 0·021

9. Find the square roots of the following numbers, giving the answers to parts (h) and (i) to the nearest third place of decimals.

(a) 1 849   (b) 44·89   (c) 7·84   (d) 4·752 4

(e) 9·424 9   (f) 0·280 9   (g) 0·052 9   (h) 813

(i) 0·054 4

10. How many significant figures has each of the following values?

(a) 437   (b) 45·42   (c) 1·730

(d) 45·054 00   (e) 23·700   (f) 0·5

(g) 0·050   (h) 0·007 010   (i) 0·007 01

11. Using the recommendations of BS 1957, round off the following values to three significant figures:

(a) 437·4   (b) 5 376·41   (c) 9·235

(d) 9·225   (e) 9·730 5   (f) 28·445

12. By applying rough checks obtain approximate values of any of the parts of the previous questions 7, 8 and 9.

13. Three boroughs A, B and C having populations of 320 000, 260 000 and 220 000 respectively are to share an amount of £72 000 in proportion to population. Determine the allocation to each borough.

14. The extension of a particular spring is proportional to the applied load. A load of 420 N causes the spring to extend by 30 mm.
    (a) What load will cause an extension of 18 mm?
    (b) What extension, to three significant figures, will occur with a load of 300 N?

15. Express the following fractions as percentages:
    (a) $\frac{1}{2}$ (b) $\frac{1}{5}$ (c) $\frac{3}{4}$ (d) $\frac{5}{8}$

16. Express the following percentages as vulgar fractions:
    (a) $33\frac{1}{3}\%$ (b) $31\frac{1}{4}\%$ (c) $14\frac{2}{7}\%$ (d) $44\frac{4}{9}\%$

17. An article purchased for £4 is resold for £5·40. What percentage of the purchase price is the profit?

18. Trains hauled by steam locomotives complete a journey between two towns 75 kilometres apart in an hour and a half. The introduction of diesel haulage raises the mean speed by 20%. What time is saved by the use of diesel locomotives?

19. Producing at 'standard rate', a firm can complete an order in 20 working days. After producing for 8 days at standard rate, the production rate is accelerated by $33\frac{1}{3}\%$. How many working days are saved by accelerating production?

20. The conversion factor between metric and British units of length is that 1 in = 25·4 mm. What is the percentage error, to two significant figures, if 1 mm is taken as 0·040 in?

21. A lighting fitment is considered to be 80% efficient when first installed. At the end of every month its efficiency has fallen by 10% of the value of efficiency at the commencement of that month. What is the efficiency of the fitment at the end of four successive months after installation?

22. A firm receives its supplies of resistors from three sources, A, B and C. Source A supplies as many as B and C together, while B supplies 3 times as many as C. Of those supplied, the acceptable articles from A, B and C are 80%, 80% and 100% respectively. If a random collection of 825 acceptable resistors is issued by the firm's stores, how many can be expected to have originated from each supplier?

23. When two resistors R_1 and R_2 are connected in parallel, the equivalent resistor R can be obtained from the formula:

$$\frac{1}{R} = \frac{1}{R_1} + \frac{1}{R_2}$$

   In a circuit, the values of R_1 and R_2 are originally 3 ohms and 6 ohms respectively. If R is to remain unchanged, what percentage change must be made to R_2 when the value of R_1 is increased by $33\frac{1}{3}\%$.

24. If C is the distance across corners of a hexagon, the approximate area of the hexagon is $0.65\ C^2$. Find the percentage reduction in volume when a circular bar of diameter 20 mm is milled into hexagonal form 20 mm across corners. (π can be taken as $\frac{22}{7}$, and give the answer to the nearest whole number.)

25. In a particular measurement of angles using a unit called a radian, four-figure mathematical tables quote one minute as being equal to $0.000\ 3$ radian. The accurate value is obtained by multiplying the angle in degrees by $\dfrac{\pi}{180}$. Taking π as 3.142, calculate, to two significant figures, the percentage inaccuracy of four-figure tables for the value of one minute in radian measure.

26. 10 kg of scrap brass having a composition of 60% copper to 40% zinc are melted down with 20 kg of a different brass having a composition of 70% of copper to 30% of zinc. Determine the percentage composition of the new alloy.

27. A contract of 4 800 articles is scheduled to be completed in twelve weeks at the rate of 400 articles per week and commences at this rate. After six weeks of production at this rate the contract is rescheduled so that completion is to be effected one week early. By what percentage must the weekly production rise if a constant rate of production is to be maintained over the last five weeks?

28. What is the percentage increase in the area of a circle when its diameter of 50 mm is increased by 8%? (The area of a circle is given by $\dfrac{\pi d^2}{4}$.)

29. Ohm's Law can be stated symbolically as:

$$V = IR$$

   If the resistance R is increased by 5% and the current I decreased by 7%, find the percentage change in the potential difference V.

30. If a lamp of power P is placed d m from a screen, the intensity of illumination I at the screen is given by:

$$I = \frac{kP}{d^2}$$

k being a constant. A lamp is initially 5 m from a screen. By what percentage is I increased if the lamp is moved 1 m nearer the screen?

31. What is the percentage change in the volume of a cylinder if its radius increases by 10% and its length decreases by 4%? (The volume of a cylinder is given by $\pi r^2 L$.)

32. The bore of a water pipe has a diameter of 20 mm. Calculate, to two significant figures, the percentage reduction in cross-sectional area when the bore becomes coated with scale to a depth of 0·3 mm. (The area of a circle is given by πr^2.)

Answers to Problems 1.4

1. (a) 2   (b) 3   (c) 2   (d) 5   (e) 6   (f) 1

2. (a) 0·5   (b) 0·25   (c) 0·75   (d) 0·625
   (e) 0·562 5   (f) 0·343 75   (g) 0·078 125   (h) 0·3
   (i) 0·16   (j) 0·74   (k) 0·29   (l) 0·029

3. (a) 0·$\dot{3}$ (b) 0·$\dot{2}$ (c) 0·8$\dot{3}$ (d) 0·41$\dot{6}$
   (e) 0·2$\dot{3}$ (f) 0·541 $\dot{6}$ (g) 0·0$\dot{7}$ (h) 0·$\dot{7}$1$\dot{4}$ 2$\dot{8}$5

4. $\frac{9}{16} > \frac{4}{7} > \frac{3}{5} > \frac{8}{13} > \frac{5}{8} > \frac{2}{3}$

5. (a) $\frac{4}{5}$ (b) $\frac{41}{50}$ (c) $\frac{3}{8}$ (d) $\frac{73}{200}$
   (e) $\frac{481}{1000}$ (f) $\frac{1}{64}$ (g) $\frac{2}{3}$ (h) $\frac{1}{6}$
   (i) $\frac{1}{30}$ (j) $\frac{10}{33}$ (k) $\frac{125}{300}$ (l) $\frac{6}{7}$

6. (a) 168·517   (b) 56·545   (c) 3 831·219   (d) 2 584·384 2

7. (a) 10 817   (b) 1 353·6   (c) 1 774·89
   (d) 22·351 2   (e) 23·339 6   (f) 0·003 354

8. (a) 23   (b) 28·02   (c) 5·4   (d) 78·2
   (e) 0·92   (f) 9·82   (g) 0·02   (h) 217·38

9. (a) 43   (b) 6·7   (c) 2·8
   (d) 2·18   (e) 3·07   (f) 0·53
   (g) 0·23   (h) 28·513   (i) 0·233

10. (a) 3   (b) 4   (c) 4   (d) 7   (e) 5
    (f) 1   (g) 2   (h) 4   (i) 3

11. (a) 437   (b) 5 380   (c) 9·24
    (d) 9·22   (e) 9·73   (f) 28·4

12. Numerical values not applicable. All answers are reasonable.

13. A receives £28 800, B £23 400 and C £19 800.

14. (a) 252 N   (b) 21·4 mm.

15. (a) 50%   (b) 20%   (c) 75%   (d) 62·5%

16. (a) $\frac{1}{3}$ (b) $\frac{5}{16}$ (c) $\frac{1}{7}$ (d) $\frac{4}{9}$

17. 35%

18. 15 min

19. 3

20. 1·6%

21. 52·5%

22. $A = 400, B = 300, C = 125$

23. R_2 decreased by $33\frac{1}{3}\%$ to 4 ohms

24. 17%

25. 3·1%

26. $66\frac{2}{3}\%$ of copper with $33\frac{1}{3}\%$ of zinc

27. 20%

28. 16·6%

29. Reduced by 2·35%

30. 56·25%

31. Increased by 16·2%

32. 5·9%

1.5
Forms of numbers other than common denary numbers

1.5.1
Numbers in standard form

A number is said to be in *indicial form* when it is expressed as a base raised to a particular power. Since $125 = 5^3$, then 5^3 is said to be the indicial form of the denary number 125. Similarly 10^2 is the indicial form of the denary number 100. Let us consider the products of numbers in indicial form, those indicial forms having the same base. In particular we will consider the base 10.

$$10^2 \times 10^3 = (10 \times 10) \times (10 \times 10 \times 10)$$
$$= 10 \times 10 \times 10 \times 10 \times 10$$
$$= \text{five tens multiplied together}$$
$$= 10^5.$$

Hence, in general, $10^a \times 10^b = 10^{(a+b)}$

$$10^6 \div 10^4 = \frac{10 \times 10 \times 10 \times 10 \times 10 \times 10}{10 \times 10 \times 10 \times 10},$$

and cancelling by 10 four times gives us $10^{6-4} = 10^2$. Hence, in general, $10^a \div 10^b = 10^{(a-b)}$. Now let us deduce what is indicated by 10^0

$$10^0 = 10^{3-3} \quad \text{or} \quad 10^{2-2} \quad \text{or} \quad 10^{x-x}$$
$$= 10^3 \div 10^3 \quad \text{or} \quad 10^2 \div 10^2 \quad \text{or} \quad 10^x \div 10^x$$
$$= \text{unity in every case}$$

Hence $10^0 = 1$.

Now let us deduce what is indicated by 10^{-a}.

$$10^{-a} = 10^{(0-a)} = 10^0 \div 10^a = 1 \div 10^a = \frac{1}{10^a}.$$

Hence, in general, $10^{-a} = \dfrac{1}{10^a}$. To recapitulate,

$$10^a \times 10^b = 10^{(a+b)}$$

$$10^a \div 10^b = 10^{(a-b)}$$

while $$10^{-a} = \frac{1}{10^a}.$$

These are three special cases of the general laws of bases and indices, to which we shall give further consideration when we proceed to algebra. Now

$$431 \cdot 764\ 326 \times 1\ 000 = 431\ 764 \cdot 326$$

and we note that if we multiply by 1 000, we move the decimal point three places to the right.

For example

$$43 \cdot 2 \times 10\ 000 = 432\ 000,$$

with an implied decimal point after the rightmost zero. Now

$$454 \cdot 624 \div 100 = 4 \cdot 546\ 24,$$

and we note that if we divide by 100, the decimal point is moved two places to the left.

In general, if we divide by 10^n, the decimal point is moved n places to the left. It may be necessary to provide extra zero digits to put the decimal point in the appropriate place. For example

$$5 \cdot 176\ 36 \div 100 = 0 \cdot 051\ 763\ 6,$$

where we have had to introduce a zero digit preceding the digit 5 to put the decimal point in the appropriate place. To conform to standard practice we provide another zero before the decimal point, but this particular zero has not been occasioned by the division process.

The division operation $456 \cdot 624 \div 100$ could have been expressed as $456 \cdot 624 \times \frac{1}{100}$ and since

$$\frac{1}{100} = \frac{1}{10^2} = 10^{-2},$$

the division operation could have been written as $456 \cdot 624 \times 10^{-2}$.

Any positive number can be expressed as a product of:

(a)   a coefficient, which is unity or greater, but is less than 10;
(b)   10 raised to an appropriate power.

A number quoted in this manner is said to be expressed in *standard form*. The coefficient is determined by moving the decimal point to leave a number which is unity or greater, but is less than ten. Thus for $0 \cdot 004\ 16$ we move the decimal point three places to the right to obtain $4 \cdot 16$. For the number 31 764 we move the decimal point (not indicated, but implied after the figure 4) four places to the left to obtain $3 \cdot 176\ 4$. The power of 10 is determined by the number of places we move the decimal point. Its arithmetical sign depends upon whether the decimal point is moved to the left or to the right. If it is moved to the left, the sign is positive. If it is moved to the right, the sign is negative. For example:

$$4\ 765 \cdot 4 = 4 \cdot 765\ 4 \times 10^3$$

(decimal point moved three places to the left)

$$0 \cdot 071\ 3 = 7 \cdot 13 \times 10^{-2}$$

(decimal point moved two places to the right)

For reasons which will be apparent as our studies develop, we shall find that on many occasions our working can be eased by using only certain powers of 10. We call this usage *preferred standard form* because we prefer to use it instead of other available methods. When numbers are expressed in preferred standard form, we use only those powers of 10 which conform to the pattern of $10^{\pm 3n}$, where n is a whole number. We also find it convenient for the multiplier of our power of 10 to range from $0 \cdot 1$ to 1 000.

Let us take, for instance, the number 101 325. In standard form it is:

$$1 \cdot 013\ 25 \times 10^5$$

because the implied decimal point was moved five places to the left. In preferred standard form, the number is either:

$$101 \cdot 325 \times 10^3 \quad \text{or} \quad 0 \cdot 101\ 325 \times 10^6$$

We must emphasize that preferred standard form can, on certain occasions, be unrealistic. With areas, powers of 10 restricted to $10^{\pm 2n}$ will be found far more logical. When we go on to consider units, we shall see the delightful connection between grouping our digits in groups of three and the preferred standard form of numbers.

1.5.2
Computations with numbers in standard form

Addition and subtraction

Addition and subtraction operations on numbers expressed in standard form can only be undertaken when they are multiples of the same power of 10. The algorithm is:

Step 1. Convert the values if necessary to multiples of the same power of 10, which should be the lowest of those stated.

Step 2. Perform the addition and/or subtraction processes on the coefficients.

Step 3. Couple with result of step 2 with the power of 10 from step 1.

Step 4. Adjust the value from step 3, if required, to give the form required in the answer.

Example:

Evaluate $4 \cdot 74 \times 10^3 + 8 \cdot 573 \times 10^2 - 21 \cdot 3 \times 10^4$, and provide the answer in standard form.

Step 1. The lowest power of 10 is 10^2

$$4 \cdot 74 \times 10^3 = 47 \cdot 4 \times 10^2$$
$$8 \cdot 573 \times 10^2 = 8 \cdot 573 \times 10^2$$
$$21 \cdot 3 \times 10^4 = 2\ 130 \times 10^2$$

Step 2.
$$47 \cdot 4 + 8 \cdot 573 - 2\ 130 = 55 \cdot 973 - 2\ 130$$
$$= -2\ 074 \cdot 027$$

Step 3. Coupling $-2\ 074 \cdot 027$ from step 2 with 10^2 from step 1 produces

$$-2\ 074 \cdot 027 \times 10^2$$

Step 4.   In standard form the modulus of the coefficient is equal or greater than unity but less than 10. The modulus of the coefficient will be 2·074 027.

Now $(-2\,074{\cdot}027)$ $= -2{\cdot}074\,07 \times 10^3.$

Hence $-2\,074{\cdot}027 \times 10^2 = -2{\cdot}074\,027 \times 10^3 \times 10^2$
$= -2{\cdot}074\,027 \times 10^5$

Answer:

$$-2{\cdot}074\,027 \times 10^5$$

Multiplication and division

The multiplication and/or division of numbers expressed in standard form follows the rules for computation previously stated in this book. The algorithm is as follows:

Step 1.   Determine the sign of the answer. A multiplication or division operation between quantities of like sign produces a positive value. A multiplication or division operation between quantities of unlike sign produces a negative value.
Step 2.   Perform the multiplication and/or division operations on the moduli of the coefficients to obtain the modulus of the answer.
Step 3.   Perform the multiplication and/or division operations on the powers of 10 to obtain the power of 10 of the answer.
Step 4.   Couple together:
the arithmetical sign from step 1
the modulus from step 2
the power of 10 from step 3
to obtain an answer.
Step 5.   Adjust the answer obtained from step 4, if necessary, to obtain the form requested by the problem.

Example:

Evaluate

$$\frac{(2{\cdot}87 \times 10^3) \times (31{\cdot}24 \times 10^{-2})}{(-3{\cdot}172 \times 10^4) \times (-4{\cdot}17)},$$

giving the answer in standard form to three significant figures.

Step 1. $\dfrac{(+) \times (+)}{(-) \times (-)} = \dfrac{(+)}{(+)} = +$

Step 2. Noting that $4\cdot17 = 4\cdot17 \times 10^0$,

$$\frac{2\cdot87 \times 31\cdot24}{3\cdot172 \times 4\cdot17}$$

by calculator, to three significant figures, is $6\cdot78$.

Step 3. Again noting that $4\cdot17 \times 10^0$

$$\frac{10^3 \times 10^{-2}}{10^4 \times 10^0} = \frac{10^{3-2}}{10^{4+0}} = \frac{10^1}{10^4} = 10^{1-4} = 10^{-3}$$

Step 4. We combine the positive sign from step 1 with the modulus $6\cdot78$ from step 2 to theoretically obtain $+6\cdot78$, which we write just as $6\cdot78$ and then multiply by the 10^{-3} obtained from step 3 to produce $6\cdot78 \times 10^{-3}$.

Step 5. This step is not required, as $6\cdot78 \times 10^{-3}$ is in the form required.

Answer:

$$6\cdot78 \times 10^{-3}.$$

A *denary* (sometimes called 'decimal') system of numbers should not be confused with a so-called metric system. A *metric* system is a system of dimensions, not numbers, associated with the metre as a measurement of length. In science we are concerned not only with magnitude, but also with what that magnitude describes, i.e. its *unit*. There is a considerable difference between 3 millimetres and 3 metres. The unit of a quantity has just as much importance as the number.

The fundamental units of science are those of length, mass and time. If we use as basic units the metre, the kilogram and the second, we have a metric system of units. If we use the centimetre, the gram and the second as our basic units, we can form a different metric system of units. Of all the metric systems which are available or could be formed, it has been internationally agreed that the preferred system will be the Système International des Unités, or *SI system of units*. In the SI system, the fundamental units of length, mass and time are the metre, the kilogram and the second respectively. As our scientific knowledge extends, we may conveniently add to what we can call the three primary fundamental units certain supplementary secondary fundamental units. At present there are three: the ampere for electricity, the candela for light, and the kelvin for heat. Others may be added in the future. Just out of interest, the degree celsius can be regarded as a unit 'associated with the SI system'.

Although the basic SI unit of length is the metre, it is absurd to presume that all lengths must be expressed in metres. Similarly, it would be completely unrealistic to express a person's age in seconds. Common sense dictates that we shall use multiples and sub-multiples of our basic units. It is extremely convenient to associate a denary system of numbers with a metric system of units, and an elegant way of doing this is to use a prefix with the basic unit to indicate the multiplier of that unit. The reader is probably aware of the prefixes kilo (meaning one thousand) and milli (meaning one thousandth) in such units as a kilogram and a millimetre. A list of multipliers is given in the following table:

Multiplier	Standard form	Name	Symbol
Million million	10^{12}	tera	T
Thousand million	10^9	giga	G
Million	10^6	mega	M
Thousand	10^3	kilo	k
Hundred	10^2	hecto	h
Ten	10^1	deca	da
Tenth	10^{-1}	deci	d
Hundredth	10^{-2}	centi	c
Thousandth	10^{-3}	milli	m
Millionth	10^{-6}	micro	μ
Thousand millionth	10^{-9}	nano	n
Million millionth	10^{-12}	pico	p
Thousand million millionth	10^{-15}	femto	f
Million million millionth	10^{-18}	atto	a

It will be observed that there are only four multipliers which do not conform to preferred standard form; these are hecto, deca, deci and centi. While it can be expected that they will be used somewhat less frequently than the others, to presume they will be completely eliminated is erroneous. The attitude that the centimetre is not a recommended SI unit and should never be used is taking purity to a ridiculous extreme. We must use those units which are suitable for a particular situation.

Just as our conventional method of indicating the number twenty-five thousand four hundred and fifty-three is

25 453 in denary form,

$2.545\,3 \times 10^4$ in standard form, and

25.453×10^3 in preferred standard form,

so we can adopt a similar convention with the presentation of units by using abbreviations. The conventions recommended by BS 1991 are our first order of preference, but if an abbreviation could lead to misinterpretation, the unit should be written out in full. The basic derived SI unit of capacity is the metre cubed. A far more practical unit is one thousandth of a metre cubed, which is a decimetre cubed. This particular unit is familiarly known all over the world as the *litre*. The litre is not an SI unit, but to presume it will not be a commonly used unit of capacity is closing one's eyes to reality. It is unfortunate that the recommended abbreviation for litre is the small letter 'el', and since this could easily be confused with the number representing unity, on occasions the unit may have to be written out in full. *Throughtout this book, to avoid confusion with unity, the small 'el' will not be used as a symbol or an abbreviation.* In abbreviations, any prefix used as a multiplier adjoins the unit without a space between them, and the abbreviation is the same for the singular and plural. For example:

5 kilometres is abbreviated to 5 km

36 millimetres is abbreviated to 36 mm

53 microseconds is abbreviated to 53 μs

Multiplication and division of units follow the normal rules of algebra, except that we use a small but distinct space (not a hyphen) when units are multiplied, whilst for convenience in typesetting, division may be indicated by an oblique stroke. Thus:

$$5\text{ m} \times 5\text{ m} \times 2\text{ m} = 50\text{ m}^3$$

$$25\text{ kg} \times 3\text{ s} = 75\text{ kg s}$$

$$6\text{ km} \div 5\text{ s} = \frac{6\text{ km}}{5\text{ s}} = 1 \cdot 2\,\frac{\text{km}}{\text{s}} = 1 \cdot 2\text{ km/s}$$

The SI system must be applied with common sense rather than used blindly in any circumstances. Recommendations and preferences mean only what they say. If they can be used naturally, they are the logical first choice. The intelligent person will depart from them when good and sufficient reasons prevail.

1.5.3
Numbers in binary form

Our usual method of counting, using natural numbers, employs the denary system of integers. The *base*, sometimes called the *radix*, of the denary system of numbers, is the value 10. The denary number

$$8\,473 = 8(10^3) + 4(10^2) + 7(10^1) + 3(10^0)$$

and the number is expressed by the coefficients of the powers of the radix 10, in descending order, terminating, for integers, with the coefficient of 10^0. We shall now proceed to another system of numbers which uses a radix of 2, which is called the binary system of numbers.

Now $15 = 8 + 4 + 2 + 1 = 2^3 + 2^2 + 2^1 + 2^0 = 1(2^3) + 1(2^2) + 1(2^1) + 1(2^0)$. In the binary system of numbers, a number is expressed by the coefficients of powers of 2, in descending order of those powers, terminating, for integers, with the coefficient of 2^0.

Thus the denary number 15 is the binary number 1111.

Similarly, since

$$14 = 1(2^3) + 1(2^2) + 1(2^1) + 0(2^0),$$

the denary number 14 is the binary number 1110.

The employment of the radix 2 for the binary system of numbers leads to every digit in a binary number being either zero or unity. The expression *binary digit* is conventionally abbreviated to *bit*. Let us express the first few natural numbers as binary numbers.

Denary		*Binary*
1	$1(2^0)$	1
2	$1(2^1) + 0(2^0)$	10
3	$1(2^1) + 1(2^0)$	11
4	$1(2)^2 + 0(2^1) + 0(2^0)$	100
5	$1(2^2) + 0(2^1) + 1(2^0)$	101
6	$1(2^2) + 1(2^1) + 0(2^0)$	110
7	$1(2^3) + 1(2^1) + 1(2^0)$	111
8	$1(2^3) + 0(2^2) + 0(2^1) + 0(2^0)$	1000
9	$1(2^3) + 0(2^2) + 0(2^1) + 1(2^0)$	1001
10	$1(2^3) + 0(2^2) + 1(2^1) + 0(2^0)$	1010
11	$1(2^3) + 0(2^2) + 1(2^1) + 1(2^0)$	1011

If we wish to convert a binary integer to its equivalent denary integer we note that the bit on the extreme right is the number of 'ones'. The next bit to the left is the coefficient of 2^1, i.e. the number of twos. The next bit is the coefficient of 2^2, i.e. the number of fours. The next bit is the coefficient of 2^3, i.e. the number of eights. And so we go on, the next bits being the numbers of sixteens, thirty-twos, sixty-fours, hundred and twenty-eights and so on. So if we wish to convert the binary number

$$1 \quad 1 \quad 0 \quad 1 \quad 1 \quad 0$$

to its denary equivalent, we can put values in circles over the bits, starting from the right-hand side, of 1, 2, 4, 8, 16, 32, 64, 128, 256, and so on, until every bit has a number in a circle above it.

In our case, we would have

$$\overset{\textstyle ㉜}{1} \quad \overset{\textstyle ⑯}{1} \quad \overset{\textstyle ⑧}{0} \quad \overset{\textstyle ④}{1} \quad \overset{\textstyle ②}{1} \quad \overset{\textstyle ①}{0}$$

and the denary equivalent is

$$(1 \times 32) + (1 \times 16) + (0 \times 8) + (1 \times 4) + (1 \times 2) + (0 \times 1)$$

and remembering the fundamental rule that multiplication takes precedence over addition, we have

$$32 + 16 + 0 + 4 + 2 + 0 = 54.$$

Hence the denary equivalent of the binary number 110110 is 54.

In the denary system of numbers we separate whole numbers from decimal fractions by a marker we call the decimal point. The syllabus for Level 1 does not include a reference to fractions in binary form, only to positive whole numbers. However, our reader will no doubt appreciate that we can use something we can refer to as a 'binary point', or merely just 'point', to separate binary whole numbers from binary fractions.

Thus, if we consider

$$110{\cdot}0101$$

to the left of the point we have the binary equivalent of the denary number 6, while to the right of the decimal point we have the equivalent of

$$0(2)^{-1} + 1(2)^{-2} + 0(2^{-3}) + 1(2^{-4})$$

i.e. of $0(\frac{1}{2}) + 1(\frac{1}{4}) + 0(\frac{1}{8}) + 1(\frac{1}{16})$

$$= \frac{1}{4} + \frac{1}{16} = \frac{4+1}{16} = \frac{5}{16}$$

and binary $110{\cdot}0101$ is equivalent to $6\frac{5}{16}$.

To convert a denary integer to a binary number, we commence by dividing the denary number by 2, being careful to add to the quotient a remainder, even if that remainder is zero. For example, in converting the denary integer 54 to its binary equivalent, we commence with

$$2 \underline{\,|\,54\,}$$
$$27 + 0$$

We now divide the quotient of 27 by 2 to obtain $13 + 1$. We then divide 13 by 2 to obtain $6 + 1$, and so on, until we arrive, as we eventually must in all cases, with $0 \div 2 = 0 + 0$. The steps can be set out as follows:

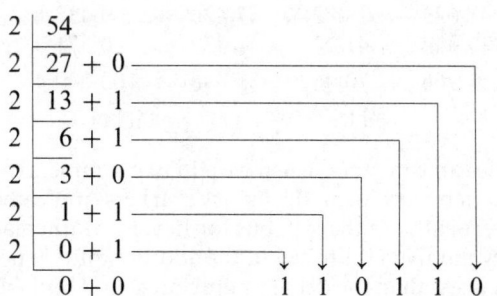

Dividing by 2, and then by 2 again, is equivalent to dividing the original dividend by 4. Dividing again by 2 is equivalent to dividing the original dividend by 8. Consequently the remainders, reading from bottom to top, are the digits of the binary equivalent, reading from right to left. The nominal indication of $0 + 0$ can be disregarded. No change is made to a binary number if a zero prefixes that number. The indication $0 + 0$ is only used as a convenient terminating arrangement. All conversions of denary integers must terminate in this manner.

Hence 54 (denary) = 110110 (binary) which may occasionally be found to be written as

$$54_{10} = 110110_2$$

indicating that 54 is given with reference to the radix 10 while 110110 is given with reference to the radix 2.

1.5.3
Computations with numbers in binary form

Since the denary number 34 is

$$1(32) + 0(16) + 0(8) + 0(4) + 1(2) + 0(1)$$
$$= 1(2^5) + 0(2^4) + 0(2^3) + 0(2^2) + 1(2^1) + 0(2^0)$$

then the denary number 34 is represented as a binary number by 100010.

In a similar manner, the denary number 25, being

$$1(2^4) + 1(2^3) + 0(2^2) + 0(2^1) + 1(2^0)$$

the denary number 25 is 11001 as a binary number.

Just as a decimal marker is implied after the last digit of a whole number in the denary system of numbers so a binary marker is implied after the last digit of a whole number in the binary system of numbers. When we add denary numbers we arrange that the decimal markers, either indicated or implied, are aligned vertically, examples being

623	28·72	57·426	99·438 8
342	41 34	1·47	0·571 4
965	70·06	58·896	100·009 4
	11		1111

In much the same manner, when we add two numbers together which are expressed in binary form, the binary markers (indicated or implied) are aligned vertically. In the syllabus for Level 1 Mathematics the addition of binary numbers is limited to the addition of two positive binary numbers, each less than the denary number 64, and both being integers, i.e. whole numbers. The implied binary marker with such binary numbers will be after the last digit and hence when we add the binary numbers the last digits are aligned vertically. In our case, adding 100010 and 11001 follows the normal practice observed with denary numbers, viz

$$100010+$$
$$11001$$
$$\overline{111011}$$

As a check, $34 + 25 = 59$

and
$$\begin{array}{cccccc} \textcircled{32} & \textcircled{16} & \textcircled{8} & \textcircled{4} & \textcircled{2} & \textcircled{1} \\ 1 & 1 & 1 & 0 & 1 & 1 = 59. \end{array}$$

Now let us consider the addition of the two denary numbers 286 and 37. Setting the problem out in the usual form, we have

$$\begin{array}{r} 286 \\ 37 \\ \hline 323 \\ \hline 11 \end{array}$$

To obtain the answer 323, we probably mentally say:

$$7 + 6 = 13$$

put down 3, carry 1,

$$1 + 3 + 8 = 12$$

put down 2, carry 1,

$$1 + 2 = 3.$$

The same general approach occurs with adding two binary numbers. When the addition of a column produces the value 2 we put down 0 and carry 1. If the total is 3 we put down 1 and carry 1.

$$\begin{array}{cccccc} & \textcircled{16} & \textcircled{8} & \textcircled{4} & \textcircled{2} & \textcircled{1} \\ \text{27 in binary form is} & 1 & 1 & 0 & 1 & 1 \\ \text{29 in binary form is} & 1 & 1 & 1 & 0 & 1 \end{array}$$

Let us add the binary numbers 11011 and 11101. We set out the problem in the usual way, viz:

$$11011+$$
$$11101$$
$$\overline{}$$

and start with the extreme right-hand column. We say $1 + 1 = 2$. This is not zero or unity, it is 2. So put down 0 and carry 1, to obtain, thus far

$$\begin{array}{r} 11011+ \\ 11101 \\ \hline 0 \\ \hline 1 \end{array}$$

Now move one column to the left. $1 + 0 + 1 = 2$, so again put down 0 and carry 1, to obtain, thus far

$$\begin{array}{r} 11011+ \\ 11101 \\ \hline 00 \\ \hline 11 \end{array}$$

So we proceed to the next column. $1 + 1 + 0 = 2$, so put down 0 and carry 1, to obtain

$$\begin{array}{r} 11011+ \\ 11101 \\ \hline 000 \\ \hline 111 \end{array}$$

Now to the next column, $1 + 1 + 1 = 3$, so this time we put down 1 and carry 1, to obtain

$$\begin{array}{r} 11011+ \\ 11101 \\ \hline 1000 \\ \hline 1111 \end{array}$$

And so the next column $1 + 1 + 1 = 3$, put down 1 and carry 1, to obtain

$$\begin{array}{r} 11011+ \\ 11101 \\ \hline 11000 \\ \hline 11111 \end{array}$$

Finally, we have to start a new column, to cater for $1 + 0$ (implied) $+ 0$ (implied) $= 1$.
  Our final result is

$$\begin{array}{r} 11011 \\ 11101 \\ \hline 111000 \end{array}$$

As a check, $27 + 29 = 56$

and

$$\frac{\overset{32}{\textcircled{32}} \quad \overset{16}{\textcircled{16}} \quad \overset{8}{\textcircled{8}} \quad \overset{4}{\textcircled{4}} \quad \overset{2}{\textcircled{2}} \quad \overset{1}{\textcircled{1}}}{1 \quad \; 1 \quad \; 1 \quad \; 0 \quad \; 0 \quad \; 0} = 56$$

The rules for adding two binary numbers are therefore quite simple. The addition is undertaken column by column commencing from the right, and:

(a) if the total is 0 or 1, write it down under the appropriate column
(b) if the total is 2, put down 0 and carry 1
(c) if the total is 3, put down 1 and carry 1.

1.5.5
The on-off mode of binary numbers

The denary system of numbers proves to be very convenient for everyday counting purposes. It requires the use of ten symbols, 0 to 9 inclusive, together with a place order of digits, in order to represent a number of any magnitude. However, other systems of numbers can be used in special circumstances if particular advantages accrue from their use.

Consider, for instance, the display of a pocket-size electronically-operated calculator. The display is in denary numbers. There are ten different values which could occur in every digit place. The display is given by arranging for one of those values to be illuminated. If this was to be obtained by simple switches, where each switch was either on or off, it may at first be thought that every symbol required a switch, so that if it was required to display the number 7, the switch associated with 7 would be 'on', and nine other switches, associated with 0 to 6 inclusive, and 8 and 9, would be 'off'.

We use the word *system* to describe a collection of individual items, called the *elements* of the system, whose properties and relationships, together with associated principles, form a coherent entity. In a more loose manner of speaking, we could say that the elements of a system naturally hang together. For instance, we talk of a system of denary numbers and of a system of binary numbers. A system does not have to be abstract, it may consist partly or wholly of actual articles, which we often colloquially refer to as the *hardware* of that system. The reliability of a system is now receiving far more attention than it did in the past. In general, a system which uses less of a particular element to perform a task is more reliable than is another system which uses more of that element to perform the same task.

Since a binary number has only two possibilities in every digit place, each digit being either zero or unity, a binary number can be used to indicate the state of a set of items, each of which can adopt one of two specific attitudes. In electrical work, this could be a simple on-off switch. The switch is either on or it is off, there is no other attitude it can adopt. We can let the digit 1 represent the switch being on and the digit 0 represent the switch being off. If we have four switches, A, B, C and D, and we use a four-digit binary number to represent the attitudes of those switches, the first digit representing switch A, the second digit switch B, and so on, then the binary number 1001 represents

> switch A is on
> switch B is off
> switch C is off
> switch D is on.

The signal received from this particular combination of switches could be used to cause a system to undertake a particular function. It could, for instance, be used in a calculator to request the display to illuminate the value 9. Since no binary number in excess of 9 needs more than four binary digits, we need only four switches to display any denary digit, instead of ten, and although a calculator displays denary numbers, the method by which this is achieved invariably employs binary numbers to cause the hardware to function. The use of the expression 'electrical switch' may well direct the thoughts of our reader to the somewhat bulky item of electrical equipment for controlling whether an illuminating device, such as a domestic light fitting, is on or off. Developments in the science of electricity, together with advancements in manufacturing methods, have produced devices similar to switches, of such minute size that they have to be viewed with magnifying aids, and a multiplicity of such switches can now quite easily be accommodated on an area no more than that of a pin head.

Furthermore, the on-off mode need not be restricted to a representation of whether an electric current is flowing or not. We could use it to represent a condition in the flow of a fluid through a hydraulic stop valve. The binary digit 0 could be used to indicate that the valve is closed and the binary digit 1 to indicate the valve is open. If a signalling device is positioned after the valve, we could let 1 indicate that a signal is displayed and 0 indicate that no signal is displayed. Alternatively, we could have a device which is pneumatic, i.e. operated by air, and let 0 indicate that no air is received by a particular item from that device, and 1 indicate that air is being received.

The use of binary digits is not restricted to the on-off states of items in systems which are operated electrically, electronically, hydraulically, pneumatically, or by combinations of these means, or other means. All the former imply the use of hardware. We can apply a concept similar to the on-off mode to an abstract concept.

Logic is the science of reasoning. We can, if we so choose, make a series of statements, each of which can either be true or false. Providing they can only be true or false we can let 1 represent a state of truth and 0 a state of falsity. A whole new science has been developed which applies mathematics to reasoning, to which our reader will be introduced should studies proceed in a particular direction. Meantime, let us return to the on-off mode of using binary digits. Consider the simple series of three electrical switches shown in fig. 1.2.

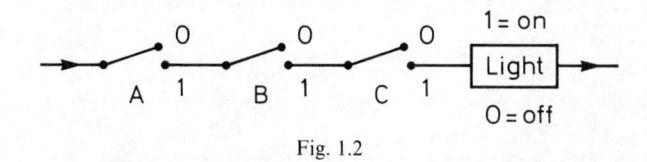

Fig. 1.2

It will be noted that the light will only come on if switch *A* AND switch *B* AND switch *C* are on.

Using the 0 or 1 mode to indicate whether a switch is open or closed, and 1 or 0 to indicate whether a signal is displayed or not, we can erect the following table.

Switch A	Switch B	Switch C	Light
1	0	0	0
0	1	0	0
0	0	1	0
1	1	1	1

A table similar to the above is called the *truth table* for the element consisting of the particular combination of three switches, and since it indicates that a signal will only be displayed if switch *A* AND switch *B* AND switch *C* are closed, it is often referred to as the truth table for an *AND* element. Now consider an element of a system which comprises of three switches arranged as in fig. 1.3.

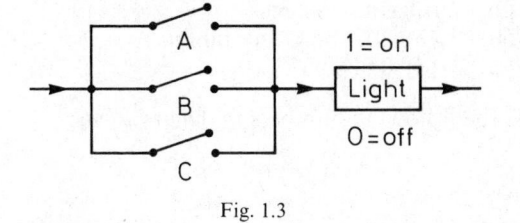

Fig. 1.3

The truth table for this element is

Switch A	Switch B	Switch C	Light
1	0	0	1
0	1	0	1
0	0	1	1
0	0	0	0

It indicates that an output will occur if switch A OR switch B OR switch C is on. It is often referred to as an *OR* element.

We are quite deliberating taking the study of the on-off mode of binary digits a little beyond the intentions of the unit, but mathematics only really becomes alive when its relevance to real situations can be demonstrated. It is hoped that this highly condensed introduction to the use of binary numbers to present information and/or commands to a system, and to interpret the output of a system, has evoked some interest in our reader. The subject is a relatively new topic in the field, when viewed against the time-span of mathematics, but its importance in the development of new technologies will be very significant.

Problems 1.5

1. Express the following quantities in standard form, by quoting the product of a coefficient which is unity or greater but less than 10, and 10 raised to an appropriate power.
   (a) 4 632   (b) one thousand
   (c) 0·762   (d) 0·800
   (e) minus one-half   (f) thirty million
   (g) five and one-half thousandths

2. Express the following numbers as a product of a coefficient which lies between 1 and 1 000, and 10 raised to the power of $3n$, where n is a positive or negative integer.
   (a) 4 960   (b) 207 thousand million
   (c) 0·562   (d) 0·036 05

3. Express the following numbers in denary form:
   (a) $6·4 \times 10^{-3}$ (b) $8·431 \times 10^2$
   (c) 8×10^{-2} (d) $3·146 \times 10^{-1}$
   (e) $3·474 \times 10^4$ (f) $1·87 \times 10^{-4}$

4. Evaluate, expressing the precise answer in standard form:
   (a) $3.62 \times 10^3 + 4.172 \times 10^3$
   (b) $476 + 23.7 \times 10^2 - 16.4 \times 10^2$
   (c) $85.7 \times 10^{-3} + 6.4 \times 10^{-2} - 437.4 \times 10^{-3}$
   (d) $4.58 - 473.1 \times 10^{-2} - 3.7 \times 10^{-1} + 0.437 \times 10^2$

5. (a) Evaluate $14.6 \times 10^{-3} + 4.74 \times 10^{-4} + 65 \times 10^{-5} - 323 \times 10^{-6}$ expressing the precise answer (i) in standard form; (ii) as a multiple of 10^{-3}.
   (b) Obtain the reciprocal of the answer in standard form to an accuracy of three significant figures.

6. Add together 3.57 km, 462 m and 41 650 cm, giving the answer as a number of metres, in standard form.

7. Evaluate, giving exact answers in standard form:
   (a) $(4.2 \times 10^2) \times (1.3 \times 10^2)$
   (b) $(45 \times 10^5) \div (9 \times 10^3)$
   (c) $0.000\ 082\ 3 \times (1.1 \times 10^5)$
   (d) $8\ 263.5 \div (7 \times 10^4)$
   (e) $(3 \times 10^7) \div (8 \times 10^8)$
   (f) $(4.8 \times 10^3) \times (5 \times 10^5) \div (6 \times 10^7)$
   (g) $(4.32 \times 10^{-2}) \div (3.6 \times 10^{-4}) \times (3.1 \times 10^2)$

8. How many millimetres are there in a kilometre? Give the answer in both standard and denary form.

9. What number, in standard form, can replace the multiplier milli ÷ micro?

10. Evaluate, giving the answers in standard form to three significant figures:
    (a) $(4.72 \times 10^4) \div (537.2 \times 10^2)$
    (b) $\sqrt{(5.67 \times 10^6)}$
    (c) $(23.92 \times 10^2)^2$
    (d) $(5.64 \times 10^3) + (58.9 \times 10^{-4}) \times (847 \times 10^3)$
    (e) $-5(292 - 292 \times 10^{-1})$

11. Express the following denary numbers as their binary equivalents:
    (a) 8   (b) 9   (c) 10   (d) 23   (e) 1
    (f) 0   (g) 37   (h) 48   (i) 57   (j) 62

12. Express the following binary numbers as their denary equivalents:
    (a) 11   (b) 101   (c) 110   (d) 1010
    (e) 1110   (f) 10101   (g) 110101   (h) 111111

13. Perform the operations indicated, giving the answers in binary form:
    (a) 10 + 01   (b) 101 + 10
    (c) 1010 + 1010   (d) 11010 + 1101
    (e) 10101 + 101   (f) 110010 + 1101

14. In the switching circuit shown in fig. 1.4, *A*, *B* and *C* are on-off switches while *D* is an indicator light. Using the on-off mode of binary numbers, complete the table, using 0 to indicate if the light is off and 1 to indicate if the light is on.

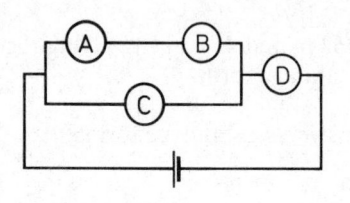

A	B	C	D
1	0	0	
1	1	0	
1	0	1	
0	0	1	
0	1	0	

Fig. 1.4

15. Complete the table shown in fig. 1.5, where *A*, *B*, *C*, *D*, *E* and *F* are on-off switches and *G* is an indicator light.

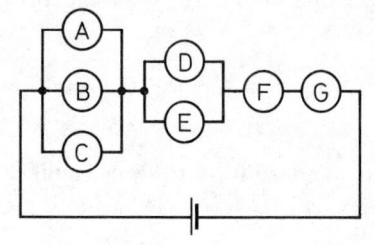

A	B	C	D	E	F	G
1	0	0	1	0	1	
1	1	1	1	1	0	
0	0	1	0	1	1	
0	1	0	0	0	1	
0	0	1	1	0	0	

Fig. 1.5

16. For each of the following statements attach a truth value of 1 if the statement is true and 0 if it is false.
    (a) E is the fifth letter of the English alphabet.
    (b) Delta is the seventh letter of the Greek alphabet.
    (c) The number 45.2×10^2 is written in standard form.
    (d) 0·33 is precisely equal to one-third.
    (e) 0·6 is precisely equal to two-thirds.
    (f) In shape the world is a perfect sphere.
    (g) Air consists of oxygen and nitrogen only.

Answers to Problems 1.5

1. (a) 4.632×10^3 (b) 1×10^3
  (c) 7.62×10^{-1} (d) 8.00×10^{-1}
  (e) -5×10^{-1} (f) 3×10^7
  (g) 5.5×10^{-3}

2. (a) 4.960×10^3 (b) 207×10^9
  (c) 562×10^{-3} (d) 36.05×10^{-3}

3. (a) $0.006\,4$ (b) 843.1
  (c) 0.08 (d) $0.314\,6$
  (e) $34\,740$ (f) $0.000\,187$

4. (a) 7.792×10^3 (b) 1.206×10^3
  (c) -2.877×10^{-1} (d) $4.317\,9 \times 10^1$

5. (a)(i) $1.540\,1 \times 10^{-2}$ (ii) 15.401×10^{-3} (b) 6.49×10^1

6. $4.448\,5 \times 10^3$ m

7. (a) $5.46 \times 10^4 = 54\,600$ (b) $5 \times 10^2 = 500$
  (c) $9.053 \times 10^0 = 9.053$ (d) $1.180\,5 \times 10^{-1} = 0.118\,05$
  (e) $3.75 \times 10^{-2} = 0.037\,5$ (f) $4 \times 10^1 = 40$
  (g) $3.72 \times 10^4 = 37\,200$

8. $10^6 = 1\,000\,000$

9. 1×10^3

10. (a) 8.79×10^{-1} (b) 2.38×10^3
  (c) 5.72×10^6 (d) 1.06×10^4
  (e) 1.31×10^3

11. (a) 1000   (b) 1001   (c) 1010   (d) 10111
  (e) 1   (f) 0   (g) 100101   (h) 110000
  (i) 111001 (j) 111110

12. (a) 3   (b) 5   (c) 6   (d) 10
  (e) 14   (f) 21   (g) 53   (h) 63

13. (a) 11   (b) 111
  (c) 10100   (d) 100111
  (e) 11010   (f) 111111

14. Column D to read, from top to bottom 0, 1, 1, 1 and 0.

15. Column G to read, from top to bottom 1, 0, 1, 0 and 0.

16. (a) 1   (b) 0   (c) 0   (d) 0
  (e) 1   (f) 0   (g) 0

1.6
Aids to computation

1.6.1
The use of digital calculators

The cost of landing a man on the moon is often thought by many people to be a somewhat unfortunate use of resources, but the technological 'spin-off' has been enormous. In particular, the development of ultra-rapid lightweight switching devices has been a tremendous boost to miniaturisation. The original calculating machines, mainly operating by mechanical means, were heavy and slow-acting. The latest electronic calculating machines are pocket size, lightweight, have virtually no moving parts to be maintained, and operate at phenomenal speed. Their reliability is incredible compared with previous types, and advances in production methods has meant that the cost of a machine of performing an immense variety of functions is now but a minute fraction of the cost of a mechanically operating machine, and in many cases, far cheaper than that computation aid which has given so much assistance in the past, the slide rule.

The number of different models now available is extremely large, and calculators are one commodity where fair competition has reacted favourably in favour of the purchaser. Some advice has been offered in the introduction of this volume regarding the purchase of a calculator applicable to our reader's present studies in mathematics. The number of different models readily available means that it is virtually impossible to provide instruction in their use which is applicable to every model. Some provide more functions than others, some have one-function keys while others have keys for two, and occasionally more, functions, some have stores or memories, others have not.

There is, however, certain advice which can be applied to the use of any calculator. The first is that the instruction book for a particular calculator should be read with care, and any examples given in that instruction book should be worked through several times to gain experience in the use of that particular calculator. The second is that instructions for one particular calculator are not necessarily valid for other models, particularly if one changes from a model with one-function keys to a model with two-function keys. Above all, a user should always adopt the practice of clearing the calculator before embarking upon a new computation. Simple processes are fed into most calculators in the logical form by which they are stated. For instance, for a simple process such as addition or subtraction or multiplication or division the input is

| Clear | First value | Sign of operation | Second value | = |

Care must be taken with 'chain' calculations to ensure whether or not a particular calculator can interpret the distributive law of mathematics, i.e. that multiplication and/or division takes precedence over addition and/or subtraction. A particular calculator may be capable of doing this for a small number of entries, but it would be most unwise to presume that it can be done so for a large number of entries. It could well be that a computation involving combinations of the four basic processes may have to be undertaken step by step and due note taken of intermediary values if the calculator does not possess a capability of storing information for subsequent recall.

It would be most advisable for our reader to develop a standard procedure which reflects the calculator which is regularly used. The author's standard practice is to use all of the digits which his calculator can provide. It is agreed that this is generally far more than is ever required for the majority of computations, but in most cases it does save the bother of determining an appropriate number of significant figures for each and every step of a long computation. As a final step, and only as a final step, the last answer is rounded-off to a reasonable number of significant figures appropriate to the data being processed. For the level of work for which this book was intended, there are very few occasions where a final accuracy of more than three significant figures will be required. In the industrial situation circumstances may be different. An example is given in Chapter 5 in reference to precise measurement. It must be emphasised that the practice used by the author reflects the capabilities of his own particular calculator. Other calculators may suggest that other more suitable practices may have to be adopted.

There is no point in proceeding to special problems to consolidate the information given previously in this article. At various points in this text book there are worked examples and problems which have been evaluated by formal arithmetic and/or with the aid of mathematical tables. Our reader is encouraged to use these worked examples and problems to gain experience in the use of a particular calculator, using the instruction manual for that calculator as a guide to its capabilities.

1.6.2
Logarithms

Numbers can be indicated in various forms. The number which in usual practice is written as 8 000 (the denary form) can also be written as:

$$8 \times 10^3 \quad \text{or} \quad 20^3 \quad \text{or} \quad 10^{3 \cdot 903 \, 1}$$

(the last value being an approximation).

The first number is written in standard form, while the other two are written in *indicial forms*. Using symbolic notation, the general case of a number in indicial form is that:

$$N = b^x$$

If two of the values of N, b and x are known, the third is automatically fixed. We say that x is the logarithm of N to the base b. *The logarithm of a number to a given base is therefore the power to which the base has to be raised to give that number.* The word logarithm is conveniently abbreviated to 'log', and the base is indicated in the following manner:

If
$$N = b^x$$

then
$$\log_b N = x$$

As examples:

since $8 = 2^3$, $\log_2 8 = 3$

similarly $25 = 5^2$, $\log_5 25 = 2$

                $4 = 16^{0.5}$, $\log_{16} 4 = 0.5$

and finally $0.1 = 10^{-1}$, $\log_{10} 0.1 = -1$

We shall shortly see in this article that logarithms to the base 10 often prove to be a useful aid to computation. Logarithms can be established for any base. The original concept of logarithms is attributed to Napier, who calculated them for a very special base. The advantage of the base 10 for computation is attributed to Briggs, a friend of Napier, and logarithms to the base 10 were at one time called Briggs' logarithms. The more usual expression nowadays is common logarithms. The original logarithms attributed to Napier were at one time called Naperian logarithms, but again modern usage tends to favour the expression natural logarithms. If common logarithms to the base 10 are used, it is not usual to indicate the base. For instance, since $10^4 = 10\,000$ we simply write log $10\,000 = 4$.* The base is only indicated if it is a base other than 10.

If our reader has not studied algebra previously, or may not appreciate the difference between algebraic terms and expressions such as $3a$, a^3 and $a + 3$, it would be advisable at this juncture to read through and take due note of the basic algebra which introduces Chapter 2.

Now let

M be a first number so that $M = 10^a$, and hence log $M = a$

and N be a second number so that $N = 10^b$, and hence log $N = b$.

---

* There has been a recent recommendation that the abbreviation for common logarith be lg. We shall use the abbreviation log until such time as it becomes necessary to consider logarithms other than common logarithms.

In a previous article, dealing with denary numbers, we established certain relationships from which can deduce that:

$$10^a \times 10^b = 10^{(a+b)}$$
$$10^a \div 10^b = 10^{(a-b)}$$
$$(10^a)^n = 10^{(a \times n)} = 10^{an}$$
$$\sqrt[n]{10^a} = 10^{(a \div n)} = 10^{a/n}$$
$$10^0 = 1$$
$$10^{-a} = \frac{1}{10^a}$$

Let us now consider the first of these relationships.

$$MN = M \times N = 10^a \times 10^b = 10^{(a+b)}$$

and if $\log M = a$ and $\log N = b$ then the log of $MN = a + b = \log M + \log N$.

Proceeding in a similar manner with the remaining relationships:

$$\log \frac{M}{N} = \log M - \log N$$

$$\log M^n = n \times \log M = n \log M$$

$$\log \sqrt[n]{M} = \frac{\log M}{n}$$

$$\log 1 = 0$$

$$\log \frac{1}{M} = -\log M.$$

Example:

If $\log x = a$, write down the expressions for:

(i) $\log (10x)$;   (ii) $\log \left(\dfrac{x}{10}\right)$;   (iii) $\log (x^2)$;   (iv) $\log (1\,000x)$

assuming all logarithms are to the base 10.   (C.G.L.I.)

Answers:

(i) $1 + a$ (ii) $a - 1$
(iii) $2a$ (iv) $3 + a$

Note: The problem said 'write down' the answers, and this means what it says. The examiner expects no working, but to show how the answers were obtained, the following are the thought processes which resulted in the answers.

$$\log 10x = \log (10 \times x) = \log 10^1 + \log x = 1 + a$$

$$\log \frac{x}{10} = \log x - \log 10 = a - 1$$

$$\log x^{\frac{1}{2}} = 2 \log x = 2a$$

$$\log 1\,000.x = \log (1\,000 \times x) = \log 1\,000 + \log x$$
$$= \log 10^3 + \log x = 3 - a$$

Example:

If $\log (xy) = a + b$ and $\log x = a$, explain why $\log y = b$. Find in terms of a and b:

(i) $\log \left(\dfrac{x}{y}\right)$;   (ii) $\log (x^2)$;   (iii) $\log (xy^2)$;   (iv) $\log \sqrt{xy}$

(C.G.L.I.)

The log of a product is the sum of the separate logs.

Hence $\log (xy) = \log x + \log y$

We are told $\log (xy) = a + b$

$\therefore \log x + \log y = a + b$

and since $\log x = a$

by subtracting these equations $\log y = b$

(i) $\log \dfrac{x}{y} = \log x - \log y = a - b$

(ii) $\log x^2 = 2 \log x = 2a$

(iii) $\log xy^2 = \log x + \log y^2 = \log x + 2 \log y$
$$= a + 2b$$

(iv) $\log (\sqrt{xy}) = \dfrac{\log x + \log y}{2} = \dfrac{a + b}{2}$

Answers:

(i) $a - b$

(ii) $2a$

(iii) $a + 2b$

(iv) $\dfrac{a + b}{2}$

Tables of logarithms

Let us presume that our reader was faced with the problem of establishing a value for

$$\frac{453{\cdot}7 \times 10{\cdot}14^2}{\sqrt[3]{141} \times 0{\cdot}08}.$$

If a calculator is available which is capable of providing an answer to the computation it would be a waste of valuable time not to use it. In particular, the result could be obtained with very great accuracy. If a calculator were not available, a slide rule could be used, but it would need very careful manipulation to obtain an answer which could be relied upon to three significant figures. If formal arithmetic were to be used, the computation would be extremely tedious. The number of individual computations would be so great that the probability of an error or errors would be quite high. If, however, we recall the rules for logarithmic computation from the previous article, the log of the answer to the above problem would be:

$$\left\{ \log 453{\cdot}7 + 2 \log(10{\cdot}14) - \left(\frac{\log 141}{3} + \log 0{\cdot}08 \right) \right\}.$$

If we can find the log of any number, and also a number corresponding to a log, then the tedious processes of multiplication, division, the determination of powers and the extraction of roots can be replaced by simple addition and subtraction together with multiplication and division by very simple numbers.

We know that we can express any number in standard form by stating a coefficient between unity and ten, together with 10 raised to a power. For revision:

$$65 = 6{\cdot}5 \times 10^1$$
$$6\,500 = 6{\cdot}5 \times 10^3$$
$$0{\cdot}065 = 6{\cdot}5 \times 10^{-2}$$
$$0{\cdot}000\,065 = 6{\cdot}5 \times 10^{-5}$$

Applying the rules of logarithms, and using common logarithms (i.e. logs to the base 10):

$$\log 10^1 = 1,\ \log 10^3 = 3,\ \log 10^{-2} = -2 \text{ and } \log 10^{-5} = -5$$

and since $$\log (ab) = \log a + \log b$$

then $$\log 65 = (\log 6{\cdot}5) + 1$$
$$\log 6\,500 = (\log 6{\cdot}5) + 3$$
$$\log 0{\cdot}065 = (\log 6{\cdot}5) - 2$$
$$\log 0{\cdot}000\,065 = (\log 6{\cdot}5) - 5$$

Hence the common logarithm of any number which consists of 65 followed by zeros, or 65 preceded by a decimal point and zeros, can be determined if we can find the common logarithm of 6·5. It follows that we can find the common logarithm of any positive number, no matter how large or how small, if we can find the common logarithm of any number between unity and 10. (The common logarithm of unity is zero.) It is for this reason that common logarithms are so convenient. In future we shall not use the full expression 'common logarithm' but the conventional abbreviation 'log'.

With the aid of a series it is possible to calculate the log of any number between 1 and 10. This has been done for us, and the information has been collected in tables. Many tables of logs have been published. Those recommended by the author for use in the initial studies of technician mathematics are *Four-figure Mathematical Tables* by Frank Castle, published by Macmillan. The title should be carefully noted, as Mr Castle has compiled several books of mathematical tables. As the title indicates, the values are given to a significance of four figures. The last figure is therefore suspect, particularly if several values are added and/ or subtracted. In general, the use of four-figure tables produces reliable answers to three significant figures. If a greater degree of accuracy is required, tables with more than four figures should be used. In engineering manufacture, the accurate location of holes often necessitates the use of seven-figure tables. For the degree of accuracy we require at present, four-figure tables will suffice. Castle's *Four-figure Mathematical Tables* are recommended because, in addition to tables of logs, they contain many other tables which we shall eventually find extremely useful.

The reader should now open the tables at the pages headed 'logarithms'. He will observe that there are no decimal points, since all we need from the tables are significant figures. The first two significant figures of the number whose log is required are given in the extreme left-hand column. There are ten columns of four-figure numbers corresponding to the third significant figure of the number, and finally nine vertical columns of small additions for the fourth significant figure of the number. The tables produce the decimal portion of a logarithm; since the logs of numbers between unity and ten lie between zero and unity, and noting that a decimal point is implied previous to every set of four figures:

$$\log 3\cdot000 = 0\cdot477\ 1$$

$$\log 3\cdot100 = 0\cdot491\ 4$$

$$\log 3\cdot170 = 0\cdot501\ 1$$

$$\log 3\cdot174 = 0\cdot501\ 7 \text{ (obtained from } 5\ 011 + 6)$$

This decimal portion is called the *mantissa* of the log, and since the log of any number between unity and ten lies between zero and unity, *the mantissa is always positive.*

Using our rules for logarithmic computation, and the mantissas given above:

$$\log 30 = \log(3 \times 10) \qquad = \log 3 + \log 10$$
$$= 0{\cdot}477\,1 + 1 \qquad = 1{\cdot}477\,1$$

$$\log 3\,100 = \log(3{\cdot}1 \times 10^3) = \log 3{\cdot}1 + \log 10^3$$
$$= 0{\cdot}491\,4 + 3 \qquad = 3{\cdot}491\,4$$

$$\log 317 = \log(3{\cdot}17 \times 10^2) = \log 3{\cdot}17 + \log 10^2$$
$$= 0{\cdot}501\,1 + 2 \qquad = 2{\cdot}501\,1$$

$$\log 31{\cdot}74 = \log(3{\cdot}174 \times 10) = \log 3{\cdot}174 + \log 10$$
$$= 0{\cdot}501\,7 + 1 \qquad = 1{\cdot}501\,7$$

Suppose we had to obtain $\log 0{\cdot}03$:

$$\log 0{\cdot}03 = \log(3 \div 100) = \log 3 - \log 100 = 0{\cdot}477\,1 - 2$$

We will now pause a moment and remember that the mantissa is positive. The whole number portion of a logarithm is called the *characteristic*, and here we have the combination of a negative characteristic and a positive mantissa. *We show that only the characteristic is negative by placing the negative sign above the characteristic*, referring to it as a bar, $\bar{2}$ being read as 'bar two'. Thus:

$$\log 0{\cdot}03 = \bar{2}{\cdot}477\,1, \text{ implying } -2 + 0{\cdot}477\,1$$

similarly:

$$\log 0{\cdot}003\,17 = \log(3{\cdot}17 \times 10^{-3}) = \log 3{\cdot}17 + \log 10^{-3}$$
$$= 0{\cdot}501\,1 + (-3) = \bar{3}{\cdot}501\,1$$

The characteristic of a log is the power of 10 when the number is expressed in standard form. It is the number of decimal places we have to traverse to produce a number between unity and ten. If we move to the right it is negative, while if we move to the left it is positive. If we try this out we shall find the following simple methods of determining characteristics:

(a)   If the number is greater than unity, the characteristic is positive and one less than the number of figures to the left of the decimal point.

(b)   If the number lies between unity and zero, the characteristic is negative and one more than the number of zeros immediately after the decimal point.

Hence

the characteristic for 3·142   is 0
                for 31·42   is 1
                for 0·031 42   is $\bar{2}$
                for 0·314 2   is $\bar{1}$
                for 0·000 314 2 is $\bar{4}$

The logs of 3·142, 31·42, 0·031 42, etc. all have the same positive mantissa of 0·497 2.

If the first two significant figures of the number lie between 10 and 20, the differences in the final columns of log tables are appreciable. In order to obtain as accurate a mantissa as possible, we use two different sets of small differences for the last significant figure. The set we use depends upon whether the line of four figures in the log tables is the upper row or the lower row. For example:

$$\log 1·236 = 0·092\ 0\ \text{(obtained from 0899 + 21)}$$

but $$\log 1·276 = 0·105\ 8\ \text{(obtained from 1038 + 20)}$$

If we require to find the log of a number which has more than four significant figures, it would be most unwise to jump to the conclusion that we should round off the number to four figures and then look up its logarithm. We should work as accurately as the tables allow, and our procedure should be decided by the steps in the small differences columns. For example:

$$\log 3·095 = 0·490\ 7$$

and $$\log 3·096 = 0·490\ 9$$

in which case log 3·095 5 could be reasoned as 0·490 8.

Our best course is to find the logs of the nearest four significant figures above and below and allocate the mantissa by proportion.

Example:

Find log 904·457.
The characteristic is 2:

$$\log 904·5 = 2·956\ 4$$
$$\log 904·4 = 2·956\ 4$$

Hence, since the logs are the same:

$$\log 904·457 = 2·956\ 4$$

Example:

Find log 14·733 2.
The characteristic is 1:

$$\log 14{\cdot}74 = 1{\cdot}168\ 5$$
$$\log 14{\cdot}73 = 1{\cdot}168\ 2$$

14·733 2 is about one-third of the way from 14·73 to 14·74.

Hence we take $\qquad \log 14{\cdot}733\ 2 = 1{\cdot}168\ 3$

Antilogarithms

By adopting the rules of logarithmic computation to obtain the value of:

$$45{\cdot}26 \times 0{\cdot}407\ 2 \times 317{\cdot}4$$

our first step is to find the logarithms of each of these numbers and add them together. We then have the logarithm of the answer. As a final step, we have to convert this logarithm to the number which is the answer. We shall be reversing the process of finding the logarithm of a number; consequently the operation is called finding an *antilogarithm*, colloquially known as an 'antilog'.

In effect, the problem is 'if $N = 10^a$, and we are given the value of a, what is the value of N?' Most 'scientific' calculators provide a facility for performing this task, in which case the task can be rapidly, and using care, accurately accomplished. However, if a calculator is not available, our reader must be aware of some other method.

Returning for a moment to common logarithms, the common logarithm of a number consists of two parts:

(a) a mantissa, in the form of a positive decimal fraction between zero and unity;
(b) a characteristic, in the form of a positive or negative whole number, or occasionally zero.

If, in finding a common logarithm, the significant figures of a number are used to decide the mantissa, then conversely, when finding an antilog, the mantissa is used to decide the significant figures of the number. The characteristic is then used to fix the position of the decimal point. In finding the significant figures corresponding to a mantissa, we could use the logarithm tables in reverse. For convenience we use a separate set of tables, so that in looking up an antilog we follow the same general method as looking up a log. This convenience is adversely balanced by the possibility of using the wrong set of tables. The reader is now advised to open Castle's *Four-figure Tables* to the antilogarithm tables and to write in very large capitals in red, across the top of each page, ANTI-LOGS.

As already indicated, the reading of antilog tables follows the same general principle as reading log tables. The reader should now check the following sets of significant figures obtained from the mantissas quoted:

$$
\begin{array}{ll}
0{\cdot}100\ 0 & 1\ 259 \\
0{\cdot}180\ 0 & 1\ 514 \\
0{\cdot}186\ 0 & 1\ 535 \\
0{\cdot}187\ 4 & 1\ 539\ (\text{obtained from } 1\ 538 + 1)
\end{array}
$$

Having obtained the significant figures, the decimal point has to be positioned according to the characteristic. Reversing the rules used to obtain characteristics, we reason:

(a)   If the characteristic is positive, the antilog is greater than unity, while the number of the figures to the left of the decimal point is one more than the magnitude of the characteristic.

(b)   If the characteristic is negative, the antilog lies between zero and unity, and the number of zeros immediately after the decimal point is one less than the magnitude of the characteristic (after the negative sign has been discarded).

Example:

Find the antilogs of:

$$
\text{(a)}\quad 2{\cdot}517\ 6 \qquad \text{(b)}\quad \bar{3}{\cdot}098\ 5
$$

(a)   From the antilog tables, the significant figures of the number are $3\ 289 + 5 = 3\ 294$. The positive characteristic tells us the antilog exceeds unity, and that there are three figures to the left of the decimal point. Hence the value required is $329{\cdot}4$.

(b)   From antilog tables, the significant figures of the answer are $1\ 253 + 1 = 1\ 254$. The negative characteristic tells us the answer lies between zero and unity, and there are two zeros after the decimal point. Hence the value required is $0{\cdot}001\ 254$.

Answers:

$$
\text{(a)}\quad 329{\cdot}4 \qquad \text{(b)}\quad 0{\cdot}001\ 254
$$

Manipulation of logarithms

The basic rules of logarithmic computation are:

$$\log abc = \log a + \log b + \log c$$

$$\log \frac{a}{b} = \log a - \log b$$

$$\log a^n = n \log a$$

$$\log \sqrt[n]{a} = \frac{\log a}{n}$$

$$\log \sqrt[n]{a^m} = \frac{m \log a}{n}$$

The manipulation of logarithms follows the same rules as ordinary arithmetic, but special care must be taken if a logarithm has a negative characteristic and positive mantissa. Let us illustrate the processes with examples.

Example:

Add 2·079 2 and $\bar{1}$·437 2.

$$
\begin{array}{r}
2 \cdot 079\ 2 \\
\bar{1} \cdot 437\ 2 \\
\hline
1 \cdot 516\ 4
\end{array}
$$

The formal method is followed for the mantissa, and since nothing is carried forward from the mantissa, adding 2 to -1 produces 1 for the characteristic.

Example:

Add $\bar{2}$·817 6, $\bar{1}$·417 8 and 3·007 4.

$$
\begin{array}{r}
\bar{2} \cdot 817\ 6 \\
\bar{1} \cdot 417\ 8 \\
3 \cdot 007\ 4 \\
\hline
1 \cdot 242\ 8 \\
\hline
1\quad 21
\end{array}
$$

Once more we commence with adding mantissas, but in this case we have to 'carry 1'. The characteristic is the addition of 1, 3, -1 and -2, which equals 1.

Example:

Subtract $\overline{2}$·509 7 from 4·959 9.

$$\begin{array}{r} 4\cdot959\ 9 \\ \overline{2}\cdot509\ 7 \\ \hline 6\cdot450\ 2 \end{array}$$

Here there are no differences from ordinary arithmetic until we come to the subtraction of characteristics, when we obey the rule 'change the sign of the bottom line and add'. $4 - (-2)$ is equal to $4 + 2$, or 6.

Example:

Subtract $\overline{3}$·470 2 from $\overline{4}$·395 7.

$$\begin{array}{r} \overline{4}\cdot395\ 7 \\ \overline{3}\cdot470\ 2 \\ \hline \overline{2}\cdot925\ 5 \end{array}$$

Commencing at the mantissas, we proceed according to formal arithmetic until we come to the point of 'borrowing one' from the top characteristic. This makes it -5, and then:

$$-5 - (-3) = -5 + 3 = -2$$

which in logarithmic notation is written $\overline{2}$.

Example:

Multiply $\overline{1}$·307 6 by 3.

$$\overline{1}\cdot307\ 6 \times 3 = \overline{3}\cdot922\ 8$$

Here there is no 'carrying forward' and we simply multiply both the mantissa and characteristic by 3.

Example:

Multiply $\overline{1}$·407 8 by 4.

$$\overline{1}\cdot407\ 8 \times 4 = \overline{3}\cdot631\ 2$$

We proceed along the mantissa until the stage '$4 \times 4 = 16$, put down 6 and carry 1'. Thence $(4 \times -1) = -4$, and since there is 1 to carry, $-4 + 1 = -3$, which is written $\overline{3}$.

Example:

Divide $\bar{3}{\cdot}507\ 2$ by 3.

$$\frac{\bar{3}{\cdot}507\ 2}{3} = \frac{\bar{3} + 0{\cdot}507\ 2}{3}$$

$$= \bar{1} + 0{\cdot}169\ 1 \text{ (rounding off)}$$

$$= \bar{1}{\cdot}169\ 1$$

The characteristic was exactly divisible by three, and we simply divided the characteristic and the mantissa by 3.

Example:

Divide $\bar{3}{\cdot}649\ 2$ by 5.

Here we have to obtain $\dfrac{\bar{3}{\cdot}649\ 2}{5}$ which can be written:

$$\frac{-3 + 0{\cdot}649\ 2}{5}$$

We want to make the characteristic divisible by 5, without altering the magnitude of $(-3 + 0{\cdot}649\ 2)$. This we can accomplish by subtracting 2 from the characteristic and adding it to the mantissa:

$$\frac{-3 + 0{\cdot}649\ 2}{5} = \frac{-5 + 2{\cdot}649\ 2}{5} = -1 + 0{\cdot}529\ 8 \text{ (rounding off)}$$

which we write as $\bar{1}{\cdot}529\ 8$.

There are many similarities between the manufacture of an article in industry and the solution of a problem in mathematics or science. In industry, the raw material is processed using the tools available, according to a planning schedule, often with intermediary checks, the article eventually being submitted for final inspection. The raw material of a problem in mathematics and science is the data provided in the question; the processing tools are the various rules of arithmetic, formulae, mathematical tables, slide rules, and so on.

Raw materials can be processed in many ways; for instance, metal can be removed by milling, shaping or turning. A workman will only use those tools that he feels competent to handle skilfully. There are many short cuts and trick methods that a competent artisan can use, but he rarely employs them unless he knows exactly what he is doing. Furthermore, if the industrial trainee uses his eyes in the workshop, it will soon become evident that the accurate, respected craftsman is the one who invariably lays out, uses and stores his tools in an efficient manner.

The same basic principles apply to the solution of problems in mathematics. An orderly manner of setting out calculations is desirable, since:

(a) it shows the reader the various steps by which the solution was obtained;

(b) it allows an easier detection of errors when such errors unfortunately occur;

(c) it encourages pride in achievement. Industry has no vacancies for untidy, slovenly and inaccurate employees.

The expression 'rough check' is in quite common use, but in some instances the adjective 'rough' is somewhat unfortunate. So-called rough checks can be surprisingly accurate. The purpose of a rough check is to give an indication of the accuracy of an answer by using approximations. If the approximations are made with care, and chosen so that the errors of the approximations tend to neutralize each other, then rough checks will most certainly indicate violent errors, such as the misplacing of a decimal point.

The use of logarithms and rough checks is demonstrated in the worked examples which follow.

Example:

The volume of a cylinder is given by the formula:

$$V = 0.785\ 4\ d^2 h$$

and 1 m³ contains 1 000 litres. What is the capacity, in litres, of a cylindrical tank of diameter 1·05 m and length 2·16 m?

$$V \text{ (in m}^3) = 0.785\ 4\ d^2 h$$

$$\text{Capacity in litres} = 1\ 000 \times 0.785\ 4\ d^2 h$$
$$= 785.4\ d^2 h$$

$$\therefore \text{ Capacity} = 785.4 \times 1.05^2 \times 2.16 = 1\ 871$$

No.	Log
785·4	2·895 1
1·05	0·021 2
1·05	0·021 2
2·16	0·334 5
1 871	3·272 0

Rough check:

$$750 \times 1 \times 1 \times 2 = 1\ 500$$

Answer:

Capacity = 1 870 litres

Example:

The time t seconds for one complete oscillation of a simple pendulum is given by the formula:

$$t = 2\pi \sqrt{\frac{L}{g}}$$

Find the value of t if $\pi = 3 \cdot 142$, $L = 0 \cdot 5$ and $g = 9 \cdot 81$.

$$t = 2\pi \sqrt{\frac{L}{g}} = (2 \times 3 \cdot 142) \times \sqrt{\frac{0 \cdot 5}{9 \cdot 81}} = 1 \cdot 419$$

No.		Log
0·5		$\bar{1}$·699 0
9·81		0·991 7
Subt		$\bar{2}$·707 3
Root	2	$\bar{1}$·353 6
2		0·301 0
3·142		0·497 2
1·419		0·151 8

Rough check:

$$6 \times \sqrt{\frac{1}{20}} = 6 \times \text{(between } \tfrac{1}{4} \text{ and } \tfrac{1}{5})$$
$$= \text{between } 1 \cdot 5 \text{ and } 1 \cdot 2$$

Answer:

Time $= 1 \cdot 42$ s

Example:

The area A of a triangle with sides of length a, b and c is given by the formula:

$$A = \sqrt{\{s(s-a)(s-b)(s-c)\}}$$

where s is the semi-perimeter, i.e.

$$s = \frac{a+b+c}{2}$$

Find the area of a triangle having sides of length 78·6, 117·4 and 154·8 mm. Give the answer in square millimetres.

We must note carefully that the formulae include additions and subtractions and these must be eliminated before logarithms are used.

$$s = \frac{a + b + c}{2} = \frac{78 \cdot 6 + 117 \cdot 4 + 154 \cdot 8}{2} = \frac{350 \cdot 8}{2} = 175 \cdot 4$$

$$s - a = 175 \cdot 4 - 78 \cdot 6 = 96 \cdot 8$$
$$s - b = 175 \cdot 4 - 117 \cdot 4 = 58 \cdot 0$$
$$s - c = 175 \cdot 4 - 154 \cdot 8 = 20 \cdot 6$$
$$A = \sqrt{(175 \cdot 4 \times 96 \cdot 8 \times 58 \cdot 0 \times 20 \cdot 6)} = 4\,504$$

No.		Log
175·4		2·244 0
96·8		1·985 9
58·0		1·763 4
20·6		1·313 9
Root	2	7·307 2
4 504		3·653 6

Rough check:

$$\sqrt{(175 \times 100 \times 60 \times 20)}$$
$$= \sqrt{14\,000\,000}$$
$$= \text{about } 4\,000$$

Answer:

Area $= 4\,500 \text{ mm}^2$

Example:

The centre distance C millimetres for helical gears with shafts at right angles is given by the formula:

$$C = \frac{Nm(1 + \sqrt[3]{R^2})^{1 \cdot 5}}{2}$$

where
$N =$ number of teeth on the pinion
$m =$ metric module, in millimetres
$R =$ gear ratio

Find the centre distance for a gear ratio of 3, the metric module being 2 mm and the pinion having 24 teeth.

$$C = \frac{24 \times 2(1 + \sqrt[3]{9})^{1 \cdot 5}}{2}$$

$\sqrt[3]{9}$ from tables (page 26) $= 2 \cdot 080$
$$C = 24(1 + 2 \cdot 080)^{1 \cdot 5} = 24(3 \cdot 080)^{1 \cdot 5}$$

Log $3 \cdot 080^{1 \cdot 5} = 1 \cdot 5 \log 3 \cdot 080 = 1 \cdot 5 \times 0 \cdot 488\ 6 = 0 \cdot 732\ 9$
Log answer $= \log 24 + 0 \cdot 732\ 9 = 1 \cdot 380\ 2 + 0 \cdot 732\ 9 = 2 \cdot 113\ 1$
Answer $= \text{antilog } 2 \cdot 113\ 1 = 129 \cdot 7$

Rough check: $24(3)^{1\cdot5} = 24 \times \sqrt[2]{3^3} \quad = 24 \times \sqrt[2]{27}$
$$= 24 \times \text{ about } 5 = \text{ about } 120$$

Answer:

Centre distance $= 130$ mm

Before we leave the subject of determining a logarithm, in effect, determining a logarithm is a solution to the problem:

'If $N = 10^a$, what is the value of a?'

Many 'scientific' calculators can provide a facility for this task, but if a calculator can provide an answer to a computation directly without the use of logarithms, it would be illogical to use the calculator to obtain logarithms and go through the computation using those logarithms.

However, if our reader is faced with the problem of determining the logarithm of a number, and no other task, it should be noticed that if a calculator is used to determine the logarithm of a positive decimal fraction of less than unity, the display is not given in the form of a wholly negative value, and the negative sign in the display should be carefully noted.

For example, with tables,

$$\log 0\cdot5 = \bar{1}\cdot699\ 0, \quad \text{which implies } (-1 + 0\cdot699\ 0)$$

but with a calculator which is capable of determining logarithms

$$\log 0\cdot5 = -0\cdot301\ 030,$$

to six decimal places, and the negative sign preceding the logarithm should be particularly noted.

Similarly, $\log 0\cdot073\ 41$, by calculator, is $-1\cdot134\ 245$, to six decimal places.

1.6.3
Mathematical tables other than logarithms

Use of tabulated data

Many students refer to books which contain tabulated data as 'log tables'. It must be admitted that considerable use is made of logarithms and antilogarithms, but the use of the expression 'log tables' often leads a student to the view that the tables of logs and antilogs are the only important tables in such a publication. The reader has been impressed with the importance of not using logarithms unless it is essential. Let us consider certain other tables in Castle's *Four-figure Mathematical Tables*.

At times during a complete Technicians' course the reader will refer to every set of tables in Castle's *Four-figure Mathematical Tables*, but our employment of certain of the tables must be left until we have studied the topics which involve their use. The various tables will be introduced at appropriate times during the course. We shall now consider tables of square roots and of reciprocals.

Tables of square roots (and of squares)

Castle's *Four-figure Mathematical Tables* include tables which are entitled 'square roots', but before these are considered in detail let us look at the interesting tables on pages 38 and 39. These will be found to be most useful if a number has only one or two significant figures. If we take as an example the number 54, the tables indicate:

$$54^2 = 2\,916 \qquad\qquad 54^3 = 157\,464$$

(exactly, not to four significant figures)

$$\sqrt{54} = 7 \cdot 348 \qquad\qquad \sqrt[3]{54} = 3 \cdot 780$$
$$\sqrt{540} = 23 \cdot 238 \qquad\qquad \sqrt[3]{540} = 8 \cdot 143$$
$$\sqrt[3]{5\,400} = 17 \cdot 544 \quad \text{and} \quad \tfrac{1}{54} = 0 \cdot 018\,52$$

If we wish to find the square root of a number with three or four significant figures, the tables shown on pages 40–43 inclusive are used. The square roots fall into 2 groups, those of numbers 1 to 10 and those of numbers 10 to 100. The square roots of numbers from 1 to 100 are extracted in a manner similar to logarithms. The reader should check the values:

$$\sqrt{8 \cdot 78} = 2 \cdot 963$$
$$\sqrt{5 \cdot 507} = 2 \cdot 346$$
$$\sqrt{42 \cdot 4} = 6 \cdot 512$$
$$\sqrt{15 \cdot 84} = 3 \cdot 980$$

The tables can be used for positive numbers outside the range of 1 to 100 by adjusting the decimal point of the result.

We write the number in the form $a \times 10^n$, where n is an *even* number (positive or negative) and a lies between 1 and 100.

Now $$\sqrt{(a \times 10^n)} = \sqrt{a} \times 10^{n/2}$$

Hence for $\sqrt{847}$, we note that $847 = 8 \cdot 47 \times 10^2$, look up the root of $8 \cdot 47$ and multiply it by 10^1. Similarly, for $\sqrt{0 \cdot 164\,2}$, we note that

$0.164\ 2 = 16.42 \times 10^{-2}$, look up $\sqrt{16.42}$ and multiply by 10^{-1}, i.e. divide by 10. As examples:

$$\sqrt{59\ 600} = \sqrt{(5.96 \times 10^4)} = \sqrt{5.96} \times 10^2$$
$$= 2.441 \times 100 = 244.1$$

and $\qquad \sqrt{0.189\ 4} = \sqrt{(18.94 \times 10^{-2})} = \sqrt{18.94} \times 10^{-1}$

$$= \frac{4.352}{10} = 0.435\ 2$$

Although the tables are headed 'square roots', we can use them to find squares by reversing the former procedure. If we wish to know the square of 5.642, we have to determine the number of which 5.642 is the square root. Using the tables, we find the nearest value below 5.642. This is 5.639, the square root of 31.8. The difference to be added is 3, which could be a last significant figure of 3 or 4. If it could be either, we may as well be consistent and compute to an even value, so we take 31.84 as the number. If the number whose square is required lies outside the range 1 to 10, we can express the number in standard form, and note:

For example
$$(a \times 10^n)^2 = a^2 \times 10^{2n}$$
$$0.021\ 74^2 = (2.174 \times 10^{-2})^2$$
$$= 2.174^2 \times 10^{-4}$$
$$= 4.724 \times 10^{-4}$$
$$= 0.000\ 472\ 4$$

(The reader should note that in this example, the last significant figure could apparently be 3, 4, 5 or 6. The mean is 4.5, which we round off evenly to 4.)

Reciprocals of numbers

The *reciprocal* of a number is its direct inversion. The number 8.47 can be written as $\dfrac{8.47}{1}$, hence its reciprocal is $\dfrac{1}{8.47}$. The reciprocal of a single number is therefore unity divided by that number. The reciprocal of a fraction such as $\frac{23}{37}$ is $\frac{37}{23}$.

Tables of reciprocals of numbers appear in Castle's *Four-figure Mathematical Tables*. On pages 38 and 39, the end column headed $\dfrac{1}{n}$ gives directly the reciprocals of whole numbers from unity to 100 inclusive. If the number has three significant figures or more, the reciprocal can be found with the aid of the tables on pages 44 and 45.

As a number tends to increase, its reciprocal tends to decrease. The reader should now open Castle's *Four-figure Mathematical Tables* to pages 44 and 45, and note that the groups of four-figure numbers tend to diminish. Consequently the mean differences in the end column have to be subtracted, and our reader is therefore advised to write plainly in large capitals in red, at the top of each table, SUBTRACT MEAN DIFFERENCES.

Example:

If resistors of magnitude R_1, R_2 and R_3 are connected in parallel, the equivalent single resistance R can be obtained from the formula:

$$\frac{1}{R} = \frac{1}{R_1} + \frac{1}{R_2} + \frac{1}{R_3}$$

Find the value of R when $R_1 = 5.37$ ohms, $R_2 = 1.714$ ohms and $R_3 = 12.3$ ohms.

$$\frac{1}{R_1} = \frac{1}{5.37} = 0.186\ 2$$

$$\frac{1}{R_2} = \frac{1}{1.714} = 0.583\ 5$$

$$\frac{1}{R_3} = \frac{1}{12.3} = 0.081\ 3$$

By addition

$$\frac{1}{R_1} + \frac{1}{R_2} + \frac{1}{R_3} = 0.186\ 2 + 0.583\ 5 + 0.081\ 3 = 0.851\ 0$$

$$\frac{1}{R} = 0.851\ 0$$

$$\therefore R = \frac{1}{0.851\ 0} = 1.175$$

Answer:

$$R = 1.18 \text{ ohms}$$

Example:

If a segment of a circle has a height of h and a chord length of w, an approximate formula for the area of the segment is:

$$A = \frac{4h^2}{3} \sqrt{\left(\frac{w^2}{4h^2} + 0.4 \right)}$$

Use this formula to find the area of a segment when $h = 0.8$ mm and $w = 4.8$ mm.

$$\frac{w^2}{4h^2} = \frac{4.8 \times 4.8}{4 \times 0.8 \times 0.8} = 9 \text{ (by cancellation)}$$

$$9 + 0.4 = 9.4$$

$$\sqrt{9.4} = 3.066 \text{ (from tables of square roots)}$$

$$A = \frac{4h^2}{3} \times 3.066 = \frac{4 \times 0.8 \times 0.8 \times 3.066}{3}$$

$$= 2.56 \times 1.022 \text{ (by simplification and cancellation)}$$

$$= 2.616\ 32 \text{ (by formal multiplication)}$$

Since we have used four-figure tables during the sequence of operations, as a final step we should now round off to three significant figures.

Answer:

$$\text{Area} = 2.62 \text{ mm}^2$$

1.6.4
The slide rule

Until as recently as 1970, the slide rule was an aid to computation which could be expected to be found in the 'tool kit' of the majority of students engaged on a course in further or higher education which included a study of mathematics. However, the introduction of portable electronic calculators, which have many advantages over the slide rule, particularly the digital display instead of the reading of scales, has led to the slide rule losing a certain amount of favour.

It has been repeatedly stated in this book that if a calculator is available which can be used for a particular computation, the use of that calculator should be a natural first choice. Traditional arithmetic can be very time consuming and the number of individual computations leads to greater possibilities of error. Time can be saved by using aids to computations such as a slide rule or mathematical tables. In general, a slide rule will provide a quicker but less accurate result than that obtained from tables.

Like any other tool used to perform tasks, the accuracy of the results obtained depends upon the accuracy of the tool itself and upon the skill of the user. An accuracy of three significant figures, provided a slide rule is used correctly, is relatively easy to obtain if the first of the significant figures is 1 or 2, but not so easily if it is 8 or 9. The results obtained from the use of a slide rule can well be thought of as approximations, but this should not be interpreted as an invitation to careless and inaccurate working.

We shall confine our attention to the use of the slide rule for multiplication, division and computing of square roots and squares. Since the principle of its use is based on logarithms, addition and subtraction processes cannot be performed on a slide rule. The slide rule consists of three basic parts, these being the *stock*, the *slide*, and a moving frame, known as a *cursor*, which carries an engraved reference line. The cursor slides in guides on the stock. There are four basic scales, known as the *A*, *B*, *C* and *D* scales. The *B* scale is uppermost on the slide, the *C* scale being the lower scale on the slide. The *A* scale is on the stock and is opposite to and identical with the *B* scale on the slide. The *D* scale is on the stock and is opposite to and identical with the *C* scale on the slide.

If the scales are observed carefully, it will be noted that divisions are not of equal width. Furthermore, one division may represent 0·01 at one part of the scale, 0·1 at another, and other values elsewhere. This is because the scales are logarithmic and not linear. The log of 1 is zero, the log of 10 is 1, and the log of 100 is 2. Hence on the *A* and *B* scales the value of 10 lies halfway along the rule. The slide rule is usually used to produce the significant figures of an answer only; the position of the decimal point is invariably found by means of a rough check.

Multiplication

The principle for multiplication on the slide rule is that $\log(ab)$ $= \log a + \log b$. Fig. 1.6 shows how this operation is performed.

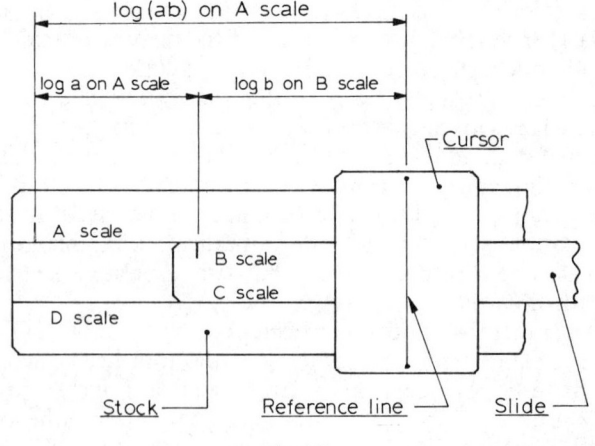

Fig. 1.6

Example:

Multiply 58·7 by 0·178.

> Move 1 on the slide opposite 5·87 on the *A* scale.
> Bring cursor over 1·78 on the *B* scale.

> Read off the answer on the *A* scale (significant figures are 1045).
> Position the decimal point by a rough check.

$$60 \times \tfrac{1}{6} = 10$$

Answer:

10·45

Division

Division is accomplished by reversing the sequence of the previous example, noting that if the 1 on the scale *B* comes out of the stock, the value on scale *A* can be found opposite 10 or 100 on scale *B*.

Example:

Divide 4·33 by 82·64.

> Move line on cursor over 4·33 on scale *A*.
> Move slide until 82·64 on scale *B* comes under cursor line.
> Answer is read on scale *A*, opposite 1, 10 or 100 of scale *B*. (In this case, opposite 100, significant figures are 524.) Find decimal point by trial.

$$4 \div 80 = \tfrac{1}{20} = 0.05$$

Answer:

0·052 4

The scales of some, but not all, slide rules include a scale for the direct determination of reciprocals, but if such a scale is not available, a reciprocal can be treated as a direct division with unity as the dividend. As an example, the reciprocal of 5·82 can be obtained from unity divided by 5·82.

Combined multiplication and division

To avoid unnecessary movements of the slide and the cursor, it is advisable to perform multiplications and divisions alternately.

Example:

Evaluate $\dfrac{58 \cdot 7 \times 1 \cdot 92 \times 0 \cdot 071}{63 \cdot 4 \times 0 \cdot 027}$.

This is obtained with the sequence:

$$58 \cdot 7 \times 1 \cdot 92 \div 63 \cdot 4 \times 0 \cdot 071 \div 0 \cdot 027$$

Move slide to bring 1 on scale *B* opposite 5·87 on scale *A*.
Move cursor line over 1·92 on scale *B*. (The cursor line indicates 5·87 × 1·92 on scale *A*.)
Move slide to bring 6·34 on scale *B* under cursor line. (The 1 on scale *B* is opposite the answer to $\dfrac{5 \cdot 87 \times 1 \cdot 92}{6 \cdot 34}$ on scale *A*.)

Move cursor line over 7·1 on scale *B*. (The answer to $\dfrac{5 \cdot 87 \times 1 \cdot 92 \times 7 \cdot 1}{6 \cdot 34}$ is now under the cursor line on scale *A*.)
Move slide so that 2·7 on scale *B* is under the cursor line. (The answer to $\dfrac{5 \cdot 87 \times 1 \cdot 92 \times 7 \cdot 1}{6 \cdot 34 \times 2 \cdot 7}$ is found on scale *A* opposite the 1 on scale *B* on the slide (significant figures are 467).)
Check on the decimal point.

$$\frac{60 \times 2 \times 0 \cdot 07}{60 \times 0 \cdot 028} = 5$$

Answer:

$$4 \cdot 67$$

Square roots and squares

If the slide is removed completely from the rule, the square roots of numbers on the *A* scale are found on scale *D* directly below. Similarly, the squares of the numbers on the *D* scale are found directly above on the *A* scale. No difficulties are caused with the determination of squares, but care should be taken in the determination of square roots that the correct range (1 to 10 or 10 to 100) is chosen on the *A* scale.

Example:

Find $\sqrt{161}$.

$$161 = 1 \cdot 61 \times 100$$
$$\sqrt{161} = \sqrt{(1 \cdot 61 \times 100)} = \sqrt{1 \cdot 61} \times 10$$

Move slide away from 1·61 on the *A* scale.
Move cursor over 1·61 on the *A* scale.
Read $\sqrt{1·61}$ under the cursor line on the *D* scale ($=1·27$).

$$\sqrt{161} = 1·27 \times 10 = 12·7$$

Answer:

12·7

Example:

Find $\sqrt{0·697}$.

$$0·697 = \frac{69·7}{100}$$

$$\sqrt{\frac{69·7}{100}} = \frac{\sqrt{69·7}}{10}$$

Move the slide away from 69·7 on the *A* scale.
Move the cursor line over 69·7 on the *A* scale.
Read $\sqrt{69·7}$ under the cursor line on the *D* scale ($=8·35$)

$$\sqrt{0·697} = \frac{8·35}{10} = 0·835$$

Answer:

0·835

Scales *C* and *D* can be used for multiplication and division in a similar manner to scales *A* and *B*, and since scales *C* and *D* are larger, theoretically greater accuracy can be obtained. On the other hand, since they only run from 1 to 10 instead of 1 to 100, there are many more occasions when a sub-answer has to be nominally multiplied or divided by 10 to keep it on the rule. The student is advised initially to sacrifice the greater accuracy obtainable with the *C* and *D* scales until he is competent in multiplying and dividing on the *A* and *B* scales.

The correct evaluation of problems involving combined multiplication and division including squares and/or square roots needs considerable care. It is advisable to perform operations one at a time rather than to run the risk of an incorrect answer due to incorrect manipulation. A slide rule is like any other engineer's tool. The skilled artisan can use a tool with dexterity only when he has practised its use over a suitable period. The reader is urged to have patience while learning the use of a slide rule. It is advisable to concentrate initially on multiplication and division, first with the A and B scales, then with the C and D scales, and to master these processes fully before proceeding further. This can then be followed by square roots, squares and combined multiplication and division. Other processes involving the use of other scales, such as reciprocals, log-log and so on, are better left to a later stage of studies.

If the reader cares to practise in the use of the slide rule, he may solve problems 1.3 with its aid, omitting questions involving powers higher than the square, and roots of a degree other than the square root.

Problems 1.2

1. P, Q, R, S are numbers whose common logarithms (i.e. logs to the base 10) are a, b, c and d respectively. Obtain in terms of a, b, c and d expressions for the common logarithms of:

(a) PQ (b) PR (c) QS

(d) $PQRS$ (e) PQS (f) $\dfrac{P}{Q}$

(g) $\dfrac{R}{S}$ (h) $\dfrac{PQ}{RS}$ (j) $\dfrac{PQR}{S}$

(k) P^2 (l) P^2Q (m) P^2Q^2

(n) P^2QR^2S (o) $\sqrt{P}$ (p) $\sqrt[4]{P}$

(q) $\sqrt[2]{(PQ)}$ (r) $\sqrt[3]{(P^2Q)}$ (s) $\sqrt[3]{(P^3QR^2)}$

(t) $\dfrac{PQR^2}{\sqrt{S}}$ (u) $\dfrac{\sqrt{P} \times (QS)}{\sqrt[3]{R}}$ (v) $\dfrac{P^4Q}{\sqrt[5]{S^2}}$

(w) $10P$ (x) $100S$ (y) $\dfrac{PQ}{100}$

(z) $\dfrac{100PQ}{\sqrt{R^3}}$ (aa) $\dfrac{R^2}{\sqrt{(100P)}}$ (ab) $\dfrac{P^3}{\sqrt[3]{(1\,000RS^2)}}$

2. Find the logs of the following numbers:

(a) 82	(b) 95	(c) 7
(d) 435	(e) 174	(f) 1 143
(g) 1 183	(h) 4·9	(j) 0·004 9
(k) 0·000 9	(l) 0·053 2	(m) 0·021 32
(n) 213·2	(o) 1·175	(p) 10^4
(q) 10^{-2}	(r) $5·372 \times 10^4$	(s) $8·168 \times 10^{-5}$
(t) 704·138	(u) 80 873	(v) 1·027 5
(w) 16·965	(x) 1·066 25	(y) 15·833 7

3. Find antilogarithms of:
   (a) 0·387 2   (b) 4·387 2   (c) 3·387 2
   (d) $\bar{1}$·387 2   (e) 0·000 0   (f) $\bar{2}$·400 0
   (g) 1·355 0   (h) 0·301 0   (j) $\bar{1}$·687 1
   (k) 2·531 1   (l) 0·137 1   (m) $\bar{4}$·590 2

The answers to the problems 4, 5, 6, 7 should be left in logarithm form.
There is no need to look up antilogarithms as a final step.

4. Perform the following additions:
   (a) 2·170 9   (b) $\bar{1}$·380 6   (c) $\bar{1}$·306 4   (d) 0·101 5
       $\bar{1}$·307 2 $\bar{1}$·070 4 $\bar{2}$·517 8 $\bar{1}$·170 6
       3·145 8 $\bar{2}$·691 5   3·508 1   2·938 4
                                                    1·001 7 $\bar{4}$·497 2

5. In the following, subtract the lower logarithm from the one above:
   (a)3·176 2   (b) 2·760 9   (c) 1·846 1   (d) $\bar{2}$·381 5
      2·548 1   3·428 1 $\bar{1}$·475 4 $\bar{4}$·778 3

6. Evaluate:
   (a) 5(0·174 8)   (b) 3(1·467 2)
   (c) 2($\bar{1}$·480 5)   (d) 4($\bar{1}$·450 6)

7. Evaluate:
   (a) $\dfrac{3·451\ 6}{4}$ (b) $\dfrac{\bar{5}·970\ 8}{5}$ (c) $\dfrac{\bar{1}·170\ 8}{4}$

   (d) $\dfrac{\bar{1}·981\ 3}{7}$ (e) $\dfrac{2(\bar{1}·497\ 2)}{3}$ (f) $\dfrac{3(\bar{2}·864\ 5)}{5}$

The reader should note that all formulae in questions 8 to 20 which
follow are commonly used in science and will gradually become of use as
studies develop. The problems below are to give him practice in the use of
logarithms, but the formulae which have been selected are not just
haphazard collections of symbols. They all have a practical application.

Our reader is reminded that when four-figure tables are used, answers
should be rounded off to three significant figures. It would be a useful
exercise for our reader to evaluate the problems using a calculator.

8. If $I = \dfrac{BD^3}{12}$, find I when $B = 5·75$ and $D = 12$.

9. If $F = \dfrac{4\pi^2 EI}{L^2}$, find F when $\pi = 3·142$, $E = 30 \times 10^6$, $I = 5·7$ and
   $L = 64$. (Give the answer in standard form.)

10. If $V = \dfrac{4\pi R^3}{3}$, find V when $\pi = 3·142$ and $R = 0·815$.

11. If $y = \dfrac{Fa^2b^2}{3EIL}$, find y when $F = 4\,480$, $a = 36$, $b = 84$, $E = 30$ $\times 10^6$, $I = 192$ and $L = 120$.

12. If $s = \dfrac{y^2 - u^2}{2a}$, find s when $v = 88$, $u = 22$ and $a = 1\cdot135$.

13. If $R_1 = \dfrac{RR_2}{R_2 - R}$, find R_1 when $R = 10\cdot54$ and $R_2 = 16\cdot23$.

14. If $D = \sqrt[3]{\dfrac{6V}{\pi}}$, and $\pi = 3\cdot142$, find D when
    (a) $V = 10\cdot4$; (b) $V = 0\cdot472$.

15. If $V = c\sqrt{mi}$, find V when $c = 120$, $m = 1\cdot508$ and $i = 3\cdot2 \times 10^{-3}$.

16. If $Q = 3\cdot09BH^{1\cdot5}$ and $B = 3\cdot5$, find the value of Q when
    (a) $H = 1\cdot414$; (b) $H = 0\cdot507$.

17. If $K = 0\cdot275\sqrt{\left(\dfrac{D^5 - d^5}{D^2 - d^2}\right)}$, find K when
    (a) $D = 2\cdot2$ and $d = 1\cdot6$; (b) $D = 0\cdot45$ and $d = 0\cdot37$.

18. If $L = \dfrac{1}{C(2\pi f)^2}$, find the value of L when $C = 0\cdot4 \times 10^{-6}$, $\pi = 3\cdot142$ and $f = 24 \times 10^3$.

19. Evaluate $H = \dfrac{M}{(d^2 + l^2)^{3/2}}$ when $M = 333$, $d = 15\cdot5$ and $l = 5\cdot2$.

20. If $P = kB^n f$, calculate by logarithms the value of P when $k = 0\cdot04$, $B = 1\cdot5$, $n = 1\cdot6$ and $f = 50$.

21. The A0 size of drawing paper is a rectangle having an area of 1 square metre with lengths of sides in the proportion of $\sqrt{2}$ to 1. Determine the lengths of the sides, in millimetres, each to three significant figures.

22. A pipe, of outside diameter D millimetres and inside diameter d millimetres, is used as a beam as part of the structure of a shipyard crane. The maximum stress σ N/mm^2 (which is incidentally equal to σ MN/m^2) for a particular loading can be determined from the formula:

$$\sigma = \frac{17\cdot6D}{\pi(D^4 - d^4)} \times 10^8 \text{ N/mm}^2$$

Find the value of σ when $D = 200$ mm and $d = 180$ mm. (Take π as $\frac{22}{7}$.)

23. If a barrel has a largest diameter of D, a diameter of the ends d and a length of L, a close approximation to its volume V is given by the formula:

$$V = \frac{\pi L}{12}(2D^2 + d^2)$$

If 1 cubic metre contains 1 000 litres, find the capacity, in litres, of a barrel where $D = 1\cdot2$ m, $d = 0\cdot8$ m and $L = 1\cdot5$ m. (Take π as $\frac{22}{7}$.)

24. A zone of a sphere reveals two circles of radii R and r, their distance apart being h. The diameter D of the sphere from which the zone was cut can be obtained from the formula:

$$\frac{D}{2} = \sqrt{\left\{ R^2 + \left(\frac{R^2 - r^2 - h^2}{2h} \right)^2 \right\}}$$

Find the value of D when $R = 75$ mm, $r = 51$ mm and $h = 28$ mm.

25. A circular blank is drawn into a circular shell in such a manner that the thickness of the base of the shell differs from the wall thickness.
If
a = outside diameter
b = inside diameter
h = depth of shell
t = thickness of base
B = approximate diameter of circular blank

then $B = \sqrt{\left\{ a^2 + (a^2 - b^2)\frac{h}{t} \right\}}$

Determine B when $a = 80$ mm, $b = 78\cdot4$ mm, $h = 54$ mm and $t = 1\cdot2$ mm.

26. A close approximation to the length L of the perimeter of an ellipse whose axes have lengths a and b is given by the formula:

$$L = \pi \sqrt{\left\{ \frac{a^2 + b^2}{2} - \frac{(a - b)^2}{8\cdot8} \right\}}$$

Find L when $a = 40$ mm, $b = 24$ mm and $\pi = 3\cdot142$.

27. In a particular turning operation, if S is the surface cutting speed in m/min and T the tool life in hours, then S and T are connected by the formula:

$$S \times \sqrt[6]{T} = \text{a constant}$$

It is known when $S = 80$ m/min, T is 45 minutes. What must be the values of S to give a tool life of:
(a) 1 hour;   (b) 30 minutes?

28. The Brinell hardness number of a material is given by the formula:

$$\frac{2F}{\pi D\{D - \sqrt{(D^2 - d^2)}\}}$$

Calculate the Brinell hardness number (to the nearest whole number) when $F = 3\ 000$, $D = 10$, $d = 3{\cdot}51$ and $\pi = 3{\cdot}142$.

29. From the results of a science experiment, the following equation was obtained:

$$\log W = \log m + 2 \log v - 0{\cdot}301\ 0$$

Deduce a formula connecting W, m and v which does not include logarithms.

It is expected that our reader will answer questions 30 to 37 which follow with the aid of mathematical tables other than logarithms, and it would be a useful exercise to check the results with the aid of a calculator.

30. Find, to three significant figures, values for the following, taking π as $3{\cdot}142$:

   (a) $\dfrac{1}{\pi}$ (b) $\dfrac{1}{2\pi}$ (c) π^2 (d) $\dfrac{1}{\pi^2}$

31. Find, to three significant figures, values for the following, taking g as $9{\cdot}81$:

   (a) $\dfrac{1}{g}$ (b) $\sqrt{(2g)}$ (c) $\dfrac{1}{\sqrt{(2g)}}$

32. A geometrical property M, of a circle, is found from the formula:

$$M = \frac{r^4}{2}$$

Find the positive value of r, to three significant figures, when $M = 37{\cdot}5$.

33. A square has a side of length 5 mm. The length of the side is reduced so that a new square is formed which is $0{\cdot}81$ of the area of the old square. Find the new length of the side.

34. Find the value, to three significant figures, of $\sqrt{(b^2 - 4ac)}$ when $b = 5{\cdot}72$, $a = 2$ and $c = 1{\cdot}6$.

35. If the longest side of a right-angled triangle is represented by c, and the sides containing the right angle by a and b, then:

$$a = \sqrt{(c^2 - b^2)}$$

Find the value of a when:
   (a) $c = 4{\cdot}1$ and $b = 0{\cdot}9$;   (b) $c = 5{\cdot}372$ and $b = 3{\cdot}828$.

36. A formula which occurs in optics is:

$$\frac{1}{v} - \frac{1}{u} = \frac{1}{f}$$

Find the value of f, to three significant figures, when $v = 2 \cdot 7$ and $u = 8 \cdot 35$.

37. When resistances R_1, R_2 and R_3 are connected in parallel, the equivalent single resistance R is found from the formula:

$$\frac{1}{R} = \frac{1}{R_1} + \frac{1}{R_2} + \frac{1}{R_3}$$

Find the value of R, to three significant figures, when $R_1 = 5 \cdot 72$ ohms, $R_2 = 8 \cdot 36$ ohms and $R_3 = 3 \cdot 17$ ohms.

Answers to Problems 1.6

1. (a) $a + b$ (b) $a + c$
   (c) $b + d$ (d) $a + b + c + d$
   (e) $a + b + d$ (f) $a - b$
   (g) $c - d$ (h) $(a + b) - (c + d)$
   (j) $a + b + c - d$ (k) $2a$
   (l) $2a + b$ (m) $2(a + b)$

   (n) $2a + b + 2c + d$ (o) $\dfrac{a}{2}$

   (p) $\dfrac{a}{4}$ (q) $\dfrac{a + b}{2}$

   (r) $\dfrac{2a + b}{3}$ (s) $a + \left(\dfrac{b + 2c}{3}\right)$

   (t) $a + b + 2c - \dfrac{d}{2}$ (u) $\dfrac{a}{2} + b + c - \dfrac{d}{3}$

   (v) $4a + b - \dfrac{2d}{5}$ (w) $1 + a$

   (x) $2 + d$ (y) $a + b - 2$

   (z) $2 + a + b - \dfrac{3c}{2}$ (aa) $2c - \left(\dfrac{2 + a}{2}\right)$

   (ab) $3a - \left(\dfrac{3 + c + 2d}{3}\right)$

2. (a) 1·913 8   (b) 1·977 7   (c) 0·845 1
   (d) 2·638 5   (e) 2·240 5   (f) 3·058 1
   (g) 3·073 0   (h) 0·690 2   (j) $\bar{3}$·690 2
   (k) $\bar{4}$·954 2   (l) $\bar{2}$·725 9   (m) $\bar{2}$·328 8
   (n) 2·328 8   (o) 0·070 0   (p) 4·000 0
   (q) $\bar{2}$·000 0   (r) 4·730 2   (s) $\bar{5}$·912 1
   (t) 2·847 7   (u) 4·907 8   (v) 0·011 8
   (w) 1·229 6   (x) 0·027 8   (y) 1·199 6

3. (a) 2·439   (b) 24 390   (c) 2 439
   (d) 0·243 9   (e) 1·000   (f) 0·025 12
   (g) 22·65   (h) 2·000   (j) 0·486 5
   (k) 339 7   (l) 1·371   (m) 0·000 389 2

4. (a) 4·623 9   (b) $\bar{3}$·142 5   (c) 2·334 0   (d) $\bar{2}$·707 7

5. (a) 0·628 1   (b) $\bar{1}$·332 8   (c) 2·370 7   (d) 1·603 2

6. (a) 0·874 0   (b) 4·401 6   (c) $\bar{2}$·961 0   (d) $\bar{3}$·802 4

7. (a) 0·862 9   (b) $\bar{1}$·194 2   (c) $\bar{1}$·792 7
   (d) $\bar{1}$·997 3   (e) $\bar{1}$·664 8   (f) $\bar{1}$·318 7

8. 828

9. 1·65 × 10^6

10. 2·27

11. 0·019 8

12. 3 200

13. 30·1

14. (a) 2·71   (b) 0·966

15. 8·34

16. (a) 18·2   (b) 3·90

17. (a) 1·17   (b) 0·115

18. 1·10 × 10^{-4}

19. 0·076 2

20. 3·83

21. 1 190 mm × 841 mm

22. 204 N/mm^2

23. 1 380 litres

24. 170 mm

25. 133 mm

26. 102 mm

27. (a) 76·3 m/min   (b) 85·6 m/min

28. 300

29. $W = \dfrac{mv^2}{2}$

30. (a) 0·318   (b) 0·159   (c) 9·87   (d) 0·101

31. (a) 0·102   (b) 4·43   (c) 0·226

32. 2·94

33. 4·5 mm

34. 4·46

35. (a) 4   (b) 3·77

36. 3·99

37. 1·64 ohms

1.7
Statistical methods

1.7.1
The arithmetical mean

If a firm is engaged on the continuous production of articles which are relatively cheap, such as safety pins, it will be extremely rare for each individual item to receive a complete inspection to determine whether or not it meets the specification, both for appearance and for dimensional accuracy. The more usual occurrence is for an inspection to be made of a *sample* of items. If the sample is reasonably representative of a much larger quantity, then a decision is made on the larger quantity based on the information obtained from the sample. As an example, suppose that at intervals throughout one shift a sample of 100 items was inspected, and two did not satisfy the specification and hence were considered defectives, it could be expected that the defectiveness of the production from the whole shift was in the vicinity of 2%. We could not state with certainty that the overall defectiveness was precisely 2%, this would require every individual item to be inspected. The deviation from 2% would depend mainly upon how closely the characteristics of the sample conformed to the characteristics of the larger quantity. In mathematical language the larger quantity from which a sample is extracted is usually called by most mathematicians a *population*, others sometimes call it a *universe*. It will be noted immediately that the mathematical usage of the word population is different to everyday usage. It is not confined solely to people.

Another typical example, this time concerning people, of the use of a sample to predict the characteristics of a population, is when an attempt is made to predict how people will vote in an election. The accuracy of the prediction will depend to some extent upon how closely the voting characteristics of the sample can be expected to conform with those of the population. The sample must be selected with care. Another consideration is the size of the sample. In general, if the sample is carefully selected, the larger the sample the more reliable the prediction.

The word *statistics*, or the term *statistical methods* to be more precise, derives from the fact that originally the science was used in connection with affairs of state. Its use has now spread to virtually all activities of everyday life, and is mainly concerned with the collecting, arranging, presenting, and the analysis of data, usually with a view to making logical predictions and decisions.

As a first excursion into statistical methods, let us assume that a firm was required to predict the quantity of items that could be produced from a line of 20 similar machines all continuously producing the same item during a single shift of eight hours. If the rate of production, per machine when continuously operating, is 75 per hour, then the maximum number of items that could be produced is hourly rate per machine × number of machines × hours per shift, i.e.

$$75 \times 20 \times 8 = 12\ 000.$$

It would be illogical to suggest that this total would ever be achieved. For a variety of reasons, at some times a machine could be idle. Typical causes of a machine being idle could be a mechanical and/or electrical breakdown, lack of availability of materials and absence of operators. The investigation would most probably commence with collecting data regarding the number of machines which were idle at a given instant. Let us presume that, either from observations or from records, the following information was obtained of the number of machines idle at

a particular instant, those instants being a reasonable representation of
what was normally occurring.

2	3	4	0	3	3	2	1	2	2
1	0	2	1	2	2	0	2	3	4
2	2	1	1	1	2	0	2	4	0
1	3	1	0	1	4	5	3	0	4
0	1	3	3	4	1	1	2	4	1
4	1	2	0	1	1	3	1	1	3
2	2	3	2	2	1	5	3	3	1
3	0	1	1	3	0	0	2	2	0
1	2	5	3	0	3	4	1	5	3
3	4	2	2	5	1	3	1	2	1

A collection of basic information such as the array of numbers above
is known by statisticians as *raw data*. It should appeal to our reader that
it would be most useful to determine the average number of machines
that are idle at any one instant. This can be done by adding all the values
in the array of numbers and dividing by the number of observations.
Our reader can accept that the answer is obtained from 200 divided by
100, which equals 2. Under normal circumstances the average number
of machines actually operating at any one instant would be 18, since 2
would be idle. In which case, it would be reasonable to suggest that a
more accurate prediction of the total output from a shift would be
obtained from hourly rate per machine × average number of machines
× hours per shift actually operating, i.e.

$$75 \times 18 \times 8 = 10\ 800$$

Thus far, we have, in a very loose manner, used the word average.
Our reader will become familiar in future studies with many different
usages of the word average, and it is most important to state with care
which particular average is under consideration. To be more precise,
in the language of statistical methods, the average we have determined
is called an *arithmetical mean*. Just as in engineering we presume that
a screw thread is right-handed unless we are specifically informed to the
contrary, if the word mean is used in statistics without any qualification,
it is always presumed to be the arithmetical mean.

We obtained the arithmetical mean by adding the values in the array
to give 200, and then divided by the number of items. We could have,
if we so wished, obtained the total of 200 by saying that the figure zero
occurred 14 times giving a sub-total of zero, the figure 1 occurred 27
times giving a sub-total of 27 and so on. Adding the sub-totals would
produce the value of 200. For reasons which will become apparent
during future studies, our reader will eventually understand why this
latter method is the more usual manner of determining an arithmetical
mean from raw data.

1.7.2
The tally diagram

Returning to our raw data, let us proceed by determining the number of occasions that no machines were idle, that one machine was idle, and so on. We can commence with the blank tabular layout which follows:

Machines idle	Occurrences	Total occurrences
0		
1		
2		
3		
4		
5		
6		
7		
8		

We now proceed systematically through the raw data, and every time a number occurs we put a vertical dash in the appropriate space. The counting of totals is facilitated if we use what would have been a fifth vertical dash as a horizontal dash through the previous four. When the task is complete, we would note that there was never more than six machines idle, and the raw data can be presented in the form which follows, which is called a *tally diagram.*

Machines idle	Occurrences	Total
0	‖‖ ‖‖ ‖‖	14
1	‖‖ ‖‖ ‖‖ ‖‖ ‖‖ ‖‖	27
2	‖‖ ‖‖ ‖‖ ‖‖ ‖‖	24
3	‖‖ ‖‖ ‖‖ ‖‖	20
4	‖‖ ‖‖	10
5	‖‖	5
6 or more		0
		100

We now have to introduce our reader to certain words and expressions commonly used in statistical methods. The thing being investigated, whose value is variable, is called a *variate*. The number of times a variate occurs is called the *frequency* of that variate. Any presentation which indicates the frequencies of different variates is called a *frequency distribution*. So, in our case, the frequency distribution of the number of machines idle can be presented in the following tabular manner:

Machines idle	0	1	2	3	4	5	6 or more
Frequency	14	27	24	20	10	5	0

Using the method indicated toward the end of the previous article, the arithmetical mean of the number of machines idle at any one instant can be found by first evaluating

$$14(0) + 27(1) + 24(2) + 20(3) + 10(4) + 5(5) + 0(6 \text{ or more})$$
$$= 0 + 27 + 48 + 60 + 40 + 25 + 0$$
$$= 200,$$

and then dividing this value of 200 by the 100 occurrences to produce the arithmetical mean of 2 machines idle. Let us proceed by indicating what was done by using a formula. If we let a variate be represented by x and its frequency by f, we have to add up the various values of (fx). (The brackets have been intentionally used to avoid a danger which could occur in future studies.) Mathematicians use the Greek letter capital sigma, Σ, to represent 'the sum of all such things as . . .'. Hence $\Sigma(fx)$ means the sum of all the individual (fx) values. The total number of items is given the symbol N, which is the sum of all the various values of f, and hence can be indicated by $\Sigma(f)$. The symbol used for an arithmetical mean, when extracted from a sample, is $\bar{x}$ (read as bar x). In passing, although it is no concern to our reader at present, the symbol μ (Greek letter small mu) is used to indicate the mean of the population from which a sample is drawn. The mean of a sample and the mean of a population could well be different, and it will be important in later studies to distinguish between them.

To recapitulate, the mean $\bar{x}$ of a frequency distribution of the type we have been considering is obtained by dividing the sum of all the values of (fx) by the number of items N. Putting this statement into the form of an algebraic formula, we have

$$\bar{x} = \frac{\Sigma(fx)}{N}, \qquad \text{where } N = \Sigma(f).$$

The determination of the arithmetical mean from a frequency distribution is facilitated by using a tabular layout, as illustrated by the worked example which follows.

Example:

The marks obtained by 50 students in a particular test provided the following information:

Mark	5	6	7	8	9	10
Frequency	2	5	20	11	8	4

Determine the arithmetical mean of the marks awarded.

We will let x represent the variate, f the frequency and $\bar{x}$ the arithmetical mean.

x	f	fx
5	2	10
6	5	30
7	20	140
8	11	88
9	8	72
10	4	40
Totals	50	380

$$\bar{x} = \frac{\Sigma(fx)}{N} = \frac{380}{50} = 7.6$$

Answer:

Arithmetical mean = 7·6 marks.

1.7.3
Discrete and continuous variates

The raw data used previously in this chapter, referring to the number of machines idle, would have been obtained by counting. The value of the variate must be either a whole number or zero. It is illogical to have a value such as '3·3 machines idle'. On the other hand, raw data may have to be obtained by measurement, for which a value cannot be precise. For example, the value quoted for the diameter of circular bar depends to a large extent upon the accuracy of the equipment being used, and to some extent upon the skill of the person making that measurement. It is often convenient to use a particular number of significant figures. A measurement of a diameter could be stated to be 33·4 mm, but what is really being said is that if values have been rounded off to an accuracy of three significant figures, the diameter of the bar lies between 33·35 and 33·45 mm.

A variate which can take any value, usually within a given range, is called a *continuous variate*. Conversely, a variate which can only take certain specific values, not necessarily whole numbers, could be referred to as a discontinuous variate, but the more commonly used expression is a *discrete variate*. In general, data for discrete variates are obtained by counting, while that for continuous variates are obtained by measurement.

It is sometimes convenient, when presenting a frequency distribution of continuous variates, to allocate items to *classes*. For example, supposing that the variate was the temperature of a sample of water at atmospheric pressure. No value could be below 0°C and no value could be above 100°C. The number of classes would depend upon the distribution of variates within that range. Ten might well suffice, 0 to 10°, 10° to 20° and so on up to 90° to 100°. However, if all the values fell between 15°C and 55°C, it would probably be better to use a smaller class width, probably 5°C. The questioning reader will immediately ask 'if we have classes of say 45°C to 50°C and 50°C to 55°C, to which class do we allocate a sample whose temperature is 50°C?' The answer is that if we are in control of the collecting of data, we would instruct the person taking the measurements to use sufficient accuracy to allocate a sample to a definite class. Sometimes, however, a statistician has to operate on data already provided. In which case class boundaries will be probably selected which produce definite allocation by classes. For instance, if the temperatures were given in whole numbers he could still use a class width of 10°C using class boundaries such as 49·5°C to 59·5°C.

The arithmetical difference between the values of the *class boundaries* is called the *class width*. It is by no means essential to use class widths of the same size, but this is the more common case. At our stage of studies we shall only use equal class widths.

If an arithmetical mean is required from a frequency distribution of continuous variates, an approximation, sufficiently accurate for our present purposes, can be obtained by using a similar method to the previous worked example. The only difference is that for the variate x we use a single value which is the mid-point of the class boundaries. What we do in effect is to convert a frequency distribution of continuous variates into an approximately equivalent frequency distribution of discrete variates. A tabular layout can still be used, since all we need to do is to add another vertical column.

Example:

Determine the approximate mean of the following frequency distribution of continuous variates.

Temperature (°C)	20 to 30	30 to 40	40 to 50	50 to 60	60 to 70
Frequency	3	9	17	13	8

Class	x	f	fx
20–30	25	3	75
30–40	35	9	315
40–50	45	17	765
50–60	55	13	715
60–70	65	8	520
Totals		50	2 390

$$\bar{x} = \frac{\Sigma(fx)}{N} = \frac{2\ 390}{50} = 47 \cdot 8$$

Answer:

Approximate mean temperature $= 47 \cdot 8°C$

Our reader may now care to solve problem 11 in the set of problems which can be found at the end of this chapter to prove that the method is an approximation, and, in particular, to form some idea of the accuracy of the approximation.

1.7.4
Relative frequency and frequency diagrams

To recall previous definitions, a variate is an item which can take a variety of values. Discrete variates can jump from one value to another, while continuous variates can take any value, usually between given limits. The frequency is the number of times a particular variate occurs. A frequency distribution is any presentation which associates particular frequencies with particular variates. The most common form of presentation is tabular, but sometimes information can be communicated better and appreciated more quickly with the aid of pictorial representations. Pictorial representations in general are given particular attention in Chapter 3 of this book, but it will be advisable at this juncture to give some consideration to the particular application of pictorial representations to frequency distributions.

The *relative frequency* of a particular variate is the ratio of the frequency of that variate to the total of the frequencies. In a previous worked example dealing with the marks of students, the variate 9 marks had a frequency of 8. The total of all the frequencies in the distribution was 50, hence the relative frequency is $8 \div 50 = 0.16$, or, as a *relative frequency percentage*, 16%.

Let us replace the original frequency distribution by a relative frequency distribution, viz:

Mark	5	6	7	8	9	10
Relative frequency	0.04	0.10	0.40	0.22	0.16	0.08

It will be observed that the total of the relative frequencies is unity. There are many ways of pictorially representing this data. Figure 1.7 shows what is known as a *bar chart*. The overall length is made a convenient size, such as 100 mm, and the lengths of the individual bars is proportional to the relative frequency. For example, the length of the bar representing 8 marks is 0.22×100 mm $= 22$ mm.

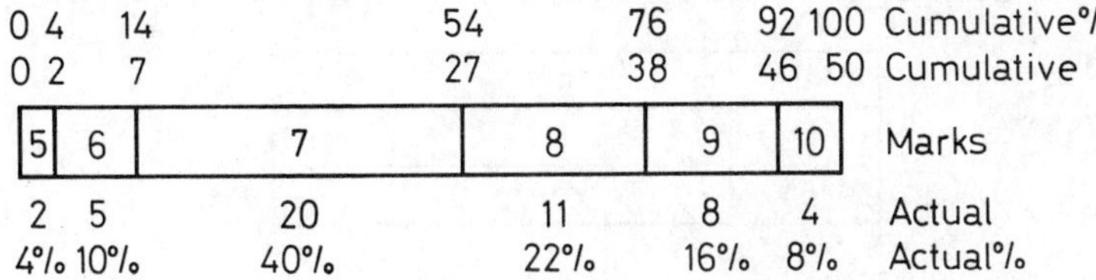

0 4	14		54		76	92	100	Cumulative%
0 2	7		27		38	46	50	Cumulative
5	6	7		8		9	10	Marks
2	5	20		11		8	4	Actual
4%	10%	40%		22%		16%	8%	Actual%

Fig. 1.7

It will be observed that two useful items which have been added is the accumulated totals of the frequencies, commencing with the first variate, called to *cumulative frequency*, together with its associated *cumulative percentage*.

A second way of pictorially representing the data is by means of the *pie chart* shown in Fig. 1.8. A 'slice' of the pie is allocated to each variate, the slice being in the form of a sector. The total of the angles subtended by all the sectors is 360°. The subtended angle for a particular variate is the relative frequency multiplied by 360°. Thus, for the '9 mark variate' having a relative frequency of 0.16, the sector angle is $0.16 \times 360° = 57.6°$. It is quite obvious from this particular pie chart that the biggest slice goes to the variate of 7 marks.

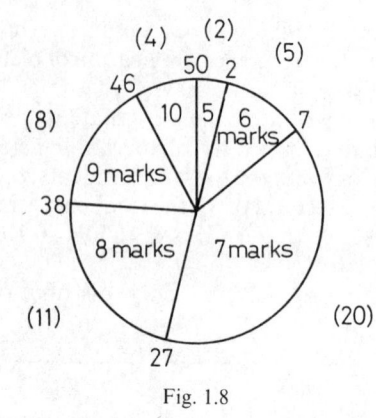

Fig. 1.8

A third way is to use a presentation sometimes referred to as a *pictogram*, which quite often has a humorous appeal to the imagination if the characters used in its construction are selected by considering their appeal to the viewer. The construction generally follows a tally diagram, but instead of tallies, some representation is used associated with what a tally represents. In our case, it is people, and so a pictogram could take a form commencing as shown in Fig. 1.9. The difficulty with a pictogram is that it may be a little difficult to indicate 'fractions of a part'. However, pictograms can be most appealing and often attract more rapid attention than some presentations.

Mark	
5	웃 웃
6	웃 웃 웃 웃 웃

Fig. 1.9

The most common form of pictorial representation of a frequency distribution uses vertical bars, one for each variate, the height being proportional to either the actual frequency. It is most important to label the axes correctly, as shown in Fig. 1.10. The diagram is called *a histogram.*

The histogram now reveals quickly some interesting features not immediately apparent from the tabular frequency distribution, from the bar chart or from the pie chart. The previous pictogram has some similarity with the histogram of Fig. 1.10, the former being aligned horizontally, the histogram vertically. It is clearly apparent that the largest frequency of 20 occurred with 7 marks, and that there was a rising tendency up to this value, and a falling tendency thereafter. In particular, the range covered 11 widths, the variate 5 being in the centre. Most of the values lay above this central value, being, as it were,

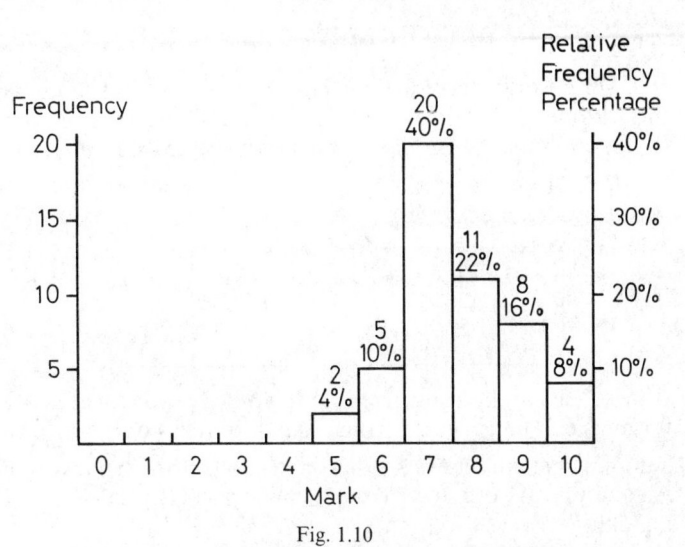

Fig. 1.10

'pushed' toward the higher values. Mathematically we would say that the histogram was 'positively skewed'. Had the marks in general, been of the order of say 0 to 6 inclusive, the histogram would have been negatively skewed.

Although there is no reference to it in TEC unit U75/005, let us go just a little further, by constructing a diagram which joins together the central points at the top of individual bars, as shown in Fig. 1.11. A diagram of this type is called a *frequency polygon*. Now if the widths became narrower the straight lines would tend to produce a curve, and if the characteristics which caused the marks to vary in the particular way they did were similar for some other statistics, such as would occur with the heights of individuals, the frequency polygon would approach a bell-shaped curve. The topic will be developed further in Book 2.

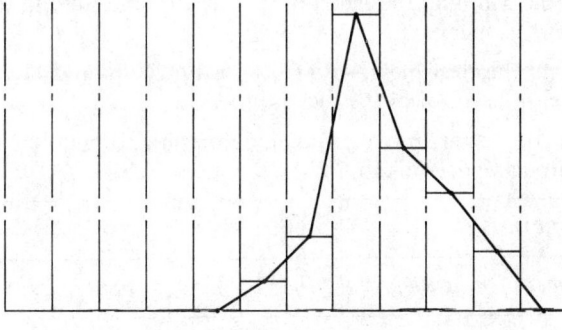

Fig. 1.11

Problems 1.7

1. If 1 370 castings have a total mass of one metric tonne, what is the arithmetical mean of the weight of a single casting, in grammes, to three significant figures? (One metric tonne = 1 000 kg.)

2. Ten forgings were extracted at random from a large batch. Their masses were stated to be:

4·8 kg	4·9 kg	5·0 kg	5·1 kg
1	4	3	2

If these ten can be considered representative, how many complete forgings can be expected from a batch of mass one metric tonne?

3. Samples of small screws are delivered to a factory in packets of 100. An analysis of the defective screws in 30 packets yielded the following data:

Defective screws	0	1	2	3	4
No. of packets	14	9	4	2	1

How many packets should be ordered to allow for 100 000 acceptable screws?

4. The production of articles from nine successive shifts was:

$$116 \quad 121 \quad 115 \quad 122 \quad 121 \quad 120 \quad 117 \quad 119 \quad 120$$

(a) What is the mean rate of production?
(b) How many are required from a tenth shift to bring the mean rate of production over all 10 shifts to 120 articles per shift?

5. The root-mean-square of a set of numbers is found by squaring them, determining the mean of the squares and finding the square root of this mean.

(a) Find the root-mean-square of 15, 12, 20, 16 and 17.
(b) Find the mean of the five numbers.

6. Fifty articles were measured on a dimension nominally 25 mm and put into categories as follows:

L (mm)	24·6	24·8	25·0	25·2	25·4
Number in this category	7	12	16	9	6

Determine the arithmetical mean of L for this particular sample of 50.

7. Four castings were chosen at random from a batch of 64 similar fettled castings and their masses were 15·7, 16·5, 15·3 and 16·7 kg. The cupola melt for the batch was 1·25 metric tonnes. Determine:

   (a) The mean mass of the four castings.
   (b) The percentage of the cupola melt that appeared in the fettled castings, assuming that the four chosen at random were representative of the whole batch.

8. The average number of articles produced in 19 successive shifts was 283. How many must be produced on the twentieth shift to raise the overall average to 285?

9. Six rivet-making machines are producing identical components. 80 different observations of the number of machines idle were as follows:

Machines idle	0	1	2	3	4	5	6
Frequency	15	15	21	18	7	3	1

   Obtain a value for the average number of machines idle, and hence determine an estimation of the total production of rivets from the group of six machines for an 8-hour shift if each machine is capable of producing 150 articles per hour when operating continually.

10. 50 samples were extracted from consignments of goods. Each sample was inspected, and the number of defectives per sample tabulated as follows:

```
3 4 2 3 4 1 1 6 1 0
0 0 2 3 6 1 6 7 0 3
1 0 3 2 7 4 7 2 4 0
3 1 1 1 0 3 4 3 1 6
4 2 7 4 2 1 2 1 7 4
```

    Compile a tally diagram, deduce a tabular frequency distribution, and determine the arithmetical mean of the number of defectives per sample.

11. The following are the production figures for 50 different shifts:

```
390 407 372 407 411 385 377 406 399 387
401 397 407 383 395 415 395 381 416 417
388 402 410 424 407 404 416 409 406 387
407 381 396 397 375 395 398 394 379 425
387 399 416 397 413 416 405 395 427 397
```

    The true arithmetical mean of the values obtained by totalling all 50 quantities and dividing by 50 is 400. Make a frequency table by putting all the values 370–379 inclusive in one cell, 380–389 in another and so on. Use the value 374·5 as the variate for the first cell, 384·5 for the next and so on. Evaluate the arithmetical mean, and determine the percentage error of the arithmetical mean obtained by this approximate method.

12. In an experiment five lines were measured, and their lengths in millimetres and inches were stated to be:

Length in millimetres	151	178	203	229	254
Length in inches	6	7	8	9	10

Determine:

(a) the average length in millimetres;
(b) the average length in inches;
(c) an 'experimental' value for the number of millimetres in an inch, to three significant figures.

13. Construct fully labelled:

(a) bar charts
(b) pie charts
(c) histograms

to illustrate the frequencies, relative frequencies, and cumulative frequencies, of the frequency distributions given in questions 2, 3, 6 and 9.

Answers to Problems 1.7

1. 730 g

2. 201

3. 1 010

4. (a) 119/shift   (b) 129

5. (a) 16·2   (b) 16·0

6. 24·98 mm

7. (a) 16·05 kg   (b) 82·2%

8. 323

9. 2 machines idle, 4 800

10.

Defectives	0	1	2	3	4	5	6	7	8 or more
Frequency	7	11	7	8	8	0	4	5	0

Arithmetical mean = 2·8 defectives

11. Mean = 399·1, error = 0·225%.

12. (a) 203 mm   (b) 8 in   (c) 25·4

13. Answers not numerical.

2 Algebra

2.1
Basic Algebraic Processes

2.1.1
Algebraic terms

Scientists often use symbols to obtain a general solution to a number of problems of a similar nature. With the aid of this general solution or *formula*, they are able to solve subsequent problems of the same type. As an example, the area of any rectangle of length L and width W is $L \times W$. If we require the area of a particular rectangle, we substitute appropriate values for L and for W.

The product of two symbols, such as x multiplied by y, is written as xy. If no arithmetical sign is indicated, a positive value is assumed. It is conventional practice with the product of symbols to write them in alphabetical order, thus:

$$c \times b \text{ is usually indicated as } bc$$

The method can be continued to more than two symbols, for example:

$$a \times d \times b \times e \times c \text{ is usually indicated as } abcde$$

The product of a number and symbols is indicated by writing the number first, thus:

$$3 \times a \times b = 3ab$$

$$4a \times 5b = (4 \times 5)(a \times b) = 20ab$$

$$5x \times 3y \times 4z = (5 \times 3 \times 4)(x \times y \times z) = 60xyz$$

The mathematical name for a multiplier is a *coefficient*. A product of numbers and symbols together with either a positive or negative sign, or implied positive sign, is called an *algebraic term*.

Hence, in the algebraic term $60xyz$, the coefficient of xyz is 60, the coefficient of yz is $60x$, and the coefficient of y is $60xz$. If no numerical coefficient is indicated, the coefficient is understood to be unity; for example, $ab = 1ab$.

2.1.2
Arithmetical signs

The rule for the determination of signs in multiplication and division, is that *a manipulation between two like signs produces a positive result, while a manipulation between two unlike signs produces a negative result.* The reader will find it safer, for the time being, to regard all symbols as positive and to carry out a separate manipulation to find the correct arithmetical sign. As examples:

$$5a \times -4b = (+ \times -)(5 \times 4)(a \times b)$$
$$= -20ab$$

and
$$-2a \times -b \times 7c = (- \times - \times +)(2 \times 1 \times 7)(a \times b \times c)$$
$$= (+ \times +)(14)(abc)$$
$$= +14abc$$
$$= 14abc$$

Division of a pair of algebraic symbols or terms is indicated similarly to a vulgar fraction. Thus:

$$a \div b \text{ is written } \frac{a}{b}$$

and
$$2x \div 3y \text{ is written } \frac{2x}{3y}$$

(In this case, the coefficient of $\frac{x}{y}$ is $\frac{2}{3}$, the coefficient of $\frac{1}{y}$ is $\frac{2x}{3}$, and so on.)

2.1.3
The Laws of indices

The product of like algebraic symbols is indicated by using that symbol with an *index* to show how many of that particular symbol are multiplied together.

$$a \times a \times a \times a = 4 \text{ '}as\text{' multiplied together}$$
$$= a^4$$

$$x \times x = 2 \text{ '}xs\text{' multiplied together}$$
$$= x^2$$

Let us now deduce some rules for the manipulation of indices.

$$x^2 \times x^3 = (x \times x) \times (x \times x \times x) = x^5$$

$$a^4 \times a = (a \times a \times a \times a) \times (a) = a^5$$

Hence, in general
$$x^m \times x^n = x^{m+n}$$

For division
$$\frac{x^5}{x^3} = \frac{x \times x \times x \times x \times x}{x \times x \times x} = x^2$$

and
$$\frac{a^6}{a^5} = \frac{a \times a \times a \times a \times a \times a}{a \times a \times a \times a \times a} = a^1 = a$$

We observe that certain of the '*as*' in the denominator cancel with some of those in the numerator, and in general:

$$\frac{x^m}{x^n} = x^{m-n}$$

For powers $(a^2)^3 = a^2 \times a^2 \times a^2$
$= (a \times a) \times (a \times a) \times (a \times a)$
$= 6$ '*as*' multiplied together
$= a^6$

and in general $(x^m)^n = x^{mn}$

The reader will observe that we have quitely introduced a new topic, a pair of brackets. This has been done to indicate quite clearly that x^m is to be raised to the nth power. A pair of brackets indicates that any operation has to be conducted on everything inside the brackets, *and that what is inside the brackets has to be treated as a complete entity.*

For roots, we can work back from the rule for products.
$\sqrt[5]{x} = x$ raised to some power so that when five of them are multiplied together they produce x^1.

$$= x^{1/5}$$

Similarly $\sqrt[3]{x} = x^{1/3}$

because $x^{1/3} \times x^{1/3} \times x^{1/3} = x^{(1/3 + 1/3 + 1/3)} = x^1 = x$

Finally $\sqrt[4]{x^3} = x^{3/4}$

because $x^{3/4} \times x^{3/4} \times x^{3/4} \times x^{3/4} = x^{(3/4 + 3/4 + 3/4 + 3/4)} = x^3$

Hence, in general $\sqrt[n]{x} = x^{1/n}$

and $\sqrt[n]{x^m} = x^{m/n}$

Before we leave the laws of indices, let us see if we can reason what is meant by x^0 and by x^{-m}

$$x^0 = x^{3-3} \quad \text{or} \quad x^{5-5} \quad \text{or} \quad x^{m-m}$$
$$= (x^3 \div x^3) \quad \text{or} \quad (x^5 \div x^5) \quad \text{or} \quad (x^m \div x^m)$$
$$= \text{unity in every case}$$

The answer of unity does not involve x, hence *anything raised to the power of zero is unity.*

Now $$\frac{1}{x^2} = x^0 \div x^2 = x^{0-2} = x^{-2}$$

and $$\frac{1}{a^3} = a^0 \div a^3 = a^{0-3} = a^{-3}$$

Hence, in general $$x^{-m} = \frac{1}{x^m}$$

Summarising the rules for indices:

$$x^m \times x^n = x^{m+n}$$

$$\frac{x^m}{x^n} = x^{m-n}$$

$$(x^m)^n = x^{mn}$$

$$\sqrt[n]{x^m} = x^{m/n}$$

and we note that $$x^{-m} = \frac{1}{x^m}$$

and that anything to the power of zero is unity.

We will now illustrate all our information on fundamental symbolic notation with some examples. They are intentionally worked out in detail. As the reader acquires skill in algebraic manipulation, he will find that it is unnecessary to write down many of the intermediary steps which are indicated. Nevertheless, it is advisable to proceed cautiously and accurately. The spending of a little extra time in clearly indicating steps towards a solution is not really as wasteful as it may seem.

Example:

Express in the simplest terms:

(a) $5a^2b \times 5a^3b^2$ (b) $42x^2y^2 \div 7xy$ (c) $27a^3b \div 2ab^3$

All symbols to have positive indices.

(a) $$\begin{aligned} 5a^2b \times 5a^3b^2 &= (5 \times 5)(a^2 \times a^3)(b \times b^2) \\ &= 25(a^{2+3})(b^{1+2}) \\ &= 25a^5b^3 \end{aligned}$$

(b) $42x^2y^2 \div 7xy = \left(\dfrac{42}{7}\right)\left(\dfrac{x^2}{x}\right)\left(\dfrac{y^2}{y}\right)$

$= 6(x^{2-1})(y^{2-1})$

$= 6x^1y^1$

$= 6xy$

(c) $27a^3b \div 2ab^3 = \left(\dfrac{27}{2}\right)\left(\dfrac{a^3}{a}\right)\left(\dfrac{b}{b^3}\right)$

$= \dfrac{27}{2}(a^{3-1})(b^{1-3})$

$= \dfrac{27}{2}a^2b^{-2}$

$= \dfrac{27}{2}a^2\left(\dfrac{1}{b^2}\right)$

$= \dfrac{27a^2}{2b^2}.$

Answers:

(a) $25a^5b^3$

(b) $6xy$

(c) $\dfrac{27a^2}{2b^2}$

If square roots are taken, the sign $\pm$, read as 'plus or minus', is applied to the numerical coefficient, together with the positive value of the symbolic term. Furthermore, it is considered quite satisfactory to leave the answers in an indicial form. If no numerical value is quoted for a root, a square root is implied.

Example:

Express in simplest terms:

(a) $\sqrt{(36x^4y^2)}$ (b) $\sqrt{(x^2y)}$ (c) $\sqrt[3]{(-27x^6y^3)}$

(a) $\sqrt{\{36x^4y^2\}} = \sqrt{\{(36)(x^4)(y^2)\}}$

$= (\pm6)(x^{4/2})(y^{2/2})$

$= (\pm6)(x^2)(y^1)$

$= \pm6x^2y$

(b)
$$\sqrt{\{x^2y\}} = \sqrt{\{(1)(x^2)(y^1)\}}$$
$$= (\pm 1)(x^{2/2})(y^{1/2})$$
$$= (\pm 1)(x^1)(y^{1/2})$$
$$= \pm xy^{1/2}$$

(c)
$$\sqrt[3]{\{-27x^6y^3\}} = \sqrt[3]{\{(-27)(x^6)(y^3)\}}$$
$$= (-3)(x^{6/3})(y^{3/3})$$
$$= (-3)(x^2)(y^1)$$
$$= -3x^2y$$

Answers:

(a) $\pm 6x^2y$

(b) $\pm xy^{1/2}$

(c) $-3x^2y$

Example:

Simplify:

(a) $\dfrac{6a^2bc}{15a^3b^2c^2} \times \dfrac{24bc^3}{3ac}$ (b) $\dfrac{-5a^2b}{4ab^3} \div \dfrac{25ab^2}{-16a^4}$

(a) $\dfrac{6a^2bc}{15a^3b^2c^2} \times \dfrac{24bc^3}{3ac} = \left(\dfrac{6 \times 24}{15 \times 3}\right)\left(\dfrac{a^2}{a^3 \times a}\right)\left(\dfrac{b \times b}{b^2}\right)\left(\dfrac{c \times c^3}{c^2 \times c}\right)$

$$= \left(\dfrac{16}{5}\right)\left(\dfrac{a^2}{a^{3+1}}\right)\left(\dfrac{b^{1+1}}{b^2}\right)\left(\dfrac{c^{1+3}}{c^{2+1}}\right)$$

$$= \left(\dfrac{16}{5}\right)\left(\dfrac{a^2}{a^4}\right)\left(\dfrac{b^2}{b^2}\right)\left(\dfrac{c^4}{c^3}\right)$$

$$= \left(\dfrac{16}{5}\right)(a^{2-4})(b^{2-2})(c^{4-3})$$

$$= \left(\dfrac{16}{5}\right)(a^{-2})(b^0)(c^1)$$

$$= \left(\dfrac{16}{5}\right)\left(\dfrac{1}{a^2}\right)(1)(c)$$

$$= \dfrac{16c}{5a^2}$$

(b) $\dfrac{-5a^2b}{4ab^3} \div \dfrac{25ab^2}{-16a^4} = \dfrac{-5a^2b}{4ab^3} \times \dfrac{-16a^4}{25ab^2}$

$$= \left(\dfrac{- \times -}{+ \times +}\right)\left(\dfrac{5 \times 16}{4 \times 25}\right)\left(\dfrac{a^2 \times a^4}{a \times a}\right)\left(\dfrac{b}{b^3 \times b^2}\right)$$

$$= \left(\dfrac{+}{+}\right)\left(\dfrac{4}{5}\right)\left(\dfrac{a^{2+4}}{a^{1+1}}\right)\left(\dfrac{b}{b^{3+2}}\right)$$

$$= (+)\left(\dfrac{4}{5}\right)\left(\dfrac{a^6}{a^2}\right)\left(\dfrac{b^1}{b^5}\right)$$

$$= \left(\dfrac{4}{5}\right)(a^{6-2})(b^{1-5})$$

$$= \left(\dfrac{4}{5}\right)(a^4)(b^{-4})$$

$$= \left(\dfrac{4}{5}\right)(a^4)\left(\dfrac{1}{b^4}\right)$$

$$= \dfrac{4a^4}{5b^4}$$

Answers:

$$\text{(a) } \dfrac{16c}{5a^2} \qquad \text{(b) } \dfrac{4a^4}{5b^4}$$

Problems 2.1

Express questions 1–7 in the simplest manner, all symbols to have positive indices:

1. (a) $3a \times 7b$ (b) $2a \times 4b \times 3c$
   (c) $2a \times -5b \times 2c$ (d) $-5c \times -3b \times -4a$
   (e) $a \times -b \times 2c \times -3d$ (f) $7x \times -2y \times -5z$

2. (a) $36xy \div 3x$ (b) $28xyz \div 4xy$
   (c) $8abc \div 4abc$ (d) $81xyz \div -3x$
   (e) $-15xy \div -3xz$ (f) $-2xy \div 8z$

3. (a) $4a \times 6b \div 8c$ (b) $6ab \times -3c \div 4d$
   (c) $8ab \div 2a \times -4c$ (d) $-4abc \div 2a \times 3d$

4. (a) $x^2 \times x^4$ (b) $x^2 \times x^3 \times x^4$
   (c) $-x \times x^5$ (d) $x^2 \times x^3 \times y^3$
   (e) $x \times x^4 \times y^2 \times y^3 \times z$ (f) $2x \times 3x^2 \times -4y$

5. (a) $a^5 \div a^2$
   (b) $a^2 \div a^5$
   (c) $-a^2 \div a$
   (d) $15a^4 \div 5a^2$
   (e) $-27x^5 \div 3x$
   (f) $-125x \div -25x^2$

6. (a) $(x^2)^3$
   (b) $\sqrt[3]{x^9}$
   (c) $(\sqrt{x})^4$
   (d) $\sqrt[2]{x^5}$
   (e) $\sqrt{(9x^4)}$
   (f) $\sqrt[3]{(-125x^6y^3)}$

7. (a) $\dfrac{27a^2b^3}{3ab^2}$
   (b) $\dfrac{-64a^4b^3}{4ab}$

   (c) $\dfrac{16a^2 \times 5b^2}{4ab}$
   (d) $\dfrac{-27a \times 4b^2}{2b \times 3a}$

   (e) $\sqrt[3]{\left(\dfrac{5a^6 \times 25a^4}{8a}\right)}$
   (f) $\left(\dfrac{2a^2 \times 3a}{4b}\right)^2$

8. Simplify:

   (a) $\dfrac{4a^2c^2}{3bd^2} \times \dfrac{9bd}{16a^3c}$
   (b) $\dfrac{9a^2b^2}{16c^4} \div \dfrac{3}{4c^2}$

9. Simplify:

   (a) $\dfrac{2a^3}{3bc} \times \dfrac{5b^2}{4ca^2} \times \dfrac{3c^2}{2ab^4}$
   (b) $\sqrt{(81a^4b^8)}$

10. Simplify:

    (a) $\dfrac{6a^2b \times 2ab^2c}{(ab)^2 \times 24abc^2}$
    (b) $x^{1/2} \times y^{1/3} \times x^{3/2} \times y^{5/3}$

11. Simplify:
    (a) $10^5 \times 10^{-2} \div 10^{-3}$
    (b) $(9x^2y^6)^{3/2}$

12. Simplify, giving answers with positive indices:

    (a) $\dfrac{10x^{-3/8}}{5x^{-3/4}}$
    (b) $\dfrac{14a^{-3}b^{-2}}{7a^2b^3}$
    (c) $\sqrt{(36p^{3/4}q^{1/2})} \times p^{-1/4}q^{1/2}$

13. Simplify, giving answers with positive indices only:

    (a) $\dfrac{4e^{-2}f^{-3/4} \times 3g^{4/3}}{18e^3f^{1/4}g^{1/3}}$
    (b) $\dfrac{\sqrt[3]{(27p^4q^5)}}{p^{-2/3}q^{-1/3}}$

14. Simplify:

    (a) $\dfrac{x^{2/3} \times y^{1/2} \times z^{3/2}}{x^{-1/3} \times y^2 \times z^{2/3}}$

15. Simplify:

$(a^{2/3}b^{2/5})^4 \div (a^{1/6}b^{1/5})^{-2}$

16. Simplify:
   (a) $(x^{3/5}y^{1/4})^2 \times (x^{2/5}y^{1/2})^{-3}$ (b) $32^{3/5}$

17. Simplify, writing answers with positive indices only:

   (a) $10^3 \times 10^{-4} \div 10^{-2}$ (b) $(a^{-2}b^{1/2})^{-3}$ (c) $\dfrac{3x^{-2}y}{y^{-2}}$

18. Simplify:
   (a) $7^{3/2} \times 28^{1/2}$ (b) $a^{-3/2} \times (4a)^{1/2}$
   (c) $9^n \times 3^{-2n}$ (d) $16^{3/2} \times 9^{1/2} \div 6$

19. Simplify the following:

   (a) $\dfrac{2a^2b}{3ab^2} \times \dfrac{7a}{5b^2}$ (b) $\sqrt{\left(\dfrac{9a^2b^2}{16c^4}\right)} \div \dfrac{3}{4c^2}$

   (c) $\dfrac{4a^2c^2}{bd^2} \times \dfrac{9bd}{16a^3c}$

Answers to Problems 2.1

1. (a) $21ab$ (b) $24abc$ (c) $-20abc$
   (d) $-60abc$ (e) $6abcd$ (f) $70xyz$

2. (a) $12y$ (b) $7z$ (c) 2

   (d) $-27yz$ (e) $\dfrac{5y}{z}$ (f) $-\dfrac{xy}{4z}$

3. (a) $\dfrac{3ab}{c}$ (b) $\dfrac{9abc}{2d}$

   (c) $-16bc$ (d) $-6bcd$

4. (a) x^6 (b) x^9 (c) $-x^6$
   (d) x^5y^3 (e) x^5y^5z (f) $-24x^3y$

5. (a) a^3 (b) $\dfrac{1}{a^3}$ (c) $-a$

   (d) $3a^2$ (e) $-9x^4$ (f) $\dfrac{5}{x}$

6. (a) x^6 (b) x^3 (c) x^2
   (d) $\pm x^{5/2}$ (e) $\pm 3x^2$ (f) $-5x^2y$

7. (a) $9ab$ (b) $-16a^3b^2$ (c) $20ab$

    (d) $-18b$ (e) $\dfrac{5a^3}{2}$ (f) $\dfrac{9a^6}{4b^2}$

8. (a) $\dfrac{3c}{4ad}$ (b) $\dfrac{3a^2b^2}{4c^2}$

9. (a) $\dfrac{5}{4b^3}$ (b) $\pm 3ab^2$

10. (a) $\dfrac{1}{2c}$ (b) x^2y^2

11. (a) 10^6 (b) $27x^3y^9$

12. (a) $2x^{3/8}$ (b) $\dfrac{2}{a^5b^5}$ (c) $\pm 6p^{1/8}q^{1/2}$

13. (a) $\dfrac{2g}{3e^5f}$ (b) $3p^2q^2$

14. $\dfrac{xz^{5/6}}{y^{3/2}}$

15. a^3b^2

16. (a) $\dfrac{1}{y}$ (b) 8

17. (a) 10 (b) $\dfrac{a^6}{b^{3/2}}$ (c) $\dfrac{3y^3}{x^2}$

18. (a) 98 (b) $\pm\dfrac{2}{a}$

19. (a) $\dfrac{14a^2}{15b^3}$ (b) $\pm ab$ (c) $\dfrac{3c}{4ad}$

2.2
Manipulation of algebraic expressions

2.2.1
Addition of algebraic expressions

The work on algebra that has been undertaken thus far has been confined to the multiplication and division of single algebraic terms, such as $3x$, $-5a^4$, $\dfrac{2\sqrt{x}}{ab^3}$, etc.

An *algebraic expression* is a collection of single terms, such as:

$$2x^4 + 3x^3 + 3x^2 - 5x$$

Algebraic expressions can be added, subtracted, multiplied and divided by methods closely similar to those used in ordinary arithmetic. Just as in arithmetical work we distinguish between units, tens, hundreds, etc. so in algebraic work we must distinguish between the different symbolic portions of algebraic expressions. In arithmetic we keep units under units, tens under tens and so on, in columns. In algebraic operations we keep x^2 terms in one column, x terms in another column and so on.

If we consider the expression:

$$3x^4 - x^3 + \frac{x^2}{4} - 5x - 15 + \frac{4}{x} - \frac{7}{x^2}$$

we observe that the power of x is decreased by one as we proceed from term to term. The expression is said to be arranged in *descending powers of x*. Until the reader becomes familiar with the layout of problems involving the manipulation of algebraic expressions, he is advised to keep as near as possible to descending powers of symbols, leaving a space for a particular term in an expression if one of the logical descending powers is omitted.

As an example, $2x^3 - x + 5$ should be spaced as:

$$2x^3 \boxed{} -x + 5$$

and $x^4 - 2x^2 + 8$ as:

$$x^4 \boxed{} -2x^2 \boxed{} +8$$

If we add 7 583 and 204, what we really do in denary arithmetic is add coefficients in the following manner, giving an answer of 7 787.

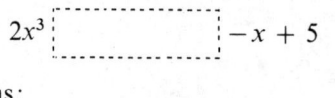

$$
\begin{array}{llll}
7(10^3) & +5(10^2) & +8(10^1) & +3(10^0) \\
        & +2(10^2) &   & +4(10^0) \\
\hline
7(10^3) & +7(10^2) & +8(10^1) & +7(10^0)
\end{array}
$$

We do exactly the same thing when adding algebraic expressions, by adding the coefficients in the columns just as we do in arthmetic.

Example:

Find the sum of $2x^2 - 3x + 5$, $x^2 + x - 11$ and $3x + 4x^2 + 2$.

$$
\begin{array}{rrr}
2x^2 & -3x & +5 \\
x^2 & +x & -11 \\
3x^2 & +4x & +2 \\
\hline
6x^2 & +2x & -4
\end{array}
$$

Answer:

$$6x^2 + 2x - 4$$

Example:

Find the sum of $x^3 + 5x$, $2x^2 - 7x + 4$, $-x^3 + 3x^2 - 5x + 2$ and $3x^3 + 2x^2 - 7$.

In this case we have descending powers of x from x^3 to the constants, which are multiples of x^0. Some of the expressions do not include all the descending powers, but columns for x^3, x^2, x and constant terms have to be included.

$$
\begin{array}{rrrr}
x^3 & & + 5x & \\
& 2x^2 & - 7x & + 4 \\
- x^3 & + 3x^2 & - 5x & + 2 \\
3x^3 & + 2x^2 & & - 7 \\
\hline
3x^3 & + 7x^2 & - 7x & - 1 \\
\end{array}
$$

Answer:

$$3x^3 + 7x^2 - 7x - 1$$

2.2.2
Subtraction of algebraic expressions

The terms are again arranged in vertical columns and we operate on the coefficients. Since 'minus minus' is plus, we can adopt the simple rule of 'change the sign of the bottom line and add'.

Example:

Subtract $3a^2 - 5a + 8$ from $5a^2 + a - 4$.

$$
\begin{array}{rrr}
5a^2 & + a & - 4 \\
3a^2 & - 5a & + 8 \\
\hline
2a^2 & + 6a & - 12 \\
\end{array}
$$

Answer:

$$2a^2 + 6a - 12$$

Example:

Subtract $5x^2 - 4x + 14$ from $2x^3 - 8x + 9$.

$$
\begin{array}{rrrr}
2x^3 & & - 8x & + 9 \\
& 5x^2 & - 4x & + 14 \\
\hline
2x^3 & - 5x^2 & - 4x & - 5 \\
\end{array}
$$

Answer:

$$2x^3 - 5x^2 - 4x - 5$$

2.2.3
Multiplication of algebraic expressions

Let us consider the multiplication of 3 457 by 745 in denary arithmetic. We can lay this out as follows:

$$
\begin{array}{r}
3\ 457 \\
745 \\
\hline
17\ 285 \\
138\ 28 \\
2\ 419\ 9 \\
\hline
2\ 575\ 465
\end{array}
$$

We follow this method in algebra, keeping the terms with like symbols in the same column. With multiplication (and division) a partial check on the accuracy is to let each symbol be unity. For multiplication, the sum of the coefficients and the constant in the answer should be the product of the sums of the coefficients and constant of the expressions that are multiplied together.

Example:

Multiply $3x^2 - 8x + 4$ by $2x^2 - 5x + 2$.

$$
\begin{array}{r}
3x^2 - 8x + 4 \\
2x^2 - 5x + 2 \\
\hline
\end{array}
$$

Multiply top line by 2: $6x^2 - 16x + 8$
Multiply top line by $-5x$: $-15x^3 + 40x^2 - 20x$
Multiply top line by $2x^2$: $6x^4 - 16x^3 + 8x^2$
Add: $6x^4 - 31x^3 + 54x^2 - 36x + 8$

Partial check by letting $x = 1$:
$$
\begin{aligned}
3 - 8 + 4 &= -1 \\
2 - 5 + 2 &= -1 \\
-1 \times -1 &= 1
\end{aligned}
$$

and $6 - 31 + 54 - 36 + 8 = 1$

Answer:

$$6x^4 - 31x^3 + 54x^2 - 36x + 8$$

Example:

Multiply $2x^3 - x + 3$ by $x^2 - 5$.

$$
\begin{array}{r}
2x^3 \qquad - x + 3 \\
x^2 \qquad\quad - 5 \\
\hline
\end{array}
$$

Multiply top line by -5: $-10x^3 \qquad + 5x - 15$
Multiply top line by x^2: $2x^5 - x^3 + 3x^2$
Add: $2x^5 - 11x^3 + 3x^2 + 5x - 15$

Partial check by letting $x = 1$:

$$2 - 1 + 3 = 4$$
$$1 - 5 = -4$$
$$4 \times -4 = -16$$

and
$$2 - 11 + 3 + 5 - 15 = -16$$

Answer:

$$2x^5 - 11x^3 + 3x^2 + 5x - 15$$

2.2.4
Division of algebraic expressions

With division, we divide the first term of the dividend by the first term of the divisor, to obtain the first term of the quotient. A remainder is established, and the next term brought down in a somewhat similar manner to ordinary arithmetic. We continue in the same manner to obtain the remaining terms of the quotient. A partial check should again be applied by summating coefficients and constants.

Example:

Divide $6a^4 - 17a^3 + 21a^2 - 27a + 10$ by $2a^2 - 5a + 2$.

$$
\begin{array}{l}
2a^2 - 5a + 2 \enclose{longdiv}{6a^4 - 17a^3 + 21a^2 - 27a + 10} \quad 3a^2 - a + 5 \\
\underline{6a^4 - 15a^3 + 6a^2} \\
\quad -2a^3 + 15a^2 - 27a \\
\quad \underline{-2a^3 + 5a^2 - 2a} \\
\qquad\qquad 10a^2 - 25a + 10 \\
\qquad\qquad \underline{10a^2 - 25a + 10}
\end{array}
$$

Partial check:
$$6 - 17 + 21 - 27 + 10 = -7$$
$$2 - 5 + 2 = -1$$
$$-7 \div -1 = 7$$

and
$$3 - 1 + 5 = 7$$

Answer:

$$3a^2 - a + 5$$

Example:

Divide $x^3 - 8$ by $\dot{x} - 2$.

In order to allow for x^2 and x terms if they should appear, the dividend of $x^3 - 8$ will be written as:

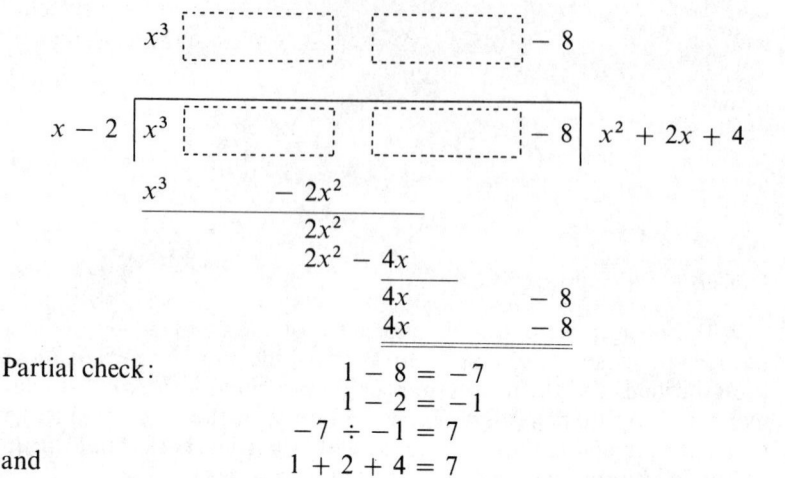

Partial check:
$$1 - 8 = -7$$
$$1 - 2 = -1$$
$$-7 \div -1 = 7$$
and
$$1 + 2 + 4 = 7$$

Answer:

$$x^2 + 2x + 4$$

2.2.5
Brackets

It was convenient to introduce brackets at an earlier stage of this volume. The reader will recall that if there are numbers and/or algebraic terms inside a pair of brackets they must be treated as a single entity. Any coefficient outside a bracket operates on every term inside the bracket. A bracketed expression should be simplified as far as possible before the coefficient operates. In simplification, we follow the rule used in arithmetic which states that multiplication and division signs have preference over addition and subtraction signs.

Examples:

$$3(a^2 + b^2 + c) = 3a^2 + 3b^2 + 3c$$
$$4a(2x^2 + y^2 - 3z) = 8ax^2 + 4ay^2 - 12az$$
$$4(x^4 - x^2 \times x) = 4(x^4 - x^3) = 4x^4 - 4x^3$$
$$-5(a^2 + b^2) = -5a^2 - 5b^2$$
$$-8x(3x^3 - 4x + 2) = -24x^4 + 32x^2 - 16x$$

It will be observed from the above examples that if the coefficient is positive, after it has operated on the bracket the signs are unchanged.

If the coefficient is negative, after it has been operated on all the signs are changed.

Remembering that an expression within brackets must be treated as a single entity, if a coefficient is itself bracketed, each term of the coefficient operates on each of the terms of the other bracketed expression.

Examples:

$$(x + y)(x + 2y) = x(x + 2y) + y(x + 2y)$$
$$= x^2 + 2xy + xy + 2y^2$$
$$= x^2 + 3xy + 2y^2$$

$$(3a - 4b)(3a + 4b) = 3a(3a + 4b) - 4b(3a + 4b)$$
$$= 9a^2 + 12ab - 12ab - 16b^2$$
$$= 9a^2 - 16b^2$$

$$(4x + 5)(2x - 7) = 4x(2x - 7) + 5(2x - 7)$$
$$= 8x^2 - 28x + 10x - 35$$
$$= 8x^2 - 18x - 35$$

$$-(3a + 2b)(a - b) = (-3a - 2b)(a - b)$$
$$= -3a(a - b) - 2b(a - b)$$
$$= -3a^2 + 3ab - 2ab + 2b^2$$
$$= -3a^2 + ab + 2b^2$$

In order to indicate precisely how entities have to be considered, there are occasions when we find it convenient to have a bracketed expression appearing with another bracket. We adopt a convention that the innermost brackets have to be removed first. In order to distinguish between sets of brackets, the innermost are shown ( ), the next { } and the next [ ]. For the scope of studies for which this volume was intended, it is not necessary to indicate more than three types of bracket sign. When brackets are removed, it is advisable to simplify before proceeding to the removal of the next pair of brackets.

Example:

Simplify $2[8x^2 - 2x\{3x - 5(2x)\}]$.

$$2[8x^2 - 2x\{3x - 5(2x)\}] = 2[8x^2 - 2x\{3x - 10x\}]$$
$$= 2[8x^2 - 2x\{-7x\}]$$
$$= 2[8x^2 + 14x^2]$$
$$= 2[22x^2]$$
$$= 44x^2$$

Answer:

$$44x^2$$

Example:

Simplify $2a[3a^2 - 4\{(a + 3b)(a - 3b) - 2b^2\} + 6]$.

$2a[3a^2 - 4\{(a + 3b)(a - 3b) - 2b^2\} + 6]$

$$= 2a[3a^2 - 4\{a^2 - 3ab + 3ab - 9b^2 - 2b^2\} + 6]$$
$$= 2a[3a^2 - 4\{a^2 - 11b^2\} + 6]$$
$$= 2a[3a^2 - 4a^2 + 44b^2 + 6]$$
$$= 2a[-a^2 + 44b^2 + 6]$$
$$= -2a^3 + 88ab^2 + 12a$$

Answer:

$$-2a^3 + 88ab^2 + 12a$$

If we have fractions such as $\dfrac{2x - 3y}{2}$ and $\dfrac{8x^2 - 3y^2}{4x}$, we could

express these as $x - \dfrac{3y}{2}$ and $2x - \dfrac{3y^2}{4x}$ respectively. It will be observed
that the denominator has been divided into each term of the numerator.
Consequently, although the dividing line of a fraction has not the
appearance of a pair of brackets, it can be regarded as a pair of brackets
for the purposes of algebraic manipulation. Thus:

$$\dfrac{2x - 3y}{2} \text{ can be indicated as } \dfrac{1}{2}(2x - 3y)$$

and $\dfrac{8x^2 - 3y^2}{4x}$ can be indicated as $\dfrac{1}{4x}(8x^2 - 3y^2)$

Problems 2.2

1. Add:
   (a) $a + 2b + 3c$, $2a + 3b + 4c$ and $3a + 4b + 5c$
   (b) $2x + 3y - z$, $5x - 3y + 2z$ and $-3x - 5y + 6z$
   (c) $x^3 + 3x^2 - 5x + 2$, $2x^3 + 7x + 5$ and $3x^2 - 2x - 7$
   (d) $a^2 + 5ab + b^2$, $-7ab + 3b^2 + 4$ and $3a^2 - b^2 - 7$

2. Subtract:
   (a) $2x^2 + 5x + 8$ from $3x^2 + 8x + 15$
   (b) $2a^2 + 3a - 5$ from $5a^2 - 6a + 2$
   (c) $2x^2 - 5x + 7$ from $-3x^3 + 2x - 15$
   (d) $2a^2 - 7ab + 5b^2$ from $3a^3 - 5a^2 + 2b^2 - 7$

3. Multiply:
   (a) $2x^2 - 5x + 6$ by $x - 3$
   (b) $a^2 + 3a - 4$ by $2a^2 + 6a - 3$

4. Multiply:
    (a) $5x^2 - 4xy + y^2$ by $3x - 7y$
    (b) $2a^2 + 3a - 4$ by $a^2 - 5a + 2$
    (c) $4x^2 - 10xy + 25y^2$ by $2x + 5y$

5. Multiply (and in each case check your result by division):
    (a) $x + 3$ by $x + 5$
    (b) $2x - 3$ by $x - 4$
    (c) $4x + 3y$ by $4x - 3y$
    (d) $x^2 + 3x - 5$ by $2x - 3$
    (e) $x^2 + xy + y^2$ by $x - y$
    (f) $3a^2 + 5a - 7$ by $2a^2 + 3a - 2$

6. Divide:
    (a) $2x^3 + 9x^2 + 5x - 12$ by $x + 3$
    (b) $2a^4 + a^3 - 20a^2 + 29a - 12$ by $2a^2 - 5a + 3$

7. Divide:
    (a) $6x^2 + 13xy - 28y^2$ by $2x + 7y$
    (b) $2x^3 - 13x^2 + 19x - 6$ by $2x - 3$
    (c) $a^3 - 64$ by $a^2 + 4a + 16$

8. Divide (and in each case check your result by multiplication):
    (a) $x^2 + 10x + 21$ by $x + 7$
    (b) $2x^2 + 5x - 3$ by $x + 3$
    (c) $x^2 - xy - 6y^2$ by $x - 3y$
    (d) $8a^3 - b^3$ by $2a - b$
    (e) $12x^4 + 28x^3 - 53x^2 - 72x + 45$ by $6x^2 - x - 15$

9. Add $2x^2 + 5x - 7$ to $3x^2 - 28x + 19$ and divide the result by $x - 4$.

10. Subtract $x^3 - 2x^2y + 3xy^2 + 4y^3$ from $2x^3 - 2x^2y + 3xy^2 + 12y^3$, and divide the result by $x + 2y$.

11. Simplify:
    (a) $3(a^2 + 4a + 7) - 2a(a - 15) - 5(-3a + 4)$
    (b) $2\{4a^2 - 5(3a - 2)\} - 3\{2a(a - 8) - 5\}$
    (c) $3[8x^3 - 3x\{x^2 - 2(4x - 3) + 7\}]$
    (d) $2a[3a^2 - 5a\{-2a + 7(a + 6) - 5(3a + 2)\}]$

12. Simplify:
    (a) $3(2x + 5) + 4(3x + 2)$
    (b) $7(3x - 2) - 5(x - 5)$
    (c) $8(2a^2 - 5a + 3) + 3(2a + 1) - 5(2a^2 - 4)$
    (d) $2\{23x - 5(4x + 2) + 7\}$
    (e) $3[8a^2 - 2\{15 - 3a(2a + 6) + 4(a + 2)\}]$

Answers to Problems 2.2

1. (a) $6a + 9b + 12c$ (b) $4x - 5y + 7z$
   (c) $3x^3 + 6x^2$ (d) $4a^2 - 2ab + 3b^2 - 3$

2. (a) $x^2 + 3x + 7$ (b) $3a^2 - 9a + 7$
   (c) $-3x^3 - 2x^2 + 7x - 22$ (d) $3a^3 - 7a^2 + 7ab - 3b^2 - 7$

3. (a) $2x^3 - 11x^2 + 21x - 18$
   (b) $2a^4 + 12a^3 + 7a^2 - 33a + 12$

4. (a) $15x^3 - 47x^2y + 31xy^3 - 7y^3$
   (b) $2a^4 - 7a^3 - 15a^2 + 26a - 8$
   (c) $8x^3 + 125y^3$

5. (a) $x^2 + 8x + 15$ (b) $2x^2 - 11x + 12$
   (c) $16x^2 - 9y^2$ (d) $2x^3 + 3x^2 - 19x + 15$
   (e) $x^3 - y^3$ (f) $6a^4 + 19a^3 - 5a^2 - 31a + 14$

6. (a) $2x^2 + 3x - 4$ (b) $a^2 + 3a - 4$

7. (a) $3x - 4y$ (b) $x^2 - 5x + 2$ (c) $a - 4$

8. (a) $x + 3$ (b) $2x - 1$ (c) $x + 2y$
   (d) $4a^2 + 2ab + b^2$ (e) $2x^2 + 5x - 3$

9. $5x - 3$

10. $x^2 - 2xy + 4y^2$

11. (a) $a^2 + 57a + 1$ (b) $2a^2 + 18a + 35$
    (c) $15x^3 + 72x^2 - 117x$ (d) $106a^3 - 320a^2$

12. (a) $18x + 23$ (b) $16x + 11$ (c) $6a^2 - 34a + 47$
    (d) $6x - 6$ (e) $60a^2 + 84a - 138$

2.3
Factors of algebraic expressions

2.3.1
Common factors

A factor of a number or an algebraic term is a quantity which will divide into that number or algebraic term without leaving a remainder. The numbers 2, 3, 5 and 7 all divide into 210 without leaving a remainder, and hence 2, 3, 5 and 7 are all factors of 210. Similarly since $(x - 1)$, $(x + 1)$ and $(x + 2)$ all divide into $x^3 + 2x^2 - x - 2$ without leaving a remainder, $(x - 1)$, $(x + 1)$ and $(x + 2)$ are all factors of $x^3 + 2x^2 - x - 2$. Although unity divides into all quantities without leaving a remainder, unity itself is not considered to be a factor.

An algebraic expression consists of two, three or more than three terms. An expression containing two terms is known as a *binomial*, one containing three terms is a *trinomial*, while one containing more than three terms is called a *polynomial*. The first step in determining the factors of any algebraic expression is to see whether or not there are common factors. If so, these should be combined into a single highest common factor.

Example:

Find the factors of $9a^2 - 12a$.

Inspection shows that each term is divisible by 3 and by a, hence 3 and a are factors, whilst $3a$ is the highest common factor. (It is true that $-3a$ is also a common factor, but it is conventional practice to consider only positive common factors.)

$$\frac{9a^2 - 12a}{3a} = 3a - 4$$

$$\therefore 9a^2 - 12a = 3a(3a - 4)$$

$3a - 4$ cannot be factorized. Hence the factors of $9a^2 - 12a$ are $3a$ and $(3a - 4)$.

Answer:

$$9a^2 - 12a = 3a(3a - 4)$$

2.3.2
Factors of binomials

Apart from a common factor, the most commonly occurring type of binomial with which we are interested in our study of factors is the difference of two perfect squares. (We will leave the sum or difference of two perfect cubes till a little later.) By multiplication:

$$(x + y)(x - y) = x^2 - xy + xy - y^2 = x^2 - y^2$$

Reversing this procedure, the factors of the difference of two perfect squares are the sum and difference of the positive values of the square roots of the terms of the binomial.

Example:

Factorise $9a^2 - 16b^2$.

Inspection reveals no common factor.
Considering positive values only, $9a^2 = (3a)^2$ and $16b^2 = (4b)^2$.
Hence the factors of $9a^2 - 16b^2$ are $(3a + 4b)$ and $(3a - 4b)$.

Answer:

$$9a^2 - 16b^2 = (3a + 4b)(3a - 4b)$$

Example:

Factorise $9y - 36y^3$.

Inspection reveals that $9y$ is the highest common factor.

$$\frac{9y - 36y^3}{9y} = 1 - 4y^2$$

A further inspection reveals that 1 and $4y^2$ are perfect squares,

$$1 = (1)^2 \quad \text{and} \quad 4y^2 = (2y)^2$$

Hence $1 - 4y^2 = (1 + 2y)(1 - 2y)$

The factors of $9y - 36y^2$ are $9y$, $(1 + 2y)$ and $(1 - 2y)$.

Answer:

$$9y - 36y^2 = 9y(1 + 2y)(1 - 2y)$$

Example:

Factorise $(5x - 2y)^2 - (4x - 2y)^2$.

The expression is the difference of two squares.

$$(5x - 2y) + (4x - 2y) = 5x - 2y + 4x - 2y = 9x - 4y$$
$$(5x - 2y) - (4x - 2y) = 5x - 2y - 4x + 2y = x$$

Answer:

$$(5x - 2y)^2 - (4x - 2y)^2 = x(9x - 4y)$$

2.3.3
Factors of trinomials

After it has been checked for common factors, a trinomial should be inspected to see whether or not it is the square of a binomial.

Now $(a + b)^2 = a^2 + 2ab + b^2$

and $(a - b)^2 = a^2 - 2ab + b^2$

The form of a trinomial which is the square of a binomial can be deduced from the above. The square of a binomial is (square of the first term) plus (twice the product of the terms) plus (square of the second term). If the second term of a trinomial is negative, the first and last terms being positive, and the trinomial is a perfect square, it is the square of the difference of two algebraic terms.

Example:

Factorise $x^2 + 6x + 9$.

There is no common factor.

$$x^2 = (x)^2 \qquad 9 = (3)^2 \qquad 6x = 2(x)(3)$$

Hence $x^2 + 6x + 9$ is the square of $x + 3$.

Answer:

$$x^2 + 6x + 9 = (x + 3)^2$$

Example:

Factorise $9a^2 - 12ab + 4b^2$.

There is no common factor.

$$9a^2 = (3a)^2 \qquad 4b^2 = (2b)^2 \qquad 12ab = 2(3a)(2b)$$

We note that the second term of the trinomial is negative. Hence $9a^2 - 12ab + 4b^2$ is the square of $3a - 2b$.

Answer:

$$9a^2 - 12ab + 4b^2 = (3a - 2b)^2$$

Example:

Factorise $75a^2b - 60ab^2 + 12b^3$.

On inspection, 3 and b are common factors; $3b$ is the highest common factor.

$$\frac{75a^2b - 60ab^2 + 12b^3}{3b} = 25a^2 - 20ab + 4b^2$$

$$25a^2 = (5a)^2 \qquad 4b^2 = (2b)^2 \qquad 20ab = 2(5a)(2b)$$

The second term of $25a^2 - 20ab + 4b^2$ is negative. Hence $25a^2 - 20ab + 4b^2$ is the square of $5a - 2b$.

Answer:

$$75a^2b - 60ab^2 + 12b^3 = 3b(5a - 2b)^2$$

Apart from common factors and perfect squares, the only other trinomials which we may be called upon to factorise at our present stage of studies are those trinomials which are the product of two binomials. In general terms:

$$(ax + b)(cx + d) = acx^2 + axd + bcx + bd$$
$$= ac(x^2) + (ad + bc)x + bd$$

What we shall be seeking will be values for a, b, c and d, with correct mathematical signs, so that:

(i) ac is the coefficient of the first term of the trinomial;
(ii) bd is the last term of the trinomial; and
(iii) $ad + bc$ is the coefficient of the second term of the trinomial.

It is now very important to proceed methodically to avoid wasted effort. As always, we first look for a common factor. Following this, on the rare occasions that the first term is negative, we can take out -1 as a temporary common factor to make the first term positive. We now look at the signs in the trinomial. If the signs of the trinomial are all positive, the factors (which are binomials) have positive signs. If the sign of the second term of the trinomial is negative while the sign of the third term is positive, the signs in the brackets are negative. If the sign of the last term of the trinomial is negative, one binomial is an addition, the other is a subtraction. In general:

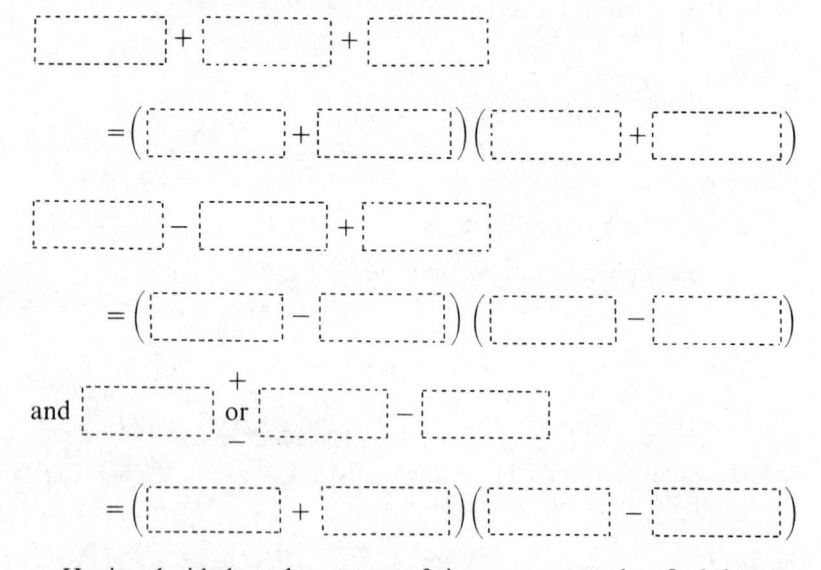

Having decided on the pattern of signs, we proceed to find the possible combinations to produce the first term, arranging that the larger factor is quoted first, and not proceeding further if the first factor becomes less than the second. For instance, for $16x^2$, we have:

$$16x \text{ and } x, \qquad 8x \text{ and } 2x, \qquad 4x \text{ and } 4x$$

and it is not necessary to consider reversals such as $2x$ and $8x$, or x and $16x$.

We now list the factors of the third term, including reversals, bearing in mind some of them can be discounted because of the sequence of the signs. We shall now find a pattern where we can patiently try every combination, without duplication, to find the correct combination to produce the second term. Let us illustrate the method with a series of examples.

Example:

Factorise $x^2 + 8x + 7$.

There is no common factor, and it is not a perfect square.

Factors of x^2		*Factors of 7*	
x	x	7	1
		1	7

(No need to consider -7 and -1, or -1 and -7, due to the pattern of signs.)

Our combination here is obvious; the factors are $(x + 7)$ and $(x + 1)$.

Answer:

$$x^2 + 8x + 7 = (x + 7)(x + 1)$$

Example:

Factorise $2x^2 - 5x + 3$.

There is no common factor, and it is not a perfect square.

Factors of $2x^2$		*Factors of $+3$*	
$2x$	x	-3	-1
		-1	-3

(The pattern of signs eliminates positive factors.)

The combinations possible are:

$$(2x - 3)(x - 1)$$

and

$$(2x - 1)(x - 3)$$

Of these, only the former produces the second term of $-5x$.

Answer:
$$2x^2 - 5x + 3 = (2x - 3)(x - 1)$$

Example:

Factorise $4x^2 + 5x - 6$.

There is no common factor, and it is not a perfect square.

Factors of $4x^2$		*Factors of* -6	
$4x$	x	6	-1
$2x$	$2x$	-6	1
		3	-2
		-3	2

We now have to try patiently all possible combinations, and the above layout assists us to list them without duplication. They are:

$$(4x + 6)(x - 1)^* \qquad (2x + 6)(2x - 1)^*$$
$$(4x - 6)(x + 1)^* \qquad (2x - 6)(x + 1)^*$$
$$(4x + 3)(x - 2) \qquad (2x + 3)(x - 2)$$
$$(4x - 3)(x + 2) \qquad (2x - 3)(x + 2)$$

We can immediately dismiss those marked with an asterisk, since the first brackets in each case would imply a common factor. Of the others, only $(4x - 3)(x + 2)$ produces the second term of $+5x$.

Answer:
$$4x^2 + 5x - 6 = (4x - 3)(x + 2)$$

2.3.4
Factors of polynomials of four terms

Having checked for common factors, and dividing by the highest common factor, should common factors occur, the polynomial should be rearranged if necessary into pairs of terms, each pair having a different common factor.

Example:

Factorise $3ac + 8bd + 6bc + 4ad$.

There is no complete common factor, but

$(3ac + 4ad)$ has a common factor of a
$(6bc + 8bd)$ has a common factor of $2b$
$$3ac + 8bd + 6bc + 4ad = (3ac + 4ad) + (6bc + 8bd)$$
$$= a(3c + 4d) + 2b(3c + 4d)$$
$$= (a + 2b)(3c + 4d)$$

Answer:

$$3ac + 8bd + 6bc + 4ad = (a + 2b)(3c + 4d)$$

Example:

Factorise $6ac - 2bd + 3bc - 4ad$.

There is no common factor.

$$(6ac + 3bc) = 3c(2a + b)$$
$$(-2bd - 4ad) = 2d(-b - 2a)$$
$$(6ac + 3bc) + (-2bd - 4ad) = 3c(2a + b) + 2d(-b - 2a)$$

We observe that the last bracketed expression can be made identical to the last but one by changing the sign of the coefficient and reversing the order of the terms.

$$(6ac + 3bc) + (-2bd - 4ad) = 3c(2a + b) - 2d(2a + b)$$
$$= (3c - 2d)(2a + b)$$

Answer:

$$6ac - 2bd + 3bc - 4ad = (2a + b)(3c - 2d)$$

2.3.5
General procedure for obtaining factors

The recommended approach to obtaining the factors of an algebraic expression is to look first for common factors. If there is a common factor, or common factors which combine to form a highest common factor, these should be divided into the original expression.

The next step is to recognise the form of the original expression, or the expression which remains after division. It will be a binomial, a trinomial or a polynomial. If it is a binomial, at our present stage of studies it will either not factorise or be the difference of two perfect squares. If it is a trinomial, the first thing to investigate is the possibility of a perfect square. If it is not, it will usually be the product of two different binomials. A polynomial of four terms will also usually be the product of two binomials.

Occasionally it may be necessary for a simplification to be performed before the form can be recognised.

Example:

Factorise $4x(x - 3y) + y(12x - 9y)$.

$$4x(x - 3y) + y(12x - 9y) = 4x^2 - 12xy + 12xy - 9y^2$$
$$= 4x^2 - 9y^2$$

This is a binomial without a common factor, and is the difference of two perfect squares.

$$4x^2 = (2x)^2 \quad \text{and} \quad 9y^2 = (3y)^2$$

Hence $4x^2 - 9y^2 = (2x + 3y)(2x - 3y)$

Answer:

$$4x(x - 3y) + y(12x - 9y) = (2x + 3y)(2x - 3y)$$

2.3.6
Functional notation and the factor theorem

If two variables are connected by a formula, say y and x, so that the value of y depends upon the value we give to x, we say that y is *a function of x*. There are various mathematical conventions for indicating the statement 'a function of x'. The most common one, which we shall use, is $f(x)$. Others are $\phi(x)$ and $F(x)$. Instead of $y = x^2 + 3x - 4$ we could state $f(x) = x^2 + 3x - 4$.

If we want to indicate the value of $f(x)$ when a specific value is given to x, such as the value of a particular function when $x = 4$, we indicate this as $f(4)$. Thus:

if $f(x) = x^2 + 3x - 4$

$$f(1) = 1 + 3 - 4 = 0$$

$$f(0) = 0 + 0 - 4 = -4$$

and $f(a) = a^2 + 3a - 4$

The following statement, known as the factor theorem, can often assist in determining factors, and is presented without proof:

$$(x - a) \text{ is a factor of } f(x) \text{ if } f(a) = 0$$

As examples:

If $f(x) = x^2 - 1, f(1) = 0$, hence $(x - 1)$ is a factor of $x^2 - 1$.
If $f(x) = x^3 - 1, f(1) = 0$, hence $(x - 1)$ is a factor of $x^3 - 1$.
If $f(x) = 4x^2 - 13x - 12, f(4) = 64 - 52 - 12 = 0$.
Hence $x - 4$ is a factor of $4x^2 - 13x - 12$.
The other factor could be found by division.

$$
\begin{array}{r}
4x + 3 \\
x - 4 \overline{\smash{\big)}\ 4x^2 - 13x - 12} \\
\underline{4x^2 - 16x} \\
3x - 12 \\
\underline{3x - 12}
\end{array}
$$

Thus $4x^2 - 13x - 12 = (x - 4)(4x + 3)$.

The use of the factor theorem can be of assistance in determining which binomials are NOT factors, by using the theorem in reverse:
If $(x - a)$ is to be a factor of $f(x)$ then $f(a) = 0$.
Thus $(x - 1)$ cannot be a factor of $x^3 + 1$, because $f(1) = 2$.
On the other hand, $(x + 1)$ is a factor, since $f(-1) = 0$.

Let us put the factor theorem to use for the factors of the sum and of the difference of two perfect cubes.

If $$f(a) = a^3 + b^3$$

then $$f(-b) = -b^3 + b^3 = 0$$

hence $\{a - (-b)\} = (a + b)$ is a factor of $a^3 + b^3$

By dividing out, we find the second factor to be:

$$a^2 - ab + b^2 = \{(a + b)^2 - 3ab\} = \{(a - b)^2 + ab\}$$

Hence
$$a^3 + b^3 = (a + b)(a^2 - ab + b^2)$$
$$= (a + b)\{(a + b)^2 - 3ab\}$$
$$= (a + b)\{(a - b)^2 + ab\}$$

Let us proceed to the difference of two cubes, by factorising $a^3 - b^3$.

If $$f(a) = a^3 - b^3$$

we note $$f(b) = b^3 - b^3 = 0$$

hence $a - b$ is a factor of $a^3 - b^3$

By dividing out, we find that the second factor is $a^2 + ab + b^2$.

$$a^2 + ab + b^2 = \{(a + b)^2 - ab\} = \{(a - b)^2 + 3ab\}$$

Hence
$$a^3 - b^3 = (a - b)(a^2 + ab + b^2)$$
$$= (a - b)\{(a + b)^2 - ab\}$$
$$= (a - b)\{(a - b)^2 + 3ab\}$$

Unless the reader has an above average memory it is advisable not to remember the factors of the sum and of the difference of two cubes but to work from first principles, remembering that one factor of $(a^3 + b^3)$ is $(a + b)$ and that one factor of $(a^3 - b^3)$ is $(a - b)$.

Example:

Factorise $8x^3 - 27y^3$.

$$\sqrt[3]{(8x^3)} = 2x \qquad \sqrt[3]{(27y^3)} = 3y$$

Hence $2x - 3y$ is a factor.

$$2x - 3y \overline{\left)\; 8x^3 \qquad\qquad - 27y^3 \right.}\; 4x^2 + 6xy + 9y^2$$
$$\underline{8x^3 - 12x^2y}$$
$$12x^2y$$
$$\underline{12x^2y - 18xy^2}$$
$$18xy^2 - 27y^3$$
$$\underline{\underline{18xy^2 - 27y^3}}$$

Answer:

$$8x^3 - 27y^3 = (2x - 3y)(4x^2 + 6xy + 9y^2)$$

2.3.7
Fractions of algebraic expressions

In article 2.2.1, the reader was shown how the processes of addition, subtraction, multiplication and division with algebraic expressions follow very closely the rules of fundamental arithmetic. We shall now proceed to a similar application of arithmetical rules to problems dealing with fractions of algebraic expressions. In the addition and subtraction of vulgar fractions we had to express fractions with a common denominator. We follow the same method with algebraic fractions by obtaining a common denominator which is the least common multiple of the various denominators.

Example:

Simplify $\dfrac{3}{2a^2} + \dfrac{4b}{3a} - \dfrac{5}{4a^2}$.

We require the L.C.M. of $2a^2$, $3a$ and $4a^2$.

$$2a^2 = 2 \times a \times a$$
$$3a = 3 \times a$$
$$4a^2 = 2 \times 2 \times a \times a$$
$$\text{L.C.M.} = (2 \times 2 \times 3) \times a^2 = 12a^2$$

$$\frac{3}{2a^2} + \frac{4b}{3a} - \frac{5}{4a^2} = \frac{6(3)}{6(2a^2)} + \frac{4a(4b)}{4a(3a)} - \frac{3(5)}{3(4a^2)}$$

$$= \frac{18 + 16ab - 15}{12a^2} = \frac{16ab + 3}{12a^2}$$

Answer:

$$\frac{16ab + 3}{12a^2}$$

Example:

Simplify $\dfrac{5x + 3}{x + 1} + \dfrac{2x + 5}{x - 1} - \dfrac{5}{2(x^2 - 1)}$

We require the L.C.M. of $x + 1$, $x - 1$ and $2(x^2 - 1)$.

Now $\qquad\qquad 2(x^2 - 1) = 2(x + 1)(x - 1)$

so $\qquad\qquad 2(x^2 - 1)$ is the L.C.M.

since $x + 1$ and $x - 1$, the other two denominators, are factors $2(x^2 - 1)$.

$$\dfrac{5x + 3}{x + 1} + \dfrac{2x + 5}{x - 1} - \dfrac{5}{2(x^2 - 1)}$$

$$= \dfrac{2(x - 1)(5x + 3)}{2(x - 1)(x + 1)} + \dfrac{2(x + 1)(2x + 5)}{2(x + 1)(x - 1)} - \dfrac{5}{2(x^2 - 1)}$$

$$= \dfrac{2(5x^2 - 2x - 3) + 2(2x^2 + 7x + 5) - 1(5)}{2(x^2 - 1)}$$

$$= \dfrac{(10x^2 - 4x - 6) + (4x^2 + 14x + 10) - (5)}{2(x^2 - 1)}$$

$$= \dfrac{14x^2 + 10x - 1}{2(x^2 - 1)}$$

This last expression cannot be simplified further.

Answer:

$$\dfrac{14x^2 + 10x - 1}{2(x^2 - 1)}$$

With multiplication and division, we have to investigate whether cancelling can take place. This usually involves factorising algebraic expressions in order to cancel factors. With division, where necessary, we follow the arithmetical rule of 'invert the divisor and multiply'.

Example:

Simplify $\dfrac{2x^2 + x - 15}{x^2 - 9}$.

$$2x^2 + x - 15 = (2x - 5)(x + 3)$$

$$x^2 - 9 = (x + 3)(x - 3)$$

$$\dfrac{2x^2 + x - 15}{x^2 - 9} = \dfrac{(2x - 5)(x + 3)}{(x + 3)(x - 3)} = \dfrac{2x - 5}{x - 3}$$

(The expression $x + 3$ cancels.)

Answer:

$$\frac{2x - 5}{x - 3}$$

Example:

Simplify $\dfrac{x^2 - 4x + 4}{2x^2 - 11x + 14} \div \dfrac{x^2 - 4}{3x^2 + 4x - 4}$

$$x^2 - 4x + 4 = (x - 2)(x - 2)$$
$$2x^2 - 11x + 14 = (x - 2)(2x - 7)$$
$$x^2 - 4 = (x + 2)(x - 2)$$
$$3x^2 + 4x - 4 = (3x - 2)(x + 2)$$

$$\frac{x^2 - 4x + 4}{2x^2 - 11x + 14} \div \frac{x^2 - 4}{3x^2 + 4x - 4}$$

$$= \frac{x^2 - 4x + 4}{2x^2 - 11x + 14} \times \frac{3x^2 + 4x - 4}{x^2 - 4}$$

$$= \frac{(x - 2)(x - 2)(3x - 2)(x + 2)}{(x - 2)(2x - 7)(x + 2)(x - 2)}$$

$$= \frac{3x - 2}{2x - 7}$$

Answer:

$$\frac{3x - 2}{2x - 7}$$

The reader is advised to use a partial check on the validity of answers by letting $x = 1$, or some other simple value if one of the expressions results in zero. In this case $x = 1$ and $x = 2$ makes certain expressions zero, so let $x = 3$. A further check could be let $x = $ zero, in which case we only check the constants.

With $x = 3$, we have $\dfrac{1}{-1} \div \dfrac{5}{35} = \dfrac{1 \times 35}{-1 \times 5} = -7$

and $\dfrac{3x - 2}{2x - 7} = \dfrac{7}{-1} = -7$

With $x = 0$, we have $\dfrac{4}{14} \div \dfrac{-4}{-4} = \dfrac{4 \times -4}{14 \times -4} = \dfrac{2}{7}$

and $\dfrac{3x - 2}{2x - 7} = \dfrac{-2}{-7} = \dfrac{2}{7}$

The reader may now care to return to the penultimate example, $\dfrac{2x^2 + x - 15}{x^2 - 9}$, and check the validity of the answer $\dfrac{2x - 5}{x - 3}$.

Problems 2.3

1. Factorise:
   (a) $x^2 - 9$
   (b) $a^2 - b^2$
   (c) $4a^2 - 9b^2$
   (d) $ay^2 - a$

   (e) $(4x - y)^2 - (2x + y)^2$
   (f) $4 - \dfrac{a^2}{16}$

2. Factorise:
   (a) $a^2 + 6a + 9$
   (b) $x^2 - 4x + 4$
   (c) $a^2 + 6ab + 9b^2$
   (d) $2x^3 + 12x^2y + 18xy^2$
   (e) $9a^2 - 3a + \frac{1}{4}$

3. Factorise:
   (a) $x^2 + 6x + 5$
   (b) $x^2 + 6x + 8$
   (c) $x^2 + 2x - 15$
   (d) $x^2 - 9x + 20$
   (e) $2x^2 + 3x + 1$
   (f) $3x^2 - 10x + 3$
   (g) $10a^2b - 7ab - 12b$
   (h) $48x^2z - 26xyz - 4y^2z$

4. Factorise:
   (a) $ac + ad + bc + bd$
   (b) $6ac + 4bd + 3bc + 8ad$
   (c) $8ax - 5by + 4ay - 10bx$
   (d) $abd + 2cd - bd - 2acd$

5. Factorise:
   (a) $5x^2 - 6x - 8$
   (b) $4a^2c - 12abc + 9b^2c$
   (c) $3ac - 8bd + 6bc - 4ad$
   (d) $x^3 - xy^2 + 2x^2y - 2y^3$

6. One of the factors of $2x^3 + 9x^2 - 11x - 30$ is $x + 5$. Find the other two.

7. Factorise:
   (a) $18 + 3a - 6a^2$
   (b) $5y + 2xy - 6x - 15$
   (c) $16x^3y^2 - 9x$

8. Factorise:
   (a) $6a^2 + 7ab - 3b^2$
   (b) $27a^2b - 3b$
   (c) $4a^2 - 4ab + b^2$
   (d) $pq - qr + 2ps - 2rs$

9. Find the four factors of:

   $$2x^4 - x^3 - 20x^2 + 13x + 30$$

   knowing that this expression can be divided by $x^2 + x - 6$ without leaving a remainder.

10. Express $\dfrac{5}{x-3} - \dfrac{3}{x-1}$ as a single algebraic fraction in its simplest form.

11. Simplify $\dfrac{1}{x+4} - \dfrac{8}{x^2-16} + \dfrac{1}{x-4}$.

12. Express as a single fraction in its simplest form:

$$\frac{3}{x+1} + \frac{2}{x-2} - \frac{2x-7}{x^2-x-2}.$$

13. Simplify:

(a) $\dfrac{x^2-1}{x^2+4x-5}$ (b) $\dfrac{a^2+6a+9}{a^2-9}$ (c) $\dfrac{x^3-1}{x^2-1}$

14. Simplify:

(a) $\dfrac{x^2+3x-10}{x^2+5x-14}$ (b) $\dfrac{bx+ay+by+ax}{cx+dy+cy+dx}$

15. Simplify $\left(\dfrac{x^2-x-12}{2x^2-9x+4}\right)\left(\dfrac{4x^2+4x-15}{2x^2+11x+15}\right)$

16. Simplify $\dfrac{a^2+8a+16}{2a^2+17a+21} \times \dfrac{2a^2+11a-21}{2a^2+5a-12} \div \dfrac{a+4}{2a-3}$.

17. If $f(a) = 2a^2 - 11a - 21$, find the values of:
    (a) $f(1)$ (b) $f(0)$ (c) $f(-7)$ (d) $f(1\cdot5)$

18. If $f(x) = 5x^2 - 7x - 24$, show that $f(3) = 0$. Hence obtain the factors of $5x^2 - 7x + 24$.

19. If $f(x) = x^3 - 8x^2 + x + 42$, show that $f(7) = 0$. Hence obtain the factors of $x^3 - 8x^2 + x + 42$.

20. If $f(a) = 125a^3 - 27b^3$, show that $f\left(\dfrac{3b}{5}\right) = 0$. Hence, or otherwise, find the factors of $125a^3 - 27b^3$.

21. Simplify:

(a) $\dfrac{x+5}{4} + \dfrac{3x+2}{6} + \dfrac{x+5}{3}$ (b) $\dfrac{2}{ab} - \dfrac{3}{bc} + \dfrac{1}{ac}$

(c) $\dfrac{5x+2}{x^2-1} + \dfrac{7}{x+1}$ (d) $\left(\dfrac{x+3}{x^2+4x+4}\right)\left(\dfrac{x+2}{x^2+x-6}\right)$

(e) $\left(\dfrac{x^2-7x+12}{2x-5}\right) \div \left(\dfrac{x^2-9}{2x^2+x-15}\right)$

Answers to Problems 2.3

1. (a) $(x + 3)(x - 3)$ (b) $(a + b)(a - b)$
   (c) $(2a + 3b)(2a - 3b)$ (d) $a(y + 1)(y - 1)$

   (e) $12x(x - y)$ (f) $\left(2 + \dfrac{a}{4}\right)\left(2 - \dfrac{a}{4}\right)$

2. (a) $(a + 3)(a + 3)$ (b) $(x - 2)(x - 2)$
   (c) $(a + 3b)(a + 3b)$ (d) $2x(x + 3y)(x + 3y)$
   (e) $(3a - \frac{1}{2})(3a - \frac{1}{2})$

3. (a) $(x + 5)(x + 1)$ (b) $(x + 4)(x + 2)$
   (c) $(x + 5)(x - 3)$ (d) $(x - 4)(x - 5)$
   (e) $(2x + 1)(x + 1)$ (f) $(3x - 1)(x - 3)$
   (g) $b(2a + 3)(5a - 4)$ (h) $2z(8x + y)(3x - 2y)$

4. (a) $(a + b)(c + d)$ (b) $(2a + b)(3c + 4d)$
   (c) $(2x + y)(4a - 5b)$ (d) $d(a - 1)(b - 2c)$

5. (a) $(5x + 4)(x - 2)$ (b) $c(2a - 3b)(2a - 3b)$
   (c) $(a + 2b)(3c - 4d)$ (d) $(x + y)(x - y)(x + 2y)$

6. $(2x + 3)$ and $(x - 2)$

7. (a) $3(2 - a)(3 + 2a)$ (b) $(2x + 5)(y - 3)$
   (c) $x(4xy + 3)(4xy - 3)$

8. (a) $(2a + 3b)(3a - b)$ (b) $3b(3a + 1)(3a - 1)$
   (c) $(2a - b)(2a - b)$ (d) $(p - r)(q + 2s)$

9. $x + 3, x - 2, 2x - 5$ and $x + 1$

10. $\dfrac{2(x + 2)}{(x - 3)(x - 1)}$ or $\dfrac{2x + 4}{x^2 - 4x + 3}$

11. $\dfrac{2}{x + 4}$

12. $\dfrac{3}{x - 2}$

13. (a) $\dfrac{x + 1}{x + 5}$ (b) $\dfrac{a + 3}{a - 3}$ (c) $\dfrac{x^2 + x - 1}{x + 1}$

14. (a) $\dfrac{x + 5}{x + 7}$ (b) $\dfrac{a + b}{c + d}$

15. $\dfrac{2x - 3}{2x - 1}$

16. $\dfrac{2a - 3}{2a + 3}$

17. (a) -30 (b) -21 (c) 154 (d) -33

18. $(x - 3)(5x + 8)$

19. $(x - 7)(x - 3)(x + 2)$

20. $(5a - 3b)(25a^2 + 15ab + 9b^2)$

21. (a) $\dfrac{13x + 39}{12}$ (b) $\dfrac{-3a + b - 2c}{abc}$

    (c) $\dfrac{12x - 5}{x^2 - 1}$ (d) $\dfrac{1}{x^2 - 4}$

    (e) $x - 4$

2.4
Simple equations

2.4.1
Solution of simple equations

An *equation* is a statement indicating an equality, and in its mathematical form is characterized by the appearance of an equals sign. With most equations an algebraic symbol occurs which represents an unknown quantity. Such an equation is said to be *solved* when a value (numerical or algebraic) is found for the unknown quantity so that when this value is substituted for the unknown quantity the state of equality is maintained. An equation can be regarded as a balance between the quantities on either side of the equals sign. Equations are solved by conducting operations which do not disturb that balance. Whatever operation is performed on one side of an equation must be performed on the other side.

A *simple equation* is one which contains the first power of the unknown quantity. If a simple equation is solved there is one value, and one value only, for the unknown quantity. A little later we shall meet equations which contain the second power of the unknown. If such an equation is solved, there will be two values for the unknown quantity. In general, an equation which contains the nth power of the unknown will have n values for the unknown.

The major proportion of the work involved in solving a simple equation is usually carried out by performing a sequence of operations so that eventually all the terms containing the unknown are brought to one side of the equation, the remaining terms appearing on the other side. The operations will vary according to the construction of the original equation. We shall illustrate the different types of operations with worked examples. When a solution has been obtained, the value should be put in the original equation to check its validity.

Example:

Solve the equation $\qquad x - 2 = 2(x - 4)$.

Remove the brackets: $\qquad x - 2 = 2x - 8$

Bring the numerical terms to one side by adding 2 to each side.

$$x - 2 + 2 = 2x - 8 + 2$$
$$x = 2x - 6$$

Bring the algebraic terms to the other side by subtracting $2x$ from each side.

$$x - 2x = 2x - 6 - 2x$$
$$-x = -6$$

The equation cannot be simplified further, and now we leave just x on the left-hand side by dividing by its coefficient, in this case -1.

$$\frac{-x}{-1} = \frac{-6}{-1}$$
$$x = 6$$

We now check the validity of the answer by substituting its value in the original equation.

Check: $\qquad\qquad x - 2 = 2(x - 4)$
$$6 - 2 = 2(6 - 4)$$
$$4 = 2(2)$$
$$4 = 4$$

The balance is maintained and hence the answer is correct. As is usual, we complete the solution by writing out the answer.

Answer:

$$x = 6$$

Let us investigate the effect of adding (or subtracting) to eliminate certain quantities.

Consider $\qquad\qquad x = 2x - 6$

and the elimination of $2x$ from the right-hand side by subtracting $2x$ from each side.

$$x - 2x = 2x - 6 - 2x$$
$$x - 2x = -6$$

If we compare this last equation with the original equation, we note that the $2x$ term has moved to the other side of the equals sign but its sign has changed from positive to negative. This indicates a useful rule which can save time in effecting solutions. Any complete term can be transferred from one side of the equals sign to the other providing its sign is changed.

Example:

Solve the equation $\dfrac{5x + 7}{4} - \dfrac{3x - 8}{7} = \dfrac{x + 9}{2}$.

When fractions appear, it is generally favourable to commence by eliminating the fractions. We should remember from previous work in this book that the lines denoting the fractions group terms into single entities. Let us never be afraid of introducing brackets to ensure that we treat them as such. We will rewrite the equation as:

$$\frac{(5x + 7)}{4} - \frac{(3x - 8)}{7} = \frac{(x + 9)}{2}$$

We can eliminate the fractions by multiplying everything by the least common multiple of 4, 7 and 2, which is 28.

$$\frac{28}{4}(5x + 7) - \frac{28}{7}(3x - 8) = \frac{28}{2}(x + 9)$$

$$7(5x + 7) - 4(3x - 8) = 14(x + 9)$$

Remove brackets:

$$35x + 49 - 12x + 32 = 14x + 126$$

Collect like terms:

$$35x - 12x - 14x = 126 - 49 - 32$$

$$9x = 45$$

$$x = 5$$

Check:
$$\frac{5x + 7}{4} - \frac{3x - 8}{7} = \frac{x + 9}{2}$$

$$\frac{25 + 7}{4} - \frac{15 - 8}{7} = \frac{5 + 9}{2}$$

$$\frac{32}{4} - \frac{7}{7} = \frac{14}{2}$$

$$8 - 1 = 7$$

$$7 = 7$$

Answer:

$$x = 5$$

Let us now consider an equation in which two fractions are equated, such as:

$$\frac{x + 2}{2} = \frac{5x - 11}{3}$$

The denominator on the left-hand side can be eliminated by multiplying both sides of the equation by 2, obtaining:

$$x + 2 = \frac{2(5x - 11)}{3}$$

The denominator 3 on the right-hand side of this latter equation can be eliminated by multiplying both sides of the equation by 3, to obtain:

$$3(x + 2) = 2(5x - 11)$$

If we compare this last equation with the original equation, we observe it to be

(denominator, right-hand side)(numerator, left-hand side)
= (denominator, left-hand side)(numerator, right-hand side)

that is, if $\frac{a}{b} = \frac{c}{d}$, then $(a \times d) = (b \times c)$.

We note that the multiplication takes place across the equals sign, hence the process is called *cross-multiplication*. We have deduced another rule which could prove to be convenient in certain cases. If an equation is formed by equating two fractions, another equation can be formed by cross-multiplication. It should be noted that any algebraic term which is not a fraction can be put in the form of a fraction by using a denominator of unity.

Example:

Solve the equation $\dfrac{7(x - 4)}{8} = 2x - 12 \cdot 5$.

Rearrange as an equation of two fractions, and use a bracket to treat $2x - 12 \cdot 5$ as a complete entity.

$$\frac{7(x - 4)}{8} = \frac{(2x - 12 \cdot 5)}{1}$$

By cross-multiplication:

$$7(x - 4) = 8(2x - 12 \cdot 5)$$

Remove brackets:

$$7x - 28 = 16x - 100$$

Collect like terms:

$$7x - 16x = -100 + 28$$
$$-9x = -72$$
$$x = \frac{-72}{-9}$$
$$x = 8$$

Check:

$$\frac{7(x - 4)}{8} = 2x - 12 \cdot 5$$
$$\frac{7(8 - 4)}{8} = 16 - 12 \cdot 5$$
$$\frac{28}{8} = 3 \cdot 5$$
$$3 \cdot 5 = 3 \cdot 5$$

Answer:

$$x = 8$$

The notes which have been put down alongside the solutions in this article were used to indicate the operations being performed, and as skill is acquired the use of such notes by students can be discarded. Now let us demonstrate how algebra can be used to help us solve practical problems.

2.4.2
Practical applications of the solution of simple equations
Example:

If a channel section has a width of W, a depth of D and a constant thickness of t, the cross-sectional area A can be found from the formula:

$$A = t(2D + W - 2t)$$

Find the value of D when $A = 600$ mm^2, $W = 70$ mm and $t = 5$ mm.

$$A = t(2D + W - 2t)$$
$$600 = 5(2D + 70 - 10)$$
$$600 = 5(2D + 60)$$
$$600 = 10D + 300$$
$$600 - 300 = 10D$$
$$300 = 10D$$
$$D = 30$$

Answer:

$$D = 30 \text{ mm}$$

Example:

The length s, measured from its lowest point, of a heavy cable suspended between two points, and the vertical height y, are connected by the equation:

$$(y + c)^2 = s^2 + c^2$$

where c is constant. If $y = 4$ when $s = 8$, show that $c = 6$. Find s when $y = 10$.

$$(y + c)^2 = s^2 + c^2$$
$$(4 + c)^2 = 8^2 + c^2$$
$$16 + 8c + c^2 = 64 + c^2$$
$$8c = 64 - 16 = 48$$
$$c = \frac{48}{8} = 6$$
$$\therefore (y + 6)^2 = s^2 + 6^2$$
$$(10 + 6)^2 = s^2 + 36$$
$$16^2 = s^2 + 36$$
$$256 = s^2 + 36$$
$$s^2 = 256 - 36 = 220$$
$$s = \sqrt{220} = \pm 14 \cdot 83$$

The negative value cannot apply $\therefore s = 14 \cdot 83$

Answer:

$$c = 6$$
$$s = 14.8 \text{ when } y = 10$$

Example:

The mass of an amount of liquid, in grams, can be found by multiplying its relative density by its volume in millilitres. The relative density of 800 millilitres of a dilute sulphuric acid is 1·12. What volume of dilute sulphuric acid of relative density 1·27 must be added to produce an acid of relative density 1·22?

Let   volume added be x millilitres
then   mass added $=$ volume $\times$ relative density
                             $= 1.27x$ grams
Mass of original acid $= 800 \times 1.12 = 896$ grams
   Mass after addition $= 896 + 1.27x$ grams
     Volume after addition $= 800 + x$

$$\text{Relative density after addition} = \frac{\text{mass}}{\text{volume}}$$

$$1.22 = \frac{896 + 1.27x}{800 + x}$$

By cross-multiplication:

$$(800 + x)\,1.22 = 896 + 1.27x$$
$$976 + 1.22x = 896 + 1.27x$$
$$976 - 896 = 1.27x - 1.22x$$
$$80 = 0.05x$$

$$x = \frac{80}{0.05} = 1\,600$$

Answer:

Volume added $= 1\,600$ millilitres

Example:

An employee produces a total of 910 articles. He commences at the rate of 20 articles per hour and, having completed a part of the total, increases his rate to 25 articles per hour. The time taken overall was 40 hours. How many articles had been made when production was changed?

Let N = number of articles made when rate changed
then $910 - N$ = number of articles made at a faster rate

$$\text{Time taken at initial rate} = \frac{N}{20}$$

$$\text{Time taken at faster rate} = \frac{910 - N}{25}$$

$$\text{Total time} = 40 \text{ hours}$$

$$\therefore \frac{N}{20} + \frac{910 - N}{25} = 40$$

L.C.M. of 20 and 25 is 100.

$$100\left(\frac{N}{20}\right) + \frac{100(910 - N)}{25} = 100(40)$$

$$5N + 4(910 - N) = 4\,000$$
$$5N + 3\,640 - 4N = 4\,000$$
$$5N - 4N = 4\,000 - 3\,640$$
$$N = 360$$

Answer:

Rate changed after 360 articles had been made.

Problems 2.4

1. Solve the following equations:

   (a) $2(2x - 5) + 3(x + 2) = 6x + 4$

   (b) $3(x + 4) - 2(x - 3) = 3x + 6$

   (c) $\dfrac{3x + 5}{4} + \dfrac{5x - 7}{6} = \dfrac{5x - 1}{3}$

   (d) $\frac{1}{2}(3x + 4) + \frac{2}{3}(x + 2) = 3x$

   (e) $3(x - 2) = \dfrac{3x + 1}{4} + 7x - 1 \cdot 5$

   (f) $\dfrac{x + 8}{x - 2} = \dfrac{x + 26}{x + 1}$

   (g) $\{2x - 3(x - 7)\} = \dfrac{3x + 2}{2}$

   (h) $5x - 3\{x - 2(2x - 4)\} = 7x - 3$

2. Certain spacing pieces are made of width either 3 mm or 5 mm. A total of 40 spacing pieces are to be used to make up an overall width of 166 mm. Determine the number of spacers of width 3 mm.

3. Find two successive positive integers whose squares differ by 15.

4. A baulk of timber of length 5·25 m is to be sawn into two pieces so that the length of one piece is three-quarters of the length of the other. Find the length of the shortest piece.

5. If $y_1 = \dfrac{3x + 12}{4}$ and $y_2 = \dfrac{26 - x}{2}$, find the value of x for which $y_1 = y_2$. Use the value of x in both equations to prove $y_1 = y_2$.

6. A first wire has a resistance of 3 ohms per metre length. A second wire has a resistance of 2 ohms per metre length. A length of the first wire is joined end-to-end (i.e. in series) with a length of the second wire so that the overall length is 50 m and the total resistance is 130 ohms. Determine the lengths of the two pieces.

7. By increasing the average speed of travel between two towns from 90 km/h to 120 km/h the journey time is reduced by one hour. Determine the distance between the two towns.

8. A rectangular sheet of metal, 23 cm × x cm, has squares of side 4 cm cut from each corner. The remainder is then folded to form a rectangular tray of depth 4 cm and capacity 1 440 cm³. Determine the value of x.

9. The outside diameter of a gear wheel is connected with the number of teeth N and a feature m of the size of the teeth so that

$$D = m(N + 2).$$

Find the value of N if $D = 95$ mm when $m = 2·5$ mm.

10. A train maintains an average speed of $0·64x$ km/h for the first hour, $0·8x$ km/h for the next 45 minutes, and x km/h for the last hour. The total distance travelled was 140 km. Determine the value of x.

11. A firm purchases goods in batches of items. The total batch purchase price is kept constant. A change in the price per item causes a change in the number of items per batch. Originally the purchase cost was £36 per item. When the price was increased by £4 per item, the batch size was decreased by 20 items. Determine the constant batch purchase price.

12. The following equation resulted from a problem dealing with mixing steam and water:

$$5\,400 + 10(100 - T) = 610(T - 20)$$

Determine the value of T.

13. An approximate relationship between the number of teeth on a milling cutter T, the diameter of cutter D and the depth cut d is:

$$T = \frac{12 \cdot 5D}{D + 4d}$$

Find the value of D when $T = 10$ and $d = 5$ mm.

14. If a shaft has a diameter D, the depth of cut d to produce a flat of width W can be obtained from the formula:

$$D - 2d = \sqrt{(D^2 - W^2)}$$

Find the value of d when $D = 34$ mm and $W = 16$ mm.

15. A keyway of width W is cut in a shaft of diameter D so that the nominal depth at the side of the keyway is h. The depth at the centre of the keyway H can be found from the formula:

$$\left(\frac{D^2 - W^2}{4}\right) = \frac{D}{2} - H + h$$

Find the value of H when $D = 26$ mm, $W = 10$ mm and $h = 4$ mm.

16. A rectangle has a length which is 4 mm greater than its breadth. By letting the length be x, obtain an expression for the perimeter of the rectangle. Hence obtain the dimensions of such a rectangle whose perimeter is 28 mm.

17. Without using tooling, articles cost £0·90 each to produce. By investing £120 in tooling, the cost is reduced to £0·30 per article. At what number of articles is the 'break even' point reached, i.e. the number of articles when the total cost by either method is identical?

18. A rectangular plate of uniform thickness originally has a length of 250 mm and a width of 120 mm. Four rectangular lightening holes are to be cut in the plate, each of length 60 mm and width x mm so that the weight of the plate is reduced by 20%. Determine the value of x.

19. A traveller starts a journey at an average speed of 30 kilometres per hour and continues at this rate for 2 hours. He rests for half an hour and then continues his journey at the rate of 40 kilometres per hour. Obtain an expression for the total distance travelled x hours after the start, when x is greater than $2\frac{1}{2}$, and hence find the total time to travel 120 kilometres.

20. The current I which flows in a simple series circuit is found by dividing the electromotive force E by the resistance R of that circuit.

The original resistance of the circuit R was not known, but when R was increased by 3 ohms, the e.m.f. had to be increased by 60% to cause the same current to flow. Find the value of R.

21. When a body passes a datum with an initial velocity u and a constant acceleration of a, the distance s from datum after a time t is given by:

$$s = ut + \frac{at^2}{2}$$

A body passes datum with a velocity of 40 m/s and after 4 seconds has travelled 200 m. Find the value of a, stating both its magnitude and its unit.

22. A brass ingot has a mass of 30 kg and contains 60% of copper. An amount of x kg of copper is melted into this ingot to produce a new ingot of mass $(30 + x)$ kg. In this new ingot there is 70% of copper. Find the amount x which was added.

23. A bin of electrical components totalling 2 000 components consists of deliveries from two suppliers A and B, of which supplier A has contributed 70%. How many components must be added, entirely from supplier B, so that his contribution is increased to 44%?

24. During an experiment using heat transfer in order to determine the specific heat c of a light mineral oil, the following equation was developed:

$$m_1 c_1 (t_2 - t_1) = (m_2 c_2 + mc)(t_1 - t_0)$$

Find the value of c given that:

$$m = 156 \qquad m_1 = 50 \qquad m_2 = 100 \qquad c_1 = c_2 = 504$$
$$t_2 = 100 \qquad t_1 = 25 \qquad t_0 = 20$$

25. Two parallel conveyor belts move in the same direction at speeds of 5 m/min. and 8 m/min. The entry points to both conveyors are at the same level. A sub-assembly enters the slower conveyor and 3 minutes later a matching sub-assembly enters the faster conveyor. How far are the sub-assemblies from the entry point when they are level?

26. If R_0 is the resistance of a particular conductor at 0°C, the resistance R_t of that conductor at a temperature of t°C is given by the equation:

$$R_t = R_0(1 + \alpha t)$$

If a conductor has a resistance of 81·6 ohms at 5°C and 88·0 ohms at 25°C, form two separate equations each containing R_0. Divide these two equations and hence find the value of α.

27. An indicator needle is oscillating about its final steady reading x. A first reading is 39, the second reading is 24, so that the first deviation is $39 - x$ and the second deviation is $x - 24$. The third reading is 34. The ratio of the first deviation to the second deviation is the same as the ratio of the second to the third. Calculate the final steady reading α.

28. An aircraft has a speed of 450 km/h in still air. On a double journey the speed on the outward leg is increased by 50 km/h due to a favourable wind. It is decreased by the same extent on the inward leg, the wind being adverse. The double journey takes 2 minutes longer than it would have been had there have been no wind. Determine the total distance of the double journey.

29. One of the values of x which satisfies the equation

$$x^2 + 0.9x = 6.3$$

is in the vicinity of $x = 2$. By substitution $2 + h$ for x in the original equation and dismissing the term which includes x^2, form a simple equation involving h. Determine the value of h, and hence find the value of x to the nearest two significant figures.

Answers to Problems 2.4

1. (a) $x = 8$ (b) $x = 6$ (c) $x = 5$ (d) $x = 4$
   (e) $x = -1$ (f) $x = 4$ (g) $x = 8$ (h) $x = 3$

2. 17

3. 7 and 8

4. 2·25 m

5. $x = 8, y_1 = y_2 = 9$

6. 20 m of 2 Ω/m, 30 m of 3 Ω/m

7. 360 km

8. 32 cm

9. 36

10. 62·5

11. £7 200

12. $T = 30$

13. $D = 80$ mm

14. $d = 2$ mm

15. $H = 5$ mm

16. $4x + 8$, 5 mm × 9 mm

17. 200

18. $x = 25$ mm

19. $40x - 40$, $x = 4$ hours

20. $R = 5$ ohms

21. 5 m/s²

22. 10 kg

23. 500

24. $c = 2\,100$

25. 40 m

26. 0·004

27. 30

28. 1 200 km

29. 2·1

2.5
Simultaneous equations

The equation $x = y + 3$ provides an infinite number of values of x, each value of x depending upon the value that is given to y. There is one, and only one, value of x, and hence only one associated value of y, which satisfies the equation $x = y + 3$, and at the same time (or *simultaneously*) satisfies an equation containing x and/or y which is not a multiple of the first equation. Let us suppose this second equation is $x + 3y = 19$.

$x = y + 3$ and $x + 3y = 19$ are a pair of *simultaneous equations*. If there are two unknowns, two separate and distinct equations will be necessary to effect a solution. Three unknowns will require three separate and distinct equations. Expressed in general terms, n unknowns require n separate and distinct equations to effect a solution. At present we are only concerned with simultaneous equations of two unknowns.

Just as with other equations, simultaneous equations can be solved by algebraic processes or by graphs. We will first consider the algebraic processes, leaving the graphical method until a later chapter. There are several methods of solution, and it takes a little experience to decide which method is quickest for a particular problem.

Example:

Solve the simultaneous equations:

$$5x = 8y + 3$$
$$x + 3y = 19$$

First method: Equating coefficients

With this method one or both equations are multiplied to produce a pair of equations where the coefficients of one of the unknowns are numerically equal. If the coefficients have the same sign, subtracting the equations eliminates one unknown and leaves a simple equation. If the coefficients have opposite signs, adding the equations will eliminate an unknown.

We first rearrange the equations, if necessary, to bring the unknowns on one side of the equals sign.

$$5x = 8y + 3$$
$$\therefore 5x - 8y = 3 \qquad \ldots \text{equation (1)}$$
$$x + 3y = 19 \qquad \ldots \text{equation (2)}$$

Equation (2) × 5: $\qquad 5x + 15y = 95$

Equation (1): $\qquad 5x - 8y = 3$

The coefficients of x are now alike, and since they both have the same sign, we shall subtract one equation from the other.

$$23y = 92$$
$$y = 4$$

Substituting this value in one of the original equations:

$$5x = 8y + 3 = 32 + 3 = 35$$
$$x = 7$$

Check: $\qquad 5x = 8y + 3, 35 = 32 + 3, 35 = 35$

$$x + 3y = 19, 7 + 12 = 19, 19 = 19$$

Answer:

$$x = 7 \qquad y = 4$$

Second method: Substitution of one equation in the other

The first step is to obtain from one equation an expression either for x or for y. In this particular problem, a value for x from the second equation is easiest.

$$x + 3y = 19$$
$$\therefore x = 19 - 3y$$

We now substitute this value in the first equation:

$$5x = 8y + 3$$
$$5(19 - 3y) = 8y + 3$$
$$95 - 15y = 8y + 3$$
$$-15y - 8y = -95 + 3$$
$$-23y = -92$$
$$y = \frac{-92}{-23} = 4, \text{ as before}$$
$$x = 19 - 3y = 19 - 12 = 7 \text{ as before}$$

Hence $x = 7$ and $y = 4$

The validity of these values has already been checked.

For most of the problems which occur in our present stage of studies, the method of equating coefficients will generally prove to be the most rapid. On some occasions the reader may prefer to arrange a substitution before the method of equating coefficients is used.

Example:

Solve the simultaneous equations:

$$\frac{5}{x} + \frac{8}{y} = 9$$
$$\frac{7}{x} - \frac{6}{y} = 4$$

Let $\dfrac{1}{x} = a$ and $\dfrac{1}{y} = b$

then $5a + 8b = 9$ ... equation (1)

and $7a - 6b = 4$ ... equation (2)

Equation (1) × 7: $35a + 56b = 63$

Equation (2) × 5: $\underline{35a - 30b = 20}$

Subtract: $86b = 43$ $b = \frac{43}{86} = \frac{1}{2}$

Substitute this value in equation (1).

$$5a + 8b = 9, 5a + 4 = 9, 5a = 9 - 4 = 5$$
$$a = 1$$

$$\frac{1}{x} = a \qquad \therefore x = \frac{1}{a} = \frac{1}{1} = 1$$

$$\frac{1}{y} = b \qquad \therefore y = \frac{1}{b} = \frac{1}{\frac{1}{2}} = 2$$

Check:
$$\frac{5}{x} + \frac{8}{y} = 9, \frac{5}{1} + \frac{8}{2} = 9, 5 + 4 = 9, 9 = 9$$

$$\frac{7}{x} - \frac{6}{y} = 4, \frac{7}{1} - \frac{6}{2} = 4, 7 - 3 = 4, 4 = 4$$

Answers:

$$x = 1 \qquad y = 2$$

Problems 2.5

1. Solve the following simultaneous equations:

   (a) $x + 2y = 11$
   $2x + y = 7$

   (b) $7x - 5y = 13$
   $2x + 3y = 17$

   (c) $3x - 2y = 4$
   $5x = 2y$

   (d) $4x + 5y = 2$

   $x - y = \dfrac{1}{20}$

   (e) $\dfrac{1}{v} - \dfrac{1}{u} = \dfrac{1}{20}$

   $\dfrac{2}{v} + \dfrac{7 \cdot 5}{u} = 2$

2. Solve the simultaneous equations:

$$3a - 4b = 0$$
$$2a + 6b = 13$$

3. Solve the simultaneous equations:

$$2x + 9y = 19$$
$$7x - 6y = 4$$

4. Determine the values of V_1 and V_2, given that:

$$5V_1 + 12V_2 = 36 \cdot 1$$
$$11V_1 + 8V_2 = 37 \cdot 1$$

5. If $W = KL + c$, find K and c given that when $L = 3$, $W = 15$ and when $L = 4$, $W = 18$.

6. Solve the simultaneous equations:

$$3x - 6y = 0$$
$$2x + 5y = 45$$

7. Solve the simultaneous equations:

$$4(x - 2) = 5(1 - y)$$
$$26x + 3y + 4 = 0$$

8. Solve the following simultaneous equations for I_1 and I_2:

$$0{\cdot}05I_1 + (I_1 + I_2) = 2{\cdot}05$$
$$0{\cdot}08I_2 + 2(I_1 + I_2) = 4{\cdot}30$$

9. Solve the simultaneous equations:

$$2R = A - \frac{1}{0{\cdot}25}$$
$$A = 3R + 1$$

10. Solve the simultaneous equations:

$$4x = 3y + 16$$
$$4y = x + 9$$

11. Solve for x and y the simultaneous equations:

$$\frac{1}{x} + \frac{1}{y} = \frac{1}{4}$$
$$\frac{5}{x} + \frac{2}{y} = 2$$

12. By writing $\dfrac{1}{x + 3y} = a$ and $\dfrac{1}{3x - y} = b$, find the values of x and y that satisfy the equations:

$$\frac{2}{x + 3y} - \frac{1}{3x - y} = 4$$
$$\frac{1}{x + 3y} + \frac{2}{3x - y} = 7$$

13. If twice the sum of two numbers exceeds four times their difference by 10 and the two numbers are in the ratio of 4:1, calculate each number.

14. The relation between the load W and the required effort E of a particular lifting machine is given by the formula:

$$E = aW + b$$

A lifting machine can lift 100 N with an effort of 15 N, and 400 N with an effort of 30 N. Determine the value of the constants a and b, and hence the value of the load when the effort is 40 N.

15. The length of a compression spring can be found from its free length L and a shortening due to the load W and stiffness S, so that the length is given by $L - (W \div S)$. Such a spring has a length of 38 mm under a load of 400 N and a length of 35·5 mm under a load of 900 N. Find the values of L and S, and hence determine the load on the spring when it has a length of 37·5 mm.

16. If a body has an initial velocity of u and has a uniform acceleration of a, its velocity v after time t is obtained from the formula:

$$v = u + at$$

Such a body has a velocity v of 40 m/s when $t = 3$ seconds and a velocity v of 100 m/s when $t = 15$ seconds. Determine the values of u and a, indicating both the magnitude and the unit in each case.

17. The selling price P of a hand tool consists of a fixed amount A plus a variable amount which is proportional to the square of the nominal size S, so that:

$$P = A + BS^2$$

where A and B are constants. If a 20 mm tool costs £0·70 and a 50 mm tool cost £3·32$\frac{1}{2}$, determine the values of A and B, and hence the cost of a 60 mm tool.

Answers to Problems 2.5

1. (a) $x = 1, y = 5$ (b) $x = 4, y = 3$
   (c) $x = -2, y = -5$ (d) $x = \frac{1}{4}, y = \frac{1}{5}$
   (e) $v = 4, u = 5$

2. $a = 2, b = 1\frac{1}{2}$

3. $x = 2, y = 1\frac{2}{3}$

4. $V_1 = 1·7, V_2 = 2·3$

5. $K = 3, c = 6$

6. $x = 10, y = 5$

7. $x = -\frac{1}{2}, y = 3$

8. $I_1 = -10, I_2 = 12 \cdot 55$

9. $A = 10, R = 3$

10. $x = 7, y = 4$

11. $x = 2, y = -4$

12. $x = \dfrac{11}{60}, y = \dfrac{1}{20}$

13. -20 and -5

14. $a = \dfrac{1}{20}, b = 10$ N; $W = 600$ N

15. $L = 40$ mm, $S = 200$ N/mm; 500 N

16. $u = 25$ m/s, $a = 5$ m/s^2

17. $A = £0 \cdot 20, B = £0 \cdot 12\frac{1}{2}; £4 \cdot 70$

2.6 Transposition of formulae

Formulae that are developed during the reader's studies are not always in a form suitable for direct use in answering a problem. For instance, the reader will soon meet a formula for the volume V of a sphere in terms of the diameter D, this being $V = \dfrac{\pi D^3}{6}$. This formula can be used directly if a problem states, 'Find the volume of a sphere of diameter 3 mm.' If a problem requires us to find the diameter of a sphere whose volume is 18 mm^3, what we require is a formula which gives the diameter in terms of the volume. In the formula $V = \dfrac{\pi D^3}{6}$, V is called the subject of the formula. What we have to do is to rearrange the equation $V = \dfrac{\pi D^3}{6}$ so that D becomes the subject instead of V. We reposition the symbols without disturbing the equality, the process being known as the *transposition of formulae*.

It must be emphasized that transposition of formulae is simply the solution of equations. The small difference when compared with previous work is that in the past most answers to equations have been numbers. In transposition of formulae the answer that is required usually consists of numbers and/or symbols.

A problem which states:

'If $V = \dfrac{\pi D^3}{6}$, make D the subject of the formula'

only means

'Solve the equation $V = \dfrac{\pi D^3}{6}$ for D.'

Consequently transposition of formulae within the scope for which this book was intended requires no further work beyond that already covered in this chapter on the solution of equations. Let us illustrate the above information by a series of typical examples, noting that in general we try to bring all the terms including the subject to one side of an equation. We will commence with formulae which equate two fractions, and we note that a single term can be made into a fraction by using a denominator of unity.

Example:

If $V = \dfrac{\pi D^3}{6}$, make D the subject of the formula.

$$V = \frac{\pi D^3}{6} \qquad \frac{V}{1} = \frac{\pi D^3}{6}$$

By cross-multiplication $\qquad \dfrac{6V}{\pi} = \dfrac{D^3}{1}$

Hence $\qquad D^3 = \dfrac{6V}{\pi} \quad \text{and} \quad D = \sqrt[3]{\dfrac{6V}{\pi}}$

Having solved an equation, the validity of the solution should be checked. Transposition of formulae, it is repeated, is only the solution of equations, and hence the answer should be checked. This is best done by finding simple values which satisfy the original formula, and then seeing if the same values satisfy the answer.

Check: In the original formula, if $D = 6$, then $V = 36\pi$.

In the answer: $\qquad D = \sqrt[3]{\dfrac{6 \times 36\pi}{\pi}} = \sqrt[3]{216} = 6$

This is apparently correct, so we now conclude the exercise by stating the answer.

Answer:

$$D = \sqrt[3]{\frac{6V}{\pi}}$$

Example:

If $T = 2\pi\sqrt{\dfrac{GJ}{IL}}$, find a formula for G.

If a root sign appears it is advisable to move everything outside the root sign to the other side of the equation.

$$T = 2\pi\sqrt{\frac{GJ}{IL}}, \qquad \frac{T}{2\pi} = \sqrt{\frac{GJ}{IL}}$$

Now square both sides to eliminate the root sign:

$$\frac{T^2}{4\pi^2} = \frac{GJ}{IL}$$

and by cross-multiplication $G = \dfrac{T^2 IL}{4\pi^2 J}$

In the original equation, if $I = 1$, $L = 2$, $J = \frac{1}{2}$ and $G = 16$, then $T = 2\pi\sqrt{4} = 4\pi$. In the final equation:

$$G = \frac{T^2 IL}{4\pi^2 J} = \frac{16\pi^2 \times 1 \times 2}{4\pi^2 \times \frac{1}{2}} = 16$$

Answer:

$$G = \frac{T^2 IL}{4\pi^2 J}$$

Example:

Make K the subject from the formula $E = \dfrac{9CK}{C + 3K}$.

We note that the line of a fraction can be interpreted as a pair of brackets.

$$\frac{E}{1} = \frac{9CK}{(C + 3K)}$$

By cross multiplication $E(C + 3K) = 9CK$

Remove brackets: $EC + 3EK = 9CK$

Bring terms involving the unknown to one side:

$$EC = 9CK - 3EK$$

Factorise R.H.S.: $\qquad EC = K(9C - 3E)$

Hence

$$\frac{EC}{(9C - 3E)} = K$$

Check: In the original equation, if $C = 2$ and $K = 1$ then:

$$E = \frac{9 \times 2 \times 1}{2 + 3} = \frac{18}{5} = 3 \cdot 6$$

In the final equation:

$$K = \frac{EC}{9C - 3E} = \frac{3 \cdot 6 \times 2}{18 - 10 \cdot 8} = \frac{7 \cdot 2}{7 \cdot 2} = 1$$

Answer:

$$K = \frac{EC}{9C - 3E}$$

Example:

If $S = \dfrac{n\sqrt{Q}}{H^{3/4}}$, make Q the subject of the formula.

Write as two fractions: $\qquad \dfrac{S}{1} = \dfrac{n\sqrt{Q}}{H^{3/4}}$

Leave root sign on one side by cross-multiplication:

$$\frac{SH^{3/4}}{n} = \frac{\sqrt{Q}}{1}$$

Square both sides: $\qquad \left(\dfrac{SH^{3/4}}{n}\right)^2 = Q$

$$\therefore Q = \frac{S^2 H^{3/2}}{n^2}$$

Check: A convenient value of H must be chosen so that $H^{3/4}(= \sqrt[4]{H^3})$ can be evaluated. Let H be 2^4 ($=16$), then $\sqrt[4]{H^3} = \sqrt[4]{(16 \times 16 \times 16)} = \sqrt[2]{(4 \times 4 \times 4)} = 2 \times 2 \times 2 = 8$. Let $Q = 4$ and $n = 3$:

then $\qquad S = \dfrac{3\sqrt{4}}{8} = \dfrac{3}{4}$

Values which fit the original formula are $n = 3$, $Q = 4$, $H = 16$ and $S = \frac{3}{4}$. Let us see whether these values fit the answer.

Check: $\quad Q = \dfrac{(\frac{3}{4})^2(16)^{3/2}}{3^2} = \dfrac{9 \times \sqrt[2]{16^3}}{16 \times 9} = \dfrac{9}{16} \times \dfrac{64}{9} = 4$

Example:

Eliminate a from the simultaneous equations:

$$P = aL^2 + K$$

$$Q = 3aL^2 + K$$

and hence find an expression for P which does not contain a.

Multiply the first equation by 3: $\qquad\qquad 3P = 3aL^2 + 3K$

Rewrite the second equation: $\qquad\qquad\quad\; Q = 3aL^2 + \;\; K$

Subtract: $\qquad\qquad\qquad\qquad\qquad \overline{3P - Q = \qquad\quad + 2K}$

$$\qquad\qquad\qquad\qquad\qquad 3P = 2K \;\; + \;\; Q$$

$$\therefore P = \dfrac{2K + Q}{3}$$

Check: Let $a = 1$, $L = 2$, $K = 3$, then $P = 7$ and $Q = 12 + 3 = 15$

$$P = \dfrac{2K + Q}{3} = \dfrac{6 + 15}{3} = \dfrac{21}{3} = 7$$

Answer:

$$P = \dfrac{2K + Q}{3}$$

Problems 2.6

In questions 1 to 23 transpose the formulae to produce formulae for the symbol stated.

1. $I = \dfrac{BD^3}{12}$ for B

2. $W = \dfrac{mv^2}{2}$ for m

3. $V = \dfrac{\pi r^2 h}{3}$ for h

4. $A = \pi r^2$ for r

5. $V = \dfrac{\pi D^3}{6}$ for D

6. $F = \dfrac{mv^2}{r}$ for v

7. $t = \sqrt{\dfrac{2s}{g}}$ for s

8. $v = c\sqrt{(mi)}$ for i

9. $R = \sqrt[3]{\dfrac{3V}{4\pi}}$ for V

10. $s = ut - \frac{1}{2}at^2$ for a

11. $K = \dfrac{2c(m+1)}{3(m-2)}$ for m

12. $S = \dfrac{n}{2}\{2a + d(n-1)\}$ for a

13. $R = \dfrac{2N}{N-n}$ for N

14. $\dfrac{1}{f} = \dfrac{1}{v} - \dfrac{1}{u}$ for v

15. $v^2 - u^2 = 2as$ for u

16. $T = 2\pi\sqrt{\dfrac{L}{g}}$ for g

17. $T = 2\pi\sqrt{\dfrac{k^2 + h^2}{gh}}$ for k

18. $N = \dfrac{1}{2L}\sqrt{\dfrac{T}{M}}$ for M

19. $\dfrac{1}{R} = \dfrac{1}{R_1} + \dfrac{1}{R_3}$ for R_1

20. $\dfrac{1}{R} = \dfrac{1}{R_1} + \dfrac{1}{R_2 + R_3}$ for R_2

21. $V = \sqrt{\dfrac{nP}{d}}$ for d

22. $A = P\left(1 + \dfrac{R}{100}\right)^n$ for R

23. $W = \dfrac{\pi LS}{4}(D^2 + d^2)$ for d

24. If $f = \dfrac{1}{2\pi\sqrt{(LC)}}$ find a formula expressing L in terms of f and C.

25. If $P = \dfrac{fa}{1 + r^2}$, express r in terms of f, a and P.

26. Make θ the subject of the formula $R = R_0(1 + \theta t)$.

27. If $A = g\left(\dfrac{1}{1 - d}\right) - f$ express d in terms of A, g and f.

28. Make b the subject of the formula:

$$x = \dfrac{mb - ad}{a + b}$$

29. If $Z = \sqrt{\{R^2 + (X_1 - X_2)^2\}}$, rearrange the formula to obtain an expression for X_2.

30. If $V = \dfrac{\pi D^2 H}{12}$ and $A = \dfrac{\pi D^2}{4}$, find an expression for V in terms of A and H.

31. If $A = \dfrac{\pi D^2}{4}$ and $C = \pi D$, find an expression for A in terms of C.

32. If $H = \dfrac{4F}{\pi d^2}$ and $D = \dfrac{0 \cdot 3 HLd}{\sqrt{V}}$, express D in terms of F, L, d and V.

Answers to Problems 2.6

1. $B = \dfrac{12I}{D^3}$

2. $m = \dfrac{2W}{v^2}$

3. $h = \dfrac{3V}{\pi r^2}$

4. $r = \sqrt{\dfrac{A}{\pi}}$

5. $D = \sqrt[3]{\dfrac{6V}{\pi}}$

6. $v = \sqrt{\dfrac{Fr}{m}}$

7. $s = \dfrac{gt^2}{2}$

8. $i = \dfrac{v^2}{c^2 m}$

9. $V = \dfrac{4\pi R^3}{3}$

10. $a = \dfrac{2(ut - s)}{t^2}$

11. $m = \dfrac{6K + 2c}{3K - 2c}$

12. $a = \dfrac{S}{n} - \dfrac{d}{2}(n - 1)$

13. $N = \dfrac{nR}{R - 2}$

14. $v = \dfrac{uf}{u + f}$

15. $u = \sqrt{(v^2 - 2as)}$

16. $g = \dfrac{4\pi^2 L^2}{T^2}$

17. $k = \sqrt{\left(\dfrac{T^2 hg}{4\pi^2} - h^2\right)}$

18. $M = \dfrac{T}{4N^2 L^2}$

19. $R_1 = \dfrac{RR_2}{R_2 - R}$

20. $R_2 = \dfrac{RR_1}{R_1 - R} - R_3$ or $\dfrac{RR_1 + RR_3 - R_1 R_3}{R_1 - R}$

21. $d = \dfrac{nP}{V^2}$

22. $R = 100\left\{\sqrt[n]{\left(\dfrac{A}{P}\right)} - 1\right\}$

23. $d = \sqrt{\left(\dfrac{4W}{\pi LS} - D^2\right)}$

24. $L = \dfrac{1}{4\pi^2 f^2 C}$

25. $r = \sqrt{\left(\dfrac{fa}{P} - 1\right)}$ or $\sqrt{\left(\dfrac{fa}{P} - P\right)}$

26. $\theta = \dfrac{R - R_0}{R_0 t}$

27. $d = 1 - \dfrac{g}{A + fg}$

28. $b = \dfrac{a(x + d)}{(m - x)}$

29. $X_2 = X_1 - \sqrt{(Z^2 - R_2)}$

30. $V = \dfrac{AH}{3}$

31. $A = \dfrac{C^2}{4\pi}$

32. $D = \dfrac{1\cdot2FL}{\pi d\sqrt{v}}$

2.7
Variation

Let us imagine we had to deduce a formula for the volume of a cylinder. The dimensions that completely specify a cylinder are the diameter and the length; hence the volume can only depend upon the diameter and the length. If we had a container with a rectangular internal cross-section and partially filled it with water, we could find the volumes of cylinders by completely submerging them in the water. The volume of the cylinder would be the apparent increase in the volume of the water. In this particular case it would be the increase in the depth of the water multiplied by the cross-sectional area of the container.

A systematic approach would be to consider the effect of changing one variable at a time. Suppose we kept the diameter of several cylinders constant, and varied only the length. We should find that if we doubled the length, we should double the volume. If we halved the length, we should halve the volume. If we represent the cylinders by subscripts 1, 2, 3, etc. then:

$$\frac{V_1}{L_1} = \frac{V_2}{L_2} = \frac{V_3}{L_3} = \text{a constant value, say } k_1$$

In general $V = k_1 L$

We would say that the volume is proportional to the length. A mathematical way of writing this is $V \propto L$. The sign between the V and the L is known as the *variation sign* and we read that it *varies as* or *is proportional to*.

Since $V \propto L$ and $V = k_1 L$, we note that we can replace the variation sign by an equals sign, provided we introduce a constant of correct magnitude.

Now let us keep the length of cylinders constant, but vary the diameter. We would find in this case that if we doubled the diameter the volume would be multiplied by four. Further experiments would show that:

$$\frac{V_1}{D_1{}^2} = \frac{V_2}{D_2{}^2} = \frac{V_3}{D_3{}^2} = \text{a constant value, say } k_2$$

In general $V = k_2 D^2$

Our formula for the volume of a cylinder must include V, D and L, and we know that our final formula must satisfy the conditions that $V = k_1 L$ and $V = k_2 D^2$. One way this can occur is when:

$$V = k_3 D^2 L \quad \text{where} \quad k_3 = k_1 k_2$$

This we can check by establishing that:

$$\frac{V_1}{D_1{}^2 L_1} = \frac{V_2}{D_2{}^2 L_2} = \frac{V_3}{D_3{}^2 L_3} = \text{a constant value, say } k_3{}^2$$

Before we proceed further, let us note particularly that if $V \propto A$ and $V \propto B$:

then $V \propto AB$

and $V = \text{a suitable constant} \times AB$

Let us now return to our experiments concerning the volume of a cylinder, having thus far established that $V = k_3 D^2 L$, and proceed to determine the magnitude of the constant k_3. We now take a particular cylinder and find its volume. Suppose that a cylinder of diameter 7 mm and length 12 mm had a volume of 462 mm³.

$$V = k_2 D^2 L, \quad \therefore k_3 = \frac{V}{D^2 L} = \frac{462 \text{ mm}^3}{49 \text{ mm}^2 \times 12 \text{ mm}} = \frac{11}{14}$$

We should note here that in this particular case the constant has no units; it is known as a *non-dimensional constant*.

We now have determined a formula for the volume of a cylinder:

$$V = \frac{11 D^2 L}{14}$$

$$\left(\text{If } \pi \text{ is taken as } \frac{22}{7}, \quad \frac{11}{14} \text{ is } \frac{\pi}{4}, \quad \text{hence } V = \frac{\pi D^2 L}{4} \right)$$

Thus far, we have met with one quantity being proportional to another, and one quantity being proportional to the square of another. The mathematicians say that they *vary directly*. In some investigations, we find that one quantity is proportional to the reciprocal of another quantity. In this case we say that one quantity *varies inversely with the other*.

For instance, if A varies inversely as B then:

$$A \propto \frac{1}{B}$$

and $\qquad A = \text{a constant value} \times \dfrac{1}{B}$

Extending this further, if A varies inversely as the square of C, then:

$$A \propto \frac{1}{C^2}$$

and $\qquad A = \dfrac{1}{C^2} \times \text{a constant value}$

Furthermore, by incorporating a point brought out earlier in this article:

if $\qquad A \propto \dfrac{1}{B}$ and $A \propto \dfrac{1}{C^2}$, then $A \propto \dfrac{1}{BC^2}$

and $\qquad A = \text{a constant value} \times \dfrac{1}{BC^2}$

or $\qquad ABC^2 = \text{a constant value}$

As a typical example of how scientific laws are deduced, let us con-sider the resistance R offered by circular wires made of a particular alloy, the temperature remaining constant. If the resistances are of the same material, the magnitude of the resistance can be changed by vary-ing the diameter d and the length L. We should find that *R varies directly as L* and that *R varies inversely as the square of d.*

Hence $$R \propto L \quad \text{and} \quad R \propto \frac{1}{d^2}$$

hence $$R \propto \frac{L}{d^2} \quad \text{and} \quad R = \frac{kL}{d^2}$$

(We will repeat that this occurs if the temperature is constant.)
Let us now see whether this constant has a unit.

$$R = \frac{kL}{d^2}, \quad \text{then} \quad k = \frac{Rd^2}{L}$$

the unit of R would be ohms
the unit of d would be metres
and the unit of L would be metres

$$\text{Unit of } k = \frac{\text{unit of } R \times \text{unit of } d^2}{\text{unit of } L}$$

$$= \frac{\Omega \times m^2}{m} = \Omega m \text{ (ohm metres)}$$

In this case, the constant is dimensional; it not only has magnitude, it also has a unit.

Example:

If a given quantity of gas is subjected to volume and temperature changes, its absolute pressure P (i.e. pressure in excess of a perfect vacuum) varies directly as the temperature in degrees celsius + 273, and inversely as the volume V.

A particular quantity of gas at 0.15 MN/m^2 absolute has a volume of 20 m^3 at 27°C. What will be its pressure at a volume of 10 m^3 and a temperature of 127°C?

$$P \propto (T + 273) \quad \text{and} \quad P \propto \frac{1}{V}$$

$$\therefore P \propto (T + 273) \times \frac{1}{V}$$

or $$P \propto \frac{T + 273}{V}$$

$$\therefore P = \frac{k(T + 273)}{V}$$

$$k = \frac{PV}{T + 273}$$

Hence $\dfrac{PV}{T + 273}$ is the same for each set of conditions.

Let subscript 1 denote the first set and subscript 2 the second set.

$$\frac{P_1 V_1}{T_1 + 273} = \frac{P_2 V_2}{T_2 + 273}$$

$$\frac{0.15 \times 20}{27 + 273} = \frac{P_2 \times 10}{127 + 273}$$

$$\frac{3}{300} = \frac{P_3 \times 10}{400}$$

$$1\,200 = 3\,000 P_2$$

$$P_2 = 0.4 \text{ (same unit as } P_1)$$

Answer:

$$\text{Final pressure} = 0.4 \text{ MN/m}^2$$

Problems 2.7

1. The force F between two magnetic poles varies directly as the strengths m_1 and m_2 and inversely as the square of the distance apart d. Write down the formula with a constant k connecting F, m_1, m_2 and d. If $F = 4$ when $m_1 = 3$, $m_2 = 7$ and $d = 2.5$, find F when $m_1 = 4$, $m_2 = 9$ and $d = 3.5$.

2. The mass M of hexagonal bars of a free machining brass commonly used in the electrical industry varies directly as the length L and directly as the square of the distance across flats W. A hexagonal bar 20 mm across flats and 50 mm long has a mass of $0.147\,2$ kg. Calculate the mass, in kilograms, of a hexagonal bar 30 mm across flats and 2 metres long.

3. The power P that can be transmitted by a motor shaft made from a particular material varies directly as the rotational speed N and directly as the cube of the diameter d. If a shaft of diameter 20 mm

can transmit 1 kW at 1 400 rev/min, what diameter shaft should be used to transmit 5 kW at 700 rev/min?

4. The mass M of a circular steel anchor ring varies directly as the mean diameter D and the square of the bar diameter d. A ring of mean diameter 80 mm made from bar of diameter 10 mm has a mass of 0·15 kg. Calculate the mass of a similar ring of mean diameter 100 mm made from bar of diameter 20 mm.

5. For a given quantity of gas, provided the temperature remains constant, the pressure P (in excess of zero) varies inversely as the volume V. A volume of 0·5 m³ of gas has a pressure of 0·104 MN/m². Determine the magnitude and unit of the constant of variation and the pressure when the volume changes to 0·4 m³ at constant temperature.

6. The power P transmitted by a particular vee-belt drive varies directly as the driving tension T and directly as the belt velocity v. A belt transmits 2 kW when the tension is 400 N and the velocity is 210 m/min. Find the power transmitted when the tension is 300 N and the belt velocity is 140 m/min.

7. The quantity of heat W generated by the passage of an electric current varies directly as the time t, directly as the square of the voltage V and inversely as the resistance R.
   (a) If $W = 500$ when $t = 2$, $V = 100$ and $R = 40$, establish the constant of variation and hence establish a formula giving W in terms of t, V and R.
   (b) Find W when $t = 0·5$, $V = 250$ and $R = 25$.

8. The intensity of illumination I provided at a point distant d from a particular light varies inversely as the square of d. The light source is originally 5 m from the point. How much nearer must the light be brought to the point to increase the intensity of illumination by $56\frac{1}{4}\%$?

9. The volume V of a particular solid varies directly as the height h and directly as the square of the base radius r. Such a solid has a base radius of 7 mm, a height of 30 mm and a volume of 1 540 mm³. Taking π as $\frac{22}{7}$, deduce a formula for V in terms of π, r and h. Hence determine the volume of a similar solid whose base radius is 3·5 mm and whose height is 24 mm.

10. The voltage drop V in a conductor varies directly as the length L of the conductor and inversely as the square of its diameter d. If L is increased by 50% and d increased by 25%, find the resulting percentage change in V. State whether V increases or decreases.

11. The electric resistance R of a wire varies directly as its length L and inversely as its cross-sectional area A. If $R = 1\,340$ when $L = 20$ and $A = 0.02$, determine R when $L = 1.5$ and $A = 0.03$.

12. A quantity Q is proportional to $\dfrac{x \times y^{1/2}}{z^2}$. If $Q = 2$ when $x = 1\frac{1}{2}$, $y = \frac{1}{9}$ and $z = \frac{1}{2}$, calculate the constant of proportionality. If x, y and z are all doubled, what will be the new value of Q?

13. (a) The electrical resistance of a copper wire of circular cross-section varies directly as its length and inversely as the square of its diameter. If two copper wires have the same resistance but the diameter of one is twice that of the other, obtain the ratio of their lengths. Hence, show that the thicker wire is sixteen times as heavy as the thinner wire.

    (b) A copper wire 10 metres long and 2 mm in diameter has a resistance of 0.054 ohms. Obtain the resistance of a second copper wire 40 metres long and 4 mm in diameter.

Answer to Problems 2.7

1. $F = \dfrac{km_1m_2}{d^2}$, $3\dfrac{171}{343}$

2. 13.2 kg

3. 43.1 mm

4. 0.75 kg

5. 52 000 Nm ($= 52$ kJ), 0.13 MN/m²

6. 1 kW

7. (a) $k = 1$, $W = \dfrac{V^2t}{R}$ (b) 1 250

8. 1 m

9. $V = \dfrac{\pi r^2 h}{3}$, 308 mm³

10. V is decreased by 4%

11. $R = 67$

12. $k = 1$, $Q = 1.41$

13. (a) $\dfrac{L_2}{L_1} = 4$ (b) 0.054 ohm

3 The Diagrammatic Representation of Relationships

3.1
Functional relationships

3.1.1
One-to-one correspondence

In a previous article, when dealing with factors, it was stated that if two variables, such as y and x, were connected by a formula so that the value of y depended upon the value we gave to x, we say that y is a function of x, and adopt the convention that

$$y = f(x)$$

is a mathematical way of expressing the relationship that the value of the *dependent variable y* depends upon the value we give to the *independent variable x*. If, in particular, we have a set of values of y and a set of values of x, and for each value of y, there is one, and only one, corresponding value of x, and conversely, for each value of x there is one, and only one, corresponding value of y, we say that a *one-to-one correspondence* exists between the two sets of values. Another expression in common use is to say that there is a *one-to-one mapping* between the set of y-values and the set of x-values.

As a typical example, consider the relationship between centimetres and inches. Since 1 inch is precisely equal to 2·54 centimetres, and we let

$$L_m = \text{a length in centimetres}$$

and

$$L_i = \text{a length in inches,}$$

then

$$L_m = 2 \cdot 54 \, L_i$$

and a one-to-one correspondence exists between the set of values of L_m and the set of values of L_i. If $L_i = 2$, then L_m must be 5·08 and no other value. Conversely, if $L_m = 6 \cdot 35$, then $L_i = 2 \cdot 5$ and no other value. On the other hand, if $y = \sqrt{x}$, and $y = 25$ then x could be either $+5$ or -5, and $y = \sqrt{x}$ is *not* a one-to-one relationship between the set of values of y and the set of the values of x.

Returning to the relationship between centimetres and inches, almost any scientific reference book will provide a set of tables which will assist conversions to be made from inches to centimetres and vice-versa. Although the author has several such reference books in his home library, his wife never makes use of them for this purpose. She uses a tape measure on which there are two parallel scales, and quickly relates one scale to another. To the accuracy she requires for dressmaking purposes it is a quick operation to translate the so-called vital statistics of 39-24-37 inches to 99-61-94 centimetres. In effect, the tape measure becomes a very practical diagrammatic representation of a one-to-one correspondence between the items of sets by using two parallel scales. Let us now proceed to other diagrammatic representations of relationships.

3.1.2
Charts, diagrams and graphs

Let us imagine a situation in industry where a firm receives batches of mouldings in a plastics material for use on electrical switchgear. The mouldings, when received from the supplier, are sent to an inspection department. They are delivered in batches of 2 000, and the first task of the inspection department is to remove the mouldings which have unsatisfactory appearance due to surface defects. Each batch has an identification number. Let us suppose that the following table indicates the results of inspecting ten successive batches.

Batch no.	21	22	23	24	25	26	27	28	29	30
Rejections	10	12	6	11	5	9	8	4	7	6

Unless the reader had some experience in interpreting tabulated data, the table does not convey a great deal of significant information.

Suppose we now represent this data on a diagram. This can be done in a variety of ways, but the diagram shown in Fig. 3.1 is how this would probably be done in industry. The diagram now rapidly conveys some significant information.

(a)   The number of rejections per batch fluctuates.
(b)   Although the number of rejections fluctuates, there is a tendency for that number to diminish.
(c)   The average of the ten batches, i.e. the average height of the bars, is about 8.

Now let us simulate another industrial situation. An estimator, as part of his duties, has to determine masses. From a reference book he

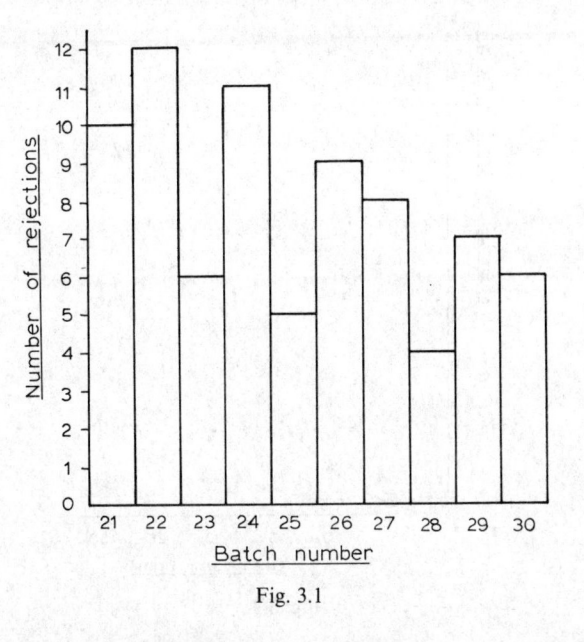

Fig. 3.1

finds the following table, giving the mass per metre run of round steel bars of a certain specification.

Diameter (mm)	5	10	15	20	25	30
Mass per metre run (kg)	0·155	0·62	1·395	2·48	3·875	5·58

Once more, unless the reader has some skill in interpreting tabulated data, there is apparently little significant information.

Let us present this information pictorially. Fig. 3.2 shows the usual manner by which this would be done.

The diagram now conveys some very interesting information not readily apparent from the tabulated data. A quite distinct trend is apparent. It is obvious that we are able to draw a continuous line which will enable us to determine the mass per metre run for other diameters not included in the table. The trend is not a directly proportional increase since the line is curved. If we double the diameter we more than double the mass per foot run.

Figures 3.1 and 3.2 are both pictorial representations of data, called *charts or diagrams*. For a pictorial representation to be a *graph* there must be some connexion, such as a formula, between the two quantities which are represented. This will be indicated by a line on the diagram.

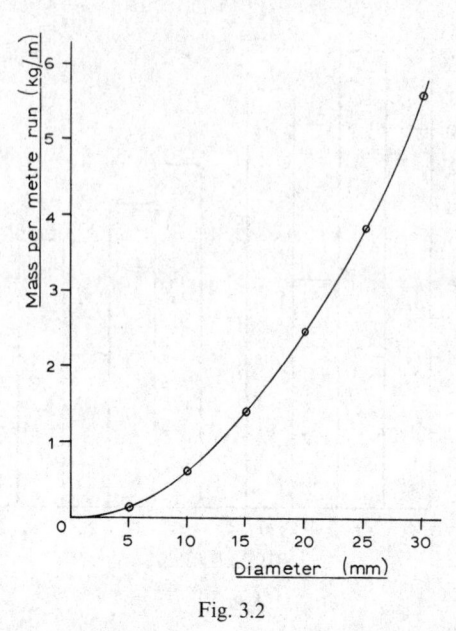

Fig. 3.2

Figure 3.1 is not a graph. For each of the two items represented, the values increase in definite steps. We cannot have such illogical values as '$5\frac{1}{4}$ rejections' or 'batch number 23·8'. The name of a diagram such as Fig. 3.1 is a *histogram*. There is no formula which connects the number of defectives with the batch number.

If a graph is drawn, the variable quantities can take any value. In Fig. 3.2, we could have a bar of diameter 23 mm, and we could have a bar with a mass of 4 kg/m run. There is a definite connexion between the mass per metre run in kilograms and the diameter in millimetres. It is actually

$$\text{Mass per metre run} = 0·006\ 2\ (\text{diameter})^2$$

We can therefore construct a graph by plotting points on a graphical field which satisfy a formula, and joining those points with a line. The line is normally continuous, and can be curved or straight. A discontinuity in the line may indicate that different formulae are applicable at different portions of the line.

Recalling definitions given in Chapter 1 the variable quantities illustrated by pictorial representations are often called *variates*. If those variates can only take specific values (such as occurred with the previously illustrated batch number and number of rejections, which must be whole numbers), the variates are said to be *discontinuous*, or *discrete*. On the other hand, the magnitude of a length or of a diameter can take any value, and such variates are said to be *continuous*. A graph can therefore be regarded as a diagrammatic representation of the connexion between continuous variates.

3.1.3
Plotting of graphs

A common (but not the only) method of graphically illustrating a formula connecting two variables, say x and y, was invented by the French mathematician Descartes. Two reference axes are positioned at right angles, it being conventional practice to position the y-axis vertically and the x-axis horizontally. These axes are the basis of *cartesian co-ordinates*, and the point where they intersect is the origin of the co-ordinates, in short, the *origin*. A particular value of x is positioned horizontally from the origin to scale. A particular value of y is positioned vertically from the origin to scale, which need not necessarily be the same scale as that used for values of x. The point $x = 3$, $y = 7$ is indicated mathematically as (3, 7), the horizontal distance being quoted first. The values of 3 and 7 are the *co-ordinates* of that particular point on the graphical field, the horizontal co-ordinate being the *abscissa* and the vertical co-ordinate being the *ordinate*.

It is conventional practice, using the origin as the starting point, to regard horizontal distances to the right as positive and horizontal distances to the left as negative. Vertical distances from the origin are positive above the origin, and negative below. Since the intersection of the x-axis and the y-axis produces four quadrants, it is useful to number them in sequence anticlockwise. The first quadrant associates positive values of x with positive values of y. The second quadrant will associate negative values of x with positive values of y, and so on.

There are several methods of indicating points on a graphical field, none of which has yet been standardised. The reader should use any method to which he has been accustomed. Many people prefer to use a small cross, the intersection of the arms of the cross being the position of the point. Others prefer to use a dot, surrounded by a circle to bring it rapidly to attention. For the time being there are no especial considerations which support any particular method but one should be adopted and adhered to. In this textbook, we shall use the 'dot and circle' convention.

Until the reader has become accustomed to plotting, the following method of plotting points on a graphical field will be found useful.

1. Draw a short faint vertical line at the value of x, the line lying in the vicinity of the value of y.
2. Draw a short faint horizontal line at the value of y, to cross the line drawn previously.
3. Put a point at the intersection of these lines.
4. Draw a small circle around the point to bring it rapidly to notice.

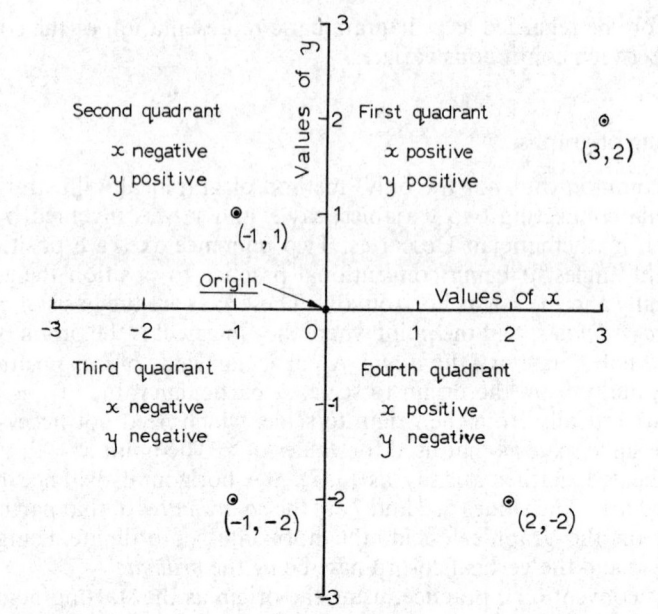

Fig. 3.3

Figure 3.3 shows the conventions employed when constructing graphs using cartesian co-ordinates. A point is indicated by a dot, but in order that its position is readily noticed, it is circled to bring it rapidly to attention. It is the dot which represents the point, not the circle. We use the circle merely for convenience in finding the point.

Some simple examples of the representation by formula on cartesian co-ordinates follow from the conventions used in Fig. 3.3. Using cartesian co-ordinates:

(a) The x-axis represents the equation $y = 0$.
(b) The y-axis represents the equation $x = 0$.
(c) The equation $y = 3$ is represented by a horizontal line through the value of 3 on the y-axis.
(d) The equation $x = -4$ is represented by a vertical line through the value -4 on the x-axis.

3.1.4
Presentation of graphs

In constructing graphs, students are advised:

(a) to label the axes;
(b) to indicate values on the axes;

(c)   where appropriate, to indicate an answer on the graph;
(d)   if there is more than one line, to state the appropriate equation near each line.

Greatest accuracy is usually obtained when the greatest area is used for the plotting. *It is not essential to include the origin on every graph.* The usual area of a graphical field used in college work is that corresponding to what is known as the international A4 size of paper. There are light rulings at 2 mm intervals, and heavier rulings at 20 mm intervals. Bearing in mind the need for labelling the axes and indicating values on them, the graphical field available for actual plotting is about 200 mm × 160 mm, i.e. about 10 large squares × 8 large squares. If the data provided gave a minimum value for y of 42·7 and a maximum values for y of 86·9 then 40 to 90 would cover all values, a range of 50. Using the 200 mm dimension would prove most convenient using a scale of 1 large square = 5 units. One small square therefore represents 0·5 unit, and decimal values are then easy to plot. If the 8 large squares were used for a range of about 50, then either 1 large square to 6 or 1 large square to 6·25 could be used, but the choice of scales such as these would lead to difficulties in plotting. Whilst it is desirable to use the greatest area of the field which is available, it should never be selected so that the greater accuracy nominally available with the larger graphical field is lost due to difficulties of accurate plotting.

Many examiners assist students by indicating suitable scales in the questions, in which case it is relatively simple to deduce whether or not the graph paper should be used with the longer side vertical.

In general, graphs have three main uses:

(a)   to convey quickly to the observer how one variable changes with respect to another;
(b)   as a rapid method of obtaining values other than those of the original data;
(c)   as a means of solving equations.

The first two of these uses will now be demonstrated with appropriate examples.

Example:

The mass of circular bars of deoxidised copper, in kilograms per metre length, varies with the diameter according to the following table.

Diameter d (mm)	10	20	30	40	50	60
Mass M (kg/m)	0·7	2·8	6·3	11·2	17·5	25·2

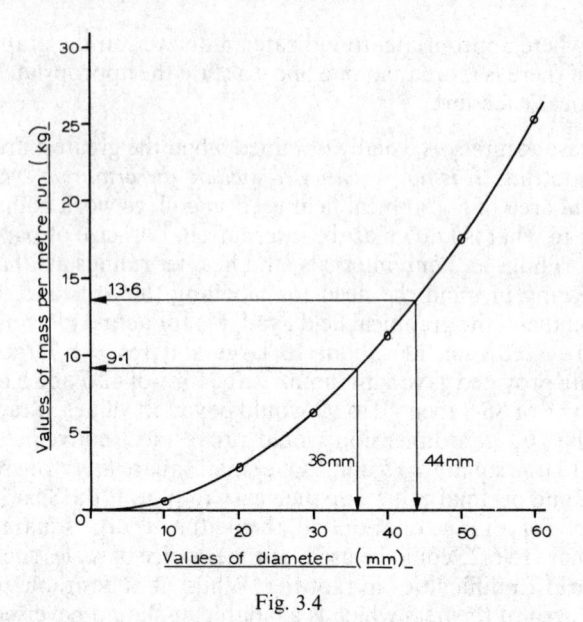

Fig. 3.4

Illustrate this data graphically, using a horizontal scale for d of 20 mm = 10 mm, and a vertical scale for M of 20 mm = 5 kg/m. Use your graph to find the weight of a tube of deoxidised copper of outside diameter 44 mm, inside diameter 36 mm and length 80 mm.

The graph is shown in Fig. 3.4. From the graph, a metre length of bar of diameter 44 mm has a mass of 13·6 kg, while a metre length of bar of diameter 36 mm has a mass of 9·1 kg.

A metre length of tube of outside diameter 44 mm and inside diameter 36 mm has a mass of 13·6 − 9·1 = 4·5 kg.

A tube of length 80 mm has a mass of $\dfrac{80}{1\,000} \times 4\cdot5$ kg = 0·36 kg.

Answer:

$$\text{Mass of tube} = 0\cdot36 \text{ kg.}$$

Example:

The intensity of illumination I provided by a particular lamp when it is at a distance d from a light meter is given by the following data.

I (units)	270	120	67·5	43·2	30·0	16·9
d (m)	1·0	1·5	2·0	2·5	3	4

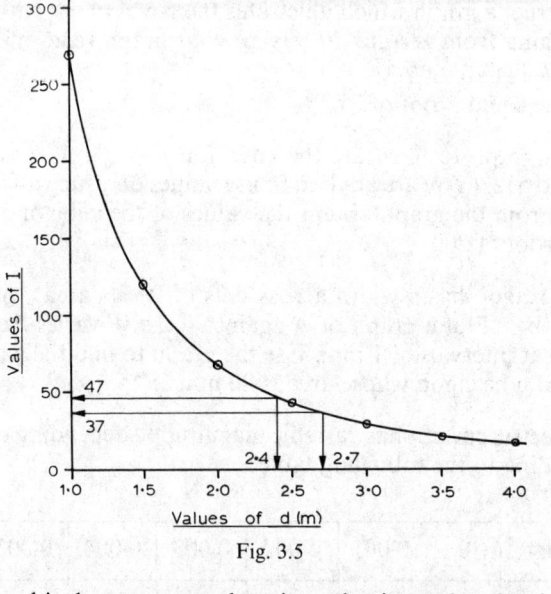

Fig. 3.5

Illustrate this data on a graph, using a horizontal scale of 20 mm = 0·5 m for d and a vertical scale of 20 mm = 50 units for I.

Use your graph to determine:

(a) The value of I when $d = 2\cdot4$ m
(b) The minimum number of lamps to provide a total value of I of 140 units when $d = 2\cdot7$ m.

The graph is shown in Fig. 3.5. From the graph $I = 47$ when $d = 2\cdot4$ and $I = 37$ when $d = 2\cdot7$.

$$\text{Number of lamps} = \frac{140}{37} = \text{over 3 but less than 4.}$$

Hence 4 lamps will be required.

Answers:

(a) $I = 47$ units
(b) 4 lamps are required

Problems 3.1

1. If $E = \dfrac{W}{20} + 10$, construct a graph that can be used to determine values of E for values of W between 0 and 100. Using the graph, obtain:

(a) the value of E when $W = 72$;
(b) the value of W when $E = 12\cdot8$.

2. Construct a graph which illustrates the law $y = x^2$, with values of x ranging from zero to 10. Using your graph read off values for
   (a) the square of 7·3;
   (b) the square root of 71.

3. Plot a graph to illustrate the equation $y = \sqrt[3]{x}$ as x varies from zero to 512. (You are advised to use values of x such as 1, 8, 27, 64, etc.) From the graph obtain the values of the cube of 7·5 and the cube root of 200.

4. If a hexagon has a width across flats of W, its area A is given by $0\cdot866\ W^2$. Plot a graph of A against W as W varies from zero to 6 mm at intervals of 1 mm. Use the graph to find the width across flats of a hexagon whose area is 20 mm^2.

5. An electric current has variable magnitude I depending on a time t, according to the following table:

t (seconds)	0	0·001	0·002	0·003	0·004	0·005	0·006
i (amperes)	64·3	81·9	94·0	99·6	98·5	90·6	76·6

Plot a graph of i against t, and from your graph determine:
   (a) the maximum value of i and the value of t at which this maximum occurs;
   (b) the first value of t for which $i = 90$ amperes.

6. A pipe of outside diameter 25 mm has a wall thickness of t so that its inside diameter is $(25 - 2t)$ mm. Plot a graph of the cross-sectional area against t as t varies from zero to 5 mm at intervals of 1 mm. Hence determine the value of t for a cross-sectional area of 225 mm^2.

7. The efficiency E of a lifting machine was determined for differing loads W, resulting in the following data:

W (N)	400	800	1 200	1 600	2 000
E (%)	25·0	33·3	37·5	40·0	41·7

Illustrate these values on a graph, and use the graph to determine the value of E when $W = 1\ 000$ N.

8. The price P of a particular type of hand tool depends upon the nominal size S according to the following table:

S (mm)	1	2	4	6	8	10
P (£)	0.30	0.45	1.05	2.05	3.45	5.25

If the firm decides to introduce a 3 mm and a 5 mm size, what prices can be expected for these two sizes?

9. A rectangular plate of length 30 mm and width 20 mm has square notches of side x cut out from each corner. The plate is then folded so as to form a rectangular tray of depth x. Show that the capacity C of the tray is given by the formula:

$$C = 4x^3 - 100x^2 + 600x$$

Plot a graph of C against x, as x varies from zero to 6 mm at intervals of 1 mm. Hence find the value of x when $C = 937.5$ mm^3.

10. A cable suspended from two pylons 100 m apart takes the form of the curve $y = \dfrac{x^2}{100} + 45$, where:

y = height in metres of a point P on the cable above the ground.
x = distance in metres of P from the lowest point on the cable.
  Plot the curve of x from -50 to 50. (Take intervals of 10.) From the graph, find:
  (a) the clearance of the cable from the ground;
  (b) the heights of the supports if the cable spans 60 m but takes the same form;
  (c) the distance between the supports to give a maximum sag of 5 m.

11. The work done per second, W joules, by an electric motor is given by:

$$W = IV - I^2R$$

where I is the current in amperes and V and R are constants. Plot a graph of W against I when $V = 200$ and $R = 100$, as the values of I range from zero to 2.4 at intervals of 0.4. From the graph determine:
  (a) the maximum value of W and the value of I at which this value occurs;
  (b) the values of I at which W is zero.

12. (a) If a cutting speed has to be kept constant at 66 m/min, take $\pi = \frac{22}{7}$ and show that the relationship between the spindle speed n rev/min and bar diameter d mm is $dn = 21\,000$.
    (b) Plot a graph to illustrate this relationship, the values of d ranging from 20 mm to 70 mm.
    (c) Read off from the graph:
       (i) the value of n when $d = 35$ mm;
       (ii) the value of d when $n = 500$ rev/min.

Answers to Problems 3.1

1. (a) 13·6   (b) 56

2. (a) 53·3   (b) 8·43

3. 422, 5·85

4. 4·8 mm

5. (a) 100 amperes when $t = 0.003\,3$ seconds
   (b) $t = 0.001\,6$ seconds

6. 3·3 mm

7. 35·7%

8. £0·70 and £1·50

9. 2·5 mm

10. (a) 45 m   (b) 54 m   (c) 44·7 m

11. (a) $W = 100$ J when $I = 1$ ampere
    (b) $W = 0$ when $I = 0$ and when $I = 2$ amperes

12. (a) and (b) require no numerical answers
    (c) (i) 600 rev/min;   (ii) 42 mm

3.2
The straight line law

3.2.1
The equation of a straight line

The gradient of a straight line on a graphical field is given by:

$$\frac{\text{the difference in the values of } y \text{ of two points on the line}}{\text{the difference in the corresponding values of } x}$$

If our reader has studied mathematics previously, the foregoing definition may have been referred to as the slope of the line. Modern usage tends to favour the use of the word slope as a general indication, such as up or down, and to associate the word gradient with specific numerical magnitudes.

Hence, if (x_1, y_1) and (x_2, y_2) are two points on the straight line, the gradient is given by:

$$\frac{y_2 - y_1}{x_2 - x_1}$$

If the straight line slopes 'up to the right' it will have a positive gradient, while if it slopes 'down to the right' it will have a negative gradient. A horizontal line has zero gradient since there is no difference in the values of y.

Let us consider the equation $y = ax + b$, where a and b are constants, and let two points on the line be (x_2, y_2) and (x_1, y_1).

If
$$y = ax + b \text{ always,}$$

then
$$y_2 = ax_2 + b$$

and
$$y_1 = ax_1 + b$$

Subtracting these equations gives:

$$y_2 - y_1 = ax_2 - ax_1$$

$$\therefore y_2 - y_1 = a(x_2 - x_1)$$

and
$$a = \frac{y_2 - y_1}{x_2 - x_1}$$

Now $\dfrac{y_2 - y_1}{x_2 - x_1}$ by definition is the gradient of the line, while a is a constant. Hence the line which represents $y = ax + b$ has a constant gradient of magnitude a. A line having a constant gradient is a straight line.

The y-axis is the line $x = 0$. If we put $x = 0$ in the general equation $y = ax + b$, then $y = b$ when $x = 0$. This indicates that the line cuts the y-axis at the value of the constant b.

The representation of the equation $y = ax + b$ by cartesian coordinates on a graphical field is a straight line of gradient a cutting the y-axis at b. The more formal way of describing where a line cuts an axis is to call it the *intercept* on that axis.

Since only two points are needed to draw a straight line, a graph of the type $y = ax + b$ can be drawn by plotting just two points and joining them by a straight line. However, the reader is recommended always to plot a third point to check the accuracy of the plotting. Conversely, the equation of a straight line can be determined if the co-ordinates of two points on the line are known. The equation is commonly referred to as the *law of the straight line*.

A straight line on a graphical field can also be specifically positioned by one point and the gradient of the line. Hence the law of the straight line can be determined if the co-ordinates of one point and the gradient of the line are known.

Example:

Find the law of the straight line that passes through the points $(2, -3)$ and $(10, 21)$.

$$y = ax + b$$

Using $x = 2, y = -3$: $-3 = 2a + b$

Using $x = 10, y = 21$: $21 = 10a + b$

By subtraction $\overline{-24 = -8a}$

$$\therefore a = \frac{-24}{-8} = 3$$

Substituting this value of a in the first equation:

$$-3 = 2(3) + b$$
$$-3 = 6 + b$$
$$-3 - 6 = b$$
$$b = -9$$

The law is $y = 3x - 9$

Check for $(2, -3)$: $-3 = 6 - 9$

$$-3 = -3$$

Check for $(10, 21)$: $21 = 30 - 9$

$$21 = 21$$

Answer:

Law is $y = 3x - 9$

Example:

Find the law of the straight line of gradient -2, passing through the point $(-2, 9)$.

$$y = ax + b$$

Since slope is -2 $a = -2$

$$\therefore y = -2x + b$$

Using $y = 9$, $x = -2$:

$$9 = (-2)(-2) + b$$
$$9 = 4 + b$$
$$9 - 4 = b$$
$$b = 5$$

The law is $y = -2x + 5$

or more conveniently $y = 5 - 2x$

Check for $(-2, 9)$: $9 = 5 - 2(-2) = 5 + 4 = 9$

Answer:

Law is $y = 5 - 2x$

3.2.2
Determination of laws from experimental data

If we conduct experiments and record values of two variables, say x and y, the results can be shown in graphical form. In many experiments in science the points will be found to lie approximately on a straight line. A typical case would be the plotting of the length of a compression spring for various amounts of loading. It is most improbable that the points would lie *exactly* on a straight line. Although the load may be known precisely, inaccuracies of observation and defects in the apparatus and in the construction of the experiment may have disturbing effects on the value recorded for the length. If the length L were plotted against the load W, the values of L would be subject to error. In another experiment, we could record the resistance R offered by a particular conductor at various temperatures T. In this experiment both R and T have to be measured and in this case the values of R and of T would each be subject to error. We can average out the errors, either in one variable or in both variables, by drawing the *best straight line* to match a particular set of plottings. If a transparent rule is used, the line can be drawn so that the errors are equally distributed, which occurs in many circumstances when as many points lie above the line as lie below. Occasionally a point or points may lie on the line.

If such a graph of a particular experiment plots R vertically and T horizontally, we know from previous work that a straight line represents the relationship $R = aT + b$. Let us suppose that a first point on the line is (T_1, R_1) and a second point on the line is (T_2, R_2).

Since $R_2 = aT_2 + b$

and $R_1 = aT_1 + b$

by subtraction

$$R_2 - R_1 = aT_2 - aT_1$$
$$= a(T_2 - T_1)$$
$$\therefore a = \frac{R_2 - R_1}{T_2 - T_1}$$

From a consideration of the above result, calculations will be eased if $T_2 - T_1$ is a convenient number for division, and greater accuracy will occur if $T_2 - T_1$ is as large as possible. Having found a value for a, the value of b can be found by substituting in the equation $R_1 = aT_1 + b$ or $R_2 = aT_2 + b$, whichever is the more convenient.

Consequently, the deduction of a straight line law of the type $y = ax + b$ can be made as follows:

1. Plot the values to the largest scales that the graph paper will allow. (*It is not necessary always to include the origin.*)
2. Draw the 'best straight line' to distribute the errors, usually by arranging as many points to lie above the line as below. Occasionally a point or points may lie on the line.
3. Select two points on the line, as far apart as possible but arranging that the distance between the x values will be a convenient divisor. This is best done by choosing the x values and then reading off the appropriate y values.
4. Substitution in $y = ax + b$ will give two simultaneous equations. By subtraction, b will be eliminated, and a can be determined. Substitution in an original equation will permit the evaluation of b.

In a few isolated circumstances the y-axis would appear, in which case the constant b can be determined immediately since it is the value of y at which the straight line crosses the y-axis. The value of a can be found by choosing one other point on the line and by substituting the (x, y) values of this point in $y = ax + b$. Since y, x and b are known, a can be determined. In order to give the greatest degree of accuracy the second point should be as far from the y-axis as possible, but allowing for a convenient value of x to simplify calculations.

Example:

For a particular nickel-chrome steel the resulting percentage reduction in area R after tempering from a temperature of $T°C$ produced the following results:

T	200	300	400	500	600
R	47·0	50·3	54·0	57·7	61·0

Plot these values, T horizontally and R vertically. Draw the best straight line through the points, and hence deduce an approximate formula giving R in terms of T.

Presuming the A4 size of graph paper, a convenient area of field for plotting is 10 large squares by 8 large squares. (A large square is 20 mm × 20 mm.) Considering the values of T, which is to be plotted horizontally, the range is $600 - 200 = 400$. The use of 8 large squares suggests itself, with 20 mm = 50°C. Plottings then conveniently occur on every other heavy line. Proceeding to the values of R, they vary from 47 to 61, a range of 14. To use all the 10 large squares would produce a difficult scale for plotting. A scale of 1 large square = 2 units uses only 7 large squares. Although the full field is not used, the values will be easier to plot accurately. (Although in this case we shall use the longer side of the paper vertically, this is not essential.) Hence the scales will be:

> 20 mm = 50°C horizontally, ranging 200 to 600
> 20 mm = 2 units vertically, ranging from 46 to 62

The graph is shown in Fig. 3.6. The best straight line has been drawn. Three points lie on the line. Of the other two, one lies above and one

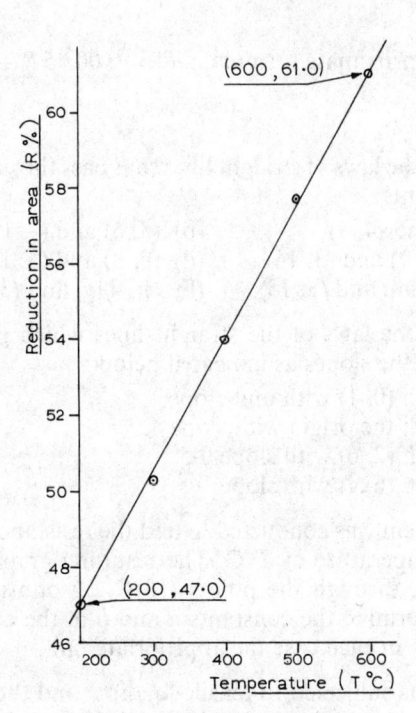

Fig. 3.6

lies below, each being displaced by one small square. Two points on the line, a convenient distance apart, are (200, 47) and (400, 61). It was pure coincidence that the two extreme points lay on the line, and are also a convenient distance apart horizontally. It should not be presumed that this always occurs.

Since $R = aT + b$

Using (600, 61): $61 = 600a + b$

Using (200, 47): $47 = 200a + b$

Subtracting: $14 = 400a$

$$a = \frac{14}{400} = 0.003\ 5$$

Substituting this value in the first equation:

$$61 = 600\ (0.003\ 5) + b$$
$$61 = 21 + b$$
$$b = 61 - 21 = 40$$

Answer:

Approximate formula is $R = 0.003\ 5T + 40$.

Problems 3.2

1. Determine the laws of straight lines that pass through the following pairs of points:

   (a) (2, 6) and (4, 8)   (b) (3, 6) and (−1, 22)
   (c) (−1, −7) and (3, 13)   (d) (0, 8) and (3, 17)
   (e) The origin and (3, 15)   (f) $(p, 4pq)$ and $(3p, -2pq)$

2. Determine the laws of the straight lines which pass through the points with the slopes as indicated below:

   (a) Through (0, 1) with unit slope.
   (b) Through the origin with slope 4.
   (c) Through (2, 6) with slope 0.5.
   (d) Through (6, 4) with slope $-\frac{2}{3}$.

3. An experiment was conducted to find the resistance R of a conductor at a temperature of $T°C$. The resulting graph was a straight line passing through the points (10°C, 52 ohms) and (40°C, 58 ohms). Determine the constants a and b in the equation $R = aT + b$, stating in each case the appropriate unit.

4. A spring was subjected to tensile loading, and the length L of the spring was plotted against the load W. The resulting graph was a

straight line passing through the points (100 N, 65 mm) and (400 N, 80 mm). Find the 'free length' of the spring, i.e. the length L when $W = 0$.

5. An experiment was conducted on a lifting machine to find the effort E required to lift a load of W. It was found that the graph of E (vertical) against W (horizontal) was a straight line passing through the points (1 000 N, 100 N) and (1 800 N, 140 N). Determine the law of the machine in the form $E = aW + b$.

6. The speed N of a flywheel that was slowing down was measured at times of t seconds after a brake was applied. The following values were obtained:

t (seconds)	20	25	30	35	40
N (rev/min)	200	175	150	125	100

Show graphically that N and t are connected by a law of the type $N = at + b$ and hence determine the speed of the flywheel when the brake was applied.

7. The following are details of a suggested relationship between the carbon content C and the ultimate tensile strength S of normalised plain carbon steels:

C (%)	0.2	0.3	0.4	0.5	0.6	0.7
S (MN/m^2)	494	566	638	710	782	854

Find graphically the law from which these values were determined in the form $S = aC + b$.

8. A channel has a level bottom with straight sides tapering outward. The width W at the water level for various depths d of water flow is given by the following data:

d (m)	1	2	3	4	5
W (m)	7.75	8.70	9.65	10.60	11.55

Show graphically that W and d are related by a formula of the type $W = ad + b$, and determine this formula.

9. In an experiment on the compression of a closely coiled helical spring, the length of the spring L was measured when varying loads W were applied. The following results were obtained:

W (N)	400	500	600	700	800	900	1 000
L (mm)	66	63·8	61	58·5	56	53·2	51

Plot these values, W horizontally and L vertically. If $L = aW + b$, find appropriate values of a and b, and hence the nominal unloaded length of the spring.

10. The resistance, in ohms, of a pure resistor depends upon temperature following the law:

$$R = r(1 + kt)$$

where R is the resistance at $t°C$, r is the resistance at $0°C$, and k is a constant. A particular resistor has a resistance of 42·4 ohms at 15°C and 45·6 ohms at 35°C. Find by a graphical method the values of r and k.

11. During an experiment on lifting tackle, the following values of the effort E to lift a load of W were recorded, the units of E and W being identical:

E	10	10·9	13	15·1	17
W	50	60	80	100	120

Plot these values, E vertically and W horizontally. Draw the best straight line for the plotting. If E and W are connected by a law of the type $E = aW + b$, determine graphically suitable values for the constants a and b.

12. The following information was obtained from the measurement of a set of lathe gears, N being the number of teeth and D being the outside diameter:

N	30	35	40	45	50
D (mm)	128	148	168	188	208

Plot these values, N horizontally and D vertically. If D and N are connected by a formula:

$$D = mN + c$$

find graphically the values of the constants m and c.

13. The following information has been extracted from a reference work, and indicates the tolerance T mm on the major diameter of a metric screw thread of pitch p mm:

Pitch p (mm)	1	1·5	2	2·5	3
Tolerance T (mm)	0·158	0·212	0·266	0·320	0·374

Plot these values, p horizontally and T vertically. Show that T and p are connected by a formula of the type $T = ap + b$. Determine the constants a and b, and hence calculate the tolerance on the major diameter of a thread of pitch 5 mm.

14. The resistance R of a length of copper wire was measured at various temperatures T, resulting in the following data:

R (ohms)	108	109·8	112·2	114	116	117·8	120·2
T (°C)	20	25	30	35	40	45	50

It was thought that R and T were connected by a formula of the type $R = aT + b$. By drawing a suitable graph find reasonable values for the constants a and b.

15. Steel samples were tested for hardness by two different methods, those being designated R and B. The values obtained were:

Sample nos.	1	2	3	4	5	6
B	382	393	402	412	421	432
R	39	40	41	42	43	44

Plot these values, R horizontally and B vertically. Show that, *over this particular range*, a law of the type $B = aR - b$ is approximately true and determine suitable values for the constants a and b.

Answers to Problems 3.2

1. (a) $y = x + 4$ (b) $y = 18 - 4x$
   (c) $y = 5x - 2$ (d) $y = 3x + 8$
   (e) $y = 5x$ (f) $y = 7pq - 3qx$

2. (a) $y = x + 1$ (b) $y = 4x$

   (c) $y = \dfrac{x}{2} + 5$ (d) $y = 8 - \dfrac{2x}{3}$

3. $a = 0\cdot2$ ohm/°C, $b = 50$ ohms

4. 60 mm

5. $E = 0\cdot005\,W + 50$ N

6. 300 rev/min, obtained from $N = 300 - 5t$

7. $S = 720\,C + 350$

8. $W = 0\cdot95\,d + 6\cdot8$

9. $a = -0\cdot025, b = 76; 76$ mm

10. $r = 40, k = 0\cdot004$

11. $a = 0\cdot1, b = 5$

12. $m = 4, c = 8$

13. $a = 0\cdot108, b = 0\cdot05; 0\cdot59$ mm

14. $a = 0\cdot4, b = 100$

15. $a = 10, b = 8$

4 Geometry

4.1
The measurement of angles

4.1.1
Sexagesimal measurement of angles

An *angle* is a measurement of an amount of turning. In article 3.1.3, dealing with graphs, the reader was introduced to the convention of two axes intersecting at right angles at the origin. Figure 4.1(a) shows axes XOX_1 and YOY_1, with the quadrants indicated.

Now let us suppose we have a line OP, and it turns in an anticlockwise direction, the end O always being coincident with the origin, as shown in Fig. 4.1(b). If the arm turns so that the end P moves through the four quadrants in numerical sequence and eventually coincides with OX, the arm has rotated through one revolution. There is no reason at all why the arm should cease turning, and consequently there is no limit to the magnitude of an angle. The basic unit is one revolution. Just as we can take a metre as the basic unit of length and subdivide it for convenience, so we can take one revolution and subdivide it into smaller quantities. One way of doing this, but not the only way, is to divide one revolution into 360 equal parts called degrees. Each of these degrees can be divided into 60 equal parts called minutes, and a minute can be further subdivided into 60 equal parts called seconds. An angle of thirty-nine degrees, twenty-seven minutes and fourteen seconds is indicated mathematically as $39° \ 27' \ 14''$.

This particular dividing is called *sexagesimal measure*. It must be emphasised that sexagesimal measure is not the only way of indicating the magnitude of an angle. There are others which we shall introduce at appropriate stages in this book.

If the arm OP moves anticlockwise from OX until it becomes coincident with OY it will have turned through one quarter of a revolution, i.e. $90°$, in which case OP is perpendicular to OX and the angle POX of $90°$ is termed a *right angle*. If the arm OP turns from OX to any angle between zero and $90°$, P will lie on the first quadrant. An angle between zero and $90°$ is called an *acute angle*. If arm OP rotates a further $90°$ it will be coincident with OX_1. POX is now a straight line, and naturally enough an angle of $180°$ is called a *straight angle*. An angle of between $90°$ and $180°$ is known as an *obtuse angle*. If OP rotates from OX by an obtuse angle, P will then fall in the second quadrant.

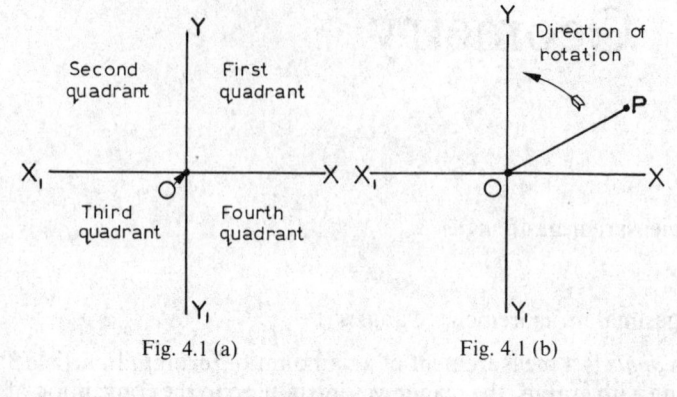

Fig. 4.1 (a)   Fig. 4.1 (b)

An angle of between 180° and 360° is called a *reflex angle*; hence if *OP* rotates from *OX* through a reflex angle, *P* will fall in the third quadrant if the angle is between 180° and 270°, and in the fourth quadrant if the angle is between 270° and 360°.

Angles which total 90° are known as *complementary angles*; for example 43° is the complementary angle of 47°. Angles which total 180° are *supplementary angles*; for example, the supplementary angle of 45° is 135°.

We have tacitly assumed a convention that the anticlockwise direction of rotation from *OX* results in a positive angle. Consequently we must accept the implication that a clockwise rotation of *OP* from *OX* results in a negative angle. Hence an angle of 330° could, if we so wished, be regarded as −30°.

A convenient method of indicating a direction is to state the nearest cardinal point (either north, east, south and west) followed by the angular deviation toward one of the adjacent cardinal points. Typical examples are N 25° W and S 10° W. Let us now assume the cardinal

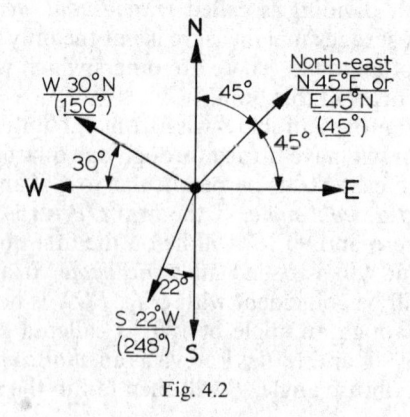

Fig. 4.2

points to be superimposed upon our four-quadrant convention. North becomes 90°, west 180°, south 270° and east 0°. If we consider the bearing N 25° W, north is 90° and a westerly deviation from north is in the anticlockwise or positive direction. N 25° W can therefore be drawn as 90° + 25° = 115° in our four-quadrant convention. In a similar manner, for the bearing S 10° W, south is 270°, a westerly deviation from south is in the clockwise or negative direction, hence S 10° W can be represented as 270° − 10° = 260°. An illustration of directional bearings is given in Fig. 4.2.

4.1.2
The centesimal measurement of angles

The word sexagesimal means 'associated with sixty', as with the sexagesimal measurement of angles and the sexagesimal sub-divisions of hours. In a similar manner, the word centesimal measurement means 'associated with one hundred'. In the *centesimal measurement of angles* the fundamental unit is the right angle and the initial subdivision is into 100 equal parts called *grades*. One centigrade is therefore equal to one-hundredth of a right angle, and for this reason it is recommended that the word centigrade be discarded when referring to temperature, and replaced by the word celsius. Thus 25°C should be read as twenty-five degrees celsius and not as twenty-five degrees centigrade. The topic of the centesimal measurement of angles has been introduced solely for the purpose of explaining the reason for the recommendation of using the word celsius instead of the word centigrade for the measurement of temperature. The centesimal measurement of angles is in very restricted use, and will not be the concern of our reader at the present stage of studies.

4.1.3
The decimal measurement of angles

The word *decimal* means 'associated with ten'. There are many circumstances where basic unit of an amount of turning should be other than one revolution, and that it will be more logical to refer a little later in this book to the *circular measurement of angles*, where the fundamental unit is one revolution divided into 2π equal parts called *radians*.

The introduction of electronic calculators has brought more clearly into focus certain difficulties associated with the sexagesimal measurement of angles, and a growing body of opinion is appearing which suggests that decimal sub-divisions of a basic unit for turning leads to greater simplicity. The point is well made, but for angles less than one revolution the use of the degree is now so firmly established that the possibilities of a change in the near future are somewhat remote. Consequently, it is suggested that a compromise could be effected where the basic unit is the degree and its sub-divisions should be wholly decimal. The purist would argue that there is no point in introducing another complication, and that angles should be measured in radians, a unit covered in article 4.3.2, with fractions being either vulgar or decimal as circumstances dictate. The mild controversy will undoubtedly go on for some considerable time. Suffice it to say that for the purposes of our reader the system of measurement that is used for angles must reflect the common usages associated with the particular set of circumstances for which a measurement of an angle is required, and that when an angle is quoted there shall be no possibility of misinterpretation. Our reader should therefore note that if no unit is associated with a numerical value when indicating the magnitude of an angle it is implied that the unit is the radian. With sexagesimal measure the symbols for degrees and/or minutes and/or seconds, as appropriate, or degrees and decimal or vulgar fractions of degrees should be clearly indicated.

To prepare our reader for any possibility of having to convert from sexagesimal measurement of angles to degrees and decimal fractions of a degree, and vice-versa, our reader should recall that 1 degree = 60 minutes and 1 minute = 60 seconds.

Example:

Convert $30.725°$ into full sexagesimal measure. ($30°$ needs no conversion.)

$$0.725° = 0.725 \times 60 \text{ minutes} = 43.5 \text{ minutes} = 43.5'$$
$$0.5' \quad = 0.5 \quad \times 60 \text{ seconds} = 30 \text{ seconds} \quad = 30''$$

Answer:

$$30.725° = 30° \ 43' \ 30''$$

Example:

Convert $30° \ 43' \ 30''$ into its decimal equivalent, the unit being the degree. ($30°$ needs no conversion.)

$$43' \ 30'' = \{(43 \times 60) + 30\} \text{ seconds}$$
$$= 2\ 580 + 30 = 2\ 610 \text{ seconds}$$

1 degree consists of 60×60 seconds $= 3\ 600$ seconds

$$30° \ 43' \ 30'' = 30\frac{2\ 610}{3\ 600} \text{ degrees} = 30.725°$$

Answer:

$$30° \; 43' \; 30'' = 30 \cdot 725°$$

4.1.4

Angular computations in sexagesimal measure

The body of opinion which favours the use of decimal notation for fractions of a degree instead of sexagesimal notation point out, with complete justification, how much easier it is to compute in decimal notation when angles have to be added and/or subtracted and/or divided, or multiplied by numbers, or any combination of these processes. However, for some considerable time to come it will be advisable for our reader to be able to perform computations involving angles in terms of sexagesimal measure.

Computations in sexagesimal measure are very similar to those undertaken with denary numbers. With the denary addition we form vertical columns of units, tens, thousands, etc. With the sexagesimal addition of angles we form vertical columns of seconds, minutes and degrees. We can 'carry one' from the total of a seconds column to the minutes column for every multiple of 60 seconds. Similarly, we can 'carry one' from the total of the minutes column to the degree column for every multiple of 60 minutes. The setting-out is very similar to denary arithmetic. For example:

$$
\begin{array}{rrr}
37° & 25' & 42'' \; + \\
 & 5' & 36'' \\
28° & 52' & \\
53° & 27' & 43'' \\
18° & 0' & 56'' \\
\hline
137° & 51' & 57'' \\
1° & 2' &
\end{array}
$$

The steps in obtaining a solution were:

$56'' + 43'' + 36'' + 42'' = 177''$

$177'' = 2' \; 57''$

Put down 57″ and carry 2′

$2'$ carried $+ \; 0' + 27' + 52' + 5' + 25' = 111'$

$111' = 1° \; 51'$

Put down 51′ and carry 1°

$1°$ carried $+ \; 18° + 53° + 28° + 37° = 137°$

The presentation of values should be noted, particularly the value of 18° 0′ 56″ and the inclusion of the value of 0′ to lend added emphasis that the number 56 applies to seconds and not to minutes. Furthermore, there is no need to put 0° in front of 5′ 36″ or 0″ behind 28° 52′.

With subtraction, if a smaller angle is to be subtracted from a larger angle, then the columns can be arranged as previously, and, if necessary we can 'borrow one' from a minutes minuend to give us 60 seconds and 'borrow one' from a degrees minuend to give us 60 minutes. For example:

$$
\begin{array}{rrr}
137° & 25' & \\
\cancel{138°} & \cancel{26'} & 17'' - \\
52° & 51' & 32'' \\
\hline
85° & 34' & 45''
\end{array}
$$

The steps in obtaining the solution were:

> $32''$ is greater than $17''$
> Borrow $1' = 60''$ from the $26'$ in the minuend
> $(60'' + 17'') - 32'' = 77'' - 32'' = 45''$

The minutes portion of the minuend now becomes $26' - 1'$ borrowed $= 25'$.

> $51'$ is greater than $25'$
> Borrow $1° = 60'$ from the $138°$ in the minuend
> $60' +$ the new 'part minuend' of $25' = 85'$
> $85' - 51' = 34'$

The degrees portion of the minuend now becomes $138° - 1°$ borrowed $= 137°$.

Finally $137° - 52° = 85°$

When subtracting an angle from a nominally smaller angle we adopt exactly the same process as in algebraic subtraction. We subtract the smaller angle from that of the larger angle, put the result in brackets and prefix with a negative sign.

Thus $17° 51' 22'' - 318° 50' 7''$

is obtained from $-(318° 50' 7'' - 17° 51' 22'')$

and our reader may care to check that this is equal to

$$-(300° 58' 45'')$$

If the result of the subtraction is to be used to indicate a particular alignment relative to the conventional four-quadrant representation, then $-(300° 58' 45'')$ is precisely the same direction as

$$360° - (300° 58' 45'')$$

which is $59° 1' 15',$

and which could have been obtained from

$$(360° + 17° 51' 22'') - (318° 50' 7'')$$
$$= (377° 51' 22'') - (318° 50' 7'')$$
$$= 59° 1' 15''$$

Multiplication by a number follows closely the method used for denary numbers, noting that 60 seconds is equal to one minute and 60 minutes is equal to one degree.

$$
\begin{array}{rrr}
22° & 16' & 21'' \times \\
 & & 7 \\
\hline
155° & 54' & 27'' \\
\hline
1° & 2' &
\end{array}
$$

The steps in obtaining the solution were:

$7 \times 21'' = 147'' = 2' 27''$
Put down 27'' and carry 2'
$7 \times 16' = 112'$
Add on the 2' carried to give $112' + 2' = 114'$
$114' = 1° 54'$
Put down 54' and carry 1°
Finally $7 \times 22° = 154°$, and add on the 1° carried to give 155°.

Division by a number similarly closely follows the method used for denary arithmetic, but every degree which is a remainder will add 60 minutes to the original minutes value, whilst every minute which is a remainder will add 60 seconds to the original seconds value. For example:

$$
\begin{array}{r|rrr}
13 & 287° & 9' & 33'' \\
\hline
 & 22° & 5' & 21''
\end{array}
$$

The steps in obtaining the solution were:

$287° \div 13 = 22°$, remainder 1°
Carry $1 \times 60 = 60'$ to the original minutes value
$60' + 9' = 69'$
$69' \div 13 = 5'$, remainder 4'
Carry $4 \times 60 = 240''$ to the original seconds value
$240'' + 33'' = 273''$
Finally $273'' \div 13 = 21''$

If an angle is divided by another angle, the quotient is a pure number. If necessary, the dividend and/or the divisor should be adjusted so that both are in terms of a single unit. It does not matter which unit is selected as long as it used both for the dividend and for the divisor. For example:

$$176° \ 20' \div 7° \ 40'$$

can be evaluated from

$$176\tfrac{1}{3}° \div 7\tfrac{2}{3}° = \frac{529}{3} \div \frac{23}{3} = \frac{529}{3} \times \frac{3}{23} = 23,$$

or, by expressing both angles in minutes, from

$$\frac{(176 \times 60) + 20}{(7 \times 60) + 40} = \frac{10\ 560 + 20}{420 + 40} = \frac{10\ 580}{460} = 23.$$

The multiplication of one angle by another will have a relevance to a much later stage of our reader's studies, and can be left until that time. It is not included in Level 1 work.

It cannot be denied that if a calculator is now to be regarded as a tool which can be expected to form part of the general kit of a technician, there is substantial evidence which favours decimal notation instead of sexagesimal notation for fractions of a degree.

4.1.5
Parallel lines and transversals

When two straight lines intersect, four angles are formed. If the lines do not intersect at right angles, two of the angles are acute, the other two are obtuse. The two acute angles are equal to each other and the two obtuse angles are equal to each other. The usual way of expressing this fact is to state that *if two straight lines intersect, vertically opposite angles are equal*.

If a straight line crosses parallel straight lines, that straight line is called a *transversal*. A transversal crossing two parallel straight lines

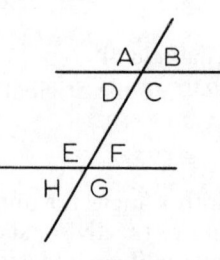

Fig. 4.3

is shown in Fig. 4.3. We have already stated:

$$\angle A = \angle C \qquad \angle B = \angle D$$
$$\angle E = \angle G \qquad \angle F = \angle H$$

In addition
$$\angle A = \angle E \qquad \angle B = \angle F$$
$$\angle C = \angle G \qquad \angle D = \angle H$$

these being known as *corresponding angles on the same side of the transversal*.

Furthermore $\quad \angle D = \angle F$ while $\quad \angle C = \angle E$

these being known as *alternate angles*.

Summarising the above:

If a transversal crosses a pair of parallel straight lines:

(a) vertically opposite angles are equal
(b) corresponding angles on the same side of the transversal are equal
(c) alternate angles are equal.

Hence, referring back to Fig. 4.3:

$$\angle A = \angle C = \angle E = \angle G$$

and
$$\angle B = \angle D = \angle F = \angle H$$

As a result of the foregoing information, parallel lines can be identified through the relationships of given pairs of angles made between those parallel lines and a transversal. For instance, if there are two straight lines crossed by a transversal, and it can be shown that corresponding angles on the same side of the transversal are equal, then the two straight lines must be parallel.

Problems 4.1

1. What angle, to the nearest second, corresponds to an amount of turning of $2\frac{1}{7}$ revolutions?

2. State whether the following angles are acute, obtuse or reflex:
   (a) 57°   (b) 125°   (c) 5°   (d) 284°   (e) 343°   (f) 179°

3. OP is an arm which rotates about the origin O of cartesian co-ordinates. The arm is originally coincident with the x-axis and rotates in an anticlockwise direction. State in which quadrant P falls when the arm rotates:
   (a) 60°   (b) 80°   (c) 120°
   (d) 170°   (e) 330°   (f) 365°
   (g) 530°   (h) 780°   (j) 4·3 revolutions

4. State which angles, in the four-quadrant convention, correspond to the following bearings:
   (a) west   (b) south-east   (c) N 10° W
   (d) W 10° N   (e) S 16° W   (f) W 20° S

5. State which bearings correspond to the following angles in the four-quadrant convention:
   (a) 25°   (b) 65°   (c) 171°   (d) 240°

6. Convert to sexagesimal measure the following angles, giving the answer to the nearest second if the resulting sexagesimal value is not precise:
   (a) $37\frac{3}{4}°$ (b) $8\frac{2}{9}°$ (c) $43\frac{5}{11}°$ (d) 82·7°
   (e) 24·312 5°   (f) 16·$\dot{4}$°

7. Convert to degrees and vulgar fractions of a degree the following sexagesimal angles:
   (a) 47° 20′   (b) 28° 54′   (c) 82° 37′ 30″
   (d) 22° 9′ $22\frac{1}{2}$″   (e) 84° 38′ $10\frac{10}{11}$″

8. Convert to degrees and decimal fractions of a degree the following sexagesimal angles, giving the answer to the nearest third place of decimals if the resulting value is not precise:
   (a) 6° 42′   (b) 18° $22\frac{1}{2}$′   (c) 16° 52′ 30″
   (d) 42° 18′ 37″   (e) 46° 0′ 23″

9. Perform the additions indicated:

   (a) 27°   4′   22″ +   (b)   6°   41′   23″ +
        3°   8′   18″   27°   18′   51″
        4°   15′   6″   106°   7′   29″
        2°   7′   11″   7°   51′   6″
   ────────────────   ────────────────

   (c) 8°   17′   21″ +   (d)   16°   41′   53″ +
        6°   5′   21′   18″
        8°   21′   6″   131°   51′   8″
              7′   14″   27°   0′   42″
   ────────────────   ────────────────

10. (a) Subtract 17° 8′ 52″ from 28° 41′ 56″.
    (b) Subtract 18° 53′ 48″ from 6° 51′ 7″.
    (c) Determine the complementary angle of 43° 8′.
    (d) Determine the supplementary angle of 51° 18′ 42″.

11. Evaluate:
    (a) 16 × 22° 30′   (b) 7 × 4° 8′ 5″
    (c) 6 × 27° 4′ 31″   (d) 13 × 16° 51′ 42″

12. Evaluate:
    (a) $32° 30' ÷ 13$ (b) $42° 13' 28'' ÷ 8$
    (c) $250° ÷ 17$, giving the answer to the nearest second.
    (d) $324° 47' 6'' ÷ 11$
    (e) $116° 21' 32'' ÷ 7$, giving the answer to the nearest second.

13. Evaluate:
    (a) $180° ÷ 11° 15'$
    (b) $48° 56' 24'' ÷ 12° 14' 6''$
    (c) $132° 21' 36'' ÷ 8° 16' 21''$
    (d) $21° 0' 43'' ÷ 189° 6' 27''$, giving the answer as a vulgar fraction.

Answers to Problems 4.1

1. $771° 25' 43''$

2. (a) Acute   (b) Obtuse   (c) Acute
   (d) Reflex   (e) Reflex   (f) Obtuse

3. (a) First   (b) First   (c) Second
   (d) Second   (e) Fourth   (f) First
   (g) Second   (h) First   (j) Second

4. (a) $180°$ (b) $315°$ (c) $100°$
   (d) $170°$ (e) $254°$ (f) $200°$

5. (a) E $25°$ N   (b) N $25°$E   (c) W $9°$ N   (d) S $30°$ W

6. (a) $37° 45'$ (b) $8° 13' 20''$ (c) $43° 27' 16''$
   (d) $82° 42'$ (e) $24° 18' 45''$ (f) $16° 26' 40''$

7. (a) $47\frac{1}{3}°$ (b) $28\frac{9}{10}°$ (c) $82\frac{5}{8}°$
   (d) $22\frac{5}{32}°$ (e) $84\frac{7}{11}°$

8. (a) $6·7°$ (b) $18·375°$ (c) $16·875°$
   (d) $42·310°$ (e) $46·006°$

9. (a) $36° 34' 57''$ (b) $147° 58' 49''$
   (c) $22° 50' 41''$ (d) $175° 55' 1''$

10. (a) $11° 33' 4''$ (b) $-(12° 2' 41'') = 347° 57' 19''$
    (c) $46° 52'$ (d) $128° 41' 18''$

11. (a) $360°$ (b) $28° 56' 35''$
    (c) $162° 27' 6''$ (d) $219° 12' 6''$

12. (a) $2° 30'$ (b) $5° 16' 41''$ (c) $14° 42' 21''$
    (d) $29° 33' 26''$ (e) $16° 37' 22''$

13. (a) 16 (b) 4
    (c) 16 (d) $\frac{1}{9}$

4.2
The types and properties of triangles

4.2.1
The types of triangles

A *polygon* is any plane figure bounded by straight sides. If the number of sides exceed three, it is assumed that no straight lines cross, otherwise there would be more than one plane figure. Names are given to polygons according to the number of sides; some of the more common polygons are given in the following list:

Number of sides	Name
Three	Triangle
Four	Quadrilateral
Five	Pentagon
Six	Hexagon
Seven	Heptagon
Eight	Octagon
Ten	Decagon
Twelve	Dodecagon

A *triangle* is a plane figure bounded by three straight lines. A triangle therefore contains three angles. The conventional way of indicating a triangle is to give each angle a capital letter.

A side is given a small letter, the same character as the opposite angle. Side *a* is opposite angle *A*, side *p* is opposite angle *P*, and so on. The convention is shown in Fig. 4.4. The longest side is opposite the largest angle and the shortest side is opposite the smallest angle.

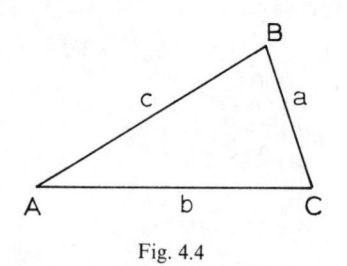

Fig. 4.4

In Fig. 4.5(a), *ABC* is a triangle, the side *AC* being continued to *D*. In Fig. 4.5(b), a line has been drawn through *C* parallel to side *AB*, and the other two sides of the triangle have been continued.

Consider first the parallel lines *AB* and *CE* cut by the transversal *BC*. Angle *B* in the triangle is equal to angle *BCE*. They are alternate angles. Now consider the parallel lines *AB* and *CE* cut by the transversal *AD*.

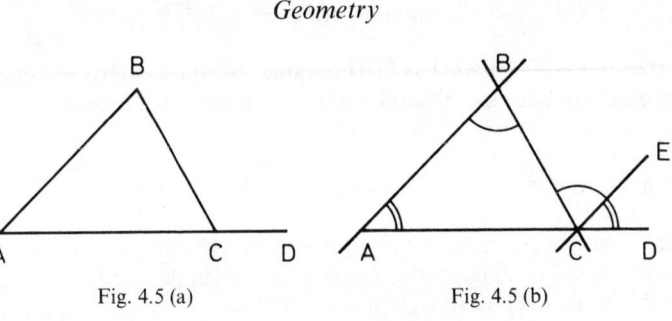

Fig. 4.5 (a)   Fig. 4.5 (b)

Angle A in the triangle is equal to angle ECD. They are corresponding angles on the same side of the transversal.

In Fig. 4.5(a), the angles A, B and C are known as the interior angles of the triangle, while angle BCD is called an exterior angle. Figure 4.5(b) shows that in any triangle an exterior angle is the sum of the opposite interior angles.

In Fig. 4.5(b), ACD is a straight angle of $180°$. Angle C is common to the triangle and to the straight angle, and totalling angle C with the equivalents of angles B and A, the figure shows that the sum of the interior angles of any triangle is $180°$.

This article was introduced by a statement which classified polygons according to the number of sides. The triangle is one particular example of the general classification of polygons. We can now take the triangle itself as a generalisation and classify it into six different kinds, according to the relationships associated with the lengths of sides and the magnitudes of angles.

An *acute angled triangle* is one in which all the angles are each less than $180°$.

An *obtuse angled triangle* is one in which one of the angles is an obtuse angle. The magnitude of an obtuse angle is between $90°$ and $180°$. Since the sum of the angles in a triangle is $180°$, if a triangle contains an obtuse angle, the other two angles must be acute. The sum of those acute angles is the supplementary angle of the obtuse angle. For instance, if the magnitude of the obtuse angle in an obtuse angled triangle is $120°$, then the sum of the other two angles, which are both acute angles, is the supplementary angle to $120°$, which is $180° - 120° = 60°$.

A *right angled triangle* is one in which one of the angles is a right angle of $90°$. The other two angles are both acute, and since the sum of the angles in a triangle is $180°$, the sum of the acute angles is $180° - 90° = 90°$. One of the acute angles is therefore the complementary angle to the other. If a right angled triangle includes an acute angle of $60°$, the other acute angle is the complementary angle to $60°$, which is $90° - 60° = 30°$. The longest side of a right-angled triangle is called the *hypotenuse*, and is the side opposite the right angle.

An *equilateral triangle* has all three sides of equal length and contains three equal angles. All the angles of an equilateral triangle are therefore equal to 180° divided by three, which is 60°. An equilateral triangle is therefore also an acute angled triangle, but it cannot possibly be either an obtuse angled triangle or a right angled triangle.

An *isosceles triangle* (isosceles from the Greek for 'equal legs') is one in which two of the sides are of equal length. The angles opposite those sides are equal in magnitude. An isosceles triangle can be either acute angled, obtuse angled, or right angled. The sum of the two equal angles, which are opposite the sides of equal length, is the supplementary angle to that angle included by the sides of equal length. For instance, if the two sides of equal length contain an angle of 80°, the sum of the two equal angles is the supplementary angle to which is 180° − 80° = 100°. Each of the equal angles is therefore equal to 50°.

A *scalene triangle* is one in which all three sides have different lengths and it contains three different angles. A scalene triangle may be acute angled, obtuse angled or right angled. It cannot possibly be either equilateral or isosceles.

Any triangle will fall into at least one of the previous six classifications, but some triangles may fall into more than one classification.

4.2.2
Similar and congruent triangles

Congruent triangles are equal in all respects. The three individual angles of one triangle are equal to the three corresponding individual angles of the other triangle, and furthermore, the three individual sides of one triangle are equal to the three individual sides of the other. Congruent triangles also have equal areas.

Similar triangles are triangles in which only the three angles of one are equal to the three angles of the other. Similar triangles are said to be *equiangular*. It should be noted that while congruent triangles are equal in area, *similar triangles are not*. There is a very important property of similar triangles which must be remembered. This property is that the ratio of corresponding sides is constant.

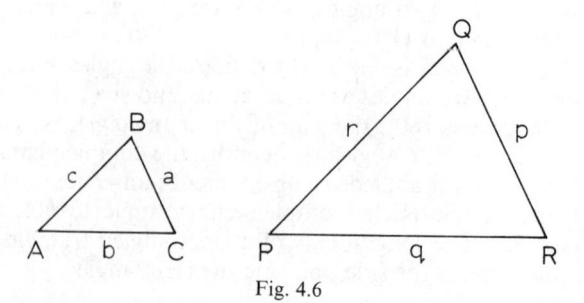

Fig. 4.6

Figure 4.6 shows two similar triangles, ABC and PQR. In the magnitude of the angles:

$$A = P, \quad B = Q \quad \text{and} \quad C = R$$

while $\dfrac{a}{p} = \dfrac{b}{q} = \dfrac{c}{r} =$ a constant for these particular triangles.

Example:

Calculate the distance X on the template shown in Fig. 4.7(a).

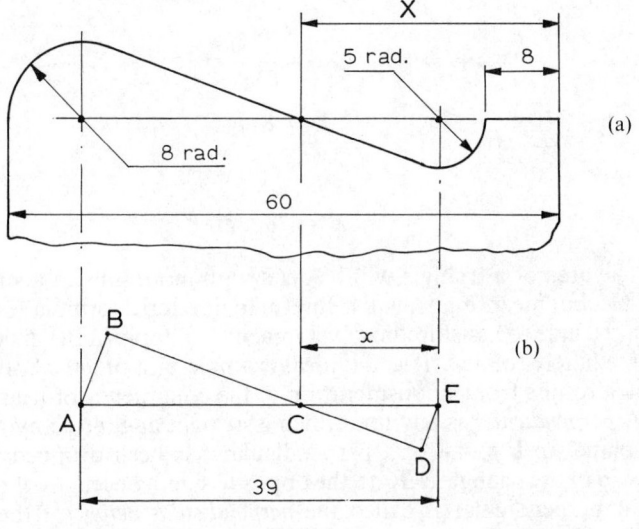

Fig. 4.7

Figure 4.7(b) shows information extracted from the drawing, identifying letters having been added to describe the triangles.

Considering triangles ABC and CDE:

1. Angle B = angle $D = 90°$, since a radius is at right angles to a tangent at a point of contact.
2. Angle C in triangle ABC is equal to angle C in triangle CDE, being vertically opposite angles.
3. From the two statements above

$$\text{Angle } A = \text{angle } E = (90° - C)$$

The three angles in triangle ABC are equal to the three angles in triangle CDE, hence triangles ABC and CDE are similar, and ratios of corresponding sides are equal.

Let $CE = x$, so that $\quad AC = 39 - x$
$$DE = 5 \text{ and } AB = 8$$

$$\frac{DE}{CE} = \frac{AB}{AC}$$

$$\frac{5}{x} = \frac{8}{39 - x}$$

$$5(39 - x) = 8x$$
$$195 - 5x = 8x$$
$$195 = 8x + 5x = 13x$$

$$x = \frac{195}{13} = 15$$

$$X = 8 + 5 + x = 28$$

Answer:

$$X = 28 \text{ mm}$$

The area of a triangle will be dealt with more fully in a subsequent article, but meantime, it will be useful to develop a formula for the area of a triangle to assist in the development of a topic in the article which immediately follows. It is a typical example of a practical application which results from a consideration of the congruency of triangles.

A *perpendicular* is any line which is at right angles to any other line or plane. In Fig. 4.8(b) a perpendicular has been dropped from the apex B of the triangle ABC to the opposite side. The length of this particular perpendicular is called the *perpendicular height* of the triangle with reference to the base b, as denoted by the dimension h. A triangle may have from one to three different perpendicular heights, but if there are two or more each one must be associated with a particular side of the triangle, called the base. We usually let the symbol b indicate the length of the appropriate base.

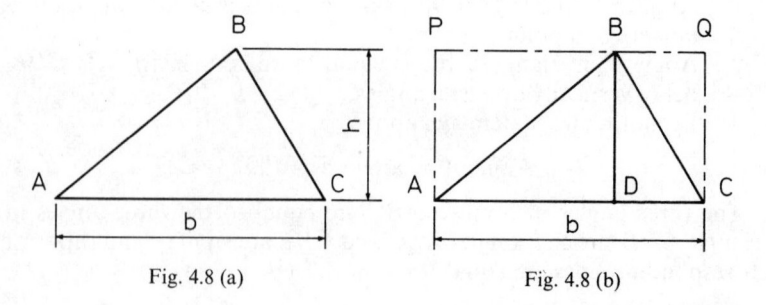

Fig. 4.8 (a)   Fig. 4.8 (b)

If we now refer to Fig. 4.8(b), lines PA and QC are perpendicular to the base b, while PBQ is a line parallel to the base b. The figure bounded by the lines AP, PQ, QC and CA is a rectangle. The area of this rectangle is the product of the lengths of two adjacent sides. In our case it can be symbolised by bh.

Now if we consider triangles APB and ABD, the three sides of one are equal in length to the three sides of the other. They are therefore congruent triangles. Since congruent triangles are equal in area, then the area of the triangle ABD is one-half of the area of its surrounding rectangle $APBD$. Similarly the area of the triangle BDC is one-half of the area of its surrounding rectangle $BQCD$. In which case, the total triangle ABC has an area of one-half of its surrounding rectangle $APQC$. The area of the triangle ABC is therefore one-half of the product of a base and its associated perpendicular height. Hence if

$$\triangle = \text{the area of a triangle}$$
$$b = \text{the length of a base of that triangle}$$
and
$$h = \text{the particular perpendicular height}$$
$$\text{associated with the base } b,$$

then
$$\triangle = \frac{bh}{2}$$

To construct a triangle from given information, our reader should recall that the conventional method of specifying a triangle is to allocate capital letters to the angles and small letters to the sides. The side opposite an angle has the letter of the same sense. For example, side a is opposite angle A, side p is opposite angle P, and so on.

To construct a triangle having definite shape and size at least one side must be given. The angles alone will only indicate shape, and any amount of similar triangles can be constructed which are equiangular. TEC unit U75/005 indicates certain specific triangles which our reader should be able to construct after having undertaken a suitable course of study. Let us consider them in the order they are given, and provide algorithms for their construction. Our reader will note that step 1 is virtually identical for each case.

Case 1. *Given three sides*

Step 1. Draw a horizontal line of length equal to one of the sides, preferably that of the longest side. Give it the appropriate letter, and allocate capital letters to its ends. For example, if the side is side c, the capital letters will be A and B.

Step 2. With compasses set to a radius equal to one of the other sides, describe an arc using an appropriate end of the line as a centre. For example, if the initial horizontal line is side c, set the compasses to side a and use B as the centre of the arc.

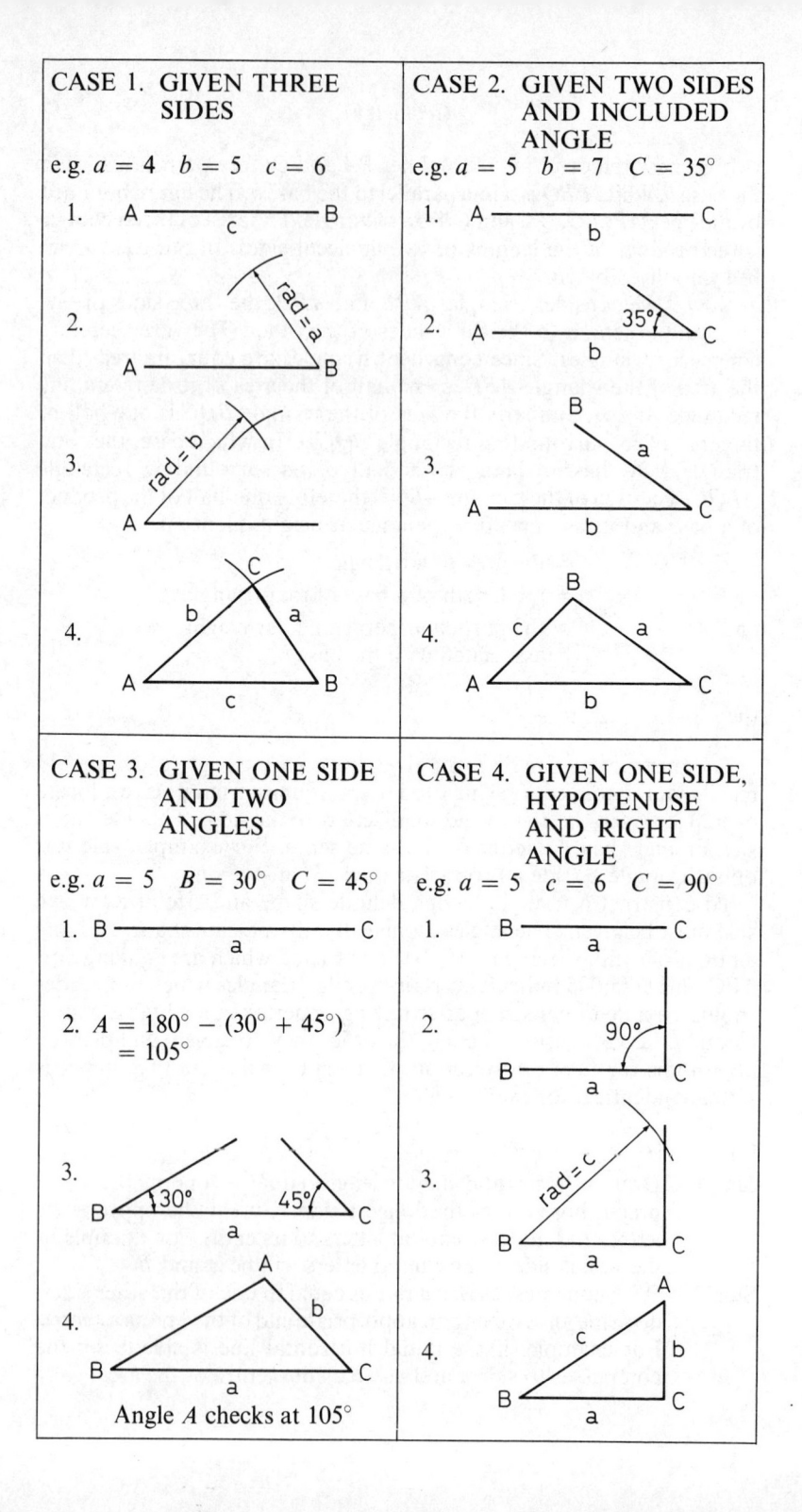

CASE 1. GIVEN THREE SIDES

e.g. $a = 4$ $b = 5$ $c = 6$

1. A ——— c ——— B

2. A ——— B   rad = a

3. A ———   rad = b

4. A ——— b, a, C, c ——— B

CASE 2. GIVEN TWO SIDES AND INCLUDED ANGLE

e.g. $a = 5$ $b = 7$ $C = 35°$

1. A ——— b ——— C

2. A ——— b ——— 35° C

3. A ——— b ——— C   B, a

4. A ——— b ——— C   c, B, a

CASE 3. GIVEN ONE SIDE AND TWO ANGLES

e.g. $a = 5$ $B = 30°$ $C = 45°$

1. B ——— a ——— C

2. $A = 180° - (30° + 45°)$
   $= 105°$

3. B 30° ——— a ——— 45° C

4. B ——— a ——— C   A, b

Angle A checks at 105°

CASE 4. GIVEN ONE SIDE, HYPOTENUSE AND RIGHT ANGLE

e.g. $a = 5$ $c = 6$ $C = 90°$

1. B ——— a ——— C

2. B ——— a ——— C   90°

3. B ——— a ——— C   rad = c

4. B ——— a ——— C   c, A, b

Step 3. Set the compasses to a radius equal to the length of the last side, and using the other end of the horizontal line as the centre, describe another arc to cut the arc described in Step 2.

Step 4. Complete the triangle by joining the ends of the horizontal line to the point of intersection of the arcs.

Case 2. Given two sides and their included angle

Step 1. Draw a horizontal line of length equal to one of the sides, preferably the longer. Give it the appropriate letter and allocate capital letters to its ends, as with the previous case.

Step 2. Construct the given angle at the appropriate end of the initial horizontal line. Using a protractor is the quickest method, but some advice is given in article 5.1.3 if a protractor is not available but mathematical tables are.

Step 3. Mark off the second side from the point of the angle. Give it its appropriate small letter and allocate the third capital letter to the end remote from the angle.

Step 4. Complete the triangle by drawing the third side to join the ends of the sides already drawn.

Case 3. Given one side and two angles

Step 1. Draw a horizontal line of length equal to the given side. Give it the appropriate small letter and allocate capital letters to its ends, as with the previous cases.

Step 2. Determine the magnitude of the third angle, knowing that the sum of the angles in any triangle is 180°.

Step 3. Construct, with the aid of a protractor, angles at the appropriate ends of the initial horizontal line.

Step 4. Complete the triangle by continuing the sloping lines until they meet, and check the size of their included angle with that either given or established from Step 2.

Case 4. Given one side, hypotenuse and a right angle

Step 1. Draw a horizontal line of length equal to the given side. Give it the appropriate small letter and allocate capital letters to its ends, as with the previous cases.

Step 2. Construct a right angle at the appropriate end of the initial horizontal line, using a protractor, a set square, or the geometrical construction for constructing a perpendicular.

Step 3. Set a pair of compasses to a radius equal to the hypotenuse. With the centre on the end of the initial horizontal line remote from the right angle, describe an arc to intersect the perpendicular.

Step 4.   Complete the triangle by drawing the hypotenuse, which is
           the straight line joining the end of the initial horizontal line
           to the intersection point on the perpendicular.

Illustrations of the preceding four cases are given on page 254.

4.2.3
The Theorem of Pythagoras

In Fig. 4.9, *ABC* is a right-angled triangle, with a square erected on
the longest side (called the *hypotenuse*). A larger square has been erected
on continuations of sides *a* and *b*. The extension of side *a* is of length *b*,

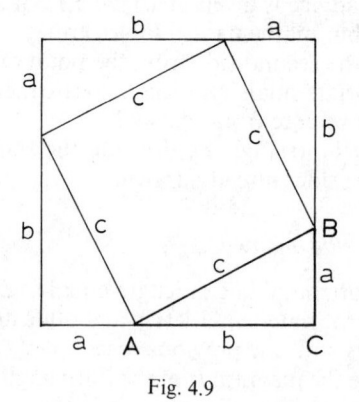

Fig. 4.9

and the extension of side *b* is of length *a*. We now have four congruent
triangles of sides *a*, *b* and *c*, surrounding a square of side *c*.

$$\text{Area of large square} = (a + b)^2$$
$$= a^2 + 2ab + b^2$$

Area of square on hypotenuse
$$= \text{area of large square} - \text{area of four triangles}$$

$$\therefore c^2 = a^2 + 2ab + b^2 - 4\left(\frac{ab}{2}\right)$$

$$c^2 = a^2 + 2ab + b^2 - 2ab$$
$$c^2 = a^2 + b^2$$

The above equation is an algebraic form of the *Theorem of Pythagoras*. This states that in a right-angled triangle the square on the
hypotenuse is equal to the sum of the squares on the other two sides.

There are certain sets of whole numbers for sides of triangles whose proportionalities produce right-angled triangles. The most well known set is 3 : 4 : 5, but some others are:

$$5 : 12 : 13 \qquad 9 : 40 : 41$$
$$7 : 24 : 25 \qquad 11 : 60 : 61$$
$$8 : 15 : 17 \qquad 12 : 35 : 37$$

Before we proceed to examples which include the Theorem of Pythagoras, let us particularly note that the Theorem only applies to right-angled triangles, *not to all triangles.*

Example:

Show that a triangle having sides of length 13 mm, 85 mm and 84 mm is right-angled.

Let c be the longest side (i.e. the hypotenuse), with a and b the other two sides. If the triangle is right-angled, using the Theorem of Pythagoras:

$$c^2 = a^2 + b^2$$

Substituting the values, working in mm units:

$$85^2 = 13^2 + 84^2$$
$$7\,225 = 169 + 7\,056 \text{ (values from tables)}$$
$$7\,225 = 7\,225$$

Answer:

The square on the hypotenuse is equal to the sum of the squares on the other two sides, hence the triangle is right-angled.

Example:

Figure 4.10 shows a link AB of length 160 mm, the rollers A and B moving in guides at right angles. Originally the centre of B was 50 mm from O. If roller B moves 10 mm nearer to O, how far does the centre of roller A move?

Let us work this example in centimetres; 160 mm = 16 cm and 50 mm = 5 cm.

According to Pythagoras:

$$(AB)^2 = (OB)^2 + (OA)^2$$
$$16^2 = (OB)^2 + (OA)^2$$
$$16^2 - (OB)^2 = (OA)^2$$
$$(OA)^2 = 256 - (OB)^2$$

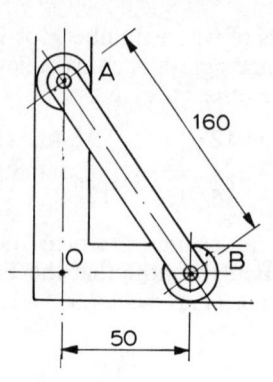

Fig. 4.10

When B is 5 cm from O:

$$(OA)^2 = 256 - 5^2 = 256 - 25 = 231$$
$$OA = \sqrt{231} = 15\cdot20$$

If B moves 1 cm nearer O, OB will be 4 cm:

$$(OA)^2 = 256 - 4^2 = 256 - 16 = 240$$
$$OA = \sqrt{240} = 15\cdot49$$

$$\text{Difference in } OA \text{ values} = 15\cdot49 \text{ cm} - 15\cdot20 \text{ cm}$$
$$= 0\cdot29 \text{ cm} = 2\cdot9 \text{ mm}$$

Answer:

Roller A moves 2·9 mm upwards

The previous worked example provides a convenient approach to a further consideration of the accuracy of answers. The answer to the problem was $\sqrt{240} - \sqrt{231}$. The answer was obtained with the aid of four-figure tables, in which the last figure is often quite suspect. The last figure of $\sqrt{240}$ is suspect, so is the last figure of $\sqrt{231}$. Due to the relative closeness of the values of $\sqrt{240}$ and $\sqrt{231}$, it would be more realistic to round-off the final answer to just one significant figure and give the answer as 3 mm. In general, if two values are obtained from four-figure tables and those values are in relatively close proximity, the difference in those values should be rounded off to a more appropriate degree of accuracy than three significant figures.

Once again the use of calculator has a distinct advantage. By calculator:

$$\sqrt{240} = 15\cdot491\ 933$$
$$\sqrt{231} = 15\cdot198\ 684$$
$$\sqrt{240} - \sqrt{231} = 0\cdot293\ 249$$

and the answer, to three significant figures, without question, is 0·293 cm which is equal to 2·93 mm.

Example:

Figure 4.11 shows a method of determining a large radius on a template. Show, by applying the Theorem of Pythagoras to triangle *EFG*, that $R = \dfrac{c^2}{8d}$

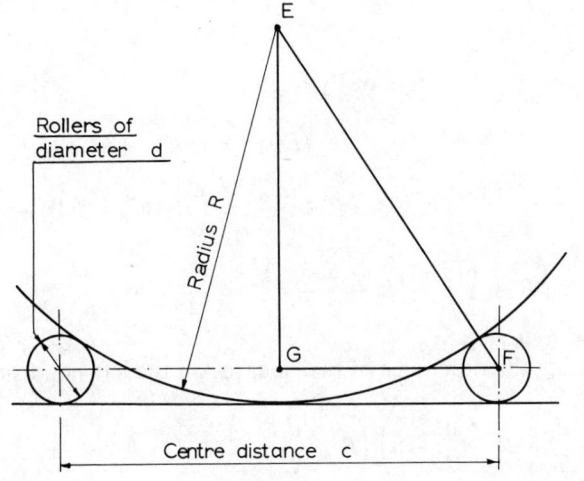

Fig. 4.11

In the right-angled triangle *EFG*:

$$GF = \frac{c}{2}, \qquad FE = R + \frac{d}{2}, \qquad EG = R - \frac{d}{2}$$

Applying the Theorem of Pythagoras:

$$(FE)^2 = (GF)^2 + (EG)^2$$

$$\therefore \left(R + \frac{d}{2}\right)^2 = \left(\frac{c}{2}\right)^2 + \left(R - \frac{d}{2}\right)^2$$

$$R^2 + Rd + \frac{d^2}{4} = \frac{c^2}{4} + R^2 - Rd + \frac{d^2}{4}$$

$$2Rd = \frac{c^2}{4}$$

whence $$R = \frac{c^2}{8d}$$

Example:

Figure 4.12 shows another method of determining a large radius. In this case, rollers of diameter d are fixed at a centre distance of C, and the height h checked with a slip gauge pile. If the height of the slip gauge pile is h, find an expression for the value of R. Find R when $C = 50$ mm, $d = 10$ mm and $h = 5$ mm.

In the right-angled triangle, EFG:

$$GF = \frac{C}{2}, \qquad FE = R + h - \frac{d}{2}, \qquad EG = R + \frac{d}{2}$$

Applying the Theorem of Pythagoras:

$$(EG)^2 = (FE)^2 + (GF)^2$$

$$\therefore \left(R + \frac{d}{2}\right)^2 = \left(R - \frac{d}{2} + h\right)^2 + \frac{C^2}{4}$$

$$\therefore \left(R + \frac{d}{2}\right)^2 - \left(R - \frac{d}{2} + h\right)^2 = \frac{C^2}{4}$$

Factorising the difference of two squares on the left-hand side:

$$\left(R + \frac{d}{2} + R - \frac{d}{2} + h\right)\left(R + \frac{d}{2} - R + \frac{d}{2} - h\right) = \frac{C^2}{4}$$

$$(2R + h)(d - h) = \frac{C^2}{4}$$

$$2R + h = \frac{C^2}{4(d - h)}$$

$$2R = \frac{C^2}{4(d - h)} - h$$

$$R = \frac{C^2}{8(d - h)} - \frac{h}{2}$$

Working in centimetres:

When $C = 5$, $d = 1$ and $h = 0.5$

$$R = \frac{5^2}{8(1 - 0.5)} - \frac{0.5}{2}$$

$$= 6.25 - 0.25 = 6 \text{ cm} = 60 \text{ mm}$$

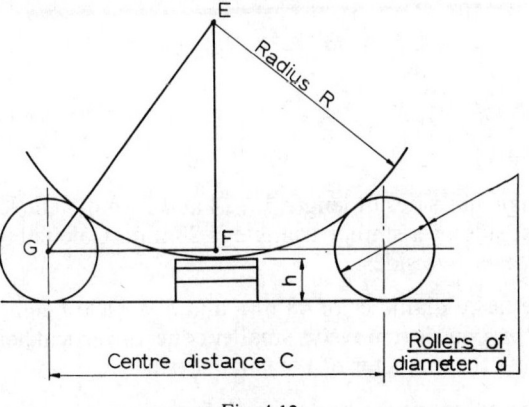

Fig. 4.12

Answer:

(a) $\quad R = \dfrac{C^2}{8(d-h)} - \dfrac{h}{2}$

(b) $\quad R = 60$ mm

Problems 4.2

1. Triangles may be classified into six types, viz: acute angled, obtuse angled, right angled, equilateral, isosceles and scalene, but a particular triangle could fall into more than one class. Construct the following triangles to a suitable scale and state into which classification or classifications the triangle falls.

(a) $a = 14, b = 16, c = 18$
(b) $a = b = 16, c = 8$
(c) $a = b = c = 10$
(d) $a = 3, b = 4, C = 90°$
(e) $a = 10, b = 10, C = 40°$
(f) $a = b = 16, C = 90°$
(g) $a = b = 15, c = 24$
(h) $a = 8, b = 12, C = 110°$
(i) $a = 12, B = 72°, C = 53°$
(j) $a = 10, b = 16, B = 90°$

2. For the following triangles, two angles A and B are given. Determine the angle C and then state into which of the six classifications given in the introduction to question 1 that the triangle falls.

(a) $A = 45°, B = 55°$
(b) $A = B = 67° \ 30'$

    (c) $A = 110°, B = 35°$
    (d) $A = 22° 17', B = 53° 52'$
    (e) $A = B = 60°$
    (f) $A = 67° 22', B = 22° 38'$
    (g) $A = B = 45°$
    (h) $A = 118° 22' 40'', B = 30° 48' 40''$

3. A triangle has sides of length 15, 24 and 36 mm. The length of the shortest side of a similar triangle is 20 mm. Calculate the lengths of the other two sides.

4. A cone has a diameter of 48 mm and a vertical height of 60 mm. From this cone is removed a smaller cone, of vertical height 20 mm. Calculate the diameter of the smaller cone.

5. A symmetrical wedge has a length from base to apex of 100 mm, the base having a width of 20 mm. If it is placed in an aperture of width 8 mm, what length of the wedge projects from the aperture?

6. A crossed belt drive connects pulleys of diameters 100 mm and 240 mm whose centres are 400 mm apart. Calculate the distance to where the belt lines cross, measured from the centre of the larger pulley.

7. Calculate the distances x and y in Fig. 4.13.

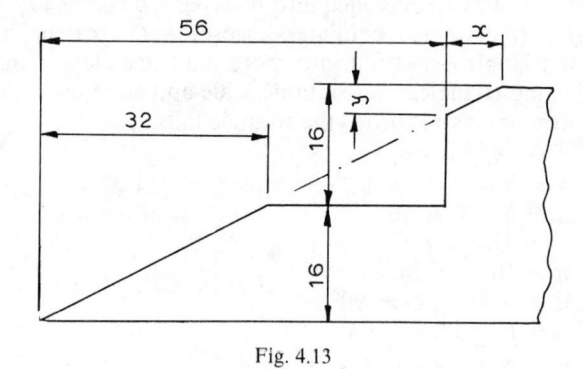

Fig. 4.13

8. The sides containing the right angle of a right-angled triangle have lengths of 12 mm and 35 mm. Calculate the length of the hypotenuse.

9. The hypotenuse of a right-angled triangle has a length of 100 mm. The other two sides are in the ratio of five to one. Determine the length of the shortest side, to three significant figures.

10. An isosceles triangle has a base of length 16 mm. Each of the other two sides has a length of 17 mm. Find the perpendicular height of the triangle.

11. Discs of diameter 27 mm are stamped out in alternate rows of five and four so that there is 3 mm between edges of adjacent holes. Determine the distance between the centre lines of parallel rows of discs, to three significant figures.

12. The profile shown in Fig. 4.14 has no straight portions. Determine the smallest width of the profile, *W*.

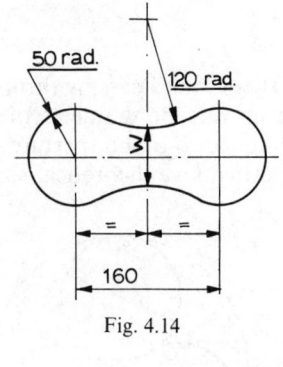

Fig. 4.14

13. *A* and *B* represent the centres of two switches on the front panel of a receiver. Measured in millimetres, the co-ordinates of *A* and *B* are (36, 71) and (60, 116) respectively. Calculate the distance, in millimetres, between the centres of *A* and *B*.

14. A field is rectangular in shape, the length being twice the breadth. The field has an area of 1 hectare.
    (a) Determine the length and breadth.
    (b) *A* and *B* represent two diagonally opposite corners. If a wire runs direct from *A* to *B* instead of along around the perimeter, determine the saving of wire, in metres (1 hectare = 10^4 m²).

15. A circular bar has a diameter of 34 mm. A cut is taken across the bar to produce a flat of width 30 mm. Calculate the depth of the cut.

16. Find the dimension *x* of the layout shown in Fig. 4.15.

17. A keyway of width 9 mm is cut in a bar of diameter 41 mm. The depth at the side of the keyway is 4 mm. Calculate its nominal depth at the centre.

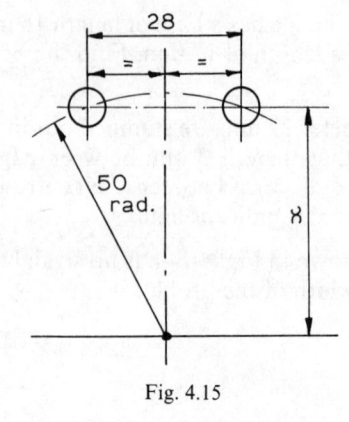

Fig. 4.15

18. Figure 4.16 shows three cables passing through a circular aperture. If the cables have an outside diameter of 8 mm, determine the smallest possible diameter D of the aperture. Give the answer to the nearest half-millimetre above theoretical size.

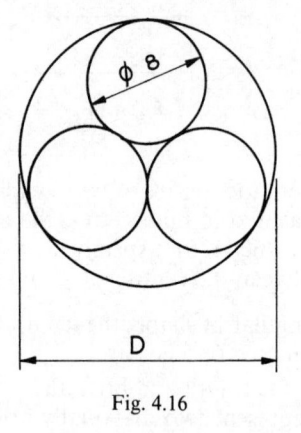

Fig. 4.16

19. Figure 4.17 shows a cable strung between three pylons.
    (a) Calculate the height of the pylon at B so that the angle of inclination remains constant.
    (b) Calculate the straight length of cable required from pylon A to pylon C.
    (c) If $3\frac{3}{4}\%$ of the theoretical length must be provided for sagging, contraction, etc. calculate the actual length of cable used.

20. A ladder of length 4·25 m rests with its foot on horizontal ground and its top against a vertical wall. Initially the foot of the ladder is

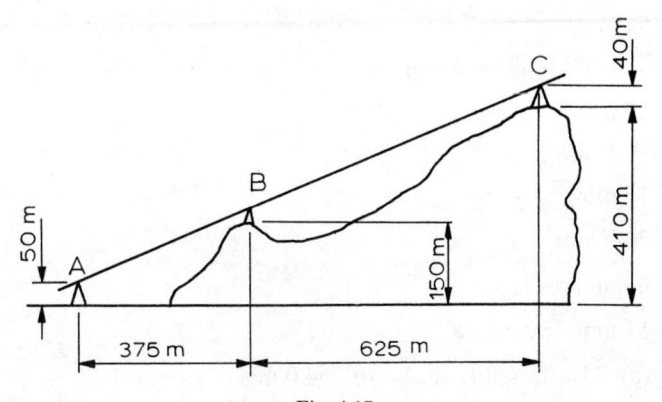

Fig. 4.17

1·19 m from the foot of the wall. How far down the wall does the top of the ladder move if the foot of the ladder is moved outward by 1·36 m?

Answers to Problems 4.2

1. (a) scalene and acute angled
   (b) isosceles and acute angled
   (c) equilateral and acute angled
   (d) right angled and scalene
   (e) isosceles and acute angled
   (f) right angled and isosceles
   (g) isosceles and obtuse angled
   (h) scalene and obtuse angled
   (i) scalene and acute angled
   (j) right angled and scalene

2. (a) 80°, scalene and acute angled
   (b) 45°, isosceles and acute angled
   (c) 35°, isosceles and obtuse angled
   (d) 103° 51′, scalene and obtuse angled
   (e) 60°, equilateral and acute angled
   (f) 90°, right angled and scalene
   (g) 45°, right angled and isosceles
   (h) 30° 48′ 40″, isosceles and obtuse angled.

3. 32 mm and 48 mm

4. 16 mm

5. 60 mm

6. 282 mm

7. $x = 8$ mm, $y = 4$ mm

8. 37 mm

9. 19·6 mm

10. 15 mm

11. 26·0 mm

12. 60 mm

13. 51 mm

14. (a) 70·7 m × 141 m   (b) 54·0 m

15. 9 mm

16. 48 mm

17. 4·5 mm

18. 17·5 mm

19. (a) 50 m   (b) 1 080 m   (c) 1 120 m

20. 0·68 m

4.3
The properties of a circle

4.3.1
Definitions relating to circles

A *circle* is a plane figure bounded by a continuous line joining an infinite set of points all of which are equidistant from a fixed point called the *centre* of the circle. The boundary, called the *circumference*, is often loosely referred to as a circle. The centre of a circle can be regarded as that point from which every part of the circumference is equidistant. A *radius* is any straight line joining the centre of the circle to the circumference. The length of this line is also loosely referred to as the radius of the circle. It is conventional practice to represent the radius of a circle by the symbol r. A *diameter* is any straight line passing through the centre of the circle which joins two points on the circumference. A diameter divides the circle into two equal plane figures called *semicircles*. An *arc* of a curve is any portion of a curve which is less than that of the whole curve, hence an arc of a circle is any portion of the circumference whose length is less than that of the whole circumference. A diameter can therefore also be considered to cut the circumference of a circle into two arcs of equal length. The diameter of a circle is represented by the symbol d, and since it can be regarded as two radii joined end-to-end in a straight line, then

$$d = 2r.$$

The length of the circumference of a circle cannot be determined absolutely in terms of its diameter, but for any circle, the ratio of the length of the circumference to the length of the diameter is of constant magnitude. This constant magnitude is used so often in mathematics that we give it the symbol π (the Greek letter small pi). Pi is the initial letter of the Greek work for perimetron, or length of the perimeter.

Hence, if

$$C = \text{the length of the circumference of a circle}$$

and $\quad d = \text{the diameter of that circle,}$

then $\quad C = \pi d$

and since $\quad d = 2r, \quad$ then $\quad C = 2\pi r.$

The determination of the value of π has a fascinating history. In early Hebrew times it was thought that the value was 3. Later, the scholar Archimedes knew that the true value lay between 220/70 and 223/71. It can be shown in branches of higher mathematics that the numerical value of π can be determined from non-terminating series. There are several of these series, such as:

$$\frac{\pi}{4} = 1 - \frac{1}{3} + \frac{1}{5} - \frac{1}{7} + \frac{1}{9} - \frac{1}{11} + \frac{1}{13} \cdots$$

$$\frac{\pi}{2} = \frac{2 \times 2}{1 \times 3} \times \frac{4 \times 4}{3 \times 5} \times \frac{6 \times 6}{5 \times 7} \times \frac{8 \times 8}{7 \times 9} \cdots$$

$$\frac{\pi}{2} = 1 + \frac{1}{3}\left(\frac{1}{2}\right) + \frac{1}{5}\left(\frac{1 \times 3}{2 \times 4}\right) + \frac{1}{7}\left(\frac{1 \times 3 \times 5}{2 \times 4 \times 5}\right) \cdots$$

If any of these series is used to establish the numerical value of π, then to ten significant figures

$$\pi = 3\cdot141\ 592\ 654.$$

The decimal equivalent of the improper vulgar fraction $\frac{355}{113}$ provides the first seven of these figures. The improper vulgar fraction $\frac{22}{7}$ has an error of approximately $0\cdot04\%$, about 1 part in 2 500. Consequently, in the past it was quite common to use the approximation of $\frac{22}{7}$ for the value of π when an accuracy of three significant figures in a final result was acceptable. The introduction of the electronic calculator with a direct input of π, however, has so simplified computation that this well-known approximation for π is now not used as often as it was in the past.

A *chord* of a circle is a straight line which joins any two points which lie on the circumference of a circle. The diameter of a circle can be considered to be one special case of a chord, being the chord of greatest length, but a chord of a circle is generally considered to be of lesser length than its diameter. A chord divides the area of a circle into two parts called *segments*. That of the greater area is called the major segment, while that of the smaller area is called the minor segment. Unless there are specific instructions to the contrary it is common practice to presume that the area contained by a chord and an arc of a circle is that of the minor segment. If two radii are drawn from the two ends of a chord, the angle contained by those radii is called the *angle subtended at the centre of the circle* by that chord. A chord actually subtends two angles, which together total 360°, and once again, unless instructions are specially given to the contrary, it is presumed that the angle subtended at the centre of a circle by a chord is the smaller of the two angles, and it will be smaller than 180°.

A *tangent line* to a circle, often called the tangent to a circle, is a straight line drawn from a point outside that circle to touch the circle at its circumference at one point only. If the straight line were to be continued beyond the point of contact it would not contact the circumference again. If a radius is drawn from the centre of the circle to the point of contact, the radius and the tangent line will lie at right angles to each other.

A *sector* of a circle is the plane figure contained by two radii and an arc of the circumference. If two radii are drawn within a circle, two sectors are formed. That of greater area is called the major sector, while that of smaller area is called the minor sector. As with segments, unless there are special instructions to the contrary, it is common practice to assume that the area contained by two radii and an arc of the circumference is the minor sector. The angle contained by the two radii is called the *included angle* of the sector. For the minor sector this included angle will be less than 180°.

The various terms used in connection with the geometry of circles are shown on page 269.

We shall proceed with certain properties of a circle which will be presented without proof.

The angle subtended at the centre of the circle is twice the angle subtended at the circumference, and all angles subtended at the circumference by the same chord are equal in magnitude.

The theorem is illustrated in Figs 4.18(a) and 4.18(b) overleaf.

In Fig. 4.18(a) obtuse angle *AOC* = twice acute angle *ABC*.
In Fig. 4.18(b), reflex angle *AOB* = twice obtuse angle *ADB*.

Let us proceed to two very important results of this theorem. The first is that if the chord is a diameter, the angle at the centre is a straight

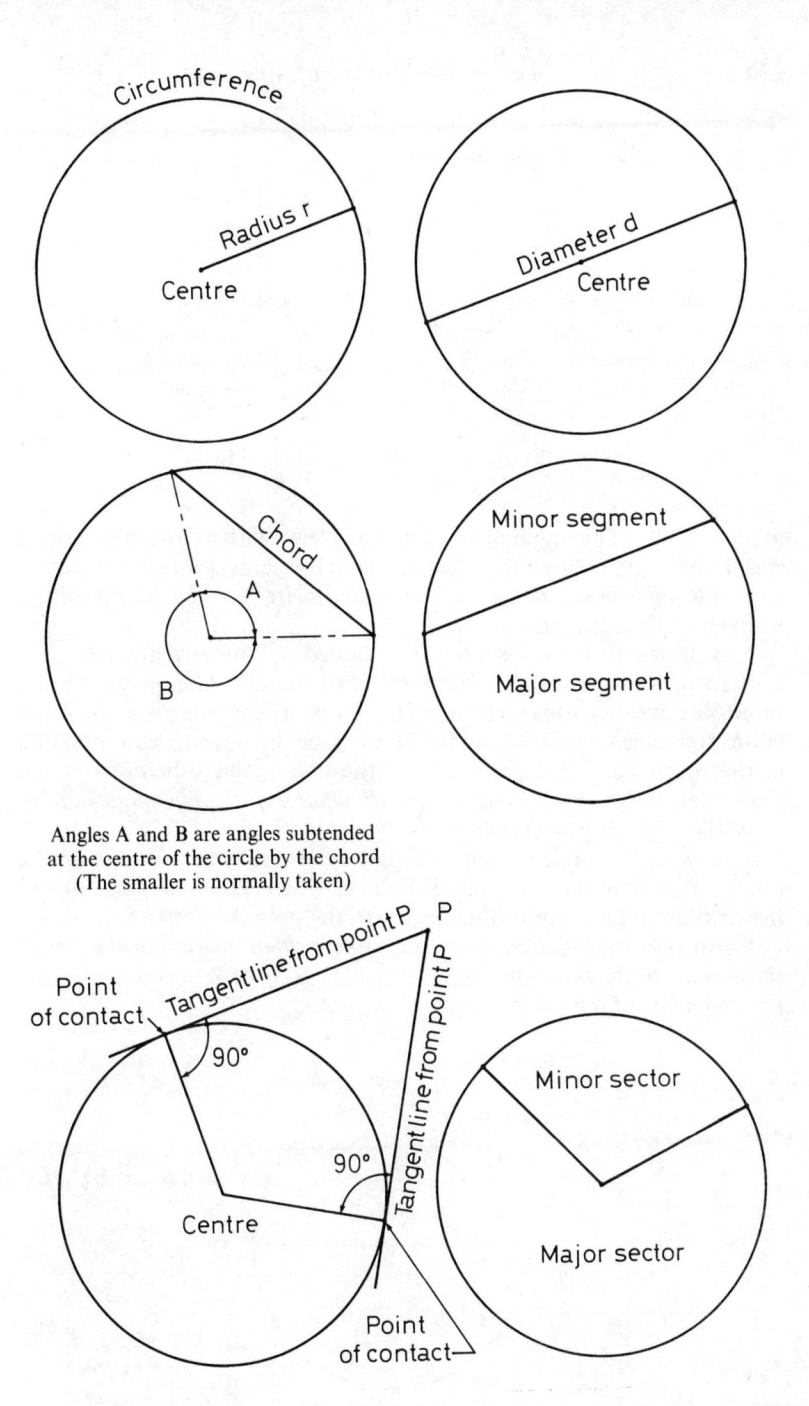

Circumference

Radius r

Centre

Diameter d

Centre

Chord

A

B

Angles A and B are angles subtended
at the centre of the circle by the chord
(The smaller is normally taken)

Minor segment

Major segment

Tangent line from point P

P

Point
of contact

Tangent line from point P

90°

90°

Centre

Point
of contact

Minor sector

Major sector

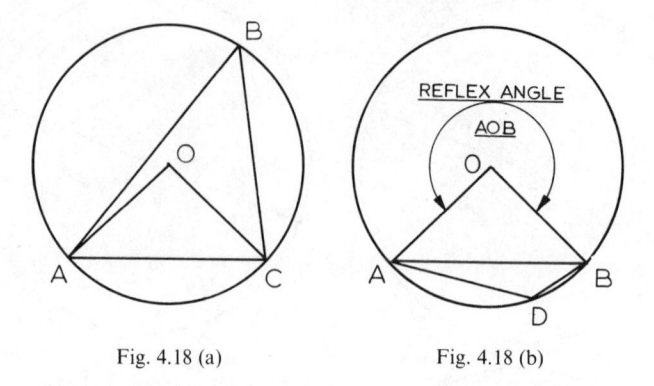

Fig. 4.18 (a)   Fig. 4.18 (b)

angle of 180°. The angle at the circumference is therefore one half of this, i.e. 90°, or a right angle. All angles in the same segment are equal, *hence the angle in a semicircle is a right angle.* In Fig. 4.19, all the angles marked *B* are right angles.

A cyclic quadrilateral is a figure bounded by four straight sides, the four corners lying on the circumference of a circle. A diagonal joining opposite corners forms a chord of the circle. If we join the ends of the chord to the centre, we can reason that since the two angles subtended at the centre total 360°, the sum of the two angles subtended at the circumference is 180°. *Hence, opposite angles of a cyclic quadrilateral total* 180°, *i.e. they are supplementary.*

A *tangent* to a circle is a line which touches the circumference at one point, but does not cut the circumference when extended. The tangent lies at right angles to a radius drawn to the point of contact.

If two chords of a circle intersect, either within or without the circle, the product of the two segments of one chord is equal to the product of the two segments of the other chord.

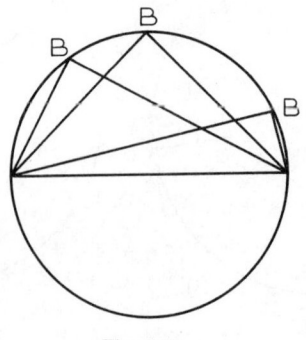

Fig. 4.19

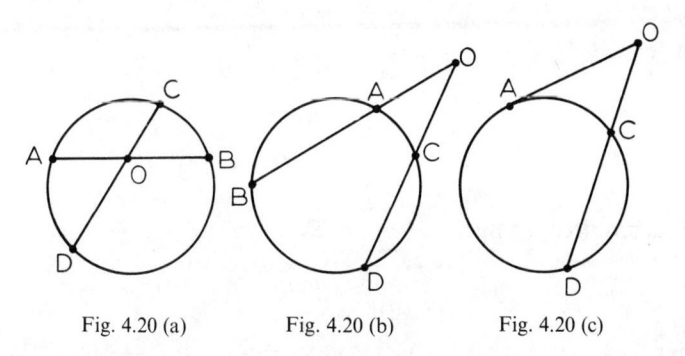

Fig. 4.20 (a)   Fig. 4.20 (b)   Fig. 4.20 (c)

Figure 4.20 shows at (a) the chords intersecting inside a circle, at (b) the chords intersecting outside a circle and at (c) the special case of one chord just touching a circle so that the chord is a tangent.

In Fig. 4.20(a) and 4.20(b), $OA \times OB = OC \times OD$.

In Fig. 4.20(c), $(OA)^2 = OC \times OD$.

An important application occurs when one chord is the diameter of the circle, the other chord intersecting the diameter at right angles within the circle, as shown in Fig. 4.21.

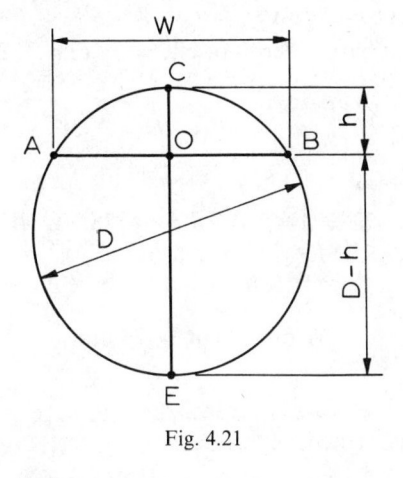

Fig. 4.21

The chord AB is bisected and $AO = OB$.

Let D = diameter of circle,

W = length of chord AB, and h = height of smaller segment.

$$AO \times OB = OE \times OC$$

$$\therefore \frac{W}{2} \times \frac{W}{2} = (D - h)h$$

and
$$\frac{W^2}{4} = Dh - h^2$$

$$\therefore W^2 = 4(Dh - h^2)$$

and
$$W = 2\sqrt{(Dh - h^2)}$$

If D is required, using
$$\frac{W^2}{4} = Dh - h^2$$

$$Dh = \frac{W^2}{4} + h^2$$

$$D = \frac{W^2}{4h} + h$$

Example:

Determine the width of the flat produced when a cut of depth 2 mm is taken across a circular bar of diameter 74 mm.

Using Fig. 4.21,
CO = depth of cut = 2 mm
OE = diameter of bar − depth of cut = 74 mm − 2 mm = 72 mm
If W = width of flat,

$$\frac{W}{2} \times \frac{W}{2} = CO \times OE,$$

$$\frac{W^2}{4} = 2 \times 72 = 144, \qquad W^2 = 576, \qquad W = \sqrt{576} = 24$$

Answer:

Width of cut = 24 mm

4.3.2
The circular measurement of angles

As our knowledge of mathematics develops, we shall find that while the degree with its subdivisions of minutes and seconds often prove to be very convenient units for the purposes of measuring an angle, there are branches of mathematics where a different unit has considerable advantages. In the circular measurement of angles we divide one complete revolution into 2π equal parts called *radians*. The abbreviation for both radian and radians is rad. One radian is therefore equivalent to $360°/2\pi$.

This value cannot be expressed precisely as a denary value. Using the sexagesimal measurement of angles one radian is approximately 57° 17′ 45″. Using the centesimal measurement of angles one radian is approximately 63·66 grades, while using the decimal measurement of angles one radian is approximately 57·296°. A geometrical appreciation of the radian can be obtained by drawing a sector of a circle whose arc has a length equal to that of the radius, as shown in Fig. 4.22.

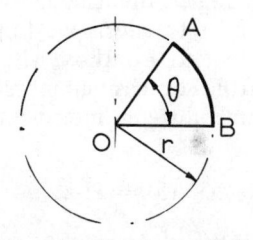

When length of arc AB is
equal to radius r , angle θ
is 1 radian

Fig. 4.22

Since a radius can be spaced along the circumference 2π times, the angle included by the two radii is $\dfrac{360°}{2\pi}$ or one radian. If the magnitude of an angle shows no indication of degrees and/or minutes and/or seconds, being given by values such as $\dfrac{\pi}{4}$, $\dfrac{2\pi}{3}$ or 0·4, it is assumed that the unit is the radian. Since 2π radians are equal to 360°, to convert radians to degrees we multiply by $\dfrac{360°}{2\pi}$. Conversely, to convert degrees to radians, we multiply by $\dfrac{2\pi}{360°}$.

As examples:

$$\frac{\pi}{3} = \frac{\pi}{3} \times \frac{360°}{2\pi} = 60°$$

$$\frac{\pi}{4} = \frac{\pi}{4} \times \frac{360°}{2\pi} = 45°$$

and $$180° = 180° \times \frac{2\pi}{360°} = \pi \text{ radians}$$

It will be observed from the last value that it is common practice when using the circular measurement of angles to express certain values as multiples of π, such as 2π, $\dfrac{3\pi}{4}$ and so on. Such values indicate the magnitude of the angles concisely. Decimal notation is normally only used if the decimal value is precise, such as with values of 0·5, 0·8, 4 and so on.

As has been previously stated in this text, if a calculator is available for conversions from the sexagesimal measurement of angles to the circular measurement of angles, that calculator should be used. It will be necessary for our reader to check the functions available of the particular calculator being used. For those with a limited range of functions, when converting from sexagesimal measure to circular measure, it may be necessary to introduce an intermediate step of the decimal measure. For example,

$$53° \ 15' = 53·25°,$$

and then the calculator is used for

$$53·25 \times 2 \times \pi \div 360$$

to obtain the equivalent in radians,

and $53° \ 15' = 0·929 \ 388$ rad, to six decimal places.

Conversely, if the value in circular measure is not a multiple of π, such as with 0·540 6, then the calculator is used for

$$0·540 \ 6 \times 360 \div 2 \div \pi$$

to obtain, to six decimal places, $30·974 \ 098°$

and
$$0·974 \ 098° = 0·974 \ 098 \times 60 \text{ minutes} = 58·445 \ 88'$$
$$0·445 \ 88' = 0·445 \ 88 \times 60 \text{ seconds} = 26·752 \ 8''$$

Hence 0·540 6 radians, to the nearest second, is

$$30° \ 58' \ 27''.$$

However, as with similar exercises in the past, it is essential for us to consider a method of procedure if a calculator is not available, if calculations made by formal arithmetic prove to be unduly time-consuming, and if determinations using a slide rule are insufficiently accurate.

Conversions can also be made using the tables on pages 32 and 33 of Castle's *Four-figure Mathematical Tables*. Values are given for intervals of one minute from zero to 89° 59′ $\left(90° \text{ is } \dfrac{\pi}{2} = 1·570 \ 8\right)$. It should be noted that since the magnitude of an angle in radians is directly proportional to its magnitude in degrees, the rows of mean differences in the end columns are constant, irrespective of the angle. Furthermore, the mean differences have to be added. For angles in excess of 90°, $\frac{\pi}{4}$ or 1·570 8 is added for every interval of 90°. The reader should now check that:

$$51° = 0·890 \ 1 \text{ radians}$$
$$80° \ 36' = 1·406 \ 7 \text{ radians}$$
$$68° \ 53' = 1·202 \ 3 \text{ radians}$$

while

$$1·309 \ 0 \text{ radians} = 75°$$
$$0·524 \ 5 \text{ radians} = 30° \ 3'$$
$$0·500 \ 0 \text{ radians} = 28° \ 39' \text{ (as accurately as tables allow)}$$

For angles in the range 90° to 360° the following procedure is normally adopted.

Example:

Convert an angle of 323° 46′ to circular measurement

$$323° \ 46'$$

Nearest whole
multiple of 90°
below value given $= \underline{270°}$ $= 3 \times 1·570 \ 8 = 4·712 \ 4$
$$\overline{53° \ 46'} \qquad\qquad = 0·938 \ 4 \text{ (from tables)}$$
$$\text{Add} \quad \overline{5·650 \ 8}$$

Answer:

$$323° \ 46' = 5·650 \ 8 \text{ radians}$$

Example:

Convert 2·753 7 radians to sexagesimal measure to the nearest minute.

$$2·753 \ 7$$

Nearest whole number
multiple of
1·570 8 below value given $= \underline{1·570 \ 8} = \quad 90°$
$$\overline{1·182 \ 9} = \quad 67° \ 46' \text{ (from tables)}$$
$$\text{Add} \quad \overline{157° \ 46'}$$

Answer:

$$2·753 \; 7 \text{ radians} = 157° \; 46'$$

The above procedure should normally be adopted within the range of zero to 360°. For very large amounts of turning, the conversion factor:

$$360° = 2\pi \text{ radians}$$

should be used, with a number of places of decimals selected for π as accuracy demands. For interest, π to ten decimal places is 3·141 592 653 6.

In certain restricted cases, when an angle is given in radians to an accuracy of two decimal places and does not exceed $\dfrac{\pi}{2}$, a very convenient table appears on page 34 of Castle's *Four-figure Mathematical Tables*. For example, 0·48 radians is given directly as 27° 30'.

It has been previously stated in this volume that the primary fundamental units of the SI system are those of mass, length and time. To these we add secondary fundamental units, one for each of certain branches of science, such as the ampere as the unit in connection with electrical science. To make the system function as a whole it is necessary to introduce other basic units which we can regard as being associated with the SI system. In particular, with the SI system, the basic unit for an angle is the radian (abbreviation rad). Consequently, using the coherence of the SI system, the basic derived unit for a rotational speed is:

$$\frac{\text{one basic unit of angle turned}}{\text{one fundamental unit of time}}$$

that is, the radian per second (abbreviation rad/s).

However, it is equally true that the unit or units used for a particular purpose should reflect the practical realities of a given situation. For many circumstances the revolution per minute is a far more realistic unit. In which case, particularly when solving problems dealing with rotation it may be necessary for our reader to convert a speed in revolutions per minute (abbreviation rev/min not rpm nor r.p.m.) to its equivalent in radians per second (abbreviation rad/s), and vice versa.

Since $1 \text{ rev} = 2\pi \text{ rad}$
and $1 \text{ min} = 60 \text{ s}$

then to convert rev/min to rad/s we multiply by $\dfrac{2\pi}{60}$, and, conversely,

to convert rad/s to rev/min we multiply by $\dfrac{60}{2\pi}$.

Example:

Convert rotational speeds of:

(a) 400 rev/min to rad/s
(b) 28·4 rad/s to rev/min,

in each case giving the answer correct to three significant figures.

(a) 400 rev/min $= 400 \times \dfrac{2\pi}{60}$ rad/s $= 41\cdot9$ rad/s

(b) 28·4 rad/s $= 28\cdot4 \times \dfrac{60}{2\pi}$ rev/min $= 271$ rev/min

Answers:

(a) 41·9 rad/s
(b) 271 rev/min

Problems 4.3

1. AB is the diameter of a circle, of length 100 mm. C is a point on the diameter so that $AC = 70$ mm. Calculate the length of the chord which is at right angles to the diameter and passes through the point C.

2. A segment of a circle has a chord of length 8 mm and a height of 2 mm. Calculate the radius of the circle.

3. A cut of depth 2 mm is taken across a circular bar of diameter 52 mm. Calculate the width of the resulting flat.

4. Two holes lie on a pitch circle, the chordal distance between their centres being 100 mm. If this chord is bisected, one of the distances from the centre of the chord to the pitch circle is 10 mm. Calculate the pitch circle diameter.

5. A circle has a diameter of 70 mm. P is a point 125 mm from the centre of the circle. Use the theorem of intersecting chords (for the special case of a tangent) to find the length of the tangent to the circle.

6. Convert:
   (a) 1·77 radians to degrees and minutes
   (b) 50° to radians

7. Convert the following to degrees:

   (a) $\dfrac{2\cdot5\pi}{12}$ radians   (b) 5·515 2 radians

8. (a) Using mathematical tables convert the following to radians:
   (i) $69° 23'$ (ii) $270°$
   (b) Using mathematical tables convert the following to degrees:
   (i) $\dfrac{5\pi}{12}$ radians   (ii) 2.775 radians

9. Evaluate $\left(100\pi t - \dfrac{\pi}{4}\right)$ when $t = 0.01$, giving the answer:
   (a) in radians, as a multiple of π (b) in degrees

10. Convert:
    (a) $2\,100$ rev/min to rad/s   (b) 44 rad/s to rev/min

Answers to Problems 4.3

1. 91.7 mm

2. 5 mm

3. 20 mm

4. 260 mm

5. 120 mm

6. (a) $101° 25'$ (b) $0.872\,7$

7. (a) $37° 30'$ (b) $316°$

8. (a) (i) $1.211\,0$; (ii) $4.712\,4$ (b) (i) $75°$; (ii) $159°$

9. (a) $\dfrac{3\pi}{4}$ (b) $135°$

10. (a) 220 rad/s   (b) 420 rev/min

4.4
The geometry of lines

4.4.1
The length of a straight line

Mensuration is the branch of mathematics which gives us the rules for finding the lengths of lines, the areas of plane figures and the volumes of solids. We will proceed with our studies in that order.

If we have two points on a graphical field, P_1 and P_2, indicated by the values (x_1, y_1) and (x_2, y_2), as shown in Fig. 4.23, then the distance between vertical lines drawn through the points is $(x_2 - x_1)$ and the distance between horizontal lines drawn through the points is $(y_2 - y_1)$.

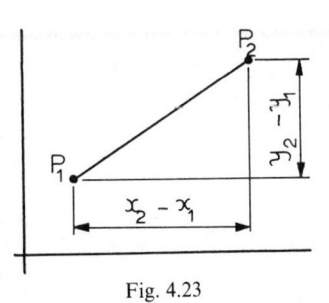

Fig. 4.23

Using the theorem of Pythagoras:

$$(P_1P_2)^2 = (x_2 - x_1)^2 + (y_2 - y_1)^2$$

hence $\qquad P_1P_2 = \sqrt{\{(x_2 - x_1)^2 + (y_2 - y_1)^2\}}$

Example:

Calculate the distance between the points $(5, 3)$ and $(-1, -5)$.

$$\begin{aligned}
\text{Distance} &= \sqrt{[\{5 - (-1)\}^2 + \{3 - (-5)\}^2]} \\
&= \sqrt{(6^2 + 8^2)} = \sqrt{(36 + 64)} \\
&= \sqrt{100} = 10 \text{ units}
\end{aligned}$$

Answer:

$$\text{Distance} = 10 \text{ units}$$

4.4.2
The length of a circular arc

In article 4.3.1 it was stated that the length C of the circumference of a circle, in terms of the radius r, was given by the formula

$$C = 2\pi r.$$

It was also stated that an arc is a portion of any curve which is less than that of the whole curve. Let us presume that the ends of the arc of a circle of radius r subtend an angle of $\theta°$ at the centre of that circle. We can imagine that the whole circumference of the circle subtends an angle of $360°$, and that the circumference is comprised of 360 identical arcs each subtending an angle of one degree. If we now collect θ of these arcs together in series, so that an arc is obtained which subtends an angle of $\theta°$, its length L can be found from

$$L = \theta \times \frac{\text{whole circumference}}{360} = \frac{\theta}{360}(2\pi r)$$

(if θ is measured in degrees).

If θ is measured in radians, the same reasoning will apply. The complete circumference of $2\pi r$ subtends 2π radians. If a circular arc subtends θ radians, the length of this circular arc can be determined from

$$L = \frac{\theta}{2\pi} \times \text{the entire circumference}$$

hence $$L = \frac{\theta}{2\pi} \times 2\pi r = r\theta.$$

(if θ is measured in radians).

Example:

Find the angle subtended at the centre of a circle of radius 12 mm by an arc of length 15 mm, giving the answer in radians and degrees.

$$L = r\theta$$

$$\therefore \theta = \frac{L}{r} = \frac{15 \text{ mm}}{12 \text{ mm}} = 1 \cdot 25 \text{ radians}$$

From tables: $1 \cdot 249\ 7$ radians $= 71° 36'$

$\underline{0 \cdot 000\ 3 \text{ radians} = \qquad 1'}$

$1 \cdot 250\ 0$ radians $= 71° 37'$

Answer:

Angle subtended $= 1 \cdot 25$ radians $= 71° 37'$

Let us consider any point on the circumference of a wheel. If that wheel rotates about its centre, the velocity of that point and similar points on the periphery is called the *peripheral speed.*

If d is the diameter of the circle, the length of the circumference is πd. If the wheel rotates once, the point travels a distance equal to πd. If the wheel makes n revolutions, the point travels a distance equal to πdn. If the n revolutions are made in unit time, then the point moves with a speed of πdn. Hence:

if $v =$ the peripheral speed
                      $d =$ the wheel diameter
                      $n =$ the rotational speed
then $v = \pi dn$

The unit of v depends upon the units given to d and to n. Using basic SI units, a speed is given in metres per second and a diameter in metres. Consequently the rotational speed to give a peripheral speed in m/s when the diameter is quoted in metres should be stated in revolutions per second.

It must be accepted that for some time to come strict SI usage cannot be expected in everyday life, and many rotational speeds will be quoted in rev/min. Just for a moment let us return to the formula $v = \pi d n$. Let us presume that the symbol n represents revolutions per second. Let us introduce another symbol ω (the small Greek letter omega), where ω represents the rotational speed in radians per second. There are 2π radians in a complete revolution, hence:

$$\omega = 2\pi n \quad \text{and} \quad n = \frac{\omega}{2\pi}$$

Substituting this value for n in $v = \pi d n$ we obtain:

$$v = \pi \times d \times \frac{\omega}{2\pi}$$

$$= d \times \frac{\omega}{2}$$

Now $\dfrac{d}{2}$ is the radius r, hence:

$$v = \omega r$$

This relationship is quite fundamental, and should be considered the preferred usage. A rotational speed, where necessary, should be converted to radians per second. If the rotational speed is quoted as N rev/min, this is equal to $2\pi N$ radians per minute, which itself is equal to $\dfrac{2\pi N}{60}$ radians per second. Hence to convert rev/min to rad/s we have to multiply by $\dfrac{2\pi}{60}$.

It should be noted from the definition of the radian that it is obtained from dividing a length by a length, and as far as the manipulation of units is concerned the magnitude of an angle in radians is a numerical value only. For example, in unit manipulation:

$$30 \text{ rad/s}^2 \text{ is regarded as } \frac{30}{s^2}.$$

The reader is strongly recommended to avoid usages such as $v = \pi d n$ or $v = \dfrac{\pi d N}{60}$. These could cause complications in later studies. The recommended usage as a first order of preference is:

$$v = \omega r$$

Example:

A drill has a diameter of 10 mm. Calculate the cutting speed at the periphery, in m/min, when the drill rotates at 210 rev/min.

$$r = \frac{d}{2} = \frac{10 \text{ mm}}{2} = 5 \text{ mm} = 0.005 \text{ m}$$

$$\omega = 210 \text{ rev/min} = 210 \times \frac{2\pi}{60} \text{ rad/s} = 21.991 \text{ rad/s}$$

$$v = \omega r = 21.991 \times 0.005 \, \frac{1}{\text{s}} \times \text{m}$$

$$= 0.109 \, 96 \text{ m/s}$$
$$= 0.109 \, 96 \times 60 \text{ m/min} = 6.597 \, 3 \text{ m/min}$$

Answer:

$$\text{Cutting speed} = 6.60 \text{ m/min}$$

Example:

The safe peripheral speed of a particular grinding wheel is stated by the manufacturers to be 1 650 m/min. Calculate the maximum angular velocity, in rev/min, of a wheel of diameter 175 mm.

$$v = 1 \, 650 \text{ m/min} = \frac{1 \, 650}{60} \text{ m/s}$$

$$r = \frac{d}{2} = \frac{175 \text{ mm}}{2} = 87.5 \text{ mm} = 0.087 \, 5 \text{ m}$$

$$v = \omega r$$

$$\therefore \omega = \frac{v}{r}$$

$$= \frac{1 \, 650}{60 \times 0.087 \, 5} \text{ rad/s}$$

To convert rad/s to rev/min we multiply by $\dfrac{60}{2\pi}$.

$$\omega = \frac{1 \, 650 \times 60}{60 \times 0.087 \, 5 \times 2\pi} = 3001.2 \text{ rev/min}$$

Answer:

$$\text{Maximum speed} = 3 \, 000 \text{ rev/min}$$

Let us review these solutions, because the reader, quite understand-ably, may feel we have made very heavy weather of what seem to be fairly simple tasks. If v is required in m/sec when d is given in millimetres and N is given in revolutions per minute, then a suitable formula is:

$$v = \frac{\pi dN}{60\ 000}$$

The use of this formula would certainly have eased the calculations in the two previous examples. While, however, the formula is true for the units used in these particular cases, for any other case where the units are different we shall require a different formula. What we really require is one single formula as an overall strategy. It is pointless to burden one's mind with a cumbersome collection of special formulae for special cases. Diameters are not necessarily quoted in millimetres, neither are speeds necessarily quoted in rev/min. It is repeated that the reader is well advised to use as the basic relationship between linear and rota-tional speeds the fundamental relationship:

$$v = \omega r$$

4.4.3
The perimeter of an ellipse

There is no precise formula which can be easily applied to find the perimeter of an ellipse. We can use approximations; the more accurate the approximation, the more involved becomes the formula. The accu-racy of an approximation usually depends on the ratio of the length of the major axis a to the length of the minor axis b.

A very rough approximation is $\frac{\pi}{2}(a + b)$.

A more accurate approximation is $\pi \sqrt{\dfrac{a^2 + b^2}{2}}$.

An even more accurate approximation, suitable for this stage of studies is:

$$\pi \sqrt{\left\{ \frac{a^2 + b^2}{2} - \frac{(a - b)^2}{8\cdot 8} \right\}}$$

There is no point in remembering these formulae; they are given for interest only. If a candidate in an examination has to determine the perimeter of an ellipse, the method of determination should be given.

Problems 4.4

1. Calculate the distance between the following points:
   (a) the origin and (5, 12)   (b) (2, 3) and (17, 11)
   (c) $(-7, -2)$ and (2, 38).

2. A straight line is plotted to the law $y = \dfrac{3x}{4} - 2$. Calculate the length of the line between the points representing values of x of 9 and 17.

3. A straight line is plotted to the law $4x - 3y + 5 = 0$. Find the length of the line between the points representing the values $x = 1$ and $x = 4$.

4. On a receiver chassis, two terminals occupy the positions indicated by the co-ordinates, in millimetres (80, 100) and (116, 148). An eyeletted connector piece connects the two terminals in a straight line. Calculate the distance between the centres of the two eyelets.

5. On an instrument panel, the centres of two rotating dials are indicated by the co-ordinates, in millimetres (47, 53) and (167, 117). The diameters of the two dials are 96 mm and 120 mm respectively. Calculate the smallest distance between the peripheries of the two dials.

6. An instrument dial has to have 50 equal divisions around its periphery. The peripheral distance between adjacent divisions is to be 2·2 mm. Determine the diameter of the dial.

7. A circular arc of radius 105 mm subtends an angle of 75°. Calculate the length of the arc.

8. Cable is wrapped around a drum at an effective diameter of 2·1 m. Through what angle (in degrees) does the drum rotate when a portion of cable of length 4·95 m is unwound from the drum?

9. A belt wraps round a motor pulley of diameter 210 mm with a contact angle of 150°.
   (a) Calculate the length of belt in contact with the pulley.
   (b) Calculate the belt velocity, in m/s, when the pulley revolves at 720 rev/min.

10. Calculate the length of the periphery of the shape shown in Fig. 4.24.

11. Cable is being unwound from a drum and rewound on to a spool. At a particular instant, cable is unwinding from the drum at an effective radius of 350 mm, the drum rotating at 60 rev/min. The

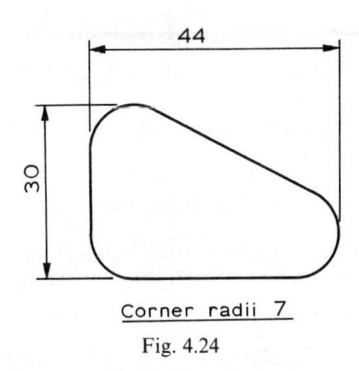

Corner radii 7

Fig. 4.24

cable is being wound on to the spool at an effective radius of 70 mm. Calculate:

(a) the linear velocity of the cable, in m/min;
(b) the rotational speed of the spool.

12. A thin-walled tube of diameter 30 mm is made from material of width 100 mm.
    (a) What is the amount of overlap?
    (b) If the strip were joined by butt welding, what would then be the diameter of the tube?

13. An oxy-cutting machine is cutting out a circular disc of diameter 2·4 m by means of a rotating arm. The flame travels around the

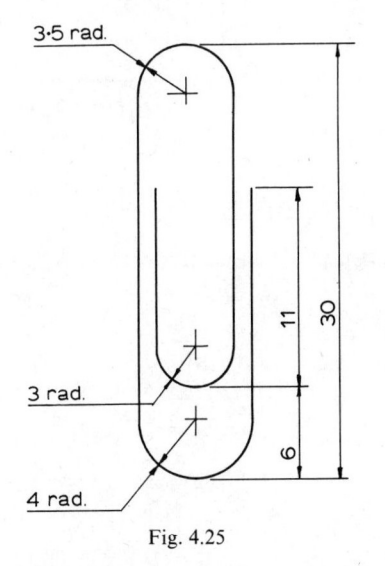

Fig. 4.25

circle at the rate of 250 mm per minute.

(a) Through what angle, to the nearest degree, does the arm turn in one minute?

(b) How long does the arm take to turn a full circle?

14. What should be the spindle speed of a lathe to provide a cutting speed of 66 m/min with a workpiece of diameter 280 mm?

15. Calculate the length of wire in the paper clip shown in Fig. 4.25. The dimensions given apply to the centre line and it may be assumed that the length of the wire is identical to the length of the centre line.

16. Figure 4.26 shows a gusset which is to be cut from mild steel plate by an oxy-acetylene flame which travels round the edge. If it takes 1 min to burn a starting hole through the plate, and then the flame moves round the form at the rate of 160 mm/min, how long altogether will it take to cut the gusset from the plate? The width of the cut may be neglected.

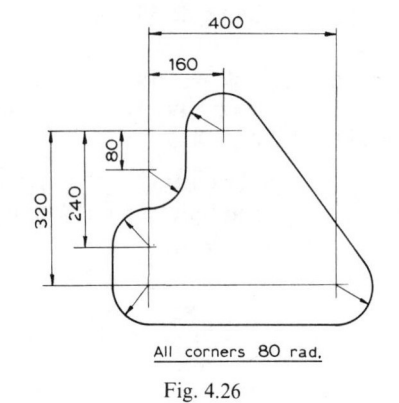

All corners 80 rad.

Fig. 4.26

Answers to Problems 4.4

1. (a) 13   (b) 17   (c) 41

2. 10

3. 5

4. 60 mm

5. 28 mm

6. 35 mm

7. 137 mm

8. 270°

9. (a) 275 mm   (b) 7·92 m/s

10. 124 mm

11. (a) 132 m/min   (b) 300 rev/min

12. (a) 5·75 mm   (b) 31·8 mm

13. (a) 12°   (b) 30·2 min

14. 75 rev/min

15. 94·0 mm

16. 11·7 min

4.5
The geometry of plane surfaces

4.5.1
The area of a triangle

The accepted mathematical convention for representing the area of a triangle is to use the symbol $\triangle$. The most commonly known formula for the area of a triangle is:

$$\triangle = \frac{\text{base} \times \text{perpendicular height}}{2}$$

This very simple formula includes a dimension of a triangle which is rarely specified, that of the perpendicular height. There are many other formulae we can use for the area of a triangle, each one being particularly suitable for a specific set of dimensions.

If the three sides of a triangle are a, b and c, then the perimeter is $a + b + c$. Let us represent the semi-perimeter by s, so that $s = \frac{a + b + c}{2}$. The reader can accept without proof that:

$$\triangle = \sqrt{\{s(s - a)(s - b)(s - c)\}}$$

This formula can be applied when the three sides are known.

Another formula which can be used to determine the area of a triangle expresses the area in terms of the product of two adjacent sides and a trigonometrical ratio of the angle included by those two sides. It is usually expressed in the form

$$\triangle = \tfrac{1}{2}ab \sin C.$$

The formula has been quoted here so that the three formulae for the area of a triangle, appropriate to Level 1 Mathematics, can be grouped together. A more detailed consideration of the last formula is given in Chapter 5, dealing with trigonometry.

Example:

A triangle has sides of length 5 mm, 12 mm and 13 mm.
(a) Prove that the triangle is right-angled.
(b) Determine its area:

$$\text{(i)} \quad \text{from} \triangle = \frac{bh}{2}$$

$$\text{(ii)} \quad \text{from} \triangle = \sqrt{\{s(s - a)(s - b)(s - c)\}}$$

(a) If the triangle is right-angled, then:

$$(\text{longest side})^2 = \text{sum of squares of other two sides}$$
$$13^2 = 12^2 + 5^2$$
$$169 = 144 + 25$$
$$169 = 169$$

The triangle is right-angled since the sizes satisfy the Theorem of Pythagoras.

(b) The sides containing the right-angle are perpendicular to each other, hence:

$$\text{(i)} \quad \triangle = \frac{bh}{2} = \frac{5 \times 12}{2} = \frac{60}{2} = 30 \text{ mm}$$

$$\text{(ii)} \quad a = 5, \quad b = 12, \quad c = 13 \quad s = \frac{a + b + c}{2} = \frac{30}{2} = 15$$

$$\triangle = \sqrt{\{s(s - a)(s - b)(s - c)\}}$$
$$= \sqrt{\{15(15 - 5)(15 - 12)(15 - 13)\}}$$
$$= \sqrt{\{15 \times 10 \times 3 \times 2\}} = \sqrt{900} = 30 \text{ mm}$$

Answer:

Area of triangle = 30 mm (by either method)

4.5.2
The area of a quadrilateral

Theoretically, a *quadrilateral* (from the Latin for four-sided) is a plane figure bounded by four straight lines. However, the term is more usually associated with a four-sided figure not included among those four-sided figures to which we give special names, and which will be referred to later in this article. It should be noted that the lengths of the four sides alone is insufficient information to specify the shape of a quadrilateral. Four rods pin-jointed at their ends can be formed into a variety of shapes. To specify the shape some means must be found to fix the position of two adjacent sides, such as the angle between them, or the distance between opposing corners of the quadrilateral.

A line joining opposite angles in a quadrilateral is called a *diagonal* of that quadrilateral. Every quadrilateral therefore has two diagonals. If they are unequal in length we can refer to them as the longer diagonal and the shorter diagonal. A diagonal divides a quadrilateral into two triangles, hence the sum of the interior angles of a quadrilateral totals twice 180°, which is 360°. Let us proceed to the names and properties of special quadrilaterals.

A *parallelogram* is a quadrilateral with two pairs of parallel sides. The parallel sides are opposite to each other and are of equal length. Opposite angles are equal to each other. Each of the diagonals divides a parallelogram into two congruent triangles. If both diagonals are drawn, the parallelogram is divided into two pairs of congruent triangles. The diagonals of a parallelogram bisect each other, that is, they cut each other in half.

If all the angles in a parallelogram are right angles, and one pair of parallel sides have different lengths to the other pair of parallel sides, the figure is more usually referred to as a *rectangle*. If all the angles in a parallelogram are right angles, and all four sides are equal in length, the figure is more usually referred to as a *square*. If a square, a rectangle, and a parallelogramic shape which is other than a square or rectangle are cut from paper and folded across a diagonal, it will be observed that while a square is symmetrical about a diagonal, the other two shapes are not.

A *rhombus* is a parallelogram having four equal sides, but no angle is a right angle. A rhombus is symmetrical about either diagonal. A diagonal divides a rhombus into two congruent isosceles triangles. Opposite angles are equal, and its diagonals bisect each other at right angles.

A *kite* is a quadrilateral which has two pairs of equal sides, but any two sides of equal length are adjacent to each other. A kite is symmetrical about one diagonal only. The diagonal of symmetry bisects the angles it joins, the other diagonal does not. The diagonals bisect each other at right angles.

A *trapezium* is a quadrilateral with one pair of parallel sides. If the non-parallel sides are equal in length, the figure is called an *isosceles trapezium*. If one of the sides is perpendicular to the parallel sides it is called a *right angled trapezium*. The words *trapezium* and *trapezoid* are often used interchangeably.

Page 290 gives diagrams illustrating the terms used in this article.

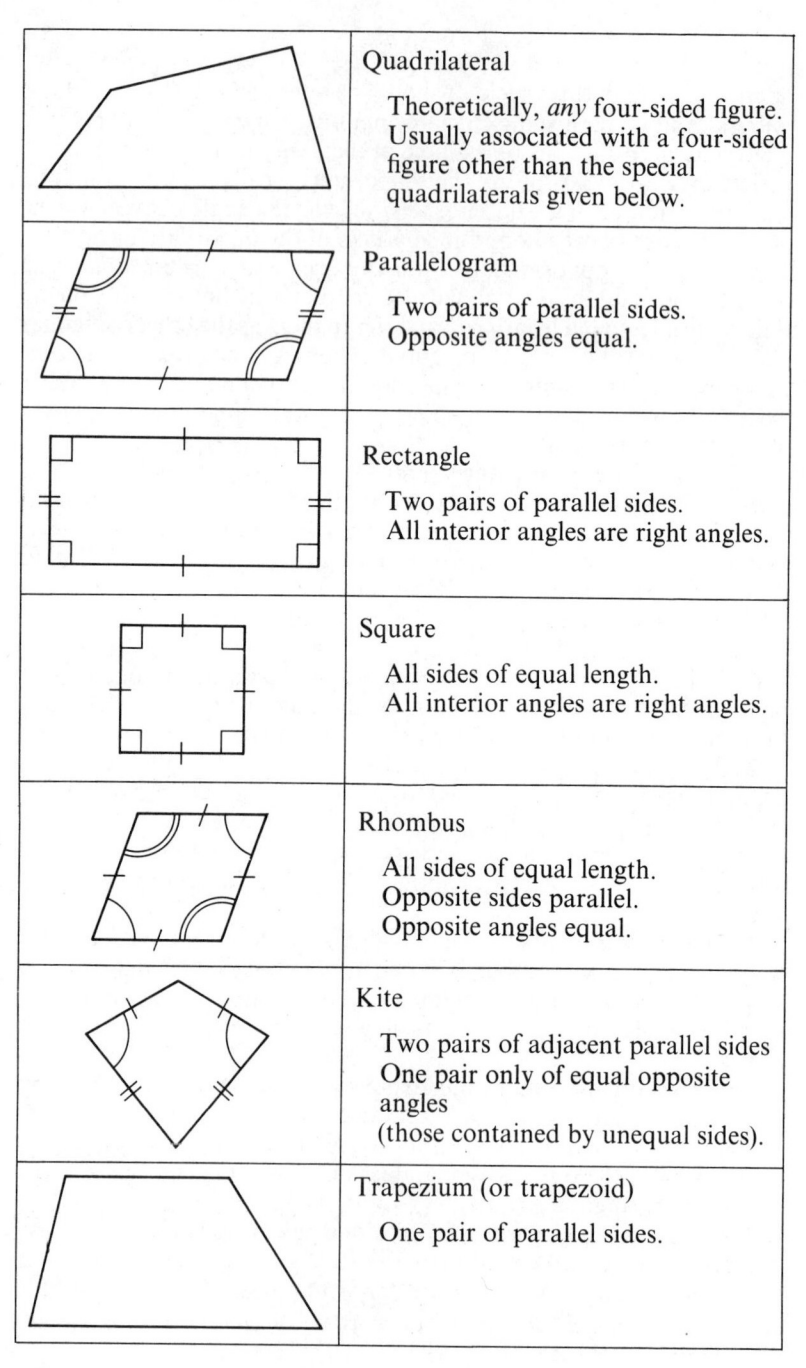

Quadrilateral

Theoretically, *any* four-sided figure. Usually associated with a four-sided figure other than the special quadrilaterals given below.

Parallelogram

Two pairs of parallel sides.
Opposite angles equal.

Rectangle

Two pairs of parallel sides.
All interior angles are right angles.

Square

All sides of equal length.
All interior angles are right angles.

Rhombus

All sides of equal length.
Opposite sides parallel.
Opposite angles equal.

Kite

Two pairs of adjacent parallel sides
One pair only of equal opposite angles
(those contained by unequal sides).

Trapezium (or trapezoid)

One pair of parallel sides.

The area of a rectangle is the product of the length and the width. If we represent the length by L and the width by W, the area of a rectangle is LW. If the rectangle has a width equal to the length, the figure is more usually referred to as a square. The dimension which usually specifies a square is the length of side L, and the area of a square is $L \times L = L^2$.

The area of any polygon can be found by dividing that polygon into triangles and summating the area of individual triangles. Let us apply this principle to special quadrilaterals.

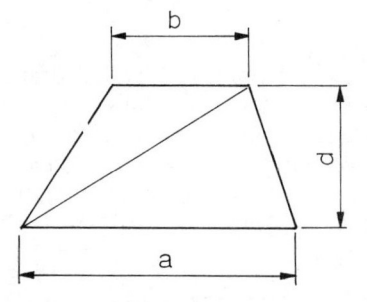

Fig. 4.27

Figure 4.27 shows a trapezium, the parallel sides being of lengths a and b and their distance apart being d. Two triangles have been formed by drawing a line between opposite corners.

$$\text{Area of a triangle} = \frac{\text{base} \times \text{perpendicular height}}{2}$$

Area of a trapezium = sum of areas of two triangles

$$= \frac{ad}{2} + \frac{bd}{2} = \frac{(a + b)}{2}d$$

i.e. mean length of parallel sides multiplied by the distance between them.

If we extend this reasoning to a parallelogram, since opposing sides are of equal length, the area of a parallelogram is the product of the length of a side and the distance between that side and its parallel side. In Fig. 4.28, the area of the parallelogram is either ad or bh.

It should be noted that since:

$$ad = bh$$

then

$$\frac{a}{b} = \frac{h}{d}$$

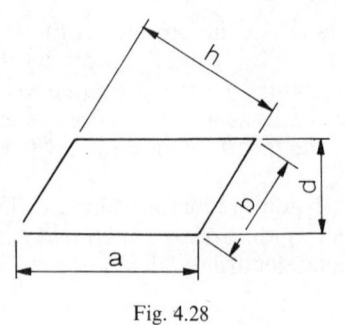

Fig. 4.28

Figure 4.29(a) shows a rhombus with a line drawn between opposite corners, this line being termed a *diagonal*.

Since the sides of a rhombus are of equal length, the diagonal divides the rhombus into two congruent isosceles triangles. In Fig. 4.29(b) the second diagonal has been added, again dividing the triangle into congruent isosceles triangles. These diagonals bisect the angles they join.

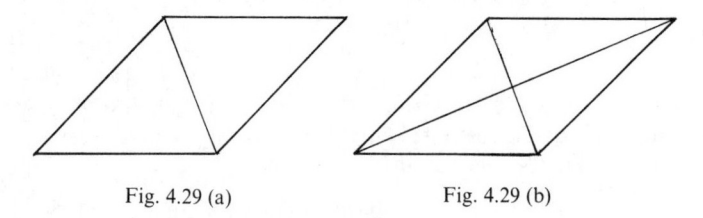

Fig. 4.29 (a)   Fig. 4.29 (b)

Since the bisector of the vertex of an isosceles triangle is at right-angles to the base, the diagonals of a rhombus intersect at right angles, and bisect each other. The diagonals of a rhombus produce four congruent right-angled triangles. The area of a rhombus is the sum of the areas of these four triangles; hence the area of a rhombus can also be obtained from half the product of the diagonals.

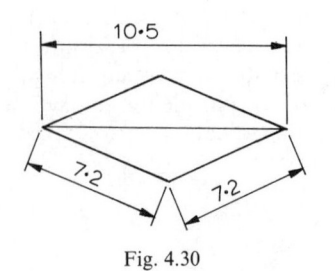

Fig. 4.30

Example:

Calculate the area of a rhombus with side 7·2 mm and the longer diagonal 10·5 mm.

First method:

The area consists of the sum of the areas of two triangles having sides 7·2, 7·2 and 10·5 mm long (Fig. 4.30).

$$a = 7·2 \qquad b = 7·2 \qquad c = 10·5$$

$$s = \frac{a + b + c}{2} = \frac{7·2 + 7·2 + 10·5}{2} = \frac{24·9}{2} = 12·45$$

$$s - a = 12·45 - 7·2 = 5·25$$

$$s - b = 12·45 - 7·2 = 5·25$$

$$s - c = 12·45 - 10·5 = 1·95$$

$$\text{Area} = \sqrt{\{s(s - a)(s - b)(s - c)\}} = \sqrt{\{(12·45)(5·25)(5·25)(1·95)\}}$$
$$= 25·87$$

$$\text{Area of rhombus} = 2 \times 25·87$$
$$= 51·74 = 51·7 \text{ mm}^2$$
$$\text{(to 3 sig. figs)}$$

Rough check:

$$12 \times 5 \times 5 \times 2 = 600$$
$$\sqrt{600} = \text{about } 25$$

No.		Log
12·45		1·095 2
5·25		0·720 2
5·25		0·720 2
1·95		0·290 0
Root	2	2·825 6
25·97		1·412 8

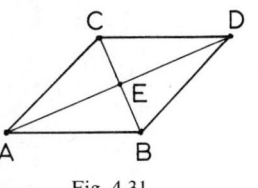

Fig. 4.31

Second method:

Diagonals AD and CB bisect each other at right angles, hence ABE is a right-angled triangle (Fig. 4.31).

$$AB = 7\cdot2 \qquad AE = \frac{10\cdot5}{2} = 5\cdot25$$

$$(AB)^2 = (EB)^2 + (AE)^2$$
$$7\cdot2^2 = (EB)^2 + 5\cdot25^2$$
$$51\cdot84 = (EB)^2 + 27\cdot56$$
$$(EB)^2 = 51\cdot84 - 27\cdot56 = 24\cdot28$$
$$EB = \sqrt{24\cdot28} = 4\cdot927$$
$$\text{Diagonal } BC = 2(EB) = 2 \times 4\cdot927 = 9\cdot854$$

$$\text{Area of rhombus} = \frac{\text{product of diagonals}}{2}$$

$$= \frac{10\cdot5 \times 9\cdot854}{2} = 51\cdot733\ 5$$

Answer:

$$\text{Area of rhombus} = 51\cdot7 \text{ mm}^2$$

4.5.3
The area of a circle, of an annulus and of an ellipse

For the time being, the reader must accept without proof that the area A of a circle of radius r is given by the formula $A = \pi r^2$, or in terms of the diameter, $A = \frac{\pi d^2}{4}$. The reader is reminded that Castle's *Four-figure Mathematical Tables* can be used to find areas of circles and used in reverse to find the diameter if the area is known.

If C is the circumference of a circle:

$$C = \pi d, \quad \text{hence} \quad d = \frac{C}{\pi} \quad \text{and} \quad \pi = \frac{C}{d}$$

Hence, if
$$A = \frac{\pi d^2}{4}$$

substituting for d gives
$$A = \frac{\pi}{4}\left(\frac{C}{\pi}\right)^2 = \frac{C^2}{4\pi}$$

and
$$A = \frac{d^2}{4} \times \frac{C}{d} = \frac{Cd}{4}$$

Here we have another example of different formulae for the area of a plane figure. The formula we use for a particular circumstance depends upon the dimensions we are given. As is our custom, we will summarise the results.

If d is the diameter of a circle, and C the circumference, then the area A is given by the formulae:

$$A = \frac{\pi d^2}{4}, \quad \text{or} \quad A = \frac{C^2}{4\pi}, \quad \text{or} \quad A = \frac{Cd}{4}$$

An *annulus* is the area which lies between two concentric circles. Once more we remind the reader that areas of circles can be found from tables. On the occasions that a formula has to be used, if D is the diameter of the larger circle and d is the diameter of the smaller circle, the area A of the annulus is given by the difference in the areas of the two circles. Hence:

$$A = \frac{\pi D^2}{4} - \frac{\pi d^2}{4}$$

$$= \frac{\pi}{4}(D^2 - d^2) = \frac{\pi}{4}(D + d)(D - d)$$

An *ellipse* is usually specified by quoting the length of the major axis a and the length of the minor axis b. The area A is then given by the formula:

$$A = \frac{\pi ab}{4}$$

Example:

Discs of steel of diameter 9 mm are stamped out from strip 38·5 mm wide so that the smallest distance between each hole or a hole and the strip edge is 0·5 mm. What percentage of the strip is waste if:

(a)   the discs are punched out in rows of four;
(b)   the discs are punched out in alternate rows of four and three?

Figure 4.32 shows the different patterns.
In the first case, a progression of 9·5 produces 4 discs.

Area fed for 4 discs = 38·5 × 9·5 = 365·75
Area of a circle 9 mm in diameter, from tables = 63·62

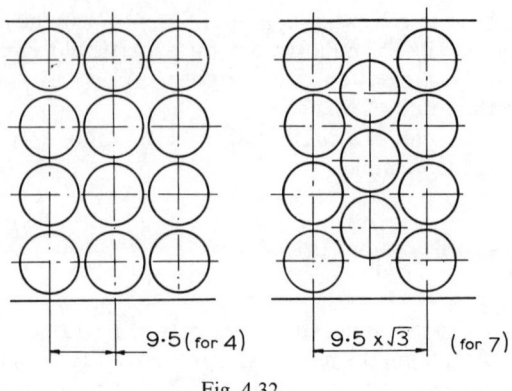

9·5 (for 4)   9·5 x √3   (for 7)

Fig. 4.32

Area of 4 discs $= 4(63·62) = 254·48$
Waste $= 365·75 - 254·48 = 111·27$

Percentage waste $= \dfrac{\text{amount of waste}}{\text{amount fed}} (100\%)$

No.	Log
11 127	4·046 4
365·75	2·563 2
30·42	1·483 2

$= \dfrac{11\ 127}{365·75} = 30·42\%$

Rough check:

$\dfrac{110}{4}\% = 27\tfrac{1}{2}\%$

In the second case, lines joining the centres of discs form equilateral triangles; hence, applying the Theorem of Pythagoras a progression of $2\left(9·5 \times \dfrac{\sqrt{3}}{2}\right) = 9·5 \times \sqrt{3}$ produces 7 discs.

Log $\sqrt{3} = 0·4771 \div 2 = 0·238\ 6$

No.	Log
38·5	1·585 5
9·5	0·977 7
√3	0·238 6
633·6	2·801 8

Area fed for 7 discs
$= 38·5 \times 9·5 \times \sqrt{3}$
$= 633·6$

Rough check:

$40 \times 10 \times 1\tfrac{3}{4} = 700$

$$\text{Area of 7 discs} = 7(63.62) = 445.34$$
$$\text{Waste} = (633.6 - 445.34) = 188.26$$

Percentage waste

No.	Log
18 826	4·274 8
633·6	2·801 8
29·72	1·473 0

$$= \frac{\text{amount of waste}}{\text{amount fed}}(100\%)$$

$$= \frac{18\ 826}{633.6}$$

$$= 29.72\%$$

Rough check:

$$18\ 000 \div 600 = 30$$

Answer:

    (a)   Waste, 4 per row $= 30.4\%$
    (b)   Waste, alternate rows $= 29.7\%$

4.5.4
The areas of sectors and of segments of a circle

    A *sector* is an area bounded by a circular arc and two radii. Inserting two radii into a circle divides that circle into two sectors. If the radii are not in line the sectors are unequal in area, and we can refer to them as the major sector (the larger in area) and the minor sector (the smaller in area). Unless we are specifically instructed to the contrary we shall consider a sector to be the minor sector, i.e. the smallest area bounded by a circular arc and two radii.

    A sector can be completely defined by the radius of the circular arc r and the included angle θ between the two radii, as shown in Fig. 4.33. The area of the complete circle of which the sector forms part is given by πr^2. We can imagine the circle to be constructed of 360 identical narrow sectors each having an included angle of one degree. The area of each of these narrow sectors is given by $\dfrac{\pi r^2}{360}$. If the included angle of a particular sector is $\theta°$, there are θ of these narrow sectors and hence the area of a sector A is given by the formula:

$$A = \frac{\theta}{360}(\pi r^2), \quad \theta \text{ being in degrees}$$

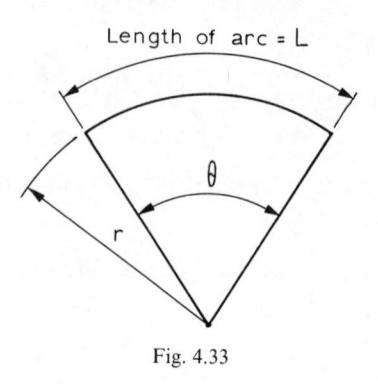

Fig. 4.33

To convert an angle in radians to an angle in degrees, we multiply by $\dfrac{360}{2\pi}$. Hence if the included angle θ is expressed in radians:

$$A = \frac{\theta}{360} \times \frac{360}{2\pi}(\pi r^2)$$

$$= \frac{r^2\theta}{2}, \quad \theta \text{ now being in radians}$$

If θ is expressed in radians, the length of the circular arc is $r\theta$. Let us represent this by L:

$$A = \frac{r^2\theta}{2} = \frac{r \times r\theta}{2} = \frac{rL}{2}$$

(note the similarity with the area of a triangle)

In a similar manner:

if $L = r\theta,$

then $r = \dfrac{L}{\theta}$

$$A = \frac{rL}{2} = \frac{L}{\theta} \times \frac{L}{2} = \frac{L^2}{2\theta}$$

We therefore have four different formulae we can use for the area of a sector; the formula we use for a particular sector depends upon which of the values r, L or θ are given. They are:

$$A = \frac{\theta}{360}(\pi r^2), \quad \theta \text{ being expressed in degrees}$$

$$A = \frac{r^2\theta}{2}, \quad \theta \text{ being expressed in radians}$$

$$A = \frac{rL}{2}, \quad L \text{ being the length of the arc}$$

$$A = \frac{L^2}{2\theta}, \quad \theta \text{ being expressed in radians}$$

Example:

Taking $\pi = \frac{22}{7}$ where necessary, calculate the areas of the following sectors:

(a) radius 7 mm, included angle 60°;
(b) length of arc 8 mm, radius = 6 mm;
(c) length of arc 10 mm, included angle 45°.

(a) $A = \dfrac{\theta}{360}(\pi r^2) = \dfrac{60}{360} \times \dfrac{22}{7} \times \dfrac{7}{1} \times \dfrac{7}{1}$

$\qquad = \dfrac{77}{3} = 25 \cdot 7 \text{ mm}^2$

(b) $A = \dfrac{rL}{2} = \dfrac{6 \times 8}{2} = 24 \text{ mm}^2$

(c) $A = \dfrac{L^2}{2\theta},$ and $45° = \dfrac{\pi}{4}$ radians

$\qquad A = \dfrac{10 \times 10 \times 4}{\pi \times 2} = \dfrac{200 \times 7}{22} = \dfrac{700}{11} = 63 \cdot 6 \text{ mm}^2$

Answers:

(a) $25 \cdot 7 \text{ mm}^2$ (b) 24 mm^2 (c) $63 \cdot 6 \text{ mm}^2$

Example:

An annulus is formed by two concentric circles of radii 6 mm and 5 mm. Calculate the area of that part of the annulus which lies between two radii inclined at 54° to one another.

From tables: area of a circle of diameter 12 mm $= 113 \cdot 1$ mm^2

                  area of a circle of diameter 10 mm $= \ \ 78 \cdot 54$ mm^2

By subtraction: area of complete annulus $= \ \ 34 \cdot 56$ mm^2

$$\text{Area of annular sector} = \frac{54}{360} \times 34 \cdot 56 = \frac{3 \times 34 \cdot 56}{20}$$

$$= \frac{103 \cdot 68}{20} = 5 \cdot 184 \text{ mm}^2$$

Answer:

$$\text{Area of annular sector} = 5 \cdot 18 \text{ mm}^2$$

A *segment* is the area contained by an arc of a circle and a chord. A chord divides a circle into two areas. Unless that chord is a diameter, one area is greater than the other. We can differentiate between the two segments by nominating the major segment to be the segment of greater area. The smaller of the two segments can be referred to as the minor segment. Unless we are specifically informed to the contrary, we shall consider the area of a segment contained by a chord and a circular arc to be that of the minor segment.

Dimensions associated with the mensuration of a minor segment are shown in Fig. 4.34. Any two of the dimensions L, r and W will enable the angle θ to be determined. The area of a minor segment can be obtained by considering it to be the area of a sector minus the area of an isosceles triangle. The area of a major segment can be considered to be the area of a complete circle minus the area of the associated minor segment.

The studies we have undertaken thus far in this book do not allow us to give a complete consideration of the area of a segment, particularly the relationship between W, r and θ. Some assistance can be offered with columns in the table on page 35 of Castle's *Four-figure Mathematical Tables*, which show the ratio of the length of a chord to a radius for

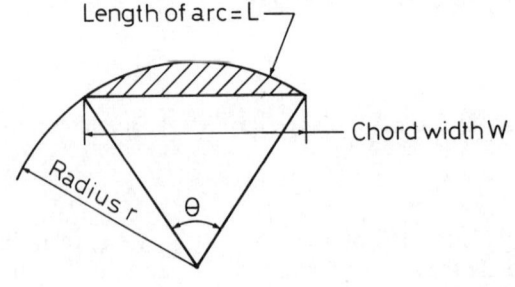

Fig. 4.34

various angles subtended at the centre of the implied circle. Unfortunately these are only given for intervals of one degree and the numerical values only to three decimal places.

A better appreciation can be obtained of the area of a segment when a study has been made of basic trigonometry. Further study of the area of a segment is best deferred at this time, and it will be resumed in article 5.2.3 in the next chapter.

4.5.5
The area of similar plane figures

If a rectangle is similar in shape to another rectangle, so that the length of a second rectangle is n times the length of the first, and the breadth of the second rectangle is n times the breadth of the first then:

$$\text{Area of first rectangle} = LB$$

and

$$\text{area of second rectangle} = nL \times nB = n^2LB$$

in which case

$$\frac{\text{area of second rectangle}}{\text{area of first rectangle}} = \frac{n^2LB}{LB} = n^2$$

If the diameter of a second circle is n times the area of first circle, then:

$$\frac{\text{Area of second circle}}{\text{Area of first circle}} = \frac{\pi(nd)^2}{4} \div \frac{4}{\pi d^2} = \frac{4\pi n^2 d^2}{4\pi d^2} = n^2$$

We could continue in the same manner for all plane figures. We should find that if a second figure has all linear dimensions n times the corresponding linear dimensions of a first figure, then:

$$\frac{\text{Area of second figure}}{\text{Area of first figure}} = n^2$$

Example:

Find the area of a circle of diameter 0·82 mm.
From tables, area of circle of diameter 8·2 mm is 52·81 mm².

$$\frac{\text{Area of circle of diameter 0·82 mm}}{\text{Area of circle of diameter 8·2 mm}} = \left(\frac{0·82}{8·2}\right)^2 = \left(\frac{1}{10}\right)^2 = \frac{1}{100}$$

$\therefore$ area of circle of diameter 0·82 mm $= \dfrac{1}{100} \times 52·81 = 0·528\,1$

Answer:

$$\text{Area} = 0·528 \text{ mm}^2$$

4.5.6
The area of irregular plane figures

If a plane figure is bounded by straight lines, or combinations of straight lines and geometrical curves, although the calculations may be very lengthy, the area of that plane figure can be determined very accurately. There are occasions when a precise determination is unnecessary, and rapidity of obtaining an approximate value of reasonable accuracy is of more importance. On other occasions, the perimeter of the figure may include portions of randomly curved lines which follow no regular geometrical pattern.

In these two latter cases the area of the irregularly shaped figure is determined approximately. There are a great many methods of determining approximately the area of an irregular figure. The simpler methods tend to be less accurate and, as greater accuracy is desired, the method adopted tends to become more complex. In our present studies we have to learn two methods, both being based on the area of a trapezium.

The irregular plane figure is divided into strips by parallel lines, called ordinates. It is conventional practice, to ease a subsequent calculation, for the strips to be of equal width. Figure 4.35 shows an irregular plane figure divided into strips of equal width by ordinates. The area of the figure is the sum of the areas of the individual strips.

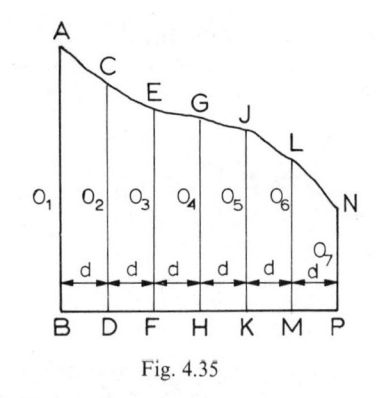

Fig. 4.35

Consider the particular strip $ABCD$. Its area is approximately equal to a trapezium of sides O_1 and O_2, whose distance apart is d.

$$\text{Area of strip } ABCD \approx \left(\frac{O_1 + O_2}{2}\right) d$$

The sign $\approx$ is read as 'is approximately equal to'.

The strip $CDEF$ has an area approximately equal to a trapezium of sides O_2 and O_3 whose distance apart is d.

$$\text{Area of strip } CDEF \approx \left(\frac{O_2 + O_3}{2}\right)d$$

We proceed across the figure in the same manner, the area of next strip being approximately equal to $\left(\dfrac{O_3 + O_4}{2}\right)d$.

The area of the figure is the sum of the areas of individual strips.

$$\text{Total area} \approx \left(\frac{O_1 + O_2}{2}\right)d + \left(\frac{O_2 + O_3}{2}\right)d + \left(\frac{O_3 + O_4}{2}\right)d \quad \text{etc.}$$

$$\approx d\left(\frac{O_1 + O_2 + O_2 + O_3 + O_3 + O_4 + O_4 \ldots}{2}\right)$$

$$\approx \frac{d}{2}(O_1 + 2O_2 + 2O_3 + 2O_4 \ldots O_L)$$

where O_L is the length of the last ordinate

$$\approx d\left(\frac{O_1 + O_L}{2} + O_2 + O_3 + O_4 \ldots\right)$$

Hence the area is given approximately by the width of a strip multiplied by sum of the mean of the first and last ordinates and the remaining ordinates. It is usually stated as:

$$\text{Area} \approx \text{width of strip} \left\{\frac{\text{first} + \text{last}}{2} + (\text{sum of other ordinates})\right\}$$

Example:

Sketch the graph of $y = 9 - x^2$ between $x = -3$ and $x = +3$. Shade the area which is bounded by the curve and the x-axis. Determine this area by the trapezoidal rule:

(a)　with six vertical strips of equal width;
(b)　with twelve vertical strips of equal width.
(c)　The area is precisely 36 units of area. Hence show that the trapezoidal rule tends to be more accurate if a greater number of strips is used.

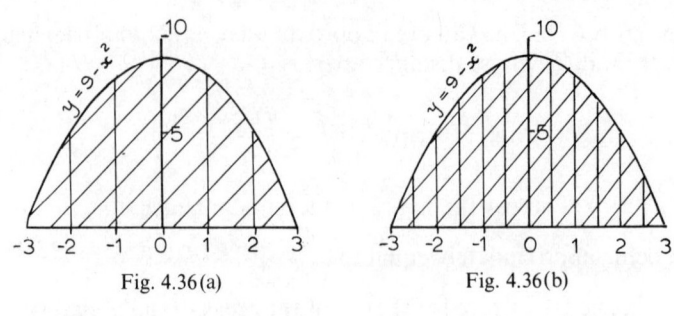

Fig. 4.36(a)   Fig. 4.36(b)

The area is in Figs. 4.36(a) and 4.36(b). Figure 4.36(a) shows six vertical strips and Fig. 4.36(b) twelve vertical strips.

(a) *Six vertical strips:*
Ordinates are 0, 5, 8, 9, 8, 5 and 0.

$$\text{Area} = \text{width of strip} \left\{ \frac{\text{first} + \text{last}}{2} + (\text{sum of others}) \right\}$$

$$= 1 \left\{ \frac{0 + 0}{2} + (5 + 8 + 9 + 8 + 5) \right\} = 1\{35\}$$

$$= 35 \text{ units of area}$$

(b) *Twelve vertical strips:*
Ordinates are 0, 2·75, 5, 6·75, 8, 8·75, 9, 8·75, 8, 6·75, 5, 2·75, 0.

$$\text{Area} = \text{width of strip} \left\{ \frac{\text{first} + \text{last}}{2} + (\text{sum of others}) \right\}$$

$$= \frac{1}{2} \left\{ \frac{0 + 0}{2} + 71 \cdot 5 \right\}$$

$$= \frac{71 \cdot 5}{2}$$

$$= 35 \cdot 75 \text{ units of area}$$

(c)   Error of six strips = 1 unit of area
            Error of twelve strips = 0·25 unit of area
Use of twelve strips produces a more accurate result.

Answer:

(a) 35 units of area
(b) 35·75 units of area
(c) Using a greater number of strips gives smaller error, and hence is more accurate.

The trapezoidal rule is the easiest method to apply of all the methods for determining approximately the area of an irregular figure. It is reasonably accurate provided at least one end of the figure is a straight line. If the ends of the figure are curved, then both the first and last ordinates are zero. This means that the first and last strips are approximated to triangles.

In most cases where the ends of the figure are curved, a closer approximation is obtained if the areas of strips are approximated to trapezia whose mean length of side is the length of another ordinate drawn midway between the existing ordinates.

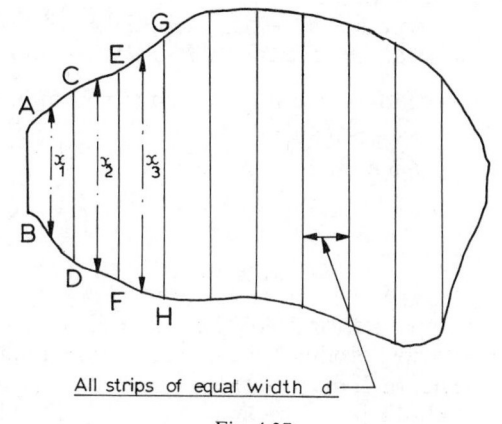

All strips of equal width   d

Fig. 4.37

Referring to Fig. 4.37, we see that chain-dotted lines have been erected to indicate lengths of lines positioned midway between the ordinates.

$$\text{Area of strip } ABCD \approx x_1 \times d$$

$$\text{that of } CDEF \approx x_2 \times d$$

$$\text{that of } EFGH \approx x_3 \times d$$

and so on.

$$\text{Total area} \approx \text{the sum of the areas of separate strips}$$

$$\approx x_1 d + x_2 d + x_3 d + x_4 d, \text{ etc.}$$

$$\approx d \text{ (the sum of the mid-ordinates)}$$

This formula is generally known as the *mid-ordinate rule*. We divide the irregular figure into strips of equal width by erecting ordinates. We measure the width of the figure in the centres of strips, and call these the mid-ordinates. The area of the figure is given by the product of the sum of the mid-ordinates and the width of one strip.

The reader should appreciate that the accuracy of either the trapezoidal rule or the mid-ordinate rule depends on the number of strips. As the strips become narrower, so the profiles of their ends become nearer to straight lines. The greater the number of strips, the greater the accuracy.

Example:

Use the mid-ordinate rule with six vertical strips to find the shaded area in the previous example.
Mid-ordinates occur at $x = -2.5, -1.5, -0.5, 0.5, 1.5$ and 2.5.
Lengths of mid-ordinates are $2.75, 6.75, 8.75, 8.75, 6.75$ and 2.75.

$$\text{Area} = \text{width of one strip} \times \text{sum of mid-ordinates}$$
$$= 1 \times 36.5$$
$$= 36.5 \text{ units of area}$$

Answer:

36.5 units of area

Whether the trapezoidal rule or the mid-ordinate rule (using the same number of strips) produces the more accurate result depends on the profile of the figure. If the profile shows no violent changes and has curves at each end, the use of the mid-ordinate rule tends to provide the more accurate answer. It should be noted, however, that on occasions the use of the trapezoidal rule with a large number of strips can produce a more accurate result than that provided by the mid-ordinate rule with a small number of strips.

Problems 4.5

1. Determine the area of an equilateral triangle with sides of length 40 mm.

2. An isosceles triangle has sides of length 40 mm, 40 mm and 20 mm.
   (a) Determine the area of the triangle.
   (b) Determine the perpendicular height of the triangle relative to the side of length 20 mm.

3. Calculate the area of a triangle with sides of length 8 mm, 11 mm and 15 mm.

4. Calculate the area of a triangle having sides of length 7.8 mm, 15.3 mm and 16.5 mm.

5. Calculate the area of a triangle with sides 3·12, 4·78 and 5·26 mm long, and hence determine the three different perpendicular heights.

6. Two circles, of diameter 78 mm and 50 mm respectively, intersect at *A* and *B*. If their common chord *AB* has a length of 30 mm, *C* being the centre of the larger circle, calculate:
   (a) the distance between their centres, *C* and *D*;
   (b) the area of triangle *CAD*.

7. Calculate the length of the side of a square whose area is equal to that of an equilateral triangle having sides of length 36 mm.

8. A rectangular plot of ground has an area of 27·2 hectares. The length is three times its width. Determine its dimensions. (1 hectare $= 10^4$ m^2).

9. Determine the area of a regular hexagon whose side has a length of 40 mm.

10. A parallelogram has two adjacent sides of length 5 mm and 8 mm, the angle between these sides being 60°. Calculate:
    (a) the area;
    (b) the distances between parallel sides;
    (c) the lengths of the diagonals.

11. Calculate the area of the quadrilateral shown in Fig. 4.38.

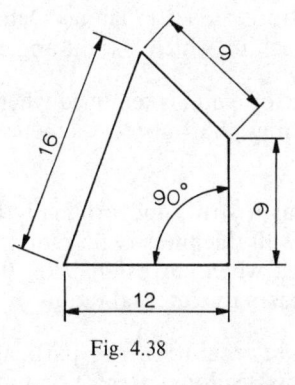

Fig. 4.38

12. Figure 4.39 is taken from a surveyor's notebook, and gives measurements taken regarding a plot of ground of the shape shown. Determine the area in ares (1 are $= 10$ m^2).

13. An irrigation channel is trapezoidal in shape. The horizontal base has a width of 1·4 m. The maximum width at the top is 2 m and the maximum depth is 600 mm. Calculate the percentage decrease in

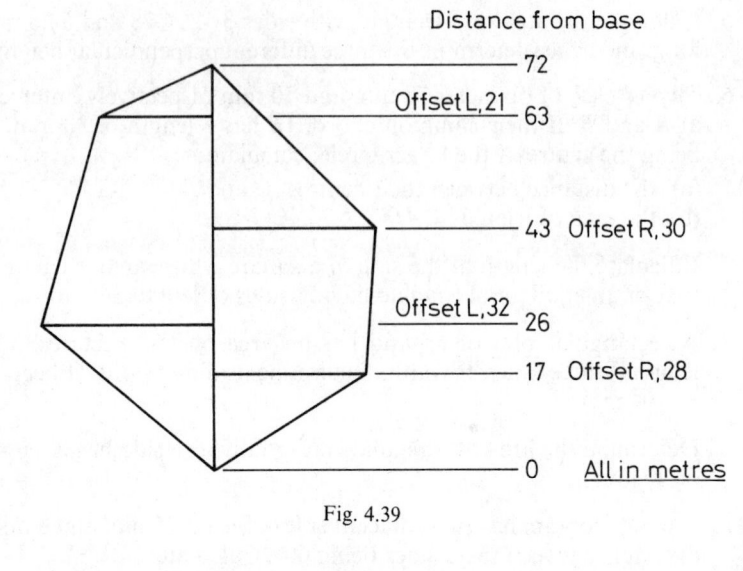

Fig. 4.39

the cross-sectional area of water flow when the depth of water decreases from 400 mm to 100 mm.

14. The parallel sides of a trapezium are in the proportion of 7 to 10. The width between the parallel sides is one-half of the longest side. The area of the trapezium is 170 mm². Determine the lengths of the sizes of the trapezium, which is right-angled.

15. What percentage of metal is removed when a cut of depth 1·5 mm is taken in turning the outside diameter of a bar of diameter 20 mm?

16. A pipe is used as a strut, and originally the outside diameter is 30 mm and the wall thickness is 2·5 mm. Calculate the percentage reduction in area when corrosion takes 0·1 mm off the external radius and increases the internal radius by the same amount.

17. It is known that in order to carry a particular load, a tubular strut must have a cross-sectional area of 550 mm². If the outside and inside diameters are to be in the proportion of 4:3, determine the outside diameter. (Take $\pi = \frac{22}{7}$.)

18. A four-way pipe junction has three inlet pipes of diameters 20 mm, 30 mm and 60 mm. Determine the diameter of the outlet pipe if its cross-sectional area is equal to the total cross-sectional area of the inlet pipes. (Take $\pi = \frac{22}{7}$ if required.)

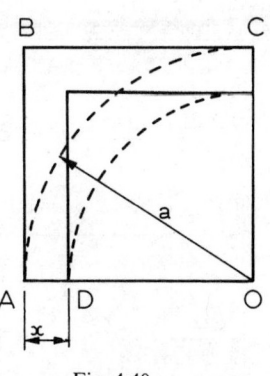

Fig. 4.40

19. An annulus has an area of 2 200 mm², the larger radius being 10 mm greater than the smaller radius. Take π as $\frac{22}{7}$ and calculate the radii.

20. Figure 4.40 shows a right-angled bend in a road with $AB = BC = a$ m (the radius of a circle with centre O). The road is to be modified as shown by the dotted lines to remove the sharp corner. If the width of the road is x m, show that the saving in road surface area of the modification is given by:

$$\left(1 - \frac{\pi}{4}\right)(2ax - x^2) \text{ m}^2$$

21. (a) The arc of a sector of a circle of radius 5 mm subtends an angle of $\frac{2\pi}{3}$ radians at the centre. Calculate the sector area.

    (b) Find the included angle of the sector of a circle whose radius is 8 mm and whose area is 50 mm².

22. A blade of an electric fan is a sector of a circle of radius 100 mm and has an area of 3 600 mm².

    (a) Calculate the angle of the sector in degrees and minutes.
    (b) Determine the speed of the fan in m/s of a point on the arc of the blade of the fan when the fan is rotating at 200 rev/min.

23. A sector of a circle radius r contains an angle θ radians at the centre of the circle. If the total perimeter of the sector is 12 mm, find a formula connecting r and θ and show that the area A of the sector is given by $A = 6r - r^2$. Find with the aid of a graph the maximum value of A and the values of r and θ corresponding to this maximum.

24. Calculate the area of the template shown in Fig. 4.41.

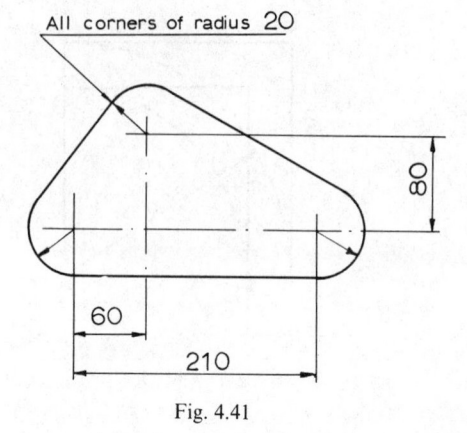

Fig. 4.41

25. On squared paper, draw accurately a semicircle on a diameter of 60 mm. Using twelve strips each of width 5 mm, so as to produce thirteen ordinates, find the length of each of the ordinates to the nearest 0·1 mm and find the area of the semicircle using the trapezoidal rule.

26. Use the information from the previous question to find the area by the mid-ordinate rule, using nominally six strips each of width 10 mm.

27. Sketch the area contained by the curve $y = x^2$, the x-axis and the line $x = 10$. Find the area using a mid-ordinate rule and five strips of width 2 units.

28. Draw an arc of a circle of radius 50 mm. Set your compasses to 60 mm, and with the point on the circle mark off and draw a chord of length 60 mm. You now have an area bounded by a circular arc of radius 50 mm and a chord of length 60 mm. Treat the chord as a base and erect ordinates to divide the figure into ten strips of width 6 mm. Obtain the area of the figure by means of the mid-ordinate rule.

29. Plot the curve, in the first quadrant, $xy = 100$. Shade the area contained by:
      (i) the x-axis   (ii) the y-axis   (iii) the line $y = 20$
      (iv) the curve $xy = 100$ (v) the line $x = 25$
  (a) Using vertical strips each of width 5 units, find the area of the shaded figure by the trapezoidal rule.
  (b) Repeat the exercise with horizontal strips of height 4 units.

30. The following table gives the depth below ground level of a cable. Plot a graph of *d* against *l*.

Horizontal distance *d* m	0	50	100	150	200	250	300	350	400
Depth of cable *l* m	1·8	2·5	3·0	3·3	3·5	3·8	3·5	2·7	1·7

Use the mid-ordinate rule to find the area under the horizontal base line and hence calculate the volume of earth, in cubic metres, to be removed in order to uncover the whole length of the cable if the trench is 1·5 m wide. Give the answer to the nearest whole number.

31. Use a scale drawing and the mid-ordinate rule with 5 strips of equal width to find the area of the template shown in Fig. 4.42.

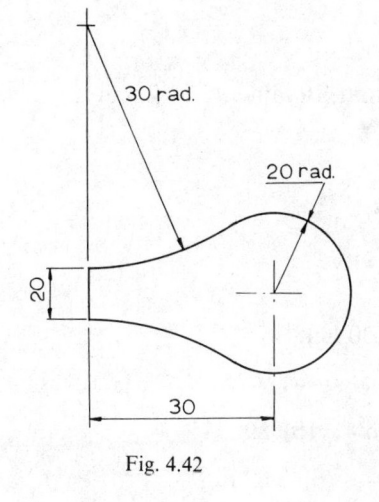

30 rad.

20 rad.

20

30

Fig. 4.42

32. Sketch the curve $y = 16 - x^2$. Shade the area which lies between the x-axis and the curve. Divide the area into 8 strips of equal width, calculate the mid-ordinates, and hence determine the shaded area with the aid of the mid-ordinate rule. If it is known that the area is $85\frac{1}{3}$ units, what is the percentage error, to 2 significant figures, caused by the adoption of the mid-ordinate rule?

Answers to Problems 4.5

1. 693 mm^2

2. (a) 387 mm^2 (b) 38·7 mm

3. 42·8 mm^2

4. 59·4 mm^2

5. 7·35 mm^2; 4·71 to 3·12 side, 3·08 to 4·78 side, 2·80 to 5·26 side

6. (a) 56 mm   (b) 420 mm^2

7. 23·7 mm

8. 903 m × 301 m

9. 4 160 mm^2

10. (a) 34·6 mm^2 (b) 4·33 mm and 6.93 mm
    (c) 11·4 mm and 7.00 mm

11. 120 mm^2

12. 29·2 ares

13. 77·3%

14. 14 mm, 20 mm, 10 mm and 11·7 mm

15. 27·8%

16. 8%

17. 40 mm

18. 70 mm

19. 40 mm and 30 mm

20. No numerical answer

21. (a) 26·2 mm^2 (b) 89° 31′

22. (a) 41° 15′   (b) 2·09 m/s

23. $A = 9$ mm^2, $r = 3$ mm, $\theta = 2$ rad $= 114°$ 36′

24. 19 300 mm^2 = 193 cm^2

25. Ordinates are 0, 16·6, 22·4, 26·0, 28·3, 29·6, 30·0, 29·6, 28·3, 26·0, 22·4, 16·6 and 0. These give an area of 1 380 mm^2.
    (The actual area is 1 414, and the error when the extremes of the figure are points producing zero ordinates should be noticed.)

26. Mid-ordinates are 16·6, 26·0, 29·6, 29·6, 26·0 and 16·6. These give an area of 1 440 mm².

27. Mid-ordinates are 1, 9, 25, 49 and 81. These give an area of 330 square units.

28. Mid-ordinates are 2·1, 5·4, 7·7, 9·2, 9·9, 9·9, 9·2, 7·7, 5·4 and 2·1. These give an area of 412 mm².

29. (a) Ordinates are 20, 20, 10, 6⅔, 5 and 4. These given an area of 268⅓ square units.
    (b) Ordinates are (from top to bottom) 5, 6¼, 8⅓, 12½, 25 and 25. These given an area of 268⅓ square units.

30. 1 800 m³

31. 1 530 mm²

32. 86 square units, 0·78%

4.6
The geometry of solids

4.6.1
The volume of a prism, of a cylinder, and of a pipe

A *prism* is a solid of constant cross-section. Unless it is specifically stated to the contrary, a prism is assumed to be a *right prism*, where the ends, having the shape of a regular polygon, lie at right angles to the axis which passes through the centres of the ends. Prisms are described by the shape of the constant cross-section, such as a triangular prism, a hexagonal prism, and so on. A square prism whose length is the same as the length of the side of the square cross-section is usually called a *cube*. A right solid of constant circular cross-section is known as a *cylinder*, while the solid has a constant annular cross-section called a *pipe*.

The volume of any right prism, or solid corresponding to a right prism, is the product of the area of the constant cross-section and the distance between the ends. A prism is usually visualised as standing on an end, when the distance between the ends is normally known as the height.

If a rectangular prism has overall dimensions L, B and H, the volume V is given by the formula:

$$V = LBH$$

(In a cube B and H are both equal to L and $V = L^3$.)

If a cylinder has a radius r and a height h, the volume is given by:

$$V = \pi r^2 h$$

or, in terms of the diameter d:

$$V = \frac{\pi d^2 h}{4}$$

If a pipe has an outside diameter D, an inside diameter d, and a height h, the volume V is given by:

$$V = \frac{\pi}{4}(D^2 - d^2)h$$

In the case of a pipe, it is more usual to have the distance between the ends much greater than a diameter, and we generally refer to the 'length' of a pipe and use the formula:

$$V = \frac{\pi}{4}(D^2 - d^2)L$$

Example:

Assuming that there is no reduction in volume during the processing, what length of wire of diameter 2·5 mm can be manufactured from a piece of material in the form of a cylinder of diameter 20 mm and length 150 mm?

Volume of wire = volume of initial cylinder.

Let the length of wire be x.

Working in millimetres:

$$\frac{\pi}{4}(2 \cdot 5)^2 \times x = \frac{\pi}{4}(20)^2 \times 150$$

Divide all through by $\frac{\pi}{4}$:

$$(2 \cdot 5)^2 \times x = (20)^2 \times 150$$
$$6 \cdot 25 \times x = 400 \times 150$$
$$x = \frac{400 \times 150}{6 \cdot 25}$$
$$\therefore x = 9\ 600$$

Answer:

Length of wire = 9 600 mm = 9·6 m

Example:

A solid roller of diameter 4 mm has a length of 20 mm. In order to reduce the mass of the roller a co-axial flat-bottomed hole is drilled in the roller of diameter 2 mm and depth 16 mm. Calculate the percentage reduction in mass.

Mass is proportional to volume, so the problem can be worked through on volumes. Working in millimetre units.

$$\text{Volume of original roller} = \frac{\pi d^2 h}{4}$$

$$= \pi(4)^2 \times 20 \div 4$$
$$= 4\pi \times 20$$
$$= 80\pi$$

$$\text{Volume of metal removed} = \pi(2^2) \times 16 \div 4$$
$$= \pi \times 16$$
$$= 16\pi$$

$$\text{Percentage reduction in volume} = \frac{\text{volume removed}}{\text{original volume}} \times 100\%$$

$$= \frac{16\pi}{80\pi} \times 100\%$$

$$= \tfrac{1}{5} \times 100\%$$
$$= 20\%$$

Answer:

$$\text{Reduction in mass} = 20\%$$

(The reader should note that since all volumes are multiples of π it was decided not to evaluate actual volumes.)

Example:

A cylindrical open-topped canister is to be pressed from sheet metal so that its diameter is equal to its height. The capacity of the canister is to be one litre ($= 10^6$ mm^3). Determine its diameter.

Let the diameter $= d$ mm
Volume of canister V = area of cross-section × height

$$V = \frac{\pi d^2}{4} \times d = \frac{\pi d^3}{4}$$

$$\frac{\pi d^3}{4} = 10^6$$

$$d^3 = \frac{4 \times 10^6}{\pi}$$

$$d = \sqrt[3]{\left(\frac{4 \times 10^6}{\pi}\right)} = 109{\cdot}2 \text{ mm}$$

No.	Log
4	0·602 1
10^6	6·000 0
Num.	6·602 1
π	0·497 2
Root 3	6·104 9
109·2	2·038 3

Rough check:

$$\sqrt[3]{1} \times \sqrt[3]{10^6} = 1 \times 10^2$$
$$= 100$$

Answer:

Diameter $= 109$ mm

4.6.2
The volume of a pyramid and of a cone

A *pyramid* is a solid which tapers uniformly from a plane figure to a point. Unless it is specifically stated to the contrary, a pyramid is assumed to be a *right pyramid*, in which a regular polygon known as the base lies at right angles to the axis joining the point to the centre of the base. A section of any pyramid taken at right angles to the base and cutting through the axis reveals a triangle. Pyramids are described by the polygon which forms the base, such as a square pyramid, a rectangular pyramid, and so on. The solid which tapers uniformly from a circular base to a point is called a *cone*.

A pyramid is usually visualised in its most stable position, standing on its base. Bearing in mind that when we say pyramids, without qualification, we are referring to right pyramids, the height of a pyramid is the distance from the point to the centre of the base. The same definition can apply to a cone.

We cannot reason formulae for the volumes of pyramids as we can with prisms. The reader must accept for the time being, although it will be proved in a later stage of the course, that the volume of any solid of pyramid form is one third of the product of the area of the base and the height.

Hence if a square pyramid has a base in the form of a square whose side has a length L, the height of the pyramid being H, then the volume V is given by:

$$V = \frac{L^2 H}{3}$$

Similarly, if a cone has a base of radius r and a height h, the volume V is given by the formula:

$$V = \frac{\pi r^2 h}{3}$$

or, in terms of the diameter d:

$$V = \frac{\pi d^2 h}{12}$$

A frustum of a pyramid is a portion cut off by two parallel planes. We shall only be concerned with frustums produced by cutting planes of *right* pyramids, that is, pyramids whose centre lines are at right angles to their bases, and also where the cutting planes are parallel to the base.

If h is the height of the frustum, A the area revealed by one cutting plane, and B the area revealed by the other cutting plane:

$$\text{Volume of the frustum} = \frac{h}{3} \{A + \sqrt{(AB)} + B\}$$

With the particular case of the frustum of a cone, if the cutting planes reveal circles of diameter R and r:

$$A = \pi R^2, \qquad B = \pi r^2$$

$$\sqrt{(AB)} = \sqrt{\{(\pi R^2)(\pi r^2)\}} = \sqrt{\{\pi^2 R^2 r^2\}} = \pi Rr$$

$$\therefore \text{Volume of a frustum of a cone} = \frac{h}{3} \{A + \sqrt{(AB)} + B\}$$

$$= \frac{h}{3} \{\pi R^2 + \pi Rr + \pi r^2\}$$

$$\text{Volume of a frustum of a cone} = \frac{\pi h}{3} \{R^2 + Rr + r^2\}$$

Example:

Figure 4.43 shows a countersunk head rivet. The rivets are made by cutting off blanks from bar of diameter 5 mm and then forging the head. At what length should the blanks be cut off?

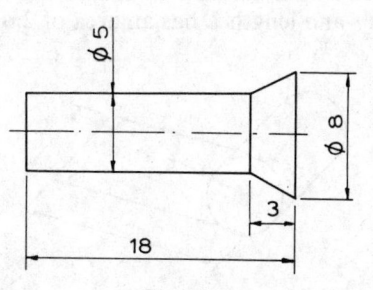

Fig. 4.43

The head is the frustum of a cone:

$$\text{Volume} = \frac{\pi h}{3}(R^2 + Rr + r^2)$$

$$= \frac{3\pi}{3}(4^2 + 4 \times 2\cdot5 + 2\cdot5^2)$$

$$= \pi(16 + 10 + 6\cdot25)$$

$$= 32\cdot25\pi$$

This is to be forged from a cylinder of diameter 5 mm. Let length of cylinder be L:

$$\frac{\pi(5^2)L}{4} = 32\cdot25\pi$$

$$L = \frac{32\cdot25\pi}{6\cdot25\pi}$$

$$= 5\cdot16 \text{ mm}$$

Increased length required $= 5\cdot16 - 3 = 2\cdot16$

$\therefore$ Length of blank $= 18 + 2\cdot16 = 20\cdot16$

Answer:

Length of blank $= 20\cdot2$ mm

4.6.3
Development of curved surfaces

If we position a cylinder of radius r and length L with its curved surface on a plane, there is a line of contact of length L between the cylinder and the plane. If we imagine the line of contact to be drawn on the cylinder, and then rotate the cylinder until the line contacts the plane once more, as in Fig. 4.44, we observe that if the curved surface of the cylinder was developed into a flat surface, it would form a rectangle with sides of length $2\pi r$ and L. Hence the curved surface of a cylinder of radius r and length L has an area of $2\pi rL$.

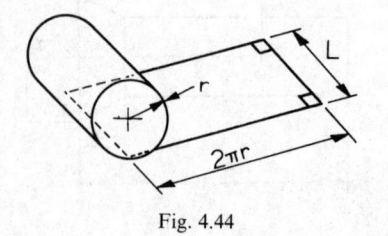

Fig. 4.44

We can develop the curved surface of a cone in exactly the same manner. If the cone has a slant height of L, the development is a sector of a circle of radius L, as shown in Fig. 4.45. The length of the arc of the sector is equal to the circumference of the base of the cone.

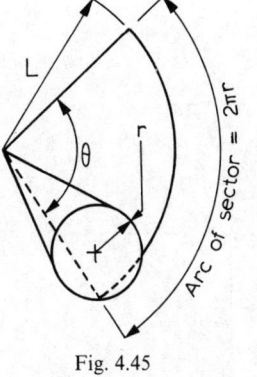

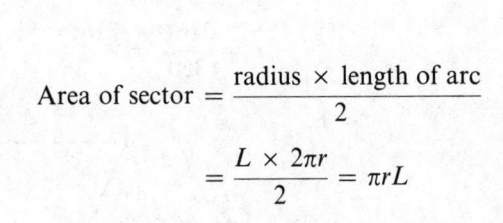

$$\text{Area of sector} = \frac{\text{radius} \times \text{length of arc}}{2}$$

$$= \frac{L \times 2\pi r}{2} = \pi r L$$

Fig. 4.45

The included angle θ of the sector, in radians, is found by dividing the length of the arc by the radius:

$$\theta \text{ (radians)} = \frac{2\pi r}{L}$$

To convert an angle in radians to degrees we multiply by $\dfrac{360}{2\pi}$.

Hence
$$\theta \text{ (degrees)} = \frac{2\pi r}{L} \times \frac{360}{2\pi} = \frac{360r}{L}$$

A *frustum* of a cone is that portion of a cone contained between two parallel planes, the planes being at right-angles to the vertical axis. A frustum can be visualised as a large cone minus a small cone. The development of the curved surface is thus the sector of a large circle minus the sector of a small circle, i.e. a sector of an annulus.

Example:

A sector of a circle has a radius of 15 mm and an included angle of 216°. The sector is formed into the curved surface of a cone, the two radii abutting without lap. Calculate:

(a) the base radius of the cone;
(b) the vertical height.

(a)   Sector angle in degrees $= \dfrac{360r}{L}$. The radius of the sector is the slant height of the cone.

$$L = 15 \text{ mm}$$

$$216° = \frac{360° \times r}{L}$$

$$r = \frac{216L}{360} = \frac{216 \times 15}{360} = 9 \text{ mm}$$

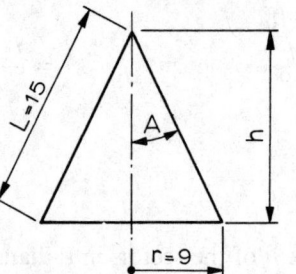

Fig. 4.46

(b)   Referring to Fig. 4.46:

$$L^2 = r^2 + h^2$$
$$h^2 = L^2 - r^2, \qquad h = \sqrt{(L^2 - r^2)}$$
$$h = \sqrt{(225 - 81)} = \sqrt{144} = 12 \text{ mm}$$

Answers:

    (a)   Base radius $= 9$ mm
    (b)   Vertical height $= 12$ mm

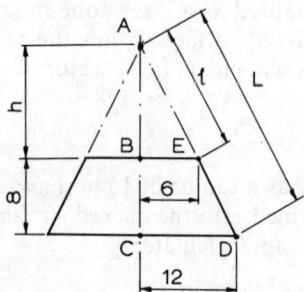

Fig. 4.47

Example:

Find the area of the curved surface of a frustum of a cone. The frustum is of height 8 mm and the radii of the top surface and the base are 6 mm and 12 mm respectively.

Figure 4.47 shows the dimensions of the frustum, which will be considered as a difference between two similar cones. Triangles ABE and ACD are similar:

$$\therefore \frac{AB}{AC} = \frac{BE}{CD} \qquad \therefore \frac{h}{h+8} = \frac{6}{12}$$

$$\therefore 12h = 6(h+8), \qquad 12h = 6h + 48$$
$$12h - 6h = 48, \qquad 6h = 48, \qquad h = 8$$
$$l = \sqrt{(6^2 + 8^2)} = \sqrt{(36 + 64)} = \sqrt{100} = 10$$
$$L = \sqrt{(12^2 + 16^2)} = \sqrt{(144 + 256)} = \sqrt{400} = 20$$

$$\begin{aligned}
\text{Area of sector of annulus} &= \pi RL - \pi rl \\
&= \pi(RL - rl) \\
&= \pi(12 \times 20 - 6 \times 10) \\
&= \pi(240 - 60) \\
&= 180\pi \\
&= 565 \cdot 56 \text{ mm}^2
\end{aligned}$$

Answer:

$$\text{Area of curved surface} = 566 \text{ mm}^2$$

4.6.4
The volume of a sphere, a spherical cap and a spherical zone

A *sphere* is a solid where any point on its surface is equidistant from a point within that sphere known as the *centre*. The distance between the centre of the sphere and its surface is called the *radius* of the sphere. If a sphere is cut by any plane, a circular plane is revealed. If the cutting plane passes through the centre the section revealed is a circle whose radius is equal to that of the sphere. The sphere would be divided into two *hemispheres*.

If a single cutting plane does not pass through the centre of the sphere, it divides the sphere into two portions of unequal volumes. The smaller of these portions is called a *spherical cap*. A *zone* of a sphere is that portion of a sphere which lies between two parallel cutting planes. At present our knowledge of mathematics is insufficient to deduce formulae connected with a sphere logically, and these must be accepted without proof.

The volume V of a sphere of radius r is given by the formula:

$$V = \frac{4\pi r^3}{3}$$

or, in terms of the diameter d

$$V = \frac{\pi d^3}{6}$$

The volume V of a spherical cap cut from a sphere of radius r, the height of the cap being h, is given by:

$$V = \pi h^2 \left(r - \frac{h}{3}\right)$$

The volume V of a spherical zone of height h, the radii of the two circles revealed by the cutting planes being r_1 and r_2 is given by the formula:

$$V = \frac{\pi h}{6}\{3(r_1{}^2 + r_2{}^2) + h^2\}$$

Example:

A hot-water tank has the form of a vertical cylinder surmounted by a hemisphere, the cylinder and hemisphere having the same diameter. If the diameter of the cylinder is 500 mm and the capacity of the tank is 200 litres, find the overall height of the tank.

$$1 \text{ m}^3 \text{ contains } 1\,000 \text{ litres, hence } V = \frac{200}{1\,000} = 0 \cdot 2 \text{ m}^3$$

$$500 \text{ mm} = 0 \cdot 5 \text{ m}$$

Let the common diameter be d and the height of the cylinder be h.

Capacity = volume of cylinder + volume of hemisphere

$$= \frac{\pi d^2 h}{4} + \frac{\pi d^3}{12}$$

$$= \frac{\pi d^2}{12}(3h + d)$$

$$0 \cdot 2 = \frac{\pi}{2 \times 2 \times 12}(3h + 0 \cdot 5)$$

$$3h + 0 \cdot 5 = \frac{0 \cdot 2 \times 2 \times 2 \times 12}{\pi} = \frac{9 \cdot 6}{\pi} = 3 \cdot 056$$

$$3h = 3 \cdot 056 - 0 \cdot 5 = 2 \cdot 556$$

$$h = \frac{2 \cdot 556}{3} = 0 \cdot 852$$

Overall height $= h + r = 0·852 + 0·25 = 1·102$

Answer:

$$\text{Overall height} = 1·10 \text{ m}$$

Example:

Figure 4.48 shows the proportions of a storage vessel having a cylindrical portion between two hemispherical ends.

(a) Taking π as $\frac{22}{7}$, deduce the simplest formula for its volume V in terms of R.

(b) Find the capacity, in litres, when R is 500 mm.

(c) Find the value of R, in millimetres, to give a capacity of 500 litres.

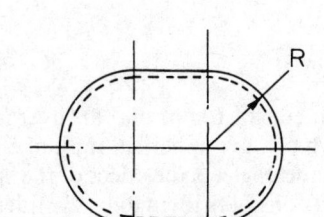

Fig. 4.48

(a) V = volume of a sphere of radius R + volume of a cylinder of radius R and length R

$$= \frac{4\pi R^3}{3} + \pi R^2 \times R = \frac{4\pi R^3}{3} + \pi R^3$$

$$= \frac{4\pi R^3 + 3\pi R^3}{3} = \frac{7\pi R^3}{3} = \frac{7 \times 22 \times R^3}{7 \times 3} = \frac{22R^3}{3}$$

(b)   500 mm $= 0·5$ m

$$V = \frac{22R^3}{3} = \frac{22 \times 0·5 \times 0·5 \times 0·5 \text{ m}^3}{3} = \frac{11}{12} \text{ m}^3$$

$1 \text{ m}^3 = 1\ 000$ litres

$$\text{Capacity} = \frac{11 \times 1\ 000}{12} \text{ litres} = 916·7 \text{ litres}$$

(c)   500 litres = 0·5 m³

$$\frac{22R^3}{3} = 0.5, \qquad R^3 = \frac{3 \times 0.5}{22} = \frac{1.5}{22} = 0.068\ 2$$

$$R = \sqrt[3]{0.068\ 2} = 0.408\ 6\ \text{m} = 408.6\ \text{mm}$$

Rough check: $0.4^3 = 0.064$

Answers:

(a)   Volume $= \dfrac{22R^3}{3}$

(b)   Capacity = 917 litres

(c)   Radius = 409 mm

Example:

A dowel has a diameter of 6 mm and an overall length of 15 mm. The dowel has ends which have spherical radii of 5 mm. One estimator in calculating the volume neglects the effect of the spherical caps on the ends of the dowels and considers them to be cylinders of diameter 6 mm and length 15 mm. A second estimator works precisely. What is the percentage error of the inaccurate estimator?

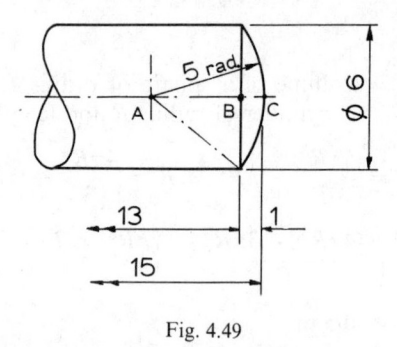

Fig. 4.49

Figure 4.49 shows the end of the dowel, with important dimensions deduced as follows: Using the 3, 4, 5 rule for a right-angled triangle, $AB = 4$ mm, $BC = 1$ mm.

A dowel consists of a cylinder of radius 3 mm and length 13 mm, with two spherical caps of radius 5 mm and height 1 mm.

Accurate method:

Volume of dowel = volume of cylinder + 2(volume of spherical cap)
$$= \pi(3^2)13 + 2\{\pi(1)^2(5 - \tfrac{1}{3})\}$$
$$= 117\pi + 9\tfrac{1}{3}\pi$$
$$= 126\tfrac{1}{3}\pi \text{ mm}^3$$

Inaccurate method:

Volume of dowel = volume of cylinder of radius 3 mm
and height 15 mm
$$= \pi(3^2)15$$
$$= 135\pi$$

Amount of error $= (135\pi - 126\tfrac{1}{3})$ mm^3
$$= 8\tfrac{2}{3}\pi \text{ mm}^3$$

Percentage inaccuracy $= \dfrac{\text{amount of error}}{\text{correct volume}} \times 100\%$

No.	Log
2 600	3·415 0
379	2·578 6
6·861	0·836 4

$$= \frac{8\tfrac{2}{3} \text{ mm}^3}{126\tfrac{1}{3} \text{ mm}^3} \times 100\%$$

$$= \left(\frac{26}{3} \times \frac{3}{379}\right) 100\%$$

$$= \frac{2\,600}{379}\% = 6\cdot 861\%$$

Rough check: $\dfrac{9}{125} = \dfrac{72}{1\,000} = \dfrac{7\cdot 2}{100} = 7\cdot 2\%$

Answer:

Inaccuracy $= 6\cdot 86\%$

4.6.5
The surface area of a sphere, of a spherical cap and of a spherical zone

A logical development of the formulae connected with the surface area of a sphere, and of parts of the surface area of a sphere, is outside the scope of this book. The formulae must be accepted without proof. The surface area A of a sphere is given by the formula:

$$A = 4\pi r^2$$

or, in terms of the diameter d:

$$A = \pi d^2$$

If a sphere is cut by a plane, the smaller portion is called a *spherical cap*. If this cap rests on its circular face, the maximum distance from that face to the curved surface is referred to as the height h of the spherical cap. The curved spherical surface area A of a spherical cap is given by the formula:

$$A = 2\pi r h$$

If a sphere is cut by two parallel planes the portion lying between those planes is called a *zone* of the sphere. If the cutting planes lie at a distance h apart, the curved spherical surface area A is given by:

$$A = 2\pi r h$$

(an identical formula to that for the curved spherical surface area of a spherical cap).

Example:

A hot-water storage tank consists of a cylinder surmounted by a hemisphere. The common radius is 400 mm and the overall height is 900 mm. Find the area, in square metres, required to cover the entire tank with lagging material.

$$400 \text{ mm} = 0.4 \text{ m}, \qquad 900 \text{ mm} = 0.9 \text{ m}$$
$$\text{Height of cylinder} = 0.9 \text{ m} - 0.4 \text{ m} = 0.5 \text{ m}$$

Total surface area = area of a circle + area of the curved surface of a cylinder + surface area of a hemisphere

$$= \pi r^2 + \pi r^2 h + \frac{4\pi r^2}{2}$$

$$= \pi r^2 + \pi r^2 h + 2\pi r^2$$
$$= \pi r^2 (1 + h + 2)$$
$$= \pi r^2 (3 + h)$$
$$= \pi r^2 (3 + 0.5)$$

$$= \frac{22 \times 0.4 \times 0.4 \times 3.5}{7} = 1.76 \text{ m}^2$$

Answer:

Area of lagging material $= 1.76 \text{ m}^2$

4.6.6
The volume of similar solids

If a first sphere has a diameter d, and a second sphere has a diameter n times larger, so that the diameter of the second sphere is nd, then:

$$\frac{\text{volume of second sphere}}{\text{volume of first sphere}} = \frac{\pi(nd)^3}{6} \times \frac{6}{\pi d^3}$$

$$= \frac{6\pi n^3 d^3}{6\pi d^3} = n^3$$

In a similar manner, if the linear dimensions of a second cone are each n times the linear dimensions of a first cone, i.e.

radius of first cone $= r$ height of first cone $= h$
radius of second cone $= nr$ height of second cone $= nh$

then
$$\frac{\text{volume of second cone}}{\text{volume of first cone}} = \frac{\pi(nr)^2(nh)}{3} \times \frac{3}{\pi r^2 h}$$

$$= \frac{3\pi n^3 r^2 h}{3\pi r^2 h} = n^3$$

We could proceed in a similar manner for all solids. We should find that if a second solid has all linear dimensions n times the corresponding linear dimensions of a first solid, then:

$$\frac{\text{volume of second solid}}{\text{volume of first solid}} = n^3$$

4.6.7
Mass calculation from drawings

The mass of an object composed of one substance only can be accurately estimated from a drawing of that object by determining its volume, and then multiplying that volume by the mass per unit volume of the substance from which the object is made.

The mass per unit volume of a substance is known as the *density* of that substance. The basic SI unit of mass is the kilogram. The basic SI unit of volume is derived from the basic unit of length, and will therefore be the metre cubed, or cubic metre. Hence the basic unit of density is the kilogram per cubic metre, the abbreviation for which is kg/m^3.

When metric units were first being formulated, about the time of the French Revolution, one gram was taken as the mass of a cubic centimetre of water. This was a reasonable basis for use at that time but, unfortunately for very precise work, water has a variable density. The variation of the density of water depends mainly upon temperature, but it also depends upon the amount of dissolved gases, this latter amount itself depending upon pressure. However, the variation in the density of water is not significantly large in terms of the accuracy required in our present studies. We can therefore assume, for the purposes of our studies, that:

$$\text{density of water} = 1 \text{ g/cm}^3$$

We are using SI units as a first order of preference, and:

$$1 \text{ g} = 10^{-3} \text{ kg} \qquad \text{while} \qquad 1 \text{ cm}^3 = 10^{-6} \text{ m}^3$$

$$\text{then density of water} = \frac{10^{-3} \text{ kg}}{10^{-6} \text{ m}^3} = 10^3 \text{ kg/m}^3$$

$$= 1\,000 \text{ kg/m}^3$$

It is interesting to note that a further convenient practical value for the density of water, since there are 1 000 litres in a cubic metre, is one kilogram per litre.

The symbol for density is the small Greek letter rho (ρ). Sometimes we find it convenient to compare the density of a given substance with the density of water. The *relative density* of a substance (symbol d) is the ratio:

$$\frac{\text{density of that substance}}{\text{density of water}}$$

As a typical example, the relative density of aluminium is about 2·70. Hence for aluminium:

$$d = \frac{\rho}{1\,000\,\text{kg/m}^3}$$

and
$$\rho = d \times 1\,000 \text{ kg/m}^3 = \text{about } 2\cdot70 \times 1\,000 \text{ kg/m}^3$$
$$= \text{about } 2\,700 \text{ kg/m}^3$$

The following tables give some approximate relative densities of some common metals and alloys.

Magnesium	1·74	Brass (60 Cu/40 Zn)	8·36
Aluminium	2·70	Brass (70 Cu/30 Zn)	8·44
Zinc	7·10	Brass (90 Cu/10 Sn)	8·78
Tin	7·30	Nickel	8·80
Cast iron	7·00–7·80	Copper	8·89
Carbon steel	7·84	Lead	11·30

At a later stage in technician studies we shall be introduced to weight. The *weight* of a body is the gravitational force exerted on the mass of that body. The SI unit of force is the newton, and hence, theoretically speaking, a weight should be expressed in newtons. In everyday life, we shall often hear people refer to a 'weight of one kilogram'. Such a statement is theoretically incorrect, but it is tolerated, somewhat reluctantly by the purist, in everyday life.

The specific weight of a substance is the weight of unit volume. The *specific gravity* of a substance is the ratio:

$$\frac{\text{specific weight of that substance}}{\text{specific weight of water}}$$

We shall see eventually that the weight of a body is directly proportional to its mass, and hence the specific gravity of a substance has precisely the same numerical value as its relative density. It seems, therefore, that it is irrelevant which expression is employed, but as a first order of preference, since we are dealing with masses, the expression relative density should be used.

Consequently, if we know the density of the material of which an article is made, and we are provided with a drawing of that article, we can determine the volume of the article and multiply it by the density to obtain the mass. Even quite complicated shapes can be broken down into a collection of simple shapes so that a volume can be obtained from a summation of smaller volumes.

With the SI system of units the fundamental unit of mass is the kilogram, while the fundamental unit of length is the metre. The basic derived unit for density is therefore the kilogram per metre cubed (kg/m^3). Consequently, if the volume of an article is evaluated in terms of cubic metres (i.e. metres cubed) and that volume is multiplied by the density in kilograms per metre cubed, the answer will be given in kilograms. Any of the original dimensions which are not in metres can be converted to metres before being substituted in a formula to determine a volume, using relationships such as $1 \text{ mm} = 10^{-3} \text{ m}$ and $1 \text{ cm} = 10^{-2} \text{ m}$. The practice of changing, if necessary, any values in the original data to equivalents in terms of fundamental SI units, is to be commended as a general principle to adopt. This is especially true if the articles are of relatively large size.

However, recommended practices should never be slavishly followed if other practices provide overriding benefits. In the particular case of articles of smaller size, whose masses are considerably less than a kilogram, a convenient unit for density is the gram per centimetre cubed because the numerical value is precisely that of the relative density (or the specific gravity). None of the materials commonly met with in everyday life has a relative density greater than 25. If a volume is determined in centimetres cubed and that volume is multiplied by the density in grams per centimetre cubed, then the mass will be given in grams. The numerical values will be generally of a low order of magnitude.

Our reader is therefore encouraged to use discretion in selecting the method to be adopted when determining the mass of an article given its dimensions. There is much to be said for determining a volume in cubic metres, but this often requires sound competence in manipulating powers of ten. If our reader prefers to determine volumes in terms of other units, there is no serious criticism that can be raised, provided always the correct numerical value and the appropriate unit for density is used.

Even quite complicated geometrical configurations can be divided into simpler geometrical shapes for which there are standard formulae for evaluating volumes. The arithmetical work involved can be considerably eased by the use of tabulated data. It is pointless to compute areas if they are readily available in works of reference, unless it is advantageous for the particular purposes of a solution to leave an area as a multiple of π. Many reference works provide excellent tables of items such as the mass per metre run of circular steel bars, the mass per metre run of flat copper bars, and so on.

Example:

A factory orders 2 metric tonnes of rectangular steel sheets 1·6 m × 1·2 m of nominal thickness 1·6 mm. If the tolerance on the thickness is 1·60 mm to 1·70 mm, what is:

(a)   the greatest number,   (b)   the least number

of complete sheets that can be expected? (1 metric tonne = 1 000 kg and take the relative density of steel to be 7·85.)

Let all the sheets be piled into a slab 1·6 m × 1·2 m and thickness t m.

$$\rho = 1\ 000\ d\ \text{kg/m}^3 = 7\ 850\ \text{kg/m}^3$$

$$\rho = \frac{m}{V} \quad \therefore V = \frac{m}{\rho} = \frac{2 \times 1\ 000}{7\ 850}\frac{\text{kg} \times \text{m}^3}{\text{kg}}$$

$$= \frac{1\ 000}{3\ 925}\ \text{m}^3 = 0\cdot254\ 8\ \text{m}^3$$

$$V = LBt, \qquad t = \frac{V}{LB} = \frac{0\cdot254\ 8\ \text{m}^3}{1\cdot6\ \text{m} \times 1\cdot2\ \text{m}} = 0\cdot132\ 7\ \text{m}$$

$$1\cdot6\ \text{mm} = 0\cdot001\ 6\ \text{m}, \qquad 1\cdot7\ \text{mm} = 0\cdot001\ 7\ \text{m}$$

(a)   Greatest number of sheets $= \dfrac{t}{\text{smallest thickness}} = \dfrac{0\cdot132\ 7\ \text{m}}{0\cdot001\ 6\ \text{m}}$

$$= 82\cdot94, \text{ i.e. 82 complete sheets}$$

(*b*) Smallest number of sheets $= \dfrac{t}{\text{largest thickness}} = \dfrac{0 \cdot 132\ 7\ \text{m}}{0 \cdot 00\ 17\ \text{m}}$

$$= 78 \cdot 65, \text{ i.e. } 78 \text{ complete sheets}$$

Answers:

    (a)   Greatest number of sheets $= 82$
    (b)   Least number of sheets $= 78$

Example:

From an engineer's reference book it is found that a metre length of steel of diameter 20 mm has a mass of 2·47 kg, and that a metre length of flat steel bar 40 mm × 10 mm has a mass of 3·14 kg. Use this information to find the mass of the lever shown in Fig. 4.50.

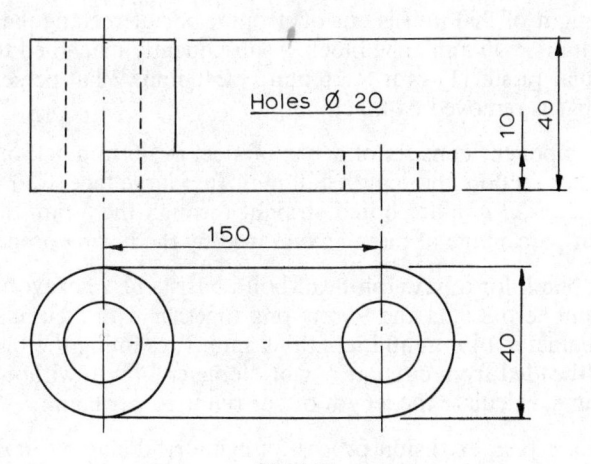

Fig. 4.50

The lever can be analysed into:

(a)   a cylinder of diameter 40 mm and length 40 mm,
    plus:
(b)   the equivalent of a length of 150 mm of 40 mm × 10 mm flat bar,
    less:
(c)   the holes, both of diameter 20 mm, totalling 50 mm in length.

The masses per metre run are proportional to the squares of their diameter.

$$\frac{\text{Mass per metre run of 40 mm bar}}{\text{Mass per metre run of 20 mm bar}} = \left(\frac{40}{20}\right)^2 = 4$$

Mass per metre run of 40 mm bar $= 4 \times 2 \cdot 47 = 9 \cdot 88$ kg

$$\text{Total mass} = \left(9 \cdot 88 \times \frac{40}{1\ 000}\right) + \left(3 \cdot 14 \times \frac{150}{1\ 000}\right) - \left(2 \cdot 47 \times \frac{50}{1\ 000}\right)$$

$$= 0 \cdot 395\ 2 \text{ kg} + 0 \cdot 471 \text{ kg} - 0 \cdot 123\ 5 \text{ kg}$$
$$= 0 \cdot 866\ 2 \text{ kg} - 0 \cdot 123\ 5 \text{ kg}$$
$$= 0 \cdot 742\ 7 \text{ kg}$$

Answer:

Mass of lever $= 0 \cdot 743$ kg

Problems 4.6

1. A length of 250 mm is cut off from a bar of rectangular section 120 mm × 40 mm. The block is subsequently machined to a rectangular prism 117 mm × 36 mm × 240 mm. What percentage of the bar is removed by machining?

2. A component consists of a bar of steel of 40 mm × 5 mm rectangular section and length 120 mm. In a large face, four holes of diameter 20 mm are bored straight through the 5 mm thickness. What percentage of metal is removed by the boring process?

3. The blank for a hexagon-head bolt consists of a hexagonal prism 22 mm across flats and 9 mm long together with a circular prism of diameter 14 mm and length 40 mm. Presuming the blanks are cold-headed from circular bar of diameter 14 mm without loss of volume, calculate the length of bar required per blank.

4. In an impact extrusion process, a cylindrical slug has a diameter of 20 mm and a length of 30 mm. A punch of hexagonal cross-section 14 mm across flats enters one end; the diameter remains at 20 mm, and when the slug is removed from the die the hexagonal hole is flat-bottomed and of depth 10 mm. Presuming that the ends of the slug remain flat, by how much has its length been increased?

5. A duct has a flat bottom of width 240 mm. The sides slope outwards so that when water flows in the duct, the cross-sectional area of flow is in the form of a trapezium. If the width of flow at the water level is 400 mm and the depth of flow is 125 mm, the velocity of flow being 2 m/s, calculate the rate of flow in litres per hour.

6. A strip of steel, of rectangular section 20 mm × 0·25 mm is wound tightly on to a former of diameter 450 mm to produce a coil of outside diameter 600 mm and thickness 20 mm. By equating volumes of the strip in its coiled and uncoiled form, calculate the length of strip, in metres, that has been wound on the former.

7. A rough casting in the form of a short length of piping has an outside diameter of 80 mm, an inside diameter of 60 mm and a length of 125 mm. During machining, 2·5 mm is removed from every surface. Determine the percentage of the original volume removed by machining.

8. Assuming no loss of volume in the process, what length of section shown in Fig. 4.51 can be extruded from a solid billet of diameter 35 mm and length 80 mm?

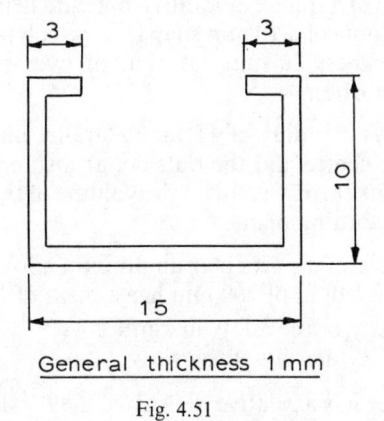

General thickness 1 mm

Fig. 4.51

9. Calculate the capacity, in litres, of a cylindrical storage tank whose internal dimensions are a diameter of 700 mm and a length of 1 200 mm.

10. A metal billet of diameter 30 mm and length 80 mm is eventually processed into discs of diameter 10 mm and thickness 1 mm. Assuming no loss of metal, calculate the number of discs produced.

11. A vertical section on the axis of a right circular cone reveals an isosceles triangle with one side of length 100 mm and two sides of length 130 mm. Calculate the vertical height and the volume of the cone.

12. The base radius R, the top radius r and the height h of the frustum of a cone are in the proportion of $7:4:7$ respectively. If the volume is 5 456 mm^3, take π as $\frac{22}{7}$ and determine the height of the frustum.

13. A pan-head rivet of diameter 15 mm has a head in the form of the frustum of a cone. The diameters of the frustum are 28 mm and 20 mm, the height being 6 mm. If the length under the head of the rivet is 40 mm, calculate the volume of material in a single rivet.

14. A vessel with a constant circular cross-section of diameter 50 mm contains water to a depth of 60 mm. If a sphere of diameter 40 mm is completely submerged in the water, how much higher does the water level rise in the vessel?

15. The volume of a hollow sphere made of thin material is often taken as the external surface area multiplied by the thickness of material. Find the percentage error if this method is used for a hollow sphere of outside diameter 50 mm and inside diameter 48 mm.

16. The head of a snap rivet of nominal diameter 10 mm can be assumed to be the cap of a sphere of radius 8 mm and height 6 mm. Calculate the total volume of a 10 mm snap rivet which holds together plates of total thickness 20 mm, i.e. that of two spherical caps and a cylindrical portion.

17. A sphere has a radius of 15 mm. Parallel planes cut the sphere between the centre and the outside, at distances of 9 mm and 12 mm from the centre. Calculate the volume of the zone of the sphere between the cutting planes.

18. A circular bar of a particular aluminium alloy with a diameter of 50 mm and a length of 200 mm has a mass of 1·1 kg. Calculate:
    (a) the density of the alloy, in kg/m^3;
    (b) the relative density of the alloy.

19. (a) If copper has a relative density of 8·89, calculate its density in kg/m^3.
    (b) Calculate the mass, in kilograms of a copper bus-bar of rectangular section 50 mm × 16 mm and length 7·5 m.

20. Taking the density of 70 Cu/30 Zn brass as 8 440 kg/m^3, construct a table that will give the mass, in kilograms per square metre, of 70 Cu/30 Zn brass sheets of thickness 1, 1·6, 2·5 and 4 mm.

21. Taking the relative density of steel to be 7·84, calculate the mass, in kilograms per metre run, of:
    (a) circular bar of diameter 50 mm;
    (b) flat bar of 50 mm × 16 mm section;
    (c) pipe of outside diameter 80 mm and wall thickness 10 mm.

22. Taking the density of steel to be 7 840 kg/m^3, calculate the mass, in kilograms, of 1 000 steel washers of outside diameter 40 mm, inside diameter 20 mm and thickness 3·2 mm.

23. Find the mass, in kilograms, of a triangular gusset plate with sides of length 180 mm, 200 mm and 260 mm, the thickness of the plate being 16 mm and the density of the material 7 840 kg/m^3.

24. A hollow float, closed at both ends, has the external appearance of a cylinder. The outside dimensions are a diameter of 200 mm and a depth of 100 mm. The constant metal thickness is 5 mm. Calculate its mass, in kilograms, if it is made of a material having a density of 8 400 kg/m^3.

25. Taking the density of steel to be 7 840 kg/m^3, what length (in metres) of steel pipe of outside diameter 40 mm and wall thickness 3 mm can be expected from one metric tonne ($= 1$ 000 kg) of such piping?

26. A rivet can be considered to be a hemisphere of radius 20 mm, together with a cylinder of diameter 25 mm and length 70 mm. Taking the density of the rivet material to be 7 840 kg/m^3, calculate the mass, in kilograms, of 100 rivets.

27. A sector of sheet metal is manipulated to form a cone of radius 50 mm and vertical height 120 mm. Assuming no allowance for seaming, and that a reference work states that the sheet metal has a mass of 0·9 kg/m^2, determine the mass of the sector, in grams.

28. An impurity in a casting can be assumed to take the form of a spherical bubble of air. The loss in mass due to the presence of the bubble is 20 g. The casting material has a relative density of 7·84. Calculate the diameter of the bubble, in millimetres.

29. Given that a metre length of steel of diameter 20 mm has a mass of 2·46 kg, construct a table that will give the mass per hundred of steel dowels 40 mm long, the diameters ranging from 5 mm to 25 mm in increments of 5 mm.

Top face 180 x 60

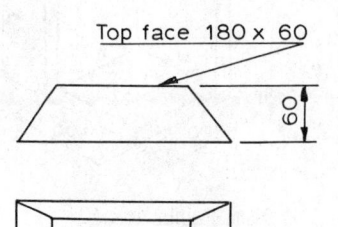

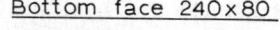

Bottom face 240x80

Fig. 4.52

30. Figure 4.52 shows the dimensions of an ingot cast in an alloy of relative density 2·8, the shape being that of the frustum of a rectangular pyramid. Calculate the number of ingots that can be made from one metric tonne (= 1 000 kg) of the alloy.

31. Find the mass, in grams, of the die casting shown in Fig. 4.53, the matherial having a relative density of 2·80. The rectangular portion is of thickness 20 mm.

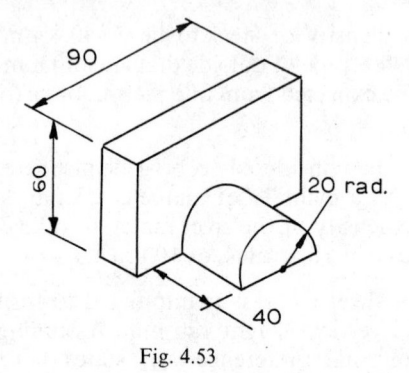

Fig. 4.53

32. Find the mass, in kilograms, of the casting shown in Fig. 4·54, the material being a grey cast iron of density 7 200 kg/m³.

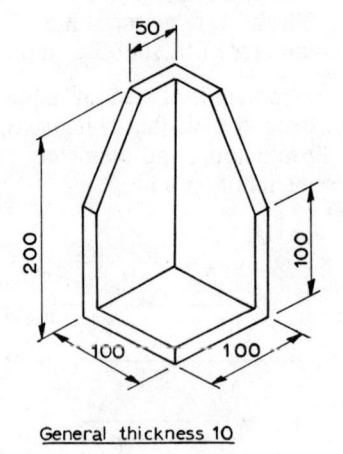

General thickness 10

Fig. 4.54

33. (a) Wire of diameter 2 mm is passed through a drawing die to reduce its diameter by 20%. If the wire enters the die at a speed of 1·6 m/s, at what speed does the wire leave the die?

(b) After leaving the die, the wire is coiled around a rotating drum. At what speed is the drum rotating, in rev/min, at the instant the wire is coiling at an effective diameter of 700 mm?

(c) What mass of wire, in kilograms, is processed per hour? Take the density of the wire to be 8 750 kg/m³.

34. Castings are to be made to the design shown in Fig. 4.55.

(a) If the material is an aluminium alloy of relative density 2·80, calculate:

  (i) the mass of a single casting, in grams;

  (ii) the number of complete castings that can be obtained from one metric tonne (= 1 000 kg) of the alloy. (Neglect scrap.)

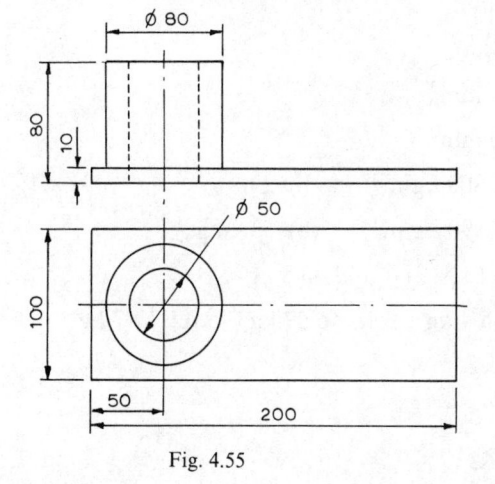

Fig. 4.55

(b) If a cast iron of relative density 7·28 were substituted for the aluminium alloy and the cast iron castings cost, by mass, one quarter the price of the aluminium alloy, calculate:

  (i) the percentage increase in mass;

  (ii) the percentage saving at cost.

Answers to Problems 4.6

1. 15·8%

2. 26·2%

3. 64·5 mm

4. 35·4 mm

5. 4 800

6. 495 m

7. 52%

8. 2 080 mm

9. 462 litres

10. 720

11. 120 mm, 314 000 mm^2

12. 14 mm

13. 4 900 mm^3

14. 17·1 mm

15. 4·11%

16. 2 930 mm^3

17. 1 080 mm^3

18. (a) 2 800 kg/m^3 (b) 2·8

19. (a) 8 890 kg/m^3 (b) 53·3 kg

20. 8·44, 13·5, 21·1 and 33·8 kg

21. (a) 15·4 kg   (b) 6·27 kg   (c) 17·2 kg

22. 23·6 kg

23. 2·25 kg

24. 4·96 kg

25. 3 670 m

26. 40·1 kg

27. 18·4 g

28. 16·8 mm

29. 0·615, 2·46, 5·64, 9·84 and 15·4 kg

30. 402

31. 373 g

32. 2·96 kg

33. (a) 2·5 m/s   (b) 68·2 rev/min   (c) 158 kg

34. (a)(i) 1 100 g   (ii) 909   (b)(i) 160%   (ii) 35%

5 Trigonometry

5.1
Basic trigonometry

5.1.1
Basic trigonometrical ratios

Trigonometry literally means the measurement of triangles. Our first excursion into trigonometry will be confined to right-angled triangles. Consider two similar right-angled triangles lettered conventionally, as shown in Fig. 5.1.

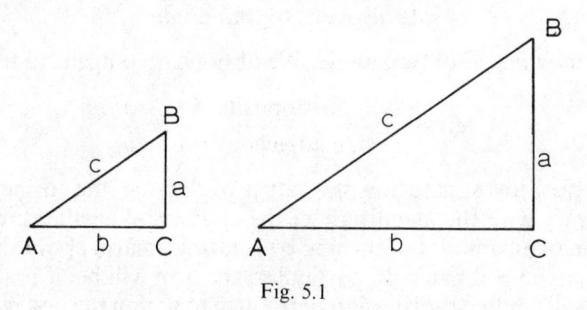

Fig. 5.1

The longest side of a right-angled triangle is the hypotenuse. We can also give names to the other sides with reference to the acute angles. For example, with reference to angle A:

side a is the 'side opposite A'

and side b is the 'side adjacent to A'

(There are actually two sides adjacent to A, but we have already nominated one as the hypotenuse.)

Since the triangles are similar, the ratios of corresponding sides are constant. The ratios $\dfrac{a}{c}, \dfrac{b}{c}$ and $\dfrac{a}{b}$ are the same for all similar right-angled triangles.

In a right-angled triangle, *and only in a right-angled triangle*, the ratio:

$$\frac{\text{side opposite an angle}}{\text{hypotenuse}}$$

339

is called the *sine* of that angle. We abbreviate sine to sin but read it as sine.

Hence
$$\sin A = \frac{\text{side opposite } A}{\text{hypotenuse}} = \frac{a}{c}$$

In a similar fashion, in a right-angled triangle, the ratio:

$$\frac{\text{side adjacent to an angle}}{\text{hypotenuse}}$$

is called the *cosine* of that angle. We abbreviate cosine to cos.

Hence
$$\cos A = \frac{\text{side adjacent to } A}{\text{hypotenuse}} = \frac{b}{c}$$

Finally, in a right-angled triangle, the ratio:

$$\frac{\text{side opposite an angle}}{\text{side adjacent to that angle}}$$

is called the *tangent* of that angle. We abbreviate tangent to tan.

Hence
$$\tan A = \frac{\text{side opposite } A}{\text{side adjacent to } A} = \frac{a}{b}$$

Let us now investigate how the values of the sine, the cosine and the tangent vary with the magnitude of the angle A. We will return to the indication of an amount of turning by rotating an arm about the origin of cartesian co-ordinates. In particular, the arm will be of unit length, and we shall confine the rotation of the arm to within the first quadrant, as shown in Fig. 5.2.

In the right-angled triangle ABC, in terms of cartesian co-ordinates:

$$a = y, \qquad b = x \qquad \text{and} \qquad c = 1$$

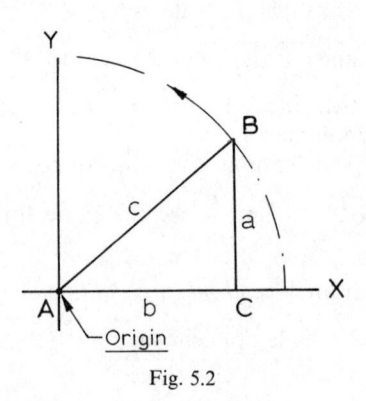

Fig. 5.2

In the first quadrant all these values are positive, and therefore ratios of these values will be positive. Hence if A lies between zero and 90°, the ratios of sin A, cos A and tan A are all positive values.

The sine of angle A is:

$$\frac{\text{side opposite } A}{\text{hypotenuse}} = \frac{a}{c} = \frac{y}{1} = y$$

When A is zero, y is zero and sin 0° = 0. As the angle A increases, y increases but not in direct proportion. y reaches its maximum value when $A = 90°$, when $y = 1$, hence sin 90°. = 1.

The cosine of angle A is:

$$\frac{\text{side adjacent to } A}{\text{hypotenuse}} = \frac{b}{c} = \frac{x}{1} = x$$

When A is zero, x is unity and cos 0° = 1. As the angle A increases, x decreases, but not in direct proportion. x reaches its minimum value of zero when $A = 90°$, hence cos 90° = 0.

It should be particularly noted that since y or x can never exceed unity, the value of the sine or the cosine can never exceed unity. When the magnitude of angle A lies between zero and 90°, the range of the sine is from zero to unity, and that of the cosine is from unity to zero.

The tangent of angle A is:

$$\frac{\text{side opposite } A}{\text{side adjacent to } A} = \frac{a}{b} = \frac{y}{x}$$

When the magnitude of angle A is zero, $y = 0$ and $x = 1$, hence tan 0° = 0. As the angle A increases, y increases but x decreases. Hence as the angle increases, the tangent increases, but not in direct proportion. When $A = 45°$, $x = y = 1$, hence tan 45° = 1. As A nears 90°, y is increasing only slightly, but x is diminishing very rapidly. At the value of 90°, $y = 1$ and $x = 0$, whence tan 90° = $\frac{1}{0}$, a value of infinite magnitude.

Let us pause in our development of basic trigonometry to recapitulate the basic trigonometrical ratios.

$$\sin A = \frac{\text{side opposite } A}{\text{hypotenuse}}$$

$$\cos A = \frac{\text{side adjacent to } A}{\text{hypotenuse}}$$

$$\tan A = \frac{\text{side opposite } A}{\text{side adjacent to } A}$$

We have already considered how these values vary as the angle increases. Let us now be more precise by calculating specific values, and plotting them on a graph. We have already calculated values of the ratios for 0° and 90°. Let us proceed by determining the values for 30°, 45° and 90°, since triangles containing these angles are simple to construct.

Figure 5.3(a) shows an equilateral triangle ABC with angle B bisected.

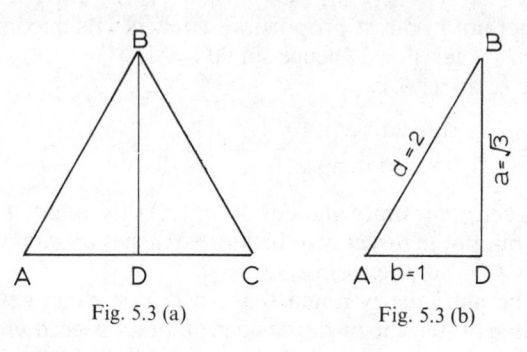

Fig. 5.3 (a)   Fig. 5.3 (b)

In the triangle ABD:

$$A = 60°, \qquad B = 30°$$
hence
$$D = 180° - (60° + 30°) = 90°$$

ABD is therefore a right-angled triangle, and is repeated for convenience in Fig. 5.3(b). If each side of the original equilateral triangle had a length of 2 units, since ABD and BDC are congruent triangles, $AD = DC = 1$ unit. Using the Theorem of Pythagoras, $(BD)^2 = (AB)^2 - (AD)^2 = 4 - 1 = 3$. Hence $BD = \sqrt{3}$ units long.

Thus in the right-angled triangle ABD:

$$A = 60° \qquad a = \sqrt{3} \qquad \text{(side opposite } A)$$
$$B = 30° \qquad b = 1 \qquad \text{(side adjacent to } A)$$
$$D = 90° \qquad d = 2 \qquad \text{(hypotenuse)}$$

$$\sin 60° = \sin A = \frac{\text{side opposite } A}{\text{hypotenuse}} = \frac{\sqrt{3}}{2} = 0.866\,0$$

$$\cos 60° = \cos A = \frac{\text{side adjacent to } A}{\text{hypotenuse}} = \frac{1}{2} = 0.500\,0$$

$$\tan 60° = \tan A = \frac{\text{side opposite } A}{\text{side adjacent to } A} = \frac{\sqrt{3}}{1} = 1.732\,1$$

$$\sin 30° = \sin B = \frac{\text{side opposite } B}{\text{hypotenuse}} = \frac{1}{2} = 0.500\,0$$

$$\cos 30° = \cos B = \frac{\text{side adjacent to } B}{\text{hypotenuse}} = \frac{\sqrt{3}}{2} = 0\text{·}866\,0$$

$$\tan 30° = \tan B = \frac{\text{side opposite } B}{\text{side adjacent to } B} = \frac{1}{\sqrt{3}} = 0\text{·}577\,4$$

Now let us construct a right-angled triangle containing an angle of 45°. The angles must be 90°, 45° and 45°, hence the triangle is isosceles, the sides containing the right angle being of equal length.

Let this length be unity, as shown in Fig. 5.4.

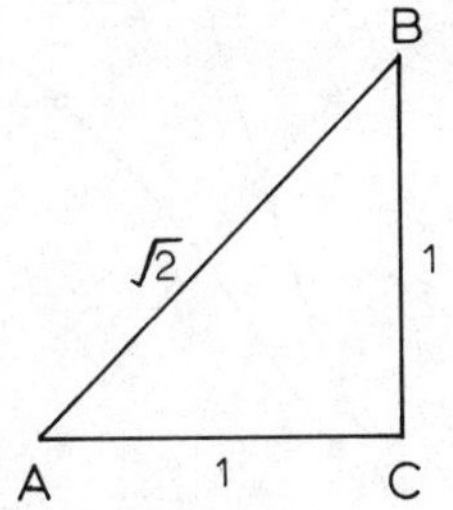

Fig. 5.4

Using the Theorem of Pythagoras:

$$c^2 = a^2 + b^2 = 1 + 1 = 2 \qquad \therefore c = \sqrt{2}$$

$$\sin 45° = \sin A = \frac{\text{side opposite } A}{\text{hypotenuse}} = \frac{1}{\sqrt{2}} = 0\text{·}707\,1$$

$$\cos 45° = \cos A = \frac{\text{side adjacent to } A}{\text{hypotenuse}} = \frac{1}{\sqrt{2}} = 0\text{·}707\,1$$

$$\tan 45° = \tan A = \frac{\text{side opposite } A}{\text{hypotenuse}} = \frac{1}{1} = 1\text{·}000\,0$$

Let us now collect all our information, and show graphically how the basic trigonometrical ratios change as the angle increases from zero to 90°.

Angle	0°	30°	45°	60°	90°
sin	0	0·500 0	0·707 1	0·866 0	1·000 0
cos	1·000 0	0·866 0	0·707 1	0·500 0	0
tan	0	0·577 4	1·000 0	1·732 1	Infinite

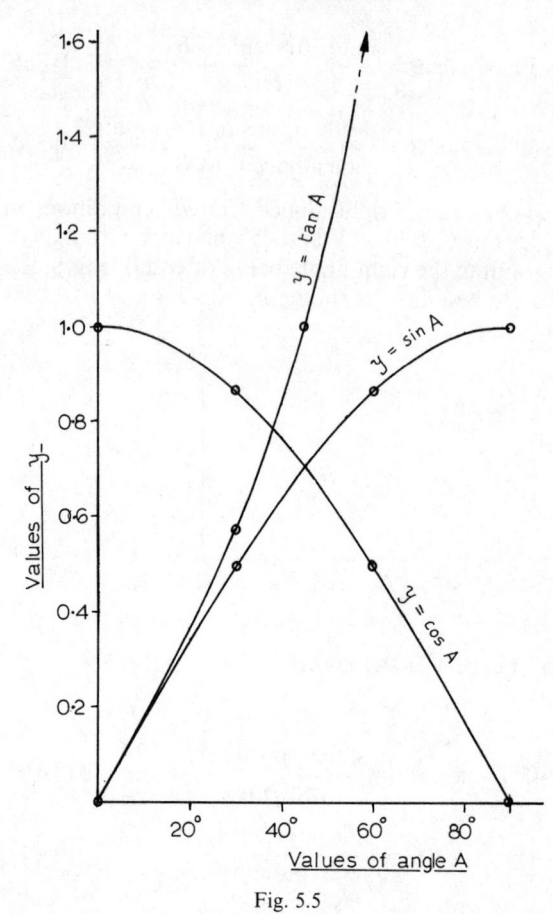

Fig. 5.5

The final graph is shown in Fig. 5.5. The graph is actually incomplete, since as the angle approaches 90° the tangent of an angle approaches a value of infinite magnitude.

In a right-angled triangle, the two angles other than the right angle total 90°. Angles which total 90° are said to be complementary to each other. Referring back to Fig. 5.1:

$$\sin A = \frac{\text{side opposite } A}{\text{hypotenuse}} = \frac{\text{side adjacent to } B}{\text{hypotenuse}} = \cos B$$

Hence the sine of an angle is the cosine of its complementary angle, and vice versa. For example:

$$\sin 60° = \cos 30°$$
$$\sin 80° = \cos 10°$$
$$\cos 25° = \sin 65°$$

5.1.2

The determination of the numerical values of trigonometrical ratios

The technician student does not study mathematics merely as an abstract topic restricted to the manipulation of signs and symbols. It serves as a very useful tool to assist with the solving of problems. In many cases the data given in a problem is used to construct a mathematical model of the problem, such as when an equation is formed. Some equations may include trigonometric ratios.

Example:

A guy-rope of length 30 mm runs from a vertical telegraph pole to the horizontal ground level. The guy-rope makes an angle of 41° to the horizontal. How far is it from the base of the pole to the ground attachment?

The details are illustrated in Fig. 5.6. We are told that the pole is vertical and the ground is horizontal, hence C is 90°. We therefore have a right-angled triangle.

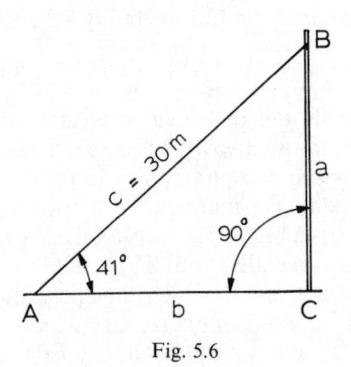

Fig. 5.6

We are told that angle A is 41° and that $c = 30$ m; this is the known side. b is the unknown side.

With problems of this kind, a useful approach is to form the fraction $\dfrac{\text{unknown side}}{\text{known side}}$, and equate this ratio to a trigonometrical ratio of one of the angles.

In this case,
$$\frac{\text{unknown side}}{\text{known side}} = \frac{b}{c} = \cos 41°$$

Hence
$$b = c \cos 41°$$
$$= 30 \text{ m} \times \cos 41°$$
$$= 22 \cdot 641\ 287 \text{ m, by calculator}$$
$$= 22 \cdot 6 \text{ m to three significant figures}$$

Answer:

Distance = 22·6 m

It is a commendable practice to construct a scale diagram, and to measure that diagram to see if any answer is reasonable. In this case, measuring the base line in Fig. 5.6 shows that the answer is quite realistic.

In obtaining the previous answer, the entry to a calculator which has 'cos' appearing as the upper indication on a two-function key, the entry will probably take the sequence:

| clear | 30 | × | 41 | ↑ | cos | pause | = |

The pause has been included because with some calculators it takes a significant amount of time, when compared with other operations, for a calculator to determine the numerical value of the trigonometrical ratio of a particular angle. The numerical value of that trigonometrical ratio should appear in the display before the equals key is depressed. On some calculators, 'cos' appears as the lower indication on a two-function key, and the entry would probably take the sequence,

| clear | 30 | × | 41 | cos | pause | = |

and once again care should be taken to ensure that the value of the cosine of 41° appears in the display before the equals key is depressed.

A calculator which can accept an entry in full sexagesimal measure, or, operating in a converse manner, can display an answer in full sexagesimal measure, is a highly specialised piece of equipment, and its high cost reflects that specialisation. The calculator which our reader will most probably use is one in which the entry and display allow for degrees and decimal fractions of degrees. In which case, the tangent of 33° 45′ is found by an entry requesting the calculator to obtain the tangent of 33·75°. The entry would probably take the sequence:

| clear | 33·75 | ↑ | tan |

(for tan appearing as an 'upper' function on a two-function key), or

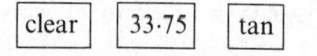

(for tan appearing as a 'lower function' on a two function key).

Some calculators provide a facility for an entry and display for either the circular measure of angles or for degrees and decimal fractions of a degree. In which case, care should be taken that the radian-degree shift is in the proper position. On certain calculators the shift consists of a simple sliding switch with the indication R for radians and D for degrees. It would be appropriate at this juncture to remind our reader that in written work, if no unit is stated, it is implied that the unit is the radian. For example, $\sin \dfrac{\pi}{5}$ means the sine of the angle equal to $\dfrac{\pi}{5}$ radians. For a calculator which only permits an entry in degrees, it is first necessary to establish the equivalent of $\dfrac{\pi}{5}$ radians, which is $\dfrac{\pi}{5} \times \dfrac{360°}{2\pi} = 36°$.

In some problems it may be necessary to determine the magnitude of an angle when a particular trigonometrical ratio is known. For example, we could be asked to find the angle whose tangent is 0·537 426. The mathematical way of asking the question is to say

'determine arctan 0·537 426'.

Similarly, the statement find 'the angle whose sine is' is written 'find arcsin', while for cosine it is 'find arccos'. The symbolic notation once used, of $\sin^{-1}$, $\cos^{-1}$ and $\tan^{-1}$, is now considered to be non-standard. With an entry of probably

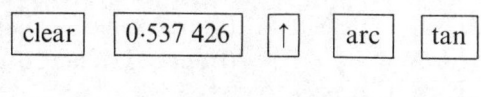

or, perhaps,

the answer is displayed as 28·254 74, to five decimal places. The display is in degrees and decimal fractions of a degree. If the answer is required in sexagesimal measure, we now subtract 28 and multiply the decimal fraction 0·254 74 by 60 to obtain 15·284 4 minutes. We now subtract 15 and multiply 0·284 4 by 60 to obtain 17·064 seconds.

Hence arctan 0·537 426 = 28° 15′ 17″ to the nearest second
= 28·254 7 to four decimal places.

As a matter of personal preference, just as the author uses all the digits that a calculator can provide during the various stages of a series of computations, and/or six decimal places, rounding-off the final answer, and the final answer only, to an appropriate number of significant figures, he normally computes trigometrical ratios using degrees and decimal fractions of a degree, only using sexagesimal measure when he is absolutely constrained to do so. For his particular purposes, sexagesimal measure is an unwarranted nuisance, but he does appreciate that in certain circumstances its use is utterly realistic.

As has been the standard practice in this textbook, our reader is urged to use a calculator if one is available and it can perform the required operation. However, it is necessary to indicate a suitable procedure if a calculator is not available. We have been able to calculate the values of certain of the trigonometrical ratios of 30°, 45° and 90°.

In higher branches of mathematics methods can be deduced by which any trigonometrical ratio of any angle can be determined. The information which we require at present has been worked out and collected in the form of tables. The reader should now open Castle's *Four-figure Mathematical Tables* at the page marked 'natural sines'. The magnitude of the sine of an angle is obtained in a manner somewhat similar to logarithms. It will be noticed that values are quoted fully at intervals of six minutes, and increments are given for intervals of one minute. The reader should now check that:

$$\sin 17° = 0.292\ 4$$
$$\sin 17° \ 12' = 0.295\ 7$$
$$\sin 61° = 0.874\ 6$$
$$\sin 87° \ 42' = 0.999\ 2$$

If we require the sine of an angle such as 21° 38', we cannot obtain this by direct reading. We proceed in the following manner:

$$\sin 21° \ 36' = 0.368\ 1 \quad \text{(nearest direct reading below)}$$
$$\underline{\text{increment for } 2' = \quad 5} \quad \text{(from end columns)}$$
$$\sin 21° \ 38' = 0.368\ 6$$

If we know the magnitude of the sine of an angle, we can find the angle by using the table of sines in a reverse manner to the method previously described. The mathematical way of abbreviating 'the angle whose sine is' is to write *arcsin*. The question 'find the angle whose sine is 0·75' can be written 'find arcsin 0·75' and the answer will be an angle. The reader may have been used to representing 'the angle whose sine is' by $\sin^{-1}$. This is now considered to be a non-standard notation.

Hence   arcsin $0.374\ 6 = 22°$
and   arcsin $0.938\ 5 = 69° \ 48'$

both these values being obtained directly.

For the case of arcsin 0·511 3, this value is not obtained directly. We proceed as follows:

$$\text{nearest direct value below} = 0.510\ 5 = \sin 30° \ 42'$$
$$\underline{\text{addition required} = \quad 8 = \text{increment for } 3'}$$
$$0.511\ 3 = \sin 30° \ 45'$$

One of the most common sources of error is in reading the wrong table. The reader is advised to write in large capitals SIN, in red, at the top of each page of the table of sines. He should now turn over to the table of cosines and write COS in large capitals, in red, at the top of each table of cosines, and then to turn over to the table of tangents and write, at the top of each table of tangents TAN, in large capitals, in red.

We are now looking at our table of tangents, which are read in a similar manner to sines. The reader should check the following values:

$$\tan 23° = 0.424\ 5$$
$$\tan 32° \ 18' = 0.632\ 2$$
$$\tan 67° \ 46' = 2.446\ 2$$

$$\arctan 0.649\ 4 = 33°$$
$$\arctan 1.118\ 4 = 48° \ 12'$$
$$\arctan 1.748\ 5 = 60° \ 14'$$

One important advantage of a calculator is that it will provide certain values quickly which take considerably more time to establish from tables. In the table of tangents it will be observed that the mean differences above 76° cease to be sufficiently accurate. The normal way of determining intermediary values is to use simple proportion.

Example:

Determine, from tables, $\tan 87° \ 14'$

$$\tan 87° \ 18' = 21.20$$
$$\tan 87° \ 12' = 20.45$$
$$\text{difference for } 6' = \overline{\quad 0.75\quad}$$

$87° \ 14'$ is 'two-sixths of the way' from $87° \ 12'$ to $87° \ 18'$

$$\text{increment to be added} = \frac{2}{6} \times 0.75 = \frac{1.50}{6} = 0.25$$

$$\tan 87° \ 14' = 20.45 + 0.25 = 20.70.$$

Another method of procedure, appropriate to Level 1 mathematics (there are others which will be introduced at later stages), can be demonstrated by reference to Fig. 5.7.

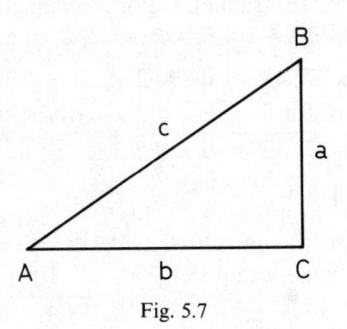

Fig. 5.7

The tangent of angle A is the ratio $\dfrac{a}{b}$. This is the reciprocal of the tangent of angle B. The angle B is the complementary angle of angle A. Hence the tangent of angle A is the reciprocal of the tangent of its complementary angle.

$$\tan 87^\circ\ 14' = \frac{1}{\tan 2^\circ\ 46'} = \frac{1}{0.048\ 4} = 20.66 \quad \text{(from reciprocal tables)}$$

Tan $87^\circ\ 14'$, by calculator, is $\tan 87.23^\circ = 20.693\ 220$ (the last two figures may differ if the calculator renders a display of eight digits).

The greater accuracy which accrues from the use of a calculator is quite apparent from the values which have been obtained for the value of the tangent of $87^\circ\ 14'$. When solving practical problems involving trigonometry our reader is urged to use a calculator where possible. The worked examples in this book must allow for a calculator not being available, and are solved with the aid of tables and the sexagesimal measure of angles. It would be most useful for our reader to check the answers by calculator.

With sines and tangents, as the angle increases, their magnitude increases, and hence the differences for extra minutes must be added. With the cosine, as the angle increases, the cosine decreases. Consequently, for extra minutes the values have to be subtracted. Returning to the table of cosines, the reader is advised now to write plainly in red, at the top of each table, S U B T R A C T M E A N D I F F E R E N C E S.

If the angle is a multiple of six minutes, the value can be read directly. For example:

$$\cos 23^\circ = 0.920\ 5$$
$$\cos 48^\circ\ 24' = 0.663\ 9$$

while in reverse
$$\text{arccos } 0.358\ 4 = 69^\circ$$
$$\text{arccos } 0.925\ 2 = 22^\circ\ 18'$$

For values which cannot be read directly, noting carefully that mean differences have to be subtracted, we proceed as follows:

To find $\cos 37^\circ\ 52'$:

$$
\begin{array}{ll}
\cos 37^\circ\ 48' = & 0.790\ 2 \\
\underline{\text{decrement for } 4' = \qquad\quad 7} & \text{(to be subtracted)} \\
\cos 37^\circ\ 52' = & 0.789\ 5
\end{array}
$$

To find arccos $0.418\ 4$:

$$
\begin{array}{lr}
\text{nearest direct value above} = 0.419\ 5 = \cos 65^\circ\ 12' \\
\underline{\text{decrement required} = \qquad 1\ 1 = \qquad\qquad 4'} \\
0.418\ 4 = \cos 65^\circ\ 16'
\end{array}
$$

This article was introduced by a reference to the establishment of the numerical value of a number multiplied by another number which was the trigonometrical ratio of an angle. Our reader was urged that if a calculator is available which can perform the operation, that calculator should be used. However, a calculator may not be available, and it has already been emphasised that a slide rule needs very careful manipulation to produce an accurate answer. Some, but certainly not all, slide rules incorporate scales of sines, of cosines and of tangents, but in normal circumstances, if a calculator is not available, our reader will use mathematical tables.

Let us take for example the establishment of a numerical value for 37·72 tan 43° 21′. The log of the answer is the log of 37·72 plus the log of the tangent of 43° 21′. There is no point in looking up tan 43° 21′ and then finding its logarithm. Pages 20 to 25 of Castle's *Four-figure Mathematical Tables* gives the values directly. One of the most common forms of error is reading the wrong table, and so our reader is recommended to print clearly LOG SIN in red at the top of pages 20 and 21, LOG COS at the top of pages 22 and 23, and LOG TAN at the top of pages 24 and 25. It is recommended that a return now be made to pages 22 and 23, and the words SUBTRACT MEAN DIFFERENCES be printed clearly in red at the top of each page.

Values are obtained in a similar manner to ordinary trigonometric ratios. Log tan 43° 21′ is $\bar{1}\cdot974\ 2 + 8 = \bar{1}\cdot975\ 0$. The calculation previously referred to can be set out

No.	Log
37·72	1·576 5
tan 43° 21′	$\bar{1}$·975 0
35·60	1·551 5

Reference to the tables will show that for log sin below 4°, for log cos above 86°, for log tan below 4° and for log tan above 86°, the mean differences are insufficiently accurate. In these regions simple proportion should be used for intermediary values.

5.1.3
The construction of an angle of any magnitude

The steps which led to the construction shown in Fig. 5.8 were:

1. A horizontal base line was drawn.
2. Two points, O and A were marked on this line.
3. A perpendicular was erected from point A.
4. With centre O, an arc of radius OA was drawn.

5. A straight line *OP* was drawn to intersect the arc and the perpendicular.
6. A point *B* was marked where the straight line intersected the arc.
7. A point *C* was marked where the straight line intersected the perpendicular.
8. A perpendicular was dropped from the point *B* to a point *D* on the horizontal base line.
9. An angle θ, angle *POA*, has been indicated.

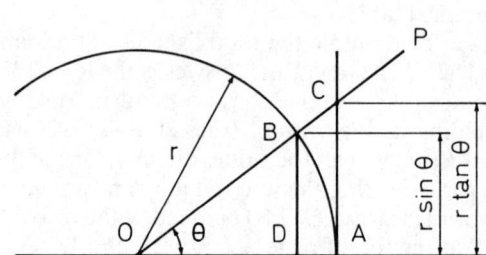

Fig. 5.8

If we denote the radius *OA* by *r*,

then
$$BD = r \sin \theta$$
$$OD = r \cos \theta$$
and
$$AC = r \tan \theta$$

The construction can be used to layout an angle of any given magnitude, and any one of the equations can be used. It is highly probable that the most convenient will be to layout the base line and mark point *O*. From *O* a distance can be set out so that *OA* is a convenient value for multiplication. For ordinary college work 100 mm (= 10 cm) is very attractive. A perpendicular from *A* can be erected, either by geometrical construction, with the aid of a protractor, or with the aid of a set square.

Let us designate the angle to be constructed by θ. If θ lies between zero and $90°$, a point *C* can be marked on the perpendicular so that $AC = OA \tan \theta$. If $OA = 100$ mm and $\theta = 36° \, 18'$, then $AC = 100 \tan 36° \, 18' = 100 \times 0.7346 =$ approximately 73·5 mm. A line joining *O* to *C* will give an angle *COA* of $36° \, 18'$.

Figure 5.9 shows how the construction can be adapted to cover other angles in the range $90°$ to $360°$. For instance, to construct an angle of $142°$ we can construct an angle of $142° - 90° = 52°$ in the first quadrant, and then use a set square or a geometrical construction to erect a perpendicular to the sloping line of the angle of $52°$.

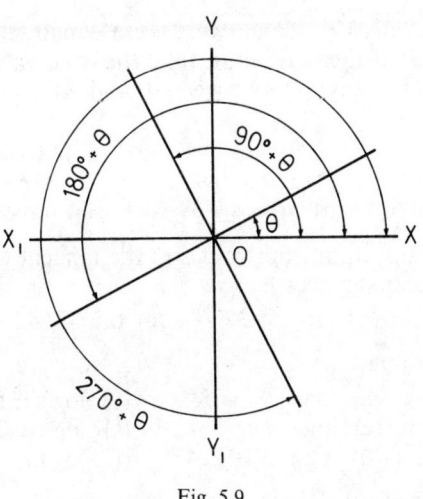

Fig. 5.9

Problems 5.1

1. Find the values of:
   (a) sin 22°   (b) sin 67°   (c) sin 8° 12′
   (d) sin 81° 18′ (e) sin 29° 53′ (f) sin 62° 3′

2. Find the values of:
   (a) cos 14°   (b) cos 53°   (c) cos 62° 48′
   (d) cos 8° 30′ (e) cos 0° 50′ (f) cos 69° 9′

3. Find the values of:
   (a) tan 81°   (b) tan 16°   (c) tan 0° 48′
   (d) tan 52° 30′ (e) tan 9° 51′ (f) tan 62° 20′

4. Find the values, where the angle lies between 0° and 90°, of:
   (a) arcsin 0·927 2 (b) arcsin 0·374 6 (c) arcsin 0·504 5
   (d) arcsin 0·917 8 (e) arcsin 0·750 0 (f) arcsin 0·380 6

5. Find the values, where the angle lies between 0° and 90°, of:
   (a) arccos 0·515 0 (b) arccos 0·970 3 (c) arccos 0·863 4
   (d) arccos 0·400 3 (e) arccos 0·449 6 (f) arccos 0·699 9

6. Find the values, where the angle lies between 0° and 90°, of:
   (a) arctan 3·270 9 (b) arctan 0·900 4 (c) arctan 2·808 3
   (d) arctan 0·240 1 (e) arctan 2·001 4 (f) arctan 0·520 2

7. In the following question, each answer should state one value only,
   but if two answers are equally appropriate from four-figure tables,

take the even value. If three answers seem equally appropriate, take the central value. If there are four, take the even value of the central two, and so on. The angles lie between 0° and 90°.

Find the values of:

(a) arcsin 0·261 4   (b) arccos 0·307 8   (c) arctan 1·590 2
(d) arcsin 0·763 2   (e) arcsin 0·987 2   (f) arccos 0·342 7
(g) arccos 0·990 6   (h) arccos 0·999 0   (j) arccos 0·996 0

8. Using simple proportion from values in the tangent tables, determine, to two decimal places:

(a) tan 85° 27'   (b) tan 85° 37'   (c) tan 87° 43'

9. Determine the values of:

(a) log sin 72°   (b) log sin 43° 54'   (c) log sin 53° 8'
(d) log cos 33°   (e) log cos 8° 36'   (f) log cos 22° 17'
(g) log tan 18°   (h) log tan 10° 24'   (i) log tan 36° 52'

10. Using simple proportion between appropriate values, determine the values, to four decimal places, of:

(a) log sin 2° 43'   (b) log cos 88° 17'
(c) log tan 3° 46'   (d) log tan 87° 8'

11. Determine, from the tables of the logarithms of trigonometrical ratios the angle whose:

(a) log sin is $\bar{1}$·938 4   (b) log tan is 0·525 2
(c) log cos is $\bar{1}$·694 1   (d) log sin is $\bar{2}$·969 7

12. Determine the values of:

(a) $\sin \dfrac{\pi}{2}$ (b) $\cos \dfrac{3\pi}{8}$ (c) $\tan \dfrac{\pi}{4}$ (d) $\tan \dfrac{\pi}{12}$

13. Determine the values, to three significant figures, of:

(a) V if $V = 400 \sin 100\pi r$ and $r = 0\cdot002$

(b) E if $E = 100 \cos\left(100\pi r - \dfrac{\pi}{2}\right)$ and $r = 0\cdot007\,5$

(c) P if $P = 20 \sin\left(120\pi r + \dfrac{\pi}{12}\right)$ and $r = \dfrac{1}{480}$

14. Using trigonometrical ratios, construct angles of:

(a) 53° 8'   (b) 43°   (c) 22° 37'   (d) 141° 30'

Answers to Problems 5.1

1. (a) 0·374 6   (b) 0·920 5   (c) 0·142 6
   (d) 0·988 5   (e) 0·498 3   (f) 0·883 3

5.2.2
The area of a regular polygon

Consider any regular polygon whose side has a length of L, the polygon having n sides. If lines are drawn from adjacent corners to the centre of the polygon, we shall obtain an isosceles triangle. A polygon of n sides with lines drawn to the centre from every corner will produce n congruent isosceles triangles. The area of the polygon will be n times the area of a single isosceles triangle. The total of the apex angles of all n congruent triangles is 360°, hence the apex angle of a single triangle is $\dfrac{360°}{n}$. One such triangle is shown in Fig. 5.11(a).

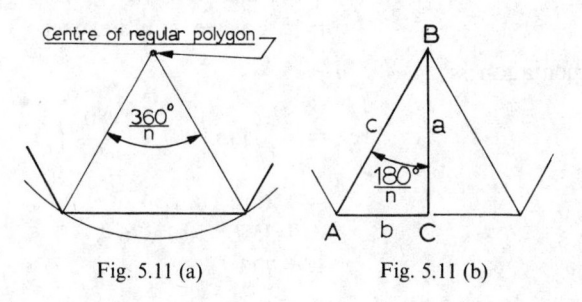

Fig. 5.11 (a)   Fig. 5.11 (b)

If we bisect the apex angle of an isosceles triangle, the bisector is at right angles to the base. Hence in Fig. 5.11(b), ABC is a right-angled triangle and angle B is half of $\dfrac{360°}{n}$, i.e. $\dfrac{180°}{n}$.

Angle A is the complementary angle to angle B, and angle $A = 90° - \dfrac{180°}{n}$.

Side b is half the length of side L, or $b = \dfrac{L}{2}$.

Perpendicular height of triangle $ABC = a$. In the right-angled triangle ABC:

$$\frac{a}{b} = \tan A, \ a = b \tan A, \ a = \frac{L}{2} \tan\left(90° - \frac{180°}{n}\right)$$

Area of large triangle $= \frac{1}{2}$ base $\times$ perpendicular height
$$= ba$$

$$= \frac{L}{2} \times \frac{L}{2} \tan\left(90° - \frac{180°}{n}\right)$$

Area of polygon having n sides = total area of n congruent triangles as Fig. 5.11(a)

$$= n\left\{\frac{L}{2} \times \frac{L}{2} \tan\left(90° - \frac{180°}{n}\right)\right\}$$

$$= \frac{nL^2}{4} \tan\left(90° - \frac{180°}{n}\right)$$

For a square: $n = 4$

Hence   area $= \dfrac{4L^2}{4} \tan\left(90° - \dfrac{180°}{4}\right)$

$$= L^2 \tan 45° = L^2 \times 1$$
$$= L^2, \text{ as we would expect}$$

For a pentagon: $n = 5$

Hence   area $= \dfrac{5L^2}{4} \tan\left(90° - \dfrac{180°}{5}\right)$

$$= 1{\cdot}25\, L^2 \tan 54°$$
$$= 1{\cdot}25\, L^2 \times 1{\cdot}376\,4$$
$$= 1{\cdot}720\, L^2$$

For a hexagon: $n = 6$

Hence   area $= \dfrac{6L^2}{4} \tan\left(90° - \dfrac{180°}{6}\right)$

$$= 1{\cdot}5\, L^2 \tan 60°$$
$$= 1{\cdot}5\, L^2 \times 1{\cdot}732\,1$$
$$= 2{\cdot}598\, L^2$$

For an octagon: $n = 8$

Hence   area $= \dfrac{8L^2}{4} \tan\left(90° - \dfrac{180°}{8}\right)$

$$= 2L^2 \tan 67° \; 30'$$
$$= 2L^2 \times 2{\cdot}414\,2$$
$$= 4{\cdot}828\, L^2$$

If the polygon has an even number of sides, opposing sides are parallel. A convenient way of indicating the size of a regular polygon with an even number of sides is to state the distance between opposing sides. In engineering this is known as the *distance across flats*.

Let us use a method similar to the previous method to obtain area of any regular polygon having an even number of sides when the distance across flats is W.

Referring back to Fig. 5.11(b), the distance a will be $\dfrac{W}{2}$ and the angle $B = \dfrac{180°}{n}$. Hence:

$$b = a \tan B = \frac{W}{2} \tan \frac{180°}{n}$$

$$\text{Area of one triangle} = \frac{\text{base} \times \text{vertical height}}{2}$$

$$= \frac{W}{2} \tan \frac{180°}{n} \times \frac{W}{2}$$

$$= \frac{W^2}{4} \tan \frac{180°}{n}$$

$$\text{Area of } n \text{ congruent triangles} = \frac{nW^2}{4} \tan \frac{180°}{n}$$

Hence the area of any regular polygon having an even number of sides n and measuring W across the flats is given by:

$$\frac{nW^2}{4} \tan \frac{180°}{n}$$

For a square: $n = 4$

Hence $\quad \text{area} = \dfrac{4W^2}{4} \tan \dfrac{180°}{n} = W^2 \tan 45°$

$$= W^2, \text{ as we would expect}$$

For a hexagon: $n = 6$

Hence $\quad \text{area} = \dfrac{6W^2}{4} \tan \dfrac{180°}{6} = \dfrac{6W^2 \tan 30°}{4}$

$$= \frac{6W^2 \times 0.577\,4}{4} = 0.866\,W^2$$

For an octagon: $n = 8$

Hence $\quad \text{area} = \dfrac{8W^2}{4} \tan \dfrac{180°}{8} = \dfrac{8W^2 \tan 22\frac{1}{2}°}{4}$

$$= 2W^2 \times 0.414\,2 = 0.828\,W^2$$

Example:

Find the area of a regular octagon whose distance across flats is 5 mm.

$$\text{Area} = \frac{nW^2}{4} \tan \frac{180°}{n}$$

$$= \frac{8(5^2)}{4} \tan \frac{180°}{8} = 50 \tan 22° \, 30'$$

$$= 50 \times 0\cdot414\,2 = 20\cdot71 \text{ mm}^2$$

Answer:

$$\text{Area of octagon} = 20\cdot7 \text{ mm}^2$$

5.2.3
The area of a segment of a circle

In article 4.4.5 our reader was requested, for the time being, to consider the area of a segment of a circle as being the difference between the area of a sector and the area of a triangle. Let us use the formula for the area of a triangle deduced in the previous article to have another look at the area of a segment. Our reader is reminded that unless instructions are given to the contrary the segment bounded by a chord and an arc is taken as the minor of the two segments into which a chord divides a circle.

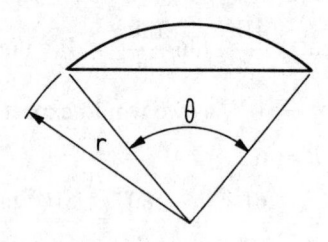

Fig. 5.12

In Fig. 5.12, we have a segment of a circle of radius r. The radii drawn to the ends of the segment contain an angle of θ.

Area of segment = area of sector − area of triangle

$$\text{Area of sector} = \frac{r^2\theta}{2}$$

Area of triangle = half product of two sides and the sine of the
included angle

$$= \tfrac{1}{2}(r)(r)\sin\theta$$

$$= \frac{r^2\sin\theta}{2}$$

Area of segment $= \dfrac{r^2\theta}{2} - \dfrac{r^2\sin\theta}{2}$

$$= \frac{r^2}{2}(\theta - \sin\theta)$$

Since no unit is quoted for θ, it is implied that the angle is in radians.
The formula requires the knowledge of the angle before a determination
of the area can be made. It would be most unusual to find a segment to
be dimensioned in such a way that it is quoted directly. In many cases
θ has to be established from other dimensions. In Fig. 5.13, the more
usual information is illustrated.

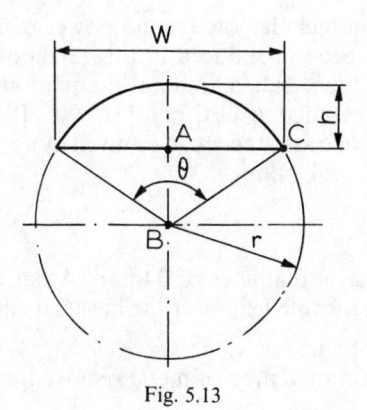

Fig. 5.13

If the length of the chord W and the radius r are known, then:

$$\frac{\dfrac{W}{2}}{r} = \sin\frac{\theta}{2} \text{ whence } \frac{\theta}{2} = \arcsin\frac{W}{2r}$$

from which θ can be established.

If two of the variables h, W and r are known, the third can be determined.

In the right-angled triangle ABC:

$$AB = r - h \qquad AC = \frac{W}{2} \qquad BC = r$$

According to Pythagoras:

$$(AB)^2 + (AC)^2 = (BC)^2$$

$$(r - h)^2 + \left(\frac{W}{2}\right)^2 = r^2$$

$$r^2 - 2rh + h^2 + \frac{W^2}{4} = r^2$$

$$2rh = h^2 + \frac{W^2}{4}$$

$$r = \frac{W^2}{8h} + \frac{h}{2}$$

This formula is easy to apply if r or L have to be determined. If the unknown is h, manipulation of the equation to bring the terms including the unknown on one side produces a further equation

$$4h^2 - 8rh + W^2 = 0$$

This latter equation includes the second power of the unknown. Our reader has not yet been introduced in this textbook to the solving of equations of this type, called quadratic equations, since quadratic equations are not included in TEC unit U75/005. If the unknown h has to be determined, some assistance can be given by applying the Theorem of Pythagoras to actual values.

Example:

A circular bar has a diameter of 34 mm. A flat of width 16 mm is machined along the whole length of the bar. Calculate:

(a)   the depth of cut;
(b)   the cross-sectional area of metal removed, in mm^2.

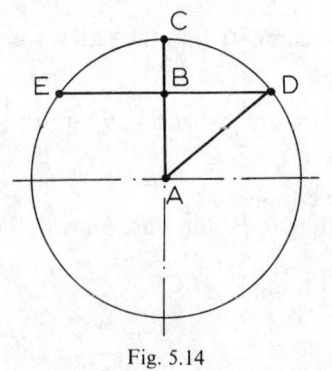

Fig. 5.14

Referring to Fig. 5.14:

$$AD = \frac{\text{diameter}}{2} = 17 \text{ mm}$$

$$BD = \text{half width of flat} = 8 \text{ mm}$$
$$(AD)^2 = (BA)^2 + (BD)^2$$
$$(BA)^2 = (AD)^2 - (BD)^2$$
$$= 17^2 - 8^2 = 289 - 64$$
$$= 225$$
$$BA = \sqrt{225} = 15$$
$$\text{Depth of cut} = CB = \text{radius } CA - BA$$
$$= 17 - 15 = 2 \text{ mm}$$

In the triangle ABD:

$$\text{Angle } A = \arcsin \tfrac{8}{17} = \arcsin 0.4706 = 28° \; 5'$$
$$\text{Angle } \theta \text{ of segment} = \text{angle } EAD = 2(28° \; 5')$$
$$= 56° \; 10'$$
$$56° \; 10' = 0.980 \; 3 \text{ radians}$$
$$\sin 56° \; 10' = 0.830 \; 6$$

$$\text{Area of segment} = \frac{r^2}{2}(\theta - \sin \theta)$$

$$= \frac{17 \times 17}{2}(0.980 \; 3 - 0.830 \; 6)$$

$$= \frac{289 \times 0.149 \; 7}{2} = 21.63 \text{ mm}^2$$

No.	Log
289	2·460 9
0·149 7	$\overline{1}$·175 2
	1·636 1
2	0·301 0
21·63	1·335 1

Rough check:

$$\frac{280 \times 0.14}{2} = 19.6$$

Answers:

Depth of cut $= 2$ mm
Cross-sectional area $= 21.6$ mm^2

The formula for the area of a segment, viz:

$$A = \frac{r^2}{2}(\theta - \sin \theta)$$

is valid for all values of θ from zero to 2π ($=360°$). However, TEC unit U75/005 does not cover the trigonometrical ratios of angles greater than $\frac{\pi}{2}$ ($=90°$). This occurs at Level 2, at which time the area of a segment will be given further consideration. At Level 1, a minor segment can be considered to be a sector minus an isosceles triangle. If the angle between the equal sides of that triangle is greater than 90°, some method other than using $\frac{r^2}{2}(\theta - \sin \theta)$ must be used to determine its area.

Problems 5.2

1. Find the area of a triangle where $a = 82$ mm, $b = 40$ mm and $C = 30°$.

2. Find the area of a triangle where $A = 73° \, 15'$, $b = 47·6$ mm and $c = 28·5$ mm.

3. The equal sides of an isosceles triangle are each of length 82·6 mm, and the angle included by those sides is 58°. Calculate the area of the triangle.

4. A rhombus has sides of length 5 mm, one of the acute angles of the rhombus being 45°. Calculate the area of the rhombus.

5. A parallelogram includes two adjacent sides of length 20 mm and 32 mm, the angle between those sides being 40°. Calculate the area of the parallelogram.

6. Calculate the area of a regular polygon having 20 sides, the distance across flats being 16 mm.

7. Calculate the area of a regular octagon:
   (a) 10 mm across flats
   (b) with side of length 5 mm.

8. Calculate the area of a regular hexagon:
   (a) 5 mm across flats
   (b) with side of length 3 mm.

9. Calculate the length of side of a regular pentagon which has an area of 100 mm².

10. Calculate the distance across flats of a regular hexagon which has an area of 140 mm².

11. A minor segment of a circle has a radius of 20 mm and an area of 50 mm². By plotting values of (θ radians $-$ sin θ) from 60° to 65° at intervals of one degree, find the angle subtended by the ends of the chord at the centre of the circle, to the nearest degree.

12. A segment is less than a semicircle and is bounded by an arc of radius 30 mm and a straight line of length 40 mm. Determine the area of the segment.

13. An open channel is semicircular in cross-section, the radius being 5 m. At a certain instant the depth of water at the centre of the channel is 2 m. Determine the cross-sectional area of the water flow.

14. What percentage of metal is removed when a flat of width 30 mm is machined on a circular bar of diameter 60 mm?

Answers to Problems 5.2

1. 820 mm²

2. 650 mm²

3. 2 890 mm²

4. 17·7 mm²

5. 411 mm²

6. 203 mm²

7. (a) 82·8 mm²   (b) 121 mm²

8. (a) 21·6 mm²   (b) 23·4 mm²

9. 7·62 mm

10. 12·7 mm

11. 64°

12. 210 mm²

13. 11·2 m²

14. 2·88%

5.3
Practical applications of trigonometry

We shall now proceed by demonstrating, with the aid of series of worked examples, how trigonometry can be used to solve a variety of problems. It must be stated at the outset that a particular problem involving trigonometry can often be solved by a variety of methods, but as far as our reader is concerned it must be assumed that our reader's knowledge of trigonometry is restricted to the contents of this book, i.e. the ratios of sine, cosine, and tangent obtained from the sides of a right-angled triangle. It could well be that some of our readers have a knowledge of trigonometry beyond the contents of this book, and could well be in a position to suggest a much quicker solution.

A typical example occurs when the angles of a triangle have to be found given the lengths of the sides. There is a specific trigonometrical rule which can be applied, but a study of this rule is not included in TEC unit U75/005. This should not cause our reader to worry unduly. It is rarely a serious fault to have to revert to basic fundamentals to solve problems. If we are given the three sides of a triangle, we can determine its area. By dropping a perpendicular from an angle to its opposite side, or a continuation of that side, we can form two right-angled triangles. The length of the perpendicular can then be found from $\triangle = \frac{1}{2}$ base $\times$ perpendicular height, and then perpendiculars can be used to determine angles.

Certain advice can be proferred to our reader:

1. A scale drawing should be made of the data provided, if a drawing is not already given.
2. If right-angled triangles are not specifically included in the data, then conditions should be investigated which produce right-angles, typical of which are:

   (a) a perpendicular bisector cuts a line at right angles;
   (b) a tangent line to a circle and a radius to the point of contact lie at right angles to each other;
   (c) a perpendicular height of a triangle lies at right-angles to an appropriate base.

3. Once an answer has been determined, the scale drawing should be measured to ensure that the answer is realistic.

Let us continue with a series of worked examples, including references to specific items in individual problems where appropriate.

Example:

Five holes lie equally spaced on a pitch circle diameter of 80 mm. Determine the chordal distance between two adjacent holes.

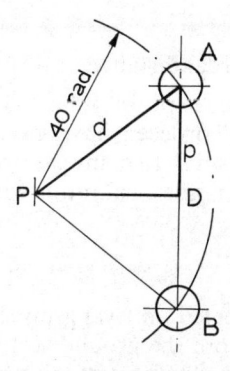

Fig. 5.15

Figure 5.15 shows the details, *P* being the centre of the circle, and *A* and *B* being two adjacent holes. We have no right-angled triangle thus far, but *PAB* is an isosceles triangle. We know that if we bisect the angle contained by the two equal sides we can form two congruent right-angled triangles. These are the triangles *PAD* and *PDB*, *D* being the mid-point of *AB*.

$$d = \text{pitch circle radius} = \frac{80 \text{ mm}}{2} = 40 \text{ mm}$$

In triangle *PAB*:

$$\text{angle } P = \frac{360°}{5} = 72°$$

Now triangle *PAD* is right-angled. In triangle *PAD*:

$$\text{angle } P = \frac{72°}{2} = 36°$$

and

$$p = \frac{AD}{2} = \text{unknown side}$$

$$\frac{AD}{AP} = \frac{\text{side opposite } P}{\text{hypotenuse}} = \sin P = \sin 36°$$

$$\therefore AD = AP \times \sin 36° = 40 \text{ mm} \times 0.587\ 8$$
$$= 23.512 \text{ mm}$$
$$AB = \text{twice } AD = 47\ 024 \text{ mm}$$

Scaling the diagram shows the value to be reasonable.

Answer:

Chordal distance = 47·0 mm

The reasoning in the above problem has been given in detail. As skill is acquired, the reader will proceed, having sketched the diagram, to an expression such as $p = d \sin P$. In many cases a general idea of the trend of the magnitude of trigonometrical ratios will often indicate which of the ratios is correct.

Example:

A telegraph pole is erected on level ground. An observer notes that the angle of elevation from the ground to the top of the pole is 32°. From a point 15 m nearer the base of the pole, the angle of elevation is 45°. Calculate the height of the pole.

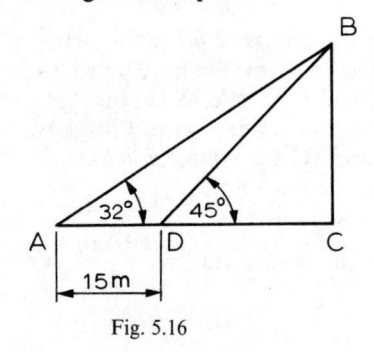

Fig. 5.16

Referring to Fig. 5.16, the pole is represented by BC.

$$\frac{BC}{DC} = \tan 45° = 1$$

$$\therefore BC = DC$$

In the triangle ABC:

$$\frac{BC}{AC} = \tan 32°$$

$$AC = AD + DC = 15 + BC, \text{ since } BC = DC$$

$$\therefore \frac{BC}{15 + BC} = \tan 32°, BC = (15 + BC) \tan 32°$$

$$BC = (15 + BC)\, 0·624\,9, \quad BC = 9·373\,5 + 0·624\,9\,BC$$

$$BC - 0·624\,9\,BC = 9·373\,5, \qquad 0·375\,1\,BC = 9·373\,5$$

$$BC = \frac{9·373\,5}{0·375\,1} = 25·00$$

No.	Log
9·373 5	0·971 9
0·975 1	$\bar{1}$·574 1
25·00	1·397 8

Rough check:

$$\frac{9\frac{3}{8}}{\frac{3}{8}} = \frac{75}{8} \times \frac{8}{3} = 25$$

Answer:

$$\text{Height of pole} = 25 \cdot 0 \text{ m}$$

Example:

Referring to the holes shown in Fig. 5.17, obtain two different expressions for the height h in terms of x. Equate these expressions and hence determine the value of x.

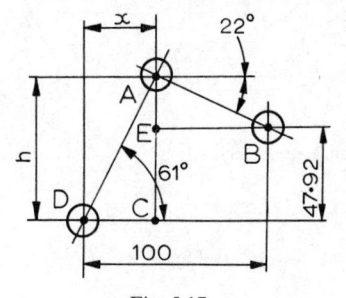

Fig. 5.17

In triangle ADC, $h = AC = DC \tan 61°$
$$= x \times 1·804\ 0$$
$$= 1·804\ 0\ x$$

In triangle ABE, $BE = 100 - x$
$$AE = BE \tan 22°$$
$$= (100 - x)\ 0·404\ 0$$
$$= 40·4 - 0·404\ 0\ x$$

$$h = AE + 47·92 = 40·4 - 0·404\ 0\ x + 47·92 = 88·32 - 0·404\ 0\ x$$

By equating values of h:

$$1\cdot804\ 0\ x = 88\cdot32 - 0\cdot404\ 0\ x$$
$$1\cdot804\ 0\ x + 0\cdot404\ 0\ x = 88\cdot32$$
$$2\cdot208\ 0\ x = 88\cdot32$$
$$x = \frac{88\cdot32}{2\cdot208} = 40$$

Answer:

$$x = 40 \text{ mm}$$

Example:

Figure 5.18 shows the cross-section of a strengthening bracket for an instrument panel. Neglecting the thickness of material, calculate the length between the two ends when the section is opened out flat.

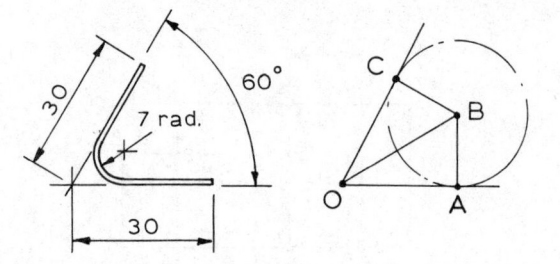

Fig. 5.18

Referring to Fig. 5.18:

$$\text{Angle } BOA = \frac{60°}{2} = 30°$$

$$\text{Angle } ABO = 90° - 30° = 60°$$
$$\text{Angle } ABC = \text{twice angle } ABO = 2 \times 60° = 120°$$

$$\frac{OA}{OB} = \tan 60°$$

$$OA = OB \tan 60° = 7 \times 1\cdot732\ 1 = 12\cdot124\ 7$$

Length of flat of each leg $= 30 - 12\cdot124\ 7 = 17\cdot875\ 3$

$$\text{Length of arc} = \frac{\theta}{360°}(2\pi r) = \frac{120°}{360°}\left(\frac{2 \times 22 \times 7}{7}\right)$$

$$= \frac{44}{3} = 14{\cdot}666\ 7$$

$$\text{Developed length} = 2(17{\cdot}875\ 3) + 14{\cdot}666\ 7$$
$$= 35{\cdot}750\ 6 + 14{\cdot}666\ 7$$
$$= 50{\cdot}417\ 3$$

Answer:

$$\text{Developed length} = 50{\cdot}4 \text{ mm}$$

Example:

In the triangle ABC, $a = 7$, $b = 13$ and $c = 10$.

(a) Calculate the area of the triangle.
(b) Determine the magnitudes of the angles A, B and C.

$$\triangle = \sqrt{s(s - a)(s - b)(s - c)}$$

$$s = \frac{a + b + c}{2} = \frac{7 + 13 + 10}{2} = \frac{30}{2} = 15$$

$$s - a = 15 - 7 = 8 \quad s - b = 15 - 13 = 2 \quad s - c = 15 - 10 = 5$$

$$\triangle = \sqrt{15 \times 8 \times 2 \times 5} = \sqrt{1\ 200} = 34{\cdot}64 \text{ (from tables).}$$

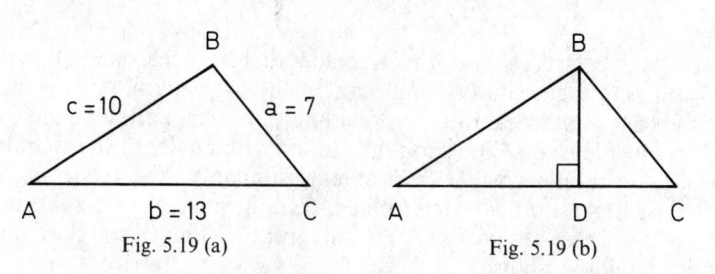

Fig. 5.19 (a)   Fig. 5.19 (b)

The triangle is shown in Fig. 5.19(a). In Fig. 5.19(b), a perpendicular has been dropped from B to the opposite side b.

$$\triangle = \frac{bh}{2} = \frac{b \times BD}{2}$$

$$BD = \frac{2 \times \triangle}{b}$$

$$BD = \frac{2 \times 34{\cdot}64}{13} = \frac{69{\cdot}28}{13} = 5{\cdot}329\ 2$$

Referring to triangle ABD,

$$\sin A = \frac{BD}{AB} = \frac{5 \cdot 329\ 2}{10} = 0 \cdot 532\ 9$$

$$A = \arcsin 0 \cdot 532\ 9 = 32° \ 12'$$

Referring to triangle BDC:

$$\sin C = \frac{BD}{BC} = \frac{5 \cdot 329\ 2}{7} = 0 \cdot 761\ 3$$

$$C = \arcsin 0 \cdot 761\ 3 = 49° \ 35'$$

$$A + B + C = 180°$$

$$B = 180° - A - C = 180° - (A + C)$$

$$A + C = 32° \ 12' + 49° \ 35' = 81° \ 47'$$

$$B = 180° - 81° \ 47' = 98° \ 13'$$

Answers:

(a)   Area of triangle $= 34 \cdot 6$

(b) $A = 32° \ 12'$, $B = 98° \ 13'$, $C = 49° \ 35'$

It would be useful to inform our reader that it is necessary at this stage of studies to ensure that an angle is acute before determining the values of the trigonometrical ratios of that angle. This can often be indicated by a scale diagram. Care should be taken with angles in the vicinity of 90°. In general terms, with problems similar to the foregoing, the smallest angle should be determined, then the next largest. Subtracting the sum of these angles from 180° will give the value of the third angle, which could be obtuse.

A useful approach to certain problems is to establish the co-ordinates of significant features from some convenient datum, as will be demonstrated in the example which follows.

Example:

Figure 5.20 shows a jig-plate. Calculate:

(a)   the checking distance x between the centres of the two holes;

(b)   the angle marked θ.

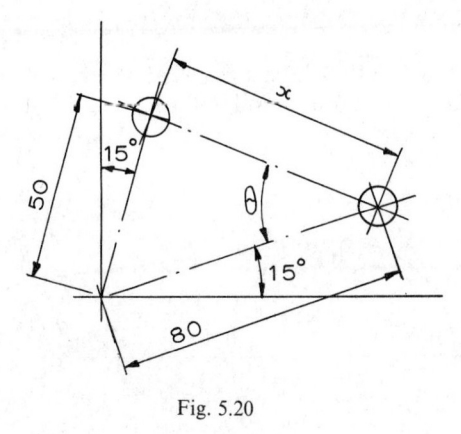

Fig. 5.20

Use the datum of dimensions in the bottom left-hand corner as the origin.

Co-ordinates of top hole = (50 sin 15°, 50 cos 15°)
= (50 × 0·258 8, 50 × 0·965 9)
= (12·940, 48·295)

Co-ordinates of bottom hole = (80 cos 15°, 80 sin 15°)
= (80 × 0·965 9, 80 × 0·258 8)
= (77·272, 20·704)

Distance between holes = $\sqrt{(77 \cdot 272 - 12 \cdot 940)^2 + (48 \cdot 295 - 20 \cdot 704)^2}$
= $\sqrt{64 \cdot 332^2 + 27 \cdot 591^2}$
= $\sqrt{4\,138 + 761 \cdot 2}$
= $\sqrt{4\,899 \cdot 2} = 70 \cdot 00$

Angle at which line joining holes lies to the horizontal

$$= \arcsin \left(\frac{\text{vertical distance between holes}}{\text{true distance between holes}} \right)$$

$$= \arcsin \frac{27 \cdot 591}{70} = \arcsin 0 \cdot 394\,2 = 23° \, 13'$$

Angle $\theta = 23° \, 13' + 15° = 38° \, 13'$

Answers:

    (a)   Distance $x = 70 \cdot 0$ mm
    (b)   Angle $\theta = 38° \, 13'$

Example:

Calculate the length of the belt of an open belt drive connecting pulleys of diameter 100 mm and 240 mm spaced at a centre distance of 250 mm.

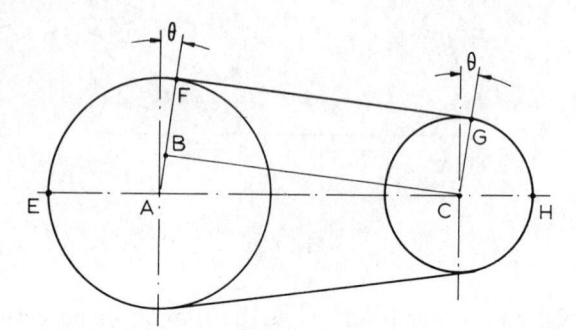

Fig. 5.21

Working throughout in centimetres, in triangle ABC (Fig. 5.21):

$$AC = 25$$
$$AB = 12 - 5 = 7$$

$$\sin \angle ACB = \frac{AB}{AC} = \frac{7}{25} = 0.28$$

$$\angle ACB = 16° \ 16'$$
$$\text{Length of belt} = 2(\text{arc } EF + FG + \text{arc } GH)$$

The angles θ are equal to $\angle ACB = 16° \ 16'$.

$$\text{Arc } EF = r\theta = 12(90° + 16° \ 16'), \text{ when the angle is}$$
$$\text{expressed in radians}$$
$$= 12(1.570\ 8 + 0.283\ 9)$$
$$= 12(1.854\ 7)$$
$$- 22.256\ 4$$
$$FG = BC = \sqrt{\{(AC)^2 - (AB)^2\}} = \sqrt{\{25^2 - 7^2\}}$$
$$= \sqrt{\{625 - 49\}} = \sqrt{576} = 24$$
$$\text{Arc } GH = r\theta = 5(90° - 16° \ 16'), \text{ when the angle is}$$
$$\text{expressed in radians}$$
$$= 5(1.570\ 8 - 0.283\ 9)$$
$$= 5(1.286\ 9)$$
$$= 6.434\ 5$$

∴ Length of belt = 2(22·256 4 + 24 + 6·434 5)
$$= 2(52·690 9)$$
$$= 105·381 8 \text{ cm} = 1\ 053·818 \text{ mm}$$

Answer:

Length of belt $= 1\ 050$ mm

Example:

Calculate the length of the belt of a crossed belt drive connecting pulleys of diameter 300 mm and 180 mm spaced at a centre distance of 400 mm.

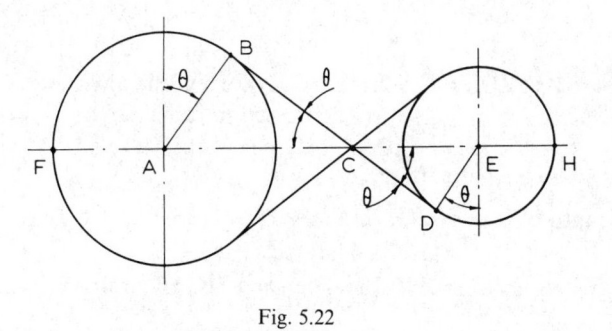

Fig. 5.22

Referring to Fig. 5.22 and working throughout in centimetres:

Triangles ABC and CDE are similar.
Let $AC = x$, then $CE = 40 - x$.

$$\frac{AB}{AC} = \frac{DE}{CE}$$

$$\therefore \frac{15}{x} = \frac{9}{40 - x}$$

$$\therefore 15(40 - x) = 9x$$
$$600 - 15x = 9x$$
$$600 = 24x$$
$$x = 25$$

In Fig. 5.22, the four angles θ are equal. In triangle ACB:

$$\sin C = \frac{15}{25} = 0.6$$

$$C = 36° \, 52'$$

i.e.
$$\theta = 36° \, 52'$$

Length of belt $= 2(\text{arc } FB + BC + CD + \text{arc } DH)$

Arc $FB = r\theta = 15(90° + 36° \, 52')$, when the angle is
expressed in radians

$$= 15(1.570\ 8 + 0.643\ 5)$$
$$= 15(2.214\ 3)$$
$$= 33.214\ 5$$

$$BC = \sqrt{\{(AC)^2 - (AB)^2\}} = \sqrt{\{25^2 - 15^2\}}$$
$$= \sqrt{\{625 - 225\}} = \sqrt{400} = 20$$

$$CD = \sqrt{\{(CE)^2 - (DE)^2\}} = \sqrt{\{15^2 - 9^2\}}$$
$$= \sqrt{\{225 - 81\}} = \sqrt{144} = 12$$

Arc $DH = 9(90° + 36° \, 52')$, when the angle is
expressed in radians

$$= 9(2.214\ 3)$$
$$= 19.928\ 7$$

Length of belt $= 2(33.214\ 5 + 20 + 12 + 19.928\ 7)$
$$= 2(85.143\ 2)$$
$$= 170.286\ 4 \text{ cm} = 1\ 702.864 \text{ mm}$$

Answer:

Length of belt $= 1\ 700$ mm

Screw threads

A *helix* is the path of a point which moves on the curved surface of
a cylinder so that the distance it moves parallel to the axis of the cylinder
is proportional to the angular rotation of the cylinder.

A particular application of a helix is a screw thread. If the cylinder
rotates one revolution, the distance moved parallel to the axis is the
lead of the helix. The pitch of a screw thread is the distance between the
centre lines of adjacent thread forms. If the helix is single-start, the
pitch is equal to the lead. If the thread is multi-start, having n starts of
pitch p, then the lead of the thread $= np$. If the helix is developed, as

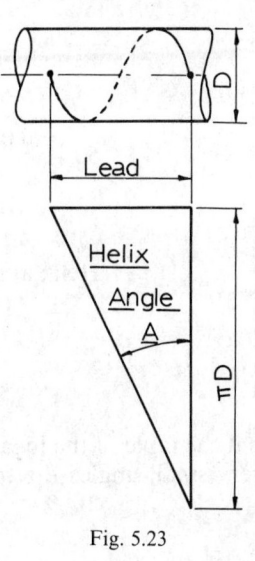

Fig. 5.23

shown in Fig. 5.23, the helix angle is given by arctan $\dfrac{\text{lead}}{\pi D}$. It should be noted that for a constant value of the lead, the value of the helix angle depends upon the value of D. Unless specifically stated to the contrary, all threads are assumed to be single-start, and the helix angle of a screw thread is assumed to correspond with the effective diameter. The reader should note that with some threads the effective diameter lies at half the thread depth. This is not true of all threads, however, particularly those of SI form.

Example:

The M16 metric thread has an outside diameter of 16 mm and a pitch of 2 mm. The effective diameter of a metric thread is given by the outside diameter minus 0·65 times the pitch. Calculate the helix angle of an M16 metric thread.

$$\text{Lead} = \text{pitch} = 2 \text{ mm}$$
$$\text{Effective diameter} = 16 - 2(0\cdot65) = 16 - 1\cdot3 = 14\cdot7$$
$$\text{Helix angle} = \arctan\left(\frac{\text{lead}}{\pi \times \text{eff. dia.}}\right) = \arctan\left(\frac{2}{\pi \times 14\cdot7}\right)$$
$$= \arctan 0\cdot043\ 3 = 2°\ 29'$$

No.	Log
π	0·497 2
14·7	1·167 3
Denom.	1·664 5
2	0·301 0
Denom.	1·664 5
0·043 3	$\bar{2}$·636 5

Rough check:

$$\frac{2}{50} = 0.04$$

Answer:

Helix angle = 2° 29′

(It should be noted that the table of the logarithms of tangents were not used because for very small angles the mean differences are insufficiently accurate.)

Tapers

The nominal dimension of a taper can be indicated either by stating its included angle or the reduction in diameter per distance of axial travel. Typical examples are 'included angle of 15°' or 'taper of 7/24'.

With the use of Fig. 5.24 the connection between the two systems can be deduced.

$$\tan \tfrac{1}{2} (\text{included angle } \theta) = \frac{BC}{AB} = \frac{\tfrac{1}{2} \text{ reduction in diameter}}{L}$$

$$\therefore \tan \frac{\theta}{2} = \frac{\text{reduction in diameter}}{2L}$$

Example:

Find the included angle corresponding to a taper of 1/8.

$$\tan \left(\frac{\text{included angle}}{2}\right) = \frac{1}{2 \times 8} = \frac{1}{16} = \arctan 3° 34′$$

$$\therefore \text{ included angle} = 3° 34′ \times 2 = 7° 8′$$

Answer:

Included angle = 7° 8′

We will conclude this book with some worked examples involving trigonometry which relate to the dimensioning of drawings, manufacturing methods and the inspection of completed work. For the accuracy

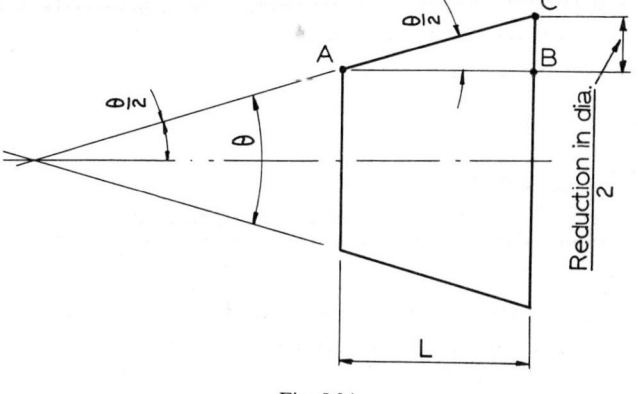

Fig. 5.24

required in these situations, four-figure tables are generally insufficiently accurate. The standard unit on an engineering drawing is the
millimetre. If other units are used express information to that fact is
given. Thus 98 is interpreted as 98 mm, but 27 metres would be indicated
by 27 m. For most cases in precision workmanship an accuracy is
desired to the nearest micrometre (often colloquially called the micron).
A micrometre is 10^{-6} metre, that is 10^{-3} mm = 0.001 mm. In the
examples which follow the answers have been determined with the aid
of a calculator, and as a final step, and only as a final step, the answers
have been rounded-off to the nearest micrometre. For angles, values
have been quoted, using decimal notation, to the nearest 0.001 of a
degree.

Use of rollers and spheres in measurement

Consider the male dovetail slide, dimensioned in millimetres and
drawn theoretically in Fig. 5.25(a). In order to machine this profile an
undercut would be necessary; an enlargment of this portion of the slide
is shown in Fig. 5.25(b).

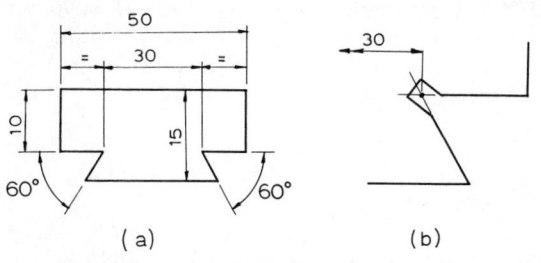

Fig. 5.25

It is not possible to measure the 30 mm dimension directly, since this dimension is between two imaginary lines in space. Problems similar to this occur repeatedly in measurement, the difficulty being overcome by measuring the 30 mm dimension indirectly. Rollers are placed in the vees and a measurement made across rollers. By the use of trigonometry the theoretical dimension of 30 mm can be checked. If the surface is internal and conical, balls are used instead of rollers.

There are two fundamental principles to remember in solving problems involving rollers (or balls) in tapers and/or vees:

1. If a roller (or ball) is placed against a flat surface, a line from the centre of the roller (or ball) to the point of contact is at right-angles to the surface and its length is equal to the radius of the roller (or ball).

2. A line from the centre of the roller (or ball) to the apex of the vee (or taper) bisects the angle of the vee (or taper).

It cannot be emphasized too much that calculations of this type invariably deal with half the included angle of the taper or vee.

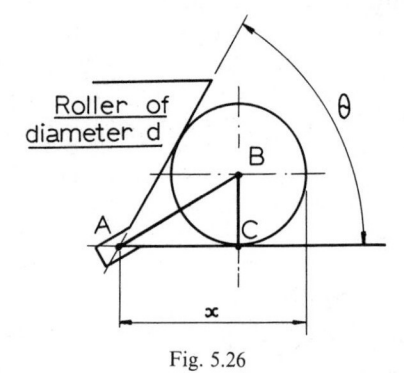

Fig. 5.26

Roller in a vee (Fig. 5.26)

Triangle ABC is right-angled, $\angle A = \dfrac{\theta}{2}$, $\angle B = 90° - \dfrac{\theta}{2}$

$$\frac{AC}{BC} = \tan B, \qquad AC = BC \tan B = \frac{d}{2} \tan\left(90° - \frac{\theta}{2}\right)$$

$$\therefore x = \frac{d}{2} + \frac{d}{2} \tan\left(90° - \frac{\theta}{2}\right)$$

whence
$$x = \frac{d}{2}\left\{1 + \tan\left(90° - \frac{\theta}{2}\right)\right\}$$

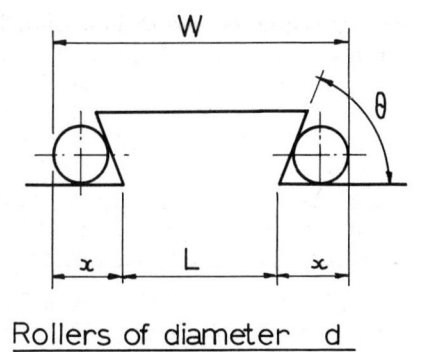

Rollers of diameter d

Fig. 5.27

In the symmetrical dovetail slide shown in Fig. 5.27:

$$W = L + 2x = L + 2\left(\frac{d}{2}\right)\left\{1 + \left(\tan\ 90° - \frac{\theta}{2}\right)\right\}$$

$$= L + d\left\{1 + \tan\left(90° - \frac{\theta}{2}\right)\right\}$$

This formula should not be applied to an asymmetrical dovetail slot. If the slot is not symmetrical, each side of the dovetail should be considered separately, as demonstrated in the example which follows.

Example:

Calculate the dimensions x and y and hence determine the checking dimension W for the asymmetrical dovetail slot shown in Fig. 5.28.

$$x = \frac{d}{2}\left\{1 + \tan\left(90° - \frac{\theta}{2}\right)\right\} = 10(1 + \tan 60°)$$

$$= 27{\cdot}320\ 508$$

$$y = \frac{d}{2}\left\{1 + \tan\left(90° - \frac{\theta}{2}\right)\right\} = 10(1 + \tan 67{\cdot}5°)$$

$$= 34{\cdot}142\ 136$$

$$W = 100 - (x + y)$$
$$= 100 - (27{\cdot}320\ 508 + 34{\cdot}142\ 136)$$
$$= 38{\cdot}537\ 356$$

Answer:

Checking distance $= 38{\cdot}537$ mm

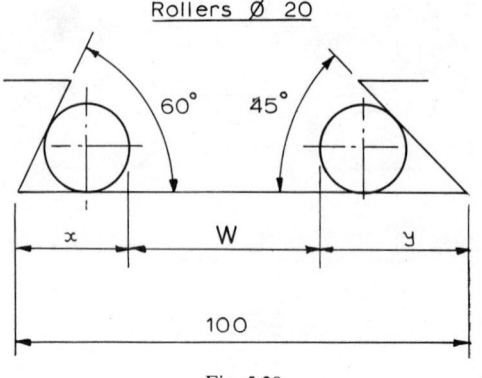

Fig. 5.28

Example:

When checking the angle θ and the nominal dimension L of the half-dovetail slot shown in Fig. 5.29, it was found that with a roller of $\varnothing 10$, the dimension W was 50·790 mm while with a roller of $\varnothing 20$, W was 21·580 mm. Calculate the angle θ and the nominal dimension L.

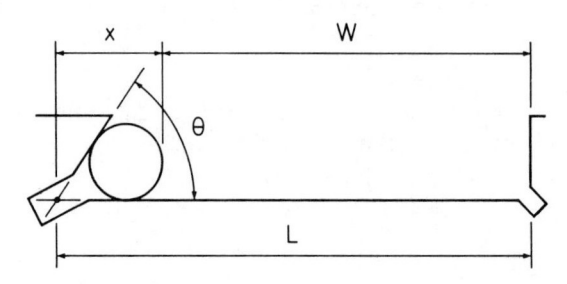

Fig. 5.29

Using the formula on page 381,

$$x = \frac{d}{2}\left\{1 + \tan\left(90° - \frac{\theta}{2}\right)\right\}$$

Let

$$1 + \tan\left(90° - \frac{\theta}{2}\right) = y,$$

then

$$L = W + dy$$

Using $W = 50{\cdot}790$, $d = 10$ $\qquad\qquad L = 50{\cdot}790 + 10y$

Using $W = 21{\cdot}580$, $d = 20$ $\qquad\qquad L = 21{\cdot}580 + 20y$

By subtraction $\qquad\qquad\qquad\qquad\qquad\overline{0 = 29{\cdot}210 - 10y}$

$$10y = 29{\cdot}210,\; y = 2{\cdot}921\,0 = 1 + \tan\left(90^\circ - \frac{\theta}{2}\right)$$

$$\tan\left(90^\circ - \frac{\theta}{2}\right) = 2{\cdot}921\,0 - 1 = 1{\cdot}921\,0$$

$$90^\circ - \frac{\theta}{2} = \arctan 1{\cdot}921\,0 = 62{\cdot}500^\circ$$

$$\frac{\theta}{2} = 90^\circ - 62{\cdot}500^\circ = 27{\cdot}500^\circ$$

$$\theta = 2 \times 27{\cdot}500^\circ = 55{\cdot}000^\circ$$

$L = 50{\cdot}790 + 10x = 50{\cdot}790 + 10(2{\cdot}921\,0) = 50{\cdot}790 + 29{\cdot}210$
$\quad = 80{\cdot}000$

Answer:

$$\theta = 55{\cdot}000^\circ,\; L = 80{\cdot}000 \text{ mm}$$

The Sine-Bar

The sine-bar consists of two rollers of equal diameter whose centres are accurately positioned on a body. In use the sine-bar is tilted, either by resting on slip gauge piles as shown in Fig. 5.30 or by being placed on an angular surface as shown in Fig. 5.31.

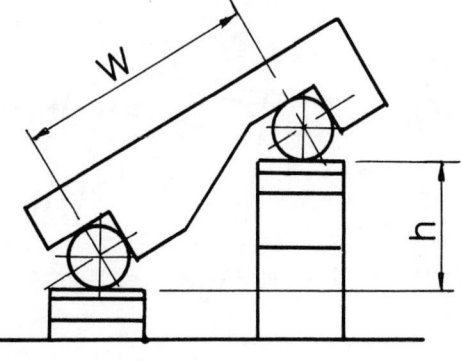

Fig. 5.30

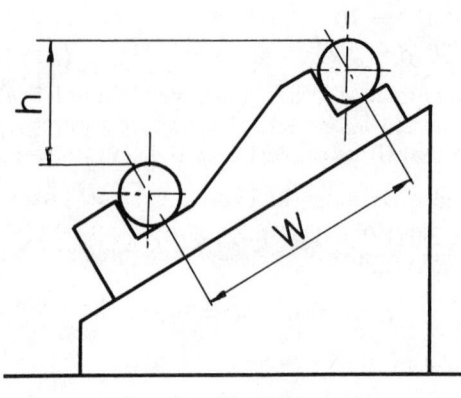

Fig. 5.31

In either case, the distance h is equal to the difference in height between the centres of the rollers.

If $\qquad\qquad W =$ centre distance of rollers
and $\qquad\qquad \theta =$ angle of tilt of sine-bar
then $\qquad\qquad h = W \sin \theta$

To facilitate calculations, W is specifically made a convenient distance, such as 100 mm or 200 mm.

As the angle of tilt approaches $90°$, the increment for a small difference of angle becomes less. Hence if an angle near $90°$ is to be checked or set using a sine-bar, greater accuracy can be obtained if use is made of a square with the sine-bar, so that the angle determined by the sine-bar is small (see Fig. 5.32).

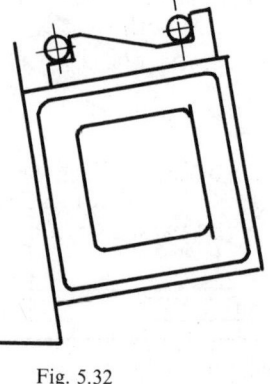

Fig. 5.32

Problems 5.3

It is expected that our reader will solve the following problems with the aid of mathematical tables. The answers should be given to the nearest third significant figure and/or the nearest minute. It would be useful practice to check the answers with the aid of a calculator.

1. In a right-angled triangle, the hypotenuse has a length of 15 mm and the shortest side a length of 9 mm. Find the length of the other side and the magnitudes of the two acute angles.

2. The distance from a point P to the centre of a circle of diameter 36 mm is 42 mm. Determine the angle between the two tangent lines drawn from P to the circle.

3. What is the radius of the largest circle that can be inscribed within an equilateral triangle having sides of length 100 mm?

4. A car climbs a hill of constant slope 4° to the horizontal. What vertical distance has been ascended by the car in travelling 800 m along the road?

5. A pylon stands on level ground. From a point 50 m from the foot of the pylon the angle of elevation to the top is 36°. Calculate the height of the pylon.

6. The legs of a pair of dividers each have a length of 160 mm. Calculate the angle between the legs when they are opened out to a distance of 58 mm.

7. The jib of a crane has a length of 10 m. When a load is attached to the jib, the jib makes an angle of 35° above the horizontal. How high has the load been lifted if the jib angle increases to 62° above the horizontal?

8. The base of an isosceles triangle has a length of 40 mm and the angles at the base are both equal to 70°. Calculate:
   (a) the length of the equal sides;
   (b) the perpendicular height of the triangle;
   (c) the area of the triangle.

9. From the top of a building of height 40 m, standing on level ground, the angles of depression of two features in line with the viewer are 11° and 27° respectively. Calculate the distance between the features.

10. A drill point has an included angle of 118°. A hole of diameter 20 mm has to be drilled so that the depth of full diameter is 48 mm. Calculate the depth to which the point of the drill has to be fed.

11. A spindle is tapered, and two measurements taken axially 160 mm apart indicate diameters of 88 mm and 64 mm. Calculate the included angle of the taper.

12. A pendulum has a length of 320 mm. The maximum swing from the centre is $18° 25'$. What is the vertical distance between the maximum and minimum heights of the moving end of the pendulum?

13. Figure 5.37 shows the details of a groove in a pulley for a vee-belt drive. Calculate the distance x.

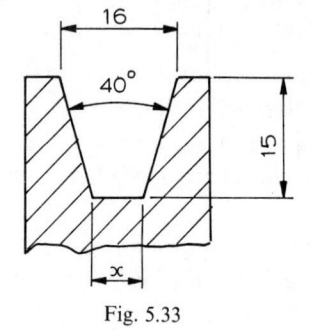

Fig. 5.33

14. Nine holes lie equally spaced on a pitch circle diameter. The direct distance between the centres of two adjacent holes is 51·3 mm. Calculate the pitch circle diameter.

15. An open belt connects two pulleys of diameters 380 mm and 200 mm, the centre distance of the pulleys being 410 mm. Calculate the included angle between the straight portions of the belt.

16. Figure 5.34 shows a corner in which a radius has to be drawn so as to be tangential at points P. Determine the magnitude of this radius, and the shortest distance x from the curve to the sharp corner.

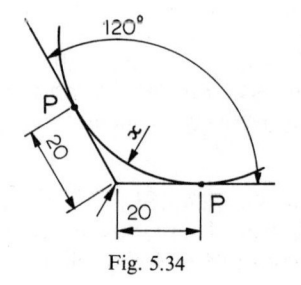

Fig. 5.34

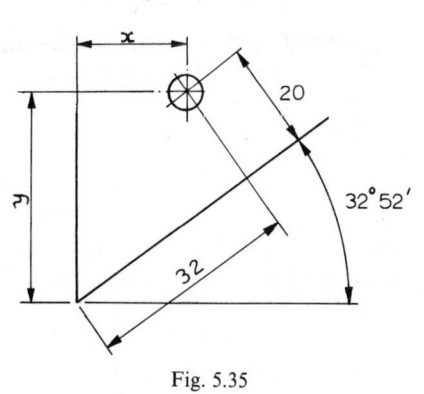

Fig. 5.35

17. Calculate the co-ordinate distances x and y of the centre of the hole shown in Fig. 5.35.

18. Two guy-ropes are attached to a vertical pole, each guy-rope being 20 m long. In plan, the guy-ropes are at right angles. Each meets the ground at an angle of 70° to the horizontal. Determine the distance between their ground attachments.

19. Calculate the distance across corners of a hexagon which measures 64 mm across flats.

20. Calculate the setting for a 200 mm sine-bar to measure an angle of 8° 43′.

21. Ten holes lie equally spaced on a pitch circle diameter of 80 mm. Calculate the chordal distance between two adjacent holes.

22. Calculate the nominal pitch circle diameter of a sprocket wheel having 36 teeth suitable for a chain of pitch 20 mm.

23. Calculate the included vertical angle and the slant height of a cone having a base diameter of 80 mm and a vertical height of 120 mm.

24. The profile of a cam consists of circles of radii 10 mm and 30 mm lying at a centre distance of 60 mm. The radii are joined by tangents. Calculate the angle between the tangents.

25. Calculate the angle A on the template shown in Fig. 5.36.

26. Calculate the checking distance W for the dovetail slot shown in Fig. 5.37.

27. Calculate the included angle of the taper of a standard milling machine spindle nose. It is designated 7/24, i.e. the diameter reduces by 7 mm for every 24 mm of axial travel.

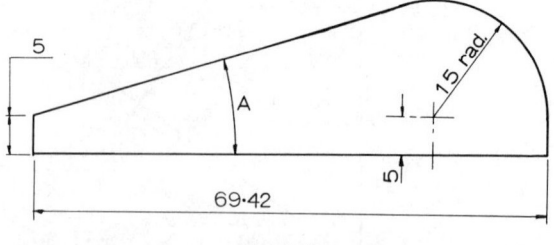

Fig. 5.36

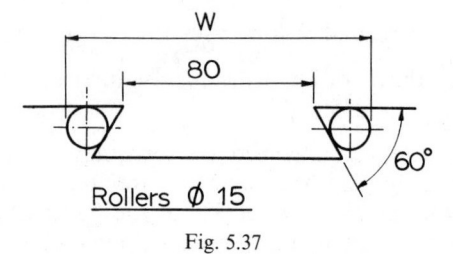

Fig. 5.37

28. Calculate the taper on diameter, in millimetres per metre of axial travel, that corresponds to an included angle of 5°.

29. Calculate the helix angle of an M36 metric thread which has an outside diameter of 36 mm and a pitch of 4 mm. The effective diameter is equal to the outside diameter minus 0·65 of the pitch.

30. A bar tapers from a diameter of 80 mm to a diameter of 40 mm over a length of 40 mm. If all of these sizes are subjected to limits of ± 0·5 mm, calculate the maximum and minimum possible included angles of the taper.

31. On an ordnance survey map using a scale of 1 : 50 000, two contour lines which representing a vertical separation of 50 m lie 8·4 mm apart. What is the average slope of the ground between those points, to the nearest minute?

32. (a) Figure 5.38 shows the cross-section of a workshop. Determine the height x to the ridge.
    (b) If the workshop has a length of 30 m and roofing costs £8.50 per square metre, find the cost of roofing.

33. From an observation point a geographical feature A lies at a distance of 5·3 km on a bearing N 27° W. From the same observa-

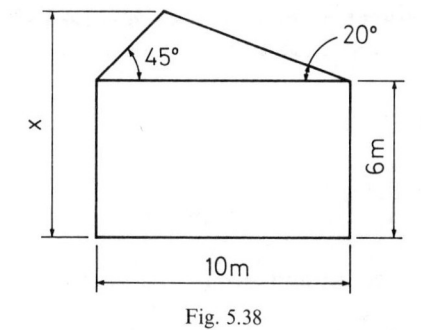

Fig. 5.38

tion point a second feature B lies at a distance of 7·2 km on a bearing N 42° E. Calculate:

(a) the distance from A to B;
(b) the bearing of B from A;
(c) the bearing of A from B.

34. A drainage channel has a horizontal base of width 1·6 m and is symmetrical about a vertical centre line. The slides slope outwards from the base with an included angle of 40°.

(a) Determine the width of the channel at any height h metres above the horizontal base, as a function of h.
(b) If water flows in the channel to a depth of 0·8 m, determine the cross sectional area of the water flow.

35. Calculate the area of the trapezium shown in Fig. 5.39.

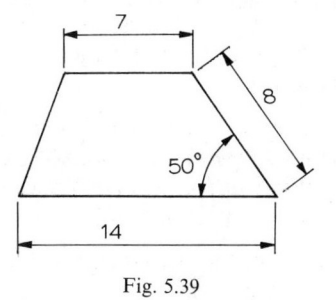

Fig. 5.39

36. A rhombus has sides of length 5 mm, one of the acute angles of the rhombus being 45°. Calculate the area of the rhombus and the length of both diagonals.

37. A parallelogram has two adjacent sides of length 5 mm and 8 mm, the angle between these sides being 40°. Calculate:
    (a) the area;
    (b) the distances between parallel sides;
    (c) the lengths of the diagonals.

38. Figure 5.40 represents the front plate of a guard used to enclose a gear train completely. Determine:
    (a) the angle θ;
    (b) the distance around the periphery, i.e. the distance around *ABCDEFA*.
    Give the answer to three significant figures.

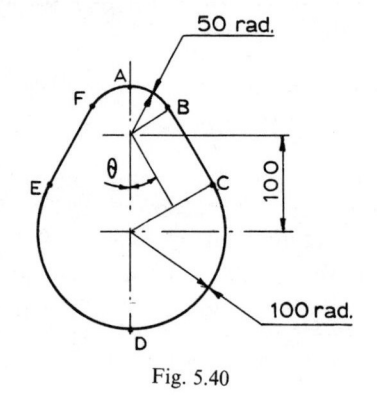

Fig. 5.40

39. Calculate the length of the centre line of the wire form shown in Fig. 5.41. (There is no straight portion on the centre line.)

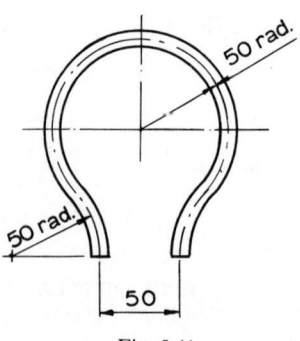

Fig. 5.41

40. Neglecting the thickness of the material, calculate the developed lengths of the brackets shown in Figs 5.42(a), 5.42(b) and 5.42(c).

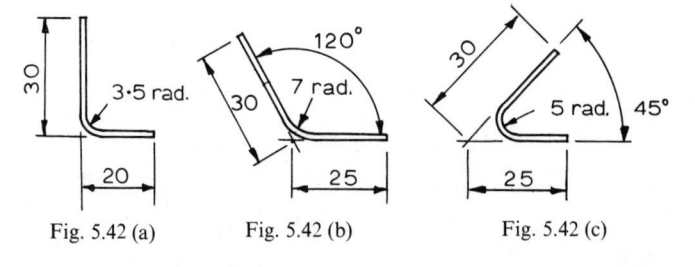

Fig. 5.42 (a)   Fig. 5.42 (b)   Fig. 5.42 (c)

41. Calculate the developed length of the cable clamp shown in Fig. 5.43. Assume that the developed length is the length along the centre line.

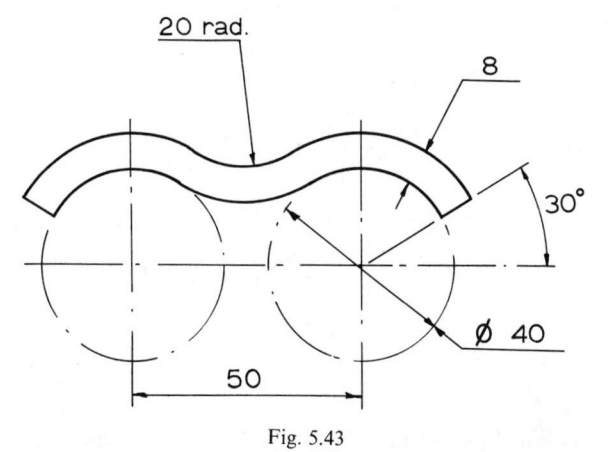

Fig. 5.43

42. Calculate the length of an open belt drive connecting pulleys of diameter 360 mm and 180 mm at a centre distance of 410 mm.

43. Calculate the length of the belt of a crossed belt drive connecting pulleys of diameter 160 mm and 320 mm at a centre distance of 510 mm.

44. Calculate the area of the template shown in Fig. 5.44.

45. A circular bar, of diameter 50 mm, is machined so that the uniform cross-section becomes a regular hexagon of the greatest possible area.
    (a) Calculate the area of this hexagon.
    (b) What percentage of the original bar is machined away?

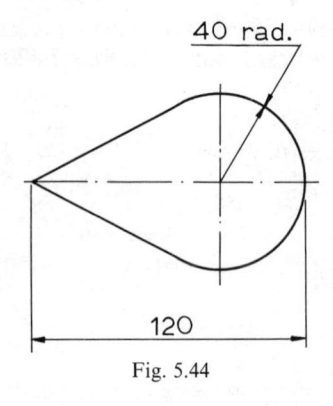

40 rad.

120

Fig. 5.44

46. Figure 5.45 shows a taper feature being checked with the aid of a disc of diameter 40 mm. Calculate the distance *x*.

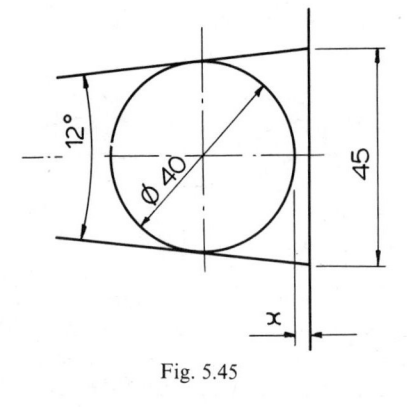

12°

Ø 40

45

x

Fig. 5.45

47. Calculate the checking distance *W* of the asymmetrical dovetail slot shown in Fig. 5.46.

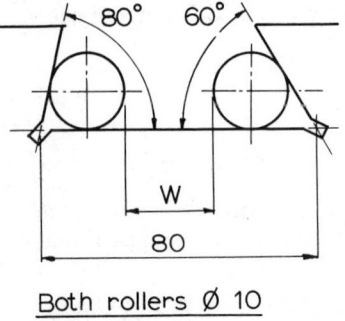

80°   60°

W

80

Both rollers Ø 10

Fig. 5.46

48. Figure 5.47 shows a portion of a gusset plate. *P* represents the centre of a drilled hole which has a diameter of 20 mm. Calculate:
    (a) the angle *AOP*;
    (b) the distance *OP*;
    (c) the thickness of metal *x*.

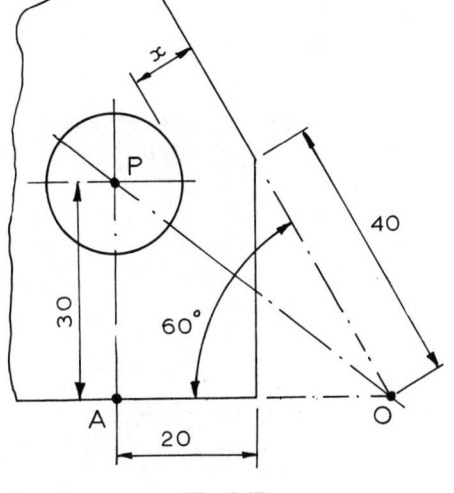

Fig. 5.47

Note. Our reader should now have sufficient knowledge of trigono-metry to solve questions 49 to 55 inclusive which now follow. However, if our reader has studied trigonometry to a later stage than is covered by this book, other methods may provide a more rapid solution. Our reader is invited to select whatever methods he or she wishes to effect solutions.

49. Figure 5.48 shows a triangle of forces diagram. Calculate the mag-nitude of the forces denoted by F_1 and F_2.

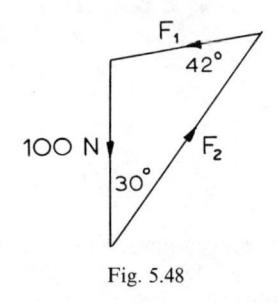

Fig. 5.48

50. In a crank and connecting rod mechanism, the crank radius is 40 mm and the connecting rod is 160 mm long. If θ denotes the angle turned by the crank from outer dead-centre, calculate the distance the slider moves as θ changes from 30° to 120°.

51. A, B and C represents the centres of three holes which lie on a pitch circle. If O represents the centre, angle $AOB = 60°$, angle $BOC = 140°$ and $AB = 100$ mm, calculate:
    (a) the pitch circle diameter;
    (b) the distances BC and CD.

52. Figure 5.50 shows one panel of a pin-jointed framework, the panel being in the form of a parallelogram. Calculate the length of the diagonal BD.

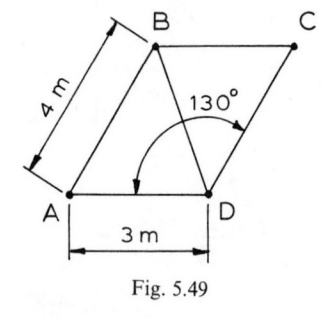

Fig. 5.49

53. In Fig. 5.50, AB represents an electricity supply, the distance AB being 5 km. C represents the position of a factory, such that $\angle CAB = 32°$ and $\angle ABC = 48°$. Calculate:
    (a) the distance BC;
    (b) the shortest distance from the factory to the supply, i.e. distance CD.

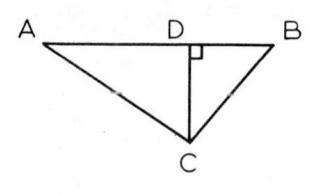

Fig. 5.50

54. Figure 5.51 shows diagrammatically a flap OP held at 45° to the horizontal by the stay QR. If $OQ = 200$ mm and $OR = 160$ mm, calculate the length QR.

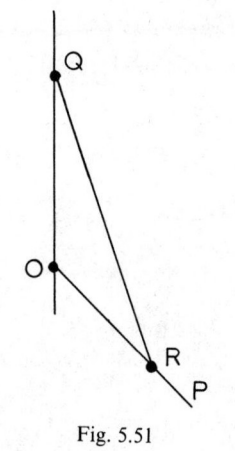

Fig. 5.51

55. In the layout shown in Fig. 5.52, $AB = 300$ mm, $BC = 160$ mm and $CA = 400$ mm. Find the co-ordinates x and y of the centre of hole B with reference to the axes shown.

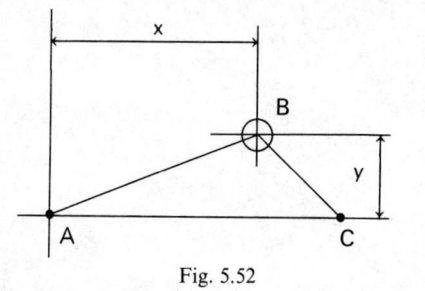

Fig. 5.52

Answers to Problems 5.3

1. 12 mm, 36° 52′, 53° 8′

2. 50° 44′

3. 28·9 mm

4. 55·8 m

5. 36·3 m

6. 20° 53′

7. 3·09 m

8. (a) 58·5 mm   (b) 55·0 mm   (c) 1 100 mm²

9. 127 m

10. 54·0 mm

11. 8° 34′

12. 16·4 mm

13. 5·08 mm

14. 150 mm

15. 25° 22′

16. 34·6 mm, 5·4 mm

17. $x = 16·0$ mm, $y = 34·2$ mm

18. 9·67 m

19. 73·9 mm

20. 30·3 mm

21. 24·7 mm

22. 229 mm

23. 36° 52′, 126 mm

24. 38° 56′

25. 16° 0′

26. 121 mm

27. 16° 36′

28. 87·4 mm

29. 2° 11′

30. 54° 52′ and 51° 25′

31. 6° 47′

32. (a) 12·7 m   (b) £2 950

33. (a) 7·25 km   (b) 4° 58′ N   (c) 4° 58′ S

34. (a) $1·6 + 0·728h$ (b) 1·51 m²

35. 64·3 mm²

36. 17·7 mm²; 3·83 mm and 9·24 mm

37. (a) 25·7 mm²   (b) 3.21 mm and 5·14 mm
    (c) 5·25 mm and 12·3 mm

38. (a) 30°   (b) 697 mm

39. 302 mm

40. (a) 48·5 mm   (b) 54·3 mm   (c) 42·6 mm

41. 103 mm

42. 1 690 mm

43. 1 890 mm

44. 6 120 mm²

45. (a) 1 620 mm²   (b) 17·3%

46. 2·74 mm

47. 55·4 mm

48. (a) 36° 52′   (b) 50 mm   (c) 9·64 mm

49. $F_1 = 74·8$ N, $F_2 = 142$ N

50. 57·1 mm

51. (a) 200 mm   (b) $BC = 182$ mm, $CA = 197$ mm

52. 3·09 m

53. (a) 2·69 km   (b) 2·00 km

54. 333 mm

55. $x = 280$ mm, $y = 106$ mm

5.4
The sine and cosine curves

Our reader will recall that when we introduced the subject of the measurement of angles we took a straight line of unit length and considered it to be a rotating arm. In particular, the pivot of the arm was coincident with the origin of cartesian co-ordinates. The rotation of the arm commenced with the arm being horizontal and lying along the positive portion of the x-axis. The arm then rotated anticlockwise to generate positive angles, the greater the amount of rotation of the arm the greater the angle. We made a special point that the arm need never stop rotating, and consequently there is no limit to the magnitude of a positive angle. If we refer to Fig. 5.54, OP is the arm and O is the origin of cartesian co-ordinates.

In Fig. 5.53, OP has rotated from the datum position through an angle of θ. The point R is a horizontal projection of the point P on to the vertical axis, i.e. the y-axis. In a similar manner, the point Q is a vertical projection of the point P on to the horizontal axis, i.e. the x-axis. OPQ is therefore a right-angled triangle.

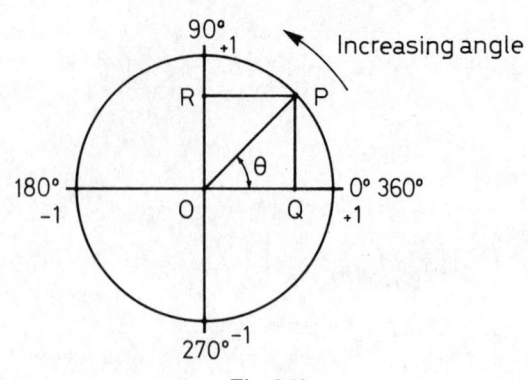

Fig. 5.53

In which case,

$$\sin \theta = \frac{PQ}{OP}$$

and if OP is of unit length, so that $OP = 1$,

then $\sin \theta = PQ$.
Now we have made $OR = PQ$,
and hence $\sin \theta = OR$.

Consequently the value of $\sin \theta$ is the value indicated by a horizontal projection of the point P on to the vertical axis, i.e. the y-axis. Let us use this statement to see what happens to the value of $\sin \theta$ when it takes any positive value. The range is not restricted to zero to 90° inclusive.

The table which follows gives certain easily determined values. The reader is asked to note that the amounts of rises and falls are not in direct proportion to changes in the magnitude of the angle.

Value of θ	Value of $\sin \theta$
ZERO	ZERO
Zero to 90°	Rising from zero to $+1$
90°	$+1$
90° to 180°	Falling from $+1$ to zero
180°	zero
180° to 270°	Falling from zero to -1
270°	-1
270° to 360°	Rising from -1 to zero
360°	zero

The arm OP has now made one complete revolution of 360°. The cycle of events from 360° to 720° is a repetition of those from zero to 360°. The cycle from 720° to 1 080° is likewise a repetition of that from zero to 360°. The cycle of events repeats itself every 360°, or, if we use the circular measure of angles, every 2π radians.

A geometric presentation which we can use to determine certain intermediate values to plot the graph of sin θ against angle θ is shown in Fig. 5.54. For convenience, only two cycles are shown, but the cycle is repeated every 360°.

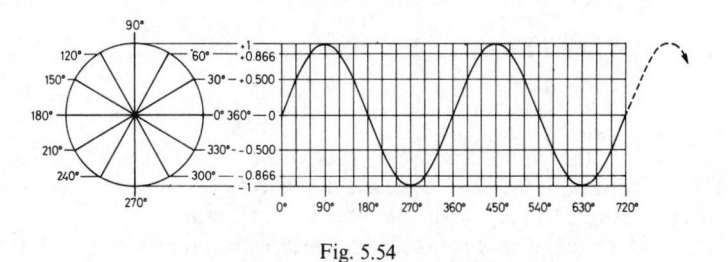

Fig. 5.54

The graph of sin θ against angle θ has the form of a wave, and consequently it is often commonly referred to as the *sine wave*. The value of sin θ is never greater than $+1$ and never less than -1, oscillating between these values about a mean value of zero. The amount of variation from a mean position of a particular oscillation, neglecting the arithmetical sign, is called the *amplitude* of that oscillation. The amplitude of a sine wave is therefore unity about a mean value of zero. The interval between a particular characteristic occurring in one cycle of events and when that characteristic occurs during an adjacent cycle, not only in magnitude but in direction of change, is called the *period* of that oscillation. The sine wave is therefore cyclic with a period of 360°, or 2π radians. In many cases the period is measured in terms of time, in which case it is then referred to as the *periodic time*.

Now let us carry out a similar procedure as we did for the sine of an angle for the cosine of an angle.

Referring back to Fig. 5.54,

$$\cos \theta = \frac{OQ}{OP}$$

and since OP has unit value, i.e. $OP = 1$,

then
$$\cos \theta = OQ,$$

and the cosine of an angle is the vertical projection of the point P on to the horizontal axis, i.e. the x-axis.

Significant values of cos θ are indicated in the table which follows:

Value of θ	Value of $\cos \theta$
Zero	$+1$
Zero to 90°	Falling from $+1$ to zero
90°	Zero
90° to 180°	Falling from zero to -1
180°	-1
180° to 270°	Rising from -1 to zero
270°	Zero
270° to 360°	Rising from zero to $+1$
360°	$+1$

Once again, the values are cyclic, repeating themselves at intervals of 360° or 2π radians. We can once more use the assistance of a geometric representation, to help with the intermediate values, but for convenience, and to assist in promoting an interesting development later in this section, we will rotate the original circle anticlockwise through 90°. The resulting cosine wave is shown in Fig. 5.55.

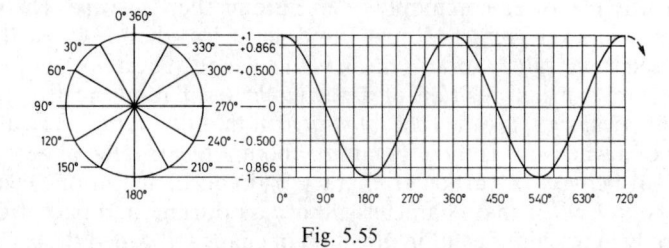

Fig. 5.55

Our reader will note that the cosine wave, like the sine wave, has an amplitude of unity about a mean position of zero, and is cyclic with a period of 360°, or 2π radians. Now let us draw the sine and cosine waves on the same graphical field over the range zero to 720°. These are shown in Fig. 5.56.

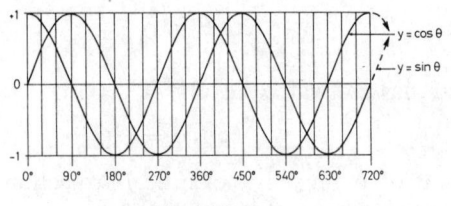

Fig. 5.56

The sine and cosine curves were deduced by taking two projections from the same rotating arm on to two axes which lie at right angles to each other. The same result can be obtained by projections on to a single axis from two rotating arms, those arms lying at right angles to each other. For the benefit of work which will be undertaken in later studies, we will put an arrow on each of the rotating arms and hence convert them into vectors, as shown in Fig. 5.57.

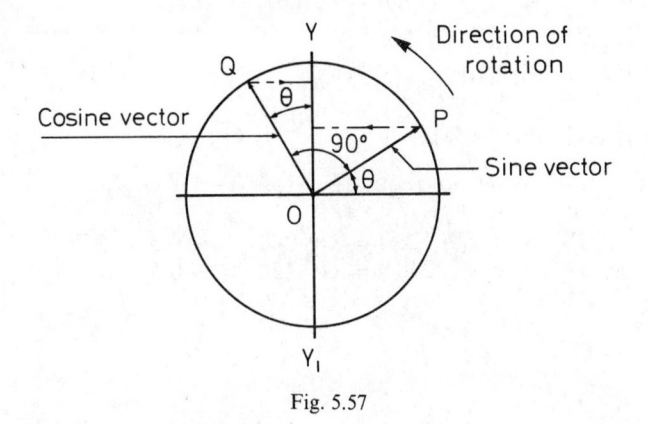

Fig. 5.57

The projection of the unit vector OP on to the vertical axis YOY_1, is $OP \sin \theta$, or $\sin \theta$ if OP is of unit length. OQ is a vector of unit length which lies at $90°$ to OP. If we consider the conventional direction of rotation to indicate the magnitude of an angle, we can consider that the vector OQ leads the vector OP by $90°$. Now the projection of OQ on to the same vertical axis YOY_1 (i.e. the axis that we used for the determination of $\sin \theta$) is $OQ \cos \theta$, or $\cos \theta$ since $OQ = OP = $ unity. Hence another consideration of the cosine wave is that it is identical with the sine wave in form, but the values lead those of the sine wave by $90°$. In the 'language' of mathematics

$$\cos \theta = \sin(\theta + 90°)$$

or, using the circular measure of angles,

$$\cos \theta = \sin\left(\theta + \frac{\pi}{2}\right).$$

We shall leave the sine and cosine curves at this point, and develop the topic in Book 2, particularly with the reference to the trigonometrical ratios of angles greater than $90°$.

Problems 5.4

1. Construct graphically, on the same field, three full waves of $\sin \theta$ and $\cos \theta$, i.e., over the range zero to $1\,080°$.

2. Although a study has not yet been undertaken of the trigonometrical ratios of angles greater than $90°$, show that the graphs constructed in the answer to question 1 indicate that:

$$\sin \theta = \cos(90° - \theta)$$

with reference to the angles:

(a) $90°$ (b) $210°$ (c) $330°$ (d) $540°$

Answers to Problems 5.4

1. Numerical answer not required
2. (a) $\sin 90° = \cos 0° = 1$
   (b) $\sin 210° = \cos(-120°) = \cos 240° = -0.5$
   (c) $\sin 330° = \cos(-240°) = \cos 120° = -0.5$
   (d) $\sin 540° = \cos(-450°) = \cos 270° = 0$

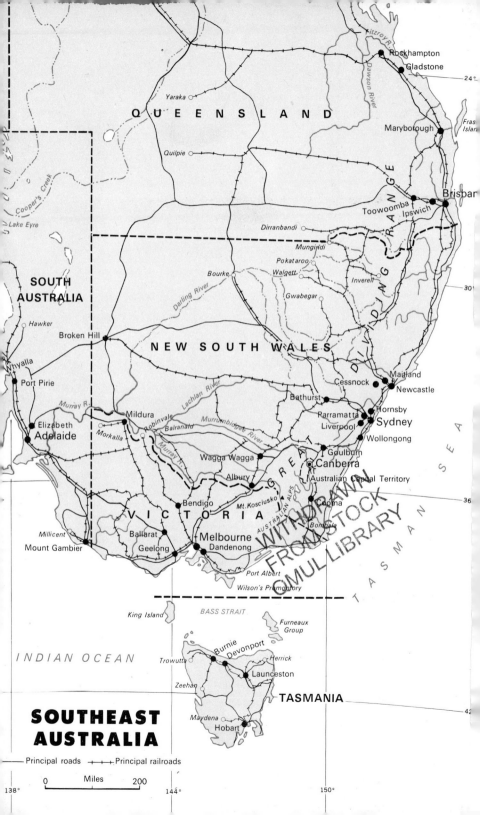

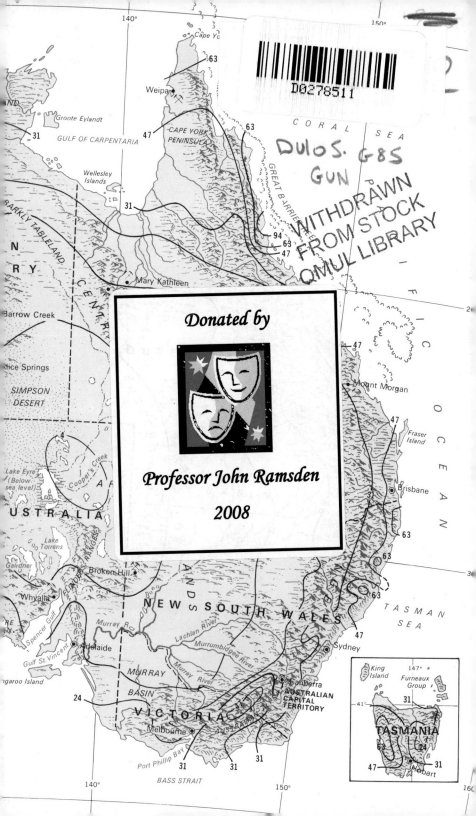

INSIDE AUSTRALIA
AND NEW ZEALAND

BOOKS BY JOHN GUNTHER

Public Affairs

INSIDE EUROPE
INSIDE ASIA
INSIDE LATIN AMERICA
INSIDE U.S.A.
INSIDE AFRICA
INSIDE RUSSIA TODAY
INSIDE EUROPE TODAY
INSIDE SOUTH AMERICA

Biography

DEATH BE NOT PROUD
ROOSEVELT IN RETROSPECT
THE RIDDLE OF MACARTHUR
EISENHOWER
TAKEN AT THE FLOOD
PROCESSION

Autobiography

THE STORY OF THE INSIDE BOOKS

Novels

THE TROUBLED MIDNIGHT
THE LOST CITY
QUARTET

Reporting

THE HIGH COST OF HITLER
D DAY
BEHIND EUROPE'S CURTAIN

Miscellaneous

DAYS TO REMEMBER (WITH BERNARD QUINT)

Travel

TWELVE CITIES

JOHN GUNTHER

Inside Australia and New Zealand

COMPLETED AND EDITED BY
WILLIAM H. FORBIS

HAMISH HAMILTON
LONDON

First published in Great Britain 1972
by Hamish Hamilton Ltd
90 Great Russell Street, London WC1

SBN 241 02180 4

Maps by Jean Paul Tremblay

TO MY FRIEND AND PUBLISHER
HAMISH HAMILTON

Printed and bound in Great Britain at
The Pitman Press, Bath

Contents

MAPS

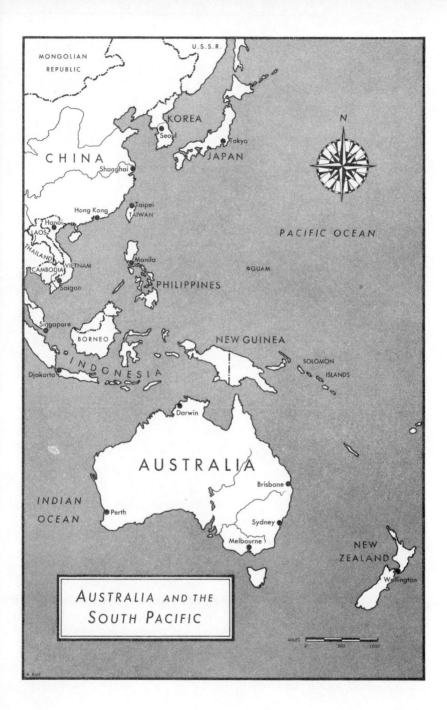

AUSTRALIA AND THE
SOUTH PACIFIC

A Note on How This
Book Was Written

DURING the five months he spent in and around Australia, John Gunther traveled everywhere, from a Sepik River village in New Guinea to the capital of each Australian state and to both the islands of New Zealand. This was in late 1969 and through January 1970, after which he returned to New York to start writing this book. As in the case of his other *Inside* books, he did not write straight through from beginning to end; it caught his fancy to commence the job by doing some of the middle passages. I never knew Gunther personally, but from what he had written it was obvious that he was attacking this project with the buoyancy and zest that characterized his other work. Suddenly, at the end of May 1970, Gunther became ill and went to the hospital. Within a week, he was dead, of a cancer that neither he nor anyone else knew he had.

His totally unexpected death left the book well begun but far from finished, and some decision had to be made about its fate. Could it, and should it, be completed at all? The assets of the project were ample: forty thick notebooks containing Gunther's interviews and observations, innumerable notes on scraps of paper (his repository for stray thoughts), documentation in the form of 150 books that Gunther had collected, mountains of clips from his well-stocked personal morgue—much of this sorted and marked—and curious memorabilia right down to the annual report of the New Zealand Cooperative Dairy Company Limited. In boxes filled with colored folders there were letters, Xeroxes, menus, photographs, briefing papers. And of course there were the chapters Gunther had written, his outline of the whole book, and, for guidance, his own book *A Fragment of Autobiography*, which describes how he put his books together. An indispensable further asset was John's

wife Jane. She had been with him on the Australia–New Zealand trip, taking part in every interview, and sharing observations with him when they returned to their hotel rooms, just as she had during their trips through other continents. She had a very specific insight into what he intended to make of the book, and was anxious to work with someone on it.

It was decided to find a writer to undertake the task. One morning in July 1970, the telephone rang in my house in Missoula, Montana, where I was living a life of sportive idleness after twenty years of work as a writer, senior editor, and correspondent for *Time* magazine. The caller was Cass Canfield, Senior Editor of Harper & Row, who had heard of me from Richard Clurman, a vice president of Time Inc. and a long-time friend of mine. Acting on Canfield's suggestion, I went to Greensboro, Vermont, where Jane Gunther has a summer house. I was strongly attracted to the job of completing the book. I admired Gunther's earlier books, particularly the one on South America; I covered that continent for *Time* for several years, and found, upon starting the assignment, that Gunther's book offered useful studies of the nine countries I had to deal with. Moreover, the thought of learning and writing about Australia and New Zealand, places that have a strong hold on the American imagination, intrigued me. It seemed to me to be a worthwhile project, much needed by many thousands of readers. While Jane Gunther was deciding that I could satisfactorily finish the book, I was deciding that the arrangement was feasible.

Soon several hundred pounds of documentation for the book arrived at the Missoula airport, and I set about completing it on the Gunther outline, with a few modifications. I also then visited Australia and New Zealand for a period of time that would have been far too short had not the reporting and interviewing for the book already been done by Gunther. The purpose of my trip was simply to get a better feel for the subject, to acquire a framework for understanding Gunther's notes, and to hear the speech and sounds of the two countries.

While I was completing the book based on Gunther's original work, and my own impressions and research, I was frequently pleased to find that whenever I needed a specific example or fact or even transition, a Gunther note would turn up to supply it. Indeed, when Australia changed Prime Ministers halfway through the writing, Gunther's notes produced an interview that he had had with the man who became the new Prime Minister. I gradually evolved a form and style for this double-authored book. For example, I at first left untouched the passages that Gunther had written in the first person, and even sometimes added an "I"—meaning Gunther—when it seemed to flow naturally and directly out of his notes. But later I thought that this

made the book sound sporadic and inconsistent, and on revision I expunged all the "I's" from the completed text. As to the book's exposition in general, I believe that I share Gunther's respect for facts, his yearning to tell them interestingly, his sense of curiosity and wonder, and his desire to be lucid. Gunther welcomed, and made use of, every fact of interest that he came across (even if the source was only a railroad timetable), though he used facts with discretion and only after rigorous checking. I tried to do the same. I made no effort to "write like Gunther," apart from adopting a few of his stylistic trademarks, using passages from his notes and relying on his methods. I like to think that the book has turned out somewhat as Gunther would have written it. Gunther was a generalist who, without apology, took the whole world as his subject, believing that specialists give too little of the spectrum to satisfy most of us; I tried to apply that principle to Australia and New Zealand. Like all the *Inside* books, this is for the reader who, knowing a little about those two countries, wants to know a thousand times more.

John Gunther's notes for this Preface observe that this is "the eighth *Inside* book (or is it the ninth?)." It is the ninth, but Gunther was in too much of a hurry to get into a new book to bother to sit down and count old ones. He wanted to bring out the "fascination and importance of Australia," and thought that "New Zealand should be of particular interest because no contemporary book exists covering it as I do." Let me add that both Gunther and I found the two countries highly agreeable, even though the reader may find some critical passages. Acknowledgments of the book's numerous debts will be found at the end, but here I wish to pay special tribute to the beautiful, discriminating, observant, cooperative, and sensitive Jane Gunther (who hates strings of adjectives). She gave me help and criticism from beginning to end, increasing the book's accuracy, clarity, relevance, and directness.

WILLIAM H. FORBIS

AUSTRALIA
Major Physical Features

CHAPTER 1

The Australasian Texture

> Australian history . . . does not read like history, but like the
> most beautiful lies. And all of a fresh new sort, no moldy old
> stale ones. It is full of surprises, and adventures, and incongrui-
> ties, and contradictions, and incredibilities; but they are all
> true, they all happened.
> —MARK TWAIN IN Following the Equator

THE WAY to start thinking about Australia is in terms of its Wild Colonial
Boys, who, as an old folk song says, "robbed the rich and helped the poor
and never hurt a good man." And this bold image brings up the convicts
sent out from England who sired the Wild Colonial Boys, and the subsequent descendants of those early settlers who fought, with the brims
of their digger hats turned up, for England in two world wars. From these
legendary figures the mind moves on to images of the inferno-like desert of
the outback, and the millions of Merino sheep that have given the coats off
their backs to keep Englishmen warm, and the kangaroo, the wallaby, and
the duck-billed platypus. This is the traditional Australia, and it can still be
sensed in the antiauthoritarian swagger of the Australians, and the exotic
flora and fauna.

But then we must swerve around and face the Australia of today: a country
that has suddenly turned out to be rich with minerals from one end to the
other, an urban country with a pair of cities each equal in population to all
but the very biggest in the United States, a country awash with recently
landed Yugoslavs and Poles and Italians and Americans, and a country that
has exchanged its former economic kinship with England for an apronstring
relationship with Japan. New Zealand shares this remolded image in a paler

form: this nation that once existed to satisfy the English appetite for butter and lamb chops is now becoming a workshop making harvesters and lumber and newsprint.

The image of the cosmopolitan, citified Australian, replacing that of the "fair dinkum digger," and of the industrial New Zealander, replacing that of the shepherd, suggests substantial changes in the last ten years. In the case of Australia, these changes are shaking the whole nation; in the case of New Zealand, they are pushing a complacent country to modernize and adapt. The more notable of the changes are these:

1. Australia and New Zealand, which for a century and a half preferred to be regarded as displaced parts of Europe, have been forced to acknowledge themselves to be parts of Asia, with urgent interests in the prosperity of their neighbors, with vital economic connections to Japan, with a crucial need to know the Oriental mind, and with a growing necessity to find a way to live with China. Militarily, Australasia,* abandoned by Britain, now finds the United States' commitment weakened by the American withdrawal from Indochina. In short, Australia and New Zealand feel that, for better or worse, they are going to wind up alone, white nations in a sea of brown.

2. The Australian economy, stimulated by fabulous discoveries of iron ore, bauxite, nickel, and other minerals, has been growing explosively, both requiring and facilitating a huge, society-changing immigration. New Zealand, without massive immigration, has managed to surpass even Australia in population increase, in proportion to size.

3. The money for Australia's growth, once derived almost solely from English investors, now comes mostly from the United States, with significant additions from Japan, which has become Australia's biggest customer. Thus many decisions vitally affecting Australia are being made in other countries; Australians are grateful for the incoming funds—and at the same time worried.

4. There has been a quantum jump in quality of life and sophistication, especially in Australia. The Australia of the beer-swilling, big-kneed man willing to "have a go" at anything is fading; an Australia of style, urbanity, education, and culture is flowering. Wellington, the wooden capital of New Zealand, is more and more becoming something of a South Seas San Francisco. In short, both countries have chucked away the dullness they once seemed not only to tolerate but to cherish.

Those are the recent changes; there remain some observations to be made about the more permanent qualities of life in these two countries.

* This somewhat flexible appellation is here construed to mean Australia, New Zealand, New Guinea, and the Bismarck Archipelago.

Australia and New Zealand are not known in familiar detail to most of the rest of the world, yet the world senses that they are worthy of greater interest. Very few Americans can name the six states of Australia—a shortcoming all the more embarrassing because millions of Australians can name most of the American states. How many Western Europeans, apart from Englishmen, can say instantly whether Auckland or Wellington is the capital of New Zealand? Yet, without being able to cite more than one or two facts about Australia or New Zealand, millions of Americans and Europeans have grasped that these antipodean countries do indeed offer a fair go, that life there might be, as one American immigrant put it, "healthier and less perplexed." A Gallup Poll says that more than *five million* Americans would like to move to Australia—enough to raise the population by 40 percent!

This hunger to be Australian comes at least in part from a strange sort of nostalgia, a yearning for the· equivalent of an earlier America, less beset by intractable problems. Australia and New Zealand have the innocence of the United States in the 1950's, when mere growth seemed a good thing in itself. Australia, at least, still has a frontier feel about it—lots of land resources untapped (if perhaps untappable), and continuing mineral discoveries that make anything seem possible. There is, of course, genuine concern about the environment in Australia and New Zealand, but there are few pollution-killed rivers to despair over, and the people seem hopeful that they can keep Los Angeles' fate from becoming theirs. Worry about families with too many babies is diluted by the plain need for more population, more workers for the farms and factories. National problems trumpeted in the newspapers frequently have a made-up air when compared to the racking plights of countries elsewhere.

Yet Australia and New Zealand are not fresh-born countries, pacing along twenty years behind the United States. Melbourne was settled long before Denver, and Sydney earlier than Melbourne. The Midland Hunt in Tasmania has been going on every year since the 1820's. A full century ago, New Zealand fought its own equivalent of the American Indian Wars. In short, these nations *have tradition*. Well over one hundred years ago, Melbourne had a theater that seated four thousand, a university, a zoo, race tracks, gaslighted streets, and balloon ascents; and its people ate such civilized delicacies as roasted black swan with port wine sauce. Australia has for ages been dotted with Stately Homes.

Another point is the importance of water. Aridity conditions all life in Australia, for even city-dwellers are conscious that they live on the perimeter of an enormous "gibber plain" where rain may not fall for years on end and no rivers run. This is country so dry that the Aborigines survive by digging

for "soaks" of wet sand that they put into their mouths to suck on and then spit out. A sign at Ceduna, on a southern stretch of the highway that rings the continent, says: "LAST RELIABLE WATER FOR 770 MILES WEST." The only sizable river, the Murray, is drawn down for irrigation water so frequently along its course that little remains to flow into the ocean—and the Australian ideal would be to conserve stream water so thoroughly that not a drop would ever reach the sea. Water could make millions of acres bloom, and vastly raise the continent's ability to support human life, and thus, perhaps, make it a target for military conquest if humanity overpopulated the rest of the earth beyond capacity. To paraphrase an old Texas joke, all that Australia needs to be a Garden of Eden is water—but that's all hell needs.

Along with dryness there is a sense of isolation and distance. A large advertising sign in Melbourne reads: "FLY QANTAS TO THE WORLD"—and the world indeed seems a far-off place. The quickest flight from New York to Sydney takes twenty-one hours and thirty-five minutes. This factor—"the tyranny of distance," as historian Geoffrey Blainey has defined it—affects all life. As an example: it was because of, not in spite of, distance that England first sent convicts to Australia. The painters of New Zealand have historically displayed a style of their own because distance kept them from tapping the trends of France or England. Isolation accounts for the fact that the flora and fauna of each nation—those kangaroos, eucalyptuses, kiwis, and tree ferns—are markedly unlike the plants and animals of the rest of the world. Similarly, distance within the country has shaped both history and modern life. A city like Perth, in Western Australia, two thousand miles from Sydney, is comparable to Honolulu in its isolation. Exploring this forbidding space took not one but many epic expeditions like that of Lewis and Clark in the United States.

Next, Australia and New Zealand are pronouncedly different from one another in half a dozen ways. Australia's climate ranges up to the hot tropical; New Zealand's ranges down to the sub-Antarctic. Australia is dry; New Zealand is plentifully watered. Australia lives in uneasy guilt with its Aborigines; New Zealand's whites and Maoris form the most successful plural society in the world. Australia, though it has supplied some of the world's best fighting men for battles abroad, has never suffered significant bloodshed on its own soil; New Zealand underwent the shattering trauma of war at home when the whites seized paramountcy from the Maoris. Australia, accepting millions of immigrants, is increasingly cosmopolitan; New Zealand, apart from its Polynesians, is quite stodgily British. Australia builds with masonry; New Zealand builds with wood. Even the birds and the trees of the two nations are quite distinct from one another. Finally, and quite simply,

an Australian feels like an Australian and a New Zealander feels like a New Zealander—two different things.

But the two countries do have in common left-leaning political philosophies, though neither is by any means socialist, despite widespread public ownership of utilities and even a few industries. The left-wing parties in both countries are against further nationalization, and the right-wing parties that have governed each country most of the time since World War II accept the existing degree of state ownership while continuing to support free enterprise—capitalism remains potent and healthy.

Here, then, are two countries well known to be stable, prosperous, and wholesome. It is a pleasure to discover that they also possess rich layers of interest, color, even excitement. Let us proceed, discussing Australia for a number of chapters, then New Zealand, and then the two together again.

CHAPTER 2

The Continent That Nature Left Unfinished

Oh! there once was a swagman camped in a billabong,
Under the shade of a coolibah tree;
And he sang as he looked at his old billy boiling,
"Who'll come a-waltzing Matilda with me?"
—FROM "WALTZING MATILDA"

BILLABONGS AND COOLIBAHS. Barramundis and wallaroos. Bandicoots and bimbils. Here is a continent where all life, and the land itself, have evolved their own individual patterns. There is a bird in Australia that builds a house and plants a lawn, and a tree that grows as tall as the spires of St. Patrick's in Manhattan. It is the world's sixth largest nation—inhabited by fewer people than Peru. You often sense in Australia that great, weird, dawn-of-history movies could be made in the long reaches of unpopulated, unpopulatable back country. You remember old theories of continental drift, and wonder where this superisland may have floated in aeons past. You are told that its rocks are the earth's oldest, to be compared with those that astronauts bring back from the moon.

*

To convey Australia's size to an American is simple: it equals the old forty-eight states minus one average-size state—say, Alabama or North Carolina. To be precise, it measures 2,967,909 square miles. It is more than twice as big as India. A man standing on Ayers Rock, in the center of the continent, would have to walk at least a thousand miles in any direction to reach the sea.

This is the only continent in the world occupied by a single nation. The main subdivisions are six states: in order of population they are New South Wales, Victoria, Queensland, South Australia, Western Australia, and Tasmania. For the most part interminable straight borders (one of them a full twelve hundred miles long) define the states in a checkerboard pattern, following lines of latitude and longitude. Except for Victoria, which is only a little bigger than Kansas, and Tasmania, the Scotland-size island off the south coast, the states are huge. California, Oregon, Washington, Idaho, Utah, Nevada, Arizona, New Mexico, Colorado, and Wyoming could all fit into Western Australia alone.

In this book each of the states is discussed in a chapter. So are (1) the Northern Territory, the prodigious square on the checkerboard that is still administered by the federal government; (2) the Australian Capital Territory, site of Canberra; and (3) the Australian, or eastern, half of the big island of New Guinea, off Australia's north coast. A curiosity is that Australia has four more territories, some none too well known even to Australians, for reasons of smallness, remoteness, or forgettableness. An example is Christmas Island* (population: 3,700), fifty-two square miles of phosphate mine in the Indian Ocean south of Java. Similarly, Australia administers the Cocos or Keeling Islands farther west in the Indian Ocean, having acquired them in the 1950's (like Christmas Island) from Britain. They produce copra and house about 630 souls. To the east is Norfolk Island, a speck of territory (three by five miles) in the Pacific 1,035 miles from Sydney. Its fifteen hundred people grow cut flowers and bean seed and pay no taxes, except that once a year the Administrator is entitled to require every male between twenty-one and fifty-five to pay a sum anywhere between ten and twenty dollars, or work for between five and ten days. Despite having no beaches—indeed, no inlets or harbors—Norfolk draws seven to eight thousand tourists a year from the Australian mainland; the round-trip air fare from Sydney is $116.

The fourth territory measures 2.4 million square miles, not far short of Australia itself. This is the Australian Antarctic Territory, a wedge of the polar continent between 45 degrees and 160 degrees east longitude. Australia runs research stations at Mawson on the Antarctic coast, at Macquarie Island off that coast, and intermittently elsewhere. Neither the United States nor the Soviet Union recognizes Australia's (or anybody else's) ownership of any part of Antarctica.

---

* Not to be confused with an even less impressive atoll called Christmas Island in the Pacific south of Hawaii, claimed by the U.S. but actually held by Britain.

*

Some points about population: (1) It is small, a mere thirteen million people, averaging out at one person for each 160 acres, or four persons per square mile. If the population were evenly spread over the country, each Australian would be standing one-half mile from the next. This is a density 1/166th that of Europe. Only twelve hundred miles from Australia is Java, which has twelve hundred persons per square mile, for a total of 65 million. (2) It is growing fast, having doubled in forty years, although high immigration and natural increase together do not achieve a total growth rate as great as many underdeveloped countries reach by births alone. (3) It is urban. Almost one-half of the whole population lives in two cities, Sydney and Melbourne, and only 15 percent of Australians live in the countryside.

Let us now get an idea of the look of the land itself—the strange plants and animals and the eerie geological formations.

Fauna

For someone used to the flora and fauna of the rest of the world, exploring the plants and animals of Australia is startling and fascinating. The key fact in explaining this exotic life is Australia's isolation. All the other continents of the world are, or were at one time, linked so that species of life were able to spread quite easily. On this great "World Continent" more than sixty million years ago the forces of evolution were shaping two kinds of mammals. One was marsupial, in which the newborn infant, after a short period of gestation, crept into a pouch on its mother's abdomen, attached itself to a teat, and completed its development outside the mother's body. The other was placental, in which the fetus remained in the womb, fed through a placenta, until relatively well developed. The marsupial mode was efficient enough at that time to have produced a big and varied population; the placentals, though basically a better model, were not yet in full production.

At this point Australia may have been briefly linked to the World Continent, allowing animals to migrate. Probably the migration route was Asia–Indonesia–New Guinea–Australia, all presumably bridged together as the result of some geological upheaval. If a short sea passage was necessary, the migrating animal may have crossed in various ways. A delightful theory, advanced by the paleontologist George Gaylord Simpson, is that some dauntless, pregnant, tree-dwelling female animal rode a tree, possibly toppled by a storm into the sea, across the water gap between the World Continent and

Australia, and proceeded to found a race. Another possibility, less credible because it depends on the unlikely theory of continental drift, is that Australia was once part of Gondwanaland, a hypothetical continent joining South America, Africa, and Antarctica.

Probably because at that time marsupials were dominant in the World Continent, Australia got an infusion of them. Then the geographical link for some reason broke, and placentals in the World Continent proceeded to evolve all the way up to Man, displacing marsupials (except for the opossum) 25 million years ago. But in Australia, marsupials had a clear track. They evolved into forms fewer in number and variety than the placentals, while retaining certain resemblances to them. Thus numbats are like anteaters, wombats like marmots, cuscus like sloths, koalas like bears, and notorcytes like moles.

The most interesting marsupials, of course, are kangaroos. They come in forty-five species, ranging in size from one foot to nine, but the most common is the red or gray kangaroo about the height and weight of a boy or man. Some big ones weigh more than two hundred pounds. The baby kangaroo, called a "joey," is one inch long at birth, following six weeks of gestation. Using the powerful forelimbs characteristic of newborn marsupials, this tiny embryo finds his way to his mother's pouch. It is astonishing that he does not lose his way. Sometimes, if he does fetch up on some other part of her body, she may distractedly scratch him away as though he were a bothersome insect. In the pouch, he clamps his mouth to a nipple and nurses unremittingly for many weeks, not leaving the pouch until six to eight months later. Even when weaned, he still uses the pouch for shelter and transportation, and may jump out to graze at the same time that his mother does.

Kangaroos walk by planting their forepaws and tail on the ground while swinging their back legs forward into position for a step. At half-speed they hop, coming to the ground on back feet and tail. At full speed, forepaws are tucked in and tail is held up as the back legs, like steel springs, thrust the animal into the air. He can travel thirty miles an hour, yet veer abruptly. The forepaws are useful for boxing—joeys box at play, and older kangaroos, suitably gloved, box human beings in circuses, occasionally scoring clever knockouts. In a fight a kangaroo uses a forepaw to pin an enemy—dingo (wild dog) or man—by the neck while rearing up on his tail and ripping the victim's belly with a hind claw.

Astonishing numbers of kangaroos are hunted and killed every year—at least a million and a half in New South Wales and Queensland alone. One hunter shot 130 kangaroos in a single night. Shooting them is easy—they bound away from any source of alarm, then pause, stock-still, to see what the danger was. Licensed hunters get two dollars per animal. The kangaroo sup-

plies meat for kangaroo-tail soup, pet food, and human consumption in Hong Kong and Japan.* It tastes like beef. The fur goes for coats, such toys as koala bears, and the leather is used for boots and baseball gloves. But much of the slaughter is simply to exterminate kangaroos as competitors for grass on Australian sheep runs. Nevertheless, this animal does not appear to be facing the fate of the American bison; perhaps as many as 100 million kangaroos of all kinds remain, owing to high birth rates. Graziers call them "noxious animals," and favor slaughtering them as the only method of control. (Somebody has said that the Australian formula for the treatment of wildlife and the environment in general is: "If it moves, shoot it. If it doesn't move, chop it down.")

Similarly plentiful are wallabies, essentially just small black scrubland kangaroos. One type of phalanger (something like a flying squirrel) is so tiny that it would fit into a slipper; its babies, the smallest form of mammal life, cannot be seen without a magnifying glass. There is a mean marsupial mouse which, if caged with a real mouse, quickly converts the real mouse into nothing but a pelt, peeled off and turned inside out. Another mean marsupial is the Tasmanian devil, raccoon-size and beady-eyed, a scavenger. Many marsupials eat meat—but the two-thumbed koala eats nothing but the tips of certain eucalyptus leaves. The pouches of all marsupials that walk on four legs open to the rear.

"Monotremes Oviparous, Ovum Meroblastic"

Even more primitive than the marsupials are the monotremes (from Greek roots that mean "one" and "hole"), the duck-billed platypus and the echidna (spiny anteater). One orifice (cloaca) serves as outlet for the intestine, kidneys, uterus and ovaries in the female, and sperm in the male. Sexual intercourse is apparently carried out cloaca to cloaca. Eggs hatch in a week or two, and a few days after that the mother's abdomen begins to sweat milk, which the babies avidly lap up. Verification of the fact that platypuses lay eggs, which had been rumored for decades, was made by a young British zoologist in 1884, who cabled the news as follows: "MONOTREMES OVIPAROUS, OVUM MEROBLASTIC."†

Australia has lizards ten feet long, which sometimes walk on their hind legs; worms ten feet long, at least when extended; dugongs, mammalian sea cows, ten feet long and half a ton in weight; and crocodiles that grow twenty feet long. There are 136 varieties of snakes, many venomous and about half

---

* Fifty-two percent of the carcass is edible, compared to 32 percent of a pig.
† Translation: "Monotremes egg-laying, egg of partial cleavage."

a dozen deadly. The most lethal is the taipan, which invariably kills its victim, and can produce enough venom at a single milking to kill two hundred sheep. The brown snake, also deadly, is noted for the wrestling matches staged by the males, who coil around one another like the strands of a rope and struggle until one is exhausted and falls away. Other killer snakes are the six-foot tiger, the black, and the death adder. But, as Alan Moorehead has pointed out, none of these snakes will attack unprovoked; instead, they run away, for the snake "is an intensely frightened and fragile creature; one blow on those delicate hairpin-thin bones will instantly break its back." Many snakes, besides being harmlessly nonvenomous, are positively beneficial; the carpet snakes and the children's python, both seven feet long, feed on rats and rabbits, and farmers hold them in esteem. Incidentally, Australia has a certain renown for its spiders. One of them measures eight inches across, and spins a web sturdy enough to catch small birds.

Bird evolution in Australia seems to have come to a dead halt for some species and reached world records of sophistication in others. The flightless, ostrich-like emu, five feet tall, can run forty miles per hour and is too stupid to stop for a paddock fence, hence crashes through. The bowerbird builds a wigwam-shaped playhouse from twigs, decorates it with colorful feathers, pebbles, shells, and broken glass, paints it with pigments of his own manufacture using a beak-held spongy brush, and plants a lawn of moss in front. The old theory was that this bower (it is not a nest) lured females for copulation. A newer idea is that the bowers, of which each male builds several, serve to stake out his territory against other birds.

Another sophisticate is the lyrebird, which improvises songs of its own, but also imitates sounds, such as a squeaky wheel or a violin or a piano chord or a baby's cry. In courting, the male climbs an earth mound of his own making, opens his fernlike tail to a lyre shape, flips it over his back so that the feather ends hang in a fringe in front of his eyes, and goes into a song-and-dance that is formidable and beautiful. In *Their Shining Eldorado* Elspeth Huxley says that the lyrebird does not sing for joy; his song warns other males away from the two to five acres that he considers his own, and is basically an instrument of hate, not pleasure.

The cassowary, another bird that does not fly, runs at express-train speed with head outstretched, armed against collision with trees by a sort of bone helmet. The male mallee fowl leads a life of desperate, panicky exhaustion keeping an even temperature in the hole-in-the-ground egg incubator that he and his wife maintain; he uses rotting compost to raise the temperature and night-cooled wet sand to lower it. Brolgas, which are cranes, dance together in groups of six to several hundred. There are also wedge-tailed eagles, a

pageant of parrots, birds of paradise, pugnacious geese, black swans, the laughing, kingfishing kookaburra—among them, 431 species of birds *unique* to Australia.

Flora

The spectacular results achieved among fauna by isolating a few species for sixty million years in a sort of continental laboratory are matched among the flora (Australia has 100,000 species of plants and animals). The most remarkable plant is the eucalyptus. We know a few forms of this tree in the U.S., transplanted from Australia to Florida and California; Americans call it gumwood and use it for pews in churches and dunnage in ships. It has also been taken to Africa and South America—it is, for example, the most prominent feature of the Andean landscape around Bogotá. But only in Australia can one see the whole exultant spread of 603 species of this tree.

The eucalyptus called karri, which flourishes in the southwest corner of Australia, grows as much as 280 feet high and 24 feet around, with trunks that shoot up without branching for 160 feet; they are comparable to many redwoods. The karri produces a tough and beautifully grained lumber, and, like a similar eucalyptus, the jarrah, is often felled by explosives. An even bigger eucalyptus is the mountain ash. By contrast, the mallee, ten to thirty feet high, covers the enormous worthless scrubland to the north and south of the central desert. Finding names for every species of eucalyptus seems to have been an exercise in inspiration: moort, bloodwood, bastard box, coolibah, red tingle, ironbark, bimbil, wandoo, and scribbly gum (from its calligraphic bark design). Despite such immense variety, the eucalyptuses share many common characteristics, as they must being of the same family. They all shed bark—the manna gum sheds fifty tons per year per acre, creating a fire hazard. The eucalyptuses' feather-shaped leaves, colored a melancholy gray-green, hang straight down, receiving sunshine on both surfaces. They stay on the tree the year round, and, when crushed in the hand, give off a camphorous odor. The name "eucalyptus" derives from Greek for "well covered," a reference to the fact that the petals fuse to cover the bud before they drop off. The brushy-looking flower consists only of stamens.

The chief vegetation of the sandy deserts (apart from plants that germinate like ghosts after a rain and then disappear) is the spinifex, a tussocky grass. In the arid margins of the desert grows mulga, a sapling-like acacia. In fact, species of acacias, or wattles, number 630, more than the eucalyptuses. Their leaf appears in the national coat of arms. Also found in the desert are saltbushes and bluebushes, with succulent leaves that absorb night moisture

and serve as sheep fodder. The semiarid savannas grow wallaby grass and kangaroo grass, making good pastoral property. In really wet places there are forests of beech, coachwood and bunya pine. The cypress pine in the mountains is thought to be the world's oldest species of conifer.

Arid Australia

One who would travel this country for pleasure would go to hell for a pastime.
                                                    —GEOGRAPHER R. T. MAURICE

Australia, which is shaped like a humped rhinoceros with its horn aimed at New Guinea, divides into three zones of differing characteristics. The *highlands* parallel the whole east coast, and although they are called the Great Dividing Range, they are actually only a series of plateaus thirty to a hundred miles wide, averaging twelve hundred feet in elevation. Next west are the *central eastern lowlands,* which traverse Australia north and south without ever rising to more than seven hundred feet above the sea. The *western plateau,* about three-quarters of the land mass, running fourteen hundred miles to the Indian Ocean, is the flat, arid, harsh, empty, and often grotesque stretch that typifies Australia in the world's imagination. Australia is mostly desert (though the word is unpopular there). More than three-quarters of the entire population live in a boomerang-shaped coastal area on the southeastern corner, where the climate is temperate and eminently habitable.

*

Alps in Australia? In the southern end of the Great Dividing Range, centered around Mount Kosciusko, there are (and the Australians drum it into you constantly) "more snowfields than in all of Switzerland." In subtropical terrain like this, the possibility of plentiful snow is difficult to grasp. But many mountains in the Australian Alps rise to six thousand feet or more, elevations sufficiently cold to guarantee skiing from June to October, and sometimes as late as Christmas. Kosciusko, Australia's highest mountain, reaches 7,310 feet. The area is within easy reach of both Sydney and Melbourne, because development of the Snowy Mountains Hydroelectric Scheme —more about that in a later chapter—has led to the construction of numerous roads. So skiers come in droves to well-equipped resorts, one of them, Thredbo, being also unashamedly expensive. The Kosciusko chalet has a chairlift three and one-half miles long; the architecture of the area is mostly Tyrolean.

The views can be both spectacular and quietly pretty—there is an awesome sense of distance. Driving through in the summer, you climb and descend and climb again, reaching alpine vegetation on the ridges and penetrating silent eucalyptus forests farther down. The inkblot waters of the lakes and ponds of the hydroelectric complex lie all around, and you are aware that inside the mountains, in blasted-out chambers, generators hum. A surprising number of parrots flap through these forests.

North of the Alps, the mountains of Australia are interesting chiefly as well-watered tablelands that support much of the country's crop farming.

The lowlands, bisecting the country along the 140th meridian of east longitude, have generally good soil but dreadfully insufficient rainfall. Oddly enough, however, it is easy to water cattle here, for the lowlands contain the world's largest artesian basin. It is necessary only to puncture the seven hundred feet of shale that seals the tops of the aquifers to get water suitable for animals (but too alkaline for irrigation). Nine thousand "bores" (as Australians call wells) make it possible for this area to support a large part of the Australian sheep-raising industry.

The southern part of the lowlands contain the Darling, Murrumbidgee, and Murray rivers, the only river *system* in the country. They are reluctant rivers. The Darling is shown on maps as a long stretch of dotted line, an indication that it ceases to flow when rain is insufficient in the headwaters. The Darling once stopped running for eighteen months, turning itself into a series of billabongs, or oxbow lakes. In short, this country, like most of Australia, is vulnerable to long and parching droughts. Lake Eyre, a feature of the lowlands, is shown as 150 miles of blue on some maps, but it is really a salt flat, 39 feet below sea level. Only a few times every century is it covered with a thin sheet of water. The rivers that flow into it are equally bogus, but technically Lake Eyre is one of the largest internal drainages in the world, more than twice as big as the Great Basin of Utah, Nevada, and California.

*

We reach now an area that seems less a geographical subdivision than an immense and forbidding Fact: the prodigious western plateau. This Fact stands stubbornly in the way of the theory that Australia can fill itself up to the brim with population. Australia's look of immensity on the map fails to show the impossibility of human life (at least with present technology) in much of the western plateau. An enormous area in the middle of it has no population at all, barring a handful of nomadic Aborigines. Almost as

underpopulated is the area stretching from this "dead heart" to the north and out to the settled strips on the east and south, where most Australians live—only one person for every *eight* square miles. This is the "outback" that city-dwellers speak of, in Sydney or Melbourne or Brisbane; this is the "back-of-beyond"; this is the "never-never."

The plateau rises a thousand feet from the lowlands that we have been investigating, and in general maintains an average elevation of twelve hundred feet. In the geographical center of the country is a scattering of low mountains, chiefly the Macdonnell and Musgrave ranges. The Macdonnells run mile after mile in geometrically parallel ridges. Both the sand and the mountains are the color of terra cotta—this is the "red center" of Australia. These tones complement the unexpected verdure of the pure-white ghost gums, desert oaks, and kurrajongs that grow there in defiance of the aridity.

Australia is sometimes called the "fossil continent" because the pedestal of the western plateau is the pre-Cambrian shield, a vast rigid block of granite a billion and a half years old. The plateau gets only four to sixteen inches of rain a year, quickly evaporated. The result is deserts, or extremely arid plains, five hundred miles wide, that sever Australia north and south, from the Eighty Mile Beach on the northwest coast to the Great Australian Bight, the concavity on the south coast. A formidable oddity among these deserts is the sterile gibber plain, many thousands of square miles in which soil and sand are blown away, leaving only loose rocks, from pebble-size to one-foot boulders. The stones can be exquisite—wind-polished chalcedony, siliceous fragments, striped and translucent quartz. These lands are not only dry on the surface; they are dry beneath. A mine in the Kalgoorlie area went down four thousand feet without hitting any water.

More desolate yet is the Nullarbor (pronounced "NULL-a-bore") Plain, a crescent of limestone adjacent to the pre-Cambrian granite on the Great Australian Bight. Any water that falls here sinks instantly into the nine-hundred-foot-deep limestone that underlies the plain. The result, as the imitation Latin name suggests, is no trees. Such is the flatness of the surface that the transcontinental railroad crosses the Nullarbor Plain in an arrow-straight line for 330 miles. The shelf terminates on the south in spectacular high cliffs, saw-tooth-edged when seen from above, that run for 120 miles unbroken, dropping vertically into the milky-green ocean.

The western plateau ends sharply in a scarp along most of the west and northwest coast. A plain at the foot of the scarp provides space for the city of Perth and the lush, well-watered shoreline to the south of that city and around the bend of Australia. On the northern edge of the western plateau lie the impressive Hamersley Range and Kimberley Mountains. Large numbers

of kangaroos, wild buffaloes, dingoes, crocodiles, and birds live here, and are all regarded as fair game for hunters.

By a road with a rather romantic conception, Australia's Highway One, a motorist can drive around the rim of the continent, barring only one gap in the north. The route is surely the world's longest: 7,664 miles. Driving it takes many weeks, because though most of it is easy going, parts are slow, rough, and even perilous. The trip starts in Cairns, on Cape York Peninsula in northeastern Australia, and sweeps around the country like the hand of a clock. Highway One goes clear to Darwin, about as far from Cairns, to stick with the clock image, as the distance from one to eleven.

It would be tidy if the road went clear around the face of the clock. But between Darwin and Cairns there are 1;150 straight-line miles, and a road would be much longer. Eventually it will be built, as a "beef road" to transport cattle that presently walk a couple of thousand miles from breeding stations in the northwest to fattening lots in the east. It could also link up with the new bauxite mines on the Gulf of Carpentaria. Otherwise it would traverse some of the world's mightiest nothingness.

*

Coolest Australia, in the south, is like warmest Europe, Spain, or southern Italy. Hottest Australia, in the north, is as torrid as equatorial Africa. The highest recorded temperature is 127.5 degrees at Cloncurry in western Queensland in 1889—which compares with the U.S. high of 134 degrees at Death Valley, California, in 1913. Except in high mountains in southeast Australia and in Tasmania, the only snow is a light freak dusting sometimes in the south. Temperature swings are not great as between day and night, and summer and winter. Sydney's mean temperatures are 71 degrees in summer (which, opposite to the Northern Hemisphere, runs from December to February) and 55 degrees in winter (from May to September).

The average annual rainfall is sixteen and one-half inches, as compared to twenty-six for the whole world. But the Australian average is compounded of coastlines in Tasmania that get one hundred inches a year weighed off against deserts that get as little as four inches a year. Even the statement that the deserts "average" four inches a year is misleading—the average is likely to be calculated from one torrential twelve-inch rainfall followed by three years of no precipitation at all. In the northwest below the monsoon belt, as much as twenty-seven inches of rain may fall in one day, to be followed by years of drought. On the other hand, the monsoons of the north and northeast are pretty reliable: during "The Wet," from November to April, twenty

to fifty inches of rain fall. Perth and the big cities of the southeast have fairly even rainfall. Cyclones and anticyclones determine the weather of Australia as they migrate through the Indian and Pacific oceans.

*

The Australian Aborigines, as we shall presently see, have survived the arrival of the white man only with great loss of numbers and culture, but their place names prevail in abundance, and give Australia much of its flavor. For the visitor, collecting these melodious, many-syllabled, and tongue-tripping words can become a mania. Herewith a well-culled list, to suggest the feel of traveling this continent: Dulkaninna, Kopperamanna, Ooroowillanie, Gringegalgona, Koolynobbing, Bimbimbie, Derribong, Oodnadatta, Nindalyup, Goondiwindi, Collymongle, Bulla Bulling. All of them have translations. Grong Grong, for example, is "very hot," and Boolaroo is "many flies."

CHAPTER 3

From Convict Colony
to Commonwealth

The jury says, "He's guilty,"
And says the judge, says he—
"For life, Jim Jones, I'm sending you
Across the stormy sea."
—OLD CONVICT BALLAD

As SOON AS men attempted to make world maps, they sensed that there had to be an Australia—an antipodes to Europe. Ptolemy called it Terra Australis Incognita on maps he made about A.D. 150, and a French cartographer in 1550, unable to contain his impatience for the actual discovery of the unknown southern world, sketched in an Australia connected to Antarctica and fitted out with camels, deer, men, and castellated fortresses. Of course, not every early authority believed the myth. St. Augustine, to whom several present-day Australian churches are forgivingly dedicated, hooted down the whole "fabulous hypothesis of men who walk a part of the earth opposite to our own, whose feet are in a position contrary to ours." But a Spanish explorer named Juan Arias, pleading with King Philip II for ships, argued that the then known land-to-sea ratio in the Northern Hemisphere had to be matched in the Southern, "according to what we are taught by sacred writ and philosophical reasoning." As late as 1769 (after many seafarers had actually touched the Australian coast) the myth-making continued. A one-time employee of the British East India Company deduced that Australia must be bigger than Asia and inhabited by fifty million people.

Considering the certainty of expectations that the "great southern continent" existed, the explorers who groped toward it, over many centuries,

seem to have been the victims of some grim cosmic joker intent on keeping them from finding it. The Spaniard Luis Vaez de Torres, sailing from South America to look for Terra Australis Incognita, managed in 1606 to shoot right through the strait (which now bears his name) between Australia and New Guinea, hardly a hundred miles wide, and never realized that his goal was almost within sight. A few decades later Abel Tasman, a Dutch East India Company skipper, discovered Tasmania and New Zealand, and died without ever knowing that Australia lay within the twelve-thousand-mile south-swinging loop he had sailed. Later Dutch ships reached Australia's west coast many times; they named the coastline New Holland, and so it was known for two hundred years.

William Dampier was the first Englishman to touch on what was finally to become a patch of British red on the map of the world. Dampier was no valiant officer seeking colonies for the Crown; he was an enterprising buccaneer who had lately seized a ship and put its captain ashore in the Philippines. Discreetly retiring from the busy trade routes, he stood south in his stolen ship and on January 4, 1688, reached the north coast of Australia. Returning to London, this amazing con man persuaded the Admiralty not only to forget his piracy but to provide him with a ship for further exploration. He managed to reach Australia briefly again before his squabbling crew forced him to return to England. Jonathan Swift took Dampier's yarn as the basis for part of *Gulliver's Travels*.

The Dutiful, Tireless, and Honorable Captain Cook

Nearly a century went by with all the colonizing nations of Europe aware that some vast land must attach to the sketched-in coast of New Holland, but none pursuing the exploration. Finally Britain saw the urgency of preserving its South Pacific interests, and in 1769 sent off Captain James Cook on the ship *Endeavour*. His ostensible mission was to observe the transit of the planet Venus from Tahiti. Turning west from Tahiti, he circumnavigated New Zealand and defined it as two islands, then sailed west again and bumped into Australia. What a picture of duty, discipline, and happy industry we get of this Yorkshire laborer's son, who is said to have invented the proverb, "An apple a day keeps the doctor away." Up the coast he went, tirelessly naming every physical feature—Point Hicks, Cape Howe, Cape Dromedary, Point Upright—of what he took to be a land of "very agreeable and pleasant aspect, diversified with hills, ridges, plains and valleys, with some few small lawns . : . we saw the smoke of fire in several places, a certain sign that the country is inhabited."

Australia was—at last—really discovered. At length Cook came to a large bay, and explored its shores for eight days, in company with the botanist Joseph Banks, discovering, among other wonders, Aborigines, oysters, and gum trees. "The great quantity of this sort of fish found in this place occasioned my giving it the name of Stingrays Harbour," he wrote in his log. Then he drew a line through that sentence, and replaced it with: "The great quantity of plants Mr. Banks and Dr. Solander found in this place occasioned my giving it the name of Botany Bay." Thus the legendary name that for a while came to signify the whole of Australia. Cook thereafter sailed along the coast eighteen hundred miles to Australia's northernmost cape, and, stepping ashore on Possession Island, gave the coast the name of New South Wales. Cook lived on, still exploring the Pacific, until 1779, when he was murdered by natives of the Hawaiian Islands.

Convicts

We come now to a touchy, paradoxical event, the second landing at Botany Bay, eighteen years after Cook. To Stone Age Australia, the continent that somehow could not even be discovered until modern history was well advanced, came its first white settlers. And who were they? Convicts from the jails and prison hulks of England. What sorts of human beings were they? How does their heritage affect Australia today?

Any American prone to condemn Australia's convict origins might well recollect a bit of history. Early in the seventeenth century American colonists induced the British Crown to send—more precisely, sell, at ten pounds each—English convicts for labor in Maryland, Virginia, and Georgia. A statute in 1666 formalized this transportation of "lewd, disorderly and lawless" persons to the American colonies, and they came at a rate of about five hundred a year for more than a century, until the American Revolution. Thus there is a not inconsiderable leaven of descendance from convicts in our own nation.

The Revolution was the operative factor in the diversion of the flow of English prisoners from our shores to Australia's. Let us get a picture of the state of crime in England at that time. The industrial revolution and the enclosure acts were tearing society apart, creating hordes of hungry, jobless, crime-prone people. Law enforcement was sketchy and primitive; criminals frequently and literally got away with murder. "Pickpockets plied their trade in the crowds that watched pickpockets being hanged at Tyburn," wrote an English historian. The frightened Establishment tried to cope by prescribing heavy exemplary punishment for those criminals who did happen to be caught. About two hundred crimes were made punishable by death,

including, incredibly, the crime of "impersonating an Egyptian." Generally the death sentence was remitted if the convict agreed to being sent out of the country. The term for this was "transportation."

Most convicts sent to Australia, or to America for that matter, were guilty of what would now be regarded as minor offenses. "The bulk of the prisoners, who comprised about one-fifth of those convicted of crime at the Assizes and Quarter Sessions of England, were the town thieves, pickpockets, and shoplifters that constituted such a serious menace to the badly policed English society of the early nineteenth century," says A. G. L. Shaw, professor of history at Monash University, Melbourne. In *The Crimes of the First Fleet Convicts*, published in 1970, historian John Cobley tracks down and describes the precise offenses of the first batch of convict colonists transported to Australia. Typically, someone stole "one sattin waistcoat, value 3s, and one corded dimity waistcoat, value 5s." To the antipodes for thefts amounting to eight shillings! But Dr. Cobley flatly denies that the convicts were "more sinned against than sinning." He also denies that any of the first convicts were transported for political crimes. What the book does not bring out is how many of the convicts were incorrigibles, transported because the specifically cited felony was only the last of many. This is probably the case with most of them. And in later shipments there were ruthless murderers and rapists, exact proportions unknown. Further as to the makeup of the convicts, of the 160,000 ultimately transported to Australia only 20,000 were women, many of them prostitutes. Of the males, a great many were boys—the Artful Dodgers and orphans and deserted children of London.

Of such components as these was the band that sailed into Botany Bay in January 1788: the First Fleet, as the Australians call it in an oddly prideful way. We cannot tarry long over their early years and hardships. The main point is that, though a raffish lot, by and large, they built a settlement just as surely and persistently as if they had landed from the *Mayflower*. The first Governor, Captain Arthur Phillip, decided at once that Botany Bay would not do as a town site.* He went up the coast in a small boat, and discovered the merits of what is now Sydney Harbor. Phillip chose for settlement a well-watered inlet that he called Sydney Cove, in honor of

---

* One of the strangest coincidences in the long hit-or-miss history of Australian exploration took place the day of the landing. Beginning in 1771, the French government interested itself for more than thirty years in South Pacific exploration. The ranking French explorer was Count Jean-François de La Pérouse, who sailed into Botany Bay just a few hours after the First Fleet arrived. Had he been first, de La Pérouse might have established a strong French claim to Australia. As it was, he and Captain Phillip greeted one another amiably, and a bit later de La Pérouse sailed off to death by shipwreck in the New Hebrides.

the Home Secretary of the time. On January 26, now Australia Day, the colonists—736 convicts and 294 marines, officers, and officials—landed there.

No Devil's Island, this. A strange aspect of the Australian penal colony was that relatively few convicts ever did time behind bars. From the first, they were more colonists than prisoners. The immediate task was to provide housing and food for themselves. Though a sorry and often drunken bunch, they—and the convicts landed by three subsequent fleets—managed in the end to build a viable community. Men once skilled only at picking pockets learned perforce how to farm. The difficulties were overwhelming —for instance, no one thought of sending the colony a plow for years. The convicts were assigned by the government as free labor to the military officers and other settlers, to work on farms or in small business enterprises. Obedience was enforced to a great extent by bribery, perquisites, or actual wages. The settlers could also get balky convicts flogged, and 42,000 of them were flogged in New South Wales between 1830 and 1837. "The way they floged them was theire armes pulled Round a large tree and their breasts squezed against the tree so the men had no power to cringe or stir," runs one vivid account of a three-hundred-lash flogging by a convict forced to witness it. "I was to leew'rd of the flogers and I protest, tho' I was two perches [rods] from them, the flesh and skin blew in my face as they shooke off the cats."

Liquidating the Convict Culture

How did Australia dissolve the convict culture, and become, in the end, a nation so much like our own? The convicts were on the whole a depraved lot. One observer called the colony "little less than an extensive Brothel." The convicts drank anything alcoholic that their betters would sell them; most of their children were bastards. But the children of the first convict settlers generally turned out to be upright and industrious. A hard-bitten lawyer sent out from London to investigate the condition of the colony in 1817 was flabbergasted to discover that "the class of inhabitants who have been born in the colony affords a remarkable exception to the moral and physical characters of their parents." There is a simple and credible explanation for the contrasting morality of the second generation. Australia provided the convicts' children with abundant opportunity; work offered better prospects than crime.

In 1813 three young explorers (of whom more later) scaled the Blue Mountain escarpment west of Sydney and changed the city from a sealed-off prison colony to a gateway to the interior. Settlers poured in from England —greengrocers, vicars' sons, sheep raisers—altering the character of the population. An enlightened Governor, Lachlan Macquarie (1810–1821),

insisted against pressure from England and from the free settlers that "emancipists," convicts who had served their terms, were in every way the equals of other men.

New South Wales resisted the landing of more convicts, and got the Crown to send them to Van Diemen's Land (as Tasmania was then called in honor of the Dutch Governor who had sent Tasman to discover it), and to the northern coast, present-day Queensland. Transportation to New South Wales ended in 1840, and the gold rush of 1852 signaled the end of convict shipment to any part of eastern Australia. "There are few English criminals who would not regard a free passage to the gold fields . . . as a great boon," wrote a governor of Van Diemen's Land, commenting on the end of transportation. Western Australia still wanted English criminals for labor, and a model system of receiving and rehabilitating convicts continued there until 1868.

Exploration's Fascination

We turn now to an intriguing question: How was a continent known only as a shoreline on nautical charts opened up and laid bare to the white man's gaze? Even after Cook's discovery of New South Wales no man knew how this eastern coast was connected to west-coast New Holland. Captain Matthew Flinders settled the riddle by finding a continuous shoreline from New Holland to New South Wales along the south coast. It was his suggestion that the continent should be known as Australia (from Latin *"australis"*—"southern"). With the known parts of the continent hooked together, the problem to the people of Sydney became: what lay back of the thousand-foot scarp thirty miles west of Sydney? Going up the valleys led only to vertical gorges and impassable waterfalls. A clever young man named Gregory Blaxland overcame this obstacle in 1813; keeping to the top of a long highland spur he was able to look down to the other side.

Blaxland figuratively had the continent at his feet, but he could not know that. Subsequent invaders, following Blaxland's route, found languid rivers, the Murrumbidgee, the Murray, and the Darling, that flowed westward for a ways and then disappeared in seas of reeds. Charles Sturt, who discovered the Darling in 1829, wrote: "Its course is involved in mystery. Does it make its way to the south coast, or exhaust itself in feeding a succession of swamps in the center of the island?" The next year, he and his party carried a whaleboat by bullock cart to the Murrumbidgee, and then, cutting themselves off from any possibility of returning the way they had come, floated down that river, discovering its junction with the Murray and a bit later with the Darling. Proceeding downstream, he passed through terrain of

which he wrote, "My eye never fell upon a country of more promising aspect or more favorable position"—words that inspired the founding and colonization of South Australia six years later. Sturt emerged at the mouth of the Murray on the south coast of Australia near present Adelaide, and thus in one bold stroke solved the riddle of the river drainage.

In the life of this Australian explorer, there remains another story touched with romance—and this time with failure. Taking note of bird-migration routes at two widely separated points in southern Australia, Sturt in 1844 triangulated a point on the empty map where the courses of the flyways theoretically had to cross. He concluded that some paradise, desirable at least to birds, had to exist there. His expedition to find it encountered brutal hardships and broke his health. He never found whatever it was the cockatoos and parakeets were flying toward. Sturt was left in despair at failing to be "the first to place my foot in the centre of this vast territory."

We can mention only two or three of the other explorers. Edward John Eyre *walked* the entire south coast along the Great Australian Bight in 1841 and found not a single watercourse in fifteen hundred miles. A Polish count named Strzelecki explored the heights of the mountains southwest of Sydney, and named Mount Kosciusko in honor of the Polish soldier-states-man who volunteered in the American Revolution and served under Washington at Yorktown. In 1860 John McDouall Stuart penetrated to the very center of Australia and planted a Union Jack as "a sign to the natives that the dawn of liberty, civilization, and Christianity is about to break upon them." A year later Robert O'Hara Burke and W. J. Wills explored Australia, south to north and back, starting from a base in New South Wales, using camels imported for the purpose. They suffered such torments and delays on the return trip that the staff at the base camp gave up hope of seeing them and pulled out just seven hours before they finally got back to the camp. Both men thereupon died of privation. Many of these explorers *navigated* their way through the trackless, empty outback; Burke and Wills, for example, followed the meridian of 140 degrees east longitude. By 1875 the exploration of Australia was basically finished, the continent known.

Australia Grows Up

The tapestry of nineteenth-century Australian history is woven of wool and touched with many colors, most excitingly gold. We see in it the extraordinary conversion of a convict colony into a rousing radical democracy, a nation put together with neither princes nor peasants.

The century divides, in the political sense, with the discovery of gold in 1851. During the first half, following the explorations we have traced, the geographical lines were drawn that defined what were then six colonies and are now six states. New South Wales was founded as a place to dump convicts. It sent out the first settlers to Tasmania to keep France from claiming that island. They in turn, after some years, sent settlers to Victoria because it seemed fertile and rich. South Australia was born of lengthy intellectual discussions on the founding of an ideal colony, spun out chiefly by an incredible theorizer, Edward Gibbon Wakefield, who was serving a term in a London jail at the time. Queensland was another convict colony resulting from New South Wales' wish to have no more prisoners. Western Australia was the consequence of British determination to colonize the west coast.

A Gallery of Scoundrels and/or Heroes

A half-dozen fabulous men gave the colonies their directions in the first half of the century. The most eccentric of them was John Macarthur, who arrived in Sydney in 1791 as a lieutenant in the New South Wales Corps. He set himself up as businessman, sheep raiser, and financier, attending to military duty in what little time he had left. When a Yankee trader sailed into Sydney Harbor with a cargo of general supplies and rum, Macarthur and some other officers bought the cargo, retailed it at monopoly prices, and even paid the troops in rum. Henceforward for years, amidst corruption and drunkenness, this "Rum Corps" peddled monopoly liquor and dominated commerce. One of Macarthur's many enemies, a man with a talent for balanced alliteration, called him "as sharp as a razor and as rapacious as a shark." Debt and rum paralyzed the community. Settlers sold out to officers.

Yet to this avaricious scoundrel belongs much of the credit for founding Australia's greatest industry, sheep raising. As early as 1797 he bred Spanish Merinos that turned out to thrive on Australian pastures and remain the top breed of Australian sheep to this day. When he was sent to England in 1801, to stand trial for dueling, Macarthur took along samples of his wool and persuaded the Privy Council to grant him five thousand acres of good pasture and thirty convicts to work it. In 1806 he and fellow officers confronted a Governor who tried to break up their rum dealing, and summarily deposed him from office.

The Governor whom Macarthur and Company deposed was none other than Captain William ("Breadfruit") Bligh, victim seventeen years earlier

of the mutiny on the *Bounty*. Set adrift by the mutineers with a few loyal men, he had sailed an open boat four thousand miles west from Tahiti to Timor, not far off the northwestern coast of Australia. Despite his reputation for brutality, Bligh actually did have a heart, and as Governor he attempted to keep the Rum Corps from exploiting the poor. So Macarthur and other officers packed him off to Tasmania.

Bligh's successor, Lieutenant Colonel Lachlan Macquarie, was such a significant figure in Australian life that ultimately *two* rivers were named for him, the Lachlan and the Macquarie. He has been described by historian Ernest Scott as a man "of thoroughly dependable character, gentlemanly in his manners, kindly but firm, with the pride of a pedigreed Highland laird, the paternal authority of a Hebrew patriarch, the masterful self-sufficiency of a Norman baron, and the personal rectitude of an English squire." Macquarie succeeded in liquidating the Rum Corps. Having established the principle that an emancipated convict had repaid his debt to society and was worthy of general respect, he proved the point by occasionally inviting former convicts to dinner.

The end of the Napoleonic Wars in 1815 freed shipping to resume the transportation of convicts, and 18,500 of them were sent to Macquarie. However, New South Wales was transformed from a prison camp to a prosperous colony under Macquarie's long regime. His imprint on the face of Sydney is all-pervasive. He laid out the streets and, employing the exceptional talents of a convict-architect, Francis Greenway, gave the city many beautiful buildings; Macquarie Lighthouse, a colonnaded hospital, St. Andrew's Cathedral, and a tower-fronted stable now used as the State Conservatorium of Music are some of them. For his work Greenway, whose crime had been forgery, was duly pardoned. One sees his face today on the Australian five-dollar bill.

The first native-born Australian of note was the flamboyant William Charles Wentworth. A town in New South Wales is named for him as is a leading hotel in Sydney. He was the son of a convict mother and a gallant surgeon who, though not actually convicted on a charge of highway robbery in London's Old Bailey, had volunteered his services for the voyage of the First Fleet. As a youth Wentworth accompanied Blaxland on his exploration over the mountains west of Sydney. Wentworth's inauspicious parentage did not prevent him from going to Cambridge, and he came back to make himself, through politics and journalism, the most famous of all Australians at mid-century. At one point early in his political career he turned so angrily anti-British that he suggested Australia should tie its destiny to the United States. His great achievement was gaining a representative legislature for New South Wales, by lobbying in the British House

of Commons. He also established the University of Sydney, the country's first, in 1852. Though the "exclusionists," free settlers and their descendants who wanted to bar anyone with convict associations from their society, snubbed Wentworth socially, by the time he died, in 1872, he was known as "the father of Australian democracy."

*

"Squatter" is a term of contempt or pity in the United States; in Australia it commands respect. The bountiful fleece of the Merino sheep, as we have noted, was seen to be Australia's salvation almost from the first. Before 1830 the government granted land free on the pastures west of Sydney; then, changing policy, it proposed to sell land for five shillings an acre, limiting purchases to areas in nineteen counties conveniently close to Sydney. Defying the policy, graziers headed west with their flocks to squat on land beyond the nineteen counties. Soon "not all the armies of England, not 100,000 soldiers scattered throughout the bush" (as one Governor put it) could have herded the squatters back. The government retreated and asked the graziers only that they pay a ten-pound-a-year license. Courts began to uphold the squatters' land claims. This successful "land grabbing" gave the squatters some status; their success in raising sheep completed the work of making the word respectable. Not that the life—generally in a fly-specked, verminous one-room hut—was in the least gratifying. "There is no romance in monotony and mutton fat," wrote S. H. Roberts in *The Squatting Age in Australia*. In only three decades Australia became the main foreign supplier of high-quality wool for the mills of England.

To go ahead a bit: by the 1850's, '60's, and '70's Australian graziers had almost universally adopted the fence, at first post-and-rail and later wire, and abandoned the European principle that flocks had to be brought to shelter each night; these two measures vastly reduced the labor needed to run sheep and gave them thicker fleeces. They also found how to get water for their sheep from artesian wells. In 1880 the sailing of the refrigerator ship *Strathleven* to London with forty tons of meat signaled the big-scale use of sheep for mutton. At the close of the century Australia was grazing 100 million sheep—shepherd to the world.

*

Now: gold. It was incredibly, joyfully, easy to find. Huge nuggets, most notably the $50,000 Welcome Stranger, which weighed 141 pounds, lay in clay only a foot or so beneath the surface. Around Bendigo and Ballarat,

both within one hundred miles of Melbourne, gold was found not in placers or in solid rock as in the U.S., but in alluvial deposits, carried there in ages past by streams. To find it, at first, took scarcely any more effort than leaning down.

The first big gold strike was made by Edward Hargraves, who had been a California forty-niner. He found gold in a creek near Bathurst, a hundred miles from Sydney. Then came the discoveries at Bendigo and Ballarat. The news went round the world. Gold-seekers rushed to the fields, at first from Adelaide and Tasmania, then from abroad. Laborers left factories; graziers deserted their sheep runs; sailors quit ships. Sydney ran out of soap and candles, for lack of anyone to make them. In 1852 in the month of September alone, nineteen thousand immigrants landed in Melbourne, plus some thousands in Sydney, and trudged to the goldfields afoot. By nationality they were mostly English, Scottish, Welsh, or Irish—followed in numbers by Americans and Germans. Then the Chinese began to arrive, forty thousand of them by 1857, most of them meekly mining the tailings from earlier diggings. The population of Australia grew from 405,000 in 1851 to 1,146,-000 in 1861. In the same period Victoria grew from 87,000 to 540,000, to surpass New South Wales as Australia's largest colony.

The Australian gold rush brought none of the lynching, vigilantism, and killing that characterized the much smaller California gold rush (where a thousand men were murdered in the early 1850's, and only one killer punished). But it did foment the only battle ever fought on Australian soil. To dig, miners paid licenses of thirty shillings a month, and mounted policemen were unleashed on "license hunts" to collect the fees. The very concept of police in Australia triggers a hostility that apparently goes clear back to convict days (and continues until now). Miners found it particularly onerous to climb a mine shaft at the peremptory call of a cop. The police grew ever more arrogant and brutal. By a familiar escalation—a murder, a riot, prosecution of the rioters—feeling heightened until the miners felt obliged to build themselves a stockade at a town named Eureka. Some of the miners, increasing their list of demands from abolition of licenses to universal manhood suffrage, called for the formation of a Republic of Victoria. Police and the military charged the stockade. In the brief skirmish twenty-two miners and six policemen were killed. This fight was the bloodiest ever fought on Australian soil.

Australians have always seemed to sense that the gold rush, climaxed by the Eureka battle, has deep national meaning. As early as 1851 W. C. Wentworth foresaw that the gold era "must in a few years precipitate us from a colony into a nation."

The Second Half of the Nineteenth Century

After the gold rush, the population tripled. The proportion of unemancipated convicts dwindled to virtually zero. The rebellion at Eureka helped inspire reforms that ultimately included male suffrage, the secret ballot, and land reform. The seeds of Australian racism—the so-called White Australia policy —were planted; miners were prejudiced against and envious of the Chinese, who "were hard-working and had the usual good fortune attending those who work hard."

Finally, the Australians came to call themselves "diggers," a word then adopted by the rest of the world.

*

When did Australia become independent from Great Britain? That is a question often asked by Americans who have our own Revolution somewhat obsessively in mind, and who in certain cases of ignorance may not concede that Australia (or Canada either) has independence even yet. Australia (as well as Canada) does, of course. But independence came imperceptibly, and vestiges of English control remain to this day.

New South Wales was the mother colony, from which were later severed Tasmania, Victoria, and Queensland. As early as 1825 it had a Legislative Council named by the Governor, consisting entirely of government officials, and empowered only to approve measures submitted by the Governor. It was impotent, but it contained the possibility of change. Three years later, the British Parliament enlarged the size of the Council and gave it the right to have "opinions." By 1842 Parliament provided for a thirty-six member Council, two-thirds of whom were locally elected. By 1853 the Council was able to change the Constitution, a prerogative now traditional in Australia (but to this day regarded as unthinkable in the United States—as though our Congress could amend the Constitution as easily as it passes a law). The new Constitution created a powerful elected Legislative Assembly. In the ensuing process of widening the electorate and increasing democracy, the Governor gave up powers gradually and gracefully, moving from virtual dictator to a mere symbol of the Crown. One Governor of New South Wales found at a certain point that he was "powerless to do good, or to prevent it."

The exact steps by which the states became independent of colonial rule varied in curious ways. Victoria, breaking off from New South Wales in 1851, gave itself self-government two years later with an Executive Council

collateral to its Legislative Council. South Australia, created a Crown Colony in 1841, had a partly elected Legislative Council by 1851 and a constitution and self-government by 1856. Tasmania, held back in the democratic process by its heavy proportion of convicts, got independence from New South Wales in 1825 but no "representative institutions" until 1853, and responsible government came only in 1856. Western Australia, first colonized in 1825, waited until 1893, the date of big gold strikes in Coolgardie and Kalgoorlie, to attempt government by an *appointed* Legislative Council and Assembly. It did not move on to elected legislatures until 1911, well after the founding of the federal government. That part of Australia which to this day does not govern itself, the Northern Territory, was originally part of South Australia. In 1911 it became a territory under the jurisdiction of Canberra.

*

The influence of the country's social complexion on politics throughout this period is fairly clear. The convict heritage dedicated the nation to the ordinary man, and squatters and free settlers who tried to imitate the English landed gentry succeeded only in being slightly ridiculous. Australian agriculture, built around the ratio of one sheep to forty acres of dry land, prohibited the formation of a European-style peasantry. By 1885, wrote Shaw, Australia had "a society with a high standard of living, and a strongly egalitarian outlook lacking that respect for birth, wealth, and intellect which was usual in nineteenth-century England. In this sense, the Australian colonies were in the van of progress, and had successfully pioneered a radical political system, part cause and part result of the high status in the community of the 'common man.' "

A part of this political system, more radical than we now remember, is the Australian ballot. In the U.S. in the middle of the nineteenth century, ballots were generally printed by political parties on differing colors of paper, making secrecy dubious. Australian states, as soon as they began to elect representatives to their councils and assemblies, specified a uniform ballot printed by the government. This simple guarantee of secrecy struck Americans as most desirable, and it came into general use in the U.S. in the 1890's.

A quite natural concomitant to the thrust for democracy was the growth of unions. Craft unions started in the 1830's, and miners were pretty thoroughly organized by 1878. The Australian sheep shearers were organized into a powerful union in 1886. Trade unionism in Australia was stronger than

in any other country at the time. However, a Great Maritime Strike in 1891, involving many unions, failed. William Lane, a labor journalist and strike leader, who had soaked up his philosophy from the Americans Henry George and Edward Bellamy, quit Australia in disillusionment and tried to found an Australian Utopia in Paraguay. By the end of the century, labor had regained its power, pioneering effective compulsory arbitration with a success not yet achieved in the U.S.

In 1893 a severe depression exposed the weakness of six separate colonies on one continent with no machinery for joint concerns. Federation, apathetically discussed for a decade, took on urgency, if for no other reason than to gain respect for Australia in the eyes of world bankers, who found the states poor risks taken one by one. The hampering separation of the states grew painful and ludicrous, notably in the customs duties that they charged one another on goods in interstate commerce.

One Tasmanian politician told his people, in urging support of federation: "If you vote for the bill you will found a great and glorious nation under the bright Southern Cross, and meat will be cheaper, and you will live to see the Australian race dominate the South Seas . . . and you will have a market for potatoes and apples." On January 1, 1901, the first day of the new century, the Commonwealth of Australia came into being.

An Epilogue on the Convict Heritage

The problem of coming to grips with the convict heritage "gave most Australian historians megrim and threatened not a few with schizophrenia," says historian Russell Ward. Some theorized that the convicts, painted as simple rabbit poachers, were basically more innocent than the harsh judges in England; this was what was taught in schools. Others skated through the topic in a hasty and gingerly way—some still do. Still others took the nobody's-perfect line, quoting Daniel Defoe on Englishmen: "The most scoundrel race that ever lived; a horrid crowd of rambling thieves and drones." There was a poignant need for rationalizations. Melbourne and Adelaide took pride in not having been settled by convicts. Descendants of convicts burned papers that connected them to their ancestors. An English clergyman wrote, "There is no word in the English language of which one requires to make a more studied use in Australia than the word 'convict.' "

Now, seen in the light of Australia's success as a country, this national skeleton in the closet has mostly become a matter for jokes and indifference. How should a book like this one touch on the convict derivation? "Exploit it to the utmost!" bellowed one leading Australian editor. People boast that

they are among the one in twenty Australians descended from . . . well, not so much convicts as colonizers. One is urged to look at the results, a great functioning nation, and to judge people by what they are, not by what their ancestors may have been. The subject is out in the open. If there is shame, it is for the wretched treatment of many of the convicts after they reached Australia.

But apparently Australia is doomed never to escape completely from having had felons for founding fathers. A year or so ago a Manila columnist, inveighing against the White Australia policy, charged that Australians "still cling to the habits they inherited from their forefathers—the convicts from England's prisons." One converses with a noted author in Adelaide, and the phrase "lousy convicts dumped on us" slips out in his speech. At dinner, there may be a half a dozen Australians boisterously putting down the whole concept of any shame in convict colonization. But there are likely to be one or two who are not making jokes, but sitting strangely silent.

CHAPTER 4

The Lucky Country

Then live Australia! Matron young and mild,
Bear still bright Mercy's banner high unfurled!
Pardon and peace for Britain's fallen child,
Refuge for all the oppress'd of all the world.
—FROM A PROPOSED AUSTRALIAN
NATIONAL ANTHEM

AUSTRALIA HAS no national anthem of its own, and in fact one does not sense much nationalism here. What one notices more are symptoms of pragmatism, hedonism, and philistinism. In a moment of irritation, former Prime Minister Harold Holt once called Australia a "lotus land." There's some truth in that. But it would be fatuous to portray all of Australia as a lazy nation; prodigies of work have been accomplished in building the nation and its industries. The country functions—the plumbing works, the trains run, the mail gets delivered, the schools educate.

Still, where many Germans or Americans, for example, tend to make a cult of work, going on and on, obsessively, the bulk of Australians seem to keep in mind a certain moment every day and every week when work stops and the enjoyment of life begins. This state of mind implies a number of corollaries. For one, money loses priority as the all-powerful incentive. Constance Lee Butler, an American woman who went to Australia many years ago and married an Australian, observes that the Australian "often would prefer to earn less in order to have more free time to go to his weekender and swim and sail or paint his house and grow roses." Through some legality, the forty-hour week here officially works out at exactly thirty-five hours and fifty-one minutes. An American corporation officer says, "The

Australian working man has a nice setup here. He likes things pretty much as they are. It's difficult to tempt him with higher wages."

Australia offers vast possibilities for pleasure—and not only the simple pleasures of sport and the outdoors, but the larger tranquillities of freedom, prosperity, stability, and very little responsibility for the plight of the far-off, troubled rest of the world. Australians like to think that Australia exists for the good of Australians, not the other way around. This is not a lack of patriotism —more about that in a minute—but rather what Australians regard as the right order of priorities.

This pleasure principle works a couple of other ways, too. A girl journalist in Sydney argues that "Australians are happy because they consciously set aside religion and philosophy and take pleasure in what they have: prosperity, indifference to the world's problems, plenty of sunshine." In cab-driver language this comes out as: "Why worry about things you can't control?" A consequence is a kind of easy optimism. "She'll be right" is an all-purpose reassurance used whenever disaster impends. Strangely, she often is right; the gods seem to be on the side of the users of this jaunty Australianism. A parallel attitude is "It's good enough," said, perhaps, of a slapdash repair job—rarely do Australians take a tough, meticulous, Teutonic stand that something has damn well *got* to be right.

Is Australia, then, an earthly paradise? Not quite. People suffer diseases, fight, hate, work at unpleasant jobs, get bored. Books pour from the presses to tell Australians that they really don't have it so good. Nevertheless, the undercurrent of stubborn optimism flows strong. In 1964 Donald Horne, editor of the weekly newsmagazine the *Bulletin*, wrote a book criticizing every aspect of Australian society, and for a final flourish of irony called it *The Lucky Country*. To his astonishment, everyone took the title in its straightforward meaning, and it has been repeated so much in speech and print as to become an Australian byword, expressive of the country's euphoria.

Things the way they are are quite good enough to keep the majority of the people from wanting to rock the boat. Some force from the outer world —the pressure to accept Asian immigration, perhaps—may eventually muss up Australian complacency. Until then, a national characteristic is what New Left critic Dennis Altman calls "extraordinarily pronounced squareness." One is told that university students routinely undergo a period of derivative rebellion, "but they get square in two years."

To many people, elsewhere in the world, the sources of Australian contentment might appear to be rather slender. It's a boring, petty kind of life where ex-servicemen have an absolute right to a government job, and, worse yet, want it. Broadly speaking, the country seems quite satisfied to be badly

dressed—the pedestrians make the streets look like Brighton on a dim Sunday. The country is solid with people who have a low threshold of contentment—unambitious, middle-class people who are indifferent about getting on. "Outside the intellectual pockets, a bare-minded philistinism prevails . . . a bronzed, back-slapping, amiable mass, for whom foreign affairs and the world of ideas might as well not exist," says Ross Terrill, an Australian political scientist.

When Pope Paul VI visited Sydney in December 1970, he chided the Australians for their materialism and isolation. He warned of the temptations of "self-centeredness, hedonism, eroticism," and he implored Australians not to "close your limited circle for the sake of selfish satisfaction." Hedonism is indeed widespread in Australia, and the "closed circle" charge apparently referred to racial immigration bars.

*

An enduring source of Australian pride is its military tradition. There is a genuine and permanent awe at how the Aussies twice went out and threw away their lives and blood for England. The country is cluttered with monuments to World Wars I and II, and Anzac Day, April 25, the anniversary of the attack on Gallipoli (of which more later), is taken far more seriously than Australia Day, January 26, anniversary of the landing of the first convicts.* In the background are other struggles where Australians served—the Boer War, the Sudan, the siege of Peking. Australians contend that their soldiers, though prone to refuse to salute their officers, are the world's toughest fighting men, the winners of the country's international reputation. Sometimes the country seems to be symbolized by the rugged bare knees of the troops. Nowadays, it must be added, one senses that the older generation of Australians trades on the martial tradition—the young hate it, and joke about it.

The convict heritage interacts with patriotism in both pride and shame. A verse from the nineteenth century says:

> Small honor have we for the mother's name,
> Who stained our birth with the brand of shame.
> We were flesh of her flesh, and bone of her bone,
> We are lords of ourselves, and our land is our own.

A spillover from convict ancestry is a healthy distrust of authority. The historian Clive Turnbull notes "an impulse to snatch from the judicial system anyone who may fall foul of it. This is manifest on the most respectable

* Prime Minister Menzies always played down Australia Day.

level in distrust of politicians of all kinds. . . . The police, in the Australian
race memory, bear the brand of 'the system.' . . . Australians hate seeing
people put in jail." Yet the convict heritage carries with it, inevitably, a
distinct note of inferiority complex. "We are the most insecure people in the
world," says one Melbourne editor.

Little touches bear this out. A boast heard again and again is "best (or
biggest, or tallest) in the Southern Hemisphere." This is a safe and fairly
meaningless comparison, since the chief competition in that hemisphere
comes only from Brazil, Argentina, or South Africa. Defensive pride also
shows through in claims like "This book was printed on Australian paper."
Sugar packets honor the memory of Captain Cook with pictures of his
sextant, compass, portrait, cottage.

The national inferiority complex comes, one is told, from that convict
background (despite all attempts to make it a virtue), from domination by
Britain in the beginnings of the country, from being looked down upon by
the world as "too rough," from the slow development of Australian culture,
and even from the blatant Australian accent. Self-criticism is a national
hobby. A government booklet for immigrants confesses that the "ugly
Australian" can be a "blusterer, often loud-mouthed and ignorant"; a leading
educator says that the average Australian "would rather run half a mile than
sit and think deeply for five minutes." Jack's as good as his master, Australians
endlessly repeat—but Jack seems to lack the confidence to go on to be anything but Jack.

*

Good will devoid of grace, a down-to-earth pragmatism that can be
aesthetically offensive, a culture that is deliberately, almost defiantly
suburban.
—THE AUSTRALIAN CHARACTER ACCORDING TO ARTHUR KOESTLER

For most of the world, the Australian myth of the outback obscures the
fact that this is one of the most urban countries in the world. Moreover,
it has been citified for a century; way back in the 1860's the proportion of
urban to rural population in Australia surpassed that of the United States
and most of Europe. Four out of five Australians live in the cities, and most
of these in the six state capitals plus Canberra and Darwin—these two are
growing at the rate of 10 percent a year.

In addition, between 70 and 80 percent of all Australians own their own
houses. The consequence is that the cities spread over land areas as big
as any in the world. "Suburban sprawl" really takes on meaning when you

fly over the major cities of Australia. And undoubtedly this makes for an existence that is more conformist than anything found in the spreading "developments" around U.S. cities. Polls show that 98 percent of Australian suburbanites never go out in the evening except Saturdays, and television sets are on for an average of four hours a day.

Yet the Australian goes at suburban life with such a zest that he half-redeems it. Gardens are important, and whole streets bloom prettily. Men conserve energy on the job in order to slave zealously in their workshops over house repair and improvement. The trying relations of landlord and tenant are evaded by ownership. "You buy a home, engage in sports, you're on your own," says one contented householder. Moreover, a bit of sophistication is creeping into suburban life—your host may serve you a very good chicken in white wine sauce. For many an Australian, life on his own 7,500 square feet of the surface of the earth is paradise enow. Australians are indeed defiantly suburban, enthusiastically, overwhelmingly, unchangeably suburban.

*

Once the first-time visitor leaves Sydney airport and heads for his hotel, the game begins: is this country a transplanted England or a transplanted America? The expressway has the familiar engineering of the U.S. freeway system: America. But traffic keeps to the left: England. The highway signs have the interstate white-on-green colors and typography: America. But they may have a note of Old World courtesy ("DRIVE SLOWLY PLEASE"): England.* On a nearby railroad track, big double-truck freight cars (American) are interspersed with little four-wheel goods wagons (English). In the city, the newer glass-and-aluminum architecture speaks of America, the older Victorian architecture speaks of England. The commercial television derives from America, the sports from England. The excitement seems American, the manners are English. Brand names everywhere are American; the style of the newspapers is English.

"Australians combine the best and worst of California and Texas," says Sheila Scotter, former editor of the Australian Vogue, a magazine that is published in America, England, and Australia. But those two states certainly have no reverence for a queen, and Australia decidedly does. Queen Elizabeth visited Australia in 1970, strolled the streets with engaging informality, and cut prorepublican sentiment in the Gallup Poll from 40 percent to 25 percent. The Gallup Poll also shows that 65 percent of Australians want to

---

* There are even signs that say "PLEASE READ THE NEXT SIGN."

recognize Prince Charles as King when Elizabeth dies.* A Londoner reports that in general Englishmen feel very much at home in Australia, and a cynical Australian says that his countrymen will accept a bogus Englishman over an honest Australian. There are still people in Australia who have never forgiven the Duke of Windsor for abdicating.

All these comparisons grate on Australian nerves. Some say that they do not like what they had, Anglicization, and do not like what they are heading for, Americanization. Many Australians find identity by being anti-British, while laughing at Americans as naïve and immature. They say that they want an authentic independence, not to be compared with Canada's, which is flawed by the proximity and power of the United States. But Australia has not bothered to cut its ties to the British Crown with the same thoroughness as Canada. The upshot of all these tugs and pulls seems to be that Australia wants to be independent—with close ties to both Britain and the U.S.

*

Something between a pal, a friend, and a fellow worker—this is the Australian "mate," pronounced "mite." Mateship is an outgrowth of the taming of the bush in the last century, an egalitarian experience. The Australians, wrote a startled English observer in the middle of the century, "present a marked contrast to the population of England and Europe generally. As a rule, every man there is, may be, or expects soon to be, his own master. Here are no conventionalities, no touching of hats." But pioneering in the sometimes terrifying outback required more dependence of one man on another than, say, breaking sod in Nebraska. Thus a man was always pleased to find a mate, someone he could trust, and mateship took on a semi-mystical construction as an Australian quality.

Mateship produces a goodly measure of trust and candor; a man is taken at face value until proved otherwise. This makes for a generally cooperative community. Every man in a pub is the next man's mate, and the cab driver opens the door with a cheery "Hop in, mate." All this leads to a certain all-male complacency, a pub-and-club society that excludes women. Whether there is any relationship between mateship and homosexuality awaits the attention of a good sociologist.

* Which was not the case in a similar occasion in the past. In 1896 John Norton, editor of the Sydney *Truth*, raised a toast to Queen Victoria's health and long life, "if only to keep her rascal of a turf-swindling, cardsharping, wife-debauching, boozing rowdy of a son, Albert Edward, Prince of Wales, off the throne."

The Japanese, dependent on Australia in many business relationships, have made a careful study of mateship. Purchasing missions are given instructions in the intricacies of the Australian national character, as to wit:

> To understand the Australian better the Japanese must appreciate that although Australia is not a socialistic country, there are deep rooted elements of "mateship" and a genuine, "Jack is as good as his master" attitude outside the place of employment. Australians in authority tend to work closer with their immediate seniors or with their executive employees as colleagues and much more is done by discussion and conference than by authoritative direction. Any action of an overseas country which affects Australia and which in any way can be interpreted as an expression of superiority will incur an automatic antagonism from most Australians and this of course would not be in the best interests of international relationships.

The Australian Language

Sidney Baker, who has labored for more than twenty years in doing for the Australian language what H. L. Mencken did for the American, has accumulated more than five thousand words or phrases that are purely Australian, or have an Australian twist, or are greatly favored by Australians. With origins in England, the convict experience, the outback, the Australian expeditionary armies, and Pidgin English, this private language has extraordinary humor and expressiveness.

"Bloody" is the Great Australian Adjective—English too, of course. "Kangabloodyroo," they say. "Bloke" similarly survived the passage from England. Everybody is a bloke. Some odd verbs are: "winge," to complain; "yabber," to chat; "barrack," to root for; "bludge," to cadge; and "bugger," to wreck. If you want to speak this language, you must know the meaning of these nouns: "wowser," bluenose; "tucker," food; "jackeroo," apprentice on a sheep or cattle station; "poofter," male homosexual; "blue," a fight; "kip," a bed. An all-purpose exclamation is "Strewth!" To praise someone, the Australian says "Goodonyer!" "Digger" has largely gone out of style, but "dinkum," meaning true and admirable, and "fair dinkum," meaning even more so, still prevail. Some Australianisms are unbloodybelievable: the expression for "to attempt to deceive" is "to come the raw prawn." Or too vivid: the word for "vomit" is "chunder."

About the Australian accent, only a few points can be made with certainty and without giving offense. The great bulk of Australians, from one end of the country to the other, speak in a distinctive twang, employing the vocabulary suggested above. At least one other large group, generally better

educated, speaks an English that sounds to Americans and Britons much less accented, closer to the universal English of radio announcers.

The catchiest tongue, of course, is that of the first group, the accent of the pubs and sports stadiums, of tradesmen and miners and football players. London Cockney formed this accent in the beginnings of the colony, but Australians grow rather inflamed if you call it Cockney now. "Lady" becomes "lydy"; "mate" becomes "mite"; "mailing" becomes "myling"; this is "Austrylian." Accents land in strange places, and pronunciations get odd twists. A dispute is a "contrOHversy"—accent on the second o. A little reminder is a "meemo." A bookstore clerk in Sydney says that she frequently gets requests for "poultry," and learned only by hard experience that the customers really wanted poetry. Inevitably the day came when a man asked for a book of "poultry" and confusion reigned until he could at last make clear that he needed instructions on how to raise chickens.

A few years ago, a professor identified as Afferbeck Lauder defined a genial new language called "Strine," which is the word "Australian" telescoped. Strine utilizes the numerous elisions of Australian speech and the way its vowels veer unpredictably into diphthongs. The results can be hilarious. "Hazzy gairt non wither mare thorgon?" turns out to be "How is he getting on with the mouth organ?" Professor Lauder isolated many useful phrases, such as "egg nishner," a machine to cool a room, and "noker parison," incomparable. "Strine" came to be a household word; an English lady, complaining of the difficulties of train travel in Australia, said recently, "Try following an announcement comprising an Aboriginal place name read by a Greek whose halting English is complicated by a Strine accent." The merry professor turned out to be a commercial artist named Alistair Morrison. You may want to decode Afferbeck Lauder, his nom de Strine. Clue: the dictionary is arranged in . . .

Beer, Bountiful Beer

Wanted, a beautiful barmaid,
  To shine in a drinking den;
To entrap the youth of a nation,
  And ruin the City men.

Australia does not lead the world in beer-drinking—Czechoslovakia does, with 227 pints per capita per year. Australia is fourth (after West Germany and Belgium), with 201 pints (Britain averages 164, and the U.S. 111). But Australia talks a great game of beer-drinking; indubitably beer is the national beverage; and individual prodigies of consumption are achieved.

One Melbourne guzzler, for example, reportedly downed six pints of beer while the post office clock was striking six, which used to be the closing time in the pubs. *The Australian* solemnly reports that the country consumes 4,899,056,000 middies of beer a year, but this statistic is none too enlightening, since a middy measures ten ounces in New South Wales, seven ounces in Western Australia, and is not used as a measure elsewhere. Lore is so plentiful that author Cyril Pearl managed, a couple of years ago, to do a whole delightful book called *Beer, Glorious Beer!*

The circumstances of Australian beer-drinking are quaint and sometimes doleful. The beer itself, as anyone disgusted with the dry, thin, pale, light Pilsners of the United States can testify, is strong and hearty. This seems to be the consequence of using sugar, as well as malt and hops, in the brewing. The strength, depending on the state, runs from 3.5 percent of alcohol by weight to 4.5 percent—not too strong.* New South Wales law requires that beer contain *not less* than 3.3 percent of alcohol by weight. The ubiquitous pubs of Australia are only now beginning to turn themselves into amenable places along the lines of English pubs or American bars. Traditionally they have been (and often still are) grubby rooms lined in white tile like lavatories. This is traceable, at least in part, to the days when the six-o'clock closing hour meant that the pub existed not for sociability, games, or human warmth, but solely for grim, rapid drinking, which of course was bound to cause a lot of vomiting—and tile is easy to clean. The publicans, who encourage the singleminded boozing because it makes for profits, are not admired, by and large, in Australia. Connected to the public bar there is often also a saloon, where the drink may be Scotch and the customers professional men, and a lounge, where women are allowed. Pubs close on Sundays, except in Queensland, Western Australia, and Canberra.

Barmaids (sympathetically dealt with in the verse above by a last-century bluenose who wished to get the girls out of the pubs) are collectively a precious Australian tradition—cheery women who josh the customers as they fill a glass, let the head subside, top it off with more beer, and serve it: "Seventeen cents, thank you, love." Their expression for a repeat is "something similar," on the indisputable grounds that you can't have the "same" drink that you just drank. Treating in Australianese is "shouting"— "I'll shout you a beer"; the best guess about the origin of the expression is that rich, expansive gold miners used to shout for everyone in the neighborhood to come to the pub and have a beer. Taxes on beer produced $338,-614,000 in government revenue in a recent year. There is much learned

* U.S. beers range from 2.89 to 4.95 percent. New Zealand's Kings Brew measures 5.64 percent.

controversy over whether each new serving of beer requires a freshly washed glass. Superstition holds that refilling makes the beer taste better. "Most men in bars, I think about seven out of ten, prefer to keep their old glass," says the president of the Australian Hotels Association.

Clubs as well as pubs serve beer, but the implications can be misleading. An American in Brisbane wangled membership in exclusive Tattersalls, and presented himself at the bar for a beer. The bar, the barmaid, the beer, and the customers all turned out to be precise replicas of those to be found in any Brisbane pub.

Assorted Manners and Morals

A brand of matches called Federal is widely preferred because the wooden box slivers nicely to provide toothpicks and fingernail cleaners.

Women's liberation in Australia centers on a group that publishes a monthly magazine called *Mejane*, a switch on the famous line in the Tarzan movies. Germaine Greer, author of *The Female Eunuch* and a leading feminist on the London–New York axis, is an Australian. Women often have to resign their jobs when they marry. But not all women feel oppressed. "Women's liberation? What for? I can even go to a prize fight alone if I want to," argues strong-minded Sheila Scotter.

Sex among the Australian middle classes, it would seem, still suffers from Victorian attitudes; to have intercourse is to "have a naughty." Contraceptives are taxed at 27.5 percent as luxuries, and cannot be advertised.

The suicide rate for women in Australia is twice as high as the rate for men; no one knows why. The total suicide rate, 15.2 per hundred thousand of population, is considerably higher than that of, for example, the U.S., which is 10.9.

Sideburns are called "sidelevers" or "sideboards." A haircut costs $1.40. Long hair on men is every bit as common as in the United States or Europe, but the wearers give the impression that they are just being conventionally unconventional.

Vacationing abroad from this remote continent is expensive and exhausting —although one young businesswoman recently managed to make two trips to both New York and London in less than eight weeks.* Singapore and Tahiti are places that draw many better-off Australians. Fiji is the foreign

---

* These flights expose the traveler to maximum risk of suffering circadian dysrhythmia, the up-to-twelve-hour wrench of the human system's internal clock away from sun time. The disturbance seems to be worse when the flight goes from west to east, such as Sydney to London.

country (it became independent from Britain in 1970) that average Australians find near enough, exciting enough, and cheap enough to visit in large numbers—they form the islands' most numerous tourists. The shops there sell duty-free goods; one can pop over from Sydney, spend a few days at the Suva TraveLodge or the Fijian, and make purchases that save enough to pay for the trip.

Radio station call letters start with a figure: 5AD in Adelaide, 2CN and 2CY in Canberra, for example.

Americans, spending Australian dollars, tend to forget after a while that they cost $1.19 in U.S. money, and that therefore goods are not as cheap as they may seem. Thus the forty-five-cent shot of Scotch served at the Canberra Rex Hotel actually costs around fifty-one cents in American cash—still a bargain.

*

Australia used to be a world leader in cushioning its citizens from life's pitfalls by social security, but years of prosperity and antisocialist government have caused the system to fall behind. Health care is provided almost entirely by private insurance companies, which are subsidized by the government. Sick people pick their own doctors, pay the bills, and collect an average of 90 percent from the insurance companies. This self-help system works well enough to prevent the kind of medical disaster that bankrupts a family, but one is told that it helps the insurance companies more than the patients. Old-age pensions, introduced in 1908, have not kept pace with world standards, and now rank with those of Saudi Arabia and Trinidad. The full pension for a man over sixty-five (or a woman over sixty) is only $17.25 a week, and this sum is reduced, through a complicated "means test," in proportion to the pensioner's private income or wealth. For example, the pension is cut by an amount equal to about one-half of any private income over $10, and therefore drops to zero when private income reaches $41. Similarly the pension goes down in proportion to the value of a person's savings or property (apart from his house, car, furniture, personal effects, and a certain amount of life insurance), cutting out when these assets reach $21,200. People have to squander saved money to qualify for pensions. The Liberal Party wants to patch up the machinery by eliminating the means test, which would cost the government $370 million a year; the Labor Party wants a new, full-fledged National Superannuation Scheme.

With virtually full employment and evenly spread prosperity, Australia does not suffer the scabs of poverty that afflict big U.S. cities and the

countryside of the poorer states, but nevertheless nearly a million Australians live on the edge, with incomes of less than thirty-seven dollars a week for families of four. The poor include most Aborigines, the chronically sick, semiskilled workers with large families, and students.

*

Australia's auto-accident death rate, per mile traveled, is the worst in the world (though the rate per 100,000 population just about equals that of the U.S.). Upwards of 3,500 people are killed each year, and about 80,000 injured. Probably one reason is the lack of expressways with divided lanes—there are only 1,325 miles of them. Most of the 560 miles between Sydney and Melbourne are merely adequate two-lane highways; Brazil, with its Rio–São Paulo superhighway, far outclasses Australia in connecting major cities. The slaughter is carried out in a rather small arena. Australia has only 105,000 miles of concrete or asphalt highway and 143,000 miles of gravel; the rest of the 560,000-mile (versus 3,152,000 in the U.S.) network is described as "formed only" or "cleared only." Road maps show dotted-line tracks bleakly labeled "natural formation." The *Sunday Australian* carries a prominent report on "Roads to Avoid." Bumpy surfaces, lack of lighting, and fast driving at night by car-crazy kids are also blamed for the death toll, although Australian drivers in general are notoriously inept and careless. The state of Victoria in 1970 became the first jurisdiction in the world to make wearing seat-belts compulsory, as part of an effort to cut down on the killing. The move met some opposition on the grounds that civil liberty includes the right to risk death, but most people cheered the new law, and New South Wales soon passed a similar law.

The White Australia Policy

Australia has encouraged the immigration of literally millions of new settlers in recent decades, and the overwhelming majority of them have white skin. Australians are not very proud of excluding the black, brown, and yellow people of the surrounding nations and of Africa. They know that this discrimination seriously undermines their moral position in world councils. But among the vast bulk of Australians, shame is easily overbalanced, not by racism per se so much as relief at avoiding a problem. Historian A. G. L. Shaw speaks for millions when he writes: "Nor can any observer of South Africa or the Southern States of the U.S.A. now fail to rejoice that the color problem at least is one which Australia has never had to face."

Surely this is the ostrich hiding its head. Preserving a continent for thirteen million whites in a world of two and a half billion people of other colors does not sound much more viable over the long run than South African apartheid. Enforcing the immigration restrictions raises many poignant problems. Australia has to insist on visas on passports* in order that consular officers can get a look at the skin color of the prospective visitor. The exhaustive visa application requires the names of close relatives who are *not* accompanying the traveler, and the names of foreign countries where the applicant may have lived in recent years. A British technician can move to Australia quite readily unless—as in the recent case of Jan Allen, British subject, electronics engineer, and Jamaican—he happens to be black. Exclusion is a highly subjective, nothing-in-writing process, going back ultimately to the discretion of the Minister of Immigration himself. Desirable white immigrants get travel subsidies; blacks and Asians, unless superlatively qualified, get bland refusals.

Late in the last century, the *Bulletin*, then as now an important Australian periodical, carried a shamelessly racist motto: "Australia for the Australians —the cheap Chinaman, the cheap nigger, and the cheap European pauper to be absolutely excluded." The operative word then was "cheap": these races would work for less than the white man, and take his job. The second Prime Minister, Alfred Deakin, was thought to be most decently tolerant when he said in a speech that the Japanese were not inferior to white men, but simply different, and perhaps superior in some respects. One Immigration Minister, speaking more bluntly, said: "Two Wongs don't make a white."

The *Bulletin*, let it quickly be said, is now the country's leading advocate of a truly open immigration policy. "All that saves Australia from being a tenth-rate outthrow of European colonialism has been the constantly enlivening effect of immigrants," says this magazine. "If it makes sense to bring Turks to Australia to do some of the heavy work, why does it not make sense to try to get a few Japanese technologists to help think up some things for the Turks to do?" In point of fact, about 35,000 nonwhites and persons of "mixed descent" have immigrated since 1966. By and large Australians of the current generation know as well as any civilized men that skin pigmentation has nothing to do with a person's character and ability, and they accept individuals on a personal basis without much regard for

* The United States, of course, also insists on visas, and restricts immigration to 176,000 a year, less, in certain years, than Australia. But priorities other than color (such as having close relatives in the U.S.) determine who gets in, and in 1969, for example, 73,621 Asians and 5,876 Africans were admitted to the U.S.

color. But here a vicious circle comes into play: Australia is underpopulated, and therefore a sizable immigration of people with black and yellow skins would change Australian society more radically than it would if the country were more populous; and yet the mere fact of underpopulation makes excluding blacks and Asians from overcrowded places all the more disgraceful. A second consideration is that Australia is a Southeast Asian country, and must get along with its neighbors; it cannot pretend to be distant and isolated. Thus the White Australia policy has been steadily modified from the time it was first made law in the Immigration Restriction Act of 1901. Now the idea is to admit some, but not too many, nonwhites, with a preference for Asians, under the terms of the 1958 Migration Act, which set up a system of temporary and permanent entry permits.

The people who get in through these cracks in the wall are scientists, engineers, university lecturers, businessmen, teachers, nurses, and some other professionals. About forty thousand people of these categories, mostly Asian, now live in Australia. An Asian trader who can prove a business turnover of $20,000 a year or more can gain admission to Australia, and even bring an assistant. In addition, fifteen thousand Asian students go to Australian universities, but they are obliged to leave Australia when their education is complete. An ironic upshot is that they often emigrate to Canada or the United States, putting Australia in the position of subsidizing the training of talented people for the benefit of those two countries, while simultaneously frustrating Australia's desire to aid its Far Eastern neighbors by providing them with education.

Thus white Australia has a little leavening of nonwhites, surprisingly visible on the streets of the cities. The dark-skinned immigrants report generally good relations. A journalist from India says that his daughter was teased by white kids who jeered that she looked like she needed a bath. But his summation was: "Once Australians get to know you and like you, they accept you totally." An American Negro, supposedly the first, was accepted for Australian citizenship late in 1970. He was Bernard Byers, who had served for four years as a teacher of French at Melbourne's posh Trinity Grammar School, after first visiting Australia in 1959 as choreographer for a Brazilian government dance company. "I haven't found the racial bit here," he says. "Once you get a job and a circle of friends, you fit in."

An area where Australian exclusion hurts badly is the lesser Pacific islands. Many have populations much too big for their coconut economies, and need to export people. Perhaps more pointedly, not even the people of New Guinea, who are legally Australian citizens, are permitted to move to the continent.

"Britain got itself into a cleft stick in years past by having to admit Jamaicans and Pakistanis and so on or be kicked out of their lovely colonies," muses Bill Russell, a Sydney journalist. The temper of Australia in matters racial is the desire to go on being a white Asian nation. The descendants of the original English, Scottish, Welsh, and Irish settlers have found it traumatic enough to absorb the great white immigration of recent years. To aspire to the unity-in-diversity of a massive nonwhite immigration is just too unsettling. An undefined "gradualist" approach is the closest thing that anyone offers as an ultimate solution to the dilemma.

The New Australians

An awareness that Australia was in painful need of more population pervaded most of the country's history, but in this century immigration ran low until 1945 because of labor's fear of job competition and because of the Depression. Such immigrants as came were almost all British, pulling out of the dismal England of the 1920's. Attempts were made in the 1930's to send Australia's few Italians back, because of the shortage of jobs. At the Evian Conference on refugees in 1937 Australia refused to accept German Jews: "As we have no real racial problem, we are not desirous of importing one."

The war convinced many Australians that they had no more than twenty years to build a bigger industrial base founded on a bigger population. Minister of Information Arthur Calwell coined a catchy slogan, "Populate or perish." Britons were still first choice, but the snotty attitude toward troubled Europeans was dropped, and 218,000 displaced persons admitted. With a cheerful disregard for the niceties of Old World geography, the Aussies lumped all these immigrants together as "Balts and Letts." Then Australia reached out to get Europeans of every kind, with the goal of raising the population to thirty million in fifty years. The device for bringing them, apart from the propaganda that attracts their attention,* is the "assisted passage" payment of most of the ship or plane fare to Australia—or all of it, in the case of immigrants younger than nineteen. An adult Englishman, for example, need contribute only twenty pounds toward his passage.

This vigorous program brought 1,850,000 immigrants to Australia between 1947 and the end of 1971. Along with them came an equal number of unassisted "intending" settlers and persons who came on their own and

* An advertisement in the Manchester *Guardian*, for example, stresses Australia's easier living, economic boom, stable democracy, and snowless climate. It urges readers to send for a booklet called *Living in Australia*, which reprints ads from Australian periodicals showing prices, land values, and so on.

stayed in Australia for a year or more; but most of these two groups, plus some of the assisted settlers, did not stay. The net gain to the population since 1947 is something like two and one-half million, or one-fifth of all Australians. Britons comprised more than half of the immigrants, followed by Italians, Greeks, and Yugoslavs. In lesser numbers, but decidedly visible among the New Australians, are Spaniards, Scandinavians, Americans, Hungarians, South Americans, Czechs, Turks, Poles, Icelanders, and Arabs.

The swallowing of this horde of humanity has gone astonishingly well. The immigrants stirred up a latent tolerance among the natives, who learned to stop saying "dago" or "Polack" and start using the rather egregious but certainly friendly term "New Australian." An Englishman, however, remains a "pom" or "pommy,"* a word more jocular than pejorative. Integration into Australian society went so well for one Englishman, Bernard Hesling, that he wrote a genial book called *The Dinkumization and Depommification of an Artful English Immigrant.* Language problems seem more comic than troublesome. Women, staying home, are sometimes slow to learn English, but men on a job pick it up fast enough, though the result may be a jolting Hungarian-Australian accent. Some areas of the cities fill up so completely with immigrants that English is hardly to be heard. A third-generation Australian living near Prahan, a Greek-dominated suburb of Melbourne, went to the infant welfare center there and for starters was asked, in loud, carefully enunciated syllables, "D-o y-o-u s-p-e-a-k E-n-g-l-i-s-h?"

Fifteen newspapers are published in languages other than English. The government-run Overseas Telecommunications Commission advertises three-minute telephone calls for eleven dollars to "talk to someone in Europe you love," and the picture shows an Italian woman saying presumably, "*Mamma mía!*" An Italian contribution to Melbourne is a branch of the Mafia.

More broadly, the immigrants shook up the self-satisfied Old Australians by working hard and releasing new energy into the system. A story is told of a Yugoslav lens-grinder whose work was so good that his employers readily agreed to sponsor the immigration of his brother, also a lens-grinder. The brother was put on the night shift, turned out to be just as skilled, and in the end, of course, was revealed to be the same man working two shifts.

A case of particular improvement in Australia that everyone notes is culi-

---

* The origin of this word cited by Sidney Baker is so extravagant that it is herewith reprinted *in toto:* "There has been a popular theory that *pommy* came from *jimmy-grant* (1845), as a rhyme on 'immigrant,' which was shortened to *jimmy,* the word then being merged by rhyme into *pomegranate* (it is suggested that the rosy-cheeked English may have given some association with the ruddy fruit) and subsequently clipped back to *pommy.*"

nary standards.* There are hardly any really native dishes, except kangaroo-tail soup. Good cheese is rare; the best is cheddar. Oddly, restaurants don't serve lamb or mutton very often. It is still impossible to find truly Australian food, but you can get anything else from dolmades or a fettucini to a knish or a piroshki. No less than 40 percent of university staffs in Australia are people from other countries—also one-third of hospital staffs.

Australia's economy has run so well in the past twenty years that immigrants have had no trouble getting work. But there are plenty of cases of mutual disillusionment between Australia and the immigrants. In 1966, for example, eighteen thousand of them went back to their native countries. Immigrants are received, upon arrival, in hostels made largely of Nissen huts from old army camps, and some of them have to stay there for as much as four years before finding housing. Social tensions arise; the bluer-blooded Old Australians resist contact with the newcomers. Professional men are often disappointed to find their degrees unacceptable, forcing them to sit for new exams or take menial jobs. A few fainthearted Australians express second thoughts about expansion-by-immigration, which requires, for example, so many new schools—forgetting that the immigrants are providing Australia with talents acquired at the expense of the nations whence they came. The immigration program is criticized more soundly for its execution. A pertinent case in point, mentioned earlier, is that the government accepts unskilled Turks and Arabs but still bars skilled Japanese and Southeast Asians.

Americans moving to Australia have lately reached six thousand a year, the motive for many being disillusionment with the troubled, violent United States, and for a few stark fear of the blacks, curable only by settling in white Australia. The Australian government even subsidizes air fare for them, up to $360. Almost all of the 5,500 letters of inquiry received by a U.S. firm that gives advice on moving to Australia mentioned social unrest and violence. Inquiries to the Australian consulates in Manhattan and San Francisco shot to record numbers after the Chicago riots of 1968. Stuffy comments abound in the explanations offered by American immigrants in Australia: "We didn't like the idea of subjecting our young son and daughter to the kind of future we saw taking shape in the cities." "We knew that in Australia there was

---

*Until the food-conscious immigrants came, Australian food was confined, pretty much everyone agrees, to steak and eggs. Cyril Pearl offers a tongue-in-cheek recipe: "Take a piece of old bullock, cut as thin as possible, throw into a frying-pan with a spoonful of sump-oil (winter grade is preferable) and burn to the ground. Fry an egg in same oil. When both are cold, toss on to a soapy plate and cover with tomato sauce."

nothing to be frightened of racially." An Australian official says, "The largest number, very frankly, are attracted by the stable social climate."

The Gallup Poll reported a while back that one out of every twenty-five American adults would like to emigrate to Australia. Many of the American New Australians, of course, have motives more admirable than flight from U.S. troubles. They sense a frontier spirit, an opportunity for business success, a chance for a sunnier, easier life. With nice irony, many go for a motive best expressed in a famous document of the land they leave: "the pursuit of happiness."

Many return to the U.S., too, let down by Australia. The chief reason is the lower standard of living. Australia is prosperous, but its per capita GNP is half that of the U.S. Prices are lower, but wages much lower. A business executive who may have been getting twenty-five thousand American dollars a year in the United States will get ten thousand Australian dollars* in Australia—and is less likely to yearn for more because high-bracket income taxes are so steep. The bulk of the immigrants accept these setbacks in return for a less competitive life. They become plumbers, journalists, schoolteachers, businessmen; they run mines, ranches, garages, small industries. Some are professors, earning from $6,000 to $15,000 a year. One American became a maker of boomerangs for tourists. The Americans blend quickly into Australian society, do not stick together in "ghettos" and soon cannot be told from old Australians except at meals: they hold their forks in their right hands. Quite often they find what a New York woman, who moved to Sydney at the age of thirty-six, calls simply "a more satisfying brand of happiness."

Just as Britons have always been favored in Australia, Australians have historically been able to migrate freely to Great Britain—but no longer. The new British immigration law makes an Australian an "alien" unless he has at least one parent or grandparent born in Britain. This makes England foreign territory to some of Australia's bluest-blooded families. One victim of this law is the son of the Queen's own Governor in New South Wales, Roden Cutler.

---

* Throughout this book, money figures are expressed in Australian or New Zealand dollars, which in terms of U.S. money rose from $1.12 to $1.19 when President Nixon devalued the American dollar. The U.S. reader should remember, when he encounters a money figure in this book, that the same figure expressed in American currency would be about one-fifth higher.

CHAPTER 5

Australian Rules—And Other Arts

> Australians of my generation grew up in a world apart. Until we went abroad we had never seen a beautiful building, hardly ever heard a foreign language spoken or been to a well-acted play, or eaten a reasonably sophisticated meal, or listened to a good orchestra; and outside two or three art galleries in Melbourne and Sydney, there was scarcely a house that contained a collection of worthwhile paintings.
>
> —ALAN MOOREHEAD

TODAY, Moorehead finds his native country improved and even transformed. The way the Australians live, what *interests* them, is made of many parts—sports, arts, books, music, the press. On the scene are eccentric painters, genteel women playing at bowls, kinetic newspapers, football stars, book-bannings, opera singers, and twelve-tone composers. A country that was placid and even stodgy ten or twenty years ago is lively, active, and interesting now. Another quality is a certain lightheartedness: Australians are spared the frustrations of Britain and the great social and military-oriented schisms of the United States. The previous chapter attempted to say something about what Australians *are*—in this one let us try to examine what they *do*, at leisure, in the arts, and on television, with a look, too, at the press.

*

"Sport to many Australians is life and the rest a shadow. . . . To many it is considered a sign of degeneracy not to be interested in it," writes Donald Horne.

And of the sports, football is king. Australians play four kinds. Rugby

Union is for amateurs; Rugby League is the main winter spectator sport in New South Wales and Queensland. Soccer is mostly played by immigrants —Italians and Yugoslavs nostalgic for the game so popular in the old countries. Australian Rules, played in Western Australia, South Australia, and Tasmania, but most fervently in Victoria, is the authentic native game. It is basically a game of flowing movement, like Rugby or soccer, rather than of formation and downs, like American football. The field is enormous—190 yards long and 120 yards wide—and oval-shaped, like the ball which is used. In the end zones are four goal posts, standing in a line at intervals of seven yards. Each side has eighteen men, who aim to advance the ball to the neighborhood of the goal posts, and kick or punch it through for six points (if it goes between the two center posts) or one point (if it goes between a center post and an outer post). To move it forward, the ball can be kicked, punched, or carried (provided that it is dribbled once every ten yards). The most spectacular play is the "mark," a successful ten-foot-high catch of a kicked ball; the marker gets a free kick to the goal posts. A ball-carrier grabbed by an opponent must get rid of the ball by kicking or punching.

More than 100,000 fans turn out regularly in Melbourne to "barrack" for their teams, of which there are a couple of dozen. The tackles are not as rough as in American football, but since the players wear no protective gear, the mayhem—black eyes, cuts, bruises, and broken bones—is considerable.

*

Australians love horse racing. There is a track in every city. TAB shops, with windows for betting on horse and greyhound races, are all over the place. The initials mean Totalisator Agency Board, the state-controlled betting organization, comparable to New York State's off-track betting system. The totalizator (U.S. spelling), or pari-mutuel machine, was invented in racing-crazy Australia, as was the photo-finish camera.

The great annual madness in Melbourne is the Melbourne Cup, run every year for 111 years on the first Tuesday of November. For a few minutes the streets are deserted, as all of Australia glues itself to the radio or television broadcast. At Flemington Racecourse, a vast arena, crowds of more than 100,000 watch. Australian troops fighting in New Guinea during World War II are supposed to have declared an unofficial cease-fire so that they could listen to the running of the Cup.

One of the most memorable figures in Australian sports history is not a man but a horse. He was Phar Lap, a "mountain of a horse," over seventeen hands high, with a stride in gallop of more than twenty-five feet. He won the Melbourne Cup in 1930. Two years later, right after winning the Aqua

Caliente Handicap in California, he died. His body was disposed of in an equitable way: the skeleton went to New Zealand, where he was born; the heart went to the Institute of Anatomy in Canberra; and the hide was duly stuffed and placed in the National Museum in Melbourne, where Phar Lap may be seen to this day.

The great summer sport is cricket; it is also much more a participatory sport. Australians are so good at sports that merely reading about their skill and energy makes one want to lie down and rest for a while. Of 178 Test Matches in cricket played with England since 1882, Australia has won 73 to England's 67, with 43 drawn. Australians won the Davis Cup in tennis fifteen times in the eighteen years between 1950 and 1967. In football, golf, swimming, and Olympic sports Australia's winning record is all out of proportion to the country's small population. Undoubtedly part of this good show derives from the stress on sport instilled in every Australian child from infancy. Another reason is that Australians generally are big, strong, physically fit people. A third might be the long evenings, year round; while the people of northern nations are forced indoors in the winter, Australians are still outside doing something strenuous.

Innumerable golf courses spangle the suburbs of the big cities, and tennis courts are everywhere. Even old ladies have a sport: bowls, although, to be exact, the game is also played by old men, plus the middle-aged of both sexes. Bowling greens, impeccably level with crew-cut grass, are to be found all over the cities, the players uniformed in white except for colored hatbands; the women look a bit like nurses. The point of the game is to roll the bowl (the bowling ball) as close as possible to the jack, a small ball resting at the other end of the green. This apparently simple endeavor is complicated by a bias cut into the bowl, which makes one hemisphere more convex than the other and causes the ball to take a curved path.

The ultimate Australian madness is surfing, done in many ways: with a surfboat (four oarsmen and a man to steer), with a sort of canoe, with inflated rubber mats, with plastic-foam floats, with ordinary surfboards, with small hand boards, and, finally, with nothing at all—a trick called "body shooting." Great green waves rolling in from the Pacific and the Indian oceans make this sport the main test of skill and virility all around the continent.

The Art Scene

Art in Australia? The average American strains his brain, and eventually the names of the painters Sidney Nolan and Sir William Dobell may float to the surface. The Englishman may recall a few others, because many Aus-

tralian artists have lived, or live, in London. Yet when seen close up, art—particularly painting—in Australia turns out to possess surprising sophistication and prosperity.

In the National Art Gallery in Melbourne, a building massively elegant, there hangs (together with a good representative selection of classical painting from Europe) the work of great Australian landscapists dating back to 1837. The effect is a sense of deeper and denser tradition than the country's short Westernization and historically thin population would seem to justify. The search for comparisons brings to mind the American experience—the paintings of the Hudson River School in the Metropolitan, perhaps—and it would seem that the Australian masters equal the Americans man for man.

Australian artists today are suddenly finding considerable confidence in themselves. No longer isolated from the great trends in art in France and New York, they now want to measure themselves on international standards rather than striving to be "the best Australian."

Art is booming, which is evident from the number and prosperity of the painters and the presence of their works on every wall. The boom may be tainted by status-seeking and speculation, but it is certainly a boom in terms of energy, awareness, and the high quality of the best work.

*

Australian painting got off to a bad start, with a European academic tradition so strong that the painters could not see what was right before their eyes, and depicted Australian eucalyptuses as though they were English oaks, leaves extended instead of hanging. At length a painter named Tom Roberts visited Europe in the early 1880's, and returned full of the liberating principles of Impressionism. He and a fellow painter, Arthur Streeton, became the country's first masters. Yet isolation continued to afflict both painters and the public. A show of modern art by French and British painters, put on by the Melbourne *Herald* in 1938, produced the same jeers that similar art had earned from philistine New Yorkers at the Armory Show a full twenty-five years before. But as with the Armory Show, the Melbourne exhibit also opened eyes, especially among artists. A couple of decades of rich creativity followed, in work by the men who are now Australia's most noted. Let us see who they are.

A Gallery of Australian Painters

SIDNEY NOLAN, Melbourne-born in 1917, kept in touch with Europe in the 1930's and learned the concepts of Picasso and Klee. Drafted in 1942, he betook his urban eye to an army camp at Dimboola, in the wheatlands of

western Victoria, and painted powerful landscapes, his personal vision stirring the Australians to recognize the beauty of their stark land. The landscape is quintessentially Australian. These pictures, and Nolan himself, were at first largely unappreciated in Australia until it was learned that London admired them. Nolan then turned from landscapes to scenes from the lives of Australian legendary heroes. Among them was Mrs. Frazer, a Scottish woman who was shipwrecked on the Queensland coast in 1836 and compelled to live among the Aborigines. Meeting an escaped convict, she promised to have him pardoned if he would guide her—she was by then naked and distraught—back to white civilization. She did not keep her word, and the convict, having saved her, had to flee back to the bush. A subsequent theme has been the gibber-strewn outback (crazy birds flying upside-down over pubs with names like Dog & Duck). Nolan now lives in London, though he goes frequently to Australia and his pictures remain Australian.

*

FREDERICK WILLIAMS lives in a converted winery, with vast vinous tubs like secret rooms beneath it. This Melbourne man, born in 1927, trained at the National Gallery Schools and in London, says he "doesn't much care for the bush." This is an extraordinary statement, because for the past four years he has painted little but the bush, at first in closeups of vertical sapling country on busy canvases about five by seven feet in dimensions, now in even bigger sizes that show, typically, a gray surface sparsely dotted with candy-cane-shaped squeezings of oil colors which stand for scrub trees. To see the landscape itself after looking at a Williams painting is to see it as limitless and desolate—compelling and perhaps fear-inspiring. He makes his canvases huge as a way of suggesting that the vast Australian countryside cannot be framed. Of his present painting style, he says, "If it's good figurative it's abstraction, and if it's good abstraction it's figurative." James Mollison, the government's Director of the National Art Gallery in Canberra, thinks very highly of Williams: "Each new work is better than the last. He could compete with the top artists in New York and London."

*

WILLIAM DOBELL, who died in 1970, was called the "reincarnated Hogarth." In his thirties, which corresponded to the 1930's, he studied and traveled in Europe. Returning to Australia, he developed a personal style of portraiture, of elongated limbs and turbulent motion. In 1943 he painted an attenuated, Greco-esque picture of a friend that won a prominent prize.

Two losers sued the judges, contending that Dobell's work was a caricature, not a proper academic portrait. In a sensational trial, a court of law decided for Dobell. The artist, who did not think of himself as a modern, found the experience traumatic. It brought him a frightening overload of commissions, and he became the country's favorite portraitist and "best painter." Art historians have pigeonholed Sir Williams as a "minor twentieth-century mannerist," but of course time will tell about that.

*

RUSSELL DRYSDALE, like Dobell a Sydney artist, started painting for the paradoxical reason that his eyes were bad—it was something to do in the hospital after a corrective operation. Trained in modern art in Europe, he returned to Australia in 1939, at twenty-seven, and began painting iron-roofed buildings on the main streets of the country towns of the outback where he grew up. A newspaper sent him to make paintings of the effects of a bad drought, and these pictures made him famous as a regionalist painter. Drysdale's work has given Australia a more precise picture of itself. He was the first major Australian painter to use the Aborigines as subjects.

*

ALBERT TUCKER is an angry painter who fought his native environment until he stormed off to Japan, London, and Paris, "a refugee from Australian culture." Preoccupied with morality and evil, he became convinced that art could produce social change. Slowly a strange image took form in his work in a series of paintings: the Antipodean Head—a Buster Keatonish physiognomy painted as though cast in bronze, angular, ax-nosed, jut-jawed— an Australian! "Psycho-landscapes" he called these paintings—reflections of "the human involvement with the Australian landscape." Tucker had returned spiritually to his origins, and thereafter went back to Australia.

*

Australian art, says Mollison, is "in the situation of American art at the close of World War II." It is moving away from the nationalism represented by Nolan, Drysdale, and Tucker. Young Australian artists now go not to London but to New York and Japan. Fred Williams hopes that the internationalization of Australian art will also inspire an immigration of foreign

painters. BRETT WHITELEY, the most visible of the internationalists, detects a "will for a new definition—not nationalism or antipodeanism, with the old soft sunny blink, but something fiercer and more delicate."

One Australian dealer, Max Hutchinson, who founded galleries in Melbourne and Sydney, now operates a Manhattan branch for "high-caliber artists of all nations but most of all Australian." A Hutchinson favorite is sculptor CLEMENT MEADMORE, a Melbourne man. His basic shape is a long, massive, iron quadrilateral, twisted like the parts of a wire puzzle, and reposing in odd balance. Many New Yorkers, including Nelson Rockefeller, own Meadmores, and a fine big example rests in the plaza of the Australian Mutual Provident Society's skyscraper in Melbourne.

*

The most spectacular museum in Australia is the National Art Gallery in Melbourne. Sydney's Art Gallery of New South Wales has lately been closed for the construction of a new wing, to be named for Captain Cook. There are many charming private galleries. Bonython's, in a terrace house on the winding streets of Paddington, Sydney's Soho, sells works by Nolan, Whiteley, and Tucker, among others. Rudy Koman, in nearby Woollahra, represents Fred Williams, Jan Senbergs, John Olsen, Clifton Pugh, and David Aspden, all prominent artists. Koman sold a painting by Leonard French to the government of New South Wales, which presented it to Pope Paul VI when the Pope visited Australia in December 1970. Patrick Sharpe, a small and delicate man with a flourishing Van Gogh beard, chooses art for the Strawberry Hill Gallery, among Greek stores and Lebanese restaurants on Elizabeth Street. This also is a terrace house, 104 years old. The artists are relatively unknown moderns, and the prices are set for beginner collectors: nothing more than $70.

The market thrives. Art students twenty years old grumble if they do not make $20,000 a year on the gallery circuits. Works by big-name painters go for $8,000 and up—way up. A good Drysdale recently brought $26,000, and one of his works sold for $45,000. In part, this is a *nouveau riche* spending spree resulting from the country's minerals boom. And what is bought must be Australian; the country is proud of its parochialism.

General affluence lets everybody buy art. Buyers of contemporary art can be adventurous young businessmen, intellectuals, and architects, or well-established old families who have been buying for so long that they are sure of their taste. The Australians' house-owning complex creates a pride that needs to be complemented with paintings on the walls—and it is.

The Performing Arts

The theaters of Brisbane, Sydney, Melbourne, Adelaide, and Perth content themselves mostly with the plays of London and New York, plus interesting classics; one week late in 1971 the main productions on the boards were *Hair, Jesus Christ Revolution, Move Over Mrs. Markham, Next* and *The Guerrilla,* and *The Au Pair Man.* Smaller houses—converted cinemas, church halls, boat sheds, comparable to Off-Broadway—get into more radical theater, sometimes tough and exciting, student works, explorations into Australian history. But new Australian plays of authentic power appear so rarely that almost everyone cites Ray Lawler's 1955 *Summer of the Seventeenth Doll,* a drama of Queensland sugar-cane cutters much produced in England and the United States, as the most recent major work. Moreover, the Elizabethan Theatre Trust, a foundation for the promotion of the performing arts, subsidizes Australian plays thought to be of value, such as *The Legend of King O'Malley* (of which we will hear more later). *Oh! Calcutta!,* incidentally, opened in Melbourne despite attempts to censor it, and ran for weeks.

Many of the talented theater people have left Australia, or wish that they could. "There are more Australian actors, writers, producers, and designers working and succeeding in England than there are in Australia," says Lance Peters, president of the Australian Writers Guild. Some well-known expatriate Australian actors are Peter Finch, Leo McKern, Diane Cilento, Judith Anderson, Zoë Caldwell, and Coral Browne.

As to film, the world's first feature, *Early Christian Martyrs,* was made in Australia by the Salvation Army in 1899–1900, and subsequently two hundred silent movies were filmed there. The talking picture decimated this industry, and now the theaters cannot fulfill their legal obligation to show Australian films at least 3 percent of the time. Still, some refreshingly different movies have come out of Australia in the last couple of years. One little noticed but original film was *Ned Kelly,* directed by Tony Richardson and starring Mick Jagger as the famous Australian outlaw (more about him later). Another was *Walkabout* set in the outback and featuring a sensitive Aboriginal actor named David Gumpilil. The rate of film production is one or two a year. A big production of Patrick White's popular novel, *Voss,* is presently under way.

In ballet the situation is bright. The Australian Ballet, a young, state-aided company, toured the United States with Rudolf Nureyev in 1971 and earned from the *New York Times* the praise of being "a promising and enjoyable company in its own right." It is co-directed by the choreographer

Sir Robert Helpmann, who is Australian though often mistaken for British because of his long career in the Royal Ballet of London. Indeed, nearly one-fourth of the dancers in the Royal Ballet are reportedly Australian.

*

Australian concern with music runs broad and deep. The country that gave the world Nellie Melba and Joan Sutherland is prolifically involved in opera, ballet, symphony, composition, and its own old folk songs.

Composers are numerous. The more interesting, possibly, are:

PERCY GRAINGER (1882-1961). Born in Melbourne and buried in Adelaide, Grainger spent most of his life studying in Germany or working in the U.S. But his themes, as he said many times, were Australian, and a lifelong project of his was the Grainger Museum that now stands in the University of Melbourne.

JOHN ANTILL. Now around sixty-seven, and an incipient grand old man, Antill wrote the first symphony based on Aboriginal music, a Stravinsky-like composition called *Corroboree*. Played by a large orchestra, augmented by clunking Aboriginal thora sticks and whirring bull-roarers, and accented by shrieking flutes and piccolos, it can make the hair stand on end.

PETER SCULTHORPE. A composer who began writing music in 1937, when he was eight years old, Sculthorpe conveys the loneliness and sadness of the Australian landscape. His *Sun Music*, which comes in a four-part suite, uses gongs and cymbals and a choir that grunts, wails, and shouts. Sculthorpe has been influenced by Balinese music.

RICHARD MEALE. A story told of Meale is that he flushed an early piano composition down a toilet, for fear that if anyone played it he would be deemed mad. Born in 1932, Meale is—or was—the rebel of Australian music. His *Homage to García Lorca* is played, in tumult sufficient to shatter glass, by *two* orchestras, using twelve-note chords. Meale sometimes uses themes from the music of Japan.

Opera is practically an Australian sport; there is a theory that a dazzling high note in singing is akin to a dazzling high catch in football, and that healthy, sunny Australia produces people good at both. Historically opera has flourished, and as theater for everyone rather than something for the rich only. Yet lately it has faltered, and the Elizabethan Trust divides its boons uncertainly between theater and opera. So the performers are apt to emigrate.

Symphony concerts, eight hundred a year, are hungrily attended in Australia. When the Australian Broadcasting Commission was founded in 1932,

to give the government a role in the improvement of radio, the charter vaguely suggested that it could form studio orchestras. Energetically accepting the hint, ABC formed orchestras in all six state capitals, and turned their performances into public concerts. Soon ABC began selling seats by subscription, and now boasts of being the largest concert-giving organization in the world. The Sydney and Melbourne symphonies, subsidized by ABC and the states, have high standards, equal to American or European orchestras. Smaller orchestras are conscientiously professional; performances are almost always sold out. The ABC network takes its musical obligations seriously. One ABC programmer, asked for extra news time on the night of the Kennedy assassination, is supposed to have replied, "Oh, no, we can't interrupt the Sibelius." The drawbacks are that ABC is fusty about programing modern music, and that it has never gotten around to frequency-modulation (FM) broadcasting.

Television's pattern is a recognizable variation of the British system. Government-owned ABC (often called "Auntie") provides, on thirty-nine stations, good drama (much of it from the British Broadcasting Corporation), news, lectures and current affairs, and absolutely no advertising. The main current-affairs program, *This Day Tonight*, holds between 20 and 25 percent of the viewing audience. ABC tries to appear independent, but it is actually sensitive to government pressure. Over all ABC-TV holds only 16 percent of the audience, because it is up against the attractions of commercial television broadcast on forty-five stations mostly owned by the four big newspaper chains. These stations show *Mod Squad*, *The Dick Van Dyke Show*, *Family Affair*, *Here's Lucy*, *Flip Wilson*, etc.—eight out of ten top shows are American. The rest of the time—Australian TV offers four hundred hours of viewing a week—is given to films and sports. The number of Australian families watching television at peak hours has dropped from an astonishing (or even appalling) 80 percent in 1960 to 70 percent now. There will be no color TV in Australia until 1974.

Books and Publishing

At the busy Robertson & Mullens bookstore in Melbourne just before Christmas in 1970, one noticeable offering was a catalogue 1,037 pages long. This turned out to be a bibliography of Australian children's books, thousands and thousands of them. The point is that this underpopulated land publishes books in extraordinary volume.

Bookstores sell—with exceptions resulting from an absurd censorship—the major works of all the world, but what is most striking is the concentration, almost obsessive, on Australian themes. The stores display titles like

these: *The Early Australian Architects, A Yankee Merchant in Goldrush Australia, The Story of the Flinders Ranges, A Pictorial History of Australia at War, Aboriginal Mission Stations in Australia, The Effluent Society: Pollution in Australia.* There are books on Australian spiders, marsupials, high-country birds, lizards, venomous wildlife, wild flowers, seashells, and gem stones.

Australians claim to read more books per capita than anyone else except New Zealanders—a tradition, supposedly, that comes from days gone by when reading was almost the only entertainment on the sheep stations. An average novel can quite easily sell five thousand copies, which is thought to be good in England, a country with more than four times as many people.

With all that, literature is not as exciting in Australia as painting. The outstanding novelist is Patrick White, author of *Voss*, a psychological investigation of the life of the explorer Ludwig Leichhardt (who traveled to the west coast of the Gulf of Carpentaria and then disappeared), and *The Vivisector*, a narrative of the life of a Sydney painter who—figuratively, in his pictures—cut up his friends and relatives alive. White, born in 1912, endured a long obscurity before Australians could accept his bleak, tragic themes. Other good novelists, also accused of too close preoccupation with Australian society, are Thomas Keneally and Hal Porter.

Australia, like Ireland, does have a way with words—perhaps because a lot of Australian writers are Irish by descent. Education and a dry, irreverent wit add to the natural talent. However, experimental novels seem to be missing, and many good writers are drawn to journalism or television.

A talented Australian writer of nonfiction is Alan Moorehead, who after covering World War II in books, and going on to write extensively about Africa, returned to Australian themes in books about Captain Cook (*The Fatal Impact*) and the exploration of the interior (*Cooper's Creek*). Poetry, too, is alive and healthy. The *New York Times* calls A. D. Hope, Australia's leading poet, "one of the master conservative poets in English in this century."

*

Australian book-banning could be called comic if censorship were not always pregnant with dangers to free speech and thought. Certainly the authorities have made themselves fools time after time, banning, in the past, such books as *Ulysses, Brave New World, Farewell to Arms, God's Little Acre, Catcher in the Rye, Appointment in Samarra, The Women of Rome, The Kama Sutra, The Postman Always Rings Twice, Forever Amber, Peyton Place,* and *Lady Chatterley's Lover.* More recent examples are *The*

Ginger Man, Last Exit to Brooklyn, Naked Lunch, Borstal Boy, Another Country, The Group, and *The Carpetbaggers; Naked Lunch,* at least, remains banned still. A wryly comic case was that of *The Trial of Lady Chatterley,* banned for import in 1965; somebody simply set it into type and printed it in Australia, a process which turned out to be perfectly legal.

The book-banners are the Ministry for Customs and Excise, and the vice squads of the various state governments. Politicians solemnly support censorship, for the most part, in deference to the bluenoses. In the public consciousness, the Minister of Customs is best known as the chief censor, but Don Chipps, the Minister appointed in 1970, a man with a rugged, deep-lined, good-humored face, has been trying to steer his customs agents away from the more ludicrous kinds of censorship. He ruled in favor of admitting a book of essays by Timothy Leary that had been kept out of Australia for four years. He also stripped away secrecy concerning precisely what books the customs men had seized, and got the list down from 136 to 80. But one still meets mature, intellectual Australians who until lately suffered the humiliation of being protected by their government from reading *Tropic of Cancer* or *Tropic of Capricorn*—and are doubly resentful at having to wait so long to discover what unimportant books they are.

Portnoy's Complaint was banned by all state governments except South Australia, whose modern-minded Premier, Don Dunstan, approved its sale while professing, for the benefit of the bluenoses, that it nauseated him. Censorship in Australia is typically quite leaky—no one really determined to get a copy of a book fails to do so. In the case of *Portnoy,* the 75,000-copy edition that the police banned was actually mostly sold before the crackdown. The vice squad raiders managed to get only the last few copies, snatching one from the hands of novelist Patrick White as he was about to pay for it. At the trial over *Portnoy,* humorist Cyril Pearl testified with a straight face that the book was in a class with Rabelais and Chaucer.

Newspapers

> To learn what great things God has done in the past, read your Bible. But to learn what he permits to be done today, read your Herald.
>
> —JOHN FAIRFAX

Ownership—that is the key word in understanding the Australian press. The British press has its lords—Lord Rothermere, Lord Thomson, the late Lord Beaverbrook. The Australian press has its knights—Sir Frank Packer, Sir Warwick Fairfax and his cousin Sir Vincent, Sir John Williams, and the late Sir Keith Murdoch. Out they went a-jousting, and when the dust settled,

Australia's press had been cut up into three empires. Unlike the lords, who still break a lance against one another from time to time, the knights provided themselves with near-monopolies. Sometimes they look quite fierce, charging out to buy some stray newspaper property that may be up for grabs. Actually they are in collusion in a number of ways, most notably in joint ownership of the commercial television industry. But in recent years, as we shall see in a moment, a young publisher has got loose among the knights and stirred up much excitement.

The major agglomerations and their flagship newspapers are:

The Melbourne *Herald* group, which has the *Herald* (circulation 500,000), the Melbourne *Sun News-Pictorial* (circulation 635,000—the nation's biggest daily), big chunks of Brisbane's *Courier-Mail*, *Telegraph*, and *Sunday Mail*, the Adelaide *Advertiser*, the Hobart *Mercury*, and Perth's *West Australian* and *Daily News*. Sir Keith Murdoch put this fiefdom together, and it is bigger by far than the others. Sir John Williams ran the company shrewdly for fourteen years, before retiring in 1970.

The Fairfax group, which has the Sydney *Morning Herald*, the Sydney *Sun*, the Sydney *Sunday Sun-Herald*, the *Financial Review*, Canberra's *Times* and *News*, and Newcastle's *Morning Herald* and *Sun*. John Fairfax, quoted above, founded this empire with the Sydney *Herald* in 1831. The Fairfax family turned it into a public company some years back, but Sir Warwick (a leading art collector) and Sir Vincent still dominate.

The Packer group (Consolidated Press), which publishes Sydney's daily and Sunday *Telegraph*, the newsmagazine the *Bulletin*, and *Women's Weekly*. Sir Frank Packer is the fabled yachtsman whose *Gretels* regularly, but so far unsuccessfully, make a challenge for the America's Cup.

In the whole country, only one leading paper stands by itself. That is the venerable Melbourne *Age*, which was long dominated by the Symes family, who started it 118 years ago. Now the Fairfax group has been buying into it, and apparently controls it. The *Age* company also runs a flock of country weeklies.

A decade ago Sir Keith Murdoch's son, Rupert, got as an inheritance the unimpressive Adelaide *News*, and set out to see what he could do by way of surpassing his father's empire. He had a flair, and before long managed to acquire a couple of sex-and-crime tabloids in Sydney, the *Mirror* and the *Sunday Mirror*. Then Murdoch got the Brisbane *Sunday Truth*, the Melbourne *Sunday Truth*, the Perth *Sunday Times*, and the Adelaide *Sunday Mail*.* Murdoch has never been embarrassed at publishing gossipy, trivial, sometimes sensational papers, as all of these are, but in 1964 he took on the

* Which has to get off the presses by Saturday midnight because of a South Australian blue law.

nervy task of starting a new, serious, national paper, *The Australian*. Published first in Canberra, it met rough competition on its own intellectual level from the Canberra *Times*, and finally had to move to Sydney, where it pushed circulation up to 150,000. Next Murdoch stormed into London, winning control of the Sunday *News of the World* in exchange for the Melbourne *Sunday Truth*; a bit later he acquired the London *Sun* (which was Hugh Cudlipp's moribund attempt to make a new paper out of the remnants of the equally moribund old *Daily Herald*). Murdoch turned it around, got circulation up from 850,000 to 1,800,000, and now lives in London, running the Australian papers by phone and frequent trips. He found breaking into London tough. "In Australia and America, you know who your friends are more quickly. There's a lot more subtlety here."

The fight of the knights to get and keep papers implies that they are highly profitable, and for most of them this is indisputably true. The circulation of all the Sydney dailies runs around 300,000 each; the Sunday papers run up as high as the 750,000 of the Sydney *Sun-Herald* ("the paper you can take into your home"). In short, Australians are avid newspaper readers; there are 50 dailies in all, plus 550 others, and total circulations equal one-half of the population. Advertising runs heavy—65 percent of the Sydney *Morning Herald*, for example. The papers do not have to fight television for advertising dollars, for publishers own the bulk of the commercial stations, and own them in interesting patterns. BTQ in Brisbane is owned by the Melbourne *Herald* group *and* Packer. NWS in Adelaide is owned by Fairfax *and* Murdoch. WIN in Wollongong and NBN in Newcastle are owned by Fairfax *and* Packer *and* Murdoch. What more could a publisher ask?

If all this sounds as though profit dominated the press to the exclusion of competence in news—not so. Most of the papers, true enough, pander to the taste for animal stories, a bit of sex, disasters, local squabbles, police-court news, comics, betting odds, and sports, sports, sports. Quite shockingly, there is not a single major paper in the country that supports Labor, the party of almost half of all Australians. A pervasive establishmentarianism enervates the press. Foreign news coverage depends to a large extent on the Reuters news service, which works in conjunction with the Australian Associated Press. On the other hand, the journalists are by and large sufficiently trained and talented, and a lot of intelligence and hard thinking shows through. The Sydney *Morning Herald* and the Melbourne *Age* are responsible papers that tackle trends and problems, and occasionally take some interest in the rest of the world. And *The Australian*, amplified in 1971 with a Sunday edition summarizing the week, is serious, well written, and comprehensive, with careful attention to the arts, books, science, plus a heavy

dose of business. Circulation is 150,000. The *Bulletin,* Packer's *Time-*like magazine, heir to a fine nineteenth-century reputation for publishing Australia's best writers of the time, is now a sprightly reporter and critic of the national life. Donald Horne, the editor, is prolific, witty, observant, meditative, and a most engaging luncheon companion. Typical *Bulletin* headline (upon the downfall of Prime Minister Gorton): "WHAT WAS A NICE COUNTRY LIKE US DOING IN A POLITICAL MESS LIKE THAT?"

Reading the press can be diverting. Headline abbreviations are occasionally cryptic: "AUST" for "Australian," "MELB" for "Melbourne," "UNI" for "university," "QLD" for "Queensland," "C'TTEE" for "committee." The typography is mostly English style, which assumes that nobody will read anything unless the type is made to perform gymnastics. One paper, the Brisbane *Courier-Mail,* may not be the best in the world but it certainly must be the widest—17½ inches (compared to the *New York Times'* 14½), *eleven* columns of type. The sum impression of the Australian press is that it writes down too much, takes itself a little too seriously, and could exchange some of its circusy qualities for more depth. Nevertheless, it is certainly as informative as all but the best in England and the United States.

CHAPTER 6

Aborigines I: Sophisticated Savages

—ABORIGINAL POET KATH WALKER
Why change our sacred myths for your sacred myths?

SHARING THE CONTINENT with the white-skinned millions, whose works and ways we have been describing, is a minority race of blacks who are troubled, downtrodden, and fascinating: the Aborigines. Including persons with 50 percent or more Aboriginal blood, this race numbers 80,000, according to the 1966 census. Other authorities, by including less-than-half-bloods and allowing for increase, estimate that there are as many as 130,000 Aborigines now. This would be one Aboriginal for every one hundred whites, which compares with one American Indian for every three hundred whites. Aborigines are increasing at a rate faster than the general population—but they are still fewer than one-half of the 300,000 who, in the estimate of Professor A. R. Radcliffe-Brown, inhabited Australia in 1788. At that time there were some 500 tribes, ranging in size from 100 to 1,500 members.

The Aborigines are most visible in the sparsely populated north. Arnhem Land, the torrid, bird-ornamented protuberance on the north-central coast, is a reserve for Aborigines. Other Aborigines live in Roman Catholic, Methodist, or Anglican missions in the Northern Territory, or work on pastoral properties there. By the census, there are 21,000 Aborigines in the Northern Territory, 19,000 in Queensland, 18,500 in Western Australia, and 14,000 in New South Wales.

These are the remnants of a destroyed civilization. Before the white man

arrived, the Aborigines had a beautifully elaborated culture, one that fitted the people to the land rather than the land to the people and was pre-eminently gratifying to human emotions. Like white Americans with regard to Red Indian culture, white Australians have historically accorded the Aboriginal heritage little or no value. The memorable first comment by a white man on the Aborigines, made by explorer William Dampier in 1688, was that "The Inhabitants of this Country are the miserablest People in the World. . . . They are long-visaged and of very unpleasing Aspect, having no one graceful Feature in their Faces."

It perhaps took Sir James Frazer's publication of *The Golden Bough* in 1890 to explode upon the white world the notion that "primitive" cultures could be qualitatively comparable to those of the whites. The astounding corollary that even the "miserablest People in the World" had worthwhile traditions had to await the work of some brilliant Australian anthropologists —chiefly Professor A. P. Elkin of Sydney University—in the 1920's and 1930's. Even today most white Australians have only an indifferent glimmering as to Aboriginal* cultures, and the American writer must tackle the subject with great circumspection.

Religion, Magic, and Sex

We are talking now of the beliefs held by Aborigines in the past and to some degree even now by all but a few thousand completely integrated city-dwellers. Two main points here: (1) Aboriginal thought is inductive, not deductive; that is, cause and effect are not ineluctably related, and this is partly because (2) the religion is totemic; that is, persons have *blood relationships* with natural spirits, and these are the main determinants of life. It is thus a culture that most satisfyingly fuses man, nature, art, belief, life and death—past, present, and future. The sacred outweighs the profane. As Professor Elkin explains it, a man begins as the spirit-child of an animal or a rock or a pool, that is incarnated by his mother (sexual intercourse, in Aboriginal belief, having only a collateral role). Raised in a profane world until puberty, he is reintroduced to the sacred life by initiation ceremonies, sustained in that life by ritual until he dies, and by the rite of death returned to the totemic center for recycling into another round of life. This spiritual life is reserved chiefly for men. The Aboriginal woman, spiritually and otherwise, is considered inferior.

This belief is built on the concept of a Dream Time that goes back long

* The Australians often use the word "abos" for "Aborigines." It has an offensive and disparaging connotation and has therefore been avoided in this book.

before the arrival of the first Europeans, to a vague and indefinite period in the past. "In Dream Time there were no blackfellow; but kangaroo, iguana, bird," one Aboriginal explained to anthropologist Phyllis M. Kaberry, in Pidgin English. "All bin walk like blackfellow. Him all the same blackfellow. After he bin turn into kangaroo, iguana, bird." In other words, the animal spirits of the men of the Dream Time eventually turned into present-day animals, and the spirits of men remained totemically related to them.

These ancestral spirits have abodes—trees or pools or caves or heaps of stone—where rituals must be performed by painting, chanting, or shedding blood. Along with all this go many myths and legends. Mythology says that a giant marsupial carved the hills, and a giant rainbow snake scored out the rivers. A typical legend explains death: the moon-man Alinda had two sons who went fishing and returned claiming that they had caught nothing. Alinda, seeing grease on their hands, took them to the sea and drowned them as liars. When his two wives inquired, Alinda in turn lied that the sons had merely gone hunting. The wives discovered that he had murdered them, and burned him to death. Alinda thereupon pronounced that henceforth death would be the fate of everyone but he himself, the man in the moon—"Except for three days every month, I will live forever." Yet to tribal Aborigines, death is not natural; the blame has to be placed. Often someone in another tribe is chosen, and revenge or compensation (a woman, sometimes) is exacted.

Magic, pure black magic, was until very recently a good way of causing death. Among the Dieri tribe, a magician had merely to point a human fibula at a victim, thus "giving him the bone," to cause him to pine away and die. In another kind of magic, kidney fat, taken from a freshly garroted hostile tribesman, was used to anoint bodies and thus increase prowess and virtue. Writer Eugene Burdick reported only a decade ago that rain makers performed their magic task by prying off a chosen victim's thumbnail with an opossum tooth while chanting, "Blood flows like a river/Rushes along like a river."

The basic concept of profane and sacred life (which in itself is perfectly comprehensible to any member of a Masonic lodge) requires elaborate manhood initiation ceremonies, including circumcision. A particular horror in these rites among many Aborigines has been the practice of subincision: some time after circumcision, a young man, perhaps out hunting, is suddenly pinioned by his elders and thrown to the ground, and the underside of his penis slit from end to end through to the urethra. The purpose of this mutilation, which in Australian slang is called "whistlecock," is as vague as the purpose of ritual circumcision. The operation does not affect the man's

sexual powers.* A much-repeated story about the Aborigines is that they do not connect copulation with conception, and apparently in the past they did not, but nowadays they are mostly a bit wiser than that. They seem to think that intercourse is necessary for conception, some holding that it takes five or six ejaculations on successive days, but sex is less a prerequisite to procreation than the entry into the fetus of its spirit. Women associate awareness of pregnancy with an animal, rock, tree, whirlwind, or whatever. This spirit, entering the mother through foot, flank, navel, or mouth (but never vagina), becomes the totem of the child. In the Kimberley district of northwestern Australia, the spirit-children are thought to inhabit pools as fish or birds, waiting to infiltrate the womb of the mother-to-be. The totem determines all the child's future relationships and behavior—much as the sign of the zodiac is believed to do in some other parts of the world.

Aborigines always marry spouses who bear some degree of consanguinity. Among some tribes, a man must marry a woman who is his mother's brother's daughter; among others, the prescribed relationship is mother's mother's brother's daughter's daughter.

The Aborigines have seven hundred tongues, apparently dialects of a common original language, and none of them written. The "dialect chain" is such that an Aboriginal who talks language A can talk to one speaking language B, who in turn can talk to one speaking language C, who in turn can talk to one speaking language D. But one who speaks D cannot talk with one who speaks A. The languages have as many as forty thousand words, including four genders of nouns declinable into eight cases. A sample of one language, taken from anthropologist A. W. Howitt's early research, looks like this, when the sounds are set down in the Roman alphabet:

Yakai! yai!	Ngata	wata	tanana	ngantyai,	tananana	turu-etya.
OH DEAR!	I	NOT	HER	LOVE,	SHE	FIRE-PERSON.

How to Kill and Cook a Kangaroo

Utilize the animal's curiosity. Send part of your hunting party on ahead; the kangaroo will follow, and the rest of the party can sneak up behind him and kill him with a boomerang. To cook: dig a kangaroo-size hole and build a roaring wood fire until the hole and surrounding edges are covered six inches deep with live coals. Heave the kangaroo

---

* In an even bloodier initiation ceremony, a number of old men successively tie off their upper arms and lance veins at the elbow, directing the stream of blood over the head and body of the young man being initiated until he becomes stiff with encrusted gore. The purpose of the ritual is to infuse courage, by showing that the sight of blood is nothing and that wounds are not to be feared.

into the hole, feet protruding upward, cover with coals from the fire's edges, then shovel sand over the coals. Roast until rare.*

How, indeed, have the Aborigines survived in their hostile environment? Few foreigners, not to mention Australians, have ever seen them living primitively, but Eugene Burdick went to the outback some years ago, and produced a vivid report. In the middle of a burning desert, crouched in the shade of a boulder, were an extremely thin six-foot man, his *lubra* (woman) and their two sons. Flies crawled across the *surface* of the man's eyeballs. All were naked, though covered, in effect, by a second skin of dirt and ashes glued on with sweat. The odor was an overwhelming smell of sex. Their entire possessions consisted of: two rubbing sticks for fire, two knife-shaped stones, a container of dried worms, a dead rodent, a boomerang, spear and woomera (spear-thrower), and a dingo dog. (By contrast, anthropologist John Greenway of the University of Colorado lists the following items as essential equipment for a white man in the same terrain: a four-wheel-drive vehicle, extra fuel and water tanks, food for twice the expected stay, two spare wheels and tires, extra battery, tools, medicine, cooking gear, fly and mosquito nets, blankets, aero maps, compass and radio to call for help.)

Burdick's guide pressed the Aboriginal for a show of skill. The sons led off, hanging the rodent in a bush and peppering it with stones flung with flat trajectories and deadly accuracy. The father announced that he would throw his boomerang at the rodent and deliberately miss it. His body coiled, snapped. The boomerang started out far to the left, grazing the ground, then rose to fifteen or twenty feet. Turning, it seemed to stop. Then continuing its return, it shot down to knee height and swooped back up to miss the rodent by an inch. The Aboriginal had to move only one step sidewise to pluck the boomerang out of the air. In a third demonstration, the man disappeared toward the horizon with spear and dingo. Burdick waited briefly; the hunter trotted back with a small dead kangaroo, already half-skinned. As Burdick left, forty bloody fingers were stuffing four bloody faces. The Aboriginal's adieu was to crunch a bone between his teeth.

In a sweep of the nearby desert, barren to a white man's eyes, the Aboriginal and his family, employing uncanny powers of observation, could quickly collect ten or twelve pounds of food. The boys, throwing rocks,

---

* "On the Lower River Murray emu was cooked in this way, with its head left out, so that when steam came from its beak it was judged to be properly done." *The World of the First Australians*, R. M. and C. H. Berndt. Another Australian (though not Aboriginal) recipe tells how to cook a galah (parrot): boil with a stone until stone is tender; throw away galah and eat stone.

would kill birds, toads, snakes, and rabbits; the woman would gather insects and edible plants; the man would kill big game. As for water, Burdick watched another group of Aborigines spot a cloud on the horizon and lope over the ground for *three hours* (while he followed in a Land Rover) to get a few swallows from the puddles they finally found. Greenway notes a case in which a group of Aborigines traced a rain cloud two days, and when it failed to provide water, they calmly dug graves and lay down to die. As it happened, a patrol officer came by with water before they expired. They casually got out of their graves and asked him for a *kapati* (cup of tea).

The Aborigines do without metals, the bow, money, pottery, the wheel, politics, arithmetic,* and writing. Most importantly, they never learned to farm. They harvest and eat grass seed, but—perhaps because of the generally inhospitable terrain, perhaps because they never hit upon a suitable crop, perhaps because they were so isolated from the example of societies that did develop agriculture—they never discovered how to plant and tend and reap. Thus the consequences of agriculture—settled life and communities—eluded them, and they have always been nomads. To travel light, families may limit their possessions to about twenty pounds. The tools they prize are stone knives, ax-chisels, clubs, spears, boomerangs, hair-string bags, pandanus-palm baskets, long wooden bowls. The tools are simple but cunningly constructed of fastidiously selected materials: greenstone for axes, sandstone for grinding them, mallee wood for spears, and *keti,* an epoxy-like powdered resin from spinifex, for fastening axheads to hafts. These materials were so vital that they were long the commodities of long-distance, intertribal commerce. Pearl shell from the east coast of Cape York Peninsula, valued for ceremonial gifts, was delivered all over Australia by Aboriginal trading. Of course the most interesting weapon is the boomerang, which so impressed the first explorers of Australia. "The boomerang is a very formidable weapon," wrote the British seaman Philip Parker King (1762–1828), giving the world one of the first accounts. "It is a short curved piece of heavy wood, and is propelled through the air by the hand in so skilful a manner, that the thrower alone knows where it will fall. . . . It is used by the natives with success in killing the kangaroo, and is, I believe, more a hunting than a warlike weapon." Incidentally, the best boomerangs nowadays are made of plastic, either in Australia or by an American firm that exports to Australia. They are used for sport, and the record throw, by the late Frank Donnellan, an Irish-Australian, was 125 yards, out and back to the thrower's hand.

* Their languages have no word for numbers beyond three, although Aborigines can express higher numbers by finger signs.

The diet of the Aboriginal is as varied as a smörgåsbord. Witchetty grubs, the cigar-size larvae of a certain huge moth, are favored. "Lightly toasted, it has quite a pleasant flavor," reports anthropologist Andrew Abbie. Ants are eaten, particularly one kind that stores a sweet liquid in its abdomen in a hard ball the size of a marble. Aborigines eat wild honey, yams and tubers, pandanus and baobab nuts, grass seeds, roots and berries. From rivers they get fish and platypuses, and from the sea shellfish* and the dugong, the extraordinary thousand-pound sea mammal.

Some Aborigines make houses, "wurlies," by planting a circle of flexible sticks, bending and tying the tops together, and covering the frame thus formed with leafy branches. More commonly, families sleep in depressions of the ground, curling up into "family balls" of human flesh for warmth if necessary, or lying between two fires and rolling into the warm ashes in the cold predawn. There is a theory, which is supposed to explain the Aborigines' nakedness, that human beings can often stay warmer without clothes, covered only with dust or ashes, and that white men do not know this because they have never given the practice a fair go. Be that as it may, Aborigines often do wear some clothes, and in the south possum-skin rugs.

In appearance, Aborigines in general share several characteristics. The skin is chocolate brown, the stature somewhat less than that of white men. The hair is straight or wavy, not fuzzy; most hair is black; the men have strong black beards and mustaches. Aboriginals have broad but not Negroid noses, and deep-set eyes usually topped by a pronounced ridge. Yet there are wide divergences among them. Some Aborigines are only four and one-half feet high; some are strongly muscled and some shockingly thin-shanked and thin-armed. Many children are blond until adolescence. Skin color, according to Abbie, comes in a range from "sienna and umber through light and medium to dark vandyke." In manner the Aborigines are said to be cautious and secretive at first meeting, then gentle and quick-spoken but still rather withdrawn, and often crafty.

Some tidbits: Aborigines are rarely shortsighted or color-blind, and tooth decay is not one of their problems. Neither was alcohol, until the white man appeared; they never invented the stuff. The Aboriginal's blood pressure tends to be lower than that of the white man, but his brain size is just the same. Aboriginal women sometimes suckle dingo pups, since these dogs are much needed in indigenous life.

* Earlier Aborigines left huge shell mounds, which served to mark the sites of small settlements, at Weipa on Cape York Peninsula and in Arnhem Land. Carbon dating places the origin of one mound in the fifth century B.C.

Antiquity of the Aborigines

The popular theory of the origin of the Aborigines is that they crossed from New Guinea during the last Ice Age, somewhere between 10,000 and 25,000 years ago, when so much of the ocean was bound up as ice that the sea level dropped and the strait between Australia and the island was exposed as a land bridge (even now it is a mere one hundred feet deep). This has a plausibility that fades upon examination, chiefly because it presupposes that these colonists had sailed from Southeast Asia to New Guinea, since there was no continuous land bridge along that route. Once at sea, they might just as well have sailed to Australia. Indeed, in the more likely event that they drifted uncontrollably along the course of the regular northwest monsoon, they would have more probably landed in Australia than in New Guinea. Furthermore, there are no Australoid* people in present-day New Guinea, which suggests that it was not a steppingstone.

If the land-bridge theory can be dispensed with, so can the time it fixes for the arrival of the Aborigines, usually put at 25,000 years ago. The tendency now is to estimate the arrival at a bit upwards of eight thousand years ago, based on (1) guesses as to the age of the oldest skull found in Australia, and (2) the homogeneity of the Australoid stock. Had the Aborigines arrived as long as 25,000 years ago, surely greater variations in their tribal characteristics would have developed. Professor Abbie suggests that the early ancestors of the Aborigines were "Proto-Caucasoids" who lived in Central Asia twenty thousand years ago and sent out migrations that became the genetically related Ainus of Japan, Veddas of India and Ceylon, and the "Aryan" Europeans.

The land-bridge theory assumed that the Tasmanian Aborigines, a fuzzy-haired group now extinct, came from a wave of immigration even earlier than that of the first Australian Aborigines. Abbie and others now believe that they probably drifted to Tasmania from some island in Melanesia, which is largely inhabited by similarly fuzzy-haired people.

The early Aborigines may have had to deal with some weird animals now extinct—three huge marsupials, including the rhinoceros-size diprotodon, two kinds of giant flightless birds, and some lesser marsupial carnivores, one the size of a leopard and the other the size of a wolf. Archaeological diggings provide some interesting key dates, all B.C.: 4850, linear engravings; 2820, spearhead points in use; 1800, tulas (flake-edge knives) in use; 1470,

* By blood group typing it is possible to define the Aborigines as Australoids, a stock distinct from Caucasoids, Negroids, and Mongoloids.

specialized bone industry active. D. J. Mulvaney, an anthropologist at the Australian Institute of Aboriginal Studies, delineates a period of prolific invention signaled by the adoption of tools with hafts (handles) about 5,500 years ago.

The arrival of the white man led to such horrendous decimation of the Aborigines that Australian historians still look upon it as a subject to be got through very quickly. At first, the Englishmen were kindly. The Sydney suburb of Manly got its name from the behavior of the Aborigines there, as observed by the admiring Captain Phillip. But in the early settlements the Aborigines were quickly reduced to paupers who did odd jobs, caught measles, scarlet fever and dysentery,* and died out. When the whites first pressured the Aborigines for land, the tribes moved closer together and fell murderously upon one another. Then, greedy for yet more land, the whites started the "dispersal," a bland euphemism for massacre. They went out to "shoot abos" for fun—a gay sport on Saturdays. The police participated, to make examples of Aborigines who stole sheep or produce from land recently theirs. At its worst, the "vermin" were given food poisoned with arsenic, or killed for their tattooed skins, Ilsa Koch–style. "I am told it is no uncommon thing for these rascals to sleep all night with a Lubra (Native Female) and if she poxes him or in any way offends him perhaps shoot her before twelve the next day," says a passage from the journal of the chronicler Niel Black. In Tasmania occurred the world's only complete genocide of an entire ethnic group. The two or three thousand Aborigines there were hunted down until only some two hundred remained, and they were deported to nearby Flinders Island. By 1847 only forty were left; the last sixteen were returned to Tasmania, and the last of them, named Trucanini, died in 1876.

Ironically, even as the slaughter continued, Aborigines were giving succor to such nearly helpless white men as the Burke-Wills expedition, the "first" men to traverse Australia north and south—provided one does not count the innumerable journeys by Aborigines over that route. English humanitarians finally blew the whistle on the killing of Aborigines, and Australian states started "protection" policies in the 1870's. In 1877 a Lutheran mission penetrated to a point in the interior seven hundred miles from the nearest railhead, made contact with the nomadic blackfellows, and planted a hospital, a school, and the Gospel. By the turn of the century many

---

* But not much syphilis. The prevalence among Aborigines of yaws, a milder form of syphilis spread by lack of hygiene, seems to have provided a cross-immunity, according to Dr. Abbie.

Aborigines were beggars, pleading for gifts of clothes, without which they were not allowed in the white man's towns.

The unlovely hypothesis that the Aborigines were a self-solving problem—that is, that they would die out—dominated official policy until 1939. Even such a sober work as the *Modern Reference Encyclopedia*, published in Melbourne that year, could blandly say, "The Aborigines, of many tribes, of primitive habits, and a low order of intelligence, are disappearing." It was the realization that they were not dying out that forced the shift of policy from protection to assimilation, which meant that they would be "civilized" and given citizenship. In 1951 Paul Hasluck, then the Minister of Territories, set up a Department of Native Welfare to prepare assimilation by training Aborigines in Western ways and educating their children.

But it took until 1964 for Parliament to replace the act that made Aborigines wards of the state with one that put them under the Social Welfare Act that applies to all citizens. A peculiar referendum in 1967 completed the legal process of equality for Aborigines: census-takers were ordered to count them along with the rest of the members of the genus Homo sapiens in Australia. Much special and presumably beneficial legislation has been passed since then. "We didn't do a bad job of killing them off, but now we are trying to build them up again," says one rueful Australian government official.

A Flowering of Art and Music

Only through art can we get outside of ourselves and know another's view of the universe which is not the same as ours.
                                                        —MARCEL PROUST

If there is one aspect of Aboriginal life that stirs empathy and understanding among average white Australians, it is Aborigine art. Here we get, in fairly communicable form, another view of the universe, with diversity, humor, tradition, and an aesthetic pleasing to any eye. Art seems to have come so naturally to the Aborigines that they had no need to invent a word for it, though they have words for the individual acts of painting, carving, singing, and dancing.

The majority of Aboriginal men can paint, carve, and incise—some artists, of course, being better than others, and only a few are entitled by age, status, or prestige to create certain sacred emblems. Their motivations are never purely the artist's own satisfaction; art is for communication to others. It teaches the laws of the tribe, explains creation, endows weapons with visible potency, keeps alive religious beliefs. The most sacred secret objects,

churungas, oval-shaped slabs of painted wood or stone, contain such powerful symbolic messages among their abstract dots, meanders, spirals, and animal tracks that in some tribes women who catch a glimpse of them are put to death.

The flourishing art is produced with distinctly primitive tools and materials, at least until recent times. The historic engraving tool is an incisor tooth from a possum, with the jawbone in which it is rooted as a handle. Brushes are twigs or human hair. Sharkskin serves for sandpaper. The common painting surfaces are rock facings, in caves or outcrops, and the insides of sheets of bark, often from the stringybark eucalyptus. Wood, beeswax, and clay are used for modeling and carving. Pigments are made of charcoal, or of red and yellow ocher (both forms of hematite, iron ore), ground in water. But some of the more sophisticated artists nowadays, with the assistance of missionaries, order paints and sable brushes from Winsor & Newton, the celebrated London dealer in art supplies.

The most extraordinary works of Aboriginal art are cave paintings, found by the thousands across northern Australia, often inaccessibly high on walls and ceilings. In the northwest are the Wandjinas, spooky, looming, gigantic physiognomies of dead white, with long-lashed black eyes and black coblike noses—but no mouths. They represent the mythical creators of the surrounding land, who supposedly painted the pictures themselves. In fact, most sacred cave paintings are thought by the Aborigines to be of "nonhuman" origin, dating back to the Dream Time. They depict mythical beings who "became a painting." Other motives for cave paintings were: hunting and fishing magic, the illustration of stories and legends, and love magic, accomplished by paintings depicting copulation. The shallow caves of Ayers Rock, the big tourist attraction of central Australia, portray mythical lizard-men and other anthropoid figures.

A fascinating branch of cave art, found in western Arnhem Land and nowhere else in the world, is X-ray painting, depictions of animals, birds, fish, and reptiles showing both the external form and the internal organs—skeleton, stomach, heart, lungs, and intestines. Another variation in the same area, and a charming one, is the Mimi painting, the work of some now extinct Aboriginal people. Mimi paintings, just a couple of inches high and done only in red ocher, show stick-figure humans running, fighting, and throwing spears with a ballet-like suggestion of movement. Near Delamere, in the Northern Territory, are the mythical Lightning Brothers painted sixty feet up on a sheer rock face. These are a pair of male figures nine feet tall, each with a penis as long as his legs, tipped in red, indicating subincision.

Bark paintings apparently originated as decorations for the interior of wet-weather shelters made of this material. A typical bark painting, done in a semiabstract style with dots and patterned lines, shows men in canoes, dolphins, and nesting seagulls—the story of an expedition to collect seagull eggs. Bark paintings are so popular nowadays among fanciers of Aboriginal art that they are turned out in the thousands by a somewhat synthetic native-art industry. The National Gallery in Melbourne displays authentic examples.

The Pukamuni burial ceremony (more on that in the next chapter) on Melville Island off the north coast of Australia requires decorated poles that constitute a minor branch of Aboriginal art. Similar poles in Arnhem Land, exhibiting skillfully carved figures, are objects as sacred as the churunga. Recent anthropological discoveries among the Aborigines are figures in the round, wood sculptures of fierce faces and so on. Aborigines also expend much artistic effort on bull-roarers, slabs of wood attached to a string so that they can be whirled in the air to make a noise supposed to be the voice of a great spirit.

Three curious forms of Aboriginal plastic art remain to be mentioned. One is ground drawing, on earth smoothed and hardened with water or blood, then painted abstractly in ocher or perhaps incised in channels lined with down. Another is body painting, almost exclusively by and for men, in designs drawn in ocher or in down, cotton, or kapok glued on with blood. The last is petroglyphs, outlines of animal figures sometimes sixty feet long cut ages ago on sandstone surfaces in the neighborhood of Sydney.

*

Up and up soars the Evening Star, hanging there
   in the sky.
Men watch it, at the Place of the Dugong, the Place
   of the Clouds, the Place of the Evening Star.
Far off, at the Place of the Mist, the Place of the
   Lilies, the Place of the Dugong.
The lotus, Evening Star, hangs there on its long
   stalk held by the Spirits.

So goes an Aboriginal song, part of a long legend about the moon. The Aborigines see to it that their lives have plenty of entertainment, in song, dance, drama, and storytelling.

The songs are short and repetitious verses that go together in long cycles and tell stories. A song about a journey will touch on the traveler's thirst, the sharp stones that hurt his feet, the wild honey he found, the water he

thirsted for, the trees and spirit people he encountered. The verses are full of vivid imagery—a sunset is "like circumcision blood."

The main musical instrument is the *didjeridu*, a stopless, end-blown wood or bamboo drone pipe four to seven feet long. Elspeth Huxley describes the sound as *"gibba-yerra, gibba-yerra, gibba-yerra"*—"something between a grunting and the far-off bellow of a bull." With it go various combinations of clapping sticks, pillow drums (possum skins stuffed with feathers), bull-roarers, rattles made of baobab nuts or shells, and rasps.

Music plus dancing equals the corroboree, an Aboriginal word long since taken into Australian English to signify a big festivity. Aborigines divided their song-dance-dramas into religious cult ceremonies celebrating the adventures of Dream Time heroes, totemic ceremonies to secure the increase and prosperity of individual totems, and "playabouts," purely for fun. They often involve the miming of a totemic animal: a man may wriggle like a crocodile between the spread legs of a row of standing men; or the troupe may stamp like a herd of buffalo. The drama may recount a dream or convey a parody. A group of anthropologists attending a corroboree at Barrow Creek a few years ago watched in puzzlement as a dancer approached the other dancers one by one, scrutinized them carefully, then went through the motions of writing in a notebook. Suddenly the anthropologists realized, red-faced, that the drama was a spoof of their own interviews of members of the same tribe some time earlier. Men dance much more than women, but in some areas women have a graceful dance performed while making cat's cradles with string. Big poofs of flame from bundles of dried grass often illuminate corroborees. The ceremonies arouse strong sexual emotions, and generally end in a copulative free-for-all.

CHAPTER 7

Aborigines II: Their Role Today

*The abo pleads nolo contendere, I do not wish to contend,
every time his eyes meet those of a white man.*
—EUGENE BURDICK

JIM HAMILTON was born in an Aboriginal settlement named Cherbourg—nobody knows how it got this Gallic name—150 miles north of Brisbane, the capital of the state of Queensland. His native language, one of 150 different Aboriginal languages in the area, representing 150 different tribes, is known as Gorang Gorang. His daily language nowadays, however, is English, for he lives in Brisbane and works as a civil servant (a status few Aborigines have achieved), employed by the Queensland government's Department of Aboriginal and Island Affairs. His career is that of service to his fellows, and he is well versed in the life of the Aborigines in this state.

Most Aborigines here, according to Jim Hamilton, are what are colloquially called "no-hopers," because they have no sense of "now, past, or future." Townsmen get equal pay for equal work at industrial and other urban jobs, but they do not understand the time clock and are apt to be undisciplined. Out in the bush they have a happier environment, but they constitute a big "problem" in the country towns because they seldom become fully integrated into the community. They have a good deal of pride in their "national" position; for instance, much resentment was expressed about reports—which proved to be false—that Aboriginal graves had been bull-dozed on Melville Island to make room for cypress plantings. The Aborigines in Queensland are somewhat distrustful of education; a young boy starts

school at five for a ten-year course, whereupon he may matriculate into the local university. There are only *two* Aboriginal students at the local university, however, out of some seventeen thousand. An Aboriginal boy will say, "All that schoolwork won't get me a sharper pick."

There are no rich Aborigines, and they do not even benefit from the discovery of mineral rights on their land. Aborigines are free to accept or reject military service on the same terms as whites in the Australian armed forces. They have the vote, and most vote Labor. Many still live as nomads, getting seasonal work "mustering" (rounding up) cattle on the stations. Except socially, there is no color bar or segregation of any kind; any Aboriginal can go to the best hotel (if he can afford it) or shop. On the other hand, invisible and mysteriously exerted pressure may make it difficult for him to buy or rent a dwelling in a location of his own choosing. More often than not the Aboriginal lives on the fringes of the towns, where water, sewage, and other public amenities are deficient, and where seasonal unemployment can be severe. Every Aboriginal shares money and all other possessions with other members of the tribe. It does not seem important to him to keep them for himself, even to improve his own housing or clothing.

*

In Darwin, modern Aborigines do not in general have a fruitful life. They live mostly in huts in outlying parts of the town, and when working often get less than the official minimum wage. However, most full Aborigines manage to look proud, and stride the streets like cowboys. A basic problem is that the Aborigines themselves are in such different and rapidly shifting stages of education, assimilation, and integration. But they lack drive (or so the whites say); they drink too much (or so the whites say); they desperately need more schooling, better housing, fuller amenities. Entering Darwin from the bush, most begin at once to deteriorate.

Social mixing with whites is at the minimum, but mixed football teams play in vigorous games of Australian Rules. At some of the new mines, like Gove, on the Gulf of Carpentaria, Aborigines complain that white workers molest their women. Occasionally marriages occur between the Aborigines and the Chinese; the local legend has it that "the genes click," and that most such unions are happy. Intermarriage between Aboriginal and white is rare; usually a white man marries an Aboriginal girl, not vice versa. There has been at least one case of the reverse in recent years, a marriage between an Aboriginal athlete and a Dutch nurse, which broke up after eighteen months.

Aborigines on Melville Island

Half an hour away by air on the northern coast off Darwin are two small islands, Melville and Bathurst, closely linked and inhabited by a curious Aboriginal people, the Tiwis. From the air, Melville, which marks off the Timor Sea from the Arafura Sea, gives the impression of searing, stinging jungle. This is a wet green madhouse. The plane puts down on a crude airstrip marked by oil drums painted white; a black ibis may sometimes be seen strolling among them.

Traveling by Toyota land cruiser—rough ground here—one passes the metal carcass, intertwined with decrepit foliage, of a Japanese aircraft shot down during the war, near Snake Bay settlement. Here, local folklore has it, the first European settlement was established in the 1820's by a Belgian bishop named Gsell, who wrote an autobiography, *The Bishop with One Hundred Wives*. The story is that, determined to wipe out polygamy, he picked a hundred youthful maidens and married them all himself, so that they could not become partners in polygamous marriages with the local natives.

The chief industry of the island is the production of wood chips exported to Japan. There are long transverse rows of cypress, and at the end of each row is the name of the cutter responsible for the logs—Black Joe, Tractor Joe, Big David. Between bouts of work the Aborigines amuse themselves by shooting crocodiles. The patriarch of the community, King Larry, about sixty-five, has a splendid white beard, thirty wives (apparently the Belgian bishop's injunctions have not been carried forward), and several hundred children; he wears intricate finery for ceremonies, and is proud of the fact that, singlehanded, he "murdered," as he puts it, an entire Japanese pearling crew in 1932.

Snake Bay is an official welfare station, operated by the Commonwealth government to give sustenance to these forlorn but stubborn islanders. The natives call it Millikapiti. The population is about 250 Tiwi and a handful of whites. The Tiwis are a complicated people. They do not throw the boomerang, or play the *didjeridu*, but they speak *seventy* different languages; to master the easiest, one is told, takes ten years. The traditional Tiwi costume includes a fiber ball, called a biting bag, swung below the neck; men and women chew on this, as Arabs fondle amber beads. In Snake Bay there is a store with retail sales running to $8,000 a month, as well as a museum containing wooden totem figures and other artifacts, and a canteen, where the residents are permitted three cans of beer per day—no hard

liquor. They must drink their three beers on the spot, under supervision. There is a hospital of sorts—twelve beds under an iron roof, and so hot that if a patient did not have a temperature going in, he would surely have one coming out.

Superintendent of the Snake Bay settlement is Ian Rodgers, an impeccably civilized Briton who served many years in Kenya as a district commissioner there. According to Rodgers, the Aborigines are good carpenters, good craftsmen, and they have little temptation to leave the station (anybody is free to go) because family ties tend to hold them there. Comparing the local Tiwis with the Kikuyu in Africa, he said first that he did not think that he really knew or understood the Tiwi mind deep down, as he had come to understand the Kikuyu, but he went on to say the Tiwi are more explosive and give quicker expression to resentment or other emotions in passionate outbursts, whereas the Kikuyu always seek to hold everything in. Yet Kenya had the Mau Mau revolt, which would be unthinkable in Australia.

The Pukamuni burial ritual is unusual and ornate among the Tiwi. On the death of a person the family assembles, and an artist in the group cuts and paints a batch of wooden poles, which are placed around the grave. These poles have a special arcane significance, each symbolizing a particular member of a family, living or dead. The corpse is hung overhead. Even if it starts to decay (which as a matter of fact it does almost at once), it is not cut down till the next full moon. Then an elaborate recital takes place describing the dead man's life and qualities; after this, the family goes through a purification ceremony, and at last the actual funeral occurs. Members of the family are not allowed, as a rule, to return to their homes until rain has fallen; hence, they may be obliged to sleep out of doors for days or even weeks. Only rain can produce the necessary purification. Moreover, a tribal taboo has it that the ceremonial poles must *not* be cut, carved, or painted by any close kin of the dead; this work is assigned to outsiders, and the cost, believe it or not, can be $500.

The Tiwis are renowned for their dancing, which may be seen at their corroborees. The celebrated dancers Ted Shawn and Ruth St. Denis came here to study it on one occasion, and a group of Tiwi dancers performed at the 1970 Osaka Fair.

Aborigines and Whites—An Uneasy Relationship

White Australians, one is told, do not ever really *see* the Aborigines, not only because they live in distant places but because they are "psychologically invisible"—though perhaps this last is not so true as it used to be.

The Aborigines fit in closest to the white economy as hired hands in stock raising. Mustering cattle through the scrub, they can wear out four horses a day, and they are willing to ride boundaries on the enormous stations for as long as three months away from the homestead. And they have been willing to work for one-fifth of the pay of a unionized pastoralist, taking pay, moreover, in food and goods, thus sparing the station manager the inconvenience that stems from giving Aborigines money—the theory being that they will spend cash on liquor and tobacco and neglect their families. Aborigines on the stations live in fly-specked huts of corrugated iron, called "humpies." Often the station manager's wife does the cooking for the "gins," the Aboriginal women, who themselves do nothing at all. Unfortunately, Aboriginal horsemen are now being replaced on a large scale by helicopters and motorcycles. Though these stockmen are reasonably Westernized, they may still disappear for months at a time on "walkabouts." There they resume, to some extent, the life of killing kangaroos, of trapping goannas (lizards), of sleeping by fires, or heeding the animal spirits. It is typical to see Aborigines hovering timidly, namelessly, at the edge of a sheep shearing; but they may also be seen relaxed and chatting on the street in front of a handsome new bank in Alice Springs. For every Aboriginal standing one-legged in the desert, in the traditional posture of one foot tucked against a knee, there is another digging Grand Funk Railroad on a transistor radio.*

Whites and Aborigines do not mix much in homes. Even on the far reaches of the stations, alone in the night around a campfire, white stockmen won't always "doss down" next to Aborigines. (Contrariwise, a civilized Aboriginal lately moved out of the West End of Adelaide because of "too bloody many Italians coming in.") In fact, some whites still profess to have lynch rights over the Aborigines; the *Bulletin* quotes a cattle station manager as boasting that "if they bothered one of my girls they would be dead men, just like that, and you would get away with it here, too."

Some ten thousand Aborigines live in Sydney; others in towns like Moree and Walgett in New South Wales mix a culture of humpies, drink-scrounging, movies, and cars. In Sydney's Greenwich Village-like King's Cross, you can find bars that cater mostly to beery, staggering Aborigines and their blowzy women. Formerly, various laws forbade the Aborigines to drink, but such bans have been repealed throughout the country. Favored drinks

---

* Greenway gives this picture of a partly detribalized Aboriginal. "I had an experience with one such fellow; he emerged from the Adelaide prison with jailhouse English, the winsome habit of breaking his wife's forearm across his knee whenever she became too 'cheeky,' and a hand-cranked phonograph with one record, *Adeste Fideles*, which he played over and over again as he squatted in the red sand with the temperature 106 degrees in the waterbag."

among them are cheap wine and methylated spirits. Black prostitution is commonplace in Australia; in such a mining boom town as Roebourne, on the northwest coast, Aboriginal girls get $40 to $50 a night.

Only one percent of Aborigines get through secondary school—but in the bauxite fields Aboriginals are making $112 a week. Aborigines have suffered depressing discrimination at swimming pools, hotels, and restaurants —which spurs protests among idealistic university students in Sydney. On the other hand, welfare officers are now required to address an Aboriginal man as "mister," not "Jackie" or· "Billie," as they used to. Last year the Aborigines of Arnhem Land sued to establish ownership of the bauxite there, just as Nabalco Proprietary Ltd. (a Swiss-Australian firm) prepared to mine it; the tribesmen said that mining would damage sacred places, and that they have secret tribal documents (which unfortunately, by the terms of their beliefs, cannot be shown to the public) proving ownership of the land. They lost the suit.

Assimilation Versus Integration

What is to become of these people, who are partially adapted to white culture and growing in numbers? The debate in Australia, which often rings of the American dilemma over white-Indian relations, centers on those two code words, assimilation and integration.

Government policy toward the Aborigines in Australia has gone, historically, from (1) complicity with attempted genocide to (2) belated application of English humanity and justice to (3) welfare and protection. The fourth and present official policy goes like this: "All Aborigines and part-Aborigines will attain the same manner of living as other Australians and will live as members of a single Australian community, enjoying the same rights and privileges, accepting the same responsibilities, observing the same customs, and influenced by the same beliefs, hopes and loyalties as other Australians." This is assimilation; it conjures up a picture of a proper Aboriginal wearing shorts, drinking beer, playing at bowls or golf, watching television, and raising roses in a suburban garden. Many well-intentioned whites approve of the government policy. They want the "natives" to have better housing, but there are inevitable qualms about accepting them as neighbors. All the squirming evasions of U.S. racial strife are here, though in pallid imitation.

The concept of integration, on the other hand, aims at the coexistence of white and Aboriginal cultures, with the bulk of the Aboriginal values surviving. It assumes that Aboriginal art and music must continue to exist,

that Aboriginal language and religion can exist side by side with English and Christianity. "Integrationists" concede that naked-nomadic Aboriginal life is finished. But their point is that Aborigines have something to give, to teach, to the whites.

Mr. Wentworth's Theory

The Australian Minister for Social Services, W. C. Wentworth, has been in a lesser incarnation also the Minister-in-Charge of Aboriginal Affairs, and the latter job seemed to be his special enthusiasm.* It would be scientifically fascinating, he believes, to know what man's society was like before farming led to storage of food, division of labor, settled cities, and today's technological civilization. And, he argues, the study can be made, if tackled by anthropologists right now. The Australian Aborigines are the only people left in the world "not contaminated with agriculture." The focus, he thinks, would be on the cohesiveness of small groups, and on their language differences, which maintain tribal unity and distinguish friends from strangers. It would reveal an ancient psychology that may have planted patterns in the race memories of us all. In Australia such a study could be "pulled out of the memories of men still alive." But the opportunity is slipping away: "A language dies every month" among the rapidly detribalizing Aborigines, says Mr. Wentworth.

A similar note was struck in an interview in Canberra with Frederick D. McCarthy, the distinguished anthropologist who heads the Australian Institute for Aboriginal Studies. "In some cases, there are only two or three survivors of a tribe left," he said. "We have to act quickly. The 'wild' Aborigines are very tense. Not many left, just the few on the eastern side of Western Australia." Typically, in South Australia, construction of the space-age Woomera rocket range uprooted a lot of Stone Age Aborigines, who moved to white settlements. A drought a decade ago forced hundreds of Aborigines into missions and settlements.

One great hope for getting Aboriginal culture down on film, tape, and paper before it disappears is the Institute itself. It has fifty projects under way. Enough languages have been recorded to guarantee considerable enlightenment about Aboriginal linguistics. Similarly, hundreds of songs have been recorded, and much work done in social anthropology. The hitch, according to Wentworth, is a shortage not of money or goodwill, but of

* Prime Minister William McMahon, after taking office in March 1971, created a new ministry. Mr. Peter Howson is the first Minister for the Environment, Aborigines, and the Arts.

interest by the anthropologists of the rest of the world in a rapidly fading opportunity.

Recognition of this situation tends to modify the official policy of forcing Aborigines to attain "the same manner of living as other Australians." The government concedes the Aborigines' right to their languages and their customs, just as it gives that right to other religious or immigrant groups. But it still proposes to promote assimilation and to discourage "any kind of compulsory and permanent apartheid."

The various forms of government aid run around $30 million a year, which averages $1,500 each for the approximately twenty thousand Aboriginal families. Federal funds are pouring into the Aboriginal reserves in the Northern Territory, where housing, food, and work are offered to any Aboriginal who asks for them. Houses are concrete platforms on which rest one-room aluminum buildings with verandas. Schools are staffed at twenty students per teacher. Aboriginal children in remote spots are taught at a particularly ingenious mobile school that can be wheeled in over roadless terrain. The teachers find the mental equipment of the Aboriginal children equal to that of white ones, but question whether at adolescence the old sex and initiation ceremonies will dominate the children's energies.

Despite these efforts, the majority of Aborigines still face a discouraging future, well described by that eminent Australian, Dr. H. C. Coombs, who is among other things chairman of the Australian Council of Aboriginal Affairs. His prognosis:

If an aboriginal baby is born today it has a much better than average chance of being dead within two years.

If it does survive it has a much better than average chance of suffering from substandard nutrition to a degree likely permanently to handicap it (a) in its physical and mental potential, (b) in its resistance to disease.

It is likely in its childhood to suffer from a wide range of diseases but particularly ear, nose and throat and respiratory infections, from gastro-enteritis, from trachoma and other eye infections.

If it reaches the teen ages it is likely to be ignorant of and lacking in sound hygienic habits, without vocational training, unemployed, maladjusted and hostile to society.

If it reaches adult age it is likely to be lethargic, irresponsible and, above all, poverty-stricken—unable to break out of the iron cycle of poverty, ignorance, malnutrition, ill health, social isolation, and antagonism: if it lives in the north it has a good chance of being maimed by leprosy and, wherever, its search for affection and companionship may well end only in the misery of venereal disease.

If it happens to be a girl it is likely to conceive a baby at an age

when her white contemporary is screaming innocent adulation at some pop star and she will continue to bear babies every twelve or eighteen months until she reaches double figures or dies of exhaustion. And so the wheel will turn.

The Aborigines' Point of View

It would obviously be interesting to know what the Aborigines want in the way of integration, but apparently no well-researched study exists. A couple of factors seem clear, however. After nearly two centuries of losing life and land to the white man, the Aboriginal is distrustful of belated attempts to square things up. Yet when offered houses, schools, and jobs in the white world, the remaining primitive nomads slowly but surely shuck off their old culture and join this new one. Beyond these elementary observations, there are, of course, some worthwhile opinions about the problem from various articulate Aborigines. A typical voice is that of Bob Maza, the president of the Victorian Aborigines Association, one of about sixty-five groups defending Aboriginal interests. He calls government policy "cultural obliteration," and says, sardonically: "Let us not be too hasty about rushing into a society that is at present foundering." The closest thing to a Black Power group in Australia is the National Tribal Council. Its leader is Bruce McGuinness, who is just past thirty years of age and would appear to take his style from American black militants. "We want our land and the wealth that Honky dug out of it," he wrote last year. "We are sick and fed up with being told that we are being helped by the Man and that the Man is being very generous in giving us this and giving us that. We know that this is a whole lot of Toro Excretia. All the Man is doing is appeasing his conscience. . . . us Black Boys ain't buying any of his con."

The National Tribal Council makes ten demands in its official platform. They include better health, better education, a law against discrimination, black administration of government programs of aid for Aborigines, and legal aid and protection—Aborigines "need more lawyers than welfare officers." They also want "reacculturation. . . . Governments must abandon the failed policy of assimilation which amounts to cultural genocide." Finally, they want recognition of their ownership of "traditional" land, and compensation for the whites' seizure of it, plus royalties from mining or other exploitation of it.

Aboriginal militance in general, however, is disorganized and weak—nothing to perturb the government or anyone else. There are only one-tenth as many Aborigines in Australia, in proportion to the total population, as

there are Negroes in the United States, and no effective rebellion can be expected from these people.

Kath Walker, a best-seller poet and Queensland State Secretary of the Federal Council for the Advancement of Aborigines and Torres Strait Islanders, contends that the real pinch in Aboriginal-white relations is land. The government, she charges, flatly refuses to listen to Aboriginal demands for the return of land, while on the other hand leasing it in Kentucky-size chunks to American and British absentee landlords at fifty cents a square mile. Kath Walker, herself a vigorous and effective Aboriginal, was born in 1920, worked as a servant, joined the Australian Army, learned shorthand, and went on to write poetry. Her totem is the carpet snake. Another Aboriginal poet is Jack Davis, who some time back picketed the United Nations building in Manhattan to call attention to the Aborigines' plight. His lament:

> The tribes are all gone,
> The spears are all broken.

*

A number of other Aborigines, though painfully few all told, have been able to come from the Stone Age to eminence in the white culture. Harold Blair turned his resonant tenor into a career in opera and in teaching at the Melbourne Conservatorium of Music. Albert Namatjira made himself one of the country's best-known landscapists, in bone-and-ocher paintings that catch the desert's torrid, barren beauty—the heat that chills. Evonne Goolagong, nineteen, won the Wimbledon Women's Singles Championship in mid-1971, upsetting Margaret Court. Sister Jeanne Marie, of the Murinbadda tribe, a few years ago became the first full-blooded Aboriginal nun, and the Rev. James Noble is one of several who are Anglican clergymen. Lionel Rose, who grew up in a humpy near Melbourne and wrapped his hands in rags in order to learn how to fight, became world champion bantamweight by beating a Japanese boxer in 1968. He is said to be the first of his race to be best in the world at anything. The Australia Day Council made Rose Man of the Year in 1969. "A hundred and eighty years ago one of my mob would have been a dead cert for this," he remarked dryly.

The first Aboriginal to get a seat in any Australian Parliament, state or federal, is Neville Bonner, who was endorsed in mid-1971 by the Queensland Parliament to be a Liberal federal Senator from that state. He had been a boomerang-maker and carpenter, earning $75 a week as compared

to the $263 paid to a Senator. Other Aborigines have become authors, inventors, cricketers, actors. A number of them served with distinction in World War II and the Korean War.

*

Only faintly bruited about in Australia today is the thought that white society might not only let the Aborigine keep his values but even adopt some of them. The possibilities, though probably remote, are tantalizing. Valuable insights might be gained by studying the Aborigines' animistic religion in the light of Jungian psychology. We might also benefit by examining the Aboriginal trait of sharing and his distrust of competition. Nowadays white men everywhere suffer doubts about the ultimate value of their computerized, war-prone, often philistine civilization, and they impose it upon other and differing societies with less confidence than in the past. Perhaps Australia will be the nation that finds a way toward a more comfortable give-and-take between Western culture and the cultures that we perhaps quite wrongly call primitive.

CHAPTER 8

Labor Versus Liberals

They call no biped lord or sir,
And touch their hat to no-man.
—HENRY LAWSON

THE TREND of Australian politics in the nineteenth century was populist, collectivist, and democratic, which should have led to an advanced experiment in socialism. Instead, the political system today is a curious preservation of some old radical values (such as every man's absolute right to a job) combined with deep caution and conservatism. A social-minded person concerned with the individual's economic rights can be reasonably happy in this system, while big business and the banks are eminently comfortable. The system has bumbled along for years on basically managerial governments, and indeed managerial governments appear to be good enough for the present. We now proceed to examine Australia's modern politics, starting with a brief look at how the government works.

*

The Australian Constitution has been called "an extraordinarily uninteresting document that does not hold any great truths to be self-evident." It has no bill of rights. But it does have some curious wrinkles. For example, any Member of Parliament can introduce legislation to *reduce* expenditure or taxation, but only ministers of state can do so to *increase* expenditure or taxation.

Australia adopted from the U.S. the constitutional concept of listing fed-

eral powers and giving all others to the states. The federal powers, in brief, are control over defense, post and telegraph, customs, trade and commerce, marriage and divorce, banking, naturalization, and external (foreign) affairs. Also roughly patterned on the U.S. model are the Senate and House of Representatives. Each of the six states elects ten Senators, an arrangement which, as in the U.S., gives the less populous states leverage against the others. The House has "as nearly as practicable" twice as many members as the Senate (or 120). To create their constituencies, the population is divided by 120 (which currently yields a quotient of somewhat more than 100,000), and the states map out electoral districts containing about this many voters. In practice this leads to much gerrymandering, usually favoring the countryside. The Senators are elected for six-year terms, with one-half retiring every three years to make it a "continuing body." Representatives are elected for a maximum of three years, but as in England the Prime Minister, if defeated on a vote of confidence, must call for an election. Thus House and Senate elections are generally out of step, which, plus the same situation in each of the states, gives Australians *lots* of elections to vote in—and they must vote, under penalty of a fine.

But because of Australia's British heritage, the Constitution does not separate the executive and the legislative powers (as the U.S. Constitution does); it gives both functions to Parliament. The leader of a political party stands for election in his own constituency, and if he and his party win a plurality, he becomes Prime Minister. He then picks a cabinet from Representatives and Senators, chiefly the former. Most decisions are taken by the cabinet sitting in secrecy, and enacted rather docilely by the House and Senate. Australian Representatives are called Members of Parliament— M.P.s, as in England or Canada. The Labor Party has pledged itself since 1919 to abolish the Senate, partly because anti-labor forces have dominated that body throughout most of Australian federal history, but is not working very hard at it. On paper, the Senate does not seem powerful; for example, it can only "request" changes in money bills sent up from the House. But it can veto House bills by simply failing to pass them, and if in practice it is less powerful than the U.S. Senate, it is vastly more powerful than the Canadian Senate.

The federal government as a whole is stronger than the federal government of the United States, for three reasons. It has the sole power to borrow money—that is, the states cannot borrow. It controls all social and welfare services. And, most important, it has the sole power to levy income taxes. The federal government shares this revenue, on its own terms, with the states, much as President Nixon was in 1971 proposing to share U.S. federal gov-

ernment revenue with the fifty states. The central government can thus dictate much policy to the states, even though 70 percent of the money is disbursed to them without specifying how it should be used.

The High Court of Australia, with a chief justice and six other justices, is the court of appeals from state courts, since Parliament has not created any lower federal courts. It has original jurisdiction in a few matters, chiefly issues between the Commonwealth and the states, and review of legislation. Here we come to one of the few tenuous remaining governmental links between Australia and Britain. For there is a court higher than Australia's High Court, to which appeals can be made on certain questions. It is Britain's Privy Council. But the High Court itself must certify, that is, permit, any such appeals. It has done so only once. And that was in 1912.

A typical case of High Court action was its invalidation of a law to dissolve the Communist Party in 1951. Until a suitable building can be put up in Canberra, the capital, the seat of the High Court remains in Melbourne.

The Political System

The play of modern politics in Australia is not as simple as Democrat versus Republican, but it can be comprehended quickly. The Labor Party would control the majority of votes except for a festering schism created by some of its more conservative members, the Democratic Labor Party. This weakness lets a minority rightist party, the Liberals, in coalition with the Country Party, rule Australia. Some consequences are: (1) that Labor, the biggest single group even though it is stymied for the moment, remains potent and likely to regain control of the government at some point; and (2) that the Country Party, vital to the Liberals, swings more weight than its small constituency merits.

The last twenty-two years have been years of gratification and self-satisfaction for the Liberals, but the more active shapers of Australia, since it became a nation, have been the politicians of the Labor Party. This country is where the concept of a Labor Party germinated. And Labor, in the phrase of Australia's great interpretive historian, Sir Keith Hancock, has usually been the "party of initiative," working against the Liberals and their predecessors as the "parties of resistance." The political story of Australia since federation shows how this has come about.

The Labor Years

The secret ballot and universal manhood suffrage came to Australia well before Disraeli introduced them in England. Thus long before federation the idea that the vote could really be made to work for change had been

planted. So when the unions lost the great strikes of 1891, aimed at better hours, better pay, accident compensation, and so on, they turned grimly but confidently to politics for redress. Labor parties in the states soon developed a discipline that required Labor politicians to vote as a bloc on major issues. The first federal parliaments were split three ways, but Labor's bloc-vote solidarity gave it decisive control in the first decade of federal government.

It was a decade that fixed some of the main lines of Australian life. A navy was begun, old-age and invalid pensions established. Tariffs were raised, with the significant condition that protected industry must pay "fair and reasonable wages"; that is, the tariffs were enacted less in the interests of the manufacturer than in those of his employees. In a test case this led an influential judge to set a firm minimum wage, a social concept that did not come into effect in the U.S. until the New Deal. With no guidelines as to what might be fair and reasonable, Justice Henry Bourne Higgins took a deep breath and ruled that the minimum wage must meet "the normal needs of the average employee regarded as a human being living in a civilized community." This definition has prevailed in Australia ever since.

*

Labor got an absolute majority in Parliament in 1910 and put through further measures that the opposition denounced as socialism, chiefly the establishment of the government-run Commonwealth Bank as a "people's bank" in competition with private banks. It was a Labor Prime Minister, Andrew Fisher, who took Australia into World War I, vowing to support Britain "to the last man and the last shilling." The first fate of the "Anzacs" (Australia and New Zealand Army Corps) was to be sacrificed in Winston Churchill's ill-executed plan to take the Dardanelles. Under shattering fire, the Australians stormed the steep hill at Gallipoli, dug in and held their position for eight months, until withdrawn. To no purpose, ten thousand Anzacs died there. The date of the landing, April 25, is now Australia's most fervently celebrated national holiday, marking the most poignant episode of a war that ultimately cost Australia 59,258 dead and 166,819 wounded.

Enlistments *increased* after the Gallipoli disaster, such was Australia's patriotic fervor, but the succeeding Labor Prime Minister, William Morris Hughes, a man of furious invective,* decided under British pressure that only conscription would ultimately raise enough troops for Australia to do her share of the fighting in France. He thereupon forced the nation through

---

* He once compared a political enemy unfavorably to Judas Iscariot, on the grounds that Judas "did not gag the man he betrayed, nor did he fail to hang himself afterward."

two wrenching referendums designed to authorize conscription. The draft was sulphurously opposed by Melbourne's Roman Catholic Archbishop, Daniel Mannix, and by thousands of Irish Catholics in the Labor Party, who were furious at British suppression of the Easter (1916) Rebellion in Dublin. The referendums failed to pass, splitting Labor and destroying its power for years. Hughes, expelled from the party, teamed up with the Liberals to form a party labeled Nationalists. They ruled, partly in coalition with the Country Party, until the Depression made the country yearn for Labor's humane touch again.

Labor's sway this time lasted only a few years, until a new split. John Thomas ("The Big Fella") Lang, Labor Premier of New South Wales, semidesperate over the suffering caused by the Depression, moved wildly against the banks, alarming conservatives, including those in his own party. They defected to join the Nationalists and form a United Australia Party. This group ruled until World War II. Fighting the slump in the 1930's with the conventional economic wisdom of the time (deflation, wage cuts, government economy), the United Australia Party managed to provide Australia with a particularly miserable Depression. The birth rate dropped, immigration stopped, development ceased, and *one-third* of all workers were jobless.

By World War II the United Australia Party was weak and squabbling, and shortly before Pearl Harbor the voters put Labor back into power. High-voltage shocks went through the nation as the Japanese sank the British warships *Prince of Wales* and *Repulse* in the Gulf of Siam, overran Singapore, the Dutch East Indies and Timor, and invaded first New Britain, off New Guinea's north coast, and then New Guinea itself. Bombs fell on Australian soil at Darwin, killing 243. Here Mr. Churchill, deviser of Gallipoli, entered the picture again. He had already won sour criticism from Australia earlier in the war by opposing the withdrawal of exhausted Australian troops who had resisted eight months of siege at Tobruk in North Africa. Now, while Australians clamored to get a couple of Aussie divisions transferred home from the Middle East for defense, Churchill coolly proposed that it was more important to send them to Burma. Churchill is, to put it dryly, not quite the hero in Australia that he is elsewhere.

This is a crucial moment in Australian history. John Curtin, the new Labor Prime Minister, made a famous speech in which he said: "Without inhibitions of any kind, I make it quite clear that Australia looks to America, free of any pangs as to our traditional links or kinship with the United Kingdom." Defying Churchill, Curtin brought the Australian divisions home to fight in the Pacific theater. Events moved swiftly. General Douglas Mac-

Arthur went to Melbourne and then Brisbane as Supreme Commander of the Allied Forces in the Southwest Pacific. By May 1942, only five months after Pearl Harbor, American and Australian ships and planes shot the Japanese out of the Coral Sea, off Australia's northeast coast. For the rest of the year, and through 1943, Americans and Aussies struggled with Japanese troops on New Guinea, winning the eastern end so that MacArthur could begin the island-hopping that led to victory in the Pacific.

The war period was a time of solid success for Labor. The party put over conscription for overseas service with hardly a murmur. Essington Lewis, brilliant manager of Broken Hill Proprietary Ltd., the country's biggest mining and steelmaking firm, took charge of the wartime economy, and Australia learned to make machine tools, sulfanilamide, optical goods, and airplanes. To finance the war effort, the federal government took from the states the power to levy income taxes, a much needed advance toward stronger federalism.

This good record inspired voters to keep Labor in power after the war ended. Curtin died in 1945, and his successor Joseph Benedict ("Ben") Chifley, a former locomotive engineer: (1) brought off a carefully planned transition to peace, establishing many veterans on farms; (2) launched Australia's car industry; (3) started the huge Snowy Mountains Scheme for irrigation and hydroelectric power; (4) opened new universities in Sydney and Canberra; and (5) began the immigration policy that has recast Australia.

Suddenly and disastrously, Labor's idyl fell apart. Chifley had tried to nationalize all banks, jangling conservative nerves. Churchill's Iron Curtain speech at Fulton, Missouri, reminded Australians that Communist influence was dangerously strong in many unions. A coal miners' strike caused power shortages. And a refurbished opposition party, operating again under the Liberal label, was waiting in the wings with an eloquent if arrogant leader, Robert Gordon Menzies, lusting to take over the government. At elections in 1949, Labor toppled, and Menzies' Liberal-Country coalition took power.

Labor has never returned to power.* To understand why, we must now peer a little more closely at Labor's membership and ideology. Obviously, Labor's main constituency is the membership of trade unions—53 percent of the entire work force of 5,535,000—but the party is large enough and disparate enough to include a large left-right spectrum. The party's traditional

---

* This account, showing that Labor ruled the federal government only three times between 1910 and the present, may not seem to bear out the party's critical importance. But it was under Labor governments that the world wars were fought and the Depression was faced. And over the whole period Labor overwhelmingly dominated all state governments except those of Victoria and South Australia.

philosophy is heavily qualified socialism. Its "objective," announced in 1921, goes like this: "The democratic socialization of industry, production, distribution and exchange—to the extent necessary to eliminate exploitation and other anti-social features in those fields." But another declaration issued simultaneously, states that the party does not seek to abolish private ownership. The party is generally pacifist, holding that wars are started by capitalists, not workers.

Not many Roman Catholics in the early part of this century cared for this leftish program. But the prelate who dominated Catholicism in Australia from 1884 until his death in 1911, Francis Patrick Cardinal Moran, was an enlightened and foresighted man, who gave his imprimatur to the Labor movement, despite the predominance in it of Protestants and freethinkers. The net of this was to attach the great mass of Irish-descended Catholics in Australia firmly to the Labor Party, whether or not their ideologies quite jibed. Historically Catholics, though only a fourth of the electorate, have provided Labor with one-third of its vote. Two-thirds of all Catholics vote Labor. Any state or federal cabinet controlled by Labor is likely to contain a *majority* of Catholics.

The issue of Communism in the Labor Party provided the divisive factor. The Communist Party is tiny, but it has managed to appeal to militant Laborites who have not forgotten the days of strikes and lockouts, and the Depression. Solidly supporting the war effort from 1940 until 1945, the Communists proceeded to win great influence in many unions: ironworkers, sheet metal workers, engineers, sailors, clerks, railwaymen, dockworkers.

To a devout young Catholic lawyer in Melbourne, B. A. Santamaria, this infiltration was profoundly alarming. Using the same discipline and persistence as the Communists themselves, he initiated a movement, operating through secret union cells, to pry loose the Communist hold on union after union. Santamaria and his cohorts became a conservative force in the Labor Party.

The inevitable effect was to split the party, and the split was then greatly aggravated by the celebrated Petrov case. The leader of the party in 1954 was Herbert Vere Evatt, intellectual, lawyer, and radical thinker, whom the world had come to know well as an influential Foreign Minister at the founding of the United Nations. Now he was contesting elections with Prime Minister Menzies and the Liberal-Country coalition. With fifteen days to go before the voting, Menzies sprung on the nation the fact that Vladimir Petrov, third secretary of the Russian Embassy, had defected and turned over to the government documents reflecting on Labor, and particularly on two associates of Evatt. The well-timed blow may have cost Evatt the election, since Aus-

tralia was in a state of anti-Communist hysteria comparable to the McCarthy era in the U.S.

Anyway, Evatt thought that it did. Evatt was by no means a Communist, but in his fury at losing he rashly decided that *all* anti-Communists were his enemies, including Santamaria and his movement. Until then the movement had remained virtually unknown to the public because it had been felt that to expose it would stir up anti-Catholicism, always a potentially explosive political issue. Evatt stripped open the story of the movement and its cells. Labor split, and fell from power in several states. At length the anti-Communist, Catholic elements of the party cut out completely, and formed the far-right Democratic Labor Party. The 10 to 20 percent of votes that the DLP removed from the Australian Labor Party (to use the full name of the traditional party) account for two decades of losing to Liberal-Country coalitions.*

Menzies and the Liberals

Thus the rest of Australia's political story belongs to the Liberals, and largely to the urbane, royalist, fear-inspiring grandson of a gold-rusher who led them, Robert Gordon Menzies.

The Liberal Party represents order, respectability, the Establishment.

---

* The mechanism that the DLP employs to deprive the ALP of votes is the excruciatingly complicated Australian preferential voting system. Suppose five candidates are running for Representative. The voter does not mark "X" in the square opposite his choice; rather, he writes numbers, from one to five, opposite all the candidates' names in the order of his preference. In practice, the count may show that no one of the five has an absolute majority of first-preference votes. In that case, the candidate with the fewest first-preference votes is eliminated, and the second-preference votes on those ballots are distributed among the remaining four candidates. If, after the second-preference votes are added to the first, there is still no candidate with a majority of the total ballots cast, the process is repeated: the ballots of the candidate trailing at this stage of the computation are divided up on the basis of second preference. If he inherited some second-preference votes from the first man eliminated, these are now redistributed on the basis of third preference. And so on. Ultimately one of the five gets an absolute majority, even if his votes are composed of first, second, third, fourth, and even fifth preferences.

Senate elections, involving the choice of not one but perhaps five candidates from a list of, say, fifteen, are, God help us, *much* more complicated, using the principle of proportional representation. Australian elections can take days to count. Voting machines cannot be used. Most Australians, especially newly arrived immigrants, do not understand the system in full detail.

The DLP operates by ordering its members to vote for DLP candidates on first preference, and for Liberal candidates on second and other preferences. Few DLP candidates ever win, but the party's votes count anyway, through second preferences that work against the Australian Labor Party.

Large business firms rely upon it and finance it. It is a private-enterprise party that may regulate, but will not nationalize, business; it fosters trade and the stock market. Conservative as it is, the Liberal Party differs only by a few degrees from the Labor Party on many of the crucial issues. They stand close to one another on anti-Communism and White Australianism, and both support large-scale immigration and social welfare. Australian voters prefer choices of emphasis rather than diametrical opposites.

Menzies—born in 1894 in the Victorian countryside, educated at the University of Melbourne, a rich barrister before entering Victoria state politics—was Prime Minister as Australia entered World War II, and he was tottering. The House of Representatives was so closely divided that a mere smidgen of rebellion in Menzies' party was enough to knock the pins from under him. He fell, rather ignominiously. "It was a terrible thing to be told that your services were no longer required, at a crucial time in the world's history," he recalled later. During Labor's wartime rule, Menzies pondered deeply and formulated a proposal for a "true revival of Liberal thought which would work for social justice and security, for national power and national progress and for the full development of the individual citizen though not through the dull and deadening process of socialism."

In 1949, amidst waning confidence in the Labor Party, Menzies came to power again. Eighteen years later, the Washington *Post* called Menzies "probably the most completely successful public man of his time." Britain's Harold Wilson compared Menzies to Churchill. Sufficient facts exist to argue this case. Between the beginning and the end of Menzies' term: (1) national income grew from $4.4 billion to $20.8 billion; (2) the population grew from 7.9 million to 11.5 million, because of well-promoted encouragement to immigration; (3) university enrollment grew from 32,000 to 80,000; (4) the percentage of homeowners increased from 60 to 76; and (5) car ownership rose from one car for every twelve persons to one car for every four. Menzies and his Liberal Party started out their long regime operating under the hard-nosed theory that a bit of unemployment, say 3 or 4 percent, was salubrious. The uproar when the jobless rate reached merely 2 percent* forced Menzies to a policy of full employment through measures that others might have called socialist but in any case worked to keep the populace contented. Great foreign investments in auto-making, petroleum, cattle, farming, mines, and chemicals poured in, contributing to euphoria (while raising worries about the pinch of foreign control). Menzies turned Canberra from country town to flourishing capital. He devised federal help for universities, and saved them from going to ruin. He developed a much-needed Insti-

---

* In 1971 joblessness in the U.S. hit 6 percent.

extracts concessions for its members—for example, loopholes that let agriculturalists dodge most income taxes. The leader of the party from 1958 until 1971 was John ("Black Jack") McEwen, horny-handed and histrionic, who had been in the House of Representatives since 1934, often serving as a minister and latterly as Deputy Prime Minister. He amplified the party, appealing to industrialists by urging tariff protection for them. His successor is a tanned and hawk-eyed farmer (fruit, dairying, pigs), forty-two years old, whose name is Douglas Anthony.

The Short Regime of Harold Holt

The succession went easily to Menzies' choice, Harold Holt, then fifty-eight, an associate of Menzies for more than thirty years, who had been standing on tiptoe for the No. 1 job for eight years. He served not quite two years, and then one day he went swimming and was never seen again.

Holt was born in Sydney, son of a theatrical impresario who upon divorcing his wife sent the boy at the age of ten to a tough private school in Melbourne. He was barely out of law school at the University of Melbourne when he joined Menzies and others in a rightist organization called Young Nationalists, and he won a seat in the House of Representatives at the age of twenty-seven. Menzies put Holt in the cabinet in 1939, but Holt resigned in a few months to enlist in the artillery. One afternoon he was performing the lowliest of army chores, cleaning latrines, when a message came from Menzies that, like it or not, Holt had to return to one of the country's highest jobs, a cabinet post. Ironically, Holt then concluded that Menzies was not good enough to be Prime Minister, and helped to shaft him on that famous day in 1941 when Menzies was forced out. Usually an unforgiving man, Menzies in this case overlooked the slight, and on return to office brought Holt to the top by giving him important assignments.

Holt proved loyal and patient, and based his political stance on the Gallup Poll. "Vacillation can be part of the technique of achievement," he said. As Minister of Immigration, he humanized some of the more egregious aspects of the White Australia policy. Australians liked him because he had "no complexes."

In office he turned out to be as pro-American as Menzies was pro-British. He stood on the White House lawn and said that he would go "all the way with LBJ." He took a hard line on Vietnam, putting over conscription to keep the army big, and sending a couple of extra battalions to fight in the war. "It's bad luck for the Vietnamese that the world power conflict is being waged on their territory," he said. This proved a popular formula. Holt

won the 1966 election easily by pinning an anti-American label on Labor (which opposed the war and the draft), getting a boost from a tour of Australia by President Johnson. But, surprisingly, Holt reversed Menzies' attitude of indifference to Australia's Asian neighbors, and toured them as far as Cambodia; he also lowered immigration bars a bit. Possessed of a very Australian respect for a man's right to leisure, he would sometimes disappear for two days at a time; once his aides could not find him to deliver an "eyes only" message from President Johnson. Holt did not run the cabinet very firmly; quasi conspiracies to bring him down got started; the Liberals lost two by-elections and a Senate election.

Holt was an athletic man, a surfer and swimmer, and one Sunday morning in December 1967, after a tiring cabinet session, he went to the beach near his seaside house fifty-nine miles from Melbourne. The waves were ten feet high, and only one of the four neighbors who were with him even dared dip into the water. Holt, perhaps out of bravado, dived in and swam strongly out, just as the tide turned. He disappeared about five hundred yards from the shore, and his body was never found. Rumors may still be heard that this was suicide, but Canberra newsmen don't believe them—he was not by any means the type, they say. They guess that he was caught in a riptide and swam out with it, as surfers often do, intending to turn and get carried back; instead, handicapped by a recently injured shoulder, he was swept away. President Johnson made the long trip to Australia again to attend the funeral of his admirer, and in Djakarta, President Suharto went to a mosque to pray for the man who had made Australia a bit more neighborly toward Indonesia. Later the American Navy christened a new destroyer escort the U.S.S. *Harold E. Holt*.

John Gorton's Rise and Fall

The rise of John Grey Gorton, the next Prime Minister, was an intricate political gavotte. Holt's sudden death left no one neatly groomed for the job, as Holt had been. William McMahon, a longtime M.P. from New South Wales raised at this juncture to treasurer and deputy leader of the Liberal Party, stood out as the ranking Liberal; and Paul Hasluck, a prominent cabinet minister since 1951, had ambitions for the job and a strong following in New South Wales. Unbeknownst to them, a coalition of forces based in Victoria decided to attempt the nomination of the Victorian Gorton, then a Senator. The first step was to eliminate McMahon, which was tidily accomplished by John McEwen, the leader of the Country Party. McEwen, serving as caretaker Prime Minister after Holt, announced,

giving no reasons, that he would not serve in a government with McMahon, who then had no choice but to withdraw. It is said that McEwen was grateful to Gorton for not running against him some years previously in a local election in Victoria. Gorton was now in a position to compete for Liberal Party leadership. He defeated Hasluck, stepped up to the prime ministership, and then elevated Hasluck to Governor-General (by the customary process of recommending him to the Queen). This was patronage on a grand scale, and full of irony, for in effect Gorton made his defeated opponent into head of state, toothless as that job is in the British-Australian system. Not every Australian liked the move—the appointment of the Governor-General is supposed to be above politics. Since the Prime Minister is expected to be a member of the House of Representatives because legislation mostly originates there, Gorton resigned from the Senate, ran for the seat left vacant by Holt, and was duly elected.

The leader that Australians were so surprised to find in power turned out to be a bull in a china shop. In manner he combines the homespun and the patrician. His father and mother—he a prosperous orange grower, she a poor Irish girl—were not married; Gorton recently demolished a critic by saying that he had "achieved a status by his own efforts which I hold through action not by me but by my parents." His caved-in countenance, the result of being shot down by the Japanese off Singapore when he was a Royal Australian Air Force pilot ("My face got rather mixed up with the instrument panel"), is a political asset both as common-man homeliness and as evidence of bravery.* His patrician component comes from the schools his father sent him to: Geelong Grammar in Victoria and Brasenose College at Oxford, where, favoring Etonians as friends, he replaced his Australian twang with a cultivated British accent. In private, incidentally, the homespun side of Mr. Gorton gives way completely to the patrician. He is alert, graceful, aware, attentive, and humorous. He is married to the former Bettina Brown, of Bangor, Maine, whom he met while she was a student at the Sorbonne and he was touring Europe with her brother, an Oxford classmate. During World War II she ran the family orchards in Victoria, packing oranges, driving a tractor, irrigating, and keeping books. Since then, at Australian National University, she has intensively studied Indonesian language and literature, and once made a broadcast in Indonesian to the women of that country.

Despite his elitist English education, Gorton says he is "Australian to the bootheels," thus distinguishing himself from Sir Robert ("I'm British to

---

* Later Mr. Gorton crashed again, on Melville Island off the north coast of Australia, and lived there for ten days, like Robinson Crusoe, on turtle eggs and swordfish.

the bootheels") Menzies. A tentative Australianism was the trademark of his term of office. He proposed to change the national anthem from "God Save the Queen" to "Waltzing Matilda." He aimed at a little less subservience to the U.S. and Britain. He tried to make foreign investment profit Australia more, by letting local investors have a share of the action, and by discouraging foreign takeovers of prosperous existing industries. He doubted that Australia's minerals should be dug up and exported as rapidly as possible, and called himself an "economic nationalist." Domestically, he tried some changes, diverting some of the country's new wealth to the betterment of health and education; he faced up to the fact that Australia has some poverty, among Aborigines and New Australians and others, and proposed to fight it. He battled the Australian Medical Association, attempting to head off a raise in doctors' fees. He started a government-run Australian Industrial Development Corporation. He proposed to make the cities more aesthetically attractive and amenable, with particular concern for the monotonous suburbs.

Australians have a saying that all their Prime Ministers are either prima donnas or larrikins—and Gorton played the role like a larrikin, roughshod and impetuous, stressing his homespun side. At the first, this served to get him votes, but then he began to appear uncertain, gawky, weak and bully, ing at the same time, slow to think through issues, unable to accept discipline and work. Speaking on television, "a mixture of four parts nitrogen, one part carbon dioxide would pass out of his lungs, through his vocal cords, but the result lacked meaning," said the *Bulletin*.

The Americans found that they did not like Mr. Gorton nearly so well as Mr. Holt. New Zealand found him confusing and difficult. Mr. Gorton began to pressure the Arbitration Commission, chief tribunal for wage settlements since the times of Justice Higgins. The Prime Minister's goal was to keep down inflation, running at 4 percent a year, but it is bad form for the federal government to strong-arm the Commission. He fought state governments over control of offshore sea-bed resources, and turned them down cold when they asked for more revenue-sharing. Gorton's private life excited gossip. He took a nineteen-year-old girl, daughter of an opposition Senator, to a midnight party at the U.S. Embassy and stayed until three o'clock talking to her and no one else. He also visited Judy Garland's daughter, Liza Minnelli, in her dressing room at a Sydney night club, which was enough to set Australian tongues wagging.

In the November 1970 Senate elections, Gorton's Liberals lost eight percentage points, fetching up with only 35 percent of the total—quite a feat considering that the Liberals had unprecedented prosperity going for them.

Party backbenchers, especially those of the right, grumbled that they not only had a "socialist" for a Prime Minister, but also a vote-loser. It took only a minor blunder, in March 1971, for Gorton's enemies, some of them in his own cabinet, to get him.

Responding to a scandal in the army (of no importance to us here), Gorton telephoned the army chief of staff, saying, "I think you'd better come over here and see me." In the rigidly hierarchical Australian cabinet system, this summons bypassed the Defense Minister, Malcolm Fraser, and Mr. Fraser became furious. Gorton's going directly to the general, Fraser felt, made him look a fool and seriously damaged his career. After steaming for a few days, and after a hostile session with Gorton, he resigned, charging that Gorton had been "disloyal" to him. The Prime Minister was staggered, and set about heading off the resignation, which would certainly bring a ballot among Liberal M.P.s and Senators on whether he should continue to lead the party.

Through all this William McMahon, the Foreign Minister, stayed quiet, sensing that his ambitions were about to be fulfilled. Gorton summoned Fraser for a final appeal, but Fraser held firm. That night the party gathered for a vote, McMahon sitting just back of Gorton. The division was 33 to 33. It fell to Gorton, presiding, to cast the deciding vote. He voted, of course, against himself. On a second ballot, McMahon was elected leader and Prime Minister, and Gorton (running against his nemesis, Malcolm Fraser) was chosen deputy leader. "Which portfolio do you want?" McMahon asked the defeated Prime Minister. "Defense," Gorton replied, and thereby Mr. Fraser lost his job. But five months later Mr. McMahon fired Mr. Gorton from the Ministry of Defense, charging that he had "breached cabinet solidarity" by writing some newspaper articles; and the next week the Prime Minister put Mr. Fraser back into the cabinet as Minister for Education and Science.

William McMahon and the Established Values

Billy McMahon, a tough little sparrow, proposes to keep Australia going, and his party running it, not by innovation but by managerial competence. He has high dedication, energy, and technical proficiency in the politician's trade. Australia, in his opinion, has gone exceedingly well under two decades of conservative government, and the way to improve is to move it further to the right. In an interview, he specifically and pointedly refused to talk about John Gorton's administration, but it is easy to surmise that Gorton was too disheveled and radical for him. Years ago Mr. McMahon

defined Liberalism rather negatively as "avoiding the socialist threat," and when he became Prime Minister, he announced that he was still "very anti-Communist and very anti-socialist."

Mr. McMahon is (1) a Sydney man, succeeding three Liberal Prime Ministers from Victoria, and (2) sixty-three years old, an advanced age for a Prime Minister to take office in Australia. He comes from a family with business connections and money, which enabled him to go to Sydney Grammar, a famous 114-year-old private school. After a law degree at the University of Sydney, and a few years of practice, he enlisted in the army in 1939 and rose to major. Next he toured England, France, the United States, and Canada, and returned to study economics at the university, reading Marx, Keynes, and Schumpeter. Thus prepared, he turned himself single-mindedly to politics. He won election from a New South Wales constituency in 1949, and got ministerial rank in only two years. He was widely disliked by other ministers—but for the right reasons: he would go to cabinet meetings not only fully informed about his own department, but quite often much better briefed on other departments than their respective ministers were. Brash, ambitious, pushy, tactless—these were the adjectives applied to him then. But his talent and intelligence were undeniable, and he rose to deputy leadership of his party.

Defeated by Gorton for the prime ministership in 1968, McMahon accepted a senior-minister job as Treasurer, and proceeded to do what he does best: apply himself to mastery of the job. He ran the Treasury so well, and seemed to be taking such satisfaction in it, that Gorton waspishly moved him over to be Minister for External Affairs—to see if Billy might break his head on that one. Instead, McMahon turned in another performance of high competence, and in the process got the colonial-sounding title of the job changed to Foreign Minister. His party picked him to replace Gorton because he was the most experienced man around.

In office, William McMahon has seemed far right and uptight. Businessmen regarded him as "better than a new nickel mine." The Labor opposition labels McMahon a "big business" Prime Minister, "the most reactionary since the 1920's." He seemed obsessed with affirming Australia's support for Chiang Kai-shek and Taiwan, leaving himself off-base for dealing with such events as President Nixon's announcement that he would visit Peking, and the subsequent inevitable admission of the People's Republic of China to the United Nations. Belatedly and grudgingly, Mr. McMahon agreed to "explore the possibilities of establishing a dialogue with the government of the People's Republic." Even farmland conservatives deplore such recalcitrance; they want to sell wheat to China.

For the *Sunday Australian*, in May of 1971, Mr. McMahon wrote out his

views on Australia's future. It will be a place where per-man output is higher and where strikes are less frequent: "Inside our community some reject the ethic of work and the established values." Traffic jams, traffic deaths, and "pockets of poverty" will be attacked and reduced. The government deplores South African apartheid, but it also deplores demonstrators who use tours of South African Rugby teams as occasions to protest apartheid—as they did so violently in mid-1971 that a tour of South African cricket players was canceled. A committee of seven ministers will tackle the problem of bringing the Aborigines into the Australian society without wrecking their culture. Immigration policy will stress fewer numbers and higher quality. McMahon's curious statement omitted any mention of education, health, Vietnam, or foreign investment policy; it bore out strikingly charges that he lacks vision.

The Prime Minister is a man who appoints committees and works through channels. One of his first acts was to raise the number of ministries from an already inflated twenty-six to twenty-seven. He believes in lengthy cabinet meetings in which every member gets his say; unlike Menzies, he also holds that backbenchers have a right to know the facts and reasons behind cabinet decisions that they are asked (or, perhaps more exactly, ordered) to approve. He is rough on subordinates, phoning them at any hour with assorted requests or demands, all expressed in a cultivated and well-modulated voice, and terminated with a half-humorous "R-r-r-ight." Many of his aides resent his constant prodding, but it results in a high quality of work, and McMahon seems genuinely pleased with them. T. M. Fitzgerald, a political writer, suggests that this may be a case of the irritation that produces the pearl in the oyster.

Six years ago, when he was fifty-seven, McMahon, until then a bachelor, married a stunning girl named Sonia Hopkins, then thirty-three, who at five feet nine is two inches taller than he. He has let his sideburns grow long, and the McMahons make a dashing pair. They have a daughter, Melinda, and a son, Julian, who have between them reduced this steel-minded, urbane bantam of a man to the customary dimensions of a proud father. He phoned them just after winning his country's highest office. "Guess what Daddy is?" he said.

Labor's Blueblood Leader

If Mr. McMahon chooses to run a mere holding operation and the Liberals can produce no one better, the government of Australia might return in one way or another to Labor. This party has gradually moved from

lunch-bucket laborism to college-level liberalism. Among Laborites, foreign affairs now compete for attention with the acts of the Arbitration Commission. This new attitude is expressed and effected by Edward Gough Whitlam, leader of the party since 1967.

Gough (pronounced "Goff") Whitlam, who was born in 1916, is a bit blue of blood for a Labor man, but of course he has his job precisely because the party wanted a leader of cultivation and broad interests. He is very tall, always a commanding asset in an aspiring statesman, and his aristocratic grooming complements an arciform nose and George III lips. He studied law at the University of Sydney and "took silk" (joined the bar). His father-in-law, a New South Wales Supreme Court Justice, had close links with the Labor governments before 1949, which led Whitlam to contest and win a seat in Parliament in 1952, whence he rose to the leadership. Old-style Labor politicians, used to invective in the Billy Hughes style, questioned Whitlam's unimpassioned, lucid approach to issues. Still, Whitlam once lost his temper sufficiently to apply a gross epithet to William McMahon on the floor of the House, and to toss a glass of water at the Liberal Minister for External Affairs of the time, Paul Hasluck.

Up against Gorton in the 1969 Parliamentary elections, Whitlam managed to gain an impressive seventeen seats, and he similarly scored in the 1970 Senate election. Whitlam and his Laborites favored withdrawal from Vietnam, an end to the draft, and recognition of Communist China. Playing on a commonplace public apprehension, Whitlam charged that the Liberals were letting foreign investors turn Australia into "one vast quarry."

When McMahon took office, Whitlam tried an immediate no-confidence motion, losing 62 to 58. Then he stung McMahon by responding fast when Premier Chou En-lai showed, by inviting American and Australian Ping-Pong players to China, that he favored Chinese contact with the West. A delegation of Labor Party bigwigs, including Whitlam, went to Peking while the Prime Minister, caught in the "China lobby" thinking of the early 1950's, continued to fulminate about Australia's immutable obligations to Taiwan.

Mr. Whitlam's most mountainous task has been to weld Labor into a unified party of the left, specifically by reform of the weak Victoria branch, and he has done it effectively—barring the fact that the rightist Democratic Labor Party remains outside the mother party's folds. Labor, after being completely out of state government leadership, came to power in South Australia and Western Australia early in 1971. Unluckily for Mr.

Whitlam, he may not reap the rewards of his work. A rival has risen. He is Robert James Lee Hawke, forty-one, president of the Australian Council of Trade Unions.

Bob Hawke is an electric-minded man with a mop of wavy black hair—wavy sideburns too, a Teddy Kennedy look. He comes from the outback, Bordertown, Western Australia, where his father was a Congregational minister. He was a Rhodes scholar, and while at Oxford (arts, law, *and* economics) made the *Guinness Book of Records* by drinking two and one-half pints of beer in twelve seconds. His thesis was a study of the Australian wage arbitration system, which made him just the man, in 1958, to join the staff of lawyers that the Council of Trade Unions keeps for arguments before the Arbitration Commission. He rose to leadership of the unions in 1970.

Hawke has acted on the principle that union goals should be social change, and that the unions have the numbers and power to bring about such change without working through Parliament. He showed how this might work by a spectacular coup in 1971. The Liberals in Parliament had long blocked Labor's efforts to outlaw price-fixing by manufacturers, which keeps prices up and kills competition. So Hawke put the Council of Trade Unions into the retailing business by buying a share of a department store, which then began to demand the right to sell at discount prices, familiar enough to Americans. Specifically, Hawke zeroed in on British-based Dunlop's, maker of tennis rackets, sneakers, mattresses, and swimming-pool equipment. When Dunlop's refused to allow its products' retail prices to be lowered, the unions struck, refusing to move Dunlop's cargo and freight. In less than twenty-four hours, the limousine of the Dunlop's chairman appeared at Hawke's office, and the firm capitulated, allowing the union store to price Dunlop's products at any discount it chose. The Liberals in Parliament, seeing how the wind was blowing, hastily enacted a bill against price-fixing. All over Australia, competition increased, prices dropped, and consumers benefited.

Hawke would like to carry union initiative much further—at least into cheaper time-payment deals, building societies, perhaps banking and insurance. He may move into politics and shake up the Labor Party the way he has shaken labor itself. Australia is quite generally disenchanted with its gray, cautious, game-playing politicians; Hawke would be a fresh and youthful face. His mother recalls that when she was pregnant with Robert, her Bible fell open one day to Isaiah 9:6, which reads, "For unto us a child is born, unto us a son is given: and the government shall be upon his shoulder."

A Pair of Governors-General

A wise and cosmopolitan Australian is Richard Gardiner Casey, who aspired to be Prime Minister and became Governor-General. Casey, grandson of a pioneer Australian doctor, was at Gallipoli and France in World War I and after the war became liaison officer for Australia at the British Foreign Office. He served a decade in Parliament with the equivalent of the Liberals, losing out in the end to Menzies for Prime Minister. Menzies then sent Casey to Washington as first Australian Minister to the U.S. When the Labor government, in mid-war, seemed cool to Casey's staying in that job, he consulted with Winston Churchill and got a place in the British Cabinet as Minister to the Middle East, living in Cairo. That led him to the governorship of Bengal, certainly one up on Menzies, then the obscure leader of the opposition. But by the time Casey returned to Australia, Menzies was leading a Liberal Party that looked a sure winner, and Casey had to settle, in subsequent years, for lesser cabinet posts. He left Parliament in 1960, and Menzies saw to it that Casey got a peerage, and then appointment as Governor-General, the third native-born Australian to hold that post. His term expired in 1968; Lord Casey is a patrician, worldly in the best sense of the word, and surprisingly enough both he and his wife can pilot a plane.

*

Paul Meerna Caedwalla Hasluck, the present Governor-General, was born in Perth in 1905, and his parents were members of the Salvation Army. In his early career he was a newsman, university teacher, and civil servant in the External Affairs Department. As head of the Australian Mission to the United Nations after World War II, he clashed with the Minister, Herbert Evatt. Hasluck thereupon ran for Parliament on the Liberal ticket, and was a minister for most of the next twenty years.

Sir Paul (he is a Knight of the Grand Cross of the Order of St. Michael and St. George) is a meticulous, cautious, intellectual man, a poet moreover, with a reputation for coolness that he resents. "Just because I don't giggle or dig people in the ribs all the time, outsiders say I haven't got a sense of humor," he growls. "You go around my electorate and you'll find out what a warmhearted old fellow I am." As Governor-General, and therefore the Queen's surrogate in Australia, he has all the nonpowers of

the Queen herself. As a scholar, he is finishing two volumes of a history of Australia during World War II. He gets $20,000 a year, plus a noble residence in Canberra with twenty servants and three cars.

What Lies Ahead in Australian Politics?

• In the next few years, Labor will return to power.
• More distantly, Australia will become a republic.
• And some time after that, the republic will elect a non-British European, the son of immigrants, as President.

CHAPTER 9

Who Owns Australia?

> *I was out with my two sons rounding up cattle when the big bay I was riding stumbled and fell. I was there rubbing my leg when I picked up a rock. I thought to myself, "Strewth, that's heavy." And no wonder. The flaming thing was all copper.*
>
> —KIMBERLEY RANGE STOCKMAN TOM WILSON, IN 1970

FOR A MOMENT we will bypass 175 million Australian sheep and talk about Australia as the land of mines. This is the continent whose very seashores provide riches—sands that yield titanium and other rare metals. Minerals have taken over from wool as principal foreign money-earners for Australia. Mining production last year exceeded $1.5 billion. In ten years Australia has jumped from a hoarder of iron ore to one of the leading suppliers of minerals to the world.

Some points about the boom:

• Minerals occur in old rocks, and Australia has very old rocks. The great western shield is pre-Cambrian granite, a billion and a half years old, and many rocks to the east are Cambrian, not much younger. A cover of dunes and soils keeps the riches lightly hidden in most places, but where the rocks lie bare many a mineral find has been made casually. The extent of Australia's mineral riches was unknown until the last decade owing to insufficient prospecting in desolate country. "By the mid-seventies Australia is likely to be the world's largest exporter of a whole range of industrial raw material—iron ore, alumina, bauxite and black coal, lead and zinc—and to be a major exporter of copper, nickel and beach sand metals," says the *Times* of London.

• The vital market that makes the mining possible is Japan, which on few resources of its own has risen to become the third-largest industrial power in the world and possessor of the world's biggest iron and steel firm. The two booms are complementary.

• Most of the giant international mining companies of the world are pouring capital into Australia. American companies mine and export Australian iron to Japanese smelters for conversion into steel that is then manufactured into products sent to America. Australians get taxes, royalties, and wages—and warn that they would like more of the profits and a bigger role in the whole process.

How the Boom Began

Australia has known rich mineral finds for more than a century. The gold rush of the mid-nineteenth century was followed by the conquest in 1883–1885 of Broken Hill, New South Wales (silver, lead, and zinc), the 1891 gold strike at Kalgoorlie, four hundred miles east of Perth, and the 1923 discovery at Mount Isa and Mount Morgan in Queensland (lead, zinc, silver, and copper)—as well as allied finds of iron ore and coal. Australia had an Institute of Mining Engineers as early as 1893, and a few years later was a recognized world leader in metallurgical skills. But all this was only a prelude. Between 1946 and 1966 various companies, prospecting in the west, found forty-two significant lodes of one metal or another. The predominance of iron ore among these discoveries relieved fears expressed in a 1938 law banning exports of the material. Landmarks of the boom were:

• The 1955 discovery of bauxite at Weipa in Queensland
• The discovery of iron ore at Mount Newman in 1957
• The 1960 discovery of Mount Tom Price, in the Hamersley Range 750 miles north of Perth
• The repeal in 1961 of the law banning the export of iron ore
• The discovery of nickel at Kambalda in 1966

*

Going with all this have been some breath-taking speculation and stock market scandals—we will say more about the market in a moment. But the boom is basically deep and solid. Iron is the hard core of it, and the

geographical site of most of the mining is "The Pilbara," which encompasses the Hamersley Range in Western Australia, and nearby points. Two major companies get iron ore out. One is Hamersley Holdings, which mines Mount Tom Price. Conzinc Riotinto of Australia Ltd. owns 54 percent of this operation, and Kaiser Steel Corporation 34.5 percent. They spent $300 million developing the mine, building towns, a railroad, a pelletizing plant, and a port—Port Dampier, on the "cyclone coast"—that can take 150,000-ton ships. The mine produced 13 million tons of ore in 1969, and will produce 37.5 million tons in 1974. It all goes to Japan.

The other giant is Mount Newman Mining Company Pty. Ltd., which we report upon in detail later. Among lesser firms is Mount Goldsworthy Mining Company (owned by Cyprus Mines Corporation, Utah Construction & Mining Company, and Consolidated Gold Fields Australia Ltd.), which ships through Mount Newman's port at Hedland, a bit up the coast from Dampier. Another major Pilbara project, still being built, is the mine, railroad, and brand-new port of Cliffs Robe River Iron Associates, which has Australian, American, and Japanese money behind it. Japanese steel mills have contracted to buy $1.2 billion worth of Robe River ore.

The iron boom adds up to glittering totals. The ore reserves may be 100 trillion tons, four hundred times as much as the known North American reserves. Japanese contracts call for the export of 510 million tons in twenty years, at a price of $4.5 billion.

Aluminum and Other Metals

On Cape York Peninsula, in Queensland, is a port called Weipa, a town of pink brick houses under tall eucalyptuses. All around it for miles, one has only to dig through three feet of overburden to reach rich red bauxite, an oceanic layer of it eight feet thick, 2.5 billion tons in all, said to be the largest patch of the stuff on earth. This mining town is feverishly felling the trees, scraping away the soil, and scooping out the bauxite to make Australia the world's largest exporter. Comalco, the corporation in charge, stores the topsoil and puts it back after mining the bauxite, preserving the environment to raise cattle and grow mahogany and teak.

At the port, a half-mile moving belt conveys the warty-looking bauxite to fifty-thousand-ton ships which take it, at a rate growing toward fifteen million tons a year in 1975, to Japan, Sardinia, and the refinery at Gladstone, three hundred miles north of Brisbane. This refinery, a pretty thing in pink, green, and silver, put up at a painful cost in terms of strikes and

balky labor, turns out 1,275,000 tons a year of the white powder concentrate called alumina (aluminum oxide). Alumina is exported, or smelted to aluminum in plants at Bluff, New Zealand, and Bell Bay, Tasmania, for use in Australia. Further huge deposits of bauxite at Gove, westward across the Gulf of Carpentaria from Weipa, are being mined at the rate of 700,000 tons a year, and a large alumina plant will go into production there in 1973.

Way out west, bauxite is also big. The Darling Range behind Perth contains big deposits, and the alumina plant at Kwinana, the industrial city near Perth, produces 1,040,000 tons a year.

*

Western Mining Corporation Ltd. was formed in 1933 to mine gold in Western Australia, but as its lodes ran out it began prospecting widely. Some ironstone outcrops near Kambalda, south of Kalgoorlie, caught the attention of company geologists, and they carefully sited a test bore. In 1966, first crack out of the box they struck nickel, a metal never before found in Australia in quantities profitable for mining.

Now Western Mining produces more than 600,000 tons of ore (4 percent nickel) a year from an underground mine of shafts and drifts. Since this sort of mining leaves supporting pillars of valuable ore, the company has developed an ingenious method of pumping sand and cement back into the mined areas, which provides sufficient support for the miners to return and remove the pillars. Side effects of Western Mining's exciting mine are these: (1) nearby Kalgoorlie, the wide-open Virginia City of Australia, slowly dying as its gold gave out, is booming again, complete with whorehouses; (2) Western Mining has built a $30-million nickel refinery at Kwinana and plans to build a smelter at Kalgoorlie; and (3) other big mining combines have arrived in the area to find and mine more nickel. Australia makes its coins out of its own nickel, but most of this metal is exported.

Main buyer? Japan.

Other minerals: *copper*, already an Australian staple, will begin to come in 1972 from the 900 million tons of ore on a mountaintop in Bougainville, in the territory of New Guinea, where $350 million is going into development; Queensland *coal* flows through Gladstone to Japan at the rate of three million tons a year; *uranium*, long produced at Mary Kathleen in Queensland and Rum Jangle in the Northern Territory, was found in considerable quantities in 1970 at Nabarlek, Ranger, and Jim Jim in the Northern Territory; Consolidated Murchison produces one-third of the

world's *antimony*; the beach sands of New South Wales and Western Australia provide rutile to turn into *titanium* for toughening steel and pigmenting paint; *gold* still comes from Kalgoorlie's Golden Mile, source since 1891 of $400 million worth; *phosphate* underlies a 358-square-mile concession in Queensland; *manganese* is mined on Groote Eylandt, an island in the Gulf of Carpentaria; there is a new "rush" for *opals* at Glengarry, New South Wales (Australia produces almost 95 percent of the world's opal).

Sober opinion is that industrious exploration in the future will find minerals worth many times more than those so far discovered.

*

The involutions of corporate ownership in the Australian mining industry are too complex to concern us, but there is one giant that cannot escape attention, because it spreads over iron, aluminum, copper, lead, and zinc. This is Conzinc Riotinto of Australia Ltd. (CRA), which is 85 percent owned by Rio Tinto–Zinc Corporation Ltd. of London. CRA owns more than half of the aforementioned Hamersley Holdings, and 45 percent of Comalco Ltd., which gives CRA a strong interest in Weipa bauxite, the Gladstone refinery, and the smelters at Bluff and Bell Bay. CRA mines lead and zinc through Australian Mining and Smelting at Broken Hill, and it mines copper through a majority interest in Bougainville Mining Ltd. Sir Maurice Mawby, who was born in Broken Hill, is chairman of this octopus, and he is a shaping force in Australia.

Broken Hill Proprietary Ltd.

Moving from mining toward manufacturing, we come to Australia's biggest corporation, surpassing CRA, and No. 31 on *Fortune* magazine's list of the world's largest companies outside the U.S. Its name is a paradox, for it owns no mines at Broken Hill. The explanation is that BHP (letters as common in Australia as GM in the United States, or ICI in Britain) was the corporate vehicle in 1885 of the fortunate finders of the bonanza at Broken Hill. But slowly lead-zinc mining went sour for BHP, partly through ill-managed conflicts with labor, and partly because the company found iron ore, intending at first to use it only as flux for smelting lead. In 1915 BHP turned from lead-zinc miner to steel manufacturer, running Australia's first blast furnace near the coalfields of Newcastle, New South Wales. Under the great engineer Essington Lewis, BHP then moved into coal, shipbuilding, shipping, heavy transport, and integrated iron and steel, and came to loom in the

Australian industrial landscape like Krupp or I. G. Farben in prewar Germany, or the Du Pont or Rockefeller empires in the U.S. Its steel, more than six million tons of it a year lately and the cheapest in the world, became the base of Australian industrial advance. Additional blast furnaces and mills were built at Kwinana, near Perth, at Whyalla, in South Australia, and at Port Kembla, just south of Sydney. Port Kembla will produce 5.4 million tons a year when a $300-million expansion program is finished in 1973. BHP also clings to its old lead smelter at Port Pirie in South Australia, which is the world's largest. The company is, furthermore, a major holder of Mount Newman Mining Company Pty. Ltd. in the Pilbara.

Strangely, this far-flung activity only partially explains BHP's current immense profitability—$69 million in the 1970–1971 financial year, $160 million projected for 1975. In fact, at the end of 1971, the company was cutting back the production of lead, because of low world prices, and of steel, because of reduced local demand. The money mostly comes from a new, rich, and lucky venture: oil.

Historically, petroleum has been an embarrassing absentee from the roster of Australian minerals. Diligent exploration had produced commercial wells only at tiny Barrow Island, off the Pilbara coast, and at Moonie in Queensland. Australian production was about one-ninetieth as much as Venezuela's. In 1968 refineries had to import $246 million worth of foreign petroleum, and pay another $90 million in freight to bring it in. The first glimmer of relief came in 1968 with natural-gas strikes from offshore drilling by BHP and Jersey Standard in Bass Strait, the ocean passage between Tasmania and the continent. Oil soon followed, in quantity. By mid-1971 Australia was producing 70 percent of its refinery input, mostly from BHP's subsidiary, Hematite Petroleum. The price was a modest $2.05 per barrel.

Oil has thus already surpassed steel in BHP's profits, but the company is expanding in other minerals, too. It has prospects in bauxite near Perth, nickel at Kalgoorlie, lateritic nickel in Queensland, more iron ore at Marillana near Mount Newman, and more coal in Queensland. BHP employs 54,000 people and is capitalized at $2 billion. The company is putting up a forty-one-story head-office building in Melbourne that will generate its own electricity from natural-gas-fired engines, the exhaust heat of which will provide the structure with heating and air-conditioning.

Oil refining, incidentally, is big business in Australia and a vexed one. All the refineries and distribution networks are dominated by foreigners except one, Ampol. In advertisements, Ampol makes the purchase of gasoline a question of patriotism. "How big a slice of Australia will Australians own when your kids grow up? In the excitement of our vast new-found natural wealth

aren't we inclined to give too much away? . . . In the oil marketing industry so vital to Australia's growth and development, Ampol is *the* Australian company. So *think* before you buy your next gallon. PUT YOUR MONEY WHERE YOUR COUNTRY IS."

A Word About the Stock Market

The Sydney Stock Exchange (Melbourne's is much soberer) has been described as "a casino where undisciplined and renegade gamblers run amok—and then weep on the croupier's shoulder when the chips are against them." Scandal after scandal hits it. Figures have sometimes been forged on the "scrip" ownership certificates issued by the exchange's computers. The exchange trades heavily in penny stocks, which skate up and down. Cores from test drilling of dubious mines are regularly produced to kite stock prices—although the "punters" (speculators) in the Exchange's galleries have a saying that the best way to kill a prospect is to drill it. The Geological Society of Australia charges that many of the drill-core assays that shoot stock prices up are made by nongeologists.

A typical case was that of Leopold Minerals—"leaping Leopold," they called it. Leopold had a "good address" among the big mines of the Pilbara region, and drilling produced a test core supposedly high in nickel. Leopold leaped. Then one of the company's directors demanded to see the core. From the management came a brief silence and finally a confession that the core had been lost. The trucker who brought it from Nullagine to Perth, it was explained, arrived when the company's offices were closed, so he left the core on the doorstep, and it disappeared. Leopold's credibility, and its stock prices, plummeted, dragging down, as usual, the rest of the market. Eight brokerages collapsed.

But no one in Sydney much cares about such scandals; the punters are looking for a "runner"—a stock that is rising, and they are indifferent as to whether by manipulation or real worth. Sometimes the volume on the Sydney Exchange is double that of Wall Street. On the other hand, the market sometimes goes into a bad slump, and such has been the case lately. A significant factor in this hurly-burly is that the government does not tax capital gains.

There is, to be sure, a reasonable justification for the highly speculative market. It brings out the hundred-to-one money, the real risk capital, that pays for all that give-it-a-try prospecting which occasionally strikes it rich. The search would go much slower otherwise. And thus the market can claim that it is doing just what it is supposed to do.

Sheep Raising

Camels are snobbish
And sheep, unintelligent.
—MARIANNE MOORE

Sheep certainly do not produce a meat that the world likes as much as beef, and their fleece must meet competition from synthetic fabrics. Fewer than 5 percent of the Australian people are pastoralists, stockmen, or farmers; the price of wool is low; government subsidies are supplanting what used to be an admirably free market. But Australia, to use the standard cliché, has "ridden the sheep's back" through all of its history until now, and relies on wool to earn foreign exchange vital to the country's balance of trade. The sheep still counts a lot. Let us go out on a big sheep run and examine the life style of this stupid, panicky, disease-prone, exploitable, submissive animal.

The scene is the property called South Burra, near Canberra, managed by Allen Sutton and owned by F. M. Diarmid. South Burra has 6,000 sheep on 2,500 acres, about two to an acre. The sheep weigh from 130 to 140 pounds, although rams go up to 220.

The critical time of the year is the shearing season. Four or five shearers, who go from property to property in groups, come to South Burra to work for a couple of weeks in the property's shearing shed, using power-driven shears. The shearer, dressed in blue undershirt and checked shorts, uses his elbows to clamp the sheep's forelegs for control, and shears it belly first, in such a way that the fleece comes in one piece, weighing an average of twelve or thirteen pounds. The sheep makes no objection. An expert shearer can shear 160 sheep a day, but the average is 125, one every three or four minutes for eight hours—from 7:30 to 5:30 P.M., broken up by an hour for lunch and half-hour rest breaks at 9:30 and 3:30. The pay is twenty dollars per hundred sheep. The number of these nomadic workmen—twelve thousand-plus in all of Australia—is declining, and sheep raisers worry about this.

Once shorn off, the fleece is graded into one of eight categories—graders travel with the shearers and have the most important and specialized jobs. The wool is pressed down in boxes, forming 320-pound bales, wrapped in jute from India and neatly stenciled.

"Do the sheep catch cold after shearing?"

"Yes, you can lose a lot of sheep if the weather is bad after shearing," says Mr. Sutton. "A sheep has an awful life." Indeed, sheep die with indecent frequency. If pushed hard while being herded, many will drop dead.

A perennial threat is foot rot, a virus that causes lameness and loss of appetite. The treatment has been to pare away the infected area and force the sheep to walk through a formalin bath. The highly contagious virus can be spread by the tires of a car, the soles of shoes, and even by waddling ducks. Sheep also have to be dosed three times a year for worms, and several times more for liver flukes. Sheep suffer from blackleg and pulpy kidney, and tend to get caught in brush fires.

Specialized breeding is by artificial insemination, the semen being captured in an artificial vagina skillfully held by a station hand while the ram mounts a ewe in expectation of normal copulation. At correct temperatures, the semen, kept in little labeled egg cups, lasts up to a week, and if frozen it can be kept six months and easily transported. The heat period for the ewes is seventeen to twenty-one days long, and it can be artificially induced by hormones. Good rams, which have their pedigree numbers tattooed in their ears, cost plenty—one brought $29,000 lately in Sydney—though the price for the average ram is more like $80 to $100. Most males are castrated, which makes them wethers. A breeding ewe costs only $14, and a good young sheep just $9. At about the age of seven sheep tend to grind their teeth down to the point where they cannot feed, and are then sold to slaughterhouses for $4.

Three-fourths of the sheep are Merinos, a breed which originated in Spain and spread to France, Germany, the U.S.S.R., the U.S., Argentina, and, most successfully of all, Australia. The first Merinos reached Australia in 1797, via the Netherlands and the Cape of Good Hope. Even the sheep not classified as Merinos usually have Merino blood, including the "comebacks," which are finer than the half-breeds. The rest are English breeds, used for fat lamb production. Merinos have fine wool, beautifully crimped, so that the fleece appears curly rather than rug-like.

*

Numerous as they are, Australia's sheep constitute only one-sixth of the world's wooled-sheep population. Yet they are so productive that they supply as much as two billion pounds of wool a year, 30 percent of the world's supply, and it is of such high quality that it constitutes 38 percent of the wool used in clothing. Fighting prices that stayed level or went down* while everything else rose, and competing against synthetics, Australian wool growers managed to *double* production between 1949 and 1969, through better breeding, disease control, use of trace elements in raising pasturage and

* The price in 1968 was 42 cents a pound, and in 1971, 28 cents.

feed, much government assistance, and high efficiency (the use of motor-cycles in herding is an example). The wool, 95 percent exported, mostly to Japan, earns around $800 million a year. In a sense, every city-bred Australian is counting on fifteen sheep busily growing wool in the outback to provide him with what he needs from abroad: industrial raw materials, munitions of war, capital goods, consumer items.

Grain and Cattle

Two notes about the other main products of the Australian land:

Beef is doing well. Traditionally, cattle raising has been an activity that went on untouched by human hands in northern Queensland. Now better cows are bred; they are transported by truck instead of being moved by weight-losing drives; they are getting grain for feed as well as grass. With the future of wool in doubt, many sheep raisers are going into cattle as a hedge. A U.S. beef-import quota provides a market for beef for hamburgers, TV dinners, and the like. Moreover, Australians themselves consume 204 pounds of beef per person every year. Incidental intelligence: a "poddy" is an unbranded calf, and a "poddy-dodger" is a rustler who steals unbranded animals.

Wheat is problematical. Exports in 1971 set a new record, because the Australian Wheat Board happened to find a good market in Egypt and other important customers in Britain, Japan, Iraq, and Iran. In past years, heavy buyers have been the U.S.S.R., the People's Republic of China, India, Indonesia, and the Philippines. But there is no one dependable market or set of markets. And the whole crop tends to fluctuate widely with the weather.

Other crops worth a mention are cotton, the production of which grew from 8,000 bales in 1959 to 145,000 in 1968; apples and dried vine fruits, of which Australia produces 7 and 11 percent, respectively, of the world's supply; and forestry, not only for lumber but for paper and pulp, in which a new technology for using short-fibered wood like eucalyptus permits the country to supply 70 percent of its paper needs.

Land devoted to agriculture in Australia exceeds a billion acres, three-fifths of the whole country—a figure misleadingly high, since more than 90 percent of it is not cropped or plowed or watered but used in its natural state. This untended land is "pastoral" country, which gets less than fifteen inches of rain a year but does support almost a third of the nation's sheep. The part of Australia that gets enough water to cultivate or graze intensively amounts to only a twentieth of the country's surface.

Other Industries

We export stainless steel products to Sheffield, tulip bulbs to Holland, spaghetti to Italy, ball bearings to the Ruhr, waffle-making machinery to America, safari suits to Zambia, and Irish stew to Scotland.

—FORMER PRIME MINISTER JOHN GORTON

Australia is as intensively industrialized as any nation in the world. Twenty-seven percent of the work force is in manufacturing—the same as the United States. The auto industry, employing one-eighth of all factory workers, turns out 425,000 cars a year. The commonest car is the Holden, a compact made by General Motors–Holden; others are adaptations of European models put out by GM (34 percent of the market), Ford* (19 percent), and Chrysler (12 percent), plus Volkswagens and Japanese cars. There is one motor vehicle for every 2.7 Australians. Large oligopolies dominate industry: one, two, or three producers account for all sales in steel, soap, sugar, newsprint, tinplate, wire, glass, industrial gases, and matches.

As Mr. Gorton points out, Australia exports a surprising amount and variety of manufactured goods. An interesting specific case comes from Favelle Mort Ltd. of Sydney, builders of construction machinery. This firm supplied the eight 170-ton "Kangaroo" cranes used in building the twin 110-story skyscrapers of the new World Trade Center in New York. The company plans to start a U.S. subsidiary. A similar case of Australian enterprise abroad is the chain of TraveLodge motels in the Western U.S. which are controlled by the TraveLodge firm in Australia. Ford Motor Company of Australia exported 23,500 cars last year.

Labor constitutes the most influential pressure on Australian industry. Wage disputes are supposed to be settled, and in fact largely are settled, by the Commonwealth Arbitration Commission, which is usually under heavy pressure from the Australian Council of Trade Unions to raise wages, and by the government and employers to hold them steady. The Commission's across-the-board award at the end of 1970 was 6 percent. The machinery for arbitration, it must be pointed out, does not stop Australian labor from striking often and capriciously, in the awareness that there is always another job somewhere else.

Such aggressive labor, it would seem, might price itself too high and bring on mass unemployment. Three factors keep this from happening: (1) the

* Which early in the century built Australia's first car, a Model T, in an old wool store in Geelong.

long-term strength and depth of the Australian boom; (2) a reliable 2.5-3 percent per annum rise in productivity per man-hour; and (3) the still-modest levels of wages, which average $4,370 a year.

Thus Australia is generally a country of full employment, or even hyper-employment—although an exception must be made for late 1971, when un-certainty over world trade brought a downturn to the economy and pushed unemployment up to 1.57 percent. (By comparison, unemployment in the U.S. was running around 6 percent.) Almost anybody can get a job. More men and women are needed every year. Immigrants can flood in and be put to work at once. The proportion of married women in the work force rose from one-fourth in 1965 to one-third in 1970. Ten thousand *children* are employed in industry. *The Australian* for November 30, 1970, carried *four hundred* detailed, pleading display advertisements seeking employees. Inflation, like unemployment, has not been a serious problem in Australia. It ran around 3 percent a year during the 1960's. Unfortunately, however, it began to creep up, and in 1971 reached 6.5 percent—approximately the same rate as that of the United States.

Economic Oddities

The entire areas of power, railroads, water supply, telecommunications, ports, and a substantial part of airlines, banking, radio and television, and coastal shipping, are left in Australia to the state and federal governments. The rail-roads have an absurd impediment: the colonial governments that built them copied gauges variously from Ireland (5 feet 3 inches), from England (4 feet 8½ inches), and from remoter places that used the cheap narrow gauge (3 feet 6 inches). For years this meant that goods sent from Brisbane to Perth had to be transshipped five times; now, slowly, tracks are being standardized at 4 feet 8½ inches. Ocean shipping is also a weakness: Australian ships carry only 1.3 percent of the country's trade with the world.

The post office's telephone division still charges only five cents for a local call, and offers an STD (subscriber trunk dialing) service equivalent to the American DDD (direct distance dialing). All mail in standard envelopes goes to any part of Australia by air if air service is available. For a little extra postage, the post office guarantees to deliver mail overnight between major cities. The insurance business is dominated by London, and criticized as monopolistic and restrictive. The biggest operator of retail stores in Aus-tralia is Woolworth's Ltd., with annual sales of half a billion dollars.

Aviation is prosperous—the number of air trips per capita is the second highest in the world. Fares average 6.4 cents per mile in first class and 5.4

cents in tourist. Government-owned Qantas (pronounced Quantas—the name is an acronym for the parental Queensland and Northern Territory Aerial Services) flies ninety thousand miles of routes to thirty countries with safety, brusque virile courtesy, and sausages for breakfast. It is the second oldest airline in the world, after KLM. The main internal airlines are the government-owned Trans-Australia Airlines and the privately owned Ansett-ANA. One is told that "this is the best country in the world for flying— low mountains, hardly any bad weather."

*

There is a palpable intelligence back of Australia's industrial and agricultural progress, and a little inquiry reveals that this intelligence is called CSIRO, for Commonwealth Scientific and Industrial Research Organization. It was founded way back in 1926, when some genius realized that small amounts invested in brainwork would return large dividends in farm and factory efficiency. CSIRO's scientists have come through spectacularly. A favored example of their work is the fight against rabbits. This pest was imported to Australia in 1859 by a Geelong farmer who thought rabbits would be good for food and sport. As rabbits will, they multiplied. In sixteen years, they occupied most of the continent. A fence was built clear across Australia in an effort to stop them. Ravenously, they consumed the grass intended for sheep; they became a classic example of the dangers of tampering with an ecology. CSIRO got wind of a highly infectious rabbit disease called myxomatosis, discovered earlier by a Brazilian doctor named H. B. Aragão, and in 1950 learned how to spread it, via host mosquitoes, in Australia. In the next decade the rabbit population plummeted—but not to zero, and those that survived became immune to the disease and must be fought in other ways, chiefly poison.

Probably CSIRO's most useful work has been in agriculture. It perfected a kind of alfalfa called Townsend lucerne which is Australia's main hay. It found ways to fight erosion, noxious weeds, animal diseases. Lately it has discovered a chemical that can be introduced into the fourth stomach of a sheep to make it grow half again as much wool, and another chemical that can be introduced into the first stomach of a cow to prevent bloat. CSIRO has also devised an effective vaccine against sheep foot rot, which has been costing pastoralists $16 million a year in treatment and production loss.

CSIRO years ago pioneered in the development of radar, and lately has been into such work as making diffraction gratings, extremely flat metal-coated glass plates with fine parallel grooves, as many as 32,000 to the inch, cut into

them. They are used in assorted optical instruments to break light into component colors. Some more typical recent projects are: the production of whiter wool and machine-washable woollen slacks, taking radio pictures of the sun, mapping bush fires from the air at night, measuring shadows. CSIRO's staff numbers about 6,500, and its budget is $60 million a year.

Who Owns Australia?

About one-fourth of Australian corporate assets are owned outright from abroad, a share that rises to two-fifths in the case of manufacturing. The degree of control, through partial ownership, goes well beyond these fractions. More than half of all mineral production is controlled by non-Australian interests. Twenty percent of the cattle land in the northern sector of Western Australia is owned by Americans. A billion dollars, more or less, from abroad pour annually into the economy, which runs to about $30 billion in Gross National Product. The British share of this investment is typically about one-third, the U.S. share about one-half. Total U.S. investment, in five hundred companies, is $3.5 billion. Japan is an increasingly massive investor in Australia.

The overwhelming benefit that all this investment has conferred on Australia is, quite simply, the boom—the very prosperity of the country, the key to its full employment and enriching immigration. "Without this stimulation there would have been no mineral boom—only at best a miserable little pop. . . . I can assure you that we are only at the beginning of a program of Cecil B. De Mille proportions," says Professor R. L. Whitmore. "We will be the top mining nation in the world in thirty years." Real growth in GNP is more than 5 percent a year and increasing. Moreover, the foreign investors bring with them essential technology, provide enormous payrolls, and pay taxes averaging 43 percent of profits plus an extra 15 percent on dividends taken out of the country. An unmeasurable but doubtless huge benefit of foreign investment to the Australian economy is efficient competition that keeps local industry on its toes. In return for these rewards, official policy is to allow almost unrestricted entry of foreign capital.

Nevertheless, Australians voice a great number of complaints about their foreign friends. Many have an uneasy feeling of being dominated, or of having had the country bought out from underneath them. Some specific worries are these: (1) It is pleasant to have a billion dollars pour in every year now, but it will be painful to the balance of payments when the investment stops arriving and the profits start going out. (2) Foreigners dominate the areas of greatest growth potential—autos, pharmaceuticals, oil refining

and marketing, minerals. And their profits are higher than they expect or get elsewhere. Sir Alan Westerman, chairman of the government-backed Australian Industrial Development Corporation, a bank that is supposed to be a counterweight to foreign investment, contends that the profits after taxes of three hundred U.S. firms have been 28 percent higher than the same firms earned in any other country. (3) Foreigners restrict Australian participation, both in investment and dividends, and in management. General Motors–Holden is wholly owned by the U.S. parent* and takes the position that the Australian who wants to invest can buy GM on Wall Street. Speaking for the incoming Japanese investors, the chairman of Mitsubishi Bank says that the Japanese, by contrast, prefer joint ventures that give Australians some of the action. Ford, British Petroleum, and even GM in 1970 named Australians as the chief executives of their Australian subsidiaries, but much of the ranking management still comes from abroad. (4) Many foreigners are not true developers, bringing capital, but merely entrepreneurs who take over existing concerns, raise money on the Australian market—and then send the profits abroad.

*

Is the Australian economy socialistic? Most decidedly not. It is a mixed economy in a country run for the past two decades by conservative governments prone to give business every possible break.

Does it work? Most decidedly yes. It is hard to quibble with a full-employment economy that provides one of the world's highest standards of living. But much of this is luck—finding mineral riches.

Will it continue to work? Not without a hitch. Wool faces a declining market just as the costs of producing it go up. The discovery of ore may outrun the discovery of markets.

Should Australians be worried? Not yet.

* For assorted reasons. Ownership gives GM all the profits, avoids antitrust problems, and gains the rewards due the company from taking the risks of starting the foreign operation—risks Australians (who in former days were timid investors) did not share. It also preserves quality control, secrecy, and techniques.

CHAPTER 10

Sydney and New South Wales

My first feeling was to congratulate myself that I was born an Englishman.

—CHARLES DARWIN, UPON VIEWING SYDNEY

SYDNEY HAS the capacity to stir loyalty, to arouse emotions, to make people endlessly question its identity. The place is less a city, say its admirers, than an experience. It is a mixture of history, seaside beauty, metropolitan bustle, carefree energy, and confidence. Sydney commands a position of pre-eminence not only in Australia but in the whole South Pacific. With one set of mental spectacles, you see in Sydney a reflection of England; with another, you see a New York of the Southern Hemisphere. Numbering three million in population, this city by itself contains almost one-fourth of all Australians. Sydney is Australia's oldest, biggest, busiest, and most effervescent city. It also has an awful lot of flies.

The city proper occupies only 11 square miles out of the 670 square miles of the metropolitan area, which is equal to Greater London. This gives room for 368 separate suburbs (as locally defined), which crowd in upon the city so closely that one can walk from central Sydney to the suburb of Woolloomooloo, for example, in hardly more time than it takes to spell the name. The city, capital of the state of New South Wales, sits on the east coast of Australia. Inland, numerous small towns and highways run for 150 miles west of Sydney. Sydney residents tend to dismiss the less urbanized areas of New South Wales beyond that as the "Far West." This state, even if one of the dinkier Australian states, covers an area of 309,433 square miles, which is 50 percent bigger than Spain.

*

The function of Sydney is really that of a port, exporting wool, wheat, flour, and meat. The city also manufactures aircraft, ships, earth-moving equipment, paper, electrical and communications equipment, agricultural implements, and railway rolling stock.

The harbor is spectacular. One tiresome routine in Sydney is listening to the citizens boast about the incomparable beauty of the natural setting. Let us settle the issue with dispatch. Naples is more picturesque, because of its volcanoes and islands and grottoes. Rio is more breath-taking, because of its sugarloaf mountains plunging into the sea. Hong Kong is more exciting. San Francisco outclasses Sydney in the size and daring of its bridges. Yet if all these ports outrank Sydney for the given reasons, Sydney outranks all of them in sheer length and crenelation of shoreline, in its thirty-six bays and coves around a deep, safe harbor. Sydney is an enchanting minglement of spits of land and tongues of water, of islets and wooded peninsulas, of crescents and canals.

The site of Sydney is a drowned river mouth. Aeons ago the Parramatta River was just a narrow stream flowing east to the Tasman Sea. Then the coast tilted down and the sea backed into the river and its tributaries, forming coves and bays and fingers. The entrance to this immense harbor is a narrow passage called the Heads, so narrow in fact that Captain Cook sailed right past it without seeing it. Once inside, one can travel thirteen miles inland on the water, and the shoreline is 152 miles long. The harbor is deep and safe, if something of a bottleneck. The lighthouse which stands on the sandstone cliff of the south "head" (point) was designed in 1816 by Francis Greenway.

The best way to see the harbor is, of course, from a boat—the ferry to Manly will do. Marinas, Mediterranean houses, ships, buildings on stilts, yachts, overwater houses, and a surprising amount of brushy uninhabited foreshore—all these mix together with much allure. Twenty miles north of downtown Sydney, the Hawkesbury River and three tributaries juxtapose wild bush and superb yachting grounds, with the white-bellied sea eagle soaring in the skies.

From any of the lookouts around downtown Sydney you can spot in the harbor an island, barely more than a rock, that is officially known as Fort Denison but invariably called Pinchgut Island. There, a century and a half ago, incorrigible convicts did time on bread and water, and were occasionally hanged. Sydney residents find this poignant reminder of the injustice of

Australia's past so compelling a place to visit that a family tour on Saturday, the most popular day, must be arranged three and a half months in advance.

Of course, the dominating feature of the whole harbor scene is the Harbor Bridge, 1,650 feet in span and, despite competition from new skyscrapers, just about the highest structure on the Sydney skyline. This bridge was designed to be the world's longest single-arch span, something for Sydney to boast about. But after construction began in 1923, the American planners of the Bayonne Bridge over the Kill Van Kull in New York designed a span just twenty-five inches longer than Sydney's, and managed to finish it four months ahead of Harbor Bridge, which finally opened in early 1932. In those Depression days, New South Wales was ruled by the flamboyant Premier John Thomas Lang, regarded by conservatives as a dangerous and demagogic leftist who should not be given the honor of opening the bridge. Just as Lang was about to cut the ribbon, a young blueblood from the wealthy suburbs spurred forward on horseback and slashed the ribbon with a sword, yelling that he acted "in the name of His Majesty the King and all decent people."

Logically enough, Sydney uses its waterways for urban transport, antiquated green ferries bustling to various beaches and suburbs. Only three hydrofoils ply the harbor, crossing to the suburb of Manly from the Circular Quay at the foot of Sydney Cove. The harbor's great possibilities as a flexible commutation system seem to be far from sufficiently exploited. Still, at commuting hours the quay is lively, with the homebound girl office workers buying a concoction of fresh fruit cored, crushed, and served in a glass in the quayside shops.

But if you can fault the ferry system, you certainly cannot fault the passion of Sydney's addiction to yachting. Thousands of boats are out on Sydney Harbor and connecting waterways each weekend in the main sailing season, September through April. They range in size from eight-foot Sabots to mighty eight-meter (twenty-six-foot) racing machines. Most spectacular of all are the eighteen-footers, whose races are so exciting that yachting clubs hire ferries to follow the action, and lots of informal (and illegal) betting is the order of the day. Racing is controlled by the Yachting Association of New South Wales, which coordinates courses so that traffic around the buoys won't be impossibly dense. Despite this planning, sailors have to race right through ferries, ocean liners, and freighters. Any day one can see a fleet of small boats in the harbor. Sydney is also home port for the America's Cup challengers *Gretel I, Dame Pattie,* and *Gretel II.*

To ship captains, all the lovely water reached by steaming through Sydney Heads is merely the commercial Port of Sydney, also called Port Jackson.

Four to five miles in from the Heads, by way of two deep channels, are the wharves—120 general-cargo berths with a total length of more than eleven miles of wharfage, all close to the business center of the city. For ship repairs, the port has the Captain Cook Graving Dock, 1,139 feet long and 148 feet wide, one of the biggest in the world.

Sydney has grown to such an extent that Botany Bay, ten miles south of the city center, discarded as the original site for the penal colony, is now part of metropolitan Sydney. This bay is the expanse of water one sees from the plane landing at Kingsford Smith Airport.

Twenty-seven miles of beaches bracket Sydney Heads to the north and south. The names are melodious—Cronulla, Clovelly, Narrabeen, Curl Curl, Dee Why. Surfing is a national sport that has been growing ever since Duke Kanahamoku introduced the surfboard to Sydney from Hawaii in 1915. Surfies wear sun-bleached hair, dress California style, and expertly ride the great green waves. They drive the roads with surfboards on cars, like skis in the U.S. in the winter. A surfie cuts such a romantic figure that boys too timid to ride the big waves have been known to bleach their hair with Ajax to fake the surfie image.

Sunbathing is a mania. Above all, swimming, in this land where the Australian crawl was invented in the 1920's, is close to the universal sport. Any resident of Sydney who cannot swim is an object of pity or contempt. Perhaps that is as it should be in a city blessed with so much sand and 342 days a year when the sun shines at least seven hours, but there are perils. "BUSY DAY FOR RESCUERS," said a casual newspaper headline one day a year or so ago, and the story went on to tell how no less than two hundred swimmers had been rescued the previous day from "rips," currents that sweep out to sea.

The beaches are beset by sharks as well, so that every beach must be guarded by expert lifesavers. In practice these men are organized in voluntary lifesaving associations. There is intense *esprit de corps*, and, according to Craig McGregor in *Profile of Australia*, "many of the characteristics of pre-war Australian society are to be found there; the masculinity; the mateship; the emphasis on physical prowess, keg parties and grog-ups; the uncertainty of relationships with women; and throughout a certain homespun philosophy of life." Inevitably, however, the lifesaving clubs and surfies clash. The Surf Life Saving Association, for example, summarily refused to recognize the short-finned Malibu surfboard as lifesaving equipment, and barred those who rode them from the club.

Usually a life guard swims out tied to a light line and holds up the drowning person while other members of the lifesaving club on shore pull both in. But the clubs are equally adept at going out in surfboats to pick

up swimmers who may yell for help from two hundred yards out or more. In 1938 three freak waves in succession carried thousands of bathers out to sea all at once; the lifesavers rescued all but five. Throughout New South Wales, in 1970, lifeguards rescued 7,318 swimmers in distress. Guarded parts of the beach are marked off by flags, and only the foolhardy swim outside the flags. Helicopters drop markers to indicate sharks.

Sydney provides the definitive answer to a question one still hears debated by supposedly informed people; do sharks really kill human beings? They do—right on the Sydney beaches. Some beaches are protected by nets, and some swimmers stick to beachside swimming pools. To protect the rest of the beaches, planes and boats patrol outside the surf line, with particular vigilance during the December-February summer season, when sharks weighing 500 to 750 pounds, "nasty brutes," come to shore. When an airplane spotter sees a shark, the boats close in and pursue him with spears. Bells or other alarms send swimmers to the shore, and only half a dozen lives have been lost in the last twenty years. To have a leg clipped off by a shark causes a hemorrhage that is usually fatal.

*

Except for the members of the Iceberg Club, and a few other similar groups, who go into the water every day, people do not swim in Sydney in the winter season, which is May to September. The city does not get quite chilly enough to require central heating, or even heaters in cars, but movie theaters often post banners announcing "IT'S WARM INSIDE." In sum, Sydney has an agreeably equable climate. Humidity can be miserably high on hot summer days, but the spell is often broken in late afternoons by the arrival of a "southerly buster." This sudden gale, hitting as much as sixty knots, bowls over yachts in the harbor, but it also instantly air-conditions the city. Another wind, dry and searing, comes from the western deserts to burn up lawns and gardens; still a third is the salty northeasterly, laden with humidity. Indian summers are gentle and enduring—and sometimes end with a bang, a storm that heaves up the seas, sends down torrents of rain, and blows away everything not bolted down.

Glancing Around Central Sydney

Downtown Sydney draws much of its architecture from Victorian-Edwardian England and Mies van der Rohe's Chicago style, and then throws in on its own behalf a generous portion of colonial classics, together with green parks and marquee-shaded sidewalks. The streets are crowded, bustling with

people. Dress is varied; men wear conservative business suits or shorts with long socks. It all comes out looking uniquely Australian.

The city center begins where the great Harbor Bridge debouches onto an expressway that curves on stilts over the Circular Quay and then dives underground. Here are half a dozen glassy skyscrapers intermingled with many low buildings that go back as much as a century. Right near the foot of the bridge is a small quiet promontory, the Rocks, redolent of olden-days crime and present-day charm. The city's chief planner would like to remodel it along the lines of San Francisco's Ghirardelli Square; it possesses the requisite old warehouses and smell of history.

On the streets the endless marquees—a Sydney idiosyncrasy—form the sidewalks into open-sided tunnels, with uniform, plank-size signs hanging from above to indicate the wares and services offered in the adjacent stores. The window displays are by and large as jumbled and dusty as those of stores in country towns. A hardware store may have punches from Spain and electrical supplies from Adelaide and drills from England chockablock with vises and hammers and screwdrivers, all dangling scrawled tags. Store windows in general have a country-bumpkin look, and you have to remind yourself that you are in a big metropolis.

The main streets run broad, but not always straight or parallel: Elizabeth, Castlereagh, George; Macquarie, where the doctors' offices are; Pitt, once a row of convict huts, then the city's brothel district, and now the address of many insurance companies. There are intriguing, angled, short streets, too. On meandering Phillip Street, barristers in tight-curled white wigs, short dapper ties, and black gowns stride out among the crowds. Five-block Martin Place has been turned into a pedestrian mall by paving it with pink granite, banishing cars, and setting out fiberglass "street furniture," trees, and flower stalls—all to give Sydney what it has long needed, a central focus.

The streets are pictures of animation; vitality infuses the center of the city. The girls who descend from the high-rise office buildings wear the world's shortest skirts—anything briefer would hardly cover the navel. Australian girls more than women anywhere simply repulsed the maxi-midi fad, and adopted the mini as a permanent choice, which of course befits the climate, their exuberant personalities, and their slim, strong legs. There is much construction in the central city, surrounded by clamor, confusion, and a particularly sturdy kind of timber scaffolding. The men in shorts are businessmen—that is perfectly acceptable attire in Sydney in the summer. One also sees a surprising number of unshaven men in dirty work clothes, sometimes with packs on their backs. They do not look as if they belonged in a city, yet there they are, even in the bookstores.

The tallest building is a forty-seven-story cylinder called Australia Square, higher than the more famous spire of the eighty-four-year-old Renaissance-style General Post Office Building. The Post Office clock tower is lighter in color than the rest of the building, and for an interesting reason: it was dismantled stone by stone and stored during World War II, for fear that Japanese bombers might use it as a landmark, and thoroughly cleaned before its restoration was completed in 1964. A major hotel is the Wentworth, a high concave pile built by Qantas Airways a few years ago for $12,320,000; advice from wide-ranging Qantas pilots went into the planning, and the result was very much like a Hilton. The State Office Building, remindful of the fact that this city is New South Wales' capital, is a darkish glass edifice nicknamed "the·black stump." Town Hall, the seat of city government, has an auditorium that can hold 2,500 people for concerts on its pipe organ. Immigrants completing naturalization come here to swear allegiance to the Crown and get a cup of tea and a cake. The New South Wales Parliament is housed in a building, lovely with age and classic in architecture, on Macquarie Street. The central part of it, with the wide column-supported verandas that characterize Australian colonial buildings, was constructed in 1811–1816 as the Rum Corps hospital.

Sydney has a nice little toy underground that connects with the suburban surface railway routes. The central railway station is a gloomy cavern, with people waiting to depart for the innumerable suburbs. One train, for example, goes to places with such names as Emu Plains, Blaxland, Warrimoo, Faulconbridge, Bullaburra, Katoomba, Medlow Bath, and Mt. Victoria. As for Sydney's buses, all that needs to be said is that ten cents will take you quite a distance, and the system continues to operate double-deckers that are, as they always were elsewhere, agreeable for sightseeing.

An Innocuous Center of Sin

King's Cross, not far from Hyde Park, houses the liveliest night life in Australia. It is likened to Soho, Chelsea, and Greenwich Village. It also emanates a whiff of Forty-second Street in New York, and North Beach in San Francisco. But it is really only a smidgen of these bawdier places. There is a bit of pathos in Sydney's pride that it at last has the strip-tease. The strip-tease?

The net effect of King's Cross—balancing off the queers, the strip-shows (including male strip-shows), the whores, and female impersonators (to titillate suburbanites) against the architectural and geographical charm, the pretty plane trees, the espresso coffee bars, the fine delicatessens, the excellent gastronomy, and the peculiar innocence of the place—is far less tawdry

than that of sin centers elsewhere. Though cheek-by-jowl with poverty-ridden Woolloomooloo and other scruffy neighborhoods on one side, King's Cross melds with genteel Elizabeth Bay on the other. One feels that this is a rather innocuous place for American soldiers to come on "R and R" (rest and recuperation), which they have done lately from Vietnam, 4,500 miles away, and during World War II from all over the Pacific. Two hundred and eighty thousand of them visited Sydney and "the Cross" between 1967 and the end of 1971, to the tune of much mutual back-patting: "I wouldn't go anywhere else for R and R." "The Americans are the most courteous and polite boys we have known."

A century ago King's Cross was a Jewish neighborhood, a circle centered on the Elizabeth Street synagogue with a radius roughly determined by the two thousand cubits that constituted the maximum Sabbath journey for the Orthodox Jew. Later it became a stronghold of artists and writers; William Dobell lived and worked there. King's Cross area has several good hotels—the Chevron, the Kingsgate, the Gazebo—and the perfectly fascinating El Alamein fountain, a huge sphere like a dandelion gone to seed, filled with sparkling spray.

An Opera House with Sails

Cities everywhere yearn for the cachet of a lavish opera house. The opera house has become the cathedral of an era no longer much interested in building impressive churches. Manhattan, West Berlin, Los Angeles, Brasília, Santa Fe, and scores of other cities have built opera houses or concert halls in the past decade or two. But beyond a doubt the famed though unfinished opera house in Sydney is the most inspired and folly-ridden adventure of them all. "We've had first to look after basic things," explained Harry F. Jensen, who was Lord Mayor of Sydney when the project started in 1957. "Now we can expand and use some of our wealth in a major cultural endeavor."

The site is inspired. The highly visible structure is set on the eastern promontory of Sydney Cove, Benelong Point, named for an Aboriginal. Yachts with striped spinnakers cruise the blue harbor close at hand. The great bony arch of Harbor Bridge looms several hundred yards away. Landward are Government House and the Botanic Gardens.

The plan was bold, unique, brilliantly chosen—and trouble—from its inception. The Sydney Opera House Committee offered a prize of $10,600 for the best design, and the next year a jury of local critics met to judge 222 entries. The noted Finnish-American architect Eero Saarinen was

invited to Sydney to advise the jury. He looked through the designs they had already rejected and chose a sketch by an unknown Dane, Jørn Utzon. Utzon's design is essentially an extravagant sculpture: a building of soaring white concrete shells suggesting breeze-filled sails. The basic repeated element of the design is a three-sided pyramid, two sides formed of ballooning, white-tiled concrete shell that partly enclose the third, which is a bronze-mullioned, curving curtain of panes of glass. There are ten such shells, some embracing others and some back-to-back, some nearly vertical and some leaning sharply, and two of them off by themselves to house a restaurant.

But Utzon's imaginative concept was not matched by engineering foresight. A London consulting firm set six hundred men to work with computers on calculations of the stresses in his projected shells. Three years later, with the foundation well along, the consultants reported that casting the shells would be technologically impossible. Then Utzon, "after three years of search for a basic geometry for the shell complex," modified his design so that the shells could be prefabricated in blocks, each curve skinned off the surface of a single sphere having a radius of 246 feet, using the same molds over and over. The blocks were raised by hammerhead cranes and assembled with steel hawsers. It was then realized that the partly built platform could not be adapted to the prefabricated shells; so Utzon ordered the dynamiting of $300,000 worth of concrete piles already set into place.*

By 1966 Sydney's patience was beginning to wear thin. The newspapers discovered that Utzon had not planned parking fields for the opera house. He answered testily that there were no parking fields at the Parthenon either. He demanded $2.8 million to "build a full-scale wooden mockup of the main hall, and see whether it falls down." The government balked; he threatened to resign; the government quickly and gratefully accepted the resignation.

Peter Hall, a highly regarded Sydney architect, took over from Utzon and found shocking new obstacles. He did not believe that the main hall could be used successfully for two different purposes; "Concerts and operas do not happen in the same sort of rooms," he said. So the main hall (seating 2,750) was redesigned for symphonies only (Sydney has an excellent orchestra with a big following). If Australia produces any more opera singers like Nellie Melba or Joan Sutherland, they will have to perform in the smaller

---

* The earth, billed as firm sandstone before the project got started, turned out to be fill, "nothing but old boots and bedsteads," and ultimately 550 three-foot-thick concrete piles had to be driven to a depth of 70 feet below sea level, to reach a solid foundation.

adjacent theater (seating 1,500), which will be used for opera and ballets. Halls on other levels under the shells will seat 600, 450, and 150 people, and handle plays, films, rehearsals, and recitals. After all the energy and expenditure, Sydney will not have a structure that is primarily an opera house.

Meanwhile, the costs have mounted astronomically and the end is not in sight. What began as a projected figure of $10 million is now estimated to be $93 million (New York's new Metropolitan Opera House cost $45 million). Most of the fund-raising has been painless; it comes from the New South Wales Opera House Lotteries, which pay out a $224,000 first prize every two weeks. Ultimately the building will account for at least $200 million worth of gambling, or an average of $15 for every Australian.

No one really knows when the Sydney Opera House will be finished. It has been under construction since 1959, and the opening was first scheduled for 1963; the current target date for completion is November 1972, to be followed by a period of getting rid of bugs, and a grand opening in March 1973. But one thing is certain: it will be—indeed, it already is—an unforgettable building, absolutely original in concept, and a landmark oddly visible from many parts of the city, not because it is so tall but because it rivets the eye. A leading architect in the rival city of Melbourne may have summed up the whole project, when, confessing admiration, he called it "ridiculous, something you would expect of Latin America but not Sydney—and altogether wonderful."

A Sea of Suburbs

Neighborhoods around Sydney divide in class and aspect as sharply as though cut with a cleaver. The shores close to the center of town, where it may cost $150,000 to buy an apartment, are elegant indeed; the old rich and the high-fee professional classes live among fine views of the harbor, good architecture, chic shops, and sidewalk cafés in Potts Point, Double Bay, Bellevue Hill, Rose Bay, and Vaucluse. One sumptuous address is Elizabeth Bay House, a colonial mansion owned by the state and divided into sixteen apartments reached by a winding staircase.

Plunked among near-slums and near-ghettos is a region once decaying and now attractively restored, like Washington's Georgetown. This is Paddington—"Paddo," they call it. Its merit was its terrace houses, one fortunate legacy from the otherwise tasteless Victorian era. The houses are narrow, two-storied, with a door and a window on the street level, and above them a balcony with a lacework cast-iron railing, made of metal brought from England as ballast. Sharing common walls, painted in pas-

tels, these houses march up and down the hills to form long repetitive rooflines embellished with chimney pots and finials. To buy one of these houses, which, renovated, may cost as much as $50,000, is the fashionable thing for artists, writers, and young professional people.

South of the city is industry and yawning tattiness, although the expressway from the airport manages to dodge the drab, and winds through golf courses and parkland. Botany Bay, though historically distinguished, is presently an eyesore of power poles, blatant signs, litter, and tourist-trap "museums."

The real suburbs are places like Waverley, which package people in at the rate of nineteen thousand to the square mile. Tumbledown Redfern gives over a couple of streets to city-living Aborigines. Taylor Square is a center for Greeks, Pendle Hill is Little Malta.

Across the harbor, to the north, is a huge domain of small suburban houses for the middle and upper classes, young business executives with two children and two cars. Some clean-lined new "Australian contemporary" houses can be seen here, and their occupants lead wholesome lives. Lights go out at ten o'clock on the north shore.

To the west of Sydney proper stretch twenty miles of bungalows, broken only by the Gothic architecture of the University of Sydney, industrial sites, and the remnants of a once-ambitious green belt that got trampled down in the rush to construct. Here is where the bulk of Sydney lives, happy in the face of monotony, often glued to the telly, and pleased with the crimson rosellas, gorgeous parakeets, that inhabit the back yards.

A Certifiably Gallant Governor

Through someone's splendid foresight, the downtown section of Sydney is provided with a vast grassy area, the Botanic Gardens and the Domain, where on Sunday soapbox orators, often in weird dress, advocate or denounce this and that. The only buildings here are the Conservatorium of Music, the Art Gallery of New South Wales, and Government House. The last of these is a crenelated, turreted castle, set among gardens, tennis courts, and a swimming pool, and against a huge rubber tree with roots that must be regularly hacked away to keep them from undermining and toppling the castle. The resplendence of the place is a reminder of the past glory of the British Empire. The Governor represents the Queen, and is today only a symbolic head of state. His power is slight—in reality, empty authority.

Sir Arthur Roden Cutler, V.C., K.C.M.G., C.B.E., K.St.J., the twenty-first occupant of the castle in the Botanic Gardens, is a tall and lanky man. On July 6, 1941, Cutler, then an artillery lieutenant in the Australian Army

fighting in Syria, was carrying a phone line under heavy machine-gun fire to a forward outpost of his troops, and suffered serious wounds. He could not be rescued for twenty-six hours, by which time one leg was so infected that surgeons had to amputate it. It was for this and a number of earlier acts of similar bravery that Cutler won the Victoria Cross. His career was mostly in diplomacy—Australian High Commissioner in New Zealand, Ceylon, and Pakistan; Minister to Egypt; Consul-General in New York. As Her Majesty's Australian Ambassador to the Netherlands, he was dining one night at The Hague when the phone rang—Buckingham Palace on the line. The Queen had made him Governor of New South Wales, and he took up the position in January 1966.

Sir Roden talks with pride and affection about New South Wales, pointing out that the state has more industry than any other, with tremendous energy concentrated in Sydney. New South Wales has no unemployment—a labor shortage, rather. The Governor contends that Sydney is in fact the money center of Australia, though Melbourne claims the title.

The Governor cannot effectively attack the state's problems, because his powers are limited to "assenting in the Queen's name to bills passed by the Parliament of the State." But he can exert a bit of pressure here and there—perhaps "government by nudge" is a good way to put it—and he is deeply concerned about New South Wales' problems. He mentioned in particular the rebate system by which the Commonwealth government shares with the states the revenues it takes in as sole recipient of income taxes. Naturally, there is much controversy on this issue, since the richer states feel that they are contributing to the poorer states funds which they would like to have for their own improvement.

Sir Roden gets a tax-free salary of $20,000, but the whole cost of running Government House, including the wages of its carpenter-caretaker, is $105,000. At the end of a great hall in Government House are spaces for the coats of arms of each Governor and Sir Roden's was new and fresh there. What happens if a Governor does not have a private coat of arms? Sir Roden suspects, he says, that in times long past several of the Governors invented their own.

Epicurean Sydney

Here are the three main points to make, gastronomically speaking, about Sydney. First, it has shot in ten years from culinary dowdiness to authentic gourmet levels, solely because big-scale immigration finally brought international standards and every kind of cuisine. Second, Sydney rock oysters are

the best, cheapest, and most plentiful in the world. Third, the possibilities for using the waterfront for restaurant sites have been scandalously neglected; there are only one or two restaurants built over water. One of these is the Caprice, set on the shoreline among the tall luxury-apartment buildings that line Rose Bay. The cooking is French, and people say that it is the best place to dine in Australia. Short of being on the water, you can get a superlative view of Sydney, plus tolerable fare, from the Summit, a revolving restaurant (the world's biggest, of course—this is a land of superlatives) on top of the Australia Square building. An underground place called Weinkeller has smörgåsbord and a choice of seventy-two Australian wines, which one picks from rack after rack right in the middle of the floor. The Argyle Tavern bills itself as "137 years behind the times and proud of it." The Chelsea in King's Cross has a sensible idea, a menu in booklet form, the size of a postcard, with eight pages of offerings. The Newcastle Pub, unluckily soon to be torn down, provides not food, but rather lots of strenuous drinking among the many Sydney writers who find it congenial and the many Sydney artists who use it as a gallery to sell their paintings. It was the first pub in Sydney to serve women.

Sydney's rock oysters are grown in the Hawkesbury River. The system is to submerge long narrow wooden slats over the breeding grounds, and the young oysters, the "splat," quickly fasten themselves to the wood. After six months the oystermen transplant the sticks to less salty water upstream, and let them grow for three or four years. It seems to be a felicitous proceeding. One Australian dollar buys a huge plate of them at Victor's, an unpretentious place on King Street; and not long ago, at the annual convention of the Oyster Farmers of New South Wales, two thousand men sat down and consumed *ten thousand dozen* oysters.

In Sydney one can see, better than anywhere else, how this hearty egalitarian nation has expanded what might be the typical American country club into a remarkable phenomenon of mass eating and drinking, entertainment, and sport. An example is the St. George Leagues Club, which has 33,000 owner-members paying about $10 a year in dues in order to enjoy $2,250,000 worth of clubrooms, cabarets, gardens, squash courts, bars, movies, fashion shows, gymnasiums, and saunas—plus hockey, cricket, golf, swimming, bowling, football, and boxing. In the main dining room, three dozen Sydney rocks, plus Prawns Thermidor, Veal Normandie, wine, *and* a smörgåsbord can be had for about $2.40. The clubs are subsidized by slot machines—called poker machines because the spinning symbols form poker hands. One smaller club in Sydney nets a profit of almost $600,000 a year, while returning to the beer-drinking gamblers an average of $3,600 in winnings per day. It is said that the clubs "make gents out of workers," and

"give every man his own Las Vegas." Perhaps this kind of thing simply reflects the indigenous uncomplicated contentment with suburbia, beaches,
sports, and what are accepted as the good things of life—the attitude that it
is, after all, a fine thing to be alive in Sydney, Australia.

*

Alcohol is the subject of much debate in Sydney, and the object of much
moralizing and lawmaking. In short, the men there drink a lot, flying in
the face of numerous (but diminishing) restrictions. The chief absurdity,
inherited from England, is that by and large bars must pretend to be merely
a licensed ancillary convenience attached to a hotel, the "public room."
Thus what everyone knows to be nothing but bars must be substantial
buildings that will actually rent rooms. In Sydney they are mostly built to
the same plan, three stories of tan masonry, with swinging doors leading to
the tile-lined pub with its U-shaped bar, and the name of the publican duly
posted outside. It is commonplace to see these hotels planted on three of
the four corners at an intersection. The names are jaunty: Captain Cook,
Bat and Ball, Royal Oak, the Hero of Waterloo.

*

Sydney tidbits: pack rape, in which groups of young men assault a young
girl by turn, is such a serious problem that a select committee of the state
Parliament conducted an investigation. The investigators were shocked at the
indifference shown by the rapists: most of them thought mass intercourse
a reasonable thing to do, and they seemed sure that the girls did not
object.

The great postwar migration to Australia is strikingly evident on the
streets of Sydney. So many languages can be heard among the throngs that
English will sometimes hit the ear as slightly bizarre, the way it does in
Paris or Rome.

The telephone book has pink pages instead of yellow, and in the directory
first names are listed by initials only. In fact, Australians prefer the use of
initials for first names in general.

The American Club, atop a small skyscraper, has a unique men's room—a
row of urinals with a view, being fitted with a fine large window. It is said
that one visiting American was so impressed that he had the view, which is
of the whole spectacular harbor, photographed, enlarged to mural size, and
installed in his bathroom at home.

Outside of Sydney in New South Wales

Eucalyptus trees, as perhaps not everyone knows, give off an oil that floats in tiny droplets in the air. These droplets refract light, in accordance with the principle of Rayleigh scattering, which has the effect of making distant objects appear blue. And blue indeed, cobalt blue, are the eucalyptus-clad Blue Mountains west of Sydney.

These mountains are part of the Great Dividing Range, which, as we have seen, separates the narrow coast of New South Wales from its immense outback. The coast, of course, has the people—in Sydney; in Newcastle, population 360,000, a hundred miles north; and in Wollongong–Port Kembla, population 200,000, fifty miles south. There are no other towns in New South Wales with more than 40,000 people. Both Newcastle and Wollongong exist by virtue of immense fields of high-grade coal. North on the coast, near the Queensland border, most of Australia's bananas are grown.

In the mountains in the north, a set of plateaus, called New England, has rich soil and fast, clear streams. The western slopes here grow as much as eighty bushels of wheat per acre when it rains—and nothing when it doesn't.

Beyond the Dividing Range is a monotony of plains all the way to South Australia. Two million sheep are the weary inhabitants. The only town of interest is Broken Hill, great lode of many metals, described in the chapter on Adelaide because its links are all with that city. Toward the south of this region is Riverina, good productive land—even rice grows here, because it can be irrigated from the Murrumbidgee and Murray rivers, which, with the Lachlan, give the area its name.

Southwest of Sydney is high country discovered in convict times by escaping prisoners who had a curiously unswervable conviction that China lay off that way about 150 miles. If they had turned south, they would have come to purlieus equally exotic, the high Australian Alps.

The Snowy Mountains Scheme

At the most spectacular stretch of the Great Dividing Range, only a hundred miles from the sea and precisely in the southeast corner of the continent, a great hydroelectric scheme is being developed. Here within the space of a hundred miles rise six rivers: the Murrumbidgee, Murray, Tumut, Tooma, Snowy, and Eucumbene. Engineers have been tantalized for eighty years by the prospect of manhandling these rivers to generate power and irrigate

land. An Act of Parliament started the Snowy Mountains Scheme in 1949. It is now better than three-quarters finished and will be done in 1974.

The project will produce 3,770,000 kilowatts, more than any existing plant except Russia's Bratsk, although, of course, the Snowy Mountains Scheme is a system of powerhouses rather than a single large hydrogenerator. Yet hydroelectric power, to be sold for $50 million a year, is only half the purpose. Diverting the course of one river—the Snowy—from the eastern to the western slopes of the Great Dividing Range will add two million acre-feet of water to the irrigation of the Riverina country. Farm produce is to rise $60 million a year in value as a result. The cost of the scheme is $800 million, $80 million from the World Bank and the rest provided by the Commonwealth government out of revenue. How's that for solvency? A doughty New Zealander, Sir William Hudson, is the engineer who headed the Snowy Mountains Authority for most of its existence, and of every ten construction workers seven are Australian immigrants. To honor them, the flags of thirty countries fly in Cooma, the project headquarters.

The scheme requires sixteen large dams and many smaller ones, seven power stations, and eighty miles of aqueducts as well as one hundred miles of tunnels.

The waters of the Snowy River were wasted in the past, flowing southward down the mountains to end up in the Tasman Sea. Modern engineering, by capturing its waters, has been able to reverse their course, causing them to flow westward, adding water to the Murray, Australia's principal river. Let us hypothesize a drop of water in the upper Murrumbidgee. Once it would have coursed down the mountains and ultimately to the southern ocean near Adelaide; now, still high in the mountains, a dam diverts it and shoots it eleven and a half miles through a tunnel to join the headwaters of the Eucumbene. There, stopped by a dam 381 feet high, this drop is sent *back* under the Dividing Range in another tunnel and rushed through power generators a thousand feet underground into the Tumut River. The Eucumbene reservoir, however, has *two* outlet tunnels; another drop of water can bob nearly fifteen miles underground to join the dammed-off Snowy River, be carried with it through a nine-mile tunnel, through generators, into the Murray. There are unbelievable complexities, the point being to collect the greatest amount of water at the greatest altitude and put it to the greatest use. Two towns, Adiminby and Jindabyne, were demolished in the process, but two new ones have been built to replace them.

All of this depends on plenty of snow, and this little corner of arid Australia has it. Easterly winds bring heavy precipitation, and the mountain elevations, though not arctic-frigid, are cold enough to keep snow flying for

five months. Much of the project lies within Kosciusko State Park, both a summer and winter resort. Residents of New South Wales are proud of the good long ski tows, and vacationing Australians have their own Gstaad to go to.

*

> . . . a huge, sprawling, turbulent city, full of bustle and a zestful materialism.
>
> —CRAIG MC GREGOR

But now, back to Sydney. The sun has risen from the Tasman Sea, on any pleasant day in December, by 6 A.M. The city does not want to awaken, does not want to go to work, but the light floods into every room. The warmth, the rising humidity, put their mark on the kind of day it is going to be in the offices and factories. Out of it all percolates a peculiar blend of inertia and energy. This may be a place that is bland, unspiritual, hedonistic, far from God—but it is not violent, doubt-torn, or vicious. Here is a city with some dazzle and fun. Skyscraper roofs look down on sparkling sights. Sophistication is not yet something to be avoided. Perhaps it could have been more romantically named Botany Bay—but in the end Sydney has become a name with its own measure of romance. "No great city in the history of the world was born under so cruel a star," John Pringle, former editor of the Sydney *Morning Herald*, points out. Yet now these wretched beginnings seem transmuted into colonial tradition, and this blends with the sea and the beneficent sun and the extrovert population to make a city of character and individuality. There are those who love it, and no wonder.

CHAPTER 11

Canberra

> In Western Australia they distrust "the East." In South
> Australia they also distrust "the East." . . . In Victoria they
> distrust New South Wales. In New South Wales they distrust
> Sydney. In Canberra, in the Australian Capital Territory, they
> distrust Australia.
>
> —DONALD HORNE IN The Next Australia

A MUSICAL PLAY, put on lately in Melbourne by Australia's Elizabethan
Theatre Trust, bore the title The Legend of King O'Malley. By no means
pure legend, O'Malley was a real figure of obscure American origins, who
moved in the 1880's to Australia, where he ran for elective office from
Adelaide, Tasmania, and Darwin. He won his first election on a platform of:
(1) legitimizing bastards if the parents later married; (2) lavatories on every
train; and (3) the abolition of barmaids. A newsman of the time wrote that
O'Malley looked like "a threefold compromise between a wild-west romantic
hero from the cattle ranches, a spruiker [Australian for spieler] from
Barnum's Circus, and a Western American statesman." His beard was
Mosaic and his hair a shock of red—though he frequently described himself
as "the baldheaded eagle of the Rocky Mountains." He was marvelously
effective, and eventually became a cabinet minister during the early decades
of this century. He fought for the populist cause of a Commonwealth Bank,
and that government bank is now a great Australian power center. O'Malley's other major cause was even more noteworthy: the establishment of the
Australian capital at Canberra.

*

To begin with, Canberra is a real capital. The balky job of actually getting
the government to move from Melbourne has been brought nearly to
completion with the leisurely transfer of the defense departments to Can-

berra in recent years. The filling up of its centrally located artificial lake has given Canberra, after half a century of waiting, a comprehensible aesthetic unity. Canberra has reached world pre-eminence as a garden city. And it has grown with more control than most master-planned cities. Some 130,000 people now live in Canberra. The city took shape through years of drumfire disapproval as a "capital village" and a waste of money; but now Canberra is there, and it functions, and it beguiles the eye.

Somehow, though, Canberra has never lost the pastoral feeling of a country town. Lively crowds fill the half-dozen good restaurants (there are no night clubs) in the evenings, but the customers are not cosmopolites— just plain Australians, civil servants mostly, informally enjoying a meal. Youngsters find the capital a sleepy place—"nothing to do at night." Traffic lights are few—a blessing, but still a small-town touch. Even to play slot machines, that beloved Australian amusement, people must go to the neighboring town of Queanbeyan. A report commissioned by Parliament found that the loneliness characteristic of cities was more serious in Canberra than elsewhere, and recommended a "European-type" Sunday, with bars left open and horse and dog racing allowed.

Only a few politicians actually have flats or houses in Canberra, most of them living in hotels or motels. The center of politico-social life, however, is the Hotel Canberra, which dates back to the origins of Canberra as the capital, and throngs with Senators and Members, lobbyists seeking their favors, and figures from the business world. Bars in this hotel buzz with what is presumably talk of important affairs. The newer Canberra Rex also attracts some of this life, and other modern hotels are springing up. But weekends, and other free times, the Members of Parliament fly off to home towns or bigger cities, and much of the politicking and electioneering that precedes legislation goes on outside the capital. Even Prime Minister McMahon, who has a comfortable official residence in Canberra called the Lodge, is frequently to be found in his Sydney home.

The insistent flavor of Canberra is that of a strong bureaucracy; one-third of the working population is in public service, and the high officials of the government either live there or spend time there. The real power of Australia is here, including all the departmental secretaries who advise the ministers, and frequently know more about a department than the minister himself. A Parliamentary press gallery, mostly seasoned and serious men, operates in Canberra, of course, and the capital has one of the country's more important and responsible newspapers, the Canberra *Times*.

Canberra itself is ruled, incidentally, by the Department of Interior of the federal government. If it is to continue evolving according to the plans of the National Capital Development Commission, this probably must be

so. As a sop, the citizenry gets to elect to the House of Representatives one member, who may speak but cannot vote. To sum up, Canberra is essentially a civil servants' city, wholesome if somewhat sterile, ornamental if somewhat artificial, functional if rather remote.

The ACT and How It Grew

The state-by-state ratification of Australia's federal Constitution in 1899 required a two-thirds vote of approval by the people at large. The voters in New South Wales withheld approval until they were given a constitutional guarantee that the capital would lie within their state. The other states agreed to this, but only if the capital were established far away from sinister Sydney. Victoria made the stipulation that Melbourne should be the seat of government until the time when the new capital would be built. So the Constitution decreed that the capital "shall be in the State of New South Wales, and be distant not less than one hundred miles from Sydney."

That turned the site selection over to individual legislators from New South Wales, who battled for years, advocating existing towns. The ultimate brilliant compromise was a nearly empty valley, averaging 1,900 feet in elevation, in the Great Dividing Range, 148 air miles southwest of Sydney and about 70 miles from the Tasman Sea. As early as 1823 graziers had settled in the area, known then as Canberry, and around 1900 it had eleven big sheep properties but no sizable habitation. The only impressive structure was St. John's Church, built in 1841 by a devout sheep raiser; it stands to this day in the middle of Canberra. Though the planners of Washington, D.C., thought a mere 69 square miles sufficient, the federal government of Australia demanded and got free from the state of New South Wales a tract of 911 square miles, to include a watershed big enough to guarantee a water supply for Canberra. Later New South Wales threw in another 28 square miles on the coast at Jervis Bay, to give the capital a seaport.

King O'Malley, as Minister for Home Affairs, rammed through the legislation for the Australian Capital Territory (ACT) and, in 1909, chose the precise location of the capital town site. He then pressed forward with a grand international competition for the design of the city. The prize was modest, $3,500, but it attracted 137 entries—all from abroad, because of a sulky boycott by Australians. The winner, partly because O'Malley identified with his nationality, was an American, thirty-six-year-old Walter Burley Griffin, trained in architecture at the University of Illinois and M.I.T., a Chicago disciple of Louis Sullivan and an associate of Frank Lloyd Wright.

Dozens of names were suggested for the new capital, the most delicious,

no doubt, being Sydmeladperbrisho, an amalgam of the first syllables of the names of the six state capitals.* Others much favored were Myola, Wheatwoolgold, and Shakespeare. The official choice was simply the area's traditional name, given to it by the Aborigines, and this time spelled as Canberra. It is supposed to mean either: (1) meeting place; or (2) bend in the Molonglo, the river that traverses the city; or (3) woman's breasts, from two cone-shaped hills on the site. At the ceremony to mark the beginning of construction, on March 13, 1913, King O'Malley laid the foundation stone, and the wife of the Governor-General, Thomas Baron Denman, officially christened the city. Seemingly unsure how to accent the word, she pronounced it "Can-bra." And so it remains—though there are still Australians who say "CanBEARa."

Griffin worked seven years as Federal Capital Director, getting in the main streets and a railroad link in spite of every kind of bureaucratic obstruction. Finally he resigned in frustration. This gentle vegetarian was reduced for a while to designing incinerators in Sydney. He went on, however, to bigger projects—the Capital Theatre in Melbourne, Newman College of the University of Melbourne, the Castlecrag suburb of Sydney. Moving to India, he built a library for the university in Lucknow, and there, in 1937, he fell from a scaffold and died.

A spurt of construction in Canberra in the twenties produced five hundred residences and a Parliament House built for the ages out of heavy stone, but considered to be temporary because a bigger and more stately one will eventually be put up. The impermanence of Parliament House is somehow supposed to explain its blocky ugliness, the product of a committee of pedestrian government architects. It is modeled on the U.S. Capitol in providing chambers for the Senate and the House of Representatives, but the colors for the decor of these halls, red for the Senate and green for the House, are derived from the British House of Lords and House of Commons. The arms of the Speaker's chair in the House are made of oak from Nelson's flagship, Victory, and one of the sights in the building is a copy of the Magna Carta dating from 1297. In front of this building on May 9, 1927, as Nellie Melba sang and Avro 504Ks droned overhead, the Duke of York (later George VI) formally installed the Australian capital in Canberra. The wives of the newly arriving civil servants complained stridently over the wood-fire stoves in their kitchens, so primitive compared to the gas flames they had left behind in Melbourne.

The Depression and World War II stunted Canberra's growth. The

* That is, Sydney, Melbourne, Adelaide, Perth, Brisbane, and Hobart.

population was a mere 5,000 in 1927, and twenty years later it was still only 16,000. During the war Australia's government was a snarl of confusion, divided among Melbourne, Canberra, and Sydney. The transformation of Canberra came in the decade of 1957–1967, when Prime Minister Menzies put his government and its pocketbook behind the National Capital Development Commission, and the population grew to 100,000. The estimated total cost, so far, is $800 million. Of course, the spending goes on, at $100 million a year, and the population is scheduled to reach 250,000 by the early 1980's. The growth rate is 8 percent a year, probably the highest in the world for major cities.

*

To Burley Griffin, the site of Canberra was a natural amphitheater. On the north, three hills, Mount Ainslie, Black Mountain, and Mount Pleasant, formed the "top galleries," and their slopes the "auditorium." The Molonglo River, running from east to west at the foot of the slopes, was, together with its flood basin, the "arena." The river's southern banks formed the "terraced stage and setting of monumental Government structures sharply defined rising tier on tier to the culminating highest internal forested hill of the Capital." Further to the south lie the Australian Alps—to an American eye more like the mountains of Vermont than anything in Switzerland, but in any case a pleasing blue backdrop. Three mountains in the ACT exceed six thousand feet.

Canberra's mountain setting is high enough to make it a rather cool place on the whole. The average temperature in the summer is only 67 degrees (though once the thermometer reached 109), and in the winter it runs around 45. A snow that lies on the ground is rare. Most houses do not have central heating, relying on oil space heaters and good insulation. On an average, only twenty-three inches of rain fall each year, and lawns turn brown if not watered.

The natural elements determined Burley Griffin's plan. In his imagination the river, dammed, became a meandering lake to give Canberra recreation, space, and definition. So it does. It is named Lake Burley Griffin, and it is seven miles long. Griffin picked a small elevation near it to be Capitol Hill, where he envisioned a building for "housing archives and commemorating Australian Achievements." Although a "Capitol" has never been built, the city is laid out, roughly, in concentric circles of drives around this focal point, with spokelike boulevards radiating from it. One spoke shoots off north across a bridge over the lake to a complex of stores, banks, office buildings,

and the main hotels.* Between this spoke and another lies the temporary Parliament House and an ample lakefront area intended as the site for the future permanent Parliament House (actually, some Members of Parliament, including Prime Minister McMahon, now favor Capitol Hill for this structure).

The glory of Canberra is its trees, *eight million* of them. Many native eucalyptuses are here, because Griffin thought that species to be the "poet's tree," worth planting in large numbers. And, partly owing to a prelate at St. John's in the last century, who planted seeds wherever he went, Canberra also has most of the more colorful trees of the rest of the world. Spring is a blast of peach and cherry blossoms, fall an explosion of scarlet and orange oaks and maples. In Canberra's ten thousand acres of parks,† one can find box elders, silver birches, cedars, Indian deodars, hawthorns, cypresses, plane trees, Lombardy poplars, weeping willows, American, Chinese, and English elms, ashes, olives, walnuts, firs, spruces, and even pepper trees from Peru. An appealing custom is the government's gift, from its enormous nurseries, of forty shrubs and ten trees to any new household. Until a few years ago, the government even clipped your hedges for you. Trees stand in parking lots, shading cars and hiding their cumulative ugliness. Trees attract birds and butterflies in great numbers—one butterfly was to be seen fluttering around the office of a government official being interviewed for this book. Trees and the lake make Canberra a pleasant place for outdoor life. The numerous picnic spots have coin-in-the-slot, gas-fired barbecue hearths. You do have to slap away flies, though; the gesture is known as the "Canberra salute." The lake is splendid for sailing and fishing—but, incredibly for a new lake, too full of sewage to swim in. A powerful fountain—the Captain Cook Bicentennial Water Jet—under the surface of the lake shoots a stream as high as a forty-five-story building. Gardens in Canberra grow roses ten inches in diameter. Also engaging is the occasional sight of a kangaroo on the Canberra golf course.

The Public Buildings of Canberra

It is the capital's public buildings, unfortunately for a much-architected town, that disappoint the visitor. None seem very monumental, as Griffin

---

* When President Johnson visited Canberra, the Secret Service, for security and flexibility, booked him into several hotels, and installed hot lines to the White House in all of them. An Australian sheep raiser, visiting Canberra, was by mistake put into one of the rooms reserved for but not occupied by Johnson. The sheepman picked up the phone, and the White House, responding, was astonished to hear him bellow, "Send up a couple of beers!"

† By comparison, Manhattan's Central Park measures 840 acres.

would have had them; Parliament House, in the words of writer Colin MacInnes, is a "meaningless box"; the Treasury, the tiny-windowed Ministry of Defense, and other new edifices are neat and unmemorable. The higher buildings seem to know they are out of place, and try to hide their height by stressing horizontal lines; yet at around ten stories they are not high enough nor numerous enough, and too scattered, to make a skyline, which might be an asset. The War Memorial, on the slope of Mount Ainslie, is a ponderous fortress containing the "world's most comprehensive display of contemporary war relics"; it seems to celebrate war rather than lament the dead. The Memorial has already rushed to put up a diorama showing Aussies fighting in Vietnam. Australians take much pride in this building, reflecting their pride in the world recognition that they won by bold fighting in two world wars.

But the Parthenon-like National Library of Australia, designed by Sydney architect Walter Banning, is rectilinearly spare and proportionate, lovely as reflected in the lake, and set off by a garden of orange poppies. Inside is stained glass by the noted Australian artist Leonard French. A particularly exciting building is the Academy of Science: simply a shallow dome with the bottom scalloped so that it appears to sit lightly on a dozen delicate feet. Just inside the scallops is a moat whose water reflects light into the building. The Melbourne architect chiefly responsible, Sir Roy Grounds, says that he ranks it among his best. Canberra also has a business center, built years ago, that employs block after block of graceful arcades, almost in the Italian manner. A number of office buildings privately owned by banks and insurance companies rent space to house entire departments of the government, which thereby avoids the cost of constructing offices.

Three fascinating places to visit in and around Canberra are the mint, which is fitted with a visitors' gallery where you can watch coins being made; the Honeysuckle Creek tracking station, thirty-five miles southwest of Canberra and used in Apollo shots (it looms high and golden and seems to be worshiping outer space); and the Australian-American Memorial, an aluminum spire 258 feet high with an eagle on top—a commemoration of Australia's gratitude to the United States for helping to fight back the Japanese in World War II. It is one more sign of the uncomplicated Australian feeling of friendship for America. One curiosity is an elaborate carillon (Canberrans use the British pronunciation, and rhyme it with "million"), played for forty minutes once a fortnight by a carilloneur who comes from Sydney; no one in Canberra knows how to play it.

Establishing a new capital naturally compels the world's nations to build new embassies, in this case with decidedly mixed results, for many of them

have attempted to bring a bit of homeland architecture to Canberra. South Africa displays Cape architecture on the curve of a prominent avenue. The Japanese Embassy looks Japanese, the French French. The American Embassy carries this notion to the ultimate with two buildings that would certainly make the DAR applaud. The chancellery could be set down anywhere in Williamsburg, Virginia, and be at home; the Ambassador's residence, grander and higher up the hill, could pass for a colonial capitol. The detailing of both buildings is meticulous, down to the red bricks imported from the States and the walnut paneling of the Ambassador's office, but the people who work there seem to yearn for at least one modern touch, airconditioning. Whether these native-style buildings are fitting or not is a matter of much controversy in Canberra. Elspeth Huxley feels that to the general hodgepodge of Canberra architecture "foreign missions have added a musical-comedy touch with embassies in national costume, like peasants posed to dress for tourists."

The American Ambassador, by the way, is Walter Rice, a government trust-buster in the 1930's and more recently a member of the board of directors of Reynolds Metals and its subsidiaries. President Nixon appointed him in 1969. He is the father, by his charming young Danish wife, of two small children.

Residential buildings, it must be added, do not add much to Canberra's architecture. The government-built houses, though mostly of brick and roofed in red tile, are small, commonplace in look, and not very modern. As in other big Australian cities, the residential areas stretch endlessly, thousands of small suburban houses, each with its garden. Some glassy new apartment buildings are clean-cut and up-to-date, though it must be depressing to live in a place labeled "Block 17, Flats 1 to 9." Housing, as might be expected, is perennially short. Single people may have to stay in government hostels for as long as ten years.

The Australian National University

Canberra is the seat of the nation's major effort in postgraduate research, the Australian National University. Its Institute for Advanced Studies, comparable to Princeton's, is devoted to medicine, biological sciences, chemistry, physical sciences, social sciences, and Pacific studies. The undergraduate School of General Studies deals in the arts, economics, law, science, and Oriental studies. The teaching staff numbers close to three hundred, the students nearly four thousand. Attached to the university are a couple of Roman Catholic schools, Ursula College and John XXIII College. ANU

is only adequately housed, having also missed the chance for architectural distinction. But the university is magnificently equipped with laboratories and a huge observatory, and it offers teachers high incentives in pay and freedom. The students, who include a couple of hundred Asians, express their antiestablishmentarianism in routine ways, such as reading lists of Vietnam war casualties.* Apparently they also smoke a lot of pot. Three of the dormitories house both men and women.

The National Library of Australia, equivalent to the Library of Congress, is run by Harold L. White, who upon being interviewed turned out to be a specialist in Americana and an admirer of things American. He spoke of his difficulties in getting across to Australians, even in the academic world, that the United States has authentic civilization and culture. "The British never encouraged the Australians to think that the U.S. is great. In any evaluation of any aspect of U.S. life, like films, the temptation is to say that they are no good." Mr. White also admires the "unparalleled" physical beauty of the U.S., and deplores the way Hollywood films fail to convey it. Like the Library of Congress, the Canberra library gets every book published in the country. It also has a collection on Byzantium in thirty languages, and in the last ten years has acquired 100,000 volumes in Japanese and Chinese. The U.S. collection totals 100,000 books. The library recently bought 12,000 books on Brazil from a Portuguese collector. Mr. White, a short, alert man with gray hair and mustache, was born on the land in Victoria, in the same house as his father and grandfather. He has been national librarian for twenty-two years.

*

What of the rest of the Australian Capital Territory, outside Canberra? It is rich agricultural country, home of 600 horses, 14,000 cattle, and 290,000 sheep. It also grows wheat. But no private person, except on one or two remote freeholds, owns any of the land. The government, in a prudent move in the years after the ACT was laid out, bought the land and lets it out only on lease, even in the case of city lots. The benefits were minimization of speculation, rental profits that helped build the capital, and, most significantly, control over the growth of Canberra. Thus, in contrast to Washington, which swelled to the northwest and left its planned center in

* When students protested the visit of Vice President Spiro Agnew in Canberra early in 1970, the police responded in novel fashion. Two cops would pinion a protestor, a third would hold up a clipboard giving his name and alleged offense, and a fourth would photograph the whole scene.

a corner, Canberra has grown symmetrically out to the planned green belt that surrounds it.

The detached part of the ACT, Jervis Bay, never did become a port. It is now the site of the Royal Australian Naval College and a navy airfield, plus some summer houses and holiday camps.

Standing beside the television tower on Black Mountain in Canberra, one can see what will become of a substantial part of the ACT nearby. Growing up six miles to the southwest, outside the green belt, is the first of Canberra's satellite cities, aimed at accommodating the capital area's eventual growth to perhaps 400,000 people. This is Woden, which presently has a population of 40,000 and will grow, by plan, to 120,000. Six miles to the northwest is the second satellite, Belconnen, which has 12,000 people and will have 120,000. The third, still in the planning stage, will lie south of Woden and be called Tuggeranong. The satellites will contain the government offices in which the residents work; thus are the principles of decentralization being carried out in the Australian Capital Territory. "Canberra is just beginning to receive world recognition for its form of planned development . . . in decentralizing by the use of satellite towns," boasts Roger Johnson, chief architect for the National Capital Development Commission. "It may, in the end, be the one Australian city that will contribute to the solution of those vast problems now faced by every city in the world."

Another striking project for the next few years was described by James Mollison, Director of the National Art Gallery, a long-haired young man on the order of Gregory Peck, with an office in the same building in Canberra as the Prime Minister. This is the Australian National Gallery, to be built by 1975 on the same stretch of lakefront that already holds the library. Mr. Mollison says that Canberra has a large collection of Australian and European painting rarely on view for lack of display space. The new gallery, designed to emphasize huge, flowing internal spaces, will provide eleven exhibition rooms connected by ramps. "Nobody really likes living here, so that the government must provide some of the attractions of the more cosmopolitan centers," said Mr. Mollison. "The National Gallery will be one of them, I hope."

*

Is Canberra a success? There would seem to be three or four ways to measure it. We can guess that it would have been better if Burley Griffin's original plan had been followed to a T. Bureaucrats in the early years forced

many modifications on him, but then neither he nor anyone else provided for harmonious and aesthetic public buildings. Brasília is a better job in that respect, and has more flair (indeed, too much flair). Perhaps the comparison is worth pursuing: Canberra took far longer to build and came out smaller (Brasília went up in ten years, and now has 400,000 people). But Brasília is quite shoddy in construction and inhuman in scale; it has high-rise apartments sitting on the vast plain, as compared to Canberra's individual homes. And Canberra really contains the capitol, whereas Brazil is still half-governed from Rio. Another measurement might be against conventional cities. It is indubitably pleasant to get away from neon, filth, clangor, power lines, and fried-chicken joints. As a consequence of such considerations, the judgments about Canberra have been various. Historian Keith Hancock found Canberra "a kind of suburban garden parcelled into plots by a network of paths which have no obvious beginning and no visible end." The Duchess of Kent remarked, "It's just a little country town, isn't it? Charming, of course—but just a little country town." Donald Horne perceived much snobbery; there is a "lot of make-believe in Canberra . . . it looks down on the rest of Australia as crude, self-interested, troublesome, ignorant." It has also been called "a city in search of its soul." But Colin MacInnes found it an "authentic federal center with a high concentration of enterprising brains."

CHAPTER 12

Melbourne and Victoria

This will be the place for the future village.
—JOHN BATMAN, UPON CHOOSING
THE SITE FOR MELBOURNE, JUNE 7, 1835

MELBOURNE DIES every night at 11:30—there's not a cat on the streets. The venerable Melbourne Club, with an atmosphere of leather and dark paneling, is the stuffiest institution in Australia, outdoing Adelaide. Financial power, class consciousness, a legacy of Britishness, above all a sense of authentic tradition somehow inconsistent in a city that aspired to be a village only 137 years ago—these are the radiations from Melbourne. "Is Sir Ian in?"; "Dinner tonight at Sir Colin's"; "Sir Roy will see you at ten": knighthood is still in flower here. The town's favorite newspaper columnist, Barry Humphries, writes: "It's the only place on earth where the visitor from abroad can close his eyes and wonder if there really is life before death."

Driving in and out of the city, one sees miles of indifferent architecture; the suburbs are monotonous; the city plan is a dull gridiron determined by the government surveyor who moved in shortly after Batman, a prosperous Tasmanian pastoralist, did. But leafy parks, sweeping freeways, numerous city-enclosed golf courses, and the absence of slums rescue the defects. *One-fourth* of Melbourne proper is parks. The two main streets, Bourke and Collins—not pretentious Parisian boulevards or grand Washingtonian avenues—but simply broad, straight, parallel, mile-long streets—set a tone of settled satisfaction, of complacent bustle, of mercantile potency. This is the hub of the Australian Establishment.

"Sydney is faced to the Pacific coast of America . . . while Melbourne is faced to Europe, our mother and safest friend," observed R. H. Horne, a nine-

teenth-century writer. The two cities could not be anything but rivals, disputing one another not only in industry, financial power, political leadership, the arts, and population, but also in living styles. Melbourne is mellowly paced, sedate, English or perhaps Bostonian in tone. Sydney, at least by comparison, is cosmopolitan, glittery, New York-like, exciting. In Sydney people take guests to restaurants and in Melbourne they invite them to dine at home. Each city boasts that its way of entertaining is more sophisticated than the other's.

Melbourne nurtured the art of painting in Australia, and most of the luminaries of recent years live there or at least started there, though Sydney, too, has a strong art colony. Good writers seem to prefer Sydney, but Melbourne has the bulk of the book publishers. Sydney has light and water and some charming old crooked streets; Melbourne has theater and a tradition of providing Australia's foremost politicians. The rivalry is intense, but the two cities share wholeheartedly a belief in the pattern of suburban living.

For all its new corporate office towers, Melbourne emanates a quaint English-Australian gentility. The coatroom at the Windsor Hotel is labeled "Cloaks." A showing of a movie is called a "session." The cab drivers are downright deferential, shutting off meters before arrival, rounding out fares in the customer's favor, expecting no tips and visibly grateful when they get an extra five cents. The fare from downtown Melbourne to the new airport at Tullamarine, ten and one-half miles, runs around $3.40.

The new Southern Cross Hotel, with a lobby as big as an airport, has not only a bell captain, U.S.-style, but also a concierge. In hotels, callers tend to come directly to one's room, as in South Africa, rather than arranging lobby meetings. Newsboys are numerous, and papers cost only four cents.

Melbourne, like all Australian cities, has numberless "hotel" pubs, each with brightly striped damp towels laid on the bar to soak up spilled beer and condensation. They serve a brand of beer called Courage. One can see some truly sodden characters in these places, and blowzy women, too. But the trend is away from "hotels." A couple of more modern Melbourne pubs are the Sentimental Bloke, with an acre and a half of carpet, and the bar in the TraveLodge, with seventeen miniskirted waitresses all less than twenty years old. These places play Muzak and cater to women as much as to men.

Dining in Melbourne is a satisfactory experience, by and large—with the accent on the large. Portions are dismayingly big. There are twenty-seven good international restaurants—Ceylon curries, New York cuts, rigatoni, lobster, German dishes, and "natural" oysters served with little dark bread-

and-butter sandwiches—in the center of the city, and many others in the residential areas. Liquor restrictions have come tumbling down in recent years, and most restaurants serve cocktails, wine, and brandy as freely as in Manhattan. If you want to pour yourself a nightcap in your hotel room, any pub will sell you a bottle of good Australian brandy for $3.05.

Shops offer women's coats of suedelike kangaroo skin for around two hundred dollars. In the Army and Navy stores one can still buy digger hats, brim turned up on one side. In St. Kilda, a beachside amusement center reachable on a rickety tram from downtown Melbourne, a long row of delicatessens and shops display marzipans, Black Forest sponge cakes, fruit, and fish; the Melbourne-edited newspapers, in Magyar, Yugoslav, and German, sold in these stores testify to the big colonies of New Australians here. Melbourne abounds in bookshops.

The Melbourne police are said to be the best in Australia; from chief to the lowest cop on the beat they wear made-to-measure uniforms. Yet discipline is somewhat lax in this place where jobs are so plentiful; it is infeasible, for example, to forbid a cop to smoke on duty, and he does. As elsewhere in this country, cops suffer from the Australians' antiauthority complex. John Pringle relates that "a few years ago a New Australian—a young Yugoslav—rescued a Melbourne policeman from the death grip of a certified lunatic while a crowd of Old Australians stood by cheering the lunatic." The police have some problems easily recognizable in the U.S.; on July 4, 1968, two thousand Melbourne University students, protesting the Vietnam war, stoned the American Consulate in Melbourne and broke every single window.

More Melbourne sights and sounds: The label "P" on a car means probationary; a new driver must display it for a year. At the airport (as in Canberra) there are doors labeled "Men," "Women," "Mothers," and "Disabled"—the facility for the latter contains a trapeze with which persons in wheelchairs can maneuver. Most all of them have showers. The presence of the New Australian is much in evidence—driving around, one is suddenly in a community that is all Italian, or all Greek. Collins Street, the Golden Mile, is the downtown finance center, and it smells of crisp banknotes. One buys tickets for Tatersall's Lottery, drawn three times a week, at a big storefront on Flinders Street. On the fringes of the city, the good Colonel Sanders, with his Kentucky Fried Chicken, is present, as everywhere in Australia. The older suburban trains, part of a vast system, might be at home in Birmingham: English-style compartments, doors on each side, green leatherette and worn wood. One also sees, as in Sydney, wrought-iron balconies on many of the older houses; once thought to be Victorian junk, these railings are now prized and preserved as antiques.

A High-Speed History

Destiny seems to have preserved Melbourne from discovery and settlement until the time was ripe; had Captain Cook been able to visit the site of Melbourne and compare it with Sydney, he might well have opted for Melbourne and its rich surrounding land. When Melbourne was belatedly settled, it proved its advantages and grew at high speed from village in 1835 to a powerful city of the world in the 1880's.

Port Phillip Bay, forty miles of almost completely enclosed water on Australia's southeastern bulge, was first entered by the English in 1802. In 1803 Sydney sent out Lieutenant Colonel David Collins with three hundred convicts and a Church of England chaplain, to found a settlement. Collins scarcely bothered to look for a good site, and quickly begged permission to withdraw to Tasmania. Thus Melbourne had a thirty-year respite from occupation and was spared from having many convicts as settlers.* On that questionable basis Collins ultimately earned the honor of having a Melbourne street named for him. Explorers coming overland from Sydney in the 1820's and early 1830's discovered the size and fertility of the Murray River plain, and described grazing land in western Victoria so rich that it was dubbed "Australia Felix."

On the strength of these reports John Batman, the Tasmanian pastoralist, made a deal with the Aborigines at Port Phillip Bay: he acquired 600,000 acres of land in return for a yearly payment of one hundred blankets, fifty knives, fifty tomahawks, fifty pairs of scissors, fifty looking glasses, twenty suits of clothing, and two tons of flour. With Batman came an extraordinary man named John Pasco Fawkner, a convict's son and bookseller in Launceston, Tasmania, who brought fruit trees, seeds, and implements, and soon thereafter opened a hotel that had not a pub but an encyclopedia as its chief attraction. Governor Richard Bourke in Sydney denounced Batman and Fawkner for trespassing on Crown property, but lacked the troops to force them out. And so he capitulated, and named the village they had founded after Lord Melbourne, British Prime Minister of the time. By 1840 the area had eight thousand people and one million sheep, and Charles Joseph La Trobe, appointed by Bourke, governed from a small timber house, still standing in the Domain.

* One convict, by the name of William Buckley, escaped from Collins and stayed at Port Phillip, living among the Aborigines, until the place was finally settled—a feat so unlikely that Australians to this day say that to have a slim chance is "to have Buckley's."

Victoria won separation from New South Wales in 1851, by an act of Parliament, and became a colony with a Legislative Council. Propitiously, gold was found the same year: the colony had offered a reward to anyone who could discover gold, and strikes followed soon after at Ballarat and Bendigo. Melbourne flourished as miners poured in from the world, including American forty-niners who gave the name California Gully to a stream bed near Bendigo. Chinese came too, soon comprising one out of every seven males. American Southerners led parties to lynch the Chinese, and a lot of race prejudice rubbed off onto Australians in the process. Some miners got so rich that before they called for drinks for the crowd at Melbourne pubs they cleared the bar by sweeping off the glasses with their whips, and had it washed down with wine. In 1853 an American entrepreneur sold ice from Boston at fifty cents a pound. Bourke Street, which had had sheep grazing on it in 1845, looked like a London thoroughfare eleven years later. Melbourne passed Sydney in size, growing from 87,345 in 1851 to 540,322 in 1861. At one point there were 187 men for every 100 women in Victoria. So short was labor that La Trobe, by then Governor of the colony, had to groom his own horse.

When the gold diminished, land dealing took over to keep the boom going. A "selection" (homestead) act made millions of acres of good Victorian land available to farmers. However, the graziers already there wanted to keep their vast sheep runs, and did their best to frustrate the intent of the act by creating dummy homesteads and buying up the homesteads of others. Building trades got the eight-hour day in 1856 (in England, the battle to get it continued until after World War I). Melbourne was a world wonder of growth and prosperity. "There is perhaps no town in the world in which the ordinary working man can do better for himself than he can in Melbourne," wrote Anthony Trollope after a visit. Free, compulsory education was begun in 1872. The Melbourne international exhibition in 1880 attracted more than a million visitors, and its awards are still celebrated on the labels of various champagnes and beers. Based on all this optimism, British capital gushed in, feeding a great land-speculation boom. Twelve-story buildings, among the world's tallest, went up; beturreted, movie-set mansions rose along St. Kilda Road.

Then it all went to hell. A widespread strike in 1890, provoked by Melbourne capitalists to break the power of labor, brought months of turmoil and wrecked the unions for years. After that came a drought, then failures by overdrawn businessmen, followed by the collapse of land values, and finally the downfall of overextended banks that had lent money against land. Victoria suffered hysteria, depression, and unemployment. The result-

ing disillusionment and trauma lasted a generation, and forced Melbourne thereafter to share its financial dominance with Sydney. The comeback, based partly on the great lead-zinc-silver mine at Broken Hill and partly on farming expansion, was slow. Ultimately, a quieter prosperity returned.

Strangely, or perhaps fittingly, the excitement of Melbourne's history died in the 1893 bust. Its twentieth-century story is one of unspectacular growth. A big event seems to have been the Centenary Celebration in 1935, with a London-Melbourne air race (winning time: 70 hours and 54 minutes). In 1942 General Douglas MacArthur, having escaped Corregidor in a torpedo boat and flown to Darwin and Alice Springs, entrained for Melbourne and was met by thousands; on April 18, taking command of the Pacific war, he issued General Order No. 1 from Melbourne. In 1956 Melbourne staged the XVI Olympics with much aplomb in a huge new stadium, now the Cricket Ground. And in 1966 Victoria, dropping her Victorian ways, put an end to the "six o'clock swill," the traditional hour of prodigious beer-drinking between office- and factory-closing time, five-o'clock, and pub-closing time, six. Most pubs now stay open until ten.

*

As he lived, let him die! a base wretch
  without feeling:
A bushel of quicklime is all that he's worth!
  —VERSE IN THE Bulletin, 1880

Here let us move back a couple of decades in time, and examine a man, more myth than flesh, who to this day affects the Australian consciousness in those places where guilt, rebellion, and pride reside. He is the great Victorian highwayman, the best remembered and most-balladed Australian of them all, Ned Kelly.

Kelly was a bushranger, an outlaw of a kind that flourished in the first half of the nineteenth century, usually escaped convicts who took to the bush and robbed travelers and isolated homesteads. After the gold rush, bushranging became prevalent in the northeastern corner of Victoria. It combined crime with romance, choosing rich English pastoralists for targets. Bushrangers became heroes to the small, mostly Irish selectors (or homesteaders), whose land was being taken away from them, in one way or another, by the richer pastoralists. The police and the government sided with the rich, as they have been known to do before and since in history. Kelly's mother, widow of an Irish dairy farmer, was a selector; so was her father. Ned, born in 1855, began tangling with the pastoralists at fifteen. He "borrowed" a horse, clobbered its protesting owner, and sent the man's wife a package

of calf's testicles as a measure of his irrational resentment. By eighteen, Ned had done almost three years in prison for horse-stealing. He became a prospector for a couple of years after that, although cattle tended to disappear from whatever neighborhood he was in, and he was adept at rebranding horses. The police issued a warrant for his arrest, and Ned went into hiding. Meanwhile there was a warrant out for Ned's brother Dan, also a cattle-duffer (Australian for rustler). A drunken constable tried to arrest Dan, failed, and to cover his alcoholic disgrace invented a yarn about being attacked that led to the imprisonment of Ned's mother (who had eleven other children) for three years. This was authentic injustice. Ned conceived a powerful hatred of the police—"a parcel," he wrote, "of big ugly fat-necked wombat headed big bellied magpie legged narrow hipped splay-footed sons of Irish bailiffs or English landlords." Ned and his brother recruited two of Dan's friends, Steve Hart and Joe Byrne, and formed a gang —a gang imbued from the beginning with a Bonnie and Clyde kind of doom.

In October 1878, at Stringybark Creek, Ned and Dan attacked a camp of four policemen sent out to find the Kelly gang, and killed two of them. A third was badly wounded; Ned thought for a long time, then shot him through the heart. "I put his cloak over him and left him as well as I could and were they my own brothers I couldn't have been more sorry for them," Ned later reported. After these killings Victoria panicked. The state Parliament passed a Felons Apprehension Act entitling anyone to kill the members of the Kelly gang on sight. The gang responded, late in 1878, by sticking up a bank in the town of Euroa, eighty miles northeast of Melbourne. Next year they moved across the border of New South Wales and held a whole town at bay for two days before robbing the bank. An anonymous ballad gives the flavor of the event and the times:

> They mustered up the servants and locked them in a room,
> Saying "Do as we command you, or death will be your doom."
> The Chinaman cook "no savvied," his face was full of fear,
> But Ned soon made him savvy with a straight left to the ear.

But as the police closed in with ever bigger forces, Ned seemed to know that he was doomed. His resolve to die hard took a weird form. With stolen plow moldboards blacksmithed over a fire in the bush, he and his gang constructed jerry-rigged armor—curved body plates held together with thongs, and for head protection a cylinder with an eye slit. The final shootout came when the gang seized a town called Glenrowan, ruling it from the hotel. Policemen in large numbers laid siege. Ned, with a high sense of tragedy, came out of the hotel in his armor, firing, and was shot in the legs. The cops

managed to hit and kill Byrne, then set fire to the hotel, thus burning Dan Kelly and Steve Hart to death.

Sir Redomand Barry, the same judge who had earlier convicted Ned's mother, was assigned to try Ned. "Edward Kelly," he said, "I hereby sentence you to death by hanging. May the Lord have mercy on your soul." Replied Ned stoutly, "Yes, I will meet you there." *Both* men died a fortnight later, the judge naturally and Ned with a rope around his neck. His last words were: "Such is life."

This whole story speaks with a profound eloquence to Australians; Ned Kelly means more to them than Jesse James or Billy the Kid do to Americans. An obvious element in the appeal of the legend is the Robin Hood quality (Kelly would not steal from those who worked hard for little money). It also expressed the resentment of ordinary, hard-working people against a political and social Establishment from which they were excluded. Kelly, as Alan Moorehead notes, made his war on the government, not on individuals.

A disturbing evocation of the legend comes from painter Sidney Nolan, in a series of scenes from Kelly's life. Kelly, though he wore his armor but once in action, is shown only as a symbol, a black rectangle with a slit through which peer, in most of the paintings, a pair of eyes. Through all the paintings looms a feeling of the terror and beauty of the land, and an intimation that Ned Kelly somehow deserved more than a bushel of quicklime.

Megalopolitan Melbourne

In sheer area, Melbourne is one of the biggest cities in the world, measuring 800 square miles in its metropolitan area, as compared to 458 square miles for the American monstrosity, Los Angeles, and 670 square miles for that other Australian overgrowth, Sydney. The population is 2.5 million, half a million short of Sydney and comparable to metropolitan Montreal. Two-thirds of the people of Victoria live in Melbourne. The distance to Sydney is only 440 miles in a straight line, but the distance to Perth is greater than that between London and Moscow. The reasons for its site and growth are twofold: the Port Phillip Bay area is the only place for a port in a thousand miles of coastline, and, lying just south of the Kilmore gap in the Great Dividing Range, it has access to the rich hinterlands of the Murray Valley. As a port, Melbourne has 109 berths and 12 miles of berthage in an area of 20 square miles, all now somewhat handicapped by a newly important limitation. The depth of water over the rock bottom at Port Phillip Bay heads is only 38 feet, not enough for the new ocean leviathans.

＊

Visitor: "Does it snow here?"

Cab driver: "Once or twice it has done."

Midsummer (January) temperature *averages* 67.6 degrees Fahrenheit; in winter (July) the average is 48.9 degrees. The sun shines an average of 5.7 hours a day, which is none too much. Melbourne has the makings for smog —factories and cars—and sometimes a great menacing cloud of polluted air hangs over the city, but strong Antarctic winds tend to blow it away. The same winds can also drop the temperature from perhaps 90 degrees in the morning to maybe 55 degrees in the afternoon.

The constructions of man in Melbourne can be both colossal, such as the terra-cotta and gray Cricket Ground capable of seating nearly one percent of the population of Australia,* and flimsy, such as a freeway bridge being built over the Yarra that collapsed late in 1970, killing thirty-five workmen (a post-mortem of the disaster said that the British designers "seemed to have worked largely from intuition"). The Royal Melbourne Hospital, seven hundred beds, is where the first casualties of the Pacific War were brought in the early 1940's. Audiences of as many as seventy thousand hear concerts at the Sidney Myer Music Bowl, an amazing canopy that appears to float above the orchestra. The cavernous new airport at Tullamarine, finished in mid-1970, is Melbourne's way of squeezing out of a humiliating bind; until it opened, Sydney handled big jets on international flights. One more thing must be said about buildings in Melbourne: banks very nearly dominate the scene. The telephone book lists about thirteen hundred banks and branches in the city.

The Royal Botanic Gardens doubtless merit Sir Arthur Conan Doyle's accolade, "the most beautiful place I have ever seen," but one intriguing aspect of them is less horticultural than historical. That is the commemorative trees, ninety-five in all, planted by personages whose names evoke more than a century of time. Among them are Charles La Trobe, Prince Albert, Nellie Melba, Lord Tennyson, Paderewski, Viscount Kitchener, Viscount Jellicoe, the Queen of Tonga, Harold Macmillan, John Diefenbaker, the Duke of Edinburgh, and the King of Thailand. The great asset of the garden —one of three in the world entitled to the adjective "Royal"—is the climate. With no frost, nearly any kind of tree, tropical or temperate, will grow. The

---

* That is, 120,000; but a new stadium being built is to hold 157,000, surpassing what is presently said to be the world's biggest, Rio de Janeiro's Maracanã.

soil is fertile and the rainfall is a reliable two inches a month the year round. There are five greenhouses ("glass houses" in Australian), where tropical plants are propagated, and exhibited. Three lakes grow water lilies and accommodate water birds, including black swans. It is *de rigueur* in the garden to visit the Separation Tree, an ancient eucalyptus under which the citizenry of Melbourne celebrated their authorization in 1950 to break away from New South Wales.

A comment about suburban sprawl. Obviously, Melbourne has it bad. The typical family here occupies eight times more living space than a London family. Once beyond the fashionable inner suburbs, like Toorak, one sees little but brick and red roof tiles for miles. A man's home is his castle in Australia, and there are hundreds of thousands of castles. Plainly, this is a better way to live than in some city tenement; rose gardens abound in testimony to pride of ownership and contentment. "All a Melburnian wants is a cream-brick triple-front* with a two-tone lawnmower," says one university professor. But no one can drive through, or even fly over, these suburbs —Cheltenham, Moorabbin, Glenhuntly, Alamein, Deepdene, Yarraville, Murrumbeena, and 190 others—without concluding that the place is overextended. It is clear that the old city cannot spread any farther without a new city being founded.

The Benign Hinterlands

The Victorian countryside does not partake of the terrible aridity of the great outback that starts at Victoria's northern and western borders and extends for 1,200 miles north and west. Only a bit smaller than Western Germany, Victoria (87,884 square miles) is Australia's smallest state, apart from the island-state Tasmania. Twenty-five inches of rain (versus the national average of 16½) fall every year on a generally fertile soil. With 3.5 million people, 27 percent of all Australians, Victoria is the most densely settled state. It is also the most industrialized. It produces citrus and other fruit, grain, dairy products, and the country's best wool.

And yet Victoria does summon up the essential down-under, as dozens of painters have proved by their obsession with it. The Murray River, which forms its northern border, is completely controlled by dams, weirs, and locks, so that it is leveled out in steps and navigable by steamers (mainly excursion ships nowadays, but once vital transportation). The river thus provides irrigation, which is the secret of the valley's prosperity. In some places, a pumping system devised in the last century by two brothers from California

* A common style of Melbourne house architecture divides the front into three planes, set back one after another, like stairs tipped over on their side.

reinforces the dams. This water and the natural countryside combine to give the land its productive look. On one hand are vineyards, wheatfields, fat sheep, orange groves, and tan-brick farmhouses with television antennas. On the other hand are plains dotted with melancholy eucalyptuses. This is haunting country.

The natives burst with pride over the Murray Valley. Camping and "caravaning" (touring with a house trailer) are popular here, and the campgrounds offer all amenities. The highways are plentiful and as well paved as, say, West Virginia's. An American tourist experiences the familiar sight of numerous motels. But these places do have some un-American graces. Rooms at the Sanctuary Park Motel in Wodonga are fitted with "breakfast hatches" —small doors through which the management pushes the breakfast that the guests ordered the night before (grilled steak and eggs, thirty cents extra). Little cafés in the area are oddly meticulous about pricing: tomato soup sixteen cents, a single crumpet twenty-three cents, tea thirteen cents, fruit juice seventeen cents.

Victoria's second city, Geelong, is only one-twentieth as big as Melbourne, and being only forty-five miles away will probably join the capital in a big conurbation soon. It is an industrial center (wool and cars), port (wool, wheat, and oil), and home of a splendid boys school, Geelong Grammar. Prince Charles was sent there for a year. Ballarat, third among Victorian cities, establishes once again the point that the bulk of the Australians live in their coastal capitals: it is, with only sixty thousand people, the *second*-biggest inland city in the nation, after Canberra. Ballarat is a gold-rush town that managed to come through the bursting of that bubble without the ugliness attending such towns in the American West, for it has begonia-filled parks, good schools, and a passable art gallery. To the west is basaltic land that grows wheat and Geelong wool, the setting where in the last century farmers and graziers did their elegant best to ape the English landed gentry.

Bendigo, population 45,000, is similar; when the mines—the Golden Square, the Extended Rustlers—shut down, the town turned itself into a place for retired people. It also has an ordnance factory with a big gear-cutting plant. It is ornate, Victorian, beflowered, substantial; its park features a lush fernery and its art gallery a collection of French Impressionists. The director, John Henderson, likes to tell of putting himself out to escort a New Australian around the museum a few years ago, struggling to communicate despite the immigrant's Slavic accent and faulty comprehension. Turned out the fellow was second secretary in the Soviet Embassy in Canberra.

A Melburnian idea of a lark is to go to see the penguins. These flightless birds, a foot high and formally dressed, live on Phillip Island, along the coast, and by day swim out to sea in search of food for their young. Nightly,

at a certain moment—Phillip Island natives can predict it with great accuracy—a wave washes up and deposits perhaps ten or fifty or one hundred penguins. They rest a moment, tired from their long swim, and then march up the beach to their burrows. Subsequent waves bring similar groups. Incidentally, the penguins' feathers are glazed so well that the birds land perfectly dry. At the burrows, they disgorge partly disgested fish and plankton into the beaks of their young, who look like balls of fluff. Besides the penguin rookery, Phillip Island also has hundreds of easily visible koalas.

More Victoriana: Healsville, near Melbourne, has a wild-life sanctuary where visitors can walk among emus and kangaroos and see the playful platypus swimming in a glass tank. Gippsland, the eastern end of Victoria, has a Ninety Mile Beach on a spit that encloses charming and little-exploited lakes; the town of Sale, here, supplies the offshore oil drilling platforms in Bass Strait and has grown in population from six to ten thousand, including many Americans. There are fast-growing towns situated on Western Port Bay, an inlet of Bass Strait, which will be able to dock big, new, deep-draft ships, which cannot get to Geelong or Melbourne. Nearby French Island is to be the site of a giant petrochemical complex. By contrast, a town doomed to disappear is Yallourn; it stands ill-situated on the deposit of brown coal (lignite), thirty miles long and five to ten miles wide, which is mined to run generators that supply 70 percent of Melbourne's electricity. Finally, the railroads through all this part of the country run on broad gauge (5 feet 3 inches) except one, the line to Sydney, which is standard gauge (4 feet 8½ inches). Until this standard-gauge track was pushed through in 1962, passengers had to change trains at the border of New South Wales.

An Alluring Gallery of Art

In his office on St. Kilda Road, part of what used to be a mansion and now sprouts drawing boards on every side, Sir Roy Grounds draws the window shade. He does not do so for privacy, or to keep out the sun, but to reveal a paper with a table of figures on it, taped to the back of the shade. "On March 15, 1977, at noon, *noon*," he says, tapping the timetable, "the Melbourne Performing Arts Center will open."

Sir Roy, who is the architect for the Center, predicted in December 1960 that Melbourne's National Art Gallery, also his work, would open on August 20, 1968. It did.

The gallery and the performing-arts halls are really Phase One and Phase Two of the same project, the Victorian Arts Center; architecturally, they will form a harmonious whole. There is about them little of the rash daring of Utzon's sculptural opera house in Sydney, but they do incorporate some

strikingly unconventional concepts. The concert halls, to cite an example, will be far underground.

The National Gallery, long since in use, is all blue basalt stone and water. In form it is a parallelepiped, 200 feet wide, 500 feet long, and 52 feet high. A moat surrounds and reflects it, with a simple bridge crossing to the round-arch entrance. One turns to either side to enter the building, for the arch has an inner wall made of a huge hanging sheet of plate glass with water running down the outside to create a waterfall blur. Light for the interior comes from clerestory eaves, rooflights, and three large courts. One court resembles a balconied Elizabethan theater. The center court is a sculpture garden, with works by Rodin and Henry Moore. The third contains fountains that shoot icicle-like jets up at an angle, chords of water that curve into parabolas and break up glittering as they fall. A Great Hall for civic gatherings goes the full height of the building; slender, tapering steel columns hold up a roof of stained glass designed by Leonard French.

The interior is carpeted in gold and fitted with easy chairs of a comfort perhaps never before felt in a museum. Such seduction is strictly intentional. Sir Roy says that the gallery was "not built for esoteric art lovers, but for people who had never been in a gallery." He has in mind the likes of the New Australian laborers who built the museum. "I used to talk to the men on the job, Yugoslavs, Italians, Greeks; they could hardly speak English. I'd say, 'This is being built for you and your wife and kids. I'm trying to get a bloody good job.' "

From St. Kilda Road, on which the gallery fronts, the Performing Arts Center will appear to be a street-level plaza north of the gallery, out of which will rise a 430-foot spire, a sort of cone tapering into a needle. An opera-ballet theater, a playhouse, and an experimental theater will nest under the plaza, to a depth of 115 feet below street level. Construction, which began in 1971, required sinking a sort of cofferdam down to bedrock and then excavating the enclosed mud and gravel to create ten million cubic feet of space for the theaters. This job proved so expensive that Sir Roy was forced to lop one entire concert hall out of the plan. A curious engineering complexity is that in order to keep the structure from floating up through the surrounding mud, it will have to be anchored to the bedrock. The advantages will be acoustical and thermal insulation.

An Animated University

A professor at Australian National University reported a few years ago that of 366 directors and executives of business firms surveyed by him a full 17 percent had attended Melbourne Grammar School. Private schools still count

heavily in Victoria; a "Grammar old boy's tie" will take a man a long way. Playing Harvard to this Groton is the University of Melbourne, a public institution like all universities in Australia, but duly venerable, having been set into motion by a twenty-three-year-old chap from Trinity College, Oxford, in 1853. With twenty thousand students on a campus packed with buildings cheek-by-jowl, Melbourne University is badly overcrowded. It is a strong research school. Conversely, a university still growing up is La Trobe, opened in 1967, with one thousand students now and ten thousand anticipated by 1976. The liveliest place, though, is Monash University, founded a decade ago and now enrolling ten thousand students.

The school is named for Sir John Monash, a very admirable Victorian. After taking degrees in arts, law, and engineering at the University of Melbourne, he went into the army and ultimately became commander of the five Australian divisions that fought along the Somme in 1917 and 1918. After the war he took charge of the Victoria State Electricity Commission, at that time opposed by private enterprise as a form of socialism, and opened up the brown coal deposits at Yallourn that now supply Victoria with most of its power.

The university was created from the ground up in three years, 1958–1961. It started out to be a science and technology school, but now has arts, education, economics and politics, medicine and law, as well as science and engineering. Lunch at the Faculty Club—which, in accordance with Australian ideas of democracy, is open to the school's gardeners and janitors as well as its teachers—can be a lively experience. Cocktails and wine are served, hardly the custom at most American universities, and the talk, which is apt to go on until 3:30 in the afternoon, is provocative. The students are lively, too, and they keep stodgy Melbourne in an uproar. One Maundy Thursday they set the city on its ear by staging a mock crucifixion. They have participated in most of the causes favored by radical American students in recent years—sending aid to the Vietcong, for example, marching on Moratorium Day, sleeping-in at the Student Union building in protest to war-industry recruiting on campus, and, last but not least, howling about inadequate space on campus to park their cars.

Some complaints students voiced in a round-table interview were: "The welfare system is no good—they promise free hospitals and then they meet you at the door shaking a can." "The Aborigines in this country live in slavery, working as stockmen two hundred miles from their homes, riding sunup to sundown for bully beef, bread, and tea, and getting paid only once a year." "We are worried because of the trend to the American way of life. There are lots of aspects of America that we don't want transplanted

here." Many observers note, however, that in their very complaints Australian students copy Americans.

The dominating building on the campus is the Robert Menzies School of Humanities, nicknamed "The Ming Wing" because some Australians still give Menzies the Scottish pronunciation of "Ming-us." The former Prime Minister's cousin, Sir Douglas Menzies, is chancellor of the university, but the operating president of Monash, as at all Australian universities, is the vice chancellor, J. A. L. Matheson. A civil engineer, he built the school from scratch, but now finds the office burdensome in the same way that American college presidents do. Some of the students are tough Maoists, some are on hard drugs, and Matheson is known to be worried that public reaction to the school's extremists might conceivably force it to close.

Monash students call the University of Melbourne "The Shop." Melbourne students call the University of Monash "The Farm."

A Gallery of Victorians

Victoria's conservative political coloration is felicitously embodied in its Premier, Sir Henry Bolte, a Liberal. He is a solid man with a cannonball head and hands the shape of a spade. Country-born in Victoria, in 1908, he grazes four thousand sheep on an eleven-hundred-acre property, owns and races horses, goes off shooting, and plays golf. It is said that he got into politics because friends were pressing him to enter public life and on the third Scotch he agreed. He has been Premier more than fifteen years. "This is a sovereign state. I am Her Majesty's Premier." He is proud of being the founder of Monash University, but his own formal education ended with graduation from the Church of England Grammar School in Ballarat.

Sir Henry accepts wholeheartedly the principle that foreign investment is vital to Victoria, and encourages it with tax incentives, advice, and assistance. He often goes abroad (seven times to the United States so far) on "Promote Victoria" missions. The results are a spate of new industry, new reservoirs, new dams, new ports, new oil and gas fields offshore in Bass Strait, increased production of electricity. The Premier boasts that Victoria's industries are the country's most modern, so that its wages are higher and savings are far greater per capita. He argues that Melbourne still leads Sydney as a financial power, having the head offices of every bank but one, the head offices of most insurance companies, and the ability to float big stock and bond issues. Bolte is quite aware that Australian premiers have much more power than American governors, and he uses his to the fullest. He must contend for conservative votes with the Country Party and the breakoff Democratic Labor

Party, but managed in the 1969 election to hold his own against an anti-Liberal trend that cost the party votes in every other state. The ministers who assemble around the huge round blackwood cabinet table (14½ feet in diameter, standing on a single steel pedestal) in the Premier's office are all Liberals—no coalition here. Sir Henry, though he swears like a trooper, is as conservative in morals as he is in politics, a powerhouse of Victorian wowserism. "I am as opposed to the legalizing of homosexuality as I am to the introducing of poker [slot] machines," he said in 1970.

*

Gray, lean, swift-talking: such is Sir Ian Potter, who aspired to be a university professor but instead founded one of Melbourne's biggest stock brokerages and now diligently attends to the cultural life of Melbourne and Australia. Australians dote on saving money; Sir Ian says that 28 percent of the national income goes into savings. Everyone has a savings account, but millions of Australians are investors, too, from "penny tremblers"—small speculators—on up. Potter still talks knowledgeably about business; he knows the mining industry well; but his chief interests now are Monash University (he is chairman of the finance committee), the Performing Arts Center (he is a member of the building committee), the Elizabethan Theatre Trust (he is chairman), and Australian opera, ballet, and symphony (he helped send the latter two on recent tours of the U.S.). And, sitting in skyscraper offices carpeted in pepper-and-salt Australian wool, he ponders the nation's problems. How big should Australia be? "When I was young, the parameters suggested ten million. Now I say fifty million. And that's enough. But we cannot bring in more immigrants than a number just about equal to one percent of the population each year, which added to the one and three-quarters percent natural increase will get us to about twenty-five million in the year 2000."

*

A Victorian who holds that pretty much everything is wrong with the picture painted by Bolte is James Ford Cairns, a Labor Member of the federal House of Representatives from the town of Lalor (named for the leader of the rebels at the Eureka Stockade). Cairns holds a Ph.D. in commerce (economic history) from the University of Melbourne, doing a thesis on the welfare state in Australia. He then taught there ten years. He went into politics "to put into practice what I had learned." He early made himself leader of the opposition to the Vietnam war, and later wrote a history

of Western intervention in Vietnam since 1847. He proposes "a searching re-examination of our foreign relations, and particularly those with the United States. We must cease to be military camp followers, if we are to have a chance of building a better society in Australia." Cairns is the last remove from a firebrand—he is slow, deliberate, well built, blondish, and youthful, and has fine strong hands. But he talks a toughly critical line. "We have swallowed the myth that this is an egalitarian society, but it isn't. The difference between the wealthiest Australians today and the poorest is greater than it was in the 1930's." On foreign investment: "We are probably paying too much for our development. The country is like a puppy lying on its back, having its tummy tickled."

Cairns occupies a position in Australian politics comparable· to that of Eugene McCarthy in American politics, with a similarly small chance of attaining the highest office, though Cairns does not shrink so moodily from power. He has an affinity for youth, and they for him. Cairns perceives the young in Australia as differing more from the Establishment than any recent generation, but in a moral rather than a doctrinaire way. He sympathizes with youth's complaints about the state of education, holding that schools are largely mere purveyors of collections of facts, aimed at ensuring equality but achieving only uniformity. On the subject of race, he says: "We ought to have more colored people," and observes that the "majority of people under thirty-five" oppose white supremacy. Discrimination against Aborigines, he notes, is much stronger in the countryside than in the cities, and is rooted in a white-purity, fear-of-contamination complex. The whites scrub themselves constantly, change their white shirts frequently, and keep Aborigines out of rural hotels and swimming pools because "they're dirty."

*

One Melburnian who proves that a single man can change history is B. A. Santamaria, a dapper, semibald man of sharp intellect, systematized and logical. The son of immigrants from an island near Stromboli, Italy, Santamaria went through law school at the University of Melbourne. In 1942, alarmed at the strength of Communism among labor unions in wartime Australia, he started a clandestine movement, mostly comprised of Roman Catholics, to break Communist power by creating cells to oppose the Communist cells. He succeeded beyond expectation, gathering such a constituency that the Australian Labor Party split and has never since been able to elect a national government. Now the Democratic Labor Party holds the balance of power in many elections, and Santamaria is the eminence behind it.

Though conservative, Santamaria shares Jim Cairns' doubts over the inflow of private capital. He wonders whether foreign investment amounts to "selling out to compensate for our lack of ability to export. We can't live by tapping our bank account forever." On White Australia he says: "We don't want an immigration policy on a racial basis, but we don't want to get into the same situation as Great Britain. We want a unitary national community with a sense of identity." To Santamaria this means that Australia must carefully control the number of incoming Asians, and keep a sharp eye out for their assimilability; Filipinos, he says, are "easy to assimilate, Indians impossible." The Japanese? "People haven't forgotten the war."

On defense, he holds that Australia needs the United States as a friend, but that "there is no such thing as an automatic friend. If you're strong, you have friends; if you're weak, you don't. Client states are treated as client states—which means that they are occasionally flattered, more generally ignored, and not infrequently betrayed." National security is what preoccupies Santamaria most. "Someday the United States will withdraw from this area. China is a problem. The Soviet Union will be a major Indian Ocean power in five years. The Japanese, a major economic power already, will be a major political power. Neutrality is not for us—we can arm, and we should start now. Also, we should have an alliance with Indonesia. Australia can defend itself for less than 7 percent of its Gross National Product—a nation not willing to spend that doesn't deserve to live. But we are," he concludes, "a nation of lotus eaters."

CHAPTER 13

Inside Tasmania

Methinks I see Australian landscapes still,
But softer beauty sits on every hill.
—KIRK HERVEY, DESCRIBING
TASMANIA IN 1825

HERE IS A STATE all out of scale, and all out of context, with the rest of Australia. A dab of an island lying 125 miles south of the continent, Tasmania is green, tidy, well watered and well tended, mountainous and sundered by gorges. You can drive from anywhere to anywhere in it, and back, in a day. The sensation, upon flying in from the continent, is like arriving in Costa Rica—all dimensions appear to have shrunk. The railroad seems a toy, with its narrow gauge and 588 miles of track.

The longest east-west distance in Tasmania is 190 miles, and the longest north-south distance is 180 miles; the whole area, including nearby islands within the jurisdiction, measures only 26,215 square miles, about equal to Scotland or West Virginia. This heart-shaped island lies just as far south of the Equator as New York lies north of it. To the west there is nothing but ocean as far as Patagonia.

The people, who number barely 400,000, like to be called Tasmanians rather than Australians, and refer to the rest of Australia as "the mainland." Like Victoria in British Columbia, Tasmania likes to think of itself as "this other England"—a salute, in this case, to its white stone cottages, winding lanes, hawthorn hedgerows, Wordsworthian daffodils, Georgian architecture, and stout British stock. But the British Admiralty Hydrographic Department describes Tasmania as "the most thoroughly mountainous island in the

world," so this terrain bears little resemblance to Old England. Equally un-English are the forests of gum trees that cover half the state.

Tasmanians fuss over issues like these: Should the state permit wine bars and night clubs? How much should trucks pay for the use of the roads? Should *Playboy* be banned? (It sometimes is.) There's a slightly archaic air to some of these concerns, and this feeling of being stuck in the past comes on strong at the Legislative Council, one of the more conservative remnants of Australia's colonial period. Its mid-Victorian chambers in Hobart, the capital, are fitted with red-plush chaise longues, a blue arched ceiling, and a full-length portrait of Queen Elizabeth. The quaint qualifications required of voters for elections to this upper house are: minimum age of twenty-five years, ownership (or leasehold) of property, and either membership in a profession or a record of service in the armed forces. Actually, this venerable institution is powerless except to hold up legislation, and its prestige may have been diminished last year when one member was arrested for shoplifting. The authentic lawmaking body is the democratically elected House of Assembly.

This picture of fustiness does not by any means hold true for all of Tasmanian society. Farming methods are modern—apples, for instance, are picked by machine and never touched by a human hand until the buyer takes them from the store in some land across the seas. The motels and hotels are generally more up-to-date than the rest of Australia. Education uses an advanced, think-for-yourself approach in its independent matriculation colleges, where high school graduates study a year before entering universities. But the University of Tasmania, where most of them enter, is, with 2,500 students, too small for the state's needs, although it has a full complement of schools, including law and medicine. It is, one professor explains, halfway between the British and American patterns—not as elitist as in Britain, not as broad as in the United States. All the Australian state universities, incidentally, are in effect campuses of a single national university, since federal support imposes standards, salaries, and uniformity.

Tasmania is the state where the convict heritage looms largest; between 1803 and 1854 it received seventy thousand prisoners, as many as New South Wales, but into a much smaller area. The state's attitude to this background is still typically ambivalent. Most Tasmanians would rather not mention that grandfather or great-grandfather was a convict, even though this is the case with a dozen leading businessmen. Old files of the Hobart *Mercury* are full of perforations where people have managed to remove references to their convict ancestors, and many convict records have disappeared from the state archives.

But, as Tasmanians are ruefully aware, the history of the convicts is of much interest to tourists, and Tasmania wants and gets many tourists, both from the rest of Australia and from overseas. The Tasmanian Museum and Art Gallery in Hobart exhibits cat-o'-nine-tails, gags, handcuffs, balls and chains, yokes, and wardens' cutlasses; visitors are fascinated.

*

The Governor of Tasmania, Sir Edric Bastyan, is one of the three remaining English-born governors of Australian states (the other two are in Victoria and Western Australia). Sir Edric, a military man with a ramrod back, occupies a big Tudor Gothic house with a white stone facing, a ballroom eighty by thirty feet, and elegant gardens in the Hobart Domain—such an imposing edifice that local antiestablishmentarians propose abolishing the governorship and using Government House as Parliament House. The roof of the solarium would have to be fixed first—it leaks.

Apples, Meat, Seaweed, and Crayfish

This fortunate little state has an economy so diversified that it is unlikely ever to suffer a knockdown blow. It is the truck garden of Australia, the orchard of England, and a treasury of minerals, too. Wool, meat, dairy products, fruits and vegetables, mining, metal extraction, and timber products contribute to the economy in roughly equal amounts, and grain and fruit preserving are just behind. This state supplies potatoes for Australia and three-fourths of Australia's overseas exports of apples—big, crisp Tasman Prides,* sent mostly to England. Some single Tasmanian trees produce more than one ton of apples.

Other products of Tasmanian farms are green peas, berries, pears, pigs, fat lambs, poultry, and hops for most of Australia's beer. The state's sheep have historically been Merinos, but in the last ten years a wool grower named Ian Downie, working with the state Department of Agriculture, has developed, from Merinos and Corriedales, a high-quality breed called the Cormo. Its wool, at auction in 1971, brought forty-five cents a pound, seventeen cents more than other wools. A Japanese textile maker bought the whole crop to manufacture it into women's suits. Seaweed is harvested mechanically on the east coast of Tasmania and made into alginate, a thickener for ice cream.

* Another Australian (though not Tasmanian) apple growing popular in the United States is the green-skinned Granny Smith. It gets its name from having been first cultivated by Maria Ann Smith, of Eastwood, New South Wales, in the 1880's.

The sea off Tasmania also produces whopping crayfish, eighteen inches long, with ten legs (but no claws; only fresh-water crayfish have claws). Fishermen catch them in wicker-basket traps; they are frozen and exported to the United States. Tasmanian fishing also includes enormous quantities of scallops, and couta, local name for barracuda. Sport fishermen catch bluefin tuna weighing up to 250 pounds.

*

Tasmania's declivitous mountains, alpine lakes, high rainfall, and numerous short, rapid rivers give this state no less than half of the water-power potential of Australia. About two dozen power stations utilize this falling water to make 1,240,000 kilowatts of electricity—and make it as cheaply as anywhere in the world. This $370-million system is the "industry behind the industries" of Tasmania. Cheap power has attracted two big electrolytic zinc smelters, a large refinery, which uses alumina sent from Queensland, and a sizable pulp and paper industry.

Tasmania also has western mountain ranges that seem to be mostly minerals. Mount Lyell, developed in 1897–1898, is one of the world's great copper mines. Elsewhere the state produces lead, zinc, silver, gold, nickel, cobalt, and tungsten. Since 1968 it has been exporting iron ore from a remote spot on the Savage River in the northeast corner of the island. In this ingenious operation, the ore, crushed and separated from rock, is mixed with water to form slurry, and pumped for fifty-three miles through a nine-inch pipe to the north coast. There a pelletizing plant removes the water, mixes the ore with additives, forms it into small balls, and bakes them hard. Conveyor belts take the ore out over the open sea and pour it into the holds of ships which are positioned by hawsers secured to the rocky sea bottom. This clever arrangement avoided the cost of building a breakwater and docks, and ships have resisted ninety-mile-an-hour gales without mishap—though they generally put to sea when storms come up.

All this ore goes to Japan, for an ultimate return of $400 million over twenty years. Also going to Japan is the product of another new Tasmanian industry, wood chips. Contracts call for the export of $350 million worth of this raw material for pulp over the next eleven to fifteen years.

The Capital

Every year, the day after Christmas, a score of big sailing yachts tack out of Sydney Harbor and turn south in a 680-mile race to Hobart, where they arrive around New Year's Day. This is a nice symbolic connection between

two of the loveliest harbors in the world. Hobart,* with 150,000 people, is a sea-minded city where people in downtown offices can see, just out their windows, the frequent arrivals of merchant ships. The harbor is the estuary of the River Derwent, twelve miles up from Storm Bay. It widens at Hobart enough to accommodate large vessels—the *Queens* were there in their better days. Hobart, founded in 1803, is the second oldest city in Australia, after Sydney. It is a pleasantly provincial place, with sandstone warehouses in the city center dating from the 1830's, and some other good, old, stone buildings—Parliament House, the Theatre Royal, built in 1834 and far more famous for its architecture than for its theater, and an old brewery. A pleasant place in this area is Cat-and-Fiddle Square. So closely are the harbor and the city intermeshed that ships' bows loom over Campbell, Argyle, and Elizabeth streets, and fishing vessels, with nets and green glass buoys on their decks, lend a travelogue look to the place.

Hobart refines zinc, founds steel, prints Chinese silk for export to many parts of the world, makes Cadbury's chocolate. A forthcoming industry, seemingly out of place in this staid town, is a casino. The legal authorization for it was a traumatic procedure, with so many Tasmanians fearing the demoralizing effects of gambling that the conservative upper house blocked the proposition for a while. The House of Assembly bucked the decision to the general public by means of a referendum, which passed. Now the Wrest Point Casino Hotel is going up, at a cost of $10 million, and Hobart hopes to attract hordes of mainlanders to the only casino in Australia. However, another company wants to build a similar, though smaller, hotel-casino in Tasmania's second city, Launceston, which is even nearer to the prospective sinners of Melbourne. This prospect scares many Tasmanians, who just knew that letting gambling in would turn the state into a kind of Nevada.

Forest-clad Mount Wellington provides a stunning backdrop for Hobart. Rising out of the sea for 4,166 feet, this mountain carries snow for several months of the year; a road goes to a lookout on the summit. You can tour Hobart by taxi for eighty cents an hour—geraniums everywhere. Cruise boats go up the Derwent to New Norfolk, an experience featuring such charms as roses, black swans, and picnic hampers. Hobart and the surrounding forested area suffered a terrible conflagration on Ash Wednesday in 1967, when a thousand houses burned down, leaving nothing but a wasteland of chimneys. A good share of the damage, said to reach $500 million, was the result of a dynamite factory's blowing up; the explosion wiped out a cannery, a brewery, and a carbide plant. In the countryside, sheep were incinerated in their paddocks. Now little remains to show that the fire ever took place.

* Named, like Sydney, for the British colonial secretary at the moment of its christening.

Looking Around from Hobart

The orientation of Tasmania from this southern port city goes something like this. To the southwest is the Huon Valley, where most of Tasmania's apples come from; to the west is an area so mountainous and densely vegetated that it is not only unpopulated but virtually unexplored. To the northwest, on the Indian Ocean, is country rich in minerals as well as lovely mountains and national parks. The northern coast of the island, on tumultuous Bass Strait, is a strip of rich flat farmland. Straight north from Hobart is the sheep-raising Midlands valley, leading to Launceston. The northeast is a lightly inhabited corner of the state with good skiing country in the inland mountains; down the east coast are resort towns and snow-colored beaches. The Tasman Peninsula, which was the site of Port Arthur and the prison colony, juts off the southeast coast.

This is mostly clement country, milder even than northern California. The temperature in Hobart averages 61 degrees Fahrenheit in January, and 46 degrees in July. The Midlands, with such English trees as oaks, elms, poplars, and willows, stage a colorful autumn. In high country, such as that of the unroaded Cradle Mountain–Lake St. Clair National Park, a wintry snow scene can look virtually Canadian. The west coast is open to wild west winds and assaulted by heavy rains.

The eucalyptus remains the basic forest tree in this much beforested land, because the conifers that the white man found there when he arrived have all been cut down, and the imported Northern Hemisphere trees are mostly decorative. Conservationists fear that the eucalyptuses are going into pulp and wood chips too rapidly—such a fate does seem ignoble for these three-hundred-foot monarchs. But Tasmania started preserving national parks a long time ago, with the Scenery Preservation Act of 1915. Moreover, much country in the west is unloggable because the terrain, though few mountains are higher than five thousand feet (Mount Ossa, the highest, reaches 5,305 feet), is rent with gorges as deep as three thousand feet. A menace in this mountainous area is a curious phenomenon called the "horizontal." Saplings grow about fifty feet high, blow over, and lie parallel to, but about twenty feet above, the ground. This creates a platform that looks like terra firma, but men walking on it strike weak spots, fall through, land in an eerie vegetable dungeon, and cannot get out.

The animals of Tasmania are few but humorous. The Tasmanian devil is a heavy marsupial, black with white patches and the size of a wolverine, who earned his name by his cry, half-scream and half-snarl. In machine-like jaws, he crunches up everything from carrion to shellfish to straying lambs,

feeding a ravenous appetite. He can be tamed if taken young, and makes a contented if ponderous pet. Plenty of these devils remain, protected in national parks. By contrast, the Tasmanian wolf (also called the Tasmanian tiger) is so nearly extinct as to seem mythical. From time to time reports are cabled to the outer world that someone has seen the footprints of a Tasmanian wolf, but no authenticated sighting has taken place in years. And this beast would be easy to recognize: he has tiger-like dark stripes running vertically down his sides and rear haunches; his spikelike tail does not wag or bend; and his jaws will open to a 90-degree angle. He lives (if he lives at all) on small animals caught by chasing them until they falter from exhaustion.

Less of a joke are Tasmanian snakes, all poisonous: the copperhead, the whip, and the tiger. Tasmania has fourteen birds found nowhere else, a mouse-size possum, and a unique marsupial anteater. In high lakes grows a mountain shrimp, straight as a nail, which dates back so far as to make it a living fossil. In the Great Lake, largest of the state's several thousand shallow alpine lakes, trout grow so big that the law requires fishermen to throw back anything shorter than fourteen inches.

The set piece of bucolic beauty in Tasmania is the Derwent Valley, up from Hobart, which raises hops. This crop grows on vines that climb strings to wires strung eighteen feet overhead; to wander among them is to feel like a mouse in a cornfield. The vines are simply pulled down to harvest the hops, which are the fragrant cones that give beer its taste. Hand harvesting used to be the occasion of family excursions to the country for work vacations; now machines do the work. Hops are dried in oast-houses, which stand here and there in the valley. These are brick-and-shingle buildings with steep-sided pyramidal roofs surmounted by chimneys like obelisks.

*

Hobart people disparage Launceston, population 63,000 and 123 miles north, as "Cretinbury-on-Cesspool," but in point of fact it is a more dynamic place than the capital. It has the largest woolen mills in the Southern Hemisphere, and also makes breakfast food, tennis rackets, and boats. It is a city of waterways, up from the coast on the Tamar River. Once big ships could not reach it because of an inconvenient island downriver, but a couple of years ago the Port of Launceston Authority quarried out the center of the island to well below water level, making it like a giant rowboat, then blasted in the sides. Garden Island disappeared, and now ships as big as 55,000 tons can sail up to Launceston.

Near Launceston is Entally, an 1820's mansion now preserved as a national monument. Its builder was Mary Reibey, a girl convicted in England of stealing a ride on a squire's horse, and transported to New South Wales.

On the convict ship she met a young naval lieutenant; they later married, became traders and shipowners, and fetched up with a great Tasmanian estate.

A few other Tasmanian cities stand out for one reason or another. Port Arthur, the old prison colony on Tasman Peninsula sixty-three miles southeast of Hobart, is a monument to misery. Guides take tourists there these days and relate its history with gusto and gory details. Flogging, long terms in solitary on bread (one-half pound in twenty-four hours) and water (one quart), and other degrading punishments were inflicted on this colony's prisoners, mainly repeaters who had committed crimes in Australia after being transported. Nevertheless they put up some lovely buildings with their forced labor. The church, long since roofless but mellowly preserved, was a graceful, light, Gothic building, copied after Glastonbury Abbey, topped with spires at all of its numerous corners. Prisoners were brought to church in blinders that allowed them to see the pulpit but not their neighbors. The old lunatic asylum is now the Port Arthur Town Hall, which leads to endless wry comment. There are ruins of a powder magazine and a hospital, built like the other structures of pink brick and stone. The peninsula is hooked to the main island by Eaglehawk Neck, an isthmus only twenty yards wide. In prison times, hungry dogs were chained on short leashes across this neck, ready to rip the flesh off any escaping convict.

Queenstown, headquarters for the Mount Lyell copper mines, is one of those ecological disasters of the heedless past. You descend on a corkscrew highway through verdant rain forest to a valley where sulphur fumes killed all vegetation until new methods of smelting were introduced in 1922. Sulphur does weird things, brightly staining the rocks in chrome and purple and eating away at tombstones in cemeteries until they look like melted candles. Now patches of green are reappearing around Queenstown, adding new hues to a place often called "hell with the fires out," because it is barren despite one hundred inches of rainfall a year. The economy of the area is picking up—good new roads permit exploitation of minerals once unreachable. Nearby is Zeehan, a lead-zinc-silver mining town which once had ten thousand people, twenty-nine hotels, and a theater where Nellie Melba sang.

A Word on the History

Tasmania was discovered before the mainland of Australia itself. On December 1, 1642, the Dutch captain Abel Tasman put some men ashore at a spot on the east coast not far from what became Port Arthur. They heard voices and saw foot notches cut in a tree with a spacing that suggested a race of giant men, so they scurried back to their ships. In keeping with the

strange, unhurried discovery of Australasia, no explorer returned to Van Diemen's Land (which, as we have seen, was the name Tasman gave the place in honor of the Dutch East Indian Governor who sent him) until 1773, when Captain Tobias Furneaux reached Adventure Bay on Bruny Island just south of present-day Hobart. Fifteen years later, Captain Bligh and the *Bounty* put in at Adventure Bay and planted Tasmania's first apple trees. Matthew Flinders, sailing in a small boat with a naval surgeon named George Bass, proved Tasmania to be an island, by going around it, in 1798–1799. In 1803 Lieutenant John Bowen, who was precisely eighteen years old, was sent out from New South Wales to found a colony of convicts. Some pretty rough men, generally the more violent of the convicts, were molded into nation-builders and did a good job. By 1825, when Van Diemen's Land became a colony separate from New South Wales, Hobart was as large as Sydney. Then two circumstances combined to hold Tasmania back. One was the colonization of Victoria, which as we have seen was staged from Tasmania, and caused many people to leave the island. The second was the furious new British determination, at the end of the Napoleonic Wars in 1815, to make Australia take its felons. Tasmania became the "jail of the Empire," and the government in London did not even assume the burden of paying for the jails and wardens and police. One of the worst aspects of the punishment euphemistically known as transportation was the transportation itself. One ship, *George III*, sank in a channel off Tasmania in 1835, and troops shot the prisoners as they tried to jump into the sea. Another, the *Neva*, with 240 women, hit a reef and only twenty-two of the passengers got ashore to start their prison terms. A specialty of Port Arthur was a prison for boys at a locality named, with cold erudition, Point Puer.

Free Tasmanians railed at the "great curse" of transportation, and in 1847 an eloquent preacher and publicist named John West started a movement to end it. By 1853 Britain had agreed. Transportation of convicts ceased, but Port Arthur continued as a prison until abandoned in 1877. On a fluke caused by the slowness and uncertainty of the mails, Tasmania in 1855 became—by a few weeks—the first of the Australian colonies to receive self-government from London. The name Tasmania, assumed at this point to replace Van Diemen's Land, had been in colloquial use since 1823.

*

The Aborigines who occupied Tasmania turned out not to be giants, as Tasman feared, but they were different from the Aborigines on the mainland. They had no dingoes or boomerangs, and their hair was fuzzy rather

than silky; possibly they were by origin Melanesians who had sailed to Tasmania from the New Guinea vicinity. The two or three thousand of them were divided into twenty tribes. They wore no clothes, and lived in caves, hollow trees, and tentlike huts. They decorated their bodies with red ocher, scars, and shell necklaces. Their food was shellfish, lily roots, wallabies, fruits, and berries.

From the first, the whites regarded the Aborigines in Tasmania as little better than game to be shot and fed to the dogs. Convicts kidnaped their women. The Aborigines fought back, robbing, burning, and murdering. In 1830 the Governor of the time decided to round them up and isolate them, once and for all. A vivid account of this famous "black line" sweep comes from Charles Darwin, who visited Tasmania six years later and wrote about it in *The Voyage of the Beagle*:

> The whole island was put under martial law, and by proclamation the whole population commanded to assist in one great attempt to secure the entire race. The plan adopted was nearly similar to that of the great hunting-matches in India: a line [of three thousand soldiers and settlers] was formed reaching across the island, with the intention of driving the natives into a *cul-de-sac* on Tasman's peninsula. The attempt failed; the natives, having tied up their dogs, stole during one night through the lines. This is far from surprising, when their practised senses, and usual manner of crawling after wild animals is considered. I have been assured that they can conceal themselves on almost bare ground, in a manner which until witnessed is scarcely credible; their dusky bodies being easily mistaken for the blackened stumps which are scattered all over the country. I was told of a trial between a party of Englishmen and a native, who was to stand in full view on the side of a bare hill; if the Englishmen closed their eyes for less than a minute, he would squat down, and then they were never able to distinguish him from the surrounding stumps. But to return to the hunting-match; the natives understanding this kind of warfare, were terribly alarmed, for they at once perceived the power and numbers of the whites. Shortly afterwards a party of thirteen belonging to two tribes came in; and, conscious of their unprotected condition, delivered themselves up in despair. Subsequently by the intrepid exertions of Mr. Robinson, an active and benevolent man, who fearlessly visited by himself the most hostile of the natives, the whole were induced to act in a similar manner. They were then removed to an island, where food and clothes were provided for them.

Even though, as we have seen earlier, a few lived to return to Tasmania, by 1876 the last of them was dead. They had simply been killed off. All that is left from the genocide of the Tasmanian Aborigines is the habitat group in the Tasmanian Museum and Art Gallery in Hobart.

Back to Tasmania Today

A magnetic figure in Tasmania is the Roman Catholic Archbishop, Guilford Young. He has, in general, stood for the reformist, Vatican II position in Australia. But this complex man lined up with the right-winger B. A. Santamaria in the big Labor Party schism, and in the last year or so he has several times disciplined an angry young priest of his diocese who headed the Tasmanian Vietnam Moratorium Committee.

Archbishop Young comes from Queensland. He got his higher education in Rome during Mussolini's time, and fled Europe for New York just before World War II began. He toured the U.S., and, returning to Australia, was consecrated at thirty-one as the youngest Catholic bishop in the world. He went to Tasmania in 1954, and made a name for himself as a church liberal as well as school-builder. In his white hillside house in Hobart, with a stack of old *Encounters* behind him on a shelf, he made a couple of challenging points. Young, departing from the hopeful and nearly universal position of the politicians, thinks that "Australia is not necessary to the United States. We know we're expendable; the U.S. will become newly isolationist after the Vietnam withdrawal." He also points out that Australia, 27 percent Catholic now, will on projection be one-third Catholic by 1975. Tasmania itself has fewer Catholics proportionately than any other state except South Australia. The Archbishop, admittedly optimistically, holds that "what is happening today is not the death throes of civilization" but rather "the travail of a new age coming to birth." The reason for this hopeful view? "There is a factor X in history, which we may call the mysterious action of God."

*

Tasmanian tidbits: A taxi driver said: "Jews here go to Melbourne for the weekend, the Catholics pray, and the Anglicans stay in bed." Shops open at 9:05 A.M., not 9. Very little marijuana, LSD, or other drugs are used here, one is told. Handsome Georgian architecture prevails among the older buildings because this is one of the older colonies and also because the Victorian style did not reach here until twenty years after it infested England. Tasmania's most famous expatriates are Errol Flynn and Merle Oberon. Not much of the wave of immigration to Australia in recent decades has come to Tasmania; in fact, by a reverse process, Tasmania's young like to leave the place, to go at least as far as Melbourne or Sydney. Only three daily newspapers are published in the whole state.

CHAPTER 14

Adelaide and South Australia

*Yes, the Old Bloke makes a good job
of them up this way.*
—A STOCKMAN COMMENTING ON
A SOUTH AUSTRALIAN SUNSET

THE SMELL OF CHANGE, the pervasive sensation of a worn-out Establishment giving way to a new politics—you get that more strongly in South Australia than in any other part of the country. Here, in what has certainly been a mossback milieu, we find a percipient, mettlesome, enterprising, young, and potent new Premier, first fresh figure in Australian politics in a wombat's age. He is Don Dunstan, a Labor politician who came to office in 1970, breaking the solid phalanx of Liberal-Country leadership that then included all the other states and the federal government.* Perhaps he compares with Canada's Pierre Trudeau. Politician enough to dominate the Labor Party machine in his state, he is also an original thinker who draws support from the Adelaide intellectuals. He is married to a lecturer in economics who is part of his brainy circle. Dunstan wears his hair longish, dresses with style, plays Chopin on the piano, and likes to cook. He was born in Fiji, where his South Australian parents were running a business, and was educated in Adelaide's "best" school, St. Peter's, and at the University of Adelaide. Thus he is essentially a scion of the Adelaide oligarchy, which he has now set himself against.

Dunstan owes his election victory to an altruistic gesture by his Liberal-Country predecessor (and current leader of the opposition), Raymond Steele Hall, a man of surprisingly similar views. Hall decided that his party should cease winning elections by virtue of gerrymanders that favored the

* Western Australia went Labor a few weeks later.

countryside electorate, and passed voting reforms that cost him his office. Thus South Australia has two leading politicians expounding radical politics —though Hall's party is divided over him, and about half of it remains attached to conservative views.

Dunstan stands for mild social reform and energetic, state-prodded development. He wants to improve schools and transit, control pollution, and protect consumers. Holding that South Australia's economy is too tightly linked to the automobile and electrical industries, he wants to diversify and export. To these ends he is creating a state development corporation and a merchant bank, pushing the movie industry, and establishing trade centers in four Asian countries. South Australia is only modestly endowed with resources—some iron ore, a south-coast swathe of good farm and vineyard land, plains suitable for wheat or wool (both of which have declining markets). On this resource base, the state historically has done well; Dunstan wants to hold it all together and advance.

The roadblock is the Legislative Council, the upper house of the South Australian Parliament, a body elected under such strict voting qualifications that sixteen of its twenty members belong to the hidebound opposition. Dunstan wants to change this archaic chamber from a "bastion of selfish privilege" to a toothless "house of review," but so far he has failed. He sent up a bill to democratize the manner of election of members of the Legislative Council, but the proposed house of review refused to review itself—and it does have considerable power to delay legislation for years. Dunstan also plans to fight for civil liberties—particularly to curb eavesdropping devices and the general warrant, which permits police to search houses for any purpose or even none. He deplores dossier-keeping and censorship—for example, as we noted earlier, he did not try to block the sale of *Portnoy's Complaint*, the object of much censorial effort elsewhere in the country.

Most of these goals, of course, are radical only in Australian terms, repairs to be made in a society more heavily encrusted with tradition and immobility than its short history would seem to warrant. This is a state that takes immense pride in being the only one with no convict origins. Adelaide boasts that it is more aristocratic and less money-minded than Melbourne. How sweet Adelaide got so staid makes a paradoxically colorful story. Let us quickly run through it.

A Bounder and a Founder

The story starts in London in 1830. The talk of the city, at one point that year, was *Letter from Sydney*, a book by a bounder (also a philosopher) named Edward Gibbon Wakefield, who had just spent three years in New-

gate Gaol for abducting an heiress to Gretna Green. His book, which was widely taken to be authentically from Sydney though Wakefield had never been there, dealt with the art of colonization. It detailed errors made earlier in New South Wales, and put forth an idealistic how-to-colonize plan that Wakefield had conceived while doing time in what must have been a rather literary kind of imprisonment. Even as London's theorists, capitalists, and speculators pondered the possibility of using Wakefield's ideas to solve some of England's social problems by sending the hungry and unemployed abroad, news came from Australia that the explorer Charles Sturt had floated down the Murray River to its mouth and found "rich and lovely valleys."

All this had much to do with shaping Adelaide and South Australia as they are today. Economic philosopher Wakefield, the son of a surveyor and land agent, had an Adam Smithian notion that the price of land was the key to successful colonization. If the price was too low, or if the land was free, he thought, penniless laborers would take it and turn themselves into hopeless, struggling peasant farmers. Conversely, if the land was priced high, it would attract purchase by capitalists, who would run it well, and the funds generated from sales could be used to pay the passage of trades-men and laborers from Britain to Australia. These people could work for the capitalists and eventually save enough to buy land of their own. Wake-field also insisted on migration by people who were respectable, certainly not convicts, and preferably married, to avoid the overly masculine society typified by New South Wales. Soon a group of Londoners excited by Wake-field's ideas formed a South Australian Association to put theory into practice.

The first 546 settlers, mostly English yeoman farmers, went out in 1836. Upon landing they were sufficiently nervous to mistake the raucous laughing of kookaburras for an attack by Aborigines. At length, under a gum tree* at a seaside spot near present-day Adelaide, they proclaimed South Australia a province of the Crown.

If Wakefield seems a bit of an eccentric, he is almost matched by Colonel William Light, the founder of Adelaide. Prophetically, Light (born 1786) was the son of the founder of another city—Penang, the first British set-tlement in Malaya—his mother being his father's Portuguese-Malay mistress. Light went to sea in the Royal Navy at fourteen, left the service at sixteen, toured France and escaped when the Napoleonic Wars resumed, visited India for a year or so, and in 1808 bought a commission in the British Army. In the Fourth Dragoons he fought in Spain under Wellington for six years, taking part in forty-five actions. Going on half-pay after Napoleon's defeat, he toured Europe sketching. In 1823 he was back in Spain as a lieutenant

* Long since dead, but still preserved by chemicals.

colonel with revolutionary forces fighting the Spanish king, and nearly died of his battle wounds. Next year he met and married Mary Bennet, illegitimate daughter of the Duke of Richmond. Light bought a yacht and they cruised the Mediterranean, ultimately meeting Mohammed Ali, Pasha of Egypt, in Alexandria. This dignitary commissioned Light to create an Egyptian navy. Light left to round up sailors and a ship, and a few years later sailed to Egypt in a new paddle steamer, the *Nile*. While he was in England, Mary, who was prone to wear gold-braided trousers and a white turban with a purple tassel, became involved in a love affair that left her pregnant, and Light chucked her. For a few months Light and a Royal Navy Captain named Joseph Hindmarsh ferried troops from Egypt to Syria. Hindmarsh then returned to England, where the South Australia fever was running high. He got himself appointed Governor of the new colony, and Light was made Surveyor-General.

Fifty years old, and in bad health, Light took command of a small brig and set out for Australia, in mid-1836, assigned to explore fifteen hundred miles of shoreline to find the best location for habitation, plan a city, and then survey 150 square miles of country for settlement. In a mere matter of weeks, he found a lovely city site on a rich plain between the sea and the mountains, laid it out in a design that remains his claim to immortality, and surveyed the main streets. Hindmarsh arrived, only to clash with Light over his choice of site, but Light prevailed. He plunged into the task of making a trigonometric survey of land for settlers, and for a while kept up with sales of plots. Then he was ordered to survey another thousand square miles, and began to fall behind. New settlers could not be put on the land for lack of boundary markers. London complained. Light patiently replied that the head office "reckoned as if this survey should be made in England or Ireland where every facility is given by roads, carriages, provisions, lodgings," whereas he had to transport "tents, provisions, and even water over a country without roads." To Wakefield he wrote: "I am harassed in mind beyond all you can conceive." He resigned, and so did his staff. He was ill and in debt, and to top it off his house burned down, destroying many sketches and his journal of thirty years, not to mention his surveying instruments. He died of tuberculosis in 1839, at the age of fifty-four.

For a few years South Australia suffered headaches that Wakefield never dreamed of. Slow surveys confined new arrivals to Adelaide, where land prices soared to as much as £2,000 an acre. News of such speculative profit boomed South Australia Company shares on the stock market, which used the funds to export more settlers, compounding the problem. Then vigorous and expensive efforts to get land surveyed put the local government so far into debt that the Colonial Office had to bail it out—which was done at the cost

of demoting South Australia from a self-reliant province to a Crown Colony. It was thereupon dressed in "Downing Street swaddling clothes," that is, given a Governor, George Grey (whom we shall meet again in a later chapter), and a Legislative Council. Yet by that time 500,000 acres had been surveyed, the 5,000 settlers were moving from city to country, wheat was beginning to grow abundantly, and 200,000 sheep were on the land. Cleverly enough, the colonizers had brought in a group of Lutherans from Brandenburg, who resented the King of Prussia's attempt to unite the Lutheran and Reformed churches. They started South Australia's famed wineries.

Adelaide lies on terrain that slopes from the Mount Lofty Range to the Gulf St. Vincent, a coastal plain 150 miles long and as much as 50 miles wide, all good grain land and accessible to the sea for transport. The colony quickly became the granary of Australia, profiting more than the miners themselves from the Victorian gold rush of the 1850's. Only a decade after settlement, a miller named John Ridley invented a mechanical reaper that came into wide use. A couple of other inventions made it possible to grow grain on land covered with mallee, a scrub eucalyptus. One was the Mullenizer, a great drum that rolled over the scrub to break it down. The other was the stump-jump plow, with springed wrought-iron plowshares that could rise over stumps and roots, and eliminate the need for grubbing them out.

In 1857, a Surveyor-General named George Goyder roughed out a go-no-further line between good grain land, which had at least ten inches of rainfall yearly, and dubious desert. Foolhardy farmers crossed the "Goyder Line" and proved—for a while—that they could raise wheat on eight inches of rain.* But in 1881 nature caught up with this perilous practice, and a drought began that lasted more or less steadily until 1904. Farms were abandoned; the Commercial Bank of South Australia failed. Ultimately science—hybrids, fallowing, and superphosphate fertilizers—restored South Australia's agricultural prosperity.

Goyder's Line marked off the economically significant part of South Australia, but that was only a small fringe of the coast. The unfavored remainder of the colony opened up slowly. Charles Sturt's tortured trek northward in 1844 brought back a report of desiccated salt lakes so frighteningly barren that the colonists were discouraged—luckily, as it turned out —from abandoning wheat raising in the south for dryland sheep raising farther north. About 1860 an intriguing concept led South Australia to offer £2,000 to any explorer who could go all the way from Adelaide to the north coast. The London-Java cable was about to be extended to Australia.

---

* The average annual rainfall over good U.S. wheat country, specifically North Dakota, is eighteen inches.

The enterprising colonial government proposed to find a suitable north coast terminus and then connect it to Adelaide by almost two thousand miles of overland telegraph. John McDouall Stuart, who had made three expeditions to the north, pushed all the way to the Timor Sea in 1862. "I dipped my feet and washed my face and hands in the sea." On his return trip, privation reduced him to a "perfect skeleton," but he got the prize. The government of South Australia used his explorations to annex the whole Northern Territory in 1863 (not relinquishing it to the federal government until 1911). South Australia also built the telegraph, which linked the major Australian cities to the world, and to instantaneous world news, in 1874.

These explorations finally opened the better parcels of land north of Goyder's Line to sheep raising, inspiring formidable characters like Sidney Kidman, who came to be called the largest landholder in world history. It is said that he could go from Adelaide to Darwin without leaving his own land. He was also, in the eyes of his employees, mean and dishonest. He reputedly fired any man caught lighting his pipe with a match instead of a firebrand, when sitting around an evening campfire, because a man that wasteful must automatically be a shiftless worker.

The Constitution that South Australia got in 1855 was the most radical of all the charters thus far negotiated with the Crown by Australian colonies, reflecting the independence and nonconformism of the free settlers. It specified an elective upper (as well as lower) house, manhood suffrage, and vote by ballot. These wild democratic notions led at first to a measure of chaos: twenty-three governments under eight premiers between 1861 and 1876. But in more recent times, one Premier, Sir Thomas Playford, "the architect of present-day South Australia," managed to rule, on the Liberal–Country League ticket, from 1938 to 1965. At the turn of the century, South Australia voted a thumping five to one in favor of federation, but by the time of the Depression of the 1930's had cause to regret this enthusiasm. The federation had authorized high tariffs that made manufactured goods costly to the farmers; freights to markets were high; so were taxes. South Australia was a "mendicant state." In response to this plight, the federal government started revenue-sharing, and the state has generally prospered ever since then.

Green-Belted Adelaide

Colonel Light wrote, at the end of his life:

> The reasons that led me to fix Adelaide where it is I do not expect to be generally understood or calmly judged of at present. My enemies, however, by disputing their validity in every particular, have done me

the good service of fixing the whole of the responsibility upon me. . . . I leave it to posterity and not to them, to decide whether I am entitled to praise or blame.

The verdict would seem to be that his was a spectacularly advanced city plan that functioned well for many decades.

Adelaide is named for the consort of King William VI. The Torrens River flows through it, and Port Adelaide is seven miles to the west. Light's central city is a square-mile grid of streets alternately wide and narrow. A big square occupies the center, and four lesser squares (one named for Light) lie near the corners. The inner city is surrounded by a green belt of parklands, which exist to this day. The Torrens runs through the northern green belt, which also contains the Adelaide Oval (for cricket), the Botanic Gardens, the zoo, and the municipal golf links. The major public buildings are all in the center of the city—Parliament House, the public library, the art gallery, the museum, the University of Adelaide, the Royal Adelaide Hospital, and even the railroad station. On higher ground, north of the parklands, is a posh residential area that includes St. Peter's Anglican Cathedral.

Oddly, the problem with Adelaide, which has a population of about 750,000 (depending on where you begin to count), is that it has both too many people and too few. The inner city, way down from its 1915 peak population of 44,000, has too many factories, warehouses, hovels, and even little truck gardens. The City Council is building new blocks of flats and terrace houses to bring back a thousand people a year. On the other hand, the suburbs are overextended, for in Adelaide as elsewhere in Australia almost everyone has his own home, garden, and workshop. As an answer to overcrowding, the South Australian Housing Trust, a state-owned but independent corporation, has constructed the city of Elizabeth, a planned satellite for a planned city, seventeen miles north of Adelaide. Its housing, too, is mostly single homes, forty models rather too cutely varied but so artfully distributed that no street has two houses alike. They are divided into six neighborhoods plus parklands, with many walkways instead of streets, and easy access to shopping and places of work. All electrical cables are underground. The chief employers are a governmental Weapons Research Plant and factories that make cars and sewing machines. Most residents are immigrants, chiefly British.

The traditional adjectives to apply to Adelaide are parochial, decorous, conservative, gracious, provincially aristocratic. The tone was set a century ago by the old pastoralist families; the city has often been compared to a large, prosperous country town; you hear mention of "fifth-generation"

Australians. The police are effective, as befits an orderly town. A couple of years ago one could see astonishing displays of hats on the ladies in the dining room of the now demolished South Australia Hotel: rose petals, pink tulle, white straw with blue ribbons, and turbans, turbans, turbans. The rest of Australia may be on the verge of turning convict descent into a point of pride, but South Australia still boasts loudly that no convicts settled here, and damn few Irish either. The state has an Establishment, and the Establishment's citadel is the green-shuttered, cedar-doored Adelaide Club. Churches are so numerous that Adelaide is sometimes called "The Holy Land." Much architecture, in liver-colored brick or maroon stone, bespeaks a dowdy complacency.

This traditional description is now getting a bit out of date. Industry, technology, and immigration are forming a new, noisy, pleasure-loving Adelaide. The oligarchs are beginning to seem ineffectual . . . so many new people have never heard of them! Managers, experts, technicians are taking over. Long white gloves are going out and barbecue pits are coming in. A symptom of the change is the Adelaide Film Festival, run even-numbered years since 1960—two weeks of new movies, late nights, wine, gaiety, and culture, sponsored by the rapidly evolving Establishment. The city "considers itself a rival to Edinburgh in a furtive sort of way," says Max Harris, the Adelaide poet, critic, and publisher. It is easier to drink, dine, dance, and watch a floor show in Adelaide nowadays than it is in Melbourne.

A typical leader in this new burst of energy is Kym Bonython, best known for his art galleries here and in Sydney. His ancestors, Cornishmen, emigrated in two waves, one to Adelaide and one to found the town of Castine, Maine. Besides running his galleries, Mr. Bonython breeds cattle, operates a speedway, plays drums, serves as a disc jockey in his thirty-third year on a national program, and promotes jazz concerts, once importing Duke Ellington—with some obstructions from White Australia. South Australia's Governor, Sir Mark Oliphant, who was born in Adelaide in 1901, is a scientist, a world-renowned authority on atomic energy. He lives in a cream-stucco, slate-roofed mansion across from Parliament House and the Adelaide Club, and he gets $15,000 a year. A noted Adelaide woman is the Honorable Justice Roma Mitchell of the state Supreme Court, Australia's first woman Queen's Counsel. Another is Mrs. Edna Ayres, the charming and amusing daughter of Sid Kidman. She recalls one of her father's properties being "as big as Holland," 260 miles from one end to the other. A prominent Adelaide author is Geoffrey Dutton, who wrote a book, *Australia and the Monarchy*, which questioned the validity of the royal institution. "We have a revolting attitude of subservience," Dutton says. "We stick to the monarch. We

don't want to hurt Mum's feelings." For this sort of thing Dutton was asked to resign from the Adelaide Club.

Sir Thomas Playford, the former Premier, often boasts that South Australia is more industrialized, per capita, than any other state. General Motors–Holden has two plants in Adelaide, and Chrysler Australia Ltd. has another. Other Adelaide industry makes pipes and tubes, chemicals, farm equipment, textiles, salt, plastics, electrical equipment, footwear, clothing, and sheet metal. Adelaide must be a good place to do business. An American named R. Donald Dervin moved to Adelaide from Washington, D.C., in 1961, introduced the soft-bun hamburger into Australia, and by 1968 was selling a million dollars' worth of them every year from fourteen restaurants in Adelaide and Perth.

Adelaidiana: Upon Colonel Light's death the Adelaide Corporation voted an appropriation of twenty dollars a year for raising an annual toast to him, and the Lord Mayor does so on the first day of the annual meeting of the Municipal Council. At the Town Hall Colonel Light is present in a portrait —red coat, gold epaulets, and fawn-colored trousers. Also there, in a glass case, are Queen Adelaide's china and slippers. Sydney oysters are cheaper here than in Sydney, and in the restaurants you can also have yabbies (a small fresh-water crayfish) in vinegar, Canadian fiddleheads, prime Stilton, and papaw-and-passion-fruit sundaes. One thousand Americans live in Adelaide.

On to Nowhere Else

Again, here is an Australian state with a first city and no second city—or third or fourth, at least on the usual scales of population gradation. Next after Adelaide in South Australia is its own satellite, Elizabeth, with 33,000 people, and then steel-making Whyalla, 23,000. But though mostly little more than villages, the remaining towns and the rural population push the state's total close to 1.2 million. They occupy an area half as big as Mexico, yet almost everyone lives in the Britain-size southeastern coastal region, and only five thousand people in the rest of the state. Two long gulfs penetrate the populated area, giving it seaports and a Mediterranean—perhaps more precisely, Greek—character. A large resort island, named Kangaroo for the vast herds early explorers saw, lies at the mouth of Gulf St. Vincent. Place names offer clues to South Australian history and character: Cape Catastrophe, Streaky Bank, Mount Remarkable, Coffin Bay, and, most touchingly, Nowhere Else.

The twelve million acres of productive land in the south are dutifully

fruitful, but the north is more interesting. The strike-it-rich towns of Andamooka and Coober Pedy mine opal, that fascinating and supposedly unlucky stone. The largest opal to date was found in November 1969, at Andamooka; it weighed 220 ounces and the Adelaide opal buyer Luka Olah paid $168,000 for it. Japan buys almost all of Australia's opal, to cut and polish the stones and export them to the United States. In a typical deal, a miner might sell some pieces to a dealer for $627, who would resell them to the exporter for $1,200, who would sell them to the Japanese importer for $2,000—when finished and retailed in the U.S., the gems could bring $6,275.

The enormous blue lakes shown on maps of the South Australian outback are, in reality, white salt flats most of the time. Donald Campbell got his racing car up to 403.1 miles per hour on Lake Eyre in 1964. The plains beyond the lakes are mostly barren sand or gibber—clear across Western Australia to the Indian Ocean. This otherwise nearly useless country is therefore ideal for the rocket range that is based at Woomera, 250 miles northwest of Adelaide. This missile-launching site, a well-planned town now twenty-four years old, with water piped from a hundred miles away, has been used lately in the development of the British Black Arrow, a three-stage rocket to put three-hundred-pound scientific satellites into orbits three hundred miles high. Woomera also tracks U.S. space ships—for example, Mariner 9 when it was orbiting Mars. In the northwest corner of the state is another scene of science in the desert, Talgarno, where the Australian-designed pilotless target plane, Jindivik, was tested. Still a third such site is Maralinga, an atomic-testing station in country so flat that the transcontinental express train can be seen approaching for almost an hour. In this vicinity are the Musgrave Ranges, famed for spectacular sunsets. The talk is all about the size of the "properties" (never "ranches"). On one, the distance from gate to house is ten miles.

We cannot leave the South Australian countryside without first having a look at the long thumb of the southeast. Located here is Mount Gambier, which is both a town and an extinct volcano in the craters of which nestle four pretty lakes. One of them, for unknown reasons, turns from blue to gray each March, and back to blue in November. The buildings of the town are made from white coralline limestone quarried nearby; it is so soft that it can be cut with a carpenter's saw. Nearby, on the coast, is Lake Bonney, so polluted by a Kimberly-Clark plant that on any given day one can tell by the color of the water whether the plant is turning out green or yellow toilet paper. Premier Dunstan, who has promised to put through an Environment Act, is particularly concerned about Lake Bonney. But not far

away, at Kingston, crayfish are caught by the ton and frozen for export to the U.S.

To the north are vast farms of pine trees, planted since 1886 to provide Australia with softwood. Much farther north is rich country irrigated from the Murray River, as in western Victoria. A diesel-powered paddle steamer, the *Coonawarra*, makes a weekly excursion for 260 miles along the river. Barns in this country are made of stone or brick, testimony to the permanence and stability of the South Australian farmer.

The Wine of the Country

Here we come to wine country—South Australia has two-thirds of the nation's vineyards. And Australia's good wines are among the nation's greatest surprises to the foreigner, a steady joy not yet sufficiently known to the rest of the world. Only a decade or so ago most Australians, fervent partisans of beer, scorned wine as "plonk,"* and drank it mostly as fortified wine intended solely to intoxicate. Now Australians drink light wines as a civilized pleasure, and consume more per capita than any other English-speaking nation, including the United States and England.

Australian wine is never as good as the best French wine, but (which may be more significant) never as bad as the deliberately cheap *vin ordinaire* that so many Frenchmen drink. Vintages are remarkably even; there are few "years" that are strikingly better than other "years." Blending of two wines into one is much commoner here than elsewhere, and generally beneficial. There are delicious dry whites, called Riesling, or sometimes Moselle. The sweeter whites are known as white Burgundy or Chablis— a good white Burgundy is Houghton's. The dryer, more astringent reds go by the name of claret or cabernet; Coonawarra Estate claret is a highly regarded wine of this type. Australia also has rich red Burgundies, much champagne, a tank-fermented sparkling wine called pearl, sherries in five gradations from dry to sweet, and dessert wines—port, muscat, and tokay.

Italians, Greeks, and other immigrants helped spread the recent vogue for drinking wines, but the growers are mostly long-established firms founded by Germans or Englishmen. Fourteen big private companies make the bulk of the wine—Seppelt's, Penfold's, Mildara, Gramp's, and Leo Buring's are names one frequently sees on the bottles. These producers are backed up by six cooperatives and any number of little, one-family wineries. Not much

* Apparently a corruption of *vin blanc,* absorbed into Australian speech in France during World War I.

of their product goes abroad—something like two million gallons a year—but it can be had in all major cities of the world. Seppelt's exports to forty countries.

Half of the fourteen major Australian vineyard districts are in South Australia, the best-known being the Barossa Valley, thirty miles from Adelaide. The road to Barossa runs through squat date palms, silver fields of wild oats on one side and hay on the other, past sheep and cottages with geraniums—all under a high clear sky and against a profile of low gray-green hills. The valley floor is a flat iridescent green, rows and rows of low vines growing in red-brown earth. Gramp's, Penfold's, and Seppelt's all have enormous wineries here.

Broken Hill

South Australia staged the continent's first mining boom, back in 1842, exploiting copper from mines at Kapunda and Burra, not too far from the Barossa Valley. Shares in Burra went from five to two hundred pounds on the London market as the mines paid 200 percent dividends each quarter. Later two mountains of top-quality ore, Iron Knob and Iron Monarch, provided the basis of the blast furnaces, steel plants, and shipyards at Whyalla, heavy-industry center of the state. But the most fabulous mine in this area is, paradoxically, not in South Australia, it is thirty miles over the line in New South Wales, at Broken Hill.

Broken Hill was, and is, so rich that it came to signify the very essence of the idea that a poor man can lean down, pick up a rock, and find himself wildly, gloriously wealthy. All by itself the mine is a substantial chunk of the Australian economy. As late as 1966 Broken Hill produced one-fourth of Australia's metals, and half of its lead, zinc, and silver. Because of its location only two hundred miles from the South Australian coast, Broken Hill has always been economically, if not politically, a part of South Australia.

The country around Broken Hill is a harsh swatch of authentic Australian outback, with nine inches of rainfall a year and seven *feet* of evaporation (that is, in the few ponds, lakes, and reservoirs where that much evaporation is possible). The 150-foot-high hill was named by explorer Charles Sturt, who thought it looked sort of swaybacked. Sheep raisers moved out to that area in the 1850's, and in 1866 James McCulloch, Premier of Victoria, leased (with partners) the sheep station called Mount Gipps, which included the hill in its 900,000 acres and had a homestead only nine miles away. A clue to the context of this time and place is that in Wilcannia, a town not too far from Broken Hill, a favorite pastime in those years was racing camels—

which seems to tell us that the area was settled enough to need recreation, dry enough to require camels, and remote enough to need pack animals.

At that time prospectors, displaced by the end of the Ballarat-Bendigo gold rush in Victoria, were "fossicking" (digging) around the region, attracted by gold-bearing white quartz. But the man most intrigued by the jagged horizon line was Charles Rasp, German by birth and chemist by education, who had mysteriously settled for being a Mount Gipps sheep station boundary rider. An observant man, he saw similarities between Broken Hill and nearby silver mines. In 1883, driving stock to Adelaide, he bought a prospector's handbook, and concluded from it that at Broken Hill he was sitting on a mountain of tin. He and six other men "pegged" (claimed) one eighty-acre and six forty-acre blocks, and chipped in to hire a miner. This fellow, after two years of digging, found not tin but silver. Another discovery of silver on the same property followed soon after. "I believe £30,000 worth of silver was taken out of this spot in a fortnight," Rasp recalled later.

Before that, two luckless members of the syndicate had sold out. Sidney Kidman, the cattleman, with similar lack of faith, unloaded a part of a share of the mine that he had received in return for some steers. Silver turned out to be the least of Broken Hill's assets; it won its fame as the world's greatest lead-zinc-silver mine. Rasp and the remaining partners thereupon founded Broken Hill Proprietary, which grew to be the largest and most ubiquitous corporation in Australia. Broken Hill's population, 30 in 1885, grew to 10,000 in 1889, to 20,000 by the early nineties, and to 30,000 in 1901; it has stayed around that figure ever since.

In 1886 Broken Hill Proprietary brought in William H. Patton, superintendent of the Consolidated Virginia Mine in Virginia City, Nevada. He introduced the Comstock Lode method of mining, shoring up the stopes with imported American timber, which had the saving grace of giving out loud warning creaks before the roof caved in. Thousands of acres of U.S. timber are buried at Broken Hill, but nevertheless "creeps"—caving and earth movement on the surface, caused by the earlier failure to use timber— often wrecked costly mine buildings on top of the lode. The mine paid a million pounds a year in the early 1890's, after the South Australian government built a railroad to get the ore out. Soon Broken Hill's wealth topped that of the entire gold rush of the 1850's. One £200 share grew in six years to a value of £1,250,000.

Then, as the mine deepened, it ran into ores combined with sulphide. The best-known technology for sulphide ore at that time permitted the company to extract only two-thirds of the lead, half the silver, and none of the zinc. Slag heaps of valuable but unextractable metals piled up at the company's

smelter in Port Pirie. Finally, in 1912, BHP perfected a bubble-flotation process for crushed ore that got out 97 percent of the lead, 94 percent of the silver, and 92 percent of the zinc. It revolutionized mining both there and elsewhere in the world.

As we noted earlier, Broken Hill Proprietary withdrew many years ago from mining in this area, to concentrate instead on iron, steel, and ship-building. The companies now working there are Broken Hill South, North Broken Hill, and Australian Mining and Smelting (controlled by Conzinc Riotinto). These companies mine various parts of the same lode, which is 170 yards wide and 4 miles long. The average grade of ore is lead, 12.3 percent; zinc, 11.7 percent; silver, 4.4 ounces per ton. Exports of lead, zinc, and their concentrates earned $140 million in 1967–1968, and a good share of them went to the U.S. Broken Hill has more minerals in good crystalline form than any other deposit on earth.

These three companies are what remain, after various shakeups, of eleven major companies that once operated in Broken Hill. The early days of their corporate careers were stormy, with strikes and violence in 1892, 1908, 1916, and 1919. A single man could cause a strike; unions were numerous and divided. Then, in 1923, Patrick Eugene O'Neill, driver of a night-cart (toilet) in the mines, conceived the idea of a union of unions, and founded it under the name of the Barrier Industrial Council. Since then, the BIC and the companies, completely ignoring the arbitration process that prevails in the rest of Australia, have set the terms of employment in Broken Hill, and the BIC has gone on to control the town right down to deciding which shopkeepers shall prosper by union blessing and which shall be ruined by boycott. As much as for its wealth, Broken Hill is known for its rigid sociology. For example, every single mine worker is forced to subscribe to the union-owned newspaper, which has the brassy name of *Barrier Daily Truth*, even if this means that three or four copies of the paper get delivered to the same address every morning. The government of the town, in effect, is in the hands of the miners' union, and it shapes every aspect of life. Paddy O'Neill was known for years, until his death, as the uncrowned king of Broken Hill. His chief accomplishment was winning for the union a "lead bonus" (production bonus) that used to run a shilling a week but now comes to $5.50.

English-born Joseph B. Keenan ("Joe is the name, sir") is the current president of the Barrier Industrial Council.

Question: "Is this a company town run by a union?"

Keenan: "Yes, more or less. We're called the Kremlin; it's not true. I'm the boss of the town, yes, but not on Kremlin terms."

Mr. Keenan obligingly gives examples of the strait-jacket conditions of

life imposed by the union. No one can get a job in Broken Hill, for example, unless he has lived there for eight consecutive years out of the last ten—this is to "keep wanderers out." Women who marry must resign their jobs in three months. "Too many girls as it is for the work available," says Mr. Keenan. "We don't want married women to take daughter's jobs." There are exceptions—subject to union control. "We allowed in 'foreigners'—Italians, Yugos, Greeks, and Maltese—good citizens, good miners." Prices in shops are controlled by the union's power to declare a store "black" and boycott it. Shops that do not advertise in the *Truth* are likely to be declared black. Union membership is nine thousand—five thousand in the mines, four thousand in aboveground jobs.

The compensation for living with these rules, Mr. Keenan points out, is the "best amenities in the world. We've fought not just for pay and better living conditions, but amenities. Hospitalization, job security, a month's free vacation at the seaside with families, a thirty-five-hour week for miners underground, early retirement with pensions." The companies will lend a miner $8,000 at 2.5 percent interest, for housebuilding. If a man is killed in the mines, no one works the next day. Miners work seven-hour shifts, 8 A.M. to 3 P.M. and 4 P.M. to 11 P.M.; another shift, midnight until 7 A.M., is for maintenance. Boys are usually hired by the mines at seventeen and begin to go underground at eighteen, but only at twenty do they graduate to the status of miners, which is defined as being capable of handling high-explosive gelignite.

The men of Broken Hill appear to enjoy their amenities hugely. They have thirteen clubs, small-town variations in the style of the St. George Leagues Club in Sydney. On a Friday they may gamble a week's wages on the slot machines. For the benefit of the second-shift miners who get off at 11 P.M., some of the pubs* stay open until midnight, which violates the letter of the written law but by analogy obeys the unwritten Australian law that a sailor should be able to get a drink when he leaves his ship. Old people are considerately treated. One street crossing is designated for "AGED PEDESTRIANS."

On Sundays, miners and their families drive down the road to Menindee, a resort town on a series of reservoirs in the Darling River. There, in the midst of the saltbush desert, they swim from beaches, and sail, and fill the air with the din of outboard motors towing water-skiers. Broken Hill's pleasures are hearty ones; miners do not read much, says Allan Coulls, chief of the local library, which circulates only ninety thousand books a year.

* In the early days Broken Hill had ninety-two bars.

Many women find Broken Hill desperately boring; they are forbidden to work, and by a fine old Australian custom, meticulously observed in Broken Hill, their men spend much of their leisure time at clubs and pubs.

*

When Mr. Keenan sits down to "meet the boss," that is, to deal with mine managers, the man he sees most frequently is J. L. Liebelt, who runs the Conzinc Riotinto interests. Mr. Liebelt is president of the Mine Managers Association, which speaks for all the companies in union negotiations. He knows mine labor problems from two angles—his father worked in the mines as a "winding engine driver," that is, windlass operator for a shaft elevator. Keenan and Liebelt do not always succeed in keeping labor peace; miners walked out for six and one-half weeks in 1968. But usually, Mr. Keenan says, "We discuss. One round of discussions lasted eighteen months. Then we in the Council report back to a mass meeting of the membership. Most of the mine managers here know us. Know the town. Have lived here all their lives." Mr. Keenan, who stands every year for election, is a mine timberman who still works underground; besides that, he is a professional musician—trumpet, sax, and clarinet—and president of the musicians' union, too. Much prestige attaches to the presidency of the Barrier Industrial Council—one holder of the job was given the Order of the British Empire.

*

Broken Hill is a town that was wisely laid out, then badly built. Its mining-town psychology gave it an impermanent air. "Anyone could knock up a shanty," recalls Mr. Keenan. Now Sir Maurice Mawby, Chairman of Conzinc Riotinto, has persuaded everyone that Broken Hill's lead, zinc, and silver are good for many decades more, and the town is beautifying itself with gardens, irrigated with water brought by a pipeline from the Darling reservoirs. Houses are still made of "wood and iron," the iron (which is ugly) being galvanized roofing, walls, and fences, but nonetheless many houses are pleasant-looking. They stand on streets metallurgically named Garnet, Chloride, Iodide, Sulfide, Kaolin, Slag, Crystal, Cobalt, and Wolfram.

A hobby of some mine managers is to "go on the land." Mr. Liebelt says, "We buy virgin land, though now it's scarce, borrow from the bank, push the scrub down, put in some sheep, and in fifteen years have a lucrative

property. I have twelve thousand sheep." But whereas in the east a grazier can run four or five sheep to an acre, here it takes about twenty acres to a sheep. Properties must cover one or two hundred thousand acres to support six to ten thousand sheep. Driving through the buff-green countryside, one is much afflicted by the dry, red, sandy dust, which penetrates ears, hair, and eyes, and gets positively overwhelming when the wind blows hard; car windows must be kept closed. Broken Hill attempts to defend itself from dust by "protective tree plantations" along its southwest boundaries.

*

A final oddity about South Australia: this state, historically so stuffy and self-righteous, is the only one in the country to experiment with a liberalized abortion law. The law, which permits abortions when two doctors certify that continued pregnancy would gravely injure a woman's physical or mental health, went into force early in 1970. In the first nine months, eight hundred abortions were performed, 38 percent of them on single girls, one of them only thirteen years old. No less than 82 percent of the abortions were justified on psychiatric grounds, which led the Roman Catholic Church to continued attacks on the law for permitting "abortion on demand." The point—once again—is the image of this empress dowager of a state leading the country in experimentation and reform.

CHAPTER 15

El Dorado in Western Australia

Mining attracts the hardier and more self-reliant people.
—SIR MAURICE MAWBY

WHO EVER THOUGHT that anything could be three times the size of Texas? Western Australia is. This prodigious state covers close to one million square miles, and contains almost one-third of the total area of Australia. The state is bigger—to drive this point home—than the entire Mediterranean Sea.

But most of this stupendous terrain is desert, and the population is almost ridiculously small—little over a million people. Imagine what the United States would be like—from the Mississippi to the Atlantic, roughly the equivalent area—if it had only a million people. Distances between communities are vast, and communications, except by air, scant. The dreary endless stretch of road from Perth, the capital, to Port Hedland in the north, is known as the "Madman's Track"—there are only three towns, Geraldton, Carnarvon, and Onslow (population 260), in the whole thousand-mile distance. Until recently no other places existed at all where a motorist could pause for a drink or a meal after 8 P.M., much less a bed. Wayside restaurants and hotels today are mostly run by Greeks, the only people who can stand the loneliness. Similarly, Greeks maintain the isolated railway stations in the Sudanese desert near Khartoum.

Western Australia maintains a Flying Doctor Service, like that at Alice Springs to be described in a subsequent chapter, whereby medical aid is sent by air to isolated communities in the outback, or bush. This organization has its own radio network, which importantly assists communications in other fields. In other realms new concepts and techniques are being

constantly advanced. As an example, the Commonwealth government in collaboration with American interests has studied the possible use of nuclear power to blast out a harbor for iron exports in the north.

The Western Australian railways, owned and operated by the government, take a good deal of kidding because their trains are so mortally slow, running to an average speed of thirty-seven miles per hour. One popular ditty, sung by friends to wayfarers as they are about to set out on a journey by train, is "I'll Walk Beside You, Mate." Travel by automobile has been hampered recently by a microorganism with the nice name Desulfovibrio Desulfuricans, which eats into bitumen and destroys road surfacing. One of these bugs is capable of consuming matter equal to its own weight every five seconds. But few roads have pavement, and of the others often the best that can be said is not that they are graded or graveled but merely that they are clearly defined.

Western Australia has one of the historic gold mining centers of the world, Kalgoorlie, 350 miles east of Perth, about which, amazingly enough, Herbert Hoover once wrote a lyrical song. Mr. Hoover came to the gold-fields in 1897 when he was a twenty-three-year-old mining engineer, and spent ten years here off and on. The first stanza goes:

Do you ever dream, my sweetheart, of a twilight long ago,
Of a park in old Kalgoorlie where the bougainvillaeas grow,
Where the moon beams on the pathways trace a shimmering brocade
And the overhanging peppers form a lovers' promenade?*

Kalgoorlie is still active today—to understate the case—and every major mining company in the world has an agent or representative on its Golden Mile. Geologists say that it will take at least another hundred years before the wealth of the area has been completely surveyed.

Distances, to repeat, are great in Western Australia. Recently a teenager in a coastal village asked permission from her mother to go to a dance. Permission was refused, because the dance was fifteen hundred miles away. This is lonely and forbidding country, although the local statisticians take comfort in such a detail as that fourteen hundred different varieties of wild flowers are known. In one community, Marble Bar, the heat exceeded 100 degrees Fahrenheit for 160 consecutive days in 1923–1924, a record. Twice in a single week, a couple of years ago, stories appeared in the local press about men missing in the nearby desert. One was a twenty-four-year-old bricklayer, who lived in what is described as "hostile sand country" near

* And so on for eight mellifluous stanzas. Mr. Hoover was supposed to be in love with a local girl.

Derby in the north. "The searchers, including two native trackers, are on horseback, and have taken pack mules with them. It is impossible to travel in this country in a vehicle. There were areas which even a horse could not negotiate." Cases like this give rise to a hearty folklore. Tales are told, as of the Sahara, of intrepid but foolhardly pioneers dying of thirst on the edges of dried-up wells or springs. In spite of its desiccation, this is thought of locally as being "romantic" country.

Western Australia faces the Indian Ocean and the Timor Sea for a fabulous distance—4,350 miles. The state has, more than any other Australian state—although these are also bounded by seas—a true ocean image. The Indian Ocean beckons. Citizens of Perth go to Mauritius or the Seychelles for holidays, broadening their scope, and islanders from the great ocean visit Perth. The United States has several important strategic installations on the Indian Ocean coast, which have particular interest because of the recent upsurge of Soviet naval activity in these waters. Near Exmouth on Northwest Cape is a U.S. Navy VLF (very low frequency) radio station that can communicate with, and guide, submarines under water. From here, in the crunch, would go orders to Polaris subs in the Pacific or Indian oceans to fire their A-3 missiles, each with a range of 2,800 miles. Carnarvon has a tracking station vital to American astronauts; one of its radio towers rises to 1,271 feet, the highest structure in Australia. The bar in the Perth airport is called the Orbit-Inne, because the astronauts passed over it on most of their orbital flights.

Broome, up the coast, an important center for mother-of-pearl fishing before the age of plastics, now concentrates on beef processing. Cattle are driven here on what are called "beef roads." The traveler should never forget that Australia as a whole and Western Australia in particular are still frontiers inhabited by a frontier society.

The Development Factor

Indeed, the principal characteristic of Western Australia today, next to its staggering size and emptiness, is development, which has been prodigious in the last four or five years, particularly in minerals. Scratch any bit of earth in this whole immense region and you have a chance at least to strike unparalleled wealth. Western Australia is a genuine boom area—today's Golconda—and this is what makes it so exciting and vital to its citizens.

This state is outdistancing the rest of the country on any of a dozen indices of growth, among them population, retail sales, personal income, production, savings, and car registrations. Tidal power is being tapped near

Broome and Derby, where what are known as "king" tides have a rise and fall of no less than thirty-nine feet at peak level. On the Ord River, in the tropical north, a $27-million irrigation system makes possible the cultivation of cotton. As far as agriculture is concerned, more than a million acres of land are being reclaimed from the desert every year; the total area of land cleared rose from 22.8 million acres in 1958 to 32.8 million ten years later. Much land hitherto barren has been made fertile by the addition of what are called trace elements (various rare minerals, such as cerium, promethium, terbium, praseodymium, et cetera) to the normal phosphate fertilizer. Wheat production doubled in the five years before 1969. There are more than thirty times as many sheep in the state as there are human beings.

As to minerals, the figures rock the imagination—even the Australian imagination. Mineral production in Western Australia went from $54 million in 1965 to $579 million in 1970. Western Australia has the world's largest known deposits of high-grade iron ore, and substantial quantities of gold, bauxite, salt, nickel, manganese, copper, coal, and tin. The state's first commercial oil field was brought in at Barrow Island, on the north coast, several years ago, with recoverable reserves estimated at 200 million barrels.

One recent discovery in Western Australia is that of rutile, a mineral sand indispensable for the manufacture of titanium, used in nuclear reactors, airplane wings, and jet engines. Another valuable beach sand is ilmenite. Prospectors tread the beaches day by day searching for these and other arcane treasures. The Perth *Independent* has been known to carry sixty solid columns of small-type classified ads with registrations of mining strikes and claims. Everything in the world of minerals is a dizzying adventure.

Iron is king in Western Australia, but the crown prince—and glamor boy—is nickel, needed to harden steel. The mining boom has, naturally enough, induced a great deal of speculation—Australians are great gamblers, anyway—and the state's nickel mines have been the shooting stars of the Stock Exchange. The highest flier in the past few years was Poseidon. The mine was found quite by accident, when a prospecting crew, drilling for water at Windarra in Western Australia, hit nickel ore. The stock, selling at eighty cents, began to quiver and strain upward. Test drilling started, and in due time Poseidon announced that it had proved out several million tons of ore. Crazy to get in on the bonanza, buyers bid the stock up to $57. The price had to fall—instead, it sprinted to $130. Gathering the cheering stockholders together, the management announced that further drilling had confirmed that Poseidon could be a major operation. Early buyers could hardly believe their luck. In two weeks the price hit $200, in eight weeks

$280. The stock came to be known as "the big P." Turbulent scenes took place on the Sydney Stock Exchange. In one melee a broker had an ear torn off. Inevitably, a few professionals got to wondering how a mine consisting of a net of test cores, and absolutely no production, could be worth such staggering amounts. In two weeks the price was back to $212; in another week, $178. A fortnight after that it thudded to $63, and a year later it was down to $37.

Another Western Australian nickel stock with a recent rambunctious history is Tasminex, which rose from $7.20 to $96 a share in a few weeks. Then, its fever gone, it crashed disastrously. "I've had enough!" its owner cried—shortly after its claim was widely photographed as being "a hole worth $180 million." Tasminex has a nickel site about seventy miles from where Poseidon has its base. Still another newly celebrated mine in Western Australia is (as we noted earlier) the one at Kambalda, near Kalgoorlie, owned by Western Mining Corporation Ltd., an omnivorous Goliath. In about 1953 Western Mining decided to expand its operations, which had been based on gold, to base metals. The Kambalda deposits, test-drilled in 1966 after geologists found traces of high-grade nickel sulphide ore in what they call "gossans," or outcroppings, yielded a bonanza. Nickel mining began in August 1967.

That the boom is hugely speculative doesn't seem to cause much concern. Several local authorities are, however, sharply disturbed by the possibly disastrous effects of the mining bonanza on the basic physiognomy of Western Australia. Little thought has been given to the survival of vegetation, birds, and animals. Conservation is neglected in the face of the "insane hurry to get things out of the ground," as one university professor said. So much is being taken out of the ground so fast, that, in the words of another authority, the result could well be the transformation of Western Australia into "a vast disused quarry," before long.

Australians in general are, as we know, markedly materialistic and commerce-minded, but another side to the picture exists as well. Most citizens are possessed of—or by—what can only be called "the Dream," and a perception of this is essential for understanding the country. The "Dream," which cloaks utilitarianism with fantasy, is that of a great fulfilled Australia —on its own, secure, prosperous—with proper attention to human values. The Western Australians in particular want not only to be rich, but to be happy getting rich and happy in their enjoyment of wealth. Moreover, in this concept they consider themselves to be the bellwether for all the rest of the Commonwealth; their strong backs will bring wealth, health, and happiness to all.

Elements in the Background

The first European ever to see Western Australia was Dirk Hartog, a Dutch explorer and merchant, who arrived at Shark Bay in 1616. A subsequent Dutchman gave a swollen little watercourse in the Perth area the name Swan River, because of its black swans; the descendants of these swans are still there. In 1829 a British naval officer, Captain Charles H. Fremantle, landed at the mouth of the Swan and took formal possession of "all that part of New Holland not part of New South Wales." Since New South Wales was already British, the whole continent now became so. At the beginning, Western Australia was known as the Swan River Settlement. Fremantle's name is commemorated by Fremantle, Perth's present lodgment on the sea and the state's most active port.

Clearly the most important event in the history of Western Australia was the discovery of gold (including nuggets weighing 71 pounds and 125 pounds at Coolgardie) and the minerals outburst in 1890–1900. A 350-mile pipeline was built to carry water across the blinding desert from Perth to Kalgoorlie. At one point the population of the goldfields reached 200,000 (versus 21,000 today). The area was known as "the land of sin, sand, sorrow, sickness, sore eyes, and Sir John Forrest" (the unpopular Governor of the time). The most interesting political development in the present century was a strong, defiant move by Western Australia to secede from the rest of the Commonwealth and become an independent political entity. This was caused by resentment at being treated like a "Cinderella" by the other states, but the proposal failed. As of today, Western Australia could not survive without the rest of the Commonwealth in spite of its commanding wealth. One curiosity is that it is the only Australian state without a coat of arms.

Western Australians descend mostly from sound British stock, and, except in the mines, English blood still predominates, along with Scottish, Welsh, and Irish. The number of convicts pitched out into this wilderness was only 9,700 all told.

The Western Australians are extremely sensitive about white supremacy and intend to maintain it at any cost. In the whole of Western Australia there is hardly a single Negro. The local attitude to the Aborigines is clement —provided the Aborigines "keep to their place." In Port Hedland whites can be seen playing pool with Aboriginal boys in the best hotel. Western Australians, asked if amelioration of the white-supremacy concept might come in time, reply defensively. They say invariably that there is no bigotry about the white attitude, no racial hysteria like that in South Africa, and

very little personal antagonism to blacks or browns. One remark: "We know that we're part of Southeast Asia, but we prefer to go down the aisle alone." Another: "There's no need for us to create a problem for ourselves. Let well enough alone." Another: "I'm against anything that causes trouble. Why import it?" Still another from a last-ditcher: "I'd rather give up our whole mineral development than give an inch to any black!"

Among white foreign communities the Italians and Greeks are most conspicuous, and a few Turks and Yugoslavs are moving in. Perth has a small Jewish community at Mount Lawley, but it plays little role in civic affairs.

Finally, there should be a word about the Western Australian character and persona. The principal thing to say is that these are typically Australian —only more so. Most men have large aggressive chins, and combine outer toughness with a somewhat tepid interior. Most women are appallingly dressed. Citizens have a considerable regard for tradition, if only because this is one thing they lack. Perhaps most are more progressive in their thinking, less provincial and suburban, than their peers in the east. Many have a deep, sincere humanity, of which they are ashamed. Most are almost embarrassingly gregarious, friendly, and hospitable. This state is, as one newspaper editor put it, "a profoundly English country populated by people remarkably like Americans." But another citizen said, "Don't ever forget— Western Australia is a part of Southeast Asia inhabited by Europeans." It is still frontier country, still shining with the "Dream." People count without ratiocination on a beneficent future in which cultural and even spiritual values will find a proper place.

Then, too, Western Australia goes a bit beyond the rest of the country in its pronunciation of English ("tin sinz" for "ten cents," "to gaow" for "to go," "nawoo" for "no," "males" for "meals," "prem*ice*" for "premise"), and in its use of startling abbreviations in newspaper headlines—"TAS" for "Tasmania" or "PARLT" for "Parliament." One other minor observation: this is certainly not an area where aircraft stewardesses are chosen for looks.

Perth, a Pleasant Small City

This may be the most agreeable of all the Australian state capitals, and it has been called the last "authentic" city in the Commonwealth. Geographically it corresponds to Sydney as San Francisco does to New York. Its population is about 650,000, out of roughly a million for the whole of Western Australia, which indicates why decentralization is a lively issue. For years, poised on the Indian Ocean between sea and desert, Perth was called the most isolated capital in the world, but aviation has made a change—all the

major airlines serving Australia stop here on their way to Asia, Europe, and beyond. One remark is that Perth is the farthest capital in the world from Washington, D.C., but that it doesn't have to be—if you could only drive a hole straight through the center of the earth. A common joke characterizing the differences between the Australian state capitals runs like this: What are the first words said to a visitor in each?

Brisbane: " 'Ave a drink, mate!"

Sydney: "What's your bank balance?"

Melbourne: "Where did you go to school?"

Adelaide: "Of what religious persuasion were your parents?"

Perth: " 'Ave a drink, mate!"

The city has eight golf courses, three race tracks, and four legitimate theaters. The audiences do not rise for "God Save the Queen" after performances, as they do elsewhere in Australia. The local university is attractive and soundly run; in seventeen years its faculty has jumped from 15 professors to 58, from 1,500 students to 7,000. Municipal problems are like those almost everywhere else in the world, such as traffic, drunken driving, and a sharply increasing venereal disease rate. This is the sunniest of state capitals, and even in midwinter the temperature. seldom goes below 50 degrees; snow is unknown. An evening breeze from the Indian Ocean is called "the Fremantle doctor"; another wind, which yachtsmen hate, is "the southerly buster."

If a person moves out of Perth, people say that he has "gone east," never defining whether the destination is Sydney, Melbourne, or what. The biggest "foreign" colony is the Italian (38,000), followed by Greek (11,000-12,000, once a strong and cohesive group, but split in recent years by the schism in Athens), Dutch (6,000-7,000, split on religious grounds), and Yugoslav (4,000-5,000, split every which way). The city is still overwhelmingly British in tone and context, but American influence is rising. There are 3,900 Americans in Western Australia as a whole. The American consular district centering on Perth is, incidentally, the largest land district in the State Department system. (The one based on Hawaii is bigger, but consists mostly of water.)

Place names within the Perth metropolitan area make a fine hodgepodge from Maida Vale, pure English, to Wanneroo, pure Australian. In the outback are Norseman, Mount Thirsty, The Cups, Overshot, Bungulluping, Mangimup, Cape Naturaliste, Haul-Off Rock, and Double Tank. In fact, Western Australia has many tank towns—double-name places that literally end with "Tank," denoting the most essential feature. Again we see the importance of water or the lack thereof.

Perth has, unexpectedly, a hotel, the Parmelia, which is not only the

best in Australia but one of the best in the world, fit to rank with the Ritz in Paris, the Gritti Palace in Venice, or even Claridge's in London. The decor—done by a Chicagoan—is striking, the amenities luxurious, and the atmosphere beguiling. Chairs are built for both looks and comfort; carpets are thick, and the hardware dazzling. Brandy snifters are heated, and the bathrooms have bidets.

Ten years ago anybody in Western Australia who drank wine was a sissy, but this is no longer the case. The good restaurants in Perth have copious wine lists nowadays, and the wines are carefully classified, plentiful, and cheap. Rarely does anything cost more than four dollars a bottle. A normal-sized beer costs seventeen cents. Food is inexpensive, too—steaks at the best restaurant in town are $2.25, about half of what they are in Sydney. The pubs are licensed till 10 P.M., and booths are not allowed—this is so that a cop, checking up, can see the whole ensemble at a glance. Even in the best hotels, the floor maids do not wear uniforms but simply ordinary frocks or dresses. Bartenders, as is true everywhere in Australia, are invariably women, and these, too, work in ordinary dress—aprons are unknown. These barmaids are formidable young women, who can get rid of an unruly male customer as smoothly as a bouncer on Third Avenue.

It is difficult to assess social patterns in general. Australians are, in one dimension at least, an extremely formal people, and what is called "society" is rather rigidly put together. Much social life centers on clubs, of which there are multitudes, and white tie is often worn. At even a small affair in a club, the host is likely to make a speech with a microphone. Women, one hears, "are put severely in their place," that is, they work in the kitchen. Nobody has servants. The servant class has utterly disappeared. An odd phenomenon, also to be noted elsewhere in Australia, is that, taking a taxi, the passenger sits with the driver in front; to sit alone in back would be an unthinkable solecism. If a couple takes a taxi, the male passenger may well sit with the driver, while the woman is relegated to the back seat alone.

Games are all-important; Western Australians (like most other Australians) work off their surplus energy in sport, or dissipate it lying prostrate on a beach. One sport little known elsewhere is bushwalking—hiking across the outback. Perth carries a note of blandness, but street scenes can be animated. There are any number of espresso bars, lots selling secondhand automobiles, and shops bursting with radio music. In older days the police used British motorcycles; now, Japanese. Buses have a brace at the rear for holding baby carriages. The newspapers are full of advertisements for "aboveground swimming pools" ("You Can't Beat Our Prices"), and for savings banks which pay 6.5 percent interest per year. But a depositor, to receive this, must give a six-month notice of withdrawal after the first $200.

*

Southeast of Perth, around the jagged elbow of Cape Leeuwin, is the remarkable town of Esperance, near a point appropriately named Cape Arid. Indeed, the hinterland served by Esperance was almost totally useless for cultivation until quite recently, when, as mentioned above, a method was devised for treating the soil with new mineral fertilizers. Here a number of United States citizens—among them Art Linkletter and David Rockefeller— have large properties, called stations. One covers several hundred thousand acres.

A man who has played a critical and creative role in developing Western Australia in the last decade or so is Charles W. M. Court, the Minister for Industrial Development and for the North West until his party fell from power in 1971. Mr. Court is a heavily set six-footer, graying, with a russet complexion and boundless energy. His articulateness is notable. Born in England in 1911, he was brought to Australia by his parents at the age of six weeks. Again we see the rise of the so-called common man—his father was a plumber.

British migrants to Australia were not assisted financially in those days, and the Court family arrived under its own steam. Young Court wanted to be a lawyer, but at that time one had to pay a premium to enter a law firm, so that, he says, he "had to settle for being a chartered accountant." During the war he rose from the ranks to lieutenant colonel, and fought on Bougainville in the Solomons. Entering politics as a Liberal in 1953, he won a blue-ribbon seat at Nedlands, and has held it ever since. He still maintains a partnership in a private accountancy firm, and is an accomplished musician, being an enthusiastic cornet player and member of a local band.

"To get off the sheep's back"—that was his early philosophy. "We're still on it, but we've given the sheep a new back," he adds heartily. His predominant aim for Western Australia is to diversify, encourage industrial development, and, above all, rationalize mineral production. "Activate, activate!" he pleads. "We must come to grips with our great natural resources, harness them to our use!" His approach is vigorously positive. "People said that Perth was the most isolated capital in the world. All that they meant was that it was far away. What we've done is bring it nearer."

Mr. Court claims that Western Australia absorbed private investment capital at the rate of $2 million per week in 1960, $2 million per day in 1970, and can take $1 million *per business hour* in 1980. Total private investment in the state during the 1970's, he contends, may reach $21

billion. "It's magical," he says. "There were three thousand people in the Pilbara area, the size of Britain, in 1959. Now, forty thousand. In 1980, two hundred thousand. We mustn't be frightened of speed. We've never learned to live under prosperity. We're too slow, too easygoing. If something doesn't get in our hair, we'll probably be the richest area on earth in fifteen years. In ten years, we've opened up ten million *arable* acres, not pastoral or grazing land, but farmland. We're putting water in, stock, fences, clearing the virgin bush. And we have found oil—to our horror! It's too much! Found it in the first well we put down. Our program in the next thirty years calls for a most vigorous thrust in metallurgy. Even today, our bauxite goes out as alumina—we're not up to producing actual aluminum as yet. Like South America, we're still largely a raw-material economy. We'll get into aluminum ingots by 1976. Bauxite brings us five dollars a ton, alumina sixty dollars, but aluminum will bring six hundred dollars."

He wrote recently: "Making money is not evil. Tiredness, gutlessness, shortsightedness—these are the cancers."

*

Another remarkable man is Sir Thomas Wardle, Lord Mayor of Perth, known as "Tom the Cheap." Mr. Wardle, who was born in Perth of humble origin, is close to sixty, dark of hair, sharp of face, smallish in stature, sleek-looking, with bright eyes, neat. Mr. Wardle is a merchant—one of the best-known in Australia. Some years ago, while working in a bank, it suddenly occurred to him that the rise of TV would make movie theaters "redundant." Promptly he took a lease on an empty theater, and converted it into a big grocery store. It caught on, largely because the canny Mr. Wardle had another idea—to sell everything at prices 5 percent less than his competitors.

Now he has stores, called "Tom the Cheap," in every Australian state except Queensland, and has become a merchant of almost mythical prowess. In Western Australia, the Wardle stores sell one-third of all groceries sold in the state. "I was very cheeky," he says. "I wouldn't let the banks mortgage me, and I sold for cash." For seven years he couldn't sell cigarettes, because the tobacco companies refused to provide them at a low enough price, but they finally capitulated. In some respects Mr. Wardle's career and methods resemble those of Henry Ford, in whom he has a vivid interest.

Tom the Cheap was elected Mayor in 1967. "I came in as a clean-skin," he says. A clean-skin, in Western Australian slang, is an unbranded calf; in other words, Tom had no commitments to anybody. He was the first mayor in the city's history who had never served previously on the Municipal

Council, and politics as such doesn't interest him at all. "I keep away from politics—I'd never like to be a politician." He belongs to no party, and supports whatever Commonwealth or state government happens to be in power.

His biggest problem is, as he sees it, "to get the best development we can." This is not as easy as might be thought. "Our development came late, so that we can't benefit from experience." Better public transport is an issue; so is more parking space for commuters. The city recently hired a Norwegian expert to work out details of a planned "growth program."

Mounted on one wall of the Mayor's office is a steel case of a four-inch shell fired from the Australian warship *Perth* before it was sunk in the Battle of Sunda Strait in March 1942. The shell case was recovered and presented recently to the city of Perth. What is interesting is that coral still clings to it, and the Mayor, showing it, seems to be indicating that, in Western Australia, a kind of life grows even on dead metal.

*

Langley George Hancock, born in 1909, stout, bespectacled, tough, explosive, ruthless, with cauliflower ears, is probably the richest man in Western Australia and certainly one of the most picturesque. He exemplifies Australian traits which are supposed to be characteristic of the country as a whole, but which are in fact not seen everywhere—drive, durability. His wealth and power derive mostly from mining, and he owns or controls Mount Tom Price, in the Hamersley Range, a fabulously rich iron ore deposit. Other mineral interests include manganese, lead, tin, copper, gold, and asbestos. His agricultural properties are vast as well, including Mulga Downs, a 750,000-acre sheep station in the Fortescue Valley and another—500,000 acres—near Hamersley. Like almost all Western Australians, he is more interested in development than the present accouterments of power.

Lang Hancock, as he is generally known, comes of pioneer stock—of course. His great-great-grandfather, George Hancock, was an early settler north of Perth when this was still part of the Swan River Colony, and his great-grandmother, Emma Withnell, was known as "the first white woman in the north." Her son, Jimmy Withnell, set off the Pilbara gold rush in 1887, when, on the family property, he picked up a stone to throw at a bird and found it to be half-gold. The present Lang Hancock similarly owes part of his fortune to an accident. A vigorous prospector, he flew his own light aircraft everywhere. He was blown off course in bad weather some years ago and was forced to make an emergency landing on what seemed to be a

dreary stretch of flat rock. It turned out to be solid iron. Later, as he tells the story, he confesses that he had "a whisper" that iron might be there.

Mr. Hancock is supposed to hold more mineral leases than any man in the world, and, together with a partner, has, as he himself says, "more iron ore than the total resources of the United States and Canada combined." A single company in Japan is his biggest customer. It did not take long to find out from his conversation that he is a fierce, fixed, and tenacious advocate of private enterprise. He resents it extremely that the government takes a large share of mining profits in royalties and taxes, "for doing nothing," and says that "no government man ever found an ounce of anything." Above all he wants to draw foreign money into Western Australian investment and development, and thus personifies the present national mood.

CHAPTER 16

Across the Blazing North

We don't complain of dust unless it clogs the cash registers.
—A BUSINESSMAN IN MOUNT NEWMAN

THE SMALL PLANE dips into the wind, and one sees the desert floor below —ocher, umber, violet, dotted with dimples as well, greenish-yellow small circles, perfectly circular, in perfectly symmetrical procession, like a pattern on a child's dress. Did some sort of vegetation once grow here? Along a cinnamon-colored road, the color of Rome, a bright white car, the only car along unending miles of road, races to meet the plane.

So one arrives at Newman in the Pilbara mining area. Lonely country! The nearest oasis is Weeli Wolli creek, with a few date palms, fifty-three miles away. The nearest community is Nullagine, a shire center seventy miles away, with four or five hundred people, many of them prospectors; here an asbestos mine once stood, now shut down. The car proceeds swiftly from the airstrip to Newman through utter emptiness, the only car on the long straight road. Yet from this car it is possible to telephone New York.

The wheels cast up plumes of maroon dust. The sky seems to be absolutely cloudless; not a stray wisp of cloud, not a shred of white, is visible from one edge of the horizon to the other. The colors of the hills nearby change with the sun—blue in the morning, grayish at noon, green-orange at dusk. A willy-willy, a small twister, or tornado, builds up behind the car. Beside the road lies a bung-arrow, or lizard, two and a half feet long—a baby, says the guide. Growing on the solid, somber hills of iron is a primitive form of vegetation, the spiky spinifex. The car crosses a river, called the Fortescue, but it is invisible—it flows underground. Once a year or so, however, it floods nicely, and the nearby community taps it—the only local source of

potable water. "It's O.K. to drink," the people say, "but we have to soften it for laundry."

The rainfall here is ten inches a year in a good year, but some years bring almost nothing. Etiolated white trees, ghost gums, waver fragilely along the roads; fauna in the neighborhood includes wild camels (used for transport on the early stations), bush turkey, and, of course, kangaroos—some ducks, too. Men hunt and shoot. The sights include a shriveled little golf course, and tennis courts where, on account of the ripping heat, play is possible only at night, under lamps. But in winter Newman has a pleasant climate; the temperature can drop to 27 degrees Fahrenheit or thereabouts; taps freeze.

The two dominant mountains are Mount Whaleback (because it is shaped like a whale if you give your imagination a little run) and Mount Newman (3,700 feet), a kind of promontory on the Ophthalmia Range, which got its peculiar name because the man who discovered it, by name Giles, came down with an eye disease called "sandy blight," which nearly blinded him. Whaleback, unnamed ten years ago, is described by the local authorities as the single most valuable real estate property in the world. Its value is said to be $250 million per square mile, and it covers four square miles.

Close to the center of town is a small billboard: "WELCOME TO NEWMAN —FOUNDED 1968." A youthful community indeed! Once this was a cattle station covering a million acres; parts of it still survive nearby. The total population of Newman, this stout little wand in the wilderness, two years after its foundation, is about twelve hundred. The mining company built an airstrip, a railroad, and the town itself, and owns everything in it except the Walkabout Motel—comfortable, sixty-four beds, expensive, always full.

The miners work aboveground (Newman is an open-cut mine) in two ten-hour shifts, six days a week; nonminers in the company's employ work a sixty-hour week as well. At the mine the shifts are 6:30 A.M. to 5 P.M. (half an hour for lunch), and 8 P.M. to 6:30 A.M., and the wage rate varies by the job; the average for sixty hours is $100 to $120 a week. Many men come here for quick money, and like to work overtime; the company has a hard time getting experienced miners or skilled workers. A secretary receives $90 a week here, as against $40 in Perth. A locomotive driver gets $12,000 a year. Workers are strongly unionized, and there are five unions, one tame, four militant.

Housing has—naturally—been a problem. At first prefab houses were brought in from Adelaide, but now houses are built here on the spot. All have steel frames, air-conditioners, electric refrigerators, kitchens, water supply, and sewage disposal. The houses rent for six dollars a month, and

furniture is thrown in free. A family moving in needs nothing but blankets, sheets, and crockery. Single men live in two-story dormitories, air-conditioned, and fully furnished; however, they are supposed to wash after their stint in "the hill" in a nearby barracks, not at home. In their dormitories the workers have a mess staffed by a French (no less) firm of "cooking contractors"; full board is fourteen dollars a week. About eleven hundred gallons of beer (sixty-two kegs) are consumed per week—which works out to roughly a gallon per person, not very much.

The Newman community (before the actual town was set up) derives from a long-time Western Australia prospector, A. S. ("Stan") Hilditch, who came out to the region in 1957 searching for manganese. This was forlorn and empty country, still largely unexplored, and he asked some local Aborigines for advice. They pointed to what later became known as Mount Whaleback. They did not know, nor did Mr. Hilditch, that this would be proved to contain 650 million tons of high-grade iron ore. Hilditch came across a kind of ore deposit known as a "banded formation"; he sent it to his partner in Perth to be assayed, and it was found to be hematite with an iron content of 68.8 percent. The highest content ever known is 69.9 percent. Hilditch knew he had a big strike, but at the moment he could do nothing with it, because the Commonwealth government had imposed an embargo on the export of iron ore. This was because the authorities of the time, unaware how much ore the country possessed, thought that Australia should keep a large proportion of its ore reserves at home for purposes of security and defense. When the extent of the iron riches became known, the embargo was lifted, and Hilditch and his partner pegged their claims.

About a decade ago they worked out a royalty agreement with a financial group. They did not get as good terms as Lang Hancock did at Mount Tom Price, but it was enough to enable them to retire nicely. As of today, the company (Mount Newman Mining Company Pty. Ltd.) is owned by the Amax Iron Corporation (United States interests—25 percent), Pilbara Iron Ltd. (a holding company for another Australian mammoth, the Colonial Sugar Refining Company—30 percent), Dampier Mining Company (a holding company for Broken Hill—30 percent), Seltrust Iron Ore Ltd. (a holding company for the Selection Trust, a British amalgam—5 percent), and Mitsui and C. Itoh (a grouping of two Japanese trading companies—10 percent). Selection Trust, to add a complication, owns a portion of Amax, which increases its influence; this is not, however, ostentatiously exerted. A small percentage is held by Australian banks and insurance companies, which marks the first time that institutions in this category have gone into mining ventures.

The Newman mine looks like a hill made of chocolate fudge. It is not, of course, a hole in the ground, but a mountain, rising above the plain. One rides up in a stout vehicle, advancing layer by layer on a series of steep terraces, suffused with dust and perhaps mildly impressed by the fact that, on top at last, one is standing on a couple of hundred million dollars' worth of iron ore. Below are three hundred square miles of terrain solid with mineral leases. Technically the operation of the mine is simple—there's nothing to do except tear the mountain apart and cart pieces of it away; blasts take place continually, and each releases 50,000 to 100,000 tons of hematite. Power shovels seize 15 tons at a bite in their huge jaws. The carting is done by yellow trucks which carry the ore down the terraces; tires on these monsters cost $2,500, and they last about a month. The road is seventy feet wide, but even so accidents occur, if the trucks, fighting the steep grade, slip and turn over, despite their elaborate power brakes. The drivers get $120 a week.

Down the Slope to Hedland

From the chartered airplane flying to Port Hedland, the terrain beneath shows not a sign of life—not a human being, not a road, not a telegraph pole, not a tree. There is perhaps no other flight over territory so limitlessly barren except the one along the coast of Chile—also mining country. A railway does, however, exist, to carry the ore 265 miles from the mines to the port; this was built by a combination of American and Canadian capital, and a subsidiary of American Metal Climax, Inc., New York, holds a 25 percent interest in it.

The Port Hedland airstrip handles a lively traffic. Until a few years ago the planes to Perth or Darwin were too small to carry more than half a dozen passengers, and it could take months to get a reservation out. A welcoming sign says, "PORT HEDLAND—ALTITUDE TWENTY FEET."

A Dutchman named Peter Hedland landed here on a two-ton cutter, the Mystery, in 1867, and settlement began in 1893. So Port Hedland is, by the standards of this part of the world, a venerable town, but still raw, rough-edged, vital. The population has grown from three hundred to about three thousand in the past few years. It is usually torrid in the daytime, but a breeze blows at night. The principal problem is, of course, water. Rainfall is about sixteen inches in a normal year; it may not rain for months, then fifteen inches will fall in a single twelve-hour burst, the result of a cyclone in the Indian Ocean passing close to shore. Port Hedland suffers so acutely from lack of water that the townsmen set out tanks to collect dew. Two one-hundredths of an inch of dew comes in a normal night, which people say is

a lot. Strips of lawn that have been watered look so unnaturally green that they seem dyed.

Port Hedland is built on a mangrove swamp, and a hook of land stretches into a spit two thousand feet offshore. But to get there, except by boat, calls for an eighteen-mile trip, because of the difficult terrain. Nearby is the Monte Bello Prohibited Area, a group of islands still radioactive as a result of British nuclear tests conducted in the vicinity. There isn't much to see in Port Hedland itself. A good many houses are built on stilts to assist ventilation, because few people can afford air-conditioning; prices are extraordinarily high on goods or services that have to be brought in; a good house costs $50,000 to $60,000. A forlorn small race track stands exposed in the smarting heat, with a dirt track and a Quonset hut for serving drinks; two meetings are held a year, two days each. It seemed quite typical that, even in this metallic Shangri-la, racing should be a formidable preoccupation. Another curiosity is the presence of forty-four-gallon steel drums cut in half; they are used to protect seedling trees. The streets and roads nearby are full of discarded beer cans—people toss the empties out of their cars as they drink and ride. A large shiny white pyramid may attract the attention of visitors—a stockpile of salt mined by a San Francisco company.

There is a well-run motel which does, however, have a defect; page calls are incessant, and can be heard in every room. Nearby is an older, more conventional hotel, which looks like something out of Somerset Maugham: the garden seethes with bougainvillaea, but the kitchen serves bacon-and-egg-burgers and fish and chips. A sign in the bar says, "GENTLEMEN ARE REQUESTED TO WEAR SHOES AND SOCKS IN THE DINING ROOM."

Port Hedland is not in the strict sense a company town; only about 400 people from Newman live here. Mining is, however, its blood. The population was about 1,500 in 1959, about 6,000 in 1969, and estimates call for 30,000 by 1973. Pace, pace! Thirty-four different nationalities are numbered on the labor force—Australians first, then British, then Italians, then Yugoslavs. A few Icelanders, believe it or not, work at Mount Tom Price nearby.

Port Hedland cops are tough. A man picked up for a misdemeanor or minor infraction of the law, not worth bringing to trial or sending to jail, is driven twenty miles out of town on the road to Perth, then pitched out and told to fend for himself. The chances are that he will be picked up by some passing car. If not, he is out of luck, because, in summer, if he cannot hitch a ride and has no water, he could die of thirst or heat exposure.

At Port Hedland's mineral processing installation, enormous yellow machines on railway tracks chew up the ore that descends from Newman into chunks or pellets of manageable size. These fantastic appurtenances are

Japanese in manufacture, and, since Japan is a major market for Hedland ore, it is pleasantly fanciful to think that they are made of what they are working on.

The ore comes in from Newman by train, long solid lines of bucket-shaped cars, 140 cars to a train, eight trains a day. "They will run every two hours when we reach forty-million-tons-a-year production," says a company employee, with Australian confidence. The towering mounds of ore are watered to keep them from blowing away. The plant handling this immense operation is so fully automated that six men can operate the whole thing, but three hundred more are needed for maintenance. One sees such incomprehensible machines as the rotary car dumper, the tertiary crusher control system, the boom station, and the bucket wheel reclaimer. Ore goes out by sea to its markets in Japan, eastern Australia, Europe, et cetera, and is handled by ships which can take loads of up to 72,000 tons. The port is to be enlarged substantially to hold ships with a 100,000-ton capacity, and Hedland will become one of the six or seven ports in the world capable of this. Meanwhile, it is these Port Hedland installations that are largely responsible for the contemporary industrialization of Japan, a fact that augments local Australian pride. Beyond this is the spectacle of Port Hedland itself—such a melange of dreariness, hope, ardor, the Australian striving for identity and fulfillment, the boom-or-bust psychology, and above all the stubborn, indomitable spirit of the pioneer and frontier.

But it's a spartan life for the women in these towns. One hears such remarks as: "If only I could go somewhere where I could wear a hat!"

CHAPTER 17

Darwin and Alice Springs

*Minimum Dres (sic) for Gentlemen in Lounge-Dining Room
is Clean Shorts, Long Socks, and Clean Shirt.*
—SIGN IN A DARWIN HOTEL

THE COMMONWEALTH OF AUSTRALIA consists not merely of six states but of
a vast region known as the Northern Territory, which has not yet attained
statehood because its population is so scant, covers 523,000 square miles,
about one-sixth of the total area of Australia, and resembles Western Australia to an extent, but is much less developed. Most of it is emptiness.
Sometimes it is called the "Deep" or "Wild" North. There are only a few
more than seventy thousand people, including twenty-one thousand Aborigines. It has several cattle stations as big as Connecticut, and American
capital controls between 60 and 70 percent of the whole of its pastoral land.
Northern Australia belonged originally to New South Wales, but was an-
nexed by South Australia in 1863; in 1911 it was transferred to the jurisdiction of the Commonwealth, where it still is.

The northern headland of the territory, Arnhem Land, has been described
as "the most primitive, unexplored, uninhabited country" left in the world.
Here thousands of buffaloes still roam near the edges of forlorn seas. Aborigines have carved a special life of their own out of the bitter inland wastes.
In spite of its limitless desolation, remoteness, and aridity, Arnhem Land has
become a kind of symbol of romance to many Australians who live nearby.
Barring the bauxite mine at Gove, the area is not yet ruined by mineral discoveries, cities, or sprouting new settlements. People have the spirit of going
out West, as in American frontier days. Even pioneers with prosperous
cattle stations in North Australia like to take expeditions into Arnhem

Land—holidays with their families—to see at first hand the spectacle of a region so remote and pure.

Capital of the Northern Territory is Darwin, the principal port of the north, which a local booster pamphlet describes as "the gateway to three million square miles," and "the newest, noisiest [sic] city in Australia." In fact, it isn't noisy at all—it is a dead-end city. But, recessed as it is, Darwin somehow manages to be attractive in the way of most Australian frontier towns.

Its isolation is formidable; no other community of consequence exists within hundreds of miles; its sister city, Alice Springs, lies a thousand miles down the "bitumen," or highway. But flight time to Alice on the new jets is only about two hours, and the traveler can get there by road in twenty-six to thirty hours, if lucky; there are rest houses and petrol stations open twenty-four hours a day. The oddly named village, Rum Jungle, lies close by. Actually, Darwin's nearest important neighbor is not on the mainland at all, but is the derelict Portuguese island of Timor four hundred miles offshore to the west, which nobody but an inflamed masochist would ever want to visit twice.

In November, peak of the North Australian summer, the town simmers at an average daytime temperature of 93 degrees, and the water in the hotel's swimming pool goes up to 110 degrees. No one uses the pool, not even children, till dusk, but from then till midnight it is full of handsome young men and women, lolling at their pleasure. The salesgirl in the drugstore, asked when the weather will turn cooler, replies, "About My."

Darwin is built flat; there are only five elevators—lifts—in the whole of the Northern Territory, three here and two in Alice Springs. In the vigorously run hotel, guests inscribe their names in a large ledger, a practice common all over Australia. Air-conditioning is present, but it is somewhat crude and apt to make things too cold—icy. "Darwin rig"—shorts, long socks, shirt— is de rigueur for the bar. The dining room has a punkah; its wooden sails give a certain amount of propulsion to the sodden atmosphere. After a day or two the headwaiter will be likely to address you by your first name. Following the almost universal Australian custom, the guest carries his own drinks from bar to table. Partly this is caused by labor shortage. Australians clean their plates meticulously at lunch and dinner; nowhere are such huge meals enjoyed so hugely. Buffalo steaks are a specialty, and milk comes in frozen all the way from Brisbane. Beer is, of course, the universal drink— and how nourishingly good it is! Darwin is, in fact, reported to be the second-biggest beer-consuming city in the world on a per capita basis, after Munich. There are few inhibitions about drinking—or about anything else. A motel bears the nice name "Hot and Cold," and is favored by young lovers. A

sign on a filling station on Darwin's main street says, "SWITCH TO CALTEX-ASTRON AND DRINK UP THE SAVINGS." But it is not true, as has been written, that the streets are littered with discarded beer bottles. Darwin is spick-and-span, well kept up, a neat city.

This is the more remarkable because, of course, the prevailing background is so lush and tropical. A marvelously colored red-orange flowering tree, the poinciana, lines some streets, as do clumps of white frangipani, radiant in the sun. At night thunder cracks out of overcast skies like a succession of revolver shots. Darwin has good beaches, but swimming in the sea is impossible for several months each year, when an ugly marine animal that can be dangerous, the sea wasp, hovers in the surf. On the beaches here seashells have been found which geologists estimate to be thirty million years old. Another oddity in a different realm is that many automobiles in Darwin and its vicinity carry a frontal grill known as a bullbar, to fend off kangaroos.

Dingoes, wild dogs, roam the environs of the city, and are pests. They never attack human beings, but they seize upon calves on the outskirts of town if they are *in extremis* from fright or hunger and can find nothing else to feed on.

Darwin, like Newman and Port Hedland, has a great rush of citizens going south at Christmastime; the planes are choked. One somewhat unexpected detail of the city's landscape is that it contains two spectacularly good modern churches, one of them Roman Catholic. Houses are built commonly on stilts, so that there is room to park automobiles underneath. New houses go up all the time, and are expensive—say, $60,000. Once more we see the characteristics of a frontier economy. The local jail is built of galvanized iron; folklore is that the authorities let prisoners out at night to go to the movies, knowing that they are certain to return, if only because there is no place to which they could possibly escape; there's no road except the long track to Alice Springs. Finally among these generalizations one should mention that Darwin shares the country's exuberant mercantilism and will to get ahead in financial terms. This is a city where a man with money can do almost anything.

*

Essentially, Darwin consists of little but a long gently curving main street flanked by residential bungalows. Beyond this it is somewhat difficult to describe what it is like physically, except perhaps to characterize it with the term "provincial" town. Like practically everything else in Australia, Darwin has spread out significantly in recent years, and is growing with the times. Nowadays it measures about fourteen miles from rim to rim, and healthy

suburbs stand where there was nothing but bush ten years ago. The population is about 29,000, including half the white population of the whole Northern Territory. The biggest communities are Greek, Italian, and Aboriginal half-breeds, about 3,500 each, followed by Chinese, full Aborigines, Germans, and a handful of Maltese. There are also knots of Filipinos and Malays, who still work—virtual slaves—under an indentured-labor system. The Chinese community dates from the 1850's; the late Mayor, and for many years the grand old man of the town, by name of Harry Chan, was Chinese. The Greeks control something like 80 percent of local business. They dominate the construction industry, avoid politics, trade with avidity, and are exemplary citizens. Darwin's leading citizen and only millionaire is Michael Paspalis, a Greek property owner.

Another distinguished citizen is Tim Bowditch, editor of the local newspapers, the Darwin *News* and the *Northern Territory Times*. He and his managing director, Rod Lever, have provided the community with a high standard of journalism. Mr. Bowditch, a lean, emphatic, and conscientious man, resembles Manchester *Guardian* editors in the old days in England—emancipated, disciplined, fixed in his convictions, unassailably liberal, and a substantial (but modest) power in the community—in fact, its conscience.

Darwin lived on pearling for a long term, but this industry collapsed when plastics replaced natural pearl. Four resources sustain the city now—copper and gold at Tennant Creek, bauxite at Gove Peninsula, manganese and iron at Groote Eylandt, and an important uranium deposit at Batchelor, sixteen miles away. Then, too, Darwin lives on its own commerce as a trading center. Politics need not concern us, but lively complaints are heard that the Commonwealth government holds the territory back by unending bureaucratic needling and too much experimentation. In essence, this community is run by Canberra, and Canberra is far away. The local administrator seemed to be somewhat isolated. His pleasant one-hundred-year-old house (no air-conditioning) was knocked about by Japanese bombardment in World War II, but reconstructed. The Japanese attacked Darwin sixty-four times, but never took it. There has never been a British administrator here—only Australians.

A Glimpse of "The Alice"

We reach now Alice Springs, an attractive small city (population about ten thousand) lying directly athwart the Tropic of Capricorn at two thousand feet of elevation. It has been called the umbilicus of Australia, and it dominates, if anything so small can be said to dominate, the central desert. Ap-

proaching Alice Springs the visitor once more becomes acutely conscious of the immense distances of Australia, its unending frugal vistas. In a peculiar way "The Alice," as people in these parts call it, seems to be a land of invention. This city with its banks, bars, gift shops, a comfortable motel, hardly seems real. Yet it has the atmosphere of a flourishing town.

Alice Springs can be extraordinarily hot in the Australian summer. The atmosphere is dry, in sharp contrast to that of Darwin, and the local word is that during Christmas and in the first months of the new year "we usually go over the century mark by three in the afternoon; the hottest period of the day comes after 4:30." Christmas presents and tinsel wreaths seem incongruous when the temperature is 104 degrees.

The Alice Springs area is lonely, rugged country. The city lies not far from what is locally said to be the biggest monolith in the world, known as Ayers Rock, two and a quarter miles long. Physically, the forces of nature operate on a large scale, and a brush fire took place not long ago with a front of four hundred miles. The isolation of the community is, as is natural, marked, and the nearest town of any size is Tennant Creek (population 1,500), 313 miles to the north. Down the southward track are Port Augusta, 800 miles away, and the splendid city of Adelaide, 1,000 miles.

Plentiful legend haunts the story of the railway which connects Adelaide and Alice Springs, through a town named Quorn. This remarkable line was built by a contingent of Afghan camel drivers imported for the occasion, and it is still called "The Ghan" after them. Some descendants of their camels still survive. Passenger trains pass up and down the line twice a week, freight three times. The tracks are not ballasted, and sometimes the berserk weather blows them away. For fourteen days a few years ago Alice Springs was cut off from provisioning—milk, meat, beer—and emergency supplies had to be flown in. The Ghan is the most exuberant of Australian trains. Its bar car carries a grand piano. The maximum speed is thirty-five miles per hour, and one traveler reports that riding through the evening he sensed the "first night out" feeling one gets on transatlantic voyages. A visitor traveling up from Adelaide often brings his automobile with him; this is deposited on an open railway flatcar, and the passenger as a rule prefers to spend most of the time of the journey in his car, not the train itself. Men and women sit out in the open cars and drink beer, tossing the empties along the track.

To repeat—a hot, scored country! Dingoes roam in the derelict highlands near Alice Springs, and dead cattle line the roads. If a calf or steer is trampled to death in an overcrowded truck, its body is simply pitched overboard; if he is merely injured, the beast is shot and pushed to the side of the road. Men get lost, and occasionally die; the average man, if lost in the summer heat, cannot expect to survive more than twelve hours without

water. One American prospector died of thirst and exhaustion late in 1969, having lost his way only a small distance from "The Alice." Another, also American, took the wrong turn on his motorbike, and ran out of petrol; he was found two miles from a water bore, still alive after three days, but he died on a rescue plane sent out to fetch him. An Australian schoolteacher crash-landed a glider in the vicinity early in 1970; he survived for several days by drinking his own urine. Another Australian bushwacker died several years ago because his companions did not bother to organize a search—they thought that he had got drunk in the next settlement and decided not to come back.

What makes Alice is, of course, the desert, with purple mountains in the background. The earth is covered with gray green brush and spinifex, small eucalyptuses, ghost gums. The whole area is radiant with the peculiar quality of desert light. Here is the romance of Australia. The city itself—built flat—has a roof line predominantly gray, not red tile as in most Australian towns, because tile is too expensive to bring in. The effect, on a miniature scale, is to remind the visitor of Ibadan, another corrugated-iron city, in Nigeria. Then, too, Alice has, as is always the case in Australia, its suburban aspects; plenty of middle-class ranch houses have sprung up. Alice lies a thousand miles from any ocean, but it has its own river, the Todd. Unluckily, this stream is usually dry, having actually flowed only five times in the last ten years. A handsome war memorial, which commands a height nearby, was built to commemorate a 1963 visit to Alice Springs by Her Majesty Queen Elizabeth II, but the Queen did not have time to have a look at it. The town has an art gallery, a gift shop selling opals, a local specialty, a good-looking modern Lutheran church, a spectacularly modern school, and a swimming pool built to take 290 people a day and used by 540—loaded with chlorine as well as kids. Whites and Aborigines play and pant under trees in starfish clusters. Cops in broad felt hats—in long trousers, not shorts—look like Texas Rangers. A race track is, of course, prominent. The local newspaper, the *Centralian Advocate,* is a curiosity in the world's journalism: it covers the local news thoroughly, but has no wire services—no international or national news whatever.

In general Alice is not particularly tidy—beer bottles and cans dot the streets. The whites say that it is impossible to train the Aborigines to use receptacles. At night, the town gives forth a note of stridency, animation, even wildness—cowboys shout in the streets, children call.

A pleasant restaurant is Luigi's, made out of a refurbished, air-conditioned Quonset hut—chic for this part of the world, and with good wines. In the liquor shops Scotch whisky is sold by the gallon. In the milk bars and pizza stations hot dogs are known as saveloys.

*

The Alice had its birth in the 1870's, when explorers first began to look over the Macdonnell Ranges, where it lies. Its name was once Stuart, and soon after its founding it became a repeater station post in the telegraph line being built north-to-south across central Australia. Sir Charles Todd, the man in charge of building the telegraph, named the town for his wife, Lady Alice. The springs were a persistent water hole in the intermittent local river, which of course is named for Sir Charles. Alice Springs is a headquarters of local government for the Commonwealth, although Darwin is the territorial capital. The local district officials are, as always in Australia, competent and devoted men. Canberra keeps a mildly close eye on affairs, but Alice is too well behaved to bear much watching. As to the economy, Alice lives mostly by being an increasingly popular resort—the weather is marvelous once the crashing inferno of early summer is past.

*

A morning in the law court proved instructive: the magistrate wore no wig (too hot), and during a long session sentenced a variety of Aborigines to minor fines or admonitions for roughhouse or vagabondage during the night—there were no white defendants. Sentences varied from a seven-dollar fine for public drunkenness to four days in jail for disorderly behavior. In general, the Aborigines here seem to be lighter-skinned than in the north, cleaner, more adept. At a nearby settlement, they were being taught the rudiments of sanitation, how to cook on a stove, to clean their teeth, to use a toilet, to bake bread, and to postpone marriage till after the age of fifteen— if possible. Many Aborigines here resent having to learn English. Their chief value to the community is that they are fantastically good trackers out in the bush.

But the most interesting installation in Alice Springs is something else, the post of the Royal Flying Doctor Service. This is a truly seminal organization, whose work has been lately copied in Alaska, Kazakstan, and similar remote, empty places. The purpose is to render medical aid and services to isolated communities or families in the outback. These are given a call number and identical medical kits, packed so that Tube No. 5, let us say, an ointment, is always to be found in the bottom left-hand corner, and No. 6, a vial of penicillin, under the lid on the right. A person afflicted with illness or accident calls in to the Flying Doctor Service at any one of four periods

during the day, or may ring a special alarm bell; the service covers 120 communities, 100 cattle stations, 7 settlements, and several missions over an area of 400,000 square miles. The technique is the following. Two doctors are in constant attendance at headquarters, and these, picking up a call, attempt to make a diagnosis and prescribe treatment on the radio. If the matter is serious, one of the covey of doctors on hand may make arrangements to fly out to the patient, and, if necessary, fetch him back to the hospital. An important ancillary function has grown up out of this service, namely, that the Flying Doctors, with their intricate system of radio wavelengths, have become the instrument of almost all radio traffic in the area, not just medical. Last year the Service handled sixty thousand messages.

An important, and secret, American military installation operates at Pine Gap, twelve miles from Alice Springs. This base reportedly is one of two (the other is on Guam) that relay information from satellites keeping watch on the Soviet missile system. Some Australians fear that this base might invite a nuclear attack, in case of war. The two-hundred-man American garrison stays as far out of sight as possible, and never wears uniforms in the town; its members do, however, carry American license plates on their cars, and the local townsmen resent the fact that those bivouacked in neat little bungalows in Alice waste precious water by using it twice a day on their lawns. Australian white supremacists sought recently to make the United States forbid the employment of American Negro troops in these stations, but the effort failed.

You can, if you wish, drive sixty miles in the corrosive heat to a cattle station on the old Hamilton-Downs property. Until lately this covered fourteen hundred square miles, but it has recently been broken up and the operator, Bill Prior, uses only about seven hundred square miles now—enough! The Priors have 4,500 cattle, mostly Shorthorns, which are sold in Adelaide; water comes from ten bores with ten dams, which are dredged-out water holes. They have a cool room, but no refrigeration. Their nearest neighbor lives sixteen miles away, but they can go every so often to Alice—and, as Mrs. Prior puts it in her cheerful voice:

"Alice is noice."

CHAPTER 18

The Splendid Entity of Queensland

I like Brisbane. It has potential, and
you can make it be what you want it to be.
—LABOR LEADER JACK EGERTON

ONCE AGAIN—interminable blank distance! This is the second-largest Australian state, covering more than 667,000 square miles or 22.5 percent of the entire continent—intensely variegated, too. Queensland is bigger than the British Isles, France, Belgium, Germany, Italy, and Greece together. Its phenomena range from the great mining complex at Mount Isa, one of the toughest industrial communities anywhere, to the marine marvels of the Great Barrier Reef, which runs for 1,250 miles along the coastline, veritably one of the seven wonders of the world. There are tropical rain forests, dams, peach groves, flat grass tablelands, mountains, scrub, and desert. Several place names give a ring—Cape Upstart, Springsure, Mary Kathleen.

The population is about 1.8 million, and this goes up by 1.8 percent a year, less than the national average. About 88 percent are Australian-born, but this includes a variety of strains—Irish in particular. Queensland is still the "Deep North," and many citizens give a sense of having been burned out by the lazy tropics.

This is the state most richly endowed by nature. Lying across the Tropic of Capricorn, half above and half below, it has the climates to grow sugar, grain, cotton, beef, coffee, peanuts, wool, tobacco, timber, apples, pineapples, macadamia nuts, and dairy products. Its mines spew up copper, coal, lead, zinc, silver, bauxite, mineral sand, gas and oil, manganese, and phosphate. There is adequate room for all this: the state is thirteen hundred miles long, from north to south, and nine hundred miles wide.

Who—or what—runs Queensland? First, the big ranches, some of whose properties are enormous, as big as in New South Wales, covering thousands of square miles in the familiar Australian pattern. Second, such a monolith as the Colonial Sugar Refining Company Ltd., which is the most powerful industrial amalgam in Australia after Broken Hill.

Queensland has a kind of persecution complex. People say, "We don't get a fair shake, we're not appreciated." "We're a branch-office state, a client state." "Yes, we're the coming state, too, but we've been a long time coming." "Blame the drought." "But there's always been a drought."

Puritanism and drinking habits preoccupy the community more acutely than bigger issues. A stipendiary magistrate a while back ordered the destruction of a number of prints by Aubrey Beardsley, and copies of the record *Hair* were seized by the police. The intellectual life of the state is unimpressive. Pubs are not open on Sunday, and motels are not given licenses for Sunday drinking, except in special circumstances. Forty miles out of Brisbane, a tourist can get a drink if he is a bona fide traveler. The temperance lobby is strong. There are three race courses in Brisbane.

Queensland has an extremely strong sense of its own identity and a strong separatist tradition. Yet, paradoxically, Queenslanders are often called the "most Australian" people in the nation—perhaps because they suffer from most typically Australian blights: bush fires, droughts, dust storms, and flies in incredible numbers. Though Queenslanders, like other Australians, are overwhelmingly city-dwellers, the state has a greater psychological attachment to its outback, which brings with it much pride in physical prowess and courage. The state has an unfortunate history of racism that makes it comparable in a way to the American South. The early settlers dealt with the Aborigines more bloodily than anywhere else except Tasmania, and the South Sea Islanders imported to cut sugar cane in the last half of the nineteenth century were treated essentially as slaves. Queensland still has more Aborigines than any other state, about nineteen thousand. When a white man takes up with a black girl, he is said to be "going combo," or "comboing."

At the Top

The Governor of Queensland is a distinguished former judge, Sir Alan Mansfield. He was the Australian representative in the UN War Crimes Commission in London at the end of the war, and the chief Australian prosecutor in the Tokyo War Crimes Trials. He became Chief Justice of the Supreme Court in Queensland in 1956. The story is told of Sir Alan that once after sentencing a man, he kept moving his lips for several minutes.

Explanation: "I knew what he was thinking, and I just kept saying, 'The same to you, the same to you.' " The Governor likes to go fishing, in baggy pants and bare feet.

His appointment to the governorship follows the established patterns of Commonwealth politics—the job is a reward for service, and is given out by a combination of influences in London and Canberra difficult to assess. One thing to note about governors is that they live as a rule in huge enchanting houses, with beautiful lawns and gardens; a really good Government House can be the best hotel in the world. The *de rigueur* first duty for a visitor is to sign the book.

Sir Alan Mansfield, who was born in Indooroopilly, a Brisbane suburb, in 1902, is one of the three native Australian governors in the six states. Three of the governors are still British. This will almost certainly not be true in the future, as most Australians today would prefer to be presided over by a native Australian rather than a Briton.

A Word on Brisbane

Brisbane (pronounced "Brizbun," population about 850,000), gives the impression of being a happily helter-skelter city, full of disputatious citizens. Street scenes are vivid—young women with babies on their backs (no servants, of course), young men swinging in and out of the coffee bars, winding through the crowded streets, fighting a fierce pedestrian traffic, pushing. A big bending slug of a river, with several sigmoid loops, traverses the community. Not far away is Surfers Paradise, a spectacular real estate development along the beach, solid with motels, self-contained flats, rooming houses ad infinitum, and detritus. J. B. Priestley called it a "cardboard Miami." A million and a quarter people spend their vacations here every year. The coastal road is as bad as any average American commercial "strip": gasoline stations, night clubs, signs, and—Australian touch—"choox bars," fried-chicken stands. One advertising spectacular is a giant revolving mug of beer, complete with foam. Surfers Paradise is so totally given over to tourism that it employs pretty meter maids not to ticket overparked cars but rather to insert coins and forfend transgression of the law. Back of this strip, in the mountains, lies genuinely attractive country, Lamington National Park, with waterfalls and Antarctic beeches three thousand years old. From a helicopter over this country, one sees terrain resembling large fungoid-green tennis balls, and sprightly watercraft making wakes like feathers or filets of sole.

Lennon's Hotel in Brisbane is a rewarding place—a landmark of the area for many years. General Douglas MacArthur, on leaving Melbourne, set up his headquarters here, and his name still awakens resounding local reverbera-

tions. More G.I.s came to know Australia in this area than any other on the continent, because it was transformed into a huge, outspreading American base. Flying Fortresses operated from a secret airfield on the Atherton Tableland. It was from Lennon's that MacArthur issued his immortal phrase, "I shall return!"

Oddly enough, Brisbane and its hinterland are reminiscent of the Soviet Union in one particular, namely, that the citizens have an abnormal diffidence about culture. They desperately want to be considered "cultured," and the Russian word "*nekulturny*," indicating the pervasive stigma attached to nonculture, would be understood at once. Yet the roots of most citizens are so tenaciously, irreversibly suburban that it is difficult to make the jump. It is not impossible, however—"Up From Suburbia" might be a national motto. Another characteristic is the tendency to stick together. Legislation was proposed a couple of years ago in the local Parliament—but defeated—making it illegal to blackball a candidate for membership in a private club! The concept of mateyness could go no further.

A publisher's salesman says that Brisbane is a hard city for bookselling compared to Sydney or Melbourne, but that it seemed to be opening up a little. Beer is inordinately popular in Brisbane, and if you ask why, beer drinkers will tell you: (1) "It's cheap." (2) "Ever tasted our soft drinks?"

It's strong, too, but served in smaller doses here; the standard is a seven-ounce glass, not eight as in New South Wales. A lad in Lennon's said, "I toike in three quarts after every football game. It's good. It makes you sweat." A nice example of Australian gentility—such a paradoxical country!—is that one fashionable restaurant here gives a man a menu with prices, and his wife one without. The local citizens are good trenchermen, and appetites enormous. Never is a trace of food left on a plate. One curiosity is that *two* potato dishes are ordinarily served with an entree.

For a city of 850,000, Brisbane occupies a vast terrain—more land than all of New York City. The architecture is rather dreary, apart from some filigreed balcony railings and some quaint old buildings on stilts. No well-planned city this—just a grid of improvidently narrow streets. The City Hall, built in 1931, Italian Renaissance style, with a clock tower 320 feet high, contains a big concert hall and a sonorous pipe organ. The suburbs are the predictable row upon row of bungalows, with never a mansion or a shack, but here they are mostly on stilts, with lattice screens to admit the breeze, and corrugated-iron roofs. Palms dot the area. The Observatory is one of the oldest buildings; convicts put it up as a windmill, but it was shortly transformed into a hominy-grinding treadmill, upon which convicts had to labor fourteen hours at a stretch.

It is hot here, in the sense of averaging out high without going above 100

degrees Fahrenheit very often. The business uniform for many men is a white shirt, tie, hat, shorts, and knee socks. You frequently see signs that say starkly, "CASKET." These have nothing to do with the rituals of death; they advertise the state lottery, oddly called the Golden Casket Art Union. This curious appellation comes from the old British-continental practice of raffling off works of art.

The Labor Picture

The first Labor government in the history of Australia was installed here in 1889, which means that Queensland has had this experience for more than eighty years, a record unprecedented for the British Commonwealth or elsewhere. From grandfather to father to son, for nearly three generations, Labor has built itself in. Much of its program has been prudent, self-reliant, and far-seeing. This has never been a reckless party. Communists are rigorously kept out, and every Labor Party member must sign and renew an annual pledge stating that he is not a Communist or Fascist or a member of a Communist or Fascist front party.

Unions are run clean as a scoured pan—no Hoffas here—and exert great power in local politics. Proselytizers, called bush brothers, penetrate the outback to sign up recruits or bring yokels, to be converted to Labor, into the town. It is all quite messianic.

Another obvious and recurrent point to stress is how neatly the Australian tradition of mateship plays into trade unionism, and, from this, Labor politics. The oldest union in Queensland, which became rich early, is the shearers'. Others powerful and prominent are the barmaids, transport, carpenters, even dentists. Nonunion creatures, known as shingle-bums, are peremptorily despised.

Incontestable leader of the labor movement in Queeensland and a powerful figure in his own right is Jack (John Alfred Roy) Egerton, a man with huge hands, a face like a red ox, and an ingratiating appetite for beer. Mr. Egerton was born in 1918 to a farm family in Rockhampton, 640 miles from Brisbane. "I'm a bushie," he says proudly. He has been president of the Queensland Trades and Labor Council since 1957. He came up through the boilermakers, but he is a radio technician by trade. He has attended various labor congresses in Geneva, and likes to play golf and read.

Tommy (Thomas James) Burns, a more pliable man, with an aware, pleased, open face, is Egerton's counterpart. Burns, entirely self-made, was born in 1931, educated at a local grammar school. He got into labor broadcasting, and then became a state organizer for the Australian Labor Party. He

derives from a fisherman's family, cut ferns on state farms, grew watermelons, and served in Korea. Articulate, dynamic, this is clearly a man who will go far.

More Men of Brisbane

Premier of Queensland is Johannes Bjelke-Petersen, who is New Zealand born. A successful farmer, an aircraft pilot, a teetotaler, he teaches Sunday school and is an effective, hardheaded businessman. Mr. Bjelke-Petersen has no university degree; as a matter of fact, in the entire Queensland cabinet only two members hold university degrees, the Minister of Justice (who is an M.D.) and the Minister of Labor and Tourism. This government is unique in Australia, a Country-Liberal coalition dominated by the Country Party, which is, of course, Mr. Bjelke-Petersen's party. Generally it is said of Mr. Bjelke-Petersen that he may be the last Premier to represent and express predominantly conservative rural interests.

Vincent Gair, Senator for Queensland since 1965 and Parliamentary leader of the Democratic Labor Party, was born in Rockhampton. He rose to become Australian Labor Party Premier of Queensland in 1952–1957, but he broke with the ALP leadership and has never reacquired control of the state. Senator Gair is a violent anti-Communist and strong Roman Catholic. The Catholic population of Queensland runs to about 25 percent of the whole.

The Deputy Premier, Gordon William Chalk, a vigorous and enlightened personality, was born in 1913 and has been Treasurer of Queensland as well as Deputy Prime Minister since 1965. He rose as Minister of Transport and likes outdoor sports and reading. His nickname is, as it would be, "Chalkie." He has no more than two years of high school, but he is sharp, gregarious, an articulate talker and by all odds the ablest administrator in the state. He wants wider, more fruitful relations with the U.S., and thinks that a predominant British attitude to Australia is that it is still "colonial." He says, "I want good solid-mattress relations with all the world!"

Clem Jones, Lord Mayor of Brisbane, is interesting and extremely able, a man spry and cunning. Born in 1918 at Ipswich and educated in the Brisbane public schools, Mr. Jones is a lively cricketer and a member of the Australian Board of Control for International Cricket. By trade he is a professional surveyor. Urban planning fascinates him, and he has been a member for some years of the Town and Country Planning Association. People say that he is something of a fixer, but he fixes most things well.

Mr. Jones has traveled widely in the United States, and has taught at the University of California at Los Angeles. He is sandy-haired, and smiling—no

Australian nutcracker jaw. Asked how he takes the load off, he replies, "I don't feel that the job is work." Development is his leading problem. An innovator, he has the idea that by doing less one may do more; hence, he has conceived the idea of building new roads to a lesser rather than a higher technical level—so that they will cover a wider network and be useful to more citizens. "Our greatest problem," Mr. Jones concluded, "is lack of sufficient economy to be able to build into it faster than we are."

The Mines of Queensland

Mount Isa Mines has its formidable plant in the western outback of Queensland. The company is owned 53 percent by American Smelting & Refining, and has forty-thousand Australian shareholders who control another 35 percent. Labor relations have been turbulent at Mount Isa, and it is described as the "most unpleasant" community in Australia. All this got started in 1923, when a prospector named John Campbell camped in the dry bed of the Leichhardt River and idly chipped a rock with his hammer. He recognized sulphide of lead, and a "lead rush" began. Ultimately, the treasure of "The Isa" turned out to be copper. The mine has been opened and closed several times as the price of the lead, zinc, silver, and copper that it produced rose or sagged. Ore is smelted here to blister copper, and sent to Townsville, on the coast, for electrolytic refining. Mount Isa's eighteen thousand people live in air-conditioned houses, and drink milk brought a thousand miles in refrigerator trucks.

Mount Morgan is another copper mine that started out to be something else. This hill near Rockhampton was decapitated in the late nineteenth century for its gold, and then turned out to be much richer in copper. After many ups and downs, it still runs, a truly awesome excavation. A sight from airplanes between Mount Isa and the coast is Mary Kathleen, a uranium town built in the mid-1950's just months before uranium mining collapsed all over the world; now Mary Kathleen sits waiting for the price of uranium to rise again.

One characteristic of Queensland mining, it would seem, is that the operators often merely have to take the topsoil off to find the riches. At Weipa, as we have seen, they stripped away a growth of eucalyptus to reveal, red as a flesh wound, an endless deposit of bauxite. In the far west, at Yelvertoft, International Minerals & Chemical Corporation in 1967 similarly found a bed of phosphate fifteen feet thick and fifteen square miles in area, under thirty-five feet of earth. And a hundred miles southwest of Gladstone, at Moura, a dragline with a bucket as big as a barn strips away coal to be sent to Japan.

Queensland, like Victoria, was born by binary fission from New South

Wales. This took place in 1859; Queensland was the last of what are now the six states to become a colony. It was named for Queen Victoria—at that imperious lady's suggestion. Previously the area was known as Moreton Bay, from the name of the original convict settlement, founded in 1824 when New South Wales refused to accept any more of England's criminals. Some of the toughest felons, including men who committed crimes after reaching Australia, were sent to Moreton Bay, but modern Queenslanders are proud to cite descent from these convict-pioneers. Even though New South Wales' Governor Thomas Brisbane (for whom, of course, the capital is named) tried to keep Moreton Bay exclusively a prison colony, free settlers moved in to occupy the most fertile land of Darling Downs. After becoming a colony, Queensland put up in England and Scotland posters reading "WANTED: YOUNG MEN FOR QUEENSLAND," and the population doubled to fifty thousand. Since then Queensland's history has proceeded rather eventlessly, except that in 1915 the state pioneered compulsory voting in Australia, and in 1924 it abolished the upper house of the legislature, making itself the only unicameral state.

Sugar Country

Australia exports more sugar than any nation except Cuba, and almost all of it is grown in Queensland. The cane cutters are all white men, Italians and Maltese for the most part nowadays. This is said to be a unique show of how whites can labor with good health in the tropics. But this has not always been the case. More than a century ago Queensland planters began importing Kanakas (the word means "man" in the Polynesian languages) from the New Hebrides and the Solomon Islands to work the cotton and cane fields. The ship captains that brought the Kanakas were called "blackbirders," and the islanders were worked so hard and paid so little and punished so brutally that ultimately an official inquiry denounced the whole system as tinged with "deceit, cruelty, treachery, deliberate kidnaping, and cold-blooded murder." Under pressure comparable to that of the Abolitionists of the American Civil War, the sugar planters of the north threatened to secede from Queensland, but nevertheless indentured labor was banned in 1904, and the Kanakas mostly sent back. Subsidies, continued to this day, compensated the planters for the extra cost of employing whites. Not that they look white—they are burned to a shade of walnut by the sun and further blackened by soot from the burning of the cane fields, an operation carried out just before the harvest to drive out snakes and clear away brush. Of course, these days machines cut more and more of the cane.

The sugar ports—Bowen, Ayr, Ingham, Cairns, and particularly Towns-

ville, Queensland's second city with 61,000 people, and biggest city in the Australian tropics—preserve a bit of the separatist tradition, disdaining Brisbane. An intriguing place to the north of the sugar ports is Cooktown. An 1873 gold rush to the nearby Palmer River attracted to Cooktown twenty thousand Chinese, who exasperated the whites by sending much gold to China, not to mention the bones of their dead. The Chinese tongs fought with one another, and everybody was set upon by the cannibalistic Myalls, fiercest of the Aborigines. These man-eaters rated the Chinese above the whites because they tasted "less salty and equal to the finest bandicoot." A cyclone flattened Cooktown in 1907, hurting no one because the gold had given out and it was a ghost town. In fact, its destruction did not become known to the world for some time. Now four hundred people live there and hold tight to the real estate on the theory that it will be Queensland's next big tourist town.

The Great Barrier Reef

With an effort, the mind can grasp that this huge reef is no less than 1,250 miles long, and that its area of atolls, keys, and underwater ledges is as vast as the area of Kansas. But it takes far greater effort for the mind to grasp that all this coral was made by tiny polyps, extracting lime from sea water to build the largest structure ever formed by living organisms. How many polyps did the job take? How many years? In the reef shelters live all sorts of amazing, grotesque, and beautiful forms of life, an incomparable phenomenon, beyond anything else on earth.

The reef parallels two-thirds of the Queensland coast—in fact, its seaward side is the official state boundary. In the north it is fairly regular and ten or twelve miles wide, but in the south it breaks into larger banks and platforms, scattered as much as two hundred miles from shore. Most of it is just under water, although low tide exposes some of the crest, and there are, in the reef and the lagoon that it encloses, three thousand islands or islets. The reef is an extraordinary canvas of natural art. The fantasy of the forms, and the colors of the coral (from the air, reddish hues from the algae that encrust it, set off by the ultramarine of the ocean and the turquoise of the lagoons) are extraordinary. For the visitor, the reef itself is quite a sufficient marvel, and he can poke about, at low tide, crunching coral underfoot, with a feeling of endless astonishment at nature's inventiveness.

Yet even more exciting is the underwater life. Coral caves and recesses harbor many hundreds of species of fish and sea animals. Perhaps the oddest is the giant clam, which weighs as much as three hundred pounds, smiles

with liverish lips, and acts as host to polyps who mistake him for solid coral. Two kinds of fish kill their enemies by expelling poison through their dorsal spines—the handsome, gaudy lionfish, and the ugly stonefish, which indeed looks like a stone. Another dangerous fish is the Queensland grouper, which can weigh eight hundred pounds and actually stalks divers. Other menaces are the sea wasp, the moray eel, the Portuguese man-of-war, and, of course, the hammerhead shark. There are also repulsively pulpy coral-based creatures: the shell-less snail called nudibranch, and the sea cucumber. Others are as beautiful as their names imply: blue butterfly fish, angelfish, parrot fish, and exotic creatures such as giant turtles and the blood-spotted crab.

The underwater coral is fantastic. It grows in a whole spectrum of colors and infinite shapes—there are staghorn corals, mushroom corals, brain corals, et cetera. Above the water, reef life includes herons, noddy terns, and the unbearable muttonbird, which spends all night engaged in a kind of raucous cooing, if such a thing is possible.

Much alarm prevails in Australia these days over the possibility that the Great Barrier Reef is doomed. Incredibly enough, the reef is under deadly attack, and not by storms or men, but by a starfish. When polyps, which in size are comparable to the eraser on a pencil, extract lime from sea water, they turn it into skeletons which attach to one another to form greater and greater masses. The skeletons are cup-shaped, and the soft, living part of the polyp, attached to the inside of the cup, can emerge far enough to find food. This it does with eight stinger-tentacles that wave plankton into its mouth. Now, for reasons unknown, innumerable starfish of the crown of thorns variety (they actually look more like a small weedy mound) have begun to devour the soft polyps by the billions, sucking them up into the hundreds of little mouths which lie on the underside of the starfish's arms.* Almost two thousand square miles of the reef are said to have already been destroyed.

It was Charles Darwin, incidentally, who first explained how coral reefs are built. The key clue is that the polyp cannot survive much deeper than 180 feet. Therefore the reefs cannot have been built up from a deep ocean floor. Yet some reefs are solid coral for thousands of feet down. Darwin believed, and his theory has been substantiated, that these reefs were started on a rock foundation that lay near the surface of the sea. This foundation then slowly sank down as the busy polyps kept adding coral to the top.

Tourists—more than a quarter of a million a year—get to the reef from four ports along the coast. Many simply swing out in launches for a look or a dive, but more serious visitors tour the reef for days in chartered boats, or

* Parrot fish and angelfish are also coral eaters.

stay at island hotels. Of these islands, Green and Heron are perhaps the best-known since they are both true coral keys. Green Island has a clever sunken observatory, right in the coral, and through portholes visitors can see the whole fishy panorama. Elsewhere, people skin-dive, or swim with goggles, or ride glass-bottom boats. December to April is a bad time to visit, because of cyclones and insects. Ampol Petroleum Company recently created an environmental crisis in Australia by proposing to drill the Great Barrier Reef for oil, but ultimately bowed to public indignation and suspended the plan.

On the Land

Queensland sometimes boasts that it is "the biggest public estate in the world." The point is that, after experiencing many different systems of land tenure from squatting to land-reformist subdivision, the state now leases much of its agricultural area. The typical lease, sized according to use, runs thirty years, after which the government can take it back and cut it up or recombine it.

Some of the land is loamy and fertile. The Darling Downs area supposedly has forty feet of topsoil, and its fields are laid out with the precision of a checkerboard. But most of the terrain is the leathery, limitless outback. Australia exports more beef than any nation except Argentina, and Queensland raises half of Australia's cattle. The favored breed is Brahman crossed with Shorthorns and Herefords. A pitiable resident of the cattle plains is the "brumbie," a once-domesticated horse who has got away and gone wild. Brumbies can outrun any horse with rider, so they cannot be captured, and since they eat more pasturage animal for animal than cattle, they are shot and killed. Kangaroos get the same treatment in Queensland; precisely 6,556,743 of them were "harvested" between 1954 and 1964.

The major pest in sheep-raising areas is the dingo, and he is fended off with a draconian device. Surrounding almost all of the eight-hundred-mile-long tongue of savannah sheep country in south-central Queensland is a wire fence six feet high, with another foot buried underground. This desperation defense cost $2 million when it was built in 1963, and measures 3,500 miles in length. The dingoes get through the dog fence, but in reduced numbers. The real secret of this arid region's success in raising cattle and sheep is the artesian well. Queensland has five thousand "bores" into the Great Artesian Basin that underlies a third of Australia. This water is insufficient in supply and too mineralized to irrigate great areas, but it serves well for watering animals and sustaining humans. Roads in this vast domain are narrow and

poorly graded even when paved, and when not paved they are equally un-navigable when dry and dusty or wet and muddy. The Queenslanders' solution is to fly everywhere; they are the most air-minded people in Australia. Sometimes, when big windstorms strike, these air travelers look down on nothing but black dust for hundreds of miles, and the state is virtually blacked out for as much as a day.

Queenslandiana: "Desert" in this state means timbered country—treeless country is called "downs." In the downs, and elsewhere, one can see "magnetic" anthills, dagger-shaped termite houses around seven feet high, with the broad sides invariably running north and south. Actually, it is thought that they are oriented not to the magnetic polarity of the earth but rather to get the morning and afternoon sun. Each hill contains about two million ants. Two botanical wonders are the giant stinging tree, with large, bellicose, velvety leaves, and the ungrateful epiphytic fig, which lodges high in a host tree and then sends down curtains of roots that enclose and strangle it. A zoological wonder is the lungfish, ceradotus, which represents an ancient link in evolution: it has a bladder which it fills with air on the surface of the water, before diving for a swim.

Thursday Island, off the tip of Cape York, is a cosmopolitan dot, inhabited by Europeans, frizzy-haired Torres Strait islanders, Aborigines, Chinese, and Japanese. The main industry is diving as deep as 240 feet for oysters which are destined to be mothers of pearls. Japanese technicians of the Nippo Pearl Company Ltd. implant bits of Mississippi Valley mussel shell, and eventually the oysters produce a cultured pearl. Why is the island called Thursday? The tongue-in-cheek answer is that the strait already had a Tuesday Island and a Wednesday Island, so the next one had to be Thursday. It used to be a fort, with naval guns dominating the strategic strait.

Queensland's little northern towns can be the epitome of gentle tropical life. They look a lot alike, the streets are wide, the air is soft, the houses stilted. In the hotels, one carries one's own luggage and comes to dinner on time or not at all. This can be lotus-eater country; business executives fleeing Sydney's wear-and-tear may settle here and indulge in some such gentle way of living as running a bar. Groups of young hippies live off the dole in beach shacks. Everywhere one sees those trademarks of the tropics: bougainvillaea, poinciana, and frangipani.

Finally, a noble place is Cathedral Cave, in the Carnarvon Ranges National Park, a region of deep gorges and overhanging cliffs. This subterrane measures 270 feet by 70 feet, and is 200 feet high. Of all the caves in Australia that the Aborigines decorated with paintings, this may be the most remarkable in terms of richness and preservation.

*

There is a frontier air about most of the coastal towns outside of Brisbane, but the citizenry bridle if you mention it. A police officer in Townsville reared back at the suggestion and said, "Hardly that—but there's twelve hundred miles of land to the west of us that *is* frontier." Promethean wastes indeed!

CHAPTER 19

New Guinea: Men and Wigs, Women and Pigs

There is a bird-shaped island lying to the north of Australia, with head turned to the west toward the Moluccas, and tail running down to the south-east. This bird-shaped island is Papua and New Guinea, the largest island in the world if you except Australia and Greenland.

—SIR HUBERT MURRAY

NEW GUINEA . . . is that *Australian*? Indeed, the eastern half is, partly an old British colony taken over by Australia, and partly a United Nations Trust Territory administered by Australia. And the western half, once Dutch New Guinea, is now Indonesia's West Irian. New Guinea, hovering north of Australia, is the strand that stopped the Japanese thrust in World War II.

The whole island, 1,200 miles long, measures 312,329 square miles in area; the Australian part alone is as big as California. New Guinea is, amazingly, part of the Himalayan chain of mountains, the place where the eastern end of that great range drowns itself in the Pacific, after having been in and out of the water the length of the Malay Archipelago. Just before the range takes its final plunge, it reaches towering heights; and so in New Guinea, only two degrees south of the Equator, there are peaks rising so far above the snow line (14,600 feet) that they actually have glaciers. The highest mountain, Djaja, in West Irian, reaches up 16,503 feet. Dozens of mountains rise above 10,000 feet, and large areas of productive land lie above 5,000 feet. Vertical country! In some places peaks are rarely seen because of the clouds that constantly surround them; lower

down there are rain forests where the precipitation is almost continuous and the moss is two feet thick. Put aside any lingering impression that this island is all fetid tropics: its high, jagged backbone gives it proportionately more mountains than any other island in the world.

But, let it quickly be said, New Guinea has its fetid tropics, too. New Guineans say (though of course it has been said of other places, too) that God created this island on Saturday night, when running out of time He hastily flung down a confusion of marshes, plateaus, volcanoes, and rivers. The southern bulge contains one of the biggest swamps in the world. The Fly River* winds through it for six hundred miles, navigable in a whaleboat, though the straight-line distance thus traversed is only about two hundred miles. A long spur of the drowning Himalayas ends in a curve of islands— New Britain, New Ireland, and Manus. Through some historical foible, Australian New Guinea also includes the biggest and westernmost of the Solomon Islands, Bougainville.

New Guinea trees adapt themselves to their elevation: mangroves and many varieties of palms in the lowland; klinkii pine, laurels, oaks, and beeches on the next level up; myrtles and rhododendrons on the high mountainsides; and grasslands in the alpine areas. Where trees don't grow, the brutal kunai grass takes over, six feet high and sharp enough to cut the skin. It makes good thatch for huts. Much of the animal life resembles the Australian marsupials—tree kangaroos, cassowaries, wallabies, and bandicoots. In the New Guinea mountaintops live at least forty-three distinctive varieties of the bird of paradise. This flamboyant bird is a kind of crow bedecked with crests, breastplates, and plumes of scarlet, moss green, saffron, white, rusty blue, and other colors. Until 1921, when the slaughter was stopped, it was hunted to beautify the hats of European women; villagers still wear bird-of-paradise feathers in various ways. The bird's call is dismayingly raucous. The island also has big crocodiles, innumerable wild pigs, and May flies swarming so densely in the Sepik River that villagers catch and eat them. In surrounding waters are fourteen hundred kinds of fish, good sport and good eating.

*

Soon after you make your acquaintanceship with the Australian part of New Guinea, the 52 percent of the island east of West Irian, individual areas begin to come into focus. The south, up to the highlands, is Papua (pronounced "PAP-oo-a"), an Australian possession, with a population of

---

* Named not for any insect life there but rather for H.M.S. *Fly*, the little British vessel that first sailed into its mouth.

700,000. The north, including New Britain, Bougainville, and other islands, is the UN Trust Territory, officially known as New Guinea, even though that is also the name of the whole island. The population of this part is 1.8 million. The various subdivisions of both parts are:

1. Port Moresby, the capital, and its environs, on the southern coast of the bird's tail, an area small in size and anthropologically rather uninteresting. Here you find the Motus, whose women tattoo their faces and whose language is one of New Guinea's lingua francas, but the place has little tradition and lore.

2. The land along the Gulf of Papua, west of Port Moresby, and the endless swamps around the Fly River—empty except for some lowland villages and the kind of white man, given half over to adventure and half to drink, that one sees in ports like Daru all over the world.

3. The highlands—a bracing combination of clean air, icy mountaintops, and fine crops.

4. Lae and Madang on the north coast. The former evokes memories of the bitterest battles of World War II, and the richest discoveries of gold in New Guinea; the latter is an area that retains vestiges of the years when the Germans colonized it, plus the influence of missions that have somewhat Westernized the former clansmen.

5. "The Sepik"—that is, the 250-mile stretch of northern swamps and mountains adjacent to West Irian, which has for an axis the muddy, coiling Sepik River, down which float tangled masses of gray-green vegetation. This area is famous for its artifacts, chiefly carvings, and the art bespeaks a violent, magical, superstitious culture.

6. The islands of New Britain and New Ireland, which surpass Port Moresby for progress, prosperity, and economic influence. The Tolai people who live in Rabaul on the Gazelle Peninsula of New Britain are the territory's best educated and most politically potent. Tolai translates to Australians as independent rebel, standing up to the white man. New Ireland, two hundred miles long and an average of seven miles wide, is a strand of surf and coconuts.

7. Bougainville, northernmost of the Solomon Islands, which was arbitrarily placed under Australian control (the other Solomons are a British Protectorate). It has a great, new, wealth-creating copper mine, estimated to be the biggest in the world—and villagers who are fractious.

Uncounted hundreds of small islands, some only dots, surround the bigger land areas. Manus, northern bulwark of the Bismarck Sea, was a major American base in World War II—built up at a cost of $156 million to the point that it sometimes sheltered six hundred ships in its harbor. Then it was abandoned with an unbelievable waste of material and supplies.

The Yam-Eaters

The 2.5 million ulotrichous, or frizzy-haired, people of New Guinea, who combine various proportions of Negroid (or Papuan), Negrito, and Melanesian blood, are descendants of invaders from Asia many thousands of years ago—not much is really known about this. Most of them live in an essentially Stone Age culture, unaffected by white men until as recently as thirty years ago—New Guinea is virtually the last part of the world to be explored. Cannibalism was not unusual until Australian bush patrols combined with Australian law made the practice difficult, and some still persists in the more remote mountains. The bodies were never boiled, in the manner of a Charles Addams cartoon, but rather baked. A trench was dug, and the bottom lined with coals and hot stones, plus a layer of twigs and leaves. The beheaded body was placed on that, covered with more hot stones, and burning coals were raked over the top. The head, however, was boiled, and a deft pull on the neck vertebrae would extract the brain as an appetizer. The meat, as is well known, tastes like pork—hence the phrase "long pig" for a comestible man. Many observers have held that its consumption, apart from appearing natural and enjoyable to the villagers, filled a sharp physiological need. Beatrice Blackwood, an anthropologist who worked in Papua in the 1930's, wrote that "protein shortage is, if not the only, at least a powerful contributary cause of cannibalism." But there was ritual cannibalism, too.

It is probably more significant, however, that the main food of the inhabitants of New Guinea was and is yams. Most people eat ten or twelve pounds of yams—or sweet potatoes, taro, or related starches—every day. It follows that they must be diligent and skillful gardeners, and they are. Men compete in growing big yams, the record being somewhere around twelve feet in length; but these are for ritual, not for food, and women are excluded from the gardens where the necessary magic is practiced to produce a superyam. The very fact of gardening brings with it settlement, which made the people of New Guinea quite different from the nomadic Aboriginal hunters of Australia.

Meat is hard to come by in New Guinea. Villagers sometimes burn patches of kunai grass and then devour the charred trapped corpses of lizards and mice. The big and stupid cassowary is hunted; this bird cannot fly, but it can rip open a man's stomach with one swipe of its large sharp claw. The favored narcotic is the betel nut, fruit of the areca palm. This nut is red inside, green outside, as long as your thumb. It is chewed together with

pepper from the betel pepper plant (hence the name), and with lime, which counteracts the laxative effect. Addicts soon get dark red gums and black teeth.

The main point to make about traditional social organization in Papua New Guinea is that it is loose, decentralized, lightly governed, and lacking in hereditary leadership. There are about five hundred clans, roughly based on common ancestry and language, and averaging four thousand members each. There are five hundred linguistically complex languages (seven hundred on the whole island). But thirty-seven of them are spoken by more than ten thousand people each, and since they are somewhat related languages, intercommunication is not completely hopeless. Multilingualism, generally including either of the two main lingua francas, Pidgin and Motu, further eases the problem. Within the clans there are subclans united around their own songs, stories, ceremonies, and family ties. Subclans are further divided into villages and hamlets. Each subclan has a rather impotent chief, and the chiefs jointly form a clan council for such purposes as allocating land. The land is communally owned, but each subclan gets a beneficiary interest in a number of plots, dispersed in little gardens here and there.

The women are "hot of eye and vigorous—they'll tear you apart," according to an Australian journalist. They supposedly know of a plant that is an effective contraceptive. Dress varies widely; women mostly wear kunai-grass skirts, or patches of cloth. Men in many parts of the country feel adequately dressed if they are wearing a *kimbung*, or phallocrypt, that sheathes the penis. Some of these, of gourd or bamboo, are startlingly long —two feet or more. Size is a symbol of status. A common greeting among Chimbu men is to touch one another's genitals—a ritual known to whites as the "Chimbu handshake." Many men have bushy black beards, lending them an uncommonly fierce look.

Warfare once was the way of life among New Guinea clans, the weapons being clubs, spears, arrows, and elaborate shields. Some parts of New Guinea have spooky clan houses full of skulls lined up on shelves to attest to the ferocity of their wars and headhunters. The Australians have mostly stopped the battling by the forthright tactic of moving into an aggressor village and burning down all its houses. The instinct for violence, and the low value placed on life, continue in effect. Writer Jens Bjerre tells of a couple of village men competing to sell potatoes to a white man. One of them casually pulled a club from his belt and with one blow murdered the other— and without a glance at the dead body continued peddling his potatoes. "Killing for them is merely a sport," the white man remarked. "The young men are not really considered worthwhile until they have killed somebody."

Mount Hagen Sing Sing

One of the most bizarre experiences a Westerner can have in the whole South Pacific is to go to a "sing sing" in the New Guinea highlands. Fairly typical was one staged by the Manembi clan, who made it an occasion to give some pigs to the Rogogais. These friendly people, it appeared, had earlier helped the Manembis in a fight, and nothing makes a better gift in this strange high land than pigs—in fact, pigs along with pearl shells are the local currency.

The sing sing began with a signal from a whistle. Forty spear-bearing warriors advanced, dancing with an undulating kneel. Every fourth or fifth man carried a drum, narrow-waisted and black, played by tapping the hide stretched across one end. The warriors' appearance was wondrous and fearsome. From their belts hung, in front, woven reed skirts, and in back fluffy fronds of leaves, known hereabouts as "ass grass." A golden shell was suspended across each man's chest, and above that a chaplet of bars of bark. These bars are marks of wealth, each representing a pig or a pearl shell (*kina*) that the wearer owns. Bodies glistened with tree oil or pig's grease. Under the chin some of the dancers wore green "beards" of a parsley-like foliage, and through the septum of the nose a white bone. Faces were blackened, with white on the nose and outlining the eyes and mouth. Topping it all off were tricorn wigs, typical of these clans, made of black human hair and decorated with a brushlike centerpiece of orange-red and blue feathers from cockatoos, hawks, and birds of paradise. Some of the men carried axes in their belts, and almost all had long spears. The onlooking women, wearing bouffant grass skirts, had painted their faces with red ocher for this ceremony.

Sing sings serve various purposes. Some are simple demonstrations of unity. Some are to reassure people after a death or a tragedy. Sing sings can last a week, stretched out by bursts of suitable oratory. The sing sing is a standard part of the ceremony called *moga*, a pig exchange. Clans trade pigs, incredibly enough, as self-discipline against fighting one another; these beloved animals serve as hostages, and belligerent young clansmen are told, "We can't fight those people—they have our pigs." Pearl shells are traded in the same manner and for the same reason. This complex scheme of debt and credit assures political stability of a sort among the clans. Repayments may go on for thirty years.

Using pigs for currency is rather more difficult than using paper dollars. This four-footed kind of money has to be fed and tended, chores that the highland men assign to their women. A common sight on the roads is a

woman walking a pig or two on a leash. Sometimes women suckle pigs. To have fifteen pigs is to be gloriously rich. The *kina* shells are only slightly more convenient as money, since they run to the size of a dinner plate. They come from the coast, and are traded as many as fifteen times before reaching the mountains. Tourists pay ten dollars for *kinas* in shops, but the highlanders buy them with salt and tree oil.

*

Mount Hagen is perhaps the most engaging place in New Guinea, sleepy, faraway, and strange, nestled in a verdant valley. The streets are quick cuts of exotic color. Here we have a topless town—and many of the men are almost naked, too. The population is about four thousand. One-fourth of the people in the town are whites. The market is an energetic scene—fresh vegetables in from the gardens, sugar cane and peanuts, men with grass garlands or circlets of flowers on their heads, betel-nut chewers chomping and spitting, a pig's head on sale for three dollars. A hospitable place to stay is the Hagen Park Motel, and there are several others, all needed to house the tourists who come to the biennial Highland Show, when thousands of villagers get together to sing sing, dance, and play drums.

This town did not exist forty years ago: the very area itself was unknown to white men—completely undiscovered country. Then a pair of gold-prospector brothers, Mick and Danny Leahy, poked their heads over a mountaintop and viewed the wide, grassy Wahgi Valley, containing, to judge from the hamlets and gardens that they saw, a large population. Soon after, they explored the valley by airplane, then walked in and built an airstrip. They never found much gold, but Danny Leahy started a big coffee and tea plantation, took on six or seven wives, fathered twenty children, and lives there to this day. The road from Lae, 350 miles to the east on the coast, was finished only eight years ago, having been built, stretch by stretch, by villagers using only picks, shovels, and bare hands. In 1956 only three buildings in Mount Hagen had iron roofs, the rest all being made of kunai thatch. Only forty-nine whites lived there then. The first shops were opened around 1962. The town's name derives from a nearby 12,392-foot mountain, christened by the Leahy brothers in honor of Count von Hagen, administrator of the German sector of New Guinea in the 1880's. Exploring his domain, Von Hagen once got within forty miles of the mountain. He died in Madang and was buried in a tomb surmounted by a handsome bronze eagle. World War II bombing toppled the eagle, which was subsequently moved to the town that now bears the Count's name.

The Western Highlands District, for which Mount Hagen is the seat of

government, covers 9,000 square miles and contains 320,000 people, 5,000 of them white—a fairly dense population. A small area north of the Lagaip River in the northwest is the home of perhaps two thousand people, with whom the government has barely made contact. The Western Highlands voters send nine members to the Parliament in Port Moresby, and rule themselves, in the pattern of the whole country, through nine locally elected government councils. The highlanders tend to vote conservatively; specifically, they do not want independence for Papua New Guinea soon. They fear that the coastal people, with more education, will dominate them. The district's problems, says the district commissioner, Mick Foley, are communications (only eight hundred miles of roads) and education (only one child in five is in a school). The education gap comes sheerly from a shortage of teachers; the people are eager for schooling and happy to provide the buildings by joint effort. Ten religious missions are presently providing the bulk of the schooling—the Seventh-Day Adventists also run a leper colony for 250 patients. Mr. Foley first went to the Western Highlands more than twenty-five years ago, and learned to admire the character of the local population. "They're not as dour or withdrawn as most Melanesian people. One can get contact. We encourage our officers to camp out. These people are more virile than the coastal people." They have no tuberculosis, and little malaria because the valley's elevations—four to nine thousand feet —are above the malaria belt.

*

Tourists are sometimes approached in Mount Hagen by a man wearing the great triangular wig so customary there, and displaying enough wooden sticks dangling from his neck to prove that he is very rich in pigs. What he ardently desires from them is any newspaper they may have. Terrific value for cigarette paper! "A newspaper makes him a millionaire," says Ron Hiatt, a district officer.

The people of this valley eat little else but sweet potatoes, grown, chiefly by the women, in manicured gardens laid out in perfect checkerboards. In the warm valleys, the sweet potato matures in six months, but above nine thousand feet it can take as much as a year. Sweet potatoes sell for two cents a pound. The diet is obviously short on protein, but the highlanders make it up to some degree by occasionally slaughtering a pig, or by buying mackerel for twenty cents a tin. The land is fertile and skillfully handled. After five years of cultivation, it is allowed to lie fallow for five or even ten years, planted with the seeds of the casuarina tree, which returns nitrogen to the

soil. When the government buys clan land in order to lease it to whites, it must get approval—a signature or at least an "X"—from every single man in the clan,* for, as we have seen, the land is communally owned. The money is given to agents of the clan, or divided up among members.

When the clans fight, they use bows and arrows, dipping arrowheads (made of human or animal bone) into a dead, rotting pig to provide a slow poison. The disputes are over (1) land, (2) women, and (3) pigs. Since in this valley the clans speak seven different languages, the lingua franca is Pidgin, which the women dislike; they say that it is "heavy on the tongue."

Men and women sleep on the ground, the men in a big central house and the women in nearby huts that also house children and pigs. The hamlets formed by a number of these houses have a central meeting point for pig exchanges, marriages, funerals, and sing sings. Men are entitled to have as many wives as they can buy, the bride price running to maybe eighty dollars,† and each wife tends a garden. In practice, men usually have one wife; women dislike polygamy and use witchcraft against competitive wives. The culture regards women as inferior beings—"a piece of furniture that does housework"; their flesh and blood are thought to be somehow contaminated. On the roads one often sees a man in a wig walking along with an umbrella, his wife following behind without one. Men will not eat with women, though they do accept food from women's hands. The men usually make love to women out of doors. They sing a ritual song, a sort of nasal chant, that serves to announce that they want privacy, and this keeps other people away. At about thirteen boys leave their mothers to live with a teacher or mentor for five years. During this period they are forbidden to touch a woman. This waiting time ends in a coming-of-age ceremony, but boys are not circumcised here as they are in—for example—the Aboriginal society of Australia.

At death bodies are placed in boxes, elevated twelve or fourteen feet, to putrefy, a malodorous process carried out by "morticians" from another clan, who get paid in pigs or shells. After a couple of months the decomposed body is put into a lower box, decorated and painted, and left there until it becomes a skeleton ready for the boneyard.

White men, and their road from the coast that brings up modern civiliza-

---

* The temptation is to call these groups tribes, but Commissioner Foley says one mustn't. His word is "culture group." The internal political structure is not sufficiently strong or definite to merit the use of "tribe."

† Bride prices have been known to go as high as $5,000. This is, of course, too much for most men, unless they buy on the installment plan. However, the husband gets full use of the wife while making payments.

tion in one-ton trucks, have quite predictably Westernized some of the high-landers. Sweet potato gardens have become plantations of coffee, tea, and pyrethrum, an insecticide derived from the head of a kind of white daisy, which is dried and processed into an oleaginous fluid in a factory near Mount Hagen. The onetime yam-growers simultaneously move from the subsistence to the cash economy, put on Western clothes, and scoot around on Japanese motorcycles (although it is also possible to see perfectly naked men riding motorcycles, carrying bows and arrows). Sorcery is in decline. People who a few years ago did not travel farther than ten miles in a life-time now go longer distances. Still, the place remains as primitive as the middle of the Congo. Shopping in the stores are girls dressed only in swatches of grass, suspended fore and aft from a belt, as they push their carts around and pick cans from the shelves. Other shoppers are nearly naked men in feather headdresses. On the roads women wear lap laps, unsewn wraps of cloth, and carry one, two, or even three *billims*, which are string-net sacks adequate to hold yams or babies. A common appurtenance is a sort of umbrella or cloak made of pandanus leaves sewed together—good for fending off rain. The people in this area have rather light skins and strong, aquiline noses. Women's cheeks are frequently tattooed.

Coffee is the big cash crop of the highlands, grown by ninety thousand native entrepreneurs in small plots and by about fifty white (mostly Aus-tralian) planters on a big scale. But tea is rapidly catching up on coffee here, under the leadership of Ivor Manton, who comes from Melbourne. His tea fields, swamps until he drained them, are so rich that he had to stop using fertilizer—the tea plants grew too fast. "I have 1,350 acres to tea and 650 lads to work it, 90 percent of them bachelors," said Mr. Manton. "We house them, and we have a soccer ground for them, and other recreation grounds, too. We pay the highest tea wages in the world for pluckers, forty-two cents a day, plus food and housing. The day is ten hours." The elevation here is 5,600 feet, which means that the quality of the glistening tea leaves is excellent, for high quality goes with high elevation. Mr. Manton has a large tea-processing plant on the premises, and says that "leaves growing now will be finished tea tomorrow."

The Sepik

From Mount Hagen to Angoram is 115 miles sideways and one mile down: here we come again to sweaty coastal country and the fascinating land of the Sepik River, where the people catch crocodiles with their bare hands. The key words here are sago, artifacts, and beer.

Sago is the staple food, almost pure starch—breakfast, lunch, and dinner for most of the people in this area. To get it, the native men fell any of several suitable varieties of palm tree, strip open the trunk, extract the pith, and pound it to bits. Women knead this stuff with water, which dissolves out the starch. The resulting fluid can be kept wet in sacks, or dried into a flour. It can be served in cakes or as a porridge. One tree may yield four hundred pounds. The river provides plentiful fish to go with the sago, and it all makes for an easy life for anyone who does not mind eating the rough equivalent of white mucilage.

Rice, copra, and coffee are cash crops. The men also get money from itinerant buyers of crocodile skins, which bring forty dollars apiece. Croc hunters find the beast by probing underwater with sticks. They deftly tie the mouth shut with a cord, and bind the lashing tail with another. Another source of cash is artifacts, prized by collectors. This primitive art is chiefly comprised of figures, which can be quite tiny or more than life-size, and masks, which may be carved of wood or woven of fiber, but in either case are decorated with shells, hair, teeth, feathers, and fur. The Sepiks also make stools, shields, headdresses, bark paintings, canoe prows, and, most impressively, great *haus tambarans* (spirit houses) with high carved entrances. Motifs are the crocodile and the hornbill bird, and there are abstract designs as well. The Sepik peoples are endowed with a complex tradition of religion that provides themes for this art, plus much talent for carving and coloring. The carvings generally tell a story of some kind, and often contain sexual symbolism. Though Angoram is certainly an end-of-the-world sort of place, quite a few tourists fly in to buy artifacts, mostly sold at a store run by a local government council; but the bulk of the artifacts are bought by dealers for resale elsewhere.

A great deal of the money thus earned goes for beer. These easygoing people like liquor, and make a potent drink out of coconuts. The European population of Angoram numbers only about thirty, mostly planters. The villagers number between two and three thousand, a floating—literally— population that disappears and reappears in canoes from up and down the wide and muddy Sepik. The old but passable Angoram Hotel, with swishing fans in its four bedrooms, sits on a little hill beside the river, a few minutes' walk from the tiny airfield. Wewak, the big city of the Sepik, is seventy miles away by land, and a road is being built. Presently most people go from Angoram to Wewak by boat fifty miles downstream to the ocean and then west along the coast—a thirty-hour voyage. Pidgin is the universal language here. Parties of bird-watchers come to Angoram from as far away as the United States; one group found ninety-eight varieties in a month.

*

*A man put a stone on the ground and shot at it. They all shot at
it with arrows and blood came out. It went up into the sky and now
it is the moon.*

—KUKUKUKU FOLK STORY,
RECORDED BY BEATRICE BLACKWOOD

Twenty minutes from Lae lies the country of the Kukukukus, among whom
killing is a fine and satisfying thing, now mostly denied them by the Australians. They have a plant whose branches can be waved before the eyes
to make the sight keener before a fight—the same branches, waved over a
log, will make it lighter to lift. They know that begonia leaves, heated and
rubbed on the skin, relieve pain. They even have a plant to chew before
going to sleep, which will eliminate the necessity of urinating in the middle
of the night—going outside to do so entails the risk of murder by a watchful
enemy. Finally, perhaps most usefully, they know a plant that can be burned
to an ash that will substitute for salt, a rare seasoning that they yearn for all
their lives.*

The Kukukukus, most dreaded—at least formerly—of all the New Guinea
tribes—live in high mountains, and grow their gardens by clearing a bit
of the forest. Their best friend, their symbiotic plant, is the bamboo. Cut to
include one joint, it makes an excellent cooking pot, the Kukukuku pressure
cooker. Mushrooms done this way are said to be equal to the best of French
cooking. The bamboo makes both bows and arrows, although the black
palm provides a better bow. The bamboo cortex makes a surprisingly sharp
knife, formerly used to butcher human corpses. Fashioned into a jew's-harp
or an aeolian harp, the bamboo provides music. It also serves as food, torch,
container, tongs, and firewood.

The Kukukukus live in scattered hamlets of six houses, typically in family
groups of one old couple and a married son per house. The principal house
has a conical roof that goes down to within a foot of the ground, a low
doorhole for an entrance, and a walled-off circular corridor inside that leads
to a door to the main central room. Kukukukus do not call themselves that,
preferring to be known as Menyamyas. They number between fifty and one
hundred thousand, and occupy a territory bigger than Haiti.

Upon the death of a relative, Kukukukus smear themselves with mud and
mourn for four or five days, simultaneously lacerating their skins until blood
flows—all in the belief that the dead man's spirit is close at hand scrutinizing

---

* Another essential of life, Vitamin D, is supplied by the high-altitude sunshine.

the sincerity of the grief. The corpse is then smoked for six or eight weeks, making it a mummy. It is carried to a mountaintop, placed in sitting position on a bench with other recent decedents, all with their heads propped up by arrows so that they can gaze over the valley they knew while alive. Ants and caterpillars eventually reduce the putrefying mummies to skeletons.

That strange, ritualistic musical instrument, the bull-roarer, favored by the Aborigines in Australia, shows up also among the Kukukuku. The *tumbu-lun*, as it is called here, is a flat pipe tied with a bark string to a stick. Swung in a circle over the head, it makes an awesome drone. This serves to keep evil spirits away from the initiation ceremonies of boys into manhood—if a woman hears it, the spirit of the *tumbu-lun* makes away with her.

Sweet potatoes, grown in a laborious digging-stick cultivation that consumes the bulk of the day's hours for the bulk of the Kukukukus, are sufficient for most of their meals. The yearning for protein, though, leads them to hunt small wallabies, anteaters, birds, and even snakes. They have traps that cause a log to fall on the animal victims. Fish are stunned and caught by throwing pounded derris root (basic ingredient of rotenone) into a pool.

The Kukukukus make a cloth of bark, as elsewhere through the South Pacific, but do not decorate their clothes or their habitations very much. They make effective stone adzes and clubs, with heads in a lethal starfish shape. If they were ever to run out of meat, they could, it would seem, find a fat anthropologist behind any tree in the vicinity; New Guinea has proved to be a fruitful place to go for material for a Ph.D. dissertation.

The Cargo Cults

New Guinea seen through Western eyes is an amazing place, but the Western world seen through New Guinean eyes is nothing less than supernatural. The outer world is the source of the wondrous materials and products— food, weapons, cars, textiles, and baubles in infinite variety—that pour from the holds of ships and planes, described by the white men as "cargo." To the villagers, the origins of these highly desirable marvels are utterly mysterious—a ship appears on the horizon, an airplane materializes from heaven. Soon envy creeps in—the Europeans possess the secret for obtaining cargo, it is said, and won't tell the New Guineans. Sorcerers arise who claim that if they are duly worshiped and rewarded they can practice the magic that will produce the cargo. From this ridiculous premise, accepted blindly by villagers again and again in New Guinea's history, comes that weird phenomenon, the "cargo cult."

The oddest of these cargo cult explosions is the one that took place in

1964 on the island of New Hanover—a place particularly susceptible to this sort of hoax because it had been so impressed, as well it might, by the stupendous amounts of cargo brought to the region by the Americans during World War II. New Hanover, like the rest of New Guinea, was about to elect its first representatives to the territorial Parliament; simultaneously, as it happened, a United States Geodetic Survey team was in the area making maps. Some whimsical American suggested to the New Hanoverians that they should vote for the President of the United States, Lyndon Johnson,— certainly he could deliver the cargo! The idea caught fire instantly. New Hanoverians admonished the electoral officers that regardless of who was on the ballot they would vote only for Lyndon Johnson. Simultaneously, as an alternative, they took up collections of money to *buy* President Johnson outright. One night the navigation lights of a French airplane appeared in the sky, and hysteria prevailed on the island. Johnson was arriving, and bring- ing the cargo! When the plane flew on without landing, the furious cultists charged that Australia had stolen it. In Washington, President Johnson, him- self preparing for an election, heard of this tribute to his leadership from an Australian diplomat, and remarked that he was too busy to go to New Hanover, but would gladly send Barry Goldwater.

The Johnson cargo cult, it is pleasant to report, had a happy upshot. A Roman Catholic missionary, the Rev. Bernard Miller of Toledo, Ohio, per- suaded the cultists that the cargo they really needed was not material goods but American know-how. On the organizational framework of the cult, its members then founded a United Farmers Association. Under its auspices, they have planted 100,000 coconut trees, started a sawmill, and provided themselves with a bulldozer to build roads and a motor vessel to transport their produce.

Usually, however, when a cargo cult leader gets a sufficient grip on a village's credibility, the people stop gardening and other productive work —why slave over a yam when bounteous food will arrive any minute? In one case cultists built a large garage to house the automobiles due in the cargo destined for them. In another they built a mock airport, complete with control tower. Australian district officers keep a sharp ear out for rumors of cargo cults, and attempt to head them off. They fear that cargo cults are dangerous because they represent envy, hopelessness, and frustration. Another dark aspect is that cultists often adopt a warped form of Christianity; one told his followers that Christ was a Papuan, and that the Europeans had suppressed this information by tearing out the first page of the Bible. The government has many times taken local leaders to Australia to show them how the contents of cargoes, manufactured goods, are unmysteriously

turned out by human hands in factories and then loaded on ships to be taken to New Guinea. But most of the cultists remain convinced that cargo has a supernatural provenance.

*

One inscrutable cult in New Guinea goes by the unlikely name of the Hahalis Welfare Society. Its field of operations is the small island of Buka, just across a narrow passage from the northern tip of Bougainville. The founder of "The Welfare," as it is commonly called, is a burly, handsome man named John Teosin, now thirty-three, the product of a good education by American Roman Catholic priests in Buka, and later in government schools in Rabaul. This training won him a position of leadership in Hahalis, a village twenty-three miles up the Buka coast from the straits, when he was only sixteen. The Welfare's first goal was to force the government to build a road to the straits over which to truck out the village's main product, copra. Teosin achieved this goal, by a perfectly flabbergasting method.

Hahalis, it seems, had a sex problem: the old men, by reason of seniority, got all the young girls, and the young men had to wait. Hahalis also had a motivation problem—no one could see any really good reason to work. Teosin's solution, in 1961, was to start what soon became famous throughout New Guinea as the Hahalis Baby Garden. The town built a two-story house big enough to contain two hundred tiny cubicles, and stocked it with a corresponding number of toothsome young local girls, aged twelve and up. In return for productive work for the Welfare, young men were authorized to disport themselves among these lovelies, and the more of them they made pregnant, the better. The children became the property of the Welfare. White blood was prized—half-breed children were "easy to educate," said Teosin—and quite a few white men helped to fertilize the Baby Garden.

The Welfare quickly became such a dominant institution that Teosin decided his people could dispense with the official local government, which seemed chiefly interested in collecting taxes. Hahalis men clashed bloodily with government patrol officers. Teosin was jailed for a while. But the Australian Administrator in Port Moresby perceived that "European civilization and the government" had been deficient in developing Hahalis, and proceeded to build the road that Hahalis wanted. Soon the Welfare's copra trucks were churning down to market. The Baby Garden was reorganized, and opened with an exhibition of sex acts for the benefit of invited guests, the members of the local government council. In fact, Teosin has several times used sex orgies performed in public (or the threat of doing so) to

bludgeon the government. But he himself, married in a Catholic mission church years ago, stays strictly away from the Baby Garden.

Such is politics in Buka.

Some Towns and Cities

PORT MORESBY was chosen for the capital of combined Papua New Guinea after the war because it was the city least flattened. Even so, Port Moresby has been mostly rebuilt since then, and it has some attractive modern architecture and decent hotels. There are about 41,000 people in all. The look of the place is agreeable for the tropics. There are hills; eucalyptuses dot the landscape; the waters of the fine harbor are predictably turquoise. The temperature is bearable, never over 97.3 degrees Fahrenheit, and the rainfall is moderate (and varied: between twenty-three and seventy-two inches a year). The New Guineans in Port Moresby now wear mostly Western clothes, but one can still see pretty laughing girls in raffia skirts with flowers in their hair—a contrast to the drained-out, thin-blooded white women. Pidgin is the language heard on every hand. There is no television here or anywhere else in New Guinea.

There is still a Papuan village on stilts over the water at Hanuabada, the main settlement in the area before the whites arrived. A lot of the houses are made of an ugly material that the Australians call fibro—fibrous asbestos cement.

The Sogeri Rubber Plantation lies twenty-eight miles from Port Moresby and eighteen hundred feet up in the hills. Lisle Johnston, the manager, says that he employs 250 local men as tappers, paying up to twelve dollars a month plus housing, blankets, matches and tobacco, and family keep. On up the road are even bigger plantations.

*

RABAUL, meaning "place of mangroves," main city of New Britain, has been the setting for tremendous violence, both natural and man-made. The harbor is a caldera, the product of some thunderous prehistoric explosion of a volcano, later breached by the sea. Subsequent eruptions threw up three of the volcanoes (now dead) that adjoin the harbor, called North Daughter, the Mother, and South Daughter (the Father stands fifty miles down the island). Newer volcanoes still threaten Rabaul. One of them, Vulcan, was born during three days of smoke, explosions, steam, and ash-falls in May 1937, as the population watched in terror. The most recent blow, however,

was not an eruption but a Force 8.4 earthquake in July 1971. Tidal waves in the area swept half a mile inland, and the Rabaul waterfront was badly knocked about.

Rabaul's fine harbor outweighs its volcanic perils; but—just in case— vulcanologists are posted on the mountainsides to listen for any rumble and shout "Get out!" if necessary. Half of New Guinea's three thousand "*kong kongs*," Chinese descendants of laborers and house servants brought by the Germans, live in Rabaul. Rabaul exports sweet-scented copra and cocoa, much of it raised by the energetic Tolais, for whom this is the big city, though the population is only 10,500. During World War II, the Japanese swiftly captured Rabaul and converted it into a fortress so strong that the Americans and Australians never even tried to retake it, but instead waited until the surrender in Tokyo. The bombproof bunker of the Japanese naval commander still stands in the center of Rabaul, now deserted after having served briefly as a movie theater. Japanese-built defense tunnels lie under the whole town, and at nearby Kokopo are the seaside caverns that the Japanese excavated as ports for their supply barges.

*

You reach WEWAK on a plane with white stewardesses serving black passengers. Visible from above are many circular pools, bomb craters blasted out during the war. The town has lovely safe beaches, on a long spit of sand, and an ocean boulevard washed by a fresh breeze. Wewak lies behind what the local people call a "rain shadow"—that is, mountains on the windward side catch most of the rainfall—so Wewak is not as wet as nearby areas. One street has an odd-sounding name: Kenny V. C. Avenue; it celebrates a private named Kenny who got the Victoria Cross for his gallant capture of a Japanese gun position here.* Wewak has: (1) the most expensive motel in New Guinea, at nineteen dollars a day; and (2) Leo Arkfeld, the Roman Catholic Bishop, who flies his own plane all over the Sepik to attend to his flock.

*

LAE took advantage of being bombed to bits to relocate, moving from the coastal flats to a bench above the Huon Gulf. Though the city is new, big trees close over above the streets, and parks and gardens dot the city.

---

* The Japanese returned here (and elsewhere in New Guinea) after the war to open the mass graves of their thousands of war dead, cremate the bones, and take the ashes back home.

The market sells such temptations as smoked flying fox. On the beach, nearly covered by sand, is the hulk of a Japanese battleship. A number of Americans live here, and the town is booming with various light industries. Lae also reaches out to the gold-rush country of the 1920's, around Bulolo, and to the land of the once-dreaded Kukukukus.

A Short and Simple History

Moresby and Murray, General MacArthur, General Adachi, and Margaret Mead—these are some of the names that attach themselves, in the Westerner's mind, to New Guinea. The Spice Islands, just west, attracted Portuguese colonizers in the early 1500's, and a governor of the Moluccas was the first to sight the New Guinea mainland. After inspecting the residents he named the island Papua, from a Malay word meaning "frizzy-headed." A Spanish explorer, coming along later, descried a coastal resemblance to African Guinea, and gave the island the name that stuck. Three centuries of eventless history then drifted over New Guinea until the Dutch, branching out from Indonesia, took the western half of the island, establishing the bisecting border at 141 degrees east longitude, which remains the boundary to this day. About that time, British Naval Lieutenant Charles Yule landed from the schooner *Bramble* a bit up the coast from present-day Port Moresby to claim the eastern end of the island for Great Britain. An air of improvisation attached to this adventure; nobody remembered to bring a Union Jack, so Yule drew one with a pencil and paper, and nailed it to a handy tree. The tars on the beach sent up three cheers, but pianissimo, "lest too much noise should attract the natives." Unluckily, the sharp-eared natives were attracted anyway. Yule prudently tried to leave the beach, but his boat overturned in the surf. A cutter was sent to the rescue. It, too, broached, leaving fifteen unarmed Englishmen on the shore to be captured. And captured they were—until the *Bramble* managed to get boats through the surf to save them. Yule courageously waited until all his men were off the beach. Then—the natives having conveniently stripped him of his clothes—he dived into the waves and swam to his ship. He blandly named the beach Cape Possession. But so slow was the pace of exploration that it was not until 1873 that Captain John Moresby found the fine port that is presently the capital of New Guinea, and named it in honor of his father.

The next colonizer was Queensland, then a British colony and now, of course, an Australian state. The imperious Queensland Premier of the time, Sir Thomas McIlwraith, was worried by German agitation for the annexation

of eastern New Guinea. He cabled London, urging England to grab New Guinea; failing to get action, he sent an agent to Port Moresby to do so. Gladstone, Prime Minister in England, scowled at the colony's attempt to colonize. Queensland was not "fit to rule," he said. The rebuff shocked all Australia. In 1886, under further pressure from the Australian colonies, Britain seized Papua while allowing the Germans to take the rest of eastern New Guinea.

Britain ran the possession listlessly, but it established the principle that the villagers' life and right to their land should be left intact. Missionaries brought their dubious boons, suffering great perils in the process; two of them were spectacularly slaughtered and consumed. When Australia confederated in 1901, Britain gladly handed the place over to the new nation. Under an intelligent and energetic administrator, Sir Hubert Murray, New Guinea was left more or less as it was. This policy is the root of present-day accusations that Australia did not get started soon enough in educating Papuans for life in world society now. But Murray was conforming to the demands of Australian unions that black labor must not be developed and brought into White Australia. On balance the welfare of the Papuans seems to have been foremost in his mind; later, retired, he fought Canberra's attempt to make Papua an "asset" to Australia, and opposed a whipping-of-natives measure proposed in Parliament. While Administrator, he set up the system of Australian *kiaps* (Pidgin for district officers), who explored the territory, cut down cannibalism and tribal warfare, and kept the missionaries from overly disrupting village customs.

A month after World War I began, German New Guinea slipped easily into Australian hands: a few troops clashed with a volunteer militia of German planters on New Britain, and overcame them at a cost of six dead—first of the sixty thousand Australians killed in that war. At Versailles, Australia's Prime Minister Hughes demanded and got a League of Nations mandate to the former German colony. Australia took it over to "promote to the utmost the material and moral well-being" of its people, but promptly announced that the territory would have to be self-supporting, and to ship all its products through Sydney. Australians took on the former German plantations and ran them with indentured laborers, "recruited" from their homes and paid threepence a day, and susceptible, in the absence of their wives, to what was shudderingly known as "sexual depravity." The only event of note in the next two decades was the discovery of gold in the mountains up from Lae; Guinea Airlines was founded to fly cargo in and out, including entire gold dredges, part by part, and was the biggest-volume cargo airline in the world up to that time.

World War II

For millions of Americans, the first day they ever thought about New Guinea was January 21, 1942. That day the Japanese bombed two towns on the island; in March they made their first landings; and soon the war communiqués made household words of such remote places as Rabaul, Lae, Wewak, Port Moresby, Buna, Kokopa, the Owen Stanley Range, and Bougainville. In May, having taken Rabaul, the Japanese assembled a fleet to sail around the eastern end of New Guinea and seize Port Moresby. As it steamed into the Coral Sea, the assault force met a big American fleet, centered around two aircraft carriers. The opposing ships in that battle did not come within sight, but their planes fought savagely. The Japanese turned back—first augury of their eventual doom.

The Japanese decided to take by land what they could not take by sea. Over a six-thousand-foot pass in the Owen Stanley Range, north of Port Moresby, they poured troops far outnumbering the defending Australians, who fell back almost to Moresby. But the Japanese had outrun their supplies, and the tide of battle turned. Papuan villagers, the Fuzzie-Wuzzie Angels, rather reluctantly carried supplies in and wounded soldiers out.* That saved Moresby for good, but the Japanese still held Rabaul with ninety thousand men, and had fifty airfields in the New Guinea jungle. They stabbed back, attempting a convoy from Rabaul to Lae that was wiped out by three days of attacks by American and Australian planes in the Bismarck Sea. Left isolated in the mountains back of Lae, the Japanese held out savagely for two years, until the middle of 1944. Meanwhile, the Americans built up their stupendous base on Manus, the island where anthropologist Margaret Mead lived while she wrote *Growing Up in New Guinea*, and, after the war, *New Lives for Old*. One million Americans were funneled through this base.

General MacArthur deemed the enemy sufficiently under control so that he could transfer his headquarters from Brisbane to Hollandia, the capital of Dutch New Guinea, on the north coast just west of the border. Soon he was landing on Leyte, in the Philippines, and the New Guinea campaign was essentially over. The Japanese at Rabaul were simply bypassed, and remained until after the war was over. Elsewhere—on Bougainville, on the parts of New Britain outside of Rabaul, and at Wewak on the north coast of New Guinea—Australia's General Thomas Blamey fought on against the

---

* The first published novel by a Papuan, *The Crocodile*, by Vincent Eri, is the story of one such carrier.

105,000 Japanese troops there. In the end, the Japanese General, Hatazo
Adachi, came out of the bush so weak that he had to be carried, and handed
over his sword to an Australian colonel. Fifteen thousand Australian troops
were killed or wounded in New Guinea, and fifteen thousand New Guineans were killed—plus one hundred thousand pigs.

After the war, the League of Nations Mandate was reconstituted as a
United Nations Trust Agreement, and in 1949 the Australian Parliament
passed the Papua–New Guinea Act, which thereby became the territory's
Constitution. It joined the northern and southern halves of eastern New
Guinea into a single administrative unit, which now has a common Parliament, Supreme Court, and civil service. The UN scrutinizes Australia's performance in the trust territory by means of a visiting mission every three
years, and prods the colony toward self-government. How this works, in terms
of developing education, the economy, and the political system, is the subject of the next chapter.

CHAPTER 20

Niugini: The Future Nation

Hark and behold our stone age is swaying and groaning,
Right beneath your mighty step of pain,
Haltingly and stubbornly resisting and frowning,
But forever and ever in vain.
— A. P. ALLAN NATACHEE, MAKEO POET

WE HAVE NOW GRASPED that the eastern half of New Guinea is Australian; we must proceed to grasp that it will not be Australian much longer. The territory is slated for independence some time in the next decade.

The idea raises many questions. An island-nation may make some sense; one sliced down the middle by a boundary that came from the whim of some forgotten Dutchman is certainly less logical. It is equally unnatural to reach east and include a couple of islands from the Solomons. And what is enclosed in these artificial political-historical borders has no particular coherence—future fragmentation seems quite possible. Furthermore, independence may well sacrifice the law and order that the Australians imposed, which might seem a fusty argument, except that this law and order replaced cannibalism, tribal warfare for the fun of it, and casual murder. The slender new nation will also expose itself to the possibility of a cold world's raising duties against its products; it will be vulnerable to losing its markets just when it should be opening up bigger ones; and Australia will be under less obligation to subsidize it, as she has been doing at a recent rate of $100 million a year.

In reflection of these considerations, powerful forces oppose independence. Almost all of the conservative highland villagers are against it. The white Australian plantation owners also object. And one can correctly

surmise that the managers of the huge Conzinc Riotinto copper mine in Bougainville would rather stay under Australian law.

Nevertheless, Papua New Guinea is inevitably going to be a nation of its own, perhaps as early as 1976, for the forces in favor of independence are much stronger than those against. The Australian government has long since committed itself to the goal of New Guinean nationhood. The United Nations presses continuously for it. The strong nationalism of most of the educated natives of the territory works for independence. The consequence of all these pressures is a slow boil of passions and politics. Let us examine the panorama in more detail.

*

Australia's problem of letting go in New Guinea is ticklish now because the government made no preparation for education and self-government during all the long decades before the end of World War II. It was almost as though Australia deliberately steered itself to its present embarrassing position of possessing the last important colony in the world*—a colony bigger in size and population than many full-fledged members of the United Nations. In the process, Australia managed to fetch up with the last significant territory administered under the UN Trusteeship Council.†

Until the early 1960's, government in Australian New Guinea lay in the hands of an all-powerful Administrator appointed by the Department of External Territories in Canberra. He ruled through district officers backed up by patrol officers, in turn backed up by village authorities called *luluais* and *tultuls*. A more or less fictitious Legislative Council, composed mostly of prominent white businessmen and missionaries, lent advice. Local government councils—black men in white shirts sitting submissively around a table—assisted the district authorities. The Minister for External Territories in Canberra kept tight control of the whole show. Then came a memorable visit, in 1962, of a United Nations Mission to New Guinea, headed by Sir Hugh Foot (now Lord Caradon), last of the great British colonial governors (Jamaica, Nigeria, Cyprus). The mission censured Australia for dragging its

---

* If one accepts the fiction that Angola and Mozambique are provinces, not colonies, of Portugal.

† Which since the war has successfully terminated the mandates in Togoland, Cameroons, Somaliland, Tanganyika, Western Samoa, and Ruanda-Urundi; the only remaining trust territory besides New Guinea is the group of Pacific islands administered by the United States.

feet in developing New Guinea,* and demanded: (1) creation of a house of representatives with one hundred members elected by universal adult suffrage; (2) education through university level of one hundred Papuans and New Guineans to be administrators for the territory; and (3) a survey of the country's economic possibilities by the World Bank.

Thus prodded, the government proceeded, with vim and great goodwill, to get an assembly elected. Patrol officers took on the prodigious task of enrolling every last discoverable tribesman. In some cases, as one might expect, the tribesmen thought that the registration was for the purpose of singling out people to be taxed or murdered, but the more general reaction was wild, uncomprehending fervor. Explaining democracy, voting, and the principle of representation was almost comically problematic, but the patrols did their best, eloquently displaying maps of the capital and pictures of the House of Assembly.

The campaigning was vigorous and inventive. In a hinterlands New Britain constituency, a candidate named Lima Larebo promised, should he be elected, to recruit the doctors of Rabaul to go to the countryside and give the villagers injections that would turn their skins white. Some white candidates took new names with more appeal to the local voters; Peter Murray became Pitamari, and John Pasquarelli became Master February. One candidate campaigned by holding up an egg and assuring voters that it contained the embryo of an American soldier—upon winning, he would hatch the egg and the Americans would take over, which was, it is inspiring to note, thought to be a good thing. Another candidate cited just one impressive qualification: he had, in his lifetime, killed nine men. This New Guinea— what a culture clash!

Patrol officers took ballot boxes to every village. Enthusiasm reigned. Some voters walked for days to cast a ballot. One man who was wanted for murder could not resist showing up to vote, and was duly arrested. The process took a month, early in 1964, and when it was over Papua New Guinea had a new House of Assembly of sixty-four (not one hundred) members, ten of them appointed by the government rather than elected. The Rt. Hon. William Philip, Viscount De L'Isle, V.C., P.C., G.C.M.G., G.C.V.O., K.St.J., and Governor-General of Australia at the time, came to open the new Parliament. For a second general election in 1968, the size of the House was raised to ninety-four, still including ten appointive members.

This House of Assembly was, and remains, a rather toothless body, gain-

* Of course, the mission had to confine itself to the trust territory, and could scarcely set foot in Papua, but the logic of its criticism and proposals obviously applied to Papua, too.

ing only a little every year in the powers allowed to it by the Australian raj. But the system of which it is a part is the framework for the advance of the territory toward self-government and independence, and the territory's leaders work mostly within it. Let us meet a few of them.

The Not-Very-Angry Mr. Guise

Described as "widely hated among the Europeans" and "one of the angriest Papuans," John Guise is a docile, charming man of fifty-seven who scared the planter class a few years back by raising the idea of nationhood, but turned out to be a force for the most moderate kind of change. Guise, one of whose grandfathers was English, was born in a village near Milne Bay (on the eastern tip of New Guinea), and schooled by an Anglican mission. At twenty-six he was "top boy" at one of the stores of the Burns Philp Mercantile company, which dominates New Guinea. During the war he was drafted for road building, which led him to join the police as a sergeant and go on to become, in 1955, a village-affairs expert for the Department of District Administration. In 1960, when the Australian government decided to let six elected Papuan and New Guinean members into the old Legislative Council, Guise won the Eastern Papua seat. By then he had been, as a visiting policeman, to Brisbane, Sydney, and London (for the coronation of Queen Elizabeth); now he went to Kenya at Jomo Kenyatta's invitation, to Pago Pago for a conference of Pacific islanders, to New York as a UN adviser, to London again for an Anglican General Synod. In the Legislative Council he preached a program of reform—more education, help for village farmers, more authority for the local government councils, equal pay for equal work, the employment of native magistrates, antidiscrimination laws. Modest enough—but it upset the mossbacks. He won election to the House of Assembly when it was formed, and at once became leader of the elected members. He pushed the case for Papuan–New Guinean unity, proposing the name "Niugini," which looks like another way of spelling New Guinea, but is really a felicitous combination of words from local languages that jointly mean "stand of coconuts."

Certain members of the House of Assembly are now ministerial members in charge of, for instance, public health, stock and fisheries, trade and industry. But Guise never took a ministerial post: instead, he was elected the first island-born Speaker of the House. Just before taking office, he announced that for the ceremony he would shun a Western jacket in favor of a white bark tapa cloth over his shoulders. The switch infuriated highland members of the House, because tapa cloth is a coastal mode—so Guise sent someone

out for bird-of-paradise feathers, much worn in the highlands, and pinned them to the costume. Adding a *kina* shell on a necklace, a black cape, and a long white Westminster wig, he unified Papua, New Guinea, and old England with this one grand symbolic gesture. As his wife, who wears the blue face tattoos of the Motu people, looked on, Guise took office asking God's guidance for them both—"simple village people," in his words. In 1971, the Papua New Guinea government made its first international contact on its own. Guise and a group of members of the House visited Djakarta, pointedly disdaining to check in at Canberra first. Guise never went past the fourth grade, but he now holds an honorary Doctor of Laws degree from the University of Papua and New Guinea.

*

Guise floated up and out of Parliamentary politics when he became Speaker, but while he was still a partisan his heart was with the only party that has so far managed to hang together in the territory, the Pangu (an acronym of Papua and New Guinea) Pati (Pidgin for party). This group joins public servants, students, teachers, intellectuals, and radicals, plus more grass roots than any other aggregation, and criticizes the government. With only ten seats out of ninety-four, its main function, however unintentional, is to provide most of the other elected members with something to oppose. Pangu's opponents, highlanders, planters, and uncommitted members of the House, banded together after the 1968 election as "The Independent Group," and in 1970 formed a party called Compass. The party's main purpose is to stall independence as long as possible.

Chairman of this conservative group is Oala Oala-Rarua, one of the few tall, heavily built men of Papua, where people tend to be short and wiry. He is the son and grandson of pastors for the London Missionary Society. Some of his nine brothers (he also has two sisters) have become pastors, too. Mr. Oala-Rarua taught for a while, managed to study and travel in the United States and Europe, and got into politics as a founder of the Pangu Pati before the 1968 election. Soon "there were disagreements in the party. I resigned. Some say I was expelled." His qualifications—among other things, he speaks English much better than most Australians—quickly levitated him into the post of assistant ministerial member for the Treasury. "I had more experience. I was a teacher by profession."

As Oala-Rarua sees it, Papua New Guinea feels "no fierce, urgent desire to toss Australia into the sea." The reasons for this are that: (1) apathy, abetted by lack of education, is the prevalent mood; (2) Australia provides

vital funds for the territory; and (3) Australia provides a good administration with no serious exploitation. What does Australia get in return for the funds it spends here? Goodwill, a market, and raw materials, says Oala-Rarua. "There is no nationalism as such, as in Africa. There is no Communism either; there have been attempts, we've kicked them out of the trade unions, but there is no organized party." As for independence: "We look forward to it, though it is not a general feeling. Quite a lot of people are against it, especially in the highlands. They fear being taken over by the city people; they feel they are protected by the Australian government; they fear all good jobs would go to the coastal people, who have contact with the outside world. And the majority of the planters want the Australians to stay."

In sum, Oala-Rarua holds that the territory is getting ahead at an acceptable speed. "My children certainly have more opportunity than I had. They learn English from the word go" (although "only about a quarter of the children in the country at large even get into primary school"). He thinks that "the Administrator is right when he says we already have a large measure of self-government."

*

A couple of impassioned young men of Rabaul are John Kaputin, twenty-nine, the territory's most rousing orator, and Oscar Tammur, also twenty-nine, leaders of the Mataungan* Association, which is a sort of political party that wants to rule by setting up a district government of its own, bypassing the official one. The issue is land. The Gazelle Peninsula, site of Rabaul, has the best volcanic land in New Guinea, much of it held by Australian planters as an inheritance from the freehold system that prevailed under the German administration. Kaputin calls the planters "white pigs who have grown fat on my country." Responding slowly to the land hunger of the exploding Tolai population, the government in mid-1970 planned to resettle some of the Tolais on 302 plots carved from vacant plantations held by the administration. Without waiting, Mataungan squatters dashed in and seized the land. Blaming the land shortage on the lack of action by the Gazelle Peninsula Local Government Council, which has some white members, Mataungans by the thousands refused to pay taxes, putting the money into escrow pending Council reforms.

Prosecuting the tax-evaders, the government fined Tammur thirty dollars for refusing to pay sixteen dollars in taxes, and jailed him when he refused

* Meaning "Be alert."

to pay the fine. Tammur, a bearded teacher and copra grower who speaks fluent English, Pidgin, Toli and Mandik, is a Pangu Pati Member of the House of Assembly, but that did not save him from jail. He is a fiery backer of independence now. Kaputin, talking with John Gorton when the former Prime Minister visited New Guinea in 1970, told him that Mataunganism is "nationalism in the raw state."

*

Other redoubtable leaders dot the New Guinea scene. Albert Maori Kiki, secretary of the Pangu Pati, says right out loud: "We have hatred for the white man in our hearts." Michael Somare, a teacher, journalist, and Parliamentary leader of the Pangu Pati, is another angry young man. One fabulous early member of the House of Assembly is a Scotsman named Ian Farley Graham Downs. He went to New Guinea as a district commissioner in the Eastern Highlands, but resigned in 1956 to become a coffee planter. In the House he spoke up vigorously for highlands interests. He campaigned by papering his district with handbills in Pidgin that read:

YUPELA HARIM. VOT LONG
ELEKSION. DOWNS NAMBA WAN

When the returns came in, Downs was indeed No. 1. A less forceful member was Handabe Tiabe, elected by the Huri people of Tari in the Southern Highlands. He spoke only Huri, and sat through four years of sessions, which were mostly in Pidgin, bewildered but smiling broadly.

*

The man who occupies the position of the world's last high-level colonial governor—here tactfully called Administrator—is Leslie W. Johnson, a veteran Australian civil servant. The House of Assembly, he says, has the power to "pass laws about anything except a few reserved subjects like defense and external affairs, but the laws are subject to veto by the Administrator and the Governor-General of Australia." In practice, Mr. Johnson points out, the veto has been used only twice in six years. In one case, the House passed a law transferring control of the public (civil) service from the Australian Ministry for External Territories to itself—plainly a patronage grab. Another, says Mr. Johnson, was a "technical matter of mining legislation." There is a touch of play-acting in such maneuvers. The House of Assembly's ministerial

members "meet in a cabinet called the administrative executive council, which formally advises the Administrator," but they are "not yet fully responsible for government action."

Mr. Johnson, a sensitive and professional high-level civil servant, is articulate and aware of the territory's touchiest problems. Land tenure—as we have seen in the case of the Mataungan rebels in New Britain—is one of them. "We operate on the system that land belongs to the natives, and that's it," he says. All the land is owned by somebody other than the government, usually communally, and if the government wants this land, it has to buy it with the consent of all the co-holders. The natives cannot sell the land to individuals or corporations, but only to the government, which can then lease it.

For the more than $100 million a year spent in New Guinea, what does Australia get in return? Mr. Johnson replies: "Most Australians would say, 'Nothing.' The white man's burden. But Australia gets defense—not such a valid point since World War II—and a stable neighbor, and a sense of humanitarian service. And of course we are under pressure from Australian companies that have property here."

The Administrator plays down the differences between Papua and the New Guinea Trust Territory. "Legally, Papua is an Australian colony, and New Guinea is a trust territory of the United Nations administered by Australia. But in human terms, they are administered as one—my writ carries equally. They have one Parliament and one public service. They have a single educational system and a single budget. There is no problem of apportionment of funds because funds are apportioned on a basis of need."

*

Dr. John T. Gunther will probably be known, when New Guinea becomes independent, as the father of the nation. The New Guinean and Papuan leaders that he has helped to form—Guise is typical—are to be found all over the territory, and they are fantastically loyal to him. Born in northern New South Wales, Gunther went to medical school at Sydney University and planned a career in industrial medicine, but the war took him to New Guinea, where he served in the Royal Australian Air Force. Entering government service, he became Director of Public Health in the territory, then rose to be Deputy Administrator. But he never moved up to Administrator because of the political string-pulling that goes on in Canberra. This may have been all to the good, for Dr. Gunther got the fulfilling task of founding the University of Papua and New Guinea, and serving as its first vice

chancellor (which in Australian terms is head man). One recent New Year's Eve he said, "I haven't seen a winter in fifteen years." Such are the diligence and creative energy of this man.

Sitting on one of the orange seats in the university auditorium, a room of rough stone walls and a ceiling in the shape of a pyramidal tent, Dr. Gunther talked for a while about his school. The university started its first classes in makeshift rooms in 1966, began construction of its airy, modern buildings in 1967, and occupied them in 1969. The staff includes twelve professors and ninety other teachers and functionaries. The enrollment in 1970 was 780, two-thirds black and one-third white; Dr. Gunther hopes for an eventual ratio of eight blacks to one white. Entry requires four years of high school *plus* one year of prematriculation study—and still 38 percent of those who enter drop out. All the blacks are on scholarships: $200 a year for tuition, $160 for board and room, $100 for pocket money. The territory has five institutions of "tertiary" education—a technology school in Lae, an administrative college, an agricultural college, a medical and dental college, and a teachers college—but only the university grants degrees. The university library had forty thousand books on shelves, forty thousand not yet catalogued, and a set of copies of busts from the British Museum— Benin bronzes and Greek-Roman classics.

Dr. Gunther is, naturally, a mine of New Guinean lore. "There is not much left of the territory that white men have not walked over, but near Angoram recently a professor from here met people who had never seen a white man. Light skin is not generally the result of intermarriage with whites; the people around Port Moresby are, it is true, often lighter in skin color and straight-haired, but that is apparently because Malay traders came here long ago and influenced the heritage. We avoid the use of the word 'native,' except as an adjective, and 'indigene.'* They call themselves black, though they range in actual color from puce to Bougainville ebony. A bit of headhunting still goes on in the remoter areas—also eating Grandma because you love her. There used to be a brain disease, *kuru*, that mystified the doctors in New Guinea for a long time. It was finally traced to cannibalism. The cannibal himself would die—not right away but two or three years later."

---

* This is certainly a vexed question in New Guinea. Sir Hugh Foot's United Nations Mission decreed that the use of the word "native . . . must be eliminated." Government publications scrupulously avoid it, and writers often try to sidestep the word by saying "local people," "villagers," and the like. Other new books take the view that "native" means what the dictionary says it means, and use it throughout. In ordinary speech, whites use "native" functionally and naturally, and the Tolais of the Gazelle Peninsula call someone from another part of New Guinea a "foreign native." "Indigene" is, of course, an absurd pedantic euphemism.

*

Early in 1971, the Minister for External Territories, Charles E. Barnes, said that Australia would accept without reservation any date for self-government that the Papua New Guinea House of Assembly might choose. On the face of it, a rosy picture indeed—here is a colonial power cheerfully bending every effort to let go gracefully. But in Canberra one quickly learns that letting go involves excruciating problems, all related to the timetable. Australia seems to have a choice of cutting New Guinea adrift either much too soon or much too late, with no sensible compromise between these extremes.

The separation will come too soon if:
• The central government is not sufficiently strong and respected to restrain secessions and divisions.
• Education is not well enough advanced to supply replacements for the seven thousand Australians who hold most of the top jobs in the public service.
• Foreign investment ($50 million a year) declines sharply out of lack of confidence.
• The House of Assembly cannot develop more seriousness and responsibility. (One of Speaker John Guise's priorities is to hold down drinking on the job by members of the House of Assembly.)
• The highlanders, half of the population, continue to oppose independence.

But separation will come too late if:
• Nationalism grows to the point of impatient and possibly bloody rebellion.
• Dissension among the parts of the territory grows to the point of messy disunity at the moment of independence.
• Delay costs Australia credibility with the United Nations and the world, and makes it look like a nation hanging onto a colonial empire.

The trick, says one official in Canberra, is to "get out with cheers as our flag goes down and theirs goes up." Some trick; the dangers of both too soon and too late are quite real. Some of the black New Guineans are talking tough; Leo Hannet, a Bougainville poet and playwright, says, "The day of the *kiap* is over. As for the planters, they will have to go. But we must be fair. We must give them back their beads and axes." Education is still deplorably deficient; 72 percent of the people ten years old or more are illiterate; only about 220,000 children go to primary schools, and only

1,600 graduate every year from high school. The government still relies on mission schools for the bulk of primary education.

The sharpest dangers of all, no doubt, involve disunity and threats of secession. When a new United Nations Mission visited the territory in 1971, several hundred angry tribesmen at Wapenamanda in the highlands nearly broke up a meeting as they shouted down suggestions of self-government— they even opposed flying the United Nations flag next to the Australian. This sort of thing gives the Papuans of coastal New Guinea a feeling that numerically strong but backward neighbors are retarding their nationhood, and the attitude of the Mataungans on New Britain is that they'll go it alone into independence, rather than wait for the mossback highlanders. Many people in Bougainville also yearn to secede; one leader openly asked the 1971 UN Mission what the UN would do about it if Bougainville pulled out. The mission chairman said that he "hoped the situation would not arise."

Former Prime Minister John Gorton, on his 1970 tour of the territory, had to argue constantly that secession would damage the seceders. "The other thing which has been spoken of is self-government now, self-government for New Britain. . . . Would it not be well for each man to sit down in his own home and think what he really wants and what advantages he can get from being a part of the territory?" Part of the thinking in Bougainville is that the single biggest industry of the territory, the $350-million Conzinc Riotinto of Australia copper mine, is located there, and by secession the island can keep the resulting prosperity to itself. Gorton countered this concept by stressing the possibility that Papua is on the brink of a big oil find, as a consequence of some discoveries in the gulf, and will have some prosperity of its own to share before long.

How to let go of New Guinea has increasingly become a political issue in Australia. The leader of the Labor Party, Gough Whitlam, made a flamboyant tour of the territory early in 1970, and won the applause of the people of Papua and New Britain by proposing a firm timetable: self-government in 1972, independence in 1976 (both territorial election years). Ten thousand cheering people mobbed Whitlam in Rabaul. It was this successful maneuver that forced Gorton to visit New Guinea six months later, but as he prepared to go, planters rushed to Canberra to plead with him not to set any dates for independence. So he confined himself to assurances that it would come eventually but positively. "I do not speak of self-government in 1972 or in any calendar that you may mention," he said, and "anyone who seeks to say so . . . is dangerously simplifying the problem." Gorton won his share of applause in the conservative highlands.

The machinery that Australia uses to guide Papua New Guinea to inde-

pendence is the Westminster Statute, employed by Britain in the decolonial-
ization of nations all over the world—Nigeria, Trinidad-Tobago, India,
even Australia itself in the last century. An appointive Legislative Council
is created, later liberalized with elected members, then supplemented or
replaced by a more representative House, which in turn is liberalized by
increasing its size and the proportion of elected members, and by giving
it more and more power. Where the territory stands now is this: The House
of Assembly, the intended seat of power, has ninety-four members. Ten
are appointed by Canberra; all of these are "European" (that is, white
Australian) officers of the administration, who are supposed to give the
house stability and know-how. Sixty-nine are from "open" electorates of
the voting public at large—sixty-one of them happen to be Papuans or New
Guineans, and eight Europeans. The remaining fifteen are from "special"
electorates, meaning that they must be educated persons. Four of these are
Papuans or New Guineans, and eleven Europeans.

Administrator Johnson says that this House of Assembly is "more than
halfway to self-government." To be precise, the House and specifically its
ministerial members have taken power (subject to veto) in the areas of
primary and secondary (but not tertiary) education, public health, posts
and telegraphs, territory revenue, coastal shipping, prisons, registration and
use of land, town planning, and some other lesser functions. Canberra
reserves final authority for the judiciary, the enforcement of law and order,
internal security, higher education, external affairs, and external and large-
scale development projects in agriculture, transport, and forestry that are
financed by Australian funds.

Quirky factors in this political entity, half-colony and half-nation, are
citizenship and customs. Residents of the trust territory are Australian
Protected Persons, while residents of the old colony of Papua are Australian
citizens. Yet not even the Papuans are allowed to move to Australia; this is
one more application of the supposedly nonexistent White Australia policy.
Papua New Guinea is also cut off from Australia by tariffs, though mostly
low and preferential ones, similar to those conceded to Canada.

How much does race matter in the territory? Socially, the bars are
dropping. Discrimination prevailed at pools, theaters, hospitals, and clubs
until the mid-1960's, but a Discriminatory Practices Ordinance now works
against this sort of thing. The biggest step, perhaps, was the 1963 abolition,
under United Nations pressure, of laws that forbade blacks to drink and
censored movies that might purvey uppity ideas. Opening the pubs to
everyone forced the more prejudiced planters and businessmen to form
private clubs, but in general race relations in the territory are easy and

relaxed. Still, it is the kind of place where you are likely to hear a white planter say, "Some of my best mates are coons."

One grinding irritant remains. A decade ago, the administration began to employ Papuans and New Guineans in the public service, hitherto a preserve of Australians. At first the pay for the blacks was the same as for the whites. But as more and more blacks were hired, the administration perceived that it was: (1) building up a high salary bill for the future independent country; (2) creating a class of people who in local terms were supersalaried; and (3) pressuring such commercial companies as Burns Philp, W. R. Carpenter Ltd., and Steamships Trading Company by vastly exceeding their low pay scales. Accordingly, the government in 1964 cut the pay of the black public servants (who now number thirteen thousand) by 60 percent, while leaving the pay of the white Australians (who now number seven thousand) just what it was. This could be rationalized: the Australians had needed skills, and would quit rather than take a cut. But the widespread effect was unequal pay for equal work. Papuan and New Guinean members of the House of Assembly howled, and it fell to the unhappy Australian officials holding the appointive seats—at that time including both Dr. Gunther and Mr. Johnson —to defend the measure. Administrator Johnson now says, "It is still a sore point, and will get more so."

The practical facts of land tenure also rub the wrong way. Put in terms of bland official totals, 96 percent of the territory is "native or ownerless," while the administration owns 1,300 square miles and leases 5,000, and only 850 square miles is given over to "expatriate freehold," meaning white planters. Put another way, however, it turns out that only half of the land under commercial agriculture is cultivated by Papuans and New Guineans. Similarly, although blacks got the majority of individual loans made in 1969 by the New Guinea Development Bank, the white planters got more than half of the $5.5 million lent that year.

None of these issues is fully ablaze. But their presence helps explain the posters one sees on the bulletin boards of the University of New South Wales in Sydney: "NIUGINI LIBERATION NEEDS YOU! For the autonomous development of New Guinea. Redistribution of land. Economic, social, and political freedom." Radical students scent that New Guinea can be a cause.

*

We have seen that education in Papua New Guinea is barely stumbling along now, that schools were left entirely to missionaries until after World War II, and that there was not a single New Guinean university graduate until lately. In the face of this bad record, Australia, on urging by the

World Bank, made a tough and controversial decision in 1964—namely, that education would still take a back seat (with a goal of universal but low-level literacy), and that the stress would be put on economic development to enable the new nation to stand on its own feet. The money would go to roads, ports, communications, airports, incentives to industry, training in trades and agriculture.

All this is now getting enormous results, the Department of External Territories in Canberra contends. The idea is to raise agricultural exports, by means of two successive five-year plans ending in 1973, to about $90 million a year. The basic product is the traditional coconut, made into copra by splitting the husks and sun-drying the meat until it can be extracted and further dried in a furnace fueled by the husks. Copra provides coconut oil for food, soap, and cosmetics. Other crops are cocoa, rubber, oil palm, and—as we saw earlier—most notably, tea. Coffee is a big item; New Guinea supplies all of Australia. But increased planting is discourged because the market is saturated.

The dense forests of the island supply fine woods for plywood, and seventy sawmills are now in operation. Beef cattle thrive in the highlands. Tapioca has been brought in from Africa; here it is called manihot. A large proportion of the world's pyrethrum comes from around Mount Hagen. Gold mining produces a dribble of money, but the Bougainville copper is the real gold of the territory; copper exports alone are approaching the value of all agricultural exports put together.

The economic development plan pushes tree planting of all kinds, and encourages villagers to increase their share of commercial agriculture, even though by 1973 their share of copra exports, for example, will be only $9 million as compared to the white planters' $14 million. In general, most of the profit from the new (and old) enterprises in New Guinea goes to Australian and foreign businessmen, not to New Guineans.

There are social consequences which go along with this. Once people living in a subsistence community come into contact with a cash community, they are committed to change. The native culture gets shoved overboard in the process. Garden plots have to be consolidated willy-nilly. A system of family land inheritance replaces that of clan ownership. With money coming in, the New Guinean will want a permanent, European-style house, rather than a shack made of *pit pit* (sugar-cane fiber). The cashcropper may buy a motorbike or a car. In short, life will change from the agricultural to the industrial, and the villagers will say good-bye to the yam and the sing sing. The government will benefit as well; it cannot tax a man in the subsistence economy, but it can certainly nail a cashcropper.

Five Open Questions About New Guinea

- How much will it cost Australia after independence?
- Will it end up as a military dictatorship?
- Any chance of a Biafra here?
- Will an independent New Guinea form an alliance with West Irian and Indonesia?
- How far can a nation go on coconuts?

Four Curious Tidbits About New Guinea

The *Papua-New Guinea Post-Courier*, published in Port Moresby, is printed in both English and Pidgin.

The shillings and pence that used to circulate in New Guinea were often pierced so that they could be strung on a necklace for easy access and safekeeping.

The U.S.-based Summer Institute of Linguistics is translating the Bible into all available Papuan and New Guinean tongues—which involves first converting the nonliterate languages into written forms. American students work at this task in pairs in remote villages, and the result counts as a Ph.D. thesis.

When a village disobeyed its leader's orders, the government provided him with a tape recorder. He then taped the same orders and played the tape at full blast throughout the village. The stratagem produced instant obedience.

*

When Papua and New Guinea join as an independent country, the name will probably be Niugini, as proposed long since by John Guise.* The House of Assembly has formally adopted a committee report supporting the name, as well as a flag of red and black, bearing a white Southern Cross and a bird of paradise. The House also adopted (over administration objections) a committee recommendation favoring a new site for the nation's future capital—a place called Arona near Kainantu, in the Eastern Highlands. Thus, amid fears and doubts, New Guinea proceeds to acquire the accouterments of national life. And the most distinctive part of the national equipment,

---

* A popularity poll a while back plumped for Pagini as the new name. It won out narrowly over Paradesia.

if adopted as predicted, will be the choice of Pidgin as Niugini's official tongue.

Is this surprising? Most Americans and Englishmen seem to think that Pidgin is a sort of baby talk for traders dealing with primitive South Sea Islanders—the word itself apparently derives from "business." But in fact, particularly in New Guinea, Pidgin has long since attained a subtle and adequate Melanesian-style grammar and a sufficient vocabulary. There is even a move afoot to rename the language "Neo-Melanesian." Pidgin is the lingua franca of the territory—your hotel is likely to provide a "Do Not Disturb" sign that reads, "*Yu No Ken Kirapim Man I Slip*"—"You Cannot Interrupt the Man [in this room], Who Is Sleeping." More importantly, Pidgin is one of the three official languages permitted in the House of Assembly, the others being English and Motu.* One radio channel in Port Moresby broadcasts in nothing but Pidgin, and the New Testament has been published in this language.

To the English-speaker, Pidgin is tantalizing—enough of the words are close enough to English to provide a drift of the sense. Your ear has to make the connection between, for instance, "*guvman*" and "government," "*bagerap*" and "bugger up," or "*samting*" and "something." Learning a basic vocabulary of connectives is useful. "*Bilong*" ("belong"), also spelled "*bolong*," "*bilog*," "*long*," and "*blong*," may be Pidgin's commonest word, denoting the whole area of possession or attachment. Thus, "*meri bilong enaderfelo man*" is "another man's wife"—"*meri*" ("Mary") being the generic word for women. "*Felo*" ("fellow"), also spelled "*pela*," is perhaps the next commonest word, and it means not only "a man" but, more broadly, "piece" or "thing" or "kind." "*Gutpela kaikai*" might be translated "good kind of food." Plurals are often made by duplication: "*ol dei ol dei*" is "every day."

Here are the Ten Commandments in Pidgin, as published by the Alexishafen Catholic Mission in 1937:

1. *Mi Master, God bilong yu, yu no ken mekim masalai end ol tambaran.*
2. *Yu no ken kolim nating nem bilong God.*
3. *Yu mast santuim sande.*
4. *Yu mast mekin gud long papamama bilong yu.*
5. *Yu no ken kilim man.*
6. *Yu no ken brukim fashin bilong marit.*
7. *Yu no ken stilim samting.*

* The Motus, the main tribal group around Port Moresby, provide another territorial lingua franca much used by police, and known as Police Motu. In Motu, "Do Not Disturb" is "*Mahuta Basio Nadikaia.*"

8. *Yu no ken lai.*
9. *Yu no ken duim meri bilong enaderfelo man.*
10. *Yu no ken laik stilim samting.*

Some odd Pidgin words (and perhaps apocryphal, because whites cannot resist inventing additions to this colorful language) are these: *"pickaninny namba wan bilong queen"*: Prince Charles; *"grass bilong hed"*: hair; *"mixmasta bilong Jesus Christ"*: helicopter; *"die bokus"*: coffin; *"bokus bilong teeth yu hitim teeth bokus i cry"*: piano.

The Other End of the Island

Just to frame the picture of Papua New Guinea a little more precisely, let us briefly examine the neighbor to the west. The Indonesian half of the New Guinea island is that country's seventeenth province, and its capital is Djajapura, née Hollandia and known for a time in the early 1960's as Sukarnapura. West Irian has between 750,000 and 1,250,000 people—nobody knows for sure. Perhaps 320,000 of them live in the highlands.

The Dutch got a tenuous claim to western New Guinea in 1714, when a contemporary sultan threw it in as part of a deal. Its jungles and declivities did not compare in wealth with Java and Sumatra, and Holland scarcely gave the place a thought. As late as 1938 there were only four hundred Dutchmen in all of Dutch New Guinea. The colony served chiefly as a prison camp for insurgent Indonesians—about fifteen thousand between 1927 and 1962. Many of them were friends of Sukarno, whom the Japanese put into power by sponsoring the Republic of Indonesia in 1945. After the war, the Netherlands tried to get the old Dutch East Indies back, but both the United States and Australia supported Sukarno, and the Dutch were left with the scraps—that is, New Guinea. Belatedly, they tried to make something of the place (as Australia was beginning to do next door), half in the hope that Sukarno would fall and the Netherlands could reclaim the Indies. A governor moved into the handsome house that General MacArthur had built for himself in Hollandia while commanding the Pacific campaign. Thirteen thousand Europeans came in to develop the colony, and for a while the Dutch spent around $25 million a year in development. To keep the Indonesians from getting their hands on New Guinea, the Dutch rather haplessly tried this and that: an exchange of technicians and students with Australia that might beef up the case for joining the two halves of the island; a New Guinea Council, elected by 100,000 tribesmen, that might lead to home rule; a promise of eventual independence. But the

Dutchmen's hearts weren't in it; Sukarno seemed to know that, and stepped up pressure to redeem what the Indonesians called West Irian.* There was some military skirmishing—Indonesian paratroopers against twelve thousand Dutch troops. By 1962 self-government in Dutch New Guinea had got to the point that seven political parties were contending at elections, and not by any means did everyone in the Dutch colony wish it to be turned into an Indonesian colony. But the American diplomat Ellsworth Bunker, serving as a United Nations mediator, studied the conflict and decreed the transfer. Australia approved, feeling that Sukarno was eventually going to have West Irian and consent would avoid bad blood. Planters in the Australian half of New Guinea howled that the move brought "Communism to the borders of Australia." By way of compromise, Bunker gave Indonesia only "administrative control" pending a plebiscite to see how the West Irians liked it. This "act of free choice" was held in 1969, and the Indonesians rounded up a number of tribal chiefs who duly voted that being an Indonesian province suited them just fine.

The impression West Irian has given in the last decade is that of a country on the skids. Bribery and corruption—a UN report even called it "plunder"—prevailed. Power plants, transport, machinery of all kinds broke down. Schools declined until in one case a single teacher had a class of 280 pupils. Food ran short. Quite a few West Irians fled across the unpatrolled border to Australian territory, where each city now has a small colony of West Irians. Some of them, to the concern of Australian authorities, were anti-Indonesian rebels belonging to the Operasi Papua Merdeka—the Free Papua Movement, Papua in this sense meaning the whole island of New Guinea. In the western district of Manokwari, twelve thousand tribesmen rose in arms, and the Indonesian Army took a couple of years to put them down. Particularly before Sukarno was toppled, Australia was sufficiently worried over its new neighbor to spend $60 million building up the territory's military defenses, including a $9-million barracks at Lae for the Pacific Islands Regiment of Papuans and New Guineans (with mostly Australian officers). Shots flew across the border now and then. Education in West Irian is said to be going tolerably well now, partly as a consequence of an effort to teach all children the Indonesian language. A university, called Tjenderawasih, meaning "bird of paradise," was founded in Hollandia-Sukarnapura-Djajapura even before eastern New Guinea founded its university. But Tjenderawasih developed poorly, and has been attended mostly by civil servants.

* Irian signifies "hot climate." In prefixing it with West, Sukarno implied that the Australian part was really East Irian, and that he might like to grab that, too.

*

*The greatest need of New Guinea is time: time to teach the young,
time to stabilize the legend, time for the young to grow old, time to
cement the scattered fabric of its evolutionary thoughts, time for
planning and reflection, time to build an ideal and a constitution, and
to eliminate the tenacious restraints of the darkness from which it is
emerging.*

—OLAF RUHEN, IN Mountains in the Clouds

But there is no time. The machinery of the modern world cannot be
stopped. Australia and the United Nations have decreed that Niugini will
be the 128th nation on earth. Anticolonial Asians and Africans in the UN,
often much to the irritation of some Australians, hammer at Australia to
speed the process. The Soviet delegate to the General Assembly once asked
sarcastically whether Australia intended "to develop the [Stone Age] natives
to, say, the Bronze Age with the same slowness as humanity generally has
progressed." As a matter of fact, former Prime Minister Robert Menzies said
a decade ago, when the independence boom began, that "if in doubt you
go sooner, not later" toward independence.

One interesting possible relationship, Australian statehood, went by the
board in the process of settling for independence. The arguments for statehood, a concept pleasing to the white planters on the island, were that
Australia needed New Guinea for defense, that independence was going to
end in failure, and that in this manner Australia could once and for all
wipe away the tarnish of the White Australia policy. In the end, it was
the race problem that Australia could not swallow. More than two million
blacks would suddenly become one-sixth of the population, with the right
to elect Senators and Representatives to the federal Parliament, and with
freedom to go to the continent in search of jobs and better lives. So
Australia never got a seventh state. Another possibility that caught fire for
a while in the late 1950's was union with Dutch New Guinea, but Indonesia,
already planning to take that colony, frowned, and Australia did not press
the point.

So independence it is—the exact shape unforeseeable. Strategically, the
new country will lie safe between stout Australia on the south and the Territory of the Pacific Islands,* an American trust, on the north. A possibility of external conflict lies along that unnatural western border, the

---

* The Marianas (including Guam), the Carolines (including Yap, Ulithi, and
Truk), and the Marshalls (including Eniwetok and Kwajalein).

141st meridian, particularly because New Guinea is likely to advance at a pace that will set up an enviable contrast with West Irian. Internally, "Niugini" may quite conceivably split into "coconut republics," perhaps with the islands of New Britain, New Ireland, Manus, and Bougainville seceding from Papua New Guinea. This could be bloody, but the present leaders seem aware of the danger. "We do not wish to see in our country the things that have happened in Africa," says John Guise. A happier prospect is that we shall see, in writer Grenfell Price's nice phrase, "how quickly head-hunters could become vote-hunters, cannibals conservatives, and sorcerers statesmen." Australia, which separated itself from England without rancor and with no zeal for breaking every last tie, might be just the country to provide a similarly agreeable experience for "Niugini." Certainly some interrelationship, perhaps slightly comparable to that of the United States and Puerto Rico, will prevail for a while. Australia will provide "foreign aid," and, by supporting the Pacific Islands Regiment, internal security. But the day will come when Australia will have to cross its fingers and launch the new nation.

CHAPTER 21

New Zealand: High Alps and Walking Birds

. . . this small country at the bottom of the world . . .
—SIR KEITH HOLYOAKE

THE CONVENTIONAL first stop for the visitor to New Zealand is the biggest city, Auckland, or perhaps the capital, Wellington, or even Christchurch, the main city of the South Island. But if you want the most startling experience that the country affords, go as quickly as possible to the Hermitage, halfway down the length of the South Island. From the delicious warmth of this gracious, government-run hotel you step out into the crisp, cold air and hear . . . the thunder of avalanches! The bottom of the world takes on a strange excitement; here we reach a land—remote, oceanic, tidily homogeneous in culture—that ranges from high-alpine to subtropical.

The Hermitage, whose bar turns out a particularly potent gimlet,* is a long, low building of stone and dark wood, with 140 rooms. Hotels run by private enterprise in many countries might take lessons from the Hermitage's scrupulous instinct for hospitality and efficient management. Its main asset, though, is its stunning view of Mount Cook, black-flanked, double-peaked, upper slopes piebald with snow. At 12,349 feet, this mountain is New Zealand's highest, but it is only one of 223 peaks on the South Island that reach 7,500 feet, part of an icy spine that makes itself a presence everywhere on the five-hundred-mile-long island.

There are South Islanders who have never seen snow close up, but have

---

* As made there, two-thirds vodka, one-third maraschino liqueur, a dash of grenadine, juice of half a lemon and half an orange—shake, add a cherry.

known it all their lives as the frosting on the range. So dominant are the mountains that in places a man can stand on a summit and see the whole girth of the island, Tasman Sea to the west and Pacific Ocean to the east. Mount Cook itself, though beautiful at sunset when the side light pinks the snow, is a mean mountain to climb. High gales scour it down to glassy ice, and climbers must employ full alpine paraphernalia—crampons, ropes, guides. Three-day blizzards and omnipresent avalanches are perils; fifty-three people have been killed so far on Mount Cook's slopes. Sir Edmund Hillary practiced on Mount Cook before he conquered Everest.

The flight in to the Hermitage aboard Mount Cook Airlines ("Fly the world's most scenic airlines," says the sign at Christchurch airport) takes you over two lakes, one of glacial water, a deep slate color, and one of rain water, a pale lucent green-blue. From the hotel, buses climb a terrifying road so placed on the razorback that it falls off steeply on *both* sides. The destination is an unearthly height where nothing grows, and rock and snow make a black-and-white landscape of chilling beauty. There are also ski planes that depart from the Hermitage, circle stupendous peaks, and then land high, high up on one of the slopes, where one steps out onto the slow flux of Ball Glacier or Tasman Glacier. Some daring sportsmen use the flight as a lift so that they can ski for *fifteen* miles back to the Hermitage. Mount Cook is called Aorangi, "cloud-piercer," by the Maoris. All this is part of 173,000-acre Mount Cook National Park, which is one of ten in New Zealand that together occupy no less than one-thirteenth of the country's land area. Skiing, from July to September, is popular on dozens of exciting snowfields the length of the Southern Alps, as this range is called.

The South Island recreates not only Switzerland but also Norway. Deep, long fiords notch the entire southwestern coast, protected from man's commercial greed by being enclosed in the three million acres of Fiordland National Park. The most spectacular of these is Milford Sound, with sheer cliffs, brilliant sky and sea, and a rain forest below (which gets two hundred inches of rain a year). A dramatic tunnel through the mountains brings the highway to the head of this channel, where Sutherland Falls drop 1,904 feet. Higher up and farther inland, the pattern of narrow water is repeated in many finger lakes, with patches of dark green forest around them.

Another mountain range in the south of the South Island is fittingly named the Remarkables, and at its feet is Queenstown, one of the island's summer resorts. From the airplane, a rough-and-ready DC-3 with the luggage stowed in the back of the cabin, the immobile sheep standing on the edge of the airstrip seem like stone markers. Queenstown stands on Lake Wakatipu, which has waters of a paralyzing coldness that seems to

deter no one from water-skiing and swimming in the summer. An inexplicable and ominous characteristic of this lake is that it rises and falls two or three inches every fifteen minutes. A substantial village of stone buildings and churches, Queenstown had ten thousand people during gold-rush days a century ago; now it has sixteen hundred year-round residents and twenty thousand summer residents.

On the road north, bracken on the hills looks from a distance like fungus or moss. Dogs herd sheep, letting them disperse and then driving them together with such skill that the pattern of movement seems like a ballet. The valley floor is multicolored green; circles of glassy rock shine in the sun. On the highway, cars tow house caravans and sometimes boats as well.

Arrowtown, the next settlement north, a semi-ghost town, has a museum that was a bank in the region's gold-rush days. Gold worth £86 million sterling went through this building, some of it mined by seven hundred Chinese who had deserted ships to join the rush. An old sign flaunted by the bank says, in Chinese, "We sell Hong Kong currency. We buy gold dust. We give money on deposit. And give good interest." More than forty thousand tourists go to Arrowtown every year, and a good many, armed with picks and pans, still find a bit of gold in the river.

*

New Zealand is an end-of-the-earth sort of place, a pair of islands and associated islets lifted to their present heights, after an adventuresome geological career, as recently as five million years ago. With Patagonia and southern Chile, it is the southernmost land area in the world, quite appropriately the base for the U.S. Navy's explorations of Antarctica, whose shores lie fifteen hundred miles to the south. Here is authentic isolation. A wry, symbolic signpost on the northernmost cape of New Zealand reads like this: "Bluff [southernmost tip of New Zealand] 745 miles, Equator 2,065 miles, Sydney 1,065 miles, Suva 1,030 miles, Panama 6,579 miles." New Zealanders say that London is halfway between New Zealand and New Zealand—in other words it is about twelve thousand miles away no matter whether you go northeast or northwest. The logistics are such that New Zealand is detached even from Australia: the natural sea route from New Zealand to England is through Panama whereas the natural sea route from Australia to England is through Suez.

New Zealand's land area is 103,736 square miles: it compares in size with Italy. Of the two main islands, the South Island is the bigger by

about fourteen thousand square miles. But whereas the South Island is roughly a long rectangle, the North Island is looped and pronged with bays and peninsulas. The combined lengths of the two islands are about a thousand miles, and the shoreline, indentations and all, measures just short of ten thousand miles. The dividing water is Cook Strait—once again a name drawn from the stalwart British captain who first closely explored these waters. It is narrow, less than twenty miles, yet with tumultuous seas. In an amiable way, North and South Islanders perpetually dispute the merits of their respective terrains. Statistically, North Island has the lead. Auckland lies in the north of that island, and Wellington, the capital, is on the southern tip of it. The North Island has more than two-thirds of New Zealand's population, which is pushing three million. It also has more than half of the nation's sixty million sheep. Incidentally, domestic animals, including cattle, outnumber human beings in New Zealand by twenty-five to one, a ratio probably not exceeded by any other country in the world.

The country as a whole bears some resemblance to an archipelago and includes a number of smaller islands. Stewart Island, a spot you could drive around in an afternoon if there were a road, lies just south of South Island, and the Chatham Islands, considerably smaller, lie 530 miles to the east. Some uninhabited islets with nice names—Three Kings, Snares, Bounty, Antipodes (which is straight through the center of the earth from London), and others—poke up through the ocean offshore to north and south. As overseas territories, New Zealand possesses the Tokelau Islands and Niue Island, dots in the Polynesian South Pacific about two thousand miles to the northeast, near Samoa. New Zealand also administers a 60-degree wedge of Antarctica, Ross Dependency, which contains the U.S. bases of McMurdo and Little America. The Cook Islands, a group of fifteen pinpoints between Samoa and Tahiti, used to be New Zealand territory, but are now self-governing, with much tutelage and protection from the mother country. Rarotonga, one of the southern tier of pretty, elevated, fertile islands of the Cooks, is the torrid capital. The northern tier consists of coral atolls. All are vulnerable to fierce hurricanes, and a New Zealand government ship *Moana Roa* visits them regularly.

*

New Zealand gives the impression of having been put together from lava, pumice, graywacke (a local gray sandstone), and rich alluvium, vegetated with grass, evergreens, and ferns. Punctuating this landscape are

the white sheep, which seem like maggots when you look down from an airplane. Man's constructions are of wood, by and large; the frame house of the American West was brought to New Zealand by gold-rushers from California a century back, although here it was generally roofed with corrugated iron. Lumber, in villas, bungalows, and country houses, still predominates heavily over masonry. The geology, too, shares Californian characteristics. New Zealand, like that state, is on the "mobile belt" around the Pacific, where the earth's crust buckles and breaks. Something like four hundred earthquakes shake the nation every year, about a hundred of them noticeable without instruments. A quake at Napier in 1931 killed 255 people.

To the Maoris, the Polynesians who won christening rights to New Zealand as its first discoverers, the islands looked variously long, bright, and cloudlike. They named it Aotearoa, a musical word meaning "land of the long white cloud."* The loose Polynesian language permits of other translations, however, and (according to A Dictionary of Maori Place Names) Aotearoa could also mean "continuously clear light," "land of abiding day," or "long white world."

The more prosaic appellation "New Zealand" is the contrivance of the Dutch East Indies governors who dispatched Abel Tasman in 1642 on the voyage that took him to the shores of Tasmania and New Zealand. His name choice for the latter was Staten Landt, but his superiors preferred Nieuw Zeeland, after a province in Holland. The melodious Maori language, which is somewhat overloaded with vowels for the newcomer to pronounce without some practice (viz., Pukengerengere, Ngawhakapakoko, Tauherenikau), provides the bulk of New Zealand place names, but British nostalgia for heros (Nelson, Wellington), and old-country cities (Hastings, Plymouth) provides hundreds of others. Assorted capes have graphic names: Cape Foulwind, Cape Turnagain, Cape Runaway, Cape Kidnappers.

*

The South Island, as we have seen, drops precipitously into the Tasman Sea on a rainy, windy, sparsely inhabited West Coast. The chief phenomenon here is Franz Josef Glacier, which glides down to only nine hundred feet above sea level, well into the relatively warm rain forest, before melting. The flatter east coast provides grass for 25 million sheep and good soil for grain. The island's charming cities are there, too.

The North Island is all juts and spits and gulfs and bays. A protuberance

---

* Unkind Australians, who are prone to consider New Zealand a bit dull, say that the word means "long white shroud."

reaches north like a strand of taffy pulled away from the bulk of the island; no spot on the length of this peninsula is more than twenty miles from the sea. At the point where it joins the rest of the North Island, harbors carved in from east and west narrow the isthmus to only three-fourths of a mile, and this nexus is the sensible and scenic setting for Auckland, New Zealand's biggest city and main port. All is fairly flat here and for a hundred miles to the south of Auckland—then the country rises to volcanically active elevations of between three and nine thousand feet, and they dominate the rest of the island.

This country is prone to convulse, rumble, boil, hiss, gurgle, seethe, tremble, and blow up. Lake Taupo, the result of an explosion a few thousand years ago, is the world's largest volcanic lake, and a warm one, too, with a hundred beaches. Pumice from that explosion still covers much of the ground. Something harrowing is always happening in these parts. A Maori village was completely buried in 1886. In 1953 Mount Ruapehu cracked to release its crater lake, which foamed down to carry away a railroad bridge just as a train approached; 151 people were killed. In May 1971 the same mountain erupted for eight days. The city of Rotorua, on whose golf course steam vents are the hazards, is the tourist center for this volcanic region. At the nearby Whakarewarewa Thermal Reserve, geysers spout up fifty to a hundred feet, and mud boils in whirling, greasy patterns. The Maoris cook food by lowering it into boiling pools of water. Houses pipe in steam for doing the laundry, and the same steam, which as George Bernard Shaw said "smells like Hades"—he might have said rotten eggs—pours forth from a thousand spout holes. A sign on the road, where steam drifts across, says, "VISIBILITY HAZARD." A quarter of a million people visit Rotorua every year, one-third of them from abroad.

Six miles from Lake Taupo, at Wairakei, which has a splendid government hotel similar to the one at Mount Cook, all this superheated natural energy is put to practical use. In this long, twisting valley at the bottom of a trough of pines, hundreds of wells—four, six, and eight inches in diameter, and 570 to 4,000 feet deep—penetrate the earth to bring up live steam under high pressure (225 pounds per square inch) that operates turbines to drive thirteen generators. Much water comes up with the steam, and it is cleverly removed by shooting the mixture through a circular loop, wherein centrifugal force separates the two components. No engineers are visible at this inferno, but white plumes of steam rise everywhere from stubby chimneys. Concrete silencers hold down the roar.

One of the wells, the Rogue Bore, is an out-of-control hellhole. Begun as a nine-inch bore in 1960, it struck steam of such force that the engineers had to try to stopper it. They poured in grout (cement and gravel), but

pressure built up so high that the well blew out thousands of tons of grout and the rig besides. Vibration and hissing steam reamed out the bore to a bigger diameter; the engineers tried to kill it by pumping in cold water, and dropping in concrete blocks. Nothing worked; the constant trembling of the earth and the explosive steam and mud made the sides keep falling in and enlarged the hole until it was two hundred feet across. Every few minutes mud shoots into the air to geyser heights. Now it would cost $250,000 to kill the Rogue—and ruin a gripping tourist attraction: this roaring, sizzling, twenty-four-hour-a-day baby earthquake.

The whole belt of volcanic convulsion runs, 20 miles wide, for about 150 miles. White and pinkish sinter—silicon and calcium from evaporated geyser water—frosts the rocks in waterfall-like terraces. Tikitere, of which Shaw said, "I would willingly have paid ten pounds not to see it," is a weird and vaporous badland that matches the translation of its Maori name, "hell's gate."

The Plants

The flora and fauna of New Zealand, like those of Australia, are the freakish products of a long, separate evolution. Here plants and animals are of yet another displaced and antediluvian type, and for the same reason—separation from the "World Continent." There is a theory that seventy million years ago Australia and New Zealand were linked through a long-sunken land mass in the Tasman Sea, which lies between them. This hypothetical continent (which, in turn, may have been linked with Asia) is called Tasmantis. Certainly New Zealand has been isolated since then. Although seeds can cross oceans, and ferns, a plant of great antiquity, grow widely throughout the world, New Zealand has ferns of varieties found nowhere else. In short, the vast majority of New Zealand's plant species are endemic to the country. There are, to be sure, tantalizing contradictions. The lancewood, a tree of supposedly New Zealand–Australian origin, is found in South America, and many fuchsias found in New Zealand are also found in America. But the eucalyptuses and acacias that are dominant among Australian trees are mysteriously missing, except where introduced by man.

The New Zealand trademark is the giant ponga (pronounced "punga") fern. In this country, as nowhere else, ferns grow as trees, reaching sixty feet. The ponga's fronds are, says the poet A. R. D. Fairburn:

> . . . shaped and curved
> like the scroll of a fiddle: fit instrument
> to play archaic tunes . . .

The soil of New Zealand is ideal for forests, and probably half of it was forested before man arrived to burn and cut it; even now, about a quarter of the country is still wooded. The native trees, running in size between ponderosa pine and California redwoods, make imposing, cathedral-like forests. The kauri, which can be sliced in planks as wide as seven feet, is an evergreen softwood that grows straight and even for sixty feet before it branches. Except for some protected stands, this North Island tree has been nearly wiped out by fire and logging. Several still exist with girths of up to forty-seven feet and ages up to twelve hundred years. Other lofty evergreens, still growing strongly in the mountains, are the podocarps, softwoods of various species—rimu, miro, matai. The tree of the higher altitudes, especially on South Island, is the southern beech, a New Zealand native hardwood that sheds a bountiful pollen.

An exotic aspect of New Zealand forestry is the cultivation of plantations of trees literally called exotics, because they came from abroad. Chief among these is the radiata pine, brought from California about fifty years ago; others are the ponderosa pine, the Corsican pine, the lodgepole pine, and the Douglas fir. During the Depression, these trees were planted at a furious rate, using men who would otherwise have been unemployed. In New Zealand soil, and with New Zealand rainfall, these foreign trees grow twice as fast as they did where they came from. Well tended by the New Zealand Forest Service, radiata pine reach 125 feet in height and 38 inches in diameter at only forty years of age. Native forest still outranks man-made forest by ten to one in area, but the exotics, growing so fast, account for more than half of lumber production. New Zealand has the largest man-made forest in the world.

One oddity in New Zealand is the cabbage tree, a slender, spike-leaved plant whose shoots can be cooked and eaten—they taste like leeks. Not much else of New Zealand's specialized plant life is interesting and edible except the Chinese gooseberry, or kiwi fruit.* This plum-sized morsel, which is exported to the U.S. among other destinations, has a skin that in appearance is like the nut of a coconut. Peeled off, it reveals a soft, sea-green core that can be eaten, black seeds and all, without alarm. It tastes like a pear, but tarter.

The Animals

Three points about New Zealand fauna:

There are *no snakes.*

Only one mammal existed in the islands before man arrived, and that

---

* Properly *Actinidia chinensis,* a native of the Yangtze Valley.

was an insignificant bat. New Zealand's isolation started before the age of mammals, or even marsupials, began.

The dominant form of native animal is the flightless bird, in amazing variety.

*

Most notable of these birds is the brown, hen-size kiwi, the country's national bird, found only in New Zealand. It is so characteristic that New Zealanders call *themselves* Kiwis. Hairy-feathered, whisker-faced, stumpy-winged, this bird clumps on solid-boned legs through the forest in the dead of night—most human Kiwis have never seen the bird except in the Auckland Zoo. The kiwi smells out grubs and worms with nostrils on the very tip of its six-inch beak, and the kiwi hen lays five-inch eggs that weigh one pound, a fourth as much as she does. Kiwis are now very rare.

A bird of the same order, Apteryges, but built more like an ostrich, was the moa, now extinct. Some of these were veritable giants, twelve feet tall, with legs like a grand piano's. Darwin, who visited New Zealand aboard the *Beagle* in 1835, concluded that moas evolved only in the absence of predatory mammals—and it was precisely a predatory mammal, the early Polynesian settler, who finally wiped them out by killing them for food.

New Zealand has, or has had, flightless or nearly flightless geese, ducks, eagles, parrots, parakeets, and rails. One of the parrots is the kakapo which hitches its way through branches with its beak and talons, and having gained sufficient height glides to another tree.

The New Zealand kea, though, is a bird of different inclinations. Nesting high in the Southern Alps, this unflappable parrot, brightly feathered, swoops down on sick sheep and rips away live flesh. To humans, the kea is friendly and insatiably curious, but a pest. It steals TV antennas, windshield wipers, shoes, and pens, and it enjoys letting the air out of tires.

A reptile of distinction is the tuatara, a lizard that died out elsewhere 135 million years ago, but survives here as a living fossil. Its body temperature averages 52 degrees Fahrenheit, and it can go an hour without taking a breath. The tuatara has no ear opening and no penis, but in partial compensation it has a light-sensitive area on the top of its head that functions as a sort of a third eye. Another oddity is the toheroa, a dubiously edible shellfish that buries itself in the sand of the North Island's Ninety Mile Beach. The public is allowed to dig toheroas, by hand only, during a two-week season in August or September, "the months of the broken

fingernails." The limit is ten per person—all this to prevent the toahera from becoming extinct. Still another oddity is the weta, a cricket with a fourteen-inch legspread.

*

This flora and fauna formed the fragile prehistoric ecology that man proceeded recklessly to upset. Forest fires, used by Polynesian moa-hunters to scare the birds into spearing range, and used by English settlers to clear land for farms, denuded half the timbered area. As for the imported fauna, "the history of animal introductions is strewn with tragic mistakes, one leading on to the other," says the *New Zealand Official Yearbook*. Rabbits, six of which can eat as much as a sheep, turned out to be capable of reducing vast areas of grassland to desert. Stoats, ferrets, and weasels, brought in to kill rabbits, decimated kiwis and other native animals instead. American elk,* Australian opossums, goats, pigs, rats, mice, and even sparrows similarly thrived and unbalanced the ecology. The red deer's voracious browsing made it such a pest that the government fights it strenuously with air-dropped poisoned carrots, and with helicopter gunships manned by trained hunters known as "meat hawks." As for rabbits, the goal is extermination, by similar methods, at an annual cost of four cents per acre of the country. Several importations, however, were harmless or beneficial. The thar (pronounced "tar," and also spelled "tahr"), a Himalayan mountain sheep, and the European chamois are good game animals in the high country. And the sheep and the cow, though erosion-makers, are the uncomplaining providers of most of New Zealand's prosperity. But the authorities remain deeply worried. What might happen if the carpenter ant reached New Zealand? Or foot-and-mouth disease? Or—shades of Eden—the snake?

*

I believe we were all glad to leave New Zealand. It is not a pleasant place.

— CHARLES DARWIN

For once the great naturalist was quite wrong; he appears to have been rather upset by watching two Maoris rub noses. The New Zealand climate is temperate and oceanic; the occasional hot winds from Australia or cold

---

* Sent by Teddy Roosevelt as a decorative animal for the southern fiordlands.

blasts from Antarctica lose their zip before arriving. The prevailing westerlies, the winds that made these latitudes known to sailing ship skippers as the Roaring Forties, give both islands breezy coasts. Gusts of sixty miles an hour or more hit Wellington an average of thirty days every year. The same winds bring heavy rainfall—more than one hundred inches a year—to the entire West Coast. Elsewhere, annual rainfall measures between twenty and sixty inches, plenty for all kinds of farming. The temperature runs around 61 degrees in January, and 43 in July. Obviously the top of Mount Cook gets very cold, and in general the cities are colder than Australian cities during their winter. The climate gets cooler as one goes south which means an occasional light frost in Auckland and Wellington, and as many as thirty-six hard frosts a year in Christchurch and Dunedin, on the South Island.

Tale of Four Cities

"Dunedin and Christchurch live on last year's income. Wellington lives on this year's income. Auckland lives on next year's income." So goes a New Zealand saying about the four main cities. Another comparison: Dunedin was founded by Presbyterians, Christchurch by Anglicans, Wellington by merchants, Auckland by pirates.

*

DUNEDIN, which is the ancient name for Edinburgh, has a statue of Robert Burns in the center of its octagonal main square, and the people go ice curling in the winter (on high mountain lakes)—none of which necessarily makes it a wee bit o' auld Scotland. The town was founded in 1848, at the end of a narrow gulf on the southeast coast, by a nephew of the poet Burns, together with three hundred settlers belonging to the Free Kirk of Scotland, a schismatic group of Presbyterians. They had adopted the colonizing principles of Edward Gibbon Wakefield, the Newgate Gaol philosopher who provided the inspiration for the colonization of South Australia. That is, Dunedin and the surrounding province of Otago were to reproduce the social pattern of English country life, with squires, doctors, lawyers, pastors, artisans, and peasants, and this social order was to be preserved by making the price of land prohibitively high, preventing democratic ownership. Moreover, the land was to be farmed, not turned into sheep runs. Some fleeting thought was even given to the banishment of Scotch whisky, but cooler heads prevailed. Although only twelve years later some of the stuffier

Scotsmen were so comfortably settled that they were calling themselves the "Old Identity" of Dunedin, the plan was doomed. The land cried out to be grazed, not cultivated; worse yet, gold rushes throughout the 1860's attracted a lot of rough Irishmen to Otago, which in turn brought roads, banks, dancehalls, commerce, gambling, industry, prosperity, and democracy. The Old Identity, as historian Keith Sinclair put it, became the New Iniquity.

Still, when Dunedin was only twenty-one years old and had only thirteen thousand residents, the good Scots duly founded New Zealand's first institution of higher learning, the University of Otago. When the gold rush ended, sheep raising was so firmly established that Dunedin could carry on as a commercial center, with woolen mills and mutton exported via the country's first refrigerator ship, which was named *Dunedin*. Up went stone and brick buildings. St. Paul's Anglican Cathedral came to rival the Presbyterian First Church in size. Nowadays, democratic suburbs, with bankers and laborers cheek-by-jowl, cover all the surrounding hills, and the population is 110,000. Yet Dunedin is something of a backwater, living in the shadow of the old Victorian buildings that its earlier prosperity built.

*

CHRISTCHURCH, two hundred miles up the East Coast from Dunedin, is a place that grows flowers until the mind boggles. The main civic organization is Christchurch Beautifying Association (Inc.), and woe to the burgher who does not surround his house with roses, azaleas, chrysanthemums, daffodils, snapdragons, and flowering trees. Factories, too, must landscape the premises with flaming gardens or suffer contumely. The idea is to make Christchurch as much as possible like the country from which it originated; as a local poet once put it, " 'Tis England where an English spirit dwells." The result is an extremely prepossessing city, particularly where the river Avon* flows through an immaculate park. A local joke is that this small stream is the source of the Pacific Ocean.

Newspaper headlines call this city "CHCH" for short; American Navy sailors, who have for years used Christchurch's Harewood Airport as a jumping-off place for Operation Deep Freeze in Antarctica, call it "Chi-chi." This is a city of imitation-English schoolboys in boaters and starched collars, of eleven hundred factories, and of a tradition-encrusted (already!) university, Canterbury, that has outgrown its Gothic buildings in the nar-

---

* Named for the Avon in northern Scotland, not the Avon of Stratford, England.

row central streets and is moving to a spacious modern campus out of town. The Christchurch *Press* is 110 years old, a record few U.S. newspapers can match.

Christchurch and the attendant Canterbury province, which dominates the South Island and runs as far west as Mount Cook, is another brainchild of the inveterate colony-founder Wakefield. Morality and practicality were intertwined in his plan to acquire land cheaply (for blankets and beads from the few resident Maoris) and sell it dear, making a profit that would build schools and cathedrals, import more God-fearing citizens, and in general avoid the lawless and unstructured society that Wakefield perceived to be growing in Australia. His collaborator in this case was the Archbishop of Canterbury, serving as president of the Canterbury Association of Settlers, which had been formed by John Robert Godley, a young graduate of Christ Church College, Oxford. One-fourth of the first fifteen hundred settlers were men of sufficient wealth to pay three pounds sterling an acre for land. The first 773, who called themselves "Canterbury Pilgrims," arrived in 1850 aboard the *Charlotte Jane*, the *Randolph*, the *Cressy*, and the *Sir George Seymour*. To this day these vessels are reverently capitalized in Christchurch newspapers as "The First Four Ships."

Dame Ngaio Marsh, now seventy-two, the mystery-story writer who created detective Roderick Alleyn, is Christchurch's most world-famous resident. She does not know precisely what her parents had in mind when after the given name "Edith" they added the middle name "Ngaio," a Maori word for "clever," or "light on the water," or "little bug." Christchurch lives from its manufacturing and industry, plus the commercial benefits of Canterbury sheep raising and export via Lyttelton,* the city's harbor, reached by rail and road tunnels under an inconvenient hill. The population is 260,000, making this New Zealand's second-biggest city, but it is still a low, ground-hugging place. Forty miles of good beaches lie to the north.

*

WELLINGTON, as the capital, draws civil servants and bankers and polticians from all over New Zealand, but the atmosphere is not so much that of a capital as of a city in its own right. There are touches of San Francisco and Rio here. Obsession with government is not all-pervasive, as it is in Washington.

---

* Captain R. F. Scott's base for his expedition to the South Pole in 1912, which ended in his death on the return leg.

This city is built along the shores of a splendid, round bay, six miles across, with a narrow entrance to Cook Strait on the south. Most of Wellington sits on the narrow western foreshore, or on adjacent fill land reclaimed from the sea, or on the slopes of the steep enclosing hills. Houses hang on cliffs. Transport is managed by frighteningly vertical streets, thigh-strengthening paths, flights of steps, and a cable car that goes through a tunnel. Only two routes lead out of town, which prevents Wellingtonians from having weekend houses to the extent that New Zealanders do elsewhere. Geography thus forces a sense of unity and coherence on Wellington, making it a city of some urbanity, intellect, and culture. Partly in consequence of the diplomatic community, one hears foreign languages in the coffeehouses. The national ballet, opera, and orchestra perform here frequently. Alexander Turnbull Library, the nation's best, is at hand. For these inducements, many people accept life in high-rent inner-city apartments, rather than the suburban houses universally preferred elsewhere.

Here again, and even more emphatically, we encounter Edward Gibbon Wakefield. Wellington was another of his projects,* and to this one he came personally and stayed; his grave is in the Bolton Street Cemetery. The colonizers cast about for industries (whaling, for example) to develop the place in the Wakefieldian mold, but in the end settled for luring the national capital away from Auckland in 1865, on grounds of centrality, and with help from the South Island cities. That brought banks, insurance companies, shipping. Wellington became the city that agitated hardest for national self-government, and the capital where the world's first welfare state was drafted. Now all political power centers here.

Squeezed by its site, blustery Wellington has spilled over into a bedroom suburb called Hutt Valley, around the bay, and between them these cities hold 310,000 people. There are more people in Upper and Lower Hutt than in Wellington proper. It adds up to a metropolis by New Zealand standards, but pretty small as world cities go. The harbor was early dubbed Port Nicholson, soon shortened to Port Nick. The Maoris pronounced it "Poneke," and named a tribe the Ngati-Poneke. Nowadays Wellington comes just after Auckland in the number of city-dwelling Maoris. This is a city of fresh air and salt spray. Houses creak a lot, from earthquakes and powerful gales. If anything stands in this town, it's because it cannot be blown down or shaken down. Ropes are strung across midtown intersections to steady people in the wind, as on a ship. But new techniques in earthquake-proof construction are letting the city start to build up instead of out.

A popular confection in Wellington is an ice cream called hokey—vanilla

* Which in this case had the support of the Duke of Wellington.

with bits of honeycomb. Crime is low—only one murder last year. There is little chance of a criminal getting away, because it is hard to get out of the country. The novelist Katherine Mansfield (1888–1923) spent girlhood summers at Day's Bay, on Port Nick's eastern shore. The ferry *Maui* sails daily from Wellington for Christchurch, arriving ten hours later. The basic architecture consists of charming, rather simple wooden houses. Parliament is a curious, Greek-colonnaded building; never finished, it flies on one wing, with a central entrance far off-center. The other wing will never be built; on its site the government is constructing an amazing steel-and-concrete building in the shape of an inverted bucket, to provide Parliament with the space that it needs. A sight that delights visitors and embarrasses Wellingtonians is the government office building, said to be the world's largest wooden structure; the local opinion is that it should be proper masonry. Also of wood, and exceedingly graceful, is St. Paul's Cathedral Church. Wellington is busy building high-rises and highways, including a six-mile downtown expressway that costs $8,830,000 a mile. Seven hundred and seventy-five more acres have been "won," as the New Zealand expression has it, from Port Nicholson by filling it with rock.

The Mayor of Wellington, Sir Francis Kitts, is in his fifth term of three years each. His hobby is visiting other Wellingtons. "There are forty-three Wellingtons in the world, and I have visited fifteen. This is the biggest. Wellington, India, has about fifty-four thousand people. There are thirteen Wellingtons in the United States, and I have visited five. Wellington, Kansas, is the largest. Whenever I feel depressed, I read a clipping about myself from a paper in Wellington, Ohio." Sir Francis has won all the recent elections by huge majorities. The Mayor is a colorful orator and a powerhouse of a politician.

*

In AUCKLAND, a pretty girl rides by on a motorbike, singing into the wind, as if to express her delight at living in this metropolis, which with 615,000 people is by far New Zealand's biggest city. Auckland is a city neither English nor Scottish in tone; it is a rambling, vigorous, sophisticated, open city of many influences. Among these are the urban Maoris, who number about sixty thousand, and, together with forty thousand other South Sea islanders, make Auckland the biggest Polynesian city in the world. Instead of some tiresome British hero, a towering Maori warrior, club in hand, is depicted in a bronze statue on Queen Street. Often, you see a Polynesian man wearing a *lava lava*, wraparound skirt, in downtown Auckland. The exhibition

of Maori artifacts in the War Memorial Museum is interesting and incomparable—it includes a full-size carved Maori meeting house.

In Cornwall Park graze sheep and cows, surrounded by the city. Telephone dials are numbered backward; the "9" is where the "1" usually is. In the South Pacific Hotel, a comfortable place that provides each room with a small refrigerator containing bottles of milk, the bellboys are stout girls, and if the guest tries to lift a bag, a girl seizes it, saying, "But I'm *paid* to do this." The valet wears golf socks and shorts, and the bar, as everywhere in New Zealand, serves the smallest whiskey since prewar Vienna, five-eighths of an ounce.

Auckland's situation is lovely enough to be compared with those of Sydney and Rio and Naples. In form it is essentially an isthmus, connecting the bulk of the North Island to the long northern peninsula. This gives Auckland a double harbor, one on the Tasman Sea, used for light-draft traffic chiefly to Australia, and the other on the Pacific Ocean, used for ships to Asia, the United States, and England, via the Panama Canal. All this is complicated by bays, necks, basins, capes, and indentations impossible to describe. Suffice it to say that one geographical juncture requires a bridge, so that this city, like Sydney or San Francisco, has a soaring span to adorn the natural setting. Though built not long ago, in 1959, its four lanes quickly proved too few for the traffic—Aucklanders said the city was "too big for its bridges." A Japanese firm, employing principles used in tanker construction, added ingenious roadways to the bridge which doubled its capacity. Aucklanders call them "the Nippon clip-on."

Waitemata Harbor, the Pacific port, and outlying Hauraki Gulf are the scene of big thousand-yacht regattas. In the gulf is Rangitoto Island, a volcano extinct only eight hundred years. The island symbolizes Auckland, and gives rise to a Wellingtonian slur that Aucklanders are, owing to their excessive bustle and enterprise, "Rangitoto Yanks." Privacy-loving residents of Waiheke Island, beyond Rangitoto, commute to work in Auckland by hydrofoil. Some of the harbor islands still license prospectors to search for gold. Fee: twenty-five cents. Beaches abound in such numbers that they are rarely crowded, even on the warmest days of the October-April season. The Tasman coast, to the west, provides immense rollers for surfing.

Low volcanic cones, not necessarily dead, rise here and there in the city proper, and they are often used as parks. Most of them are curiously terraced, the work of Maoris who lived on them, and used them as fortresses, two centuries ago. Auckland housing is doggedly surburban: wooden bungalows set on quarter-acre plots—or, latterly, fifth-acre plots. A house on a promontory, such as the side of a volcano, has great status. There are splendid

villas in the high, chic parts of town. Central heating is not needed—electric radiators, run on cheap current, will do for cool spells. The quarter-acre obsession has forced Auckland into an awesome spread; it measures about twenty by thirty miles. A beflowered showplace is Ellerslie race track.

Many downtown streets are narrow, because an early governor had done some time in Egypt and concluded that narrow streets would reduce the glare and heat of the sun. On Queen Street, one might surmise that the main business of the city was department stores—or, if not that, bookshops. The sidewalks have a civilized innovation, a center line to divide pedestrian traffic. You must remember to keep to the left, like the street traffic. An ambiguous street sign says, "NO STANDING AT ALL TIMES." Once each cycle, traffic lights stop vehicles in all directions, and pedestrians, upon the sound of a buzzer, can cross the intersection diagonally if they wish.

What keeps Auckland going? Meat freezing, wool storage, and the docks that load these commodities are the main traditional industries. Near the city is a steel works and a rolling mill. Other plants manufacture fertilizer, used in enormous quantities in New Zealand. In essence, the sheep keeps Auckland going.

CHAPTER 22

New Zealand: Pakeha and Maori

I went to New Zealand, but it was closed.
—BBC JOURNALIST CLEMENT FREUD

THE STATE-RUN LOTTERY in New Zealand decided a few years ago to try a special inducement, a drawing with a grand prize of $120,000. To the puzzlement of the lottery's organizer, the tickets sold rather slowly. At length he found out why: the average New Zealander couldn't conceive of having or spending a sum as large as $120,000.

This is a nation of low-burning ambition, wholesome lives, limited intellectuality, dogged egalitarianism. New Zealand is not much given to hero worship. "Never mind" is the national motto, afternoon tea is the national ceremony, a jolly weekend is the national goal. "All of our problems can be painted in water colors, compared to the stronger colors of other countries," says W. A. Pyatt, the Anglican bishop of Christchurch. An Australian editor says that his idea of a fitting punishment for a foul felon is to "sentence him to New Zealand without option of a fine!" In 1970 the arrivals of "immigrants intending permanent residence" were outnumbered, 26,825 to 29,822, by "New Zealand residents departing permanently."

But nothing could be more erroneous than to picture this as a nation of mediocrity. Its accomplishments are many. It has built a conservative, working welfare state on an agricultural economy—specifically on the Romney sheep. It has won the world's highest standard of living after the United States and Canada. It has a stable, if static, society. And it joins Pakeha (white*) and Maori (Polynesian brown) in a degree of harmony that all other biracial nations can envy.

* The word comes from the Maori language, and also means "goblin" or "spook."

*

Few exercises are more hazardous than singling out "national character-istics," but we must make a try. Certainly the New Zealander is a person who counts on the government to make his life secure. He stresses society and cooperation over individualism and competition. New Zealanders form themselves into committees as easily and automatically as a platoon falling into step. Government is a tool to solve social problems—but if this is socialism, it is an undoctrinal, un-Marxist kind. Capitalism prevails in much of the economy, but it is said to be an ever "nicer" form of capitalism. The result is a minutely worked-out welfare state, which we shall observe in detail presently.

With this stress on a cooperative society goes a fierce egalitarianism. Social classes are visible, but not on the multilayered British model. Some pastoralists in Canterbury attempt to set themselves off as a rural aristocracy; prominent men take pride in being knighted by the Queen. But the only suggestion of an Establishment is a fitful collusion of the Federated Farmers, the con-servative National Party, and the Returned Services Association, a war veterans group. Official links between church and state, vital for a classic Establishment, are missing. "Old boys" from the private schools do not club together to dominate one field or another. Birth is not sufficient to establish status. Servants are virtually unknown. Neighborhoods mix artisans, laborers, businessmen, and professionals. A man calls his boss by his first name. "We are an egalitarian society, though very conservative," explains Bishop Pyatt. "We all know each other so well." Democracy is carried to the ultimate: almost anyone with a smidgen of a reason can get in to see the Prime Minister personally, and he can see the other ministers even more readily. Egalitarianism also discourages what New Zealanders call "ritual tipping"— they tip only when they receive some special, or especially good, service. Student representation in the administration of a university, an idea newly arrived in the U.S., is old practice in New Zealand, and a partial explanation for the infrequency of student protests and riots.

Living as they do a jillion miles from anywhere, New Zealanders are inevitably somewhat insular. A newspaperman in Wellington explains that "we haven't got a continent—we live in an ocean." The outside world has historically consisted of England—just England; New Zealand was the "most loyal" dominion, tranquil, prim, a "corner of an English vicarage." Sometimes the Britishness of New Zealand astounds even Britons—although, as in Australia, New Zealand has its quota of young rebels who will not

rise when "God Save the Queen" is played in the theaters. One solid reason for insularity is the difficulty of travel. To conserve foreign exchange, the government limits the expenditures of vacation travelers abroad to $600 a month, up to a maximum of $1,800. In any case, New Zealanders, operating in a low-price economy, find travel excruciatingly expensive, except to Australia, and of course airline distances are great. Overseas newspapers take six weeks to reach New Zealand by ship mail; local newspapers do not take much interest in international news.

A corollary of insularity is what Colin MacInnes calls New Zealand's "non-love" affair with Australia, the nearest civilization. One member of the New Zealand Parliament explains the relationship this way: "New Zealand feels, first, envious; second, somewhat frightened—they're slicker than we are, and they'll do us in—and, third, somewhat inferior. Sydney is to us what New York is to the Canadian. We think of it as another kind of universe where one has a sense of living dangerously—murder, auto accidents." It is this ambivalent feeling which possibly lures some New Zealanders to Australia, perhaps to live in Sydney's King's Cross section. But New Zealanders hold their own with Australians in sports, satisfying their need to be "one up on the Aussies." The New Zealand–Australia relationship is not like that of Canada and the U.S. Both down-under countries produce meat and wool, so that they compete with one another for markets, and Australia does not dominate the New Zealand economy as the U.S. does Canada's.

New Zealanders simply will not tolerate high-density housing; every man has his quarter-acre. They're a house-proud people, the furnishings chosen with care and taste. And although the average person lives modestly, his house and well-treated garden can be charming. The sputter of the lawnmower is the standard New Zealand noise. Going with house and garden is a car—one out of three persons owns one. Fifty thousand automobiles are assembled every year in no less than eighteen plants; but prices are so high that many people have to drive vintage cars. There are even some 1924 Maxwells around, and owners tend to keep the car that they buy for ten, twenty, or even thirty years. A related obsession is the beach or mountain weekend house, called a "bach" (pronounced "batch"). This is all part of the outdoor life that New Zealanders are mad about—although one environmentalist, writing in the *Official Yearbook*, damns the "pathetic weekend escape to the countryside, which chokes the roads and spreads a rash of substandard huts and garbage around every beauty spot."

Finally, there seems to be a troubled side to the New Zealand character. A humorless prudishness frequently prevails. When a traveling art exhibit

brought Marcel Duchamps' sculptures, *Fountain*, made of a urinal, and *Please Touch*, a foam-rubber breast, to Christchurch, Councillor P. J. Skellerup kept them from public view, saying, "I don't mind a bit of good, clean fun in the art world, but you have to draw the line somewhere." Movies are censored, though not harshly; *Ulysses* was shown on the suggestive condition that the two sexes must view it separately. The public serving of liquor has suffered all the tiresome restrictions of Australia and more; "licensed" means that an establishment may serve wine only, and even "fully licensed" restaurants, though they may serve hard liquor, have to insist that the customer order a full meal before they will bring a cocktail. The abortion laws are strict, and so rigorously enforced that girls go to Australia to get abortions. Prisons are harsh, with occasional riots and singularly tough solitary punishment for the "intractables."

The reality of New Zealand contrasts with its Victorian assumptions. Gambling is endemic: $210 million a year goes into race-track betting and lotteries. Drunkenness is a problem referred to five times in the *Official Yearbook*. Among beer drinkers, New Zealand stands fifth in the world, at 194 pints per head per year (and the beer drinkers bellow, "That's not good enough! We must drink our way to first!"). The little town of Westport once had eighty-two pubs in a single mile. One in eight children is born out of wedlock, and in many more cases the bride is pregnant on her wedding day.

One-third of all New Zealanders require sleeping pills, and the total bill for tranquilizers and mood pills runs to ten dollars per person per year. The former Governor-General, Sir Arthur Porritt, who is also an eminent surgeon, finds the amount of pill-popping "horrible." Psychologists speculate that New Zealanders are neurotic for lack of incentive in the welfare state. The suicide rate, however, is not high: 8.8 per hundred thousand of population, as compared to 10.9 in the U.S. and England, 19.6 in Sweden, and 29.3 in Hungary. Some adjustments are being made to reduce the hypocrisy of blue laws whose intent is wildly contradicted by practice. Bars may now stay open until 10 P.M., eliminating the vomitous "six o'clock swill" that used to take place daily in New Zealand (as in Australia) when bars closed at that hour. And (as in Australia) a government-approved Totalisator Agency Board runs off-track betting in place of the former illegal bookies.

Zubin Mehta, the conductor, found much to deplore on a visit to New Zealand, which in his view is "living in the England of the last century. If you want to go out on Sunday, forget it! The food is awful. Lamb is a delicious meat if cooked well, but the New Zealanders ruin it. In New Zealand, you have to go out of town to find Indians. They and the Maoris all live

in suburban slums. I took the Israel Philharmonic on tour there. We had to fill out forms that were insulting. There was a place marked 'Race?' The orchestra has a lot of Polish-Jewish violinists—the best violinists come from Poland. They filled in the question about race with 'Human.' "

And yet, to keep perspective, these shortcomings are essentially the normal contradictions of any country. On balance New Zealand is benevolent, responsible, one of the least corrupt nations of the world—modest, too. New Zealanders are proud of their moral position, and feel that their influence on foreign affairs is larger than their size because of it.

The Historical Background

New Zealand's prehistory is a maddening mixture of firm facts, probabilities, and contradictory mysteries. A probability is that the Polynesians were originally Asian Caucasoid, modified by Mongoloid and Negroid blood as they moved eastward through Indonesia and Melanesia to the Pacific islands of Hawaii, Tahiti, and Easter Island. Certain it is that they became good canoemen, capable of sailing at least a few hundred miles, well out of sight of land, in canoes that could carry 250 people. Certain it is, too, that their canoes began to take them to New Zealand more than a thousand years ago. But there is much mystery as to whether, on those voyages, they were lost, or deliberately seeking new lands to colonize (which might explain their bringing women), or perhaps even consciously sailing to islands already discovered and reported upon by an earlier traveler.

It is established, from carbon dating, that Polynesians lived in various places along the New Zealand coast by A.D. 1000. But the Maoris (rhymes with "dowries") have a tantalizing legend that their ancestors arrived in a number of big outrigger canoes about twenty-four generations ago, a date calculated to be 1350. It is known that there were two Polynesian civilizations in prehistoric New Zealand, one a Stone Age culture based on hunting the moa, and the other the more advanced stage of civilization that the first Europeans found here. The temptation is to equate the first scattered arrivals with the moa hunters, and the big canoes with higher Maori culture, but new scholarship casts doubt on it, chiefly questioning the concept of a specific fleet of canoes in 1350. The thought now is that the landings were occasional but continuous, and that the moa-hunter culture evolved into the Maori culture.

The pre-Pakeha civilization of the Maoris, a handsome people of light brown skin, was the highest in Polynesia. They were an agricultural people, artistic, religious, and warlike. For adzes and clubs, they crossed the Southern

Alps to the West Coast for a kind of translucent jade known in New Zealand as greenstone. They built, on numerous hilltops, a kind of conical fortress called a *pa*. They found flax growing in New Zealand, and learned to make it into baskets and clothing. For meeting houses and the prows of canoes, they practiced wood carving that ranks high in primitive art.

Day-to-day life revolved around the cultivation of the sweet potato, the taro root, and the yam. For other food they ate fern roots, birds, fish, and one another. The Maoris were notorious cannibals, keeping people in pens to fatten them up, and eating all shipwrecked sailors. Tribal organization was based on the family, then the subtribe, with a common ancestor, and finally the tribe, with a set of distant common ancestors who, according to legend, would have all arrived in the same canoe. The canoes, incidentally, all have names, and to this day it is common for a Maori to speak of descent from the Tainui canoe, or the Te Arawa canoe, or the Tokomaru canoe. The characteristics of the religion were, first, many gods but no supreme god, and secondly, a powerful sense of *tapu* (the origin of the English word "taboo"), which served to enforce customs and the law.

Above all, the Maoris loved warfare. Charles Darwin reports, in *The Voyage of the Beagle*, the case of a missionary who after much argument had nearly persuaded a certain Maori chief not to go to war with a nearby tribe. Then the chief discovered that his barrel of gunpowder was going bad, and would be worthless if not used at once. He immediately declared war. It is a likely speculation that English intervention, far from leading the Maoris to extermination, was what saved them from exterminating one another.

*

A great exterminator in the first half of the nineteenth century, when the white colonization took place, was Te Rauparaha, chief of a tribe seated in an area fifty miles south of Auckland. In 1818 he led his tribe south on a bloody trail of conquest, and established them on Kapiti Island, in the Tasman Sea near Wellington. There Te Rauparaha traded with American and English whaling ships for muskets, the weapon that was turning Maori battles from club fights into highly lethal warfare. Having terrorized the tribes of the North Island, Te Rauparaha sailed off, in his canoes with fearsome carved figureheads, to dominate the South Island, too. The battles ended with feasting on the bodies of the vanquished. In 1829, when South Islanders at Akaroa killed one of his best lieutenants, Te Rauparaha chartered a British sailing vessel called the *Elizabeth*, and with his warriors sailed for Akaroa. There the local chief was lured aboard by the ship's captain, who promised to

sell him guns for gunpowder. Te Rauparaha seized the chief and sent his men to sack the town and capture its leaderless tribesmen. They were stuffed into the *Elizabeth*'s hold and taken to Kapiti on a bloody voyage during which many of them were tortured and killed.

The scandal of this exploit and pressure from Christian missionaries forced the reluctant British government to take action. In 1840 the Treaty of Waitangi was signed formalizing British rule. Te Rauparaha was one of the chiefs who signed it. The chiefs apparently thought that they were getting the protection of the distant British Queen in return for allowing the Crown to buy Maori land if the tribes wished to sell. The chiefs had held a two-day debate and were not entirely naïve—word had reached them of the British extermination of the Aborigines in Tasmania; but in the end they decided there was no turning back the clock—the British were there. Within three years Te Rauparaha began to have regrets, as he watched the white colonizers grab land under their interpretation of the treaty. He became so defiant that Governor George Grey arrested him and imprisoned him aboard naval ships for eighteen months. Then Grey released him, provided him with the uniform of a British naval officer and an American Army sword and scabbard, and returned him to his tribe. It is to be feared, judging from a drawing, that Te Rauparaha looked a bit ridiculous, with his epaulets, his hawk-nosed wizened face, painted in whorls, and his rakish naval cap. He died, at seventy-nine, a year later.

We have already met Governor Grey, briefly, as the redoubtable official who forced the colonists of Adelaide from the huddled protection of the town to the productive rigors of farming the land. That episode over, in 1845 Grey was assigned to shape up New Zealand in similar style.

Grey, born in 1812, son of a lieutenant colonel in the 30th Foot, was a Governor who governed. From Sandhurst he had gone into exploring in Western Australia, where the Aborigines speared him in the hip. He ruled New Zealand despotically but efficiently from 1845 to 1853, buying Maori land for the Crown cheaply and selling it dear to the settlers, but nevertheless winning Maori admiration by building hospitals and schools and by taking such an intellectual interest in them that even today the best collection and translation of Maori proverbs is one of his published works. He virtually stopped intertribal warfare, and turned many Maoris to the commercial cultivation of potatoes and wheat for export to Australia. In 1852 he suddenly promulgated a New Zealand Constitution more democratic than those of the Australian colonies, giving the country self-government only a dozen years after the British takeover at the Treaty of Waitangi. Then Grey charged energetically off to South Africa to be Governor of Cape Town (where he

thwarted a formidable Kaffir rebellion), returning to New Zealand in 1861 when London (with Grey concurring) deemed him vital to win the Maori Wars.

The Maori tribes, astonished at how fast their land was going into the hands of the whites (by pressured purchases and all manner of legal trumpery), had in 1858 ominously joined together in their first nation, and elected a king. This potentate reigned from the area south of Auckland that is known to this day as the King Country. The place where the Maoris most ferociously resisted the land buyers and surveyors, however, was the area called Taranaki, south of the King Country. There in 1860 the Maoris burned the British settlers' farmhouses. The war had started. Three thousand Imperial troops and colonial militiamen responded by devastating a Maori village. A Maori war party from the King Country joined the Taranaki fighting; the white forces attacked Maori *pas*; there were a couple of pitched battles; but neither side won and an armed truce began.

Grey, taking charge of the war, tried threats and conciliation, building a menacing military road from Auckland to King Country, and at the same time ruling for the return of some Maori land. The Maoris, uncertain of his intentions, attacked again in Taranaki in 1863, and soon suffered a sharp counterattack. Grey decided to press from the north. The British General, Duncan Alexander Cameron, moved into King Country aboard small gunboats on the Waikato River. He attacked a *pa*, turning his howitzers and Enfield rifles against the Maoris' flintlocks, sporting rifles, and spears. The Maoris held out three days, until they were down to firing wooden bullets, and then charged out. One hundred and fifty of them were killed, but at least as many more escaped.

This equivocal victory for the Pakehas should have ended the war, except for an outlandish conseqence of the earlier Christianization of the Maoris. In 1862 a certain Maori Christian claimed that the Angel Gabriel had appeared before him in a vision leading him to found a religion that was both Christian and Maori. It combined belief in a hierarchy of angels with cannibalism, and worship of the Trinity with an incantation—*hau hau*—that would render the members of the sect invulnerable to British bullets. At their services, the *hau hau* people danced around a sort of Maypole surmounted by an Englishman's severed head. An Anglican missionary who attempted to stop this heresy was killed and his blood drunk in his own church; they even ate the missionary's eyes.

Bitterness increased when the New Zealand Parliament, in 1863, authorized confiscation of Maori land, thus loosing on the scene speculators who had a vested interest in taking over as much Maori territory as possible. The war bumbled on, but was virtually over. The British troops, who admired Maori

courage and chivalry, disliked fighting for the settlers' land hunger, and, except for a regiment, were withdrawn by 1866. Thousands of Maoris still held out in the King Country, excluding settlers and booze, but on such amiable terms that they continued to grow wheat and trade it for European goods. By 1868 Grey, anxious to end the conflict, had indulged in such excesses of zeal (he himself leading an attack on a *hau hau* village) that the Colonial Office dismissed him from the governorship.

The Maori Wars were over by 1872, the epic story of New Zealand, a bit remindful of the American Indian Wars. The result was that the settlers got the bulk of the land of New Zealand, and they proceeded to put it to production of wool, while waiting with a certain hypocritical sympathy for the Maoris to die out. About two thousand Maoris had been killed, and about one thousand Englishmen. The Maoris called it "the white man's anger" and "the fire in the fern."

After the wars, George Grey spent a dozen or so years in London, a doughty advocate of imperialism, and then retired to the island of Kawau, off Auckland. In 1876, the central government abolished the provincial governments (Otago, Canterbury, Taranaki were examples of these) that Grey had written into his Constitution of 1852, and he came storming out to run for Parliament. By the next year this lonely, forceful, eloquent figure was again leading New Zealand, as Prime Minister. His program was radical for the time: manhood suffrage, universal education, triennial election of Parliament, legal recognition for trade unions, compulsory government purchase of large estates, a tax on land, and popular election of the Governor (then as now a Crown appointment). He achieved some of these reforms, and all but the last were later passed. He also founded the reformist Liberal Party, later to be a main shaper of the country. But land profiteers managed to bring him down from the prime ministry. He served, respected but inconspicuous, as a plain M.P. until 1894, and died in London in 1898.

*

We now begin to scent the origins of the New Zealand welfare state, as a shrewd, gargantuan, genuinely reformist politician named Richard John Seddon comes into view. The politics of New Zealand before and after the Maori Wars were the politics of land tenure. As might be expected, the land had fallen into too few hands; access to land was the name of the problem.

Seddon, a Lancashireman, went out to Australia to work in the Melbourne ailway shops, and caught gold fever when the great Bendigo gold rush began. This in turn led him to gold mining on New Zealand's West Coast, and

thence to election as the area's Member of Parliament in 1879. He became one of George Grey's Liberals, and in 1892, after consulting with that old man, he accepted the prime ministership. He and his cabinet were strongly influenced by the ideas of John Stuart Mill, Henry George, and Edward Bellamy, ideas also current in Australia at the time.

The Liberals' first answers to the country's inequities were graduated taxes on income and unimproved land, accompanied by the removal of property taxes on everything but land. Then, instead of selling Crown land, the government began to lease it, on the theory that freehold led to speculation and high prices that froze out small farmers. For small farmers, the Liberals passed cheap-loan acts that freed them from having to pay interest, which ran up to 15 percent. For labor, the government devised an Industrial Conciliation and Arbitration Act that put disputes with management into a court and removed the need for strikes; not a single one occurred from 1894 to 1906. The conditions of labor were reformed right down to guaranteeing that shopgirls could have chairs to sit in.

A Legislative Council, the upper house, typically opposed the reforms, but during the 1890's it was reduced to impotence by assorted maneuvers, leading finally to its abolition in 1950; New Zealand now has one of the world's few unicameral legislatures. Seddon established penny postage, and—rather reluctantly—made New Zealand the first country in the world to give women the vote.* In a ninety-hour session of Parliament in 1898, Seddon put over governmental old-age pensions, long before the U.S. Social Security Act of 1935. And all the while he skillfully fought off New Zealand's bluenoses, who pressed for prohibition on the drinking of alcohol. Seddon died in 1906.

All this reform met with singular success. Foreigners came to New Zealand and admired it, and the fame of its prosperity and high standard of living spread round the world. To be sure, two economic factors, quite unrelated to the country's political organization, were in the background. One was the breeding of the New Zealand Corriedale sheep, a cross of the Merino and the Lincoln, and a prolific producer of both wool and mutton (though now largely supplanted by the Romney). The other was refrigeration, at the packing plants and on ships, that opened sales of New Zealand lamb, mutton, beef, and butter to the British market. One more point about this period: it turned New Zealand from a colony into a nation. In the 1890's New Zealand shared political characteristics with New South Wales, Queensland, and Victoria, and even considered federating with Australia; by 1907 it was clearly on its own as a British Dominion.

* Two lesser jurisdictions already had woman suffrage: Pitcairn Island and Wyoming.

*

The turn-of-the-century euphoria ended just before World War I with the Liberals breaking away to right and to left. The right prevailed from the war until the Depression, a period of eroding prosperity that left the standard of living about where it had been in the 1880's. By 1935 New Zealand was ready for the left-wing alternative, the Labor Party that had been growing steadily for two decades.

No single figure—no Roosevelt—stands out in the turbulent epoch of reform that followed, although Walter Nash proved brilliant as Minister of Finance, and then first New Zealand Minister to Washington, and Peter Fraser turned out to be a politically skillful wartime (1940–1949) Premier. The measures that restored prosperity (along with the rising price of wool and mutton, and later the effects of the war) were nonrevolutionary, Keynesian repairs to the system. The Reserve Bank was nationalized and given the power to lend to the government. Farmers got subsidies, and the state took over the marketing of their produce. Union membership was made compulsory. A Social Security Act, with improved old-age pensions, family allowances, sickness and unemployment benefits, and so on, was passed. The Department of Health started a medical services system. And some beginnings were made toward economic diversification.

New Zealand's experience in World War II was a strangely sour one. With utter loyalty to Britain, she sent her 2nd Division, under General Bernard Freyberg, to Egypt, and it fought with bravery and skill in Greece, Crete, North Africa, and Italy. Tempted to bring the troops back to the Pacific after Japan attacked Pearl Harbor, New Zealand nevertheless stayed in Europe, defending Britain, while the U.S. was in the South Pacific defending New Zealand—and while New Zealand was single-mindedly growing food to feed Britain. In 1943 the frustrating feeling developed that the 2nd Division had been too long in Europe, but that it was too late for it to join the Pacific fighting. The soldiers were then brought home on long furloughs, during which they developed resentments against the large number of men busy growing and shipping foodstuffs, whom they considered shirkers. Many troops refused to go back to the distant European theater, and in the end New Zealand was left without a role there or a voice in the Pacific.

After the war, in 1947, Parliament ratified the Statute of Westminster, passed in 1931 by the British Parliament to grant complete autonomy to members of the British Commonwealth. New Zealand has simply never been in a hurry to cut itself off from England, though accused of "prolonged adolescence" and addiction to "maternalism." It accepted English governors

long after Canada, Australia, and South Africa insisted that their governors be from their own countries. Not until 1968 did the National Party remove the Union Jack from the platform at party gatherings.

The Welfare State

The standard old-age pension paid in New Zealand (mostly without a means test, such as Australia requires) is $715 a year—a figure that raises a lot of questions. Is this piddling pension the measure of the New Zealand welfare state's prosperity?

The answer is that the pension is only the beginning of the benefits. To begin with, a married couple would get almost double $715. A dependent child would add about $500. A cheap, government-subsidized mortgage, probably paid off earlier, would guarantee home ownership for most New Zealanders, and property taxes are likely to be only $50 to $100 a year. For those without houses, apartments rent for as little as $40 a month. The main scourge of old age, medical bills, is removed by free hospitalization and state-guaranteed doctors' fees. A pensioner, or any other New Zealander for that matter, can get the services of a general practitioner for $1.50. "The fear of illness has been removed," says Dr. Wilton Henley, a surgeon at Auckland General Hospital. If the pensioner has sufficient savings or small investments or part-time work to push his income to $2,000 a year, he will have an income equal to that of a worker in a dairy processing plant—enough to live on in New Zealand. The aged and infirm get other benefits. If incapacitated, they are cared for in hospitals, not nursing homes, until death if necessary. If they can stay at home, Meals on Wheels brings food. "The care of old people is fantastic in New Zealand," says Sir Arthur Porritt.

Yet the cushioning of old age is only the beginning of the boons provided to New Zealanders by their government. The family benefit program gives mothers $1.50 a week for each child, and allows for borrowing against child benefits in order to finance a house on a low-interest government loan. Veterans (depending partly on whether or not they were wounded) get car loans, clothing allowances, incapacity allowances, and free passes, first-class, on the New Zealand Railways. A gallantry award brings in $1 a week, and the Victoria Cross $200 a year. Farmers get loans, cheap fertilizer and pesticides, transport subsidies, and dispensation from the standard 2 percent payroll tax. Other boons of the welfare state are profit-sharing, university educations at trifling cost, government-ordered wage boosts to keep pace with inflation, a commitment to total employment that works in practice,* subsidized life insurance, and numerous government jobs (one-fifth of all workers

---

* In 1969 only 2,082 persons collected unemployment benefits.

are in the civil service). But it is not quite a cradle-to-grave system: there is no funeral allowance.

The money cost of this comes from taxes, and the rates are steep. Typical income taxes for a family with two children run from $484 on a total income of $3,500, to $4,171 for a single person with a total income of $10,500. The top bracket (50 percent) on taxable income for a single person is reached at only $12,000, as compared to $22,000 in the United States. To most people, however, these tax rates are no particular hardship; only one million people pay any taxes at all, and they rarely get into the upper brackets because wages and salaries are low. The basic wage of a lathe operator, for instance, is $2,500, while even the Prime Minister gets only $17,000. The class called "Monied Farmers" clears $6,000 to $12,000 a year. Per capita income is $1,440, which is the seventh highest in the world. Low income, in turn, is no particular hardship because prices average out so low. The stiff $370 for a black-and-white television set is balanced out by a loaf of bread at 16 cents, lamb at 32 cents a pound, butter at 30 cents, sugar at 9 cents, and a pint of milk at 4 cents. No wonder New Zealanders eat an incredible amount of food per person per day, 3,358 calories, the highest in the world, including 242 pounds of meat per person every year.

Such is the balance in material things. On the philosophical scales, the welfare state wins stout support from the majority. The people come first in this country, they say; the economy is subordinate to their welfare. They feel that they have an unquestionable right to be well off, to have not only full employment but *more* jobs than jobholders. All of New Zealand was profoundly shocked in the bad year of 1968, when unemployment reached seven-tenths of one percent. The welfare state has eliminated slums, unnecessary disease, sweatshop labor, monopoly, and poverty. The country has shown itself to be admirably ready to experiment, using the government as a utility. Yet the government has not become an ogre; electorates are small, single votes still count, and Members of Parliament are accessible to individual voters. The idea of human dignity as a right—of decent employment and adequate health care—pervades the philosophy. Is this regimentation? "No," says a rich industrialist, "we are more individualistic than any other dominion." Sir Arthur Porritt concurs: "At first initiative was stifled, but now the individualism of the more enterprising people is as strong here as in countries without the welfare state." The new Prime Minister, John Marshall, argues that the government's objective is merely "to remove the insecurity caused by ill health, old age, and unemployment. When you take care of a man's medical care and unemployment, the risk of his losing his money and falling on his face is less acute. A man can look at life here without fear. So a man can dare to have more initiative, be more experimental." Sir Keith Holyoake,

Marshall's predecessor, says: "When I was a young man in Parliament, I, a conservative, of a conservative family, was worried by the possibility that social welfare would stifle initiative. But I have come to the conclusion that it has not damaged private initiative in New Zealand."

However, one minister, Duncan MacIntyre of Lands, Maori Affairs, and various other portfolios, holds that high income taxes are a definite curb on incentive. "The concept of social security was originally to 'build a floor,' so nobody would be destitute. Now, to finance the floor, we have to put on a ceiling, that is, heavy taxation. So there is no incentive to work." Doctors, who often feel themselves the victims of a welfare state, have generally fought against this welfare state, too, but New Zealand medicine has nevertheless managed to gain pre-eminence in heart surgery and neurosurgery, and the country pioneered in intrauterine transfusions of fetuses for Rh incompatibility or anemia. Doctors come from abroad to learn certain operations in these fields.

The argument against this utopia contends that only luck, specifically the idiosyncratic English hunger for lamb and butter that has prevailed until now, holds it together. Tough times would reveal its weaknesses. And, this argument continues, the cushioned life takes a toll in lack of individual incentive. With everyone, but most prominently labor and farmers, appealing to the government for support and protection, some other force—namely, the market—ultimately has to be allowed to make the economic decisions. There is supposed to be a demoralizing me-firstism about people coddled by a welfare state; if a person dies on Friday night, it is said, his body will be left until Monday, when the workers' weekend is over.

"I don't think socialism holds us back, but we don't display much zest," Bishop Pyatt admits. "As for social security, we are not the world's leader. We were when it began, but other people have overreached us. We have been a success because we're small, and because the bureaucrats are not dead-handed."

One criticism is directed at the indignity of the compulsory nature of health care, which seems to assume that the average New Zealander will shoot his wages on the race track and leave his sick children untended. Compulsory unionism is another target, for it is said to concentrate power in the hands of union officials who would be more responsible to their members if they had to struggle to enlist them and keep the union going.

The Maoris Today

The Maoris, as we have seen, were supposed to die out after the wars of the 1860's, and the population duly sank as a result of European tuberculosis, measles, and typhoid, until it hit bottom at 42,000 in 1896. Then, perhaps

owing to heightened immunity to disease, Maori numbers began to rise until now there are 220,000* of them—more, probably, than in the palmiest times of prehistory. They form 8 percent of all New Zealanders, and with a birth rate of forty-four per thousand they are increasing twice as fast as the Pakehas.

Once again, the central question here is the preservation of the culture and identity of the original occupants of a country against the seductions of white technology and urbanism. The Maori has a preservable culture, for it is based on agriculture. The Maori defends his heritage with spirit, born perhaps of the honorable fight he put up in the Maori Wars. Thus, though the word "Maori" means merely "ordinary," the Maoris are in fact an extraordinary people, respected for their courtesy, virility, assertiveness, grace, wit, dignity, and athletic prowess. Many New Zealanders claim more Maori blood than they really have in order to be identified as Maoris instead of as whites.

Maoris feel strong kinship obligations—every Maori is "cousin" to another. They value the spirit of *aroha* (affection or love or pity for others), and they are proud of their mellifluous language, though most also speak English, the language used in schools. The King Country survives, a paper kingdom within the nation; currently reigning is Queen Te Ata, wife of a prosperous Maori dairy farmer. With it goes a powerful church, Ratana, Christian-Maori plus faith-healing.† But Maori nationalism is negligible. Tribalism has declined to the point that many enthusiastic Maoris do not know their own tribal descents, but tribal customs are strong enough that even radical young Maoris treat their elders with much gentleness and respect. Hospital patients' beds are always thronged with visitors, but there are few Maoris in geriatric hospitals; the dying aged go back to their villages for loving attention.

On the other hand, the Europeanization of the Maori has gone so far that all of them—in strong contrast to the Aborigines of Australia, or the New Guineans—are in the money economy. Most work at unskilled and menial jobs, but there are also many big and small farmers, schoolteachers, mechanics, sheep shearers (the finest), nurses, and heavy-machinery operators, and quite a few doctors, lawyers, accountants, and dentists. Long since, certain Maoris have risen to the top of New Zealand society. Sir Peter Buck (whose Maori name was Te Rangi Hiroa) was a medical doctor and politican who turned to Polynesian anthropology and became, before his death in 1951, a professor at Yale and head of the Bishop Museum in Honolulu.

---

* Defined as persons with one-half or more Maori blood. The total rises to 350,000 when persons with less than one-half Maori blood are included.

† Mormons are also winning many Maori converts.

He is only one of several Maoris knighted for distinguished professional careers.

Of Maori houses, 73.8 percent have washing machines, 80 percent have refrigerators, 78.5 percent have flush toilets, and 61.2 percent have TV sets. The proportion of Maoris who live in cities has risen from 10 percent in 1936 to more than 50 percent, the biggest concentration being in Auckland, which on projection will be 29 percent Polynesian by 1990. Even the common Maori diseases nowadays seem European: gout, obesity, hypertension, and diabetes.

Putting Pakeha and Maori together, one gets a very peculiar national combination: one the reserved, British-descended white, concerned with butter, skiing, pensions, and gardening; the other the warm, unbuttoned, brown Polynesian, concerned with sex, song, relatives, and the land that he lost. Yet the harmony of this relationship, though overadvertised, is indisputable. Pakeha and Maori mingle physically, mentally, and culturally. They josh and argue in the pubs. The Maori collection in Auckland's War Memorial Museum, with its huge canoe, carved storehouses, greenstone adzes, and fishnets a mile long, is a monument of Pakeha pride in Maori culture, and no one likes to visit it more than the citified Maoris. New Zealand is unquestionably the beneficiary of two great traditions. Intermarriage is commonplace, Maori men to Pakeha girls as well as the other way around. In the Wellington area, more than half of Maori marriages are to Europeans. The percentage of persons with Maori blood is rising steadily.

Blatant racial prejudice is definitely bad form throughout the country. "Only a few far-out individuals practice discrimination here," says Matt Te Hau, Maori lecturer at the University of Auckland. Yet there can be no doubting who is up and who is down in New Zealand. A gentle condescension, or at least a naïve sentimentality, prevails among the whites. A visitor may hear a Pakeha lady gush, "Have you seen our Maoris?" Pakehas tell one another stories of Maori ignorance such as that Maori mothers hold their children's noses between their knees to sharpen them, and that they bury bloody rags because fire is for cooking and it's bad to use sacred fire to destroy filthy objects.

Prominent doctor: "Our Maoris are wonderful with their hands, proud of being Maoris, delightful people."

"Do you ever invite any of them to dinner?"

"Actually, very few of them are culturally adequate."

One serious judgment is that most of the credit for harmonious relations belongs to the Maoris, for being willing to put up with these attitudes. But let it be quickly added that most white New Zealanders know more about, and care more about, Maori culture than Americans do about Indian culture.

They have respected Maori virtues and power for more than a century and are proud of their relations with the Maoris. The essential tolerance springs from New Zealand's classlessness and basic decency. Many Pakeha high school students, for example, take the time to learn at least a smattering of the Maori language, and to appreciate its poetry.

*

Maoris get their say in Parliament through four members elected by more-than-half-blood Maoris. To assure this quota, the country is divided into four special Maori electoral districts which have no relation to the Pakeha districts. In addition, Maoris can, and do, stand for and win seats from Pakeha districts, though paradoxically they cannot vote for themselves in these cases. Currently seven or eight part-Maoris hold Pakeha seats. The Maori Sir James Carroll ran for office at various times on both rolls and once served as Acting Prime Minister, and pipe-smoking Sir Peter Buck rose to be Minister of Health. Presently, a Maori woman, Mrs. Tirikatene-Sullivan, daughter of an illustrious (and knighted) M.P., is the member for the Southern Maori District. She is a strikingly handsome woman, with shoulder-length black hair and beautiful hands. Zealous and articulate, she got into politics as her father's secretary, and won by 81 percent of the vote. Maori M.P.s have always been members of the Labor Party, and Maoris in general vote 70 percent pro-Labor. This segregated system of special Maori electorates, in use since 1867, is under constant attack. Critics say it is racist and tokenist, and makes Maoris second-class citizens. The most pragmatic proposal for improvement is to increase the number of Maori districts, because the Maori population has grown so swiftly since the 1867 apportionment. Dr. I. H. Kawharu, a social anthropologist at the University of Auckland and a Maori, says: "It never occurs to a Maori to be anything but a Maori. If he can't make a contribution as a Maori, he wants to know what's wrong with the system that does this to Maoris. So he wants more Maoris in Parliament, not more whites *talking* for Maoris. He doesn't want to lose consciousness of his identity."

Sometimes the Maori M.P.s are dismissed as the "little brown mandate," and the laws passed by Parliament for the benefit of Maoris have an air of being gifts. The Maori Social and Economic Advancement Act of 1945 financed Maori development in farming, put Maoris fully into the Social Security System, and wiped out legislation discriminating against Maoris. The word "native" was abolished from official lingo in 1947. Yet the Department of Maori and Island Affairs cannot seem to avoid a paternal "us and them" tone in official statements. It employs 782 wardens to curb "ex-

cessive drinking" among the Maoris. Various relationships are proposed for the indigenous people and their Western successors: integration, symbiosis, assimilation, coexistence. The Maoris seem to want to be both in and out of the white culture, symbolically integrating the washing machine with the *marae*, the still-prevalent traditional place for meetings and oratory. Sentimental Pakehas purport to value the Maori culture, but the official policy of integration is, at bottom, that Maoris should be like whites. The stress is on apprenticeship for trades, and the beneficial effects of YWCAs, and the like.

As a consequence, articulate Maoris seem more than a little irritated over their role in New Zealand, while admitting that basically they are treated better than any other minority in the world. They complain of everything from "alienation in their own land" to occasional discrimination in barbershops and movies to "mutilation of Maori place names" by announcers on the radio. Mrs. Tirikatene-Sullivan cites the case of a Maori girl who tried to register at ten hotels and was refused by three of them. But Mrs. Tirikatene-Sullivan also boasts that "we are the world's most successful plural society, or multiracial society, except possibly Hawaii." A great periodic outcry arises over how to overcome South African racial prejudice when New Zealand's Rugby team, with many Maori teammates, goes to Cape Town to play. Recently South Africa agreed to classify the Maoris as "honorary whites" for visa purposes, which made the Maoris more furious than outright exclusion would have. Racism, perhaps, is increasing a bit as Maoris change from distant farmers to city-dwellers, often in ghettos. One-third of all persons in prison are Maoris, which would seem to be an indication of educational shortcomings and a lack of good job opportunities.

The Pakeha land grabs of the past century still rankle. Of New Zealand's sixty-six million acres, Maoris now own only four million, less than they should even in proportion to numbers. A land tenure map of the North Island shows scrappy patches of Maori land indicated in yellow amidst a sea of Pakeha freehold land in green. A new land act, passed in 1967, makes it easier for Maoris to sell to Pakehas, and Maori radicals darkly interpret this as another land grab, this time, they say, on behalf of American companies that want it for oil and other minerals.

Maoris also suffer deficient education, partly because they tend to leave school at fifteen, the end of the compulsory period, and partly because 27 percent of them attend special, segregated Maori schools that are inferior to the state schools (the Maori schools are now being rapidly integrated into the state system). Out of ten thousand students at the University of Auckland, only one hundred are Maoris, one-eighth of what the ratio should be.

Note on the Maori Language

Here is a language ingratiating to the ear, confusingly homonymic, rich in syllables, a source of place names. You hear it frequently on the streets of Auckland. It is uniform throughout New Zealand, and all Maoris speak it. A surprising aspect of Maori, which like Hawaiian is a dialect of Polynesian, is that it gets along quite well without any of the sounds indicated by *b, d, j, l, q, s, v, x,* and *z.* Owing to the scholarly efforts of the missionaries, Maori can be—and is—written and printed. Strangely, the sound that the missionaries denoted as *"wh"* is more or less like *"f."* The name of the aforementioned Dr. Kawharu is pronounced "KAFaroo." Maori is by no means Pidgin, but it has adopted and transliterated innumerable English words. "Tin" is *"tini,"* "pork" is *"poaka,"* "broom" is *"puruma,"* "Christmas" is *"Kirihimete."* *"Wiwi"* comes from French, but it means "Frenchman," not "yes-yes." Thousands of Maori words are homonyms. *"Te-Hokioi,"* for example, called "the most mysterious word in the Maori language," means "symbol of love," "symbol of guidance," "symbol of war," and "I shall return." *"Rangi"* means "arrange in rows," "heaven," "period of time," "sky," "tune," and "weather."

CHAPTER 23

New Zealand: Politics, Grass,
and Potters

*They grizzle and groan, but they've got nothing to grizzle
and groan about.*

—AN AUCKLAND TAXI DRIVER

SIR KEITH HOLYOAKE, Prime Minister from 1960 until early 1972, ran New
Zealand longer than any other leader except the reformer Seddon. He
made this record partly because he possesses two characteristics that
New Zealanders greatly admire* in a politician: broad, strong shoulders, and
an education achieved on his own, without benefit of college. The first proves
that he has "used the shears"—that he knows how to raise sheep; the second,
that he is not too "airy-fairy" to understand the needs of the ordinary people.

These barefoot-boy traits, of course, quite misrepresent Sir Keith's character taken as a whole. In office, he proved able, powerful, sound and
markedly adept at steering away from mistakes. His ancestry goes back
to a headmaster at Rugby in 1687, but a more pertinent forebear may
have been an English great-granduncle who was a Chartist, an Owenite,
and a Secularist—a mid-Victorian Socialist. Some of this spirit of social
revolt crossed the ocean with almost every New Zealand settler.

Because his father was ill, Keith Jacka Holyoake, third of seven children,
had to leave school and go to work on the North Island family farm at the
age of twelve, taking full adult responsibilitiy for running the place. Self-
reliance and faith in free enterprise were bred into him by this experience.
For further education, he got tutoring from his mother, who had been a

* In addition to a first name that his countrymen seem to prefer above all others
to give their male children.

teacher, and took correspondence courses. He became, and remains, an important farmer (dairy products and tobacco). Farmer-organization politics led him into national politics, and election to Parliament in 1932 as its youngest member. In 1936 he and other politicians hooked together the National Party from the remnants of the right wing of the old Liberal Party, and in 1949 they beat Labor and came to power. Mr. Holyoake served until 1957 as Minister of Agriculture—a long run in what is regarded as a deathtrap post—and then took over as party leader. The Labor Party regained power from 1957 until 1960, with Walter Nash as Prime Minister. Then the National Party, under Keith Holyoake, won the government again.

At one point Sir Keith studied elocution, discarding a New Zealand accent in the process, and his oratory is so persuasive that on one occasion a Speaker of the House was stirred to drop his decorum and cry, "Hear, hear!" Holyoake also has a talent for mimicry. In public he is courtly, even pompous, and well groomed; he has a fine, haughty mouth. In private he is warm, hearty, and self-deprecatory all at once. Queen Elizabeth knighted him in June 1970, the first New Zealand Prime Minister to be so honored. His wife, a former girl athlete, shares his political zest. They have five children. He is a man with few other interests than politics, but after he underwent a prostate operation in 1970 it was generally assumed that he would step down to let a younger man lead the National Party in the November 1972 elections. And he did, resigning a few days before his sixty-eighth birthday, February 11, 1972. His successor, inevitably, was Deputy Prime Minister John Marshall, fifty-nine, a man of the same political cut, with a record of twenty-five years in Parliament, nineteen of them in the cabinet.

In office, Sir Keith operated with beguiling informality. He used to walk to work without a bodyguard, and often stopped at the Chinese greengrocer's on the way home. In his childhood, to avoid confusion with a cousin of the same name, he took the nickname "Kiwi," and still has it. Because he is over sixty-five, he collects a social security pension like anyone else. His home phone number, 44-797, is listed in the Wellington book, and anyone who dials it can speak directly to him. "I hope I'm not disturbing you," said a caller during Holyoake's last year in office. "Not at all," he replied. "Just working away at some cabinet papers." To some American visitors in his office, through the windows of which one could see seagulls sitting on a flagpole, he said, "Thank you for coming to such a teeny-weeny country."

*

New Zealand accepts the British Queen as her own, and the Queen exerts her largely fictitious powers through a Governor-General. For practical purposes, all sovereignty rests with the New Zealand Parliament, which can

make laws that bind itself with constitutional force, and then quite blandly amend them. At bottom, government in New Zealand rests simply on a great public consensus, rather than a coherent written constitution.

As we have seen, the upper house was quietly buried in 1950, the provinces, equivalent of states, having been abolished long before. The country is simply too small to need much subdivision. Jurisdiction drops directly from the federal level to the counties, of which there are 106, abetted by boroughs for urban areas and town districts like townships. A borough with twenty thousand or more inhabitants becomes a city. Backing this up is a labyrinthian system of no less than 657 local authorities—pest destruction boards, underground water authorities, electricity and gas boards—many of which require elected officials and entail endless elections. Voting is not compulsory (though registration is), but in important national elections turnouts often reach 80 or 90 percent.

Bureaucracy is rampant and paper work endless. "It takes three times as long to get anything done by a government office here as it does in the States," says an American resident of Wellington. The outstanding exception is the Inland Revenue Department, which provides simple tax forms, help and cooperation in filling them out, and fast tax refunds.

The Governor-General until lately has been Sir Arthur Espie Porritt, appointed in 1967 as the first native-born New Zealander to hold that job. He was born in 1900, educated in medicine at the University of Otago in Dunedin and at Oxford. He captained the New Zealand Olympic team in 1924 in Paris, where he won a bronze medal, and in 1928 in Amsterdam. As a surgeon he rose to president of the British Medical Association and of the Royal College of Surgeons. He served as surgeon to King George VI from 1946 to 1952, and as Sergeant-Surgeon to the Queen since then. In office, he toured New Zealand from the Tokelau and Cook Islands to the South Pole. Sir Arthur was succeeded at the beginning of 1972 by Sir Denis Blundell, also a New Zealander, who had been serving as his country's High Commissioner in London.

The Leader of the Opposition

Norman Kirk weighs twenty stone, 280 pounds, quite enough to displace John Marshall, and that is what he aspires to do as leader of the Labor Party. As far as shunning the more uppity kinds of education, the best he can boast is that he left school at fourteen. While pursuing further knowledge by reading seven books a week, he earned his living operating a factory engine. He got into politics as Mayor of Kaiapoi, near Christchurch, which

gave him a shot at running for Parliament from Lyttelton, Christchurch's port, in 1957. He won the seat, and by 1965 was Labor Party president and Leader of the Opposition—a fast rise, due in part to his eloquence as a television speaker. He is forty-eight years old, married, and has five children.

Labor has been in power only three years in the past twenty-two, but it remains strong because it devised the brilliant reforms of the late 1930's and because most workers belong to it. A three-seat switch in the 1969 election would have given the government to Labor. In a general sort of way, Labor is supposed to speak for the "socialist alternative," but in practice it fully accepts the free-enterprise component of the New Zealand mixed-economy system. It has specifically dropped nationalization of industry as a goal. The National Party, in theory, is a rightist supporter of private initiative over government regulation, but it in turn accepts state operation of part of the economy. Pragmatically, this means it has the support of most farmers—in fact, one-half of the Nationalist Members of Parliament are farmers.

Given this consensus on over-all goals, elections have to be fought over the question of how to achieve the goals. Kirk attacks the National Party administration for running the country by rigid "cabinet rule." The ministers, meeting privately, set policy and make decisions, without consulting the backbenchers, who are then marshaled into line to vote for what the cabinet wants, in short and uninformative sessions of Parliament. The cost, he says, is low prestige for Parliament—which to an outsider, listening to New Zealanders "grizzle and groan," in the Auckland taxi driver's phrase, seems to be the case. Kirk's remedy is longer sessions and freer debate. As to the economy, Kirk holds that reliance on the traditional customer, Britain, to buy wool and food is too risky; New Zealand "must rethink her policy in the light of her position in the Southwest Pacific and of changing patterns of trade inside the Commonwealth." Kirk contends that the country "should be exporting carpets, textiles, and garments instead of raw wool, prepacked processed foods instead of carcass meat, leather goods instead of hide and skins." Yet he puts reforms in the welfare and educational systems ahead of immediate economic development. "We are more nationalistic than the National Party," says Mr. Kirk. "We put more emphasis on the redistribution of wealth. The responsibility of government is to create equality of opportunity, regardless of wealth." The Labor Party's problem, he says, is that it historically stood for the right of the blacksmith's son to become an architect, but after he did so, no incentive was provided by Labor for the architect to stay in the party. "We've had to rebuild the party."

No doctrinaire is Mr. Kirk, nor was Sir Keith Holyoake either. Astute and

dominant, through cabinet rule, Holyoake balanced pressures success-fully for a decade—pressures that come, for example, from the Federated Farmers, the Federation of Labor, the Manufacturers Federation, the Asso-ciation of Chambers of Commerce, and the Public Service Association. For a time Holyoake had a Minister of Labor so seductive that he almost stole the unions out from under the Labor Party. The Nationalists' close call in the 1969 election, though, seems to indicate that something is amiss. The farmers are worried, naturally enough, by the prospective loss of British markets, and sometimes feel that their party is not doing enough for them; the Taranaki branch of the Federated Farmers even called upon Holyoake to dissolve his government. One notable National Party man, incidentally, is Sir Leslie Munro, former President of the United Nations General Assembly, now serving as a humble backbencher. Another is Robert Muldoon, the Finance Minister, a forceful character who aspires ultimately to be Prime Minister, but had to settle for election as deputy party leader in the early 1972 caucus that chose Marshall for the top job.

The universal criticism of the New Zealand political system is that it puts a premium on excessive caution. A perennial example is the question of prohibition. The politicians considered this to be such a hot potato that they tossed it to the public in the form of a referendum held every three years, in which the voters get to choose among the outright banishment of booze, the present style of licensed pubs, or complete government control. (Usually about 30 percent of the voters choose prohibition, but this proportion is falling slowly.) Individual members of Parliament cannot try bold initiatives, but must loyally vote with their parties. The passion for ordinary men, rather than experts, in politics provides just that—homespun types who take care of their constituents' requests and attempt no leadership. Submission to pressure groups is commonplace. One issue that did stir emotions was Viet-nam. Sir Keith sent New Zealand troops there. Labor soon began to demand that they be withdrawn, and in August 1971 Sir Keith agreed to get them out by the end of the year.

The Ombudsman

Like the four countries of Scandinavia, New Zealand has an ombudsman, a watchdog to make sure that private citizens get treated fairly in their rela-tionships with their government. By the nature of their jobs, a certain per-centage of bureaucrats anywhere in the world turn into petty tyrants—dictatorial welfare workers, red-tape artists, discourteous postal employees, overzealous tax collectors. New Zealand, using a system that basically origi-

nated in Sweden one hundred years ago, gives the man with a grievance a chance to get justice without hiring a lawyer and making a costly court case out of it. He can go to the office of Sir Guy Powles, the Ombudsman, whose jurisdiction covers not only the various branches of the state bureaucracy but also the utilities run by the government, such as the railways, the telephones, and the electric power supply.

The system works like this: A complainant must set down his gripe in writing, and accompany it with a two-dollar fee. The Ombudsman is entitled to reject any complaint he considers to be unwarranted. On complaints that he accepts, the Ombudsman runs an investigation that is kept private and deliberately informal, although he can summon and examine persons on oath, and has access to government premises and records. He must inform a government department that he is investigating it, and give it a chance to defend itself. He has guaranteed access to cabinet ministers, and the right to go to the Prime Minister if his recommendation is not acted upon. "We've never quite got to the crunch as yet," says Sir Guy, the pressures of public opinion being sufficient always to get a remedy at a lower level. Most department heads say, "Goodness, did we do that?" and make amends.

The Ombudsman can rule against any illegal act by a civil servant, and, beyond that, against any decision that appears to him to be unreasonable, unjust, or oppressive. If such a decision is made in conformity with a law, the Ombudsman can deem the law improper. Finally, and even more broadly, he can simply pronounce a decision to be wrong, an "elevated use of a plain old English word," as Sir Guy puts it.

Sir Guy, a tanned, sophisticated man with white-gray eyebrows and hazel eyes, operates with delightful simplicity, employing a staff of six that costs only $28,000 a year, and writing his reports with much candid use of the first person singular. Most of the investigation, he finds, can be done by mail. In a typical year, 1969, he received 620 complaints, many from Maoris, and investigated 285 of them. He found 66 to be justified, obtained satisfaction for the complainant in 56 of these by the mere act of investigating, made recommendations in the cases of 6 others, and dropped 4. The biggest number of justified complaints was against the Social Security Department, with the Post Office next, and the State Services Commission (industrial accidents) third; only one was against the police.

The complaints are over the sort of thing that makes people furious with bureaucracy everywhere. Two businessmen exploded when asked to answer the question "Race?" on a form as they departed from a New Zealand airport. Sir Guy agreed (on the basis of his own travels) that the question was not asked by other countries, and that it could be offensive, but accepted the

government's argument that the answer was needed for statistical reasons. A woman sought to bring to New Zealand her mother, an Anglo-Burmese who also had children living in England. New Zealand immigration policy requires that in such cases all of the children have to be living in New Zealand before the parent can be let in. Sir Guy persuaded the authorities to make an exception.

The White New Zealand Policy

To go from Australia to New Zealand is to go from the cosmopolitan to the provincial, and the reason quickly dawns on the traveler: Australia is a rich melting pot of European cultures, and New Zealand, barring the graceful Maoris, is almost wall-to-wall British—not a bad stock, of course, but in this case frightfully homogeneous. During the two decades when Australia was absorbing its polyglot New Australians and their pizzas and paellas and potato pancakes, New Zealand was assisting a few thousand Britons a year to join the earlier immigrants from the same place. The only cautious exception was the Dutch, 26,000 in all, plus pockets of Austrians, Americans, Hungarians, Germans, as well as ten thousand Chinese who immigrated in considerable numbers in the last century, a sizable colony of Indians, and a number of Yugoslavs who make wine. There are conspicuously few Greeks and Italians. The Dutch, by the way, run many restaurants, markedly improving New Zealand's choice of foods. In the last years of the 1960's, as we have seen, the country suffered a net outflow of population of up to six thousand a year (mostly to Australia). It must be added at once, however, that with a high local birth rate New Zealand managed not only to keep up with Australia in population growth rate, but even a little ahead, from 1951 to 1966. This is a fast-growing country!

Now New Zealand has decided to bring people in rapidly, but with much more racial trepidation than Australia. Manufacturers are crying for workers, and an easier labor market would slow inflation. Americans seem to be particularly desirable, provided they are not black, or Hawaiians of dark skin. With a giddy feeling of experimentation, New Zealand is also subsidizing the fares of workers from France, Belgium, Italy, West Germany, and Switzerland, as well as Holland. Sir Keith Holyoake says that they will "add a new dimension of strength and vitality to New Zealand."

Decisions as to who may immigrate are made, one by one, on the administrative level, by issuing entry permits. Formerly persons of "British birth" were not required to have permits, British birth being defined as descent from native-born (not naturalized) British subjects, and any British subject

who was "an Aboriginal native or the descendant of an Aboriginal native" was specifically excluded. This was a little too blatant; anyway, the permit system is sufficiently broad and arbitrary. Anyone denied a permit can appeal to the Minister of Immigration, who is not required to explain his decisions. In general, though, the decisions are based on "assimilability"—and New Zealanders are notorious for their ability to distinguish people who are "not like us." Prime Minister Marshall, a soft-spoken, sun-bronzed man who put in many years in the touchy job of Immigration Minister (and was simultaneously Minister of Labor, Minister of Overseas Trade, and Attorney General), says: "We seek to maintain a homogeneous society."

The color bar has one gaping breach. Cook Islanders, whose Polynesian homelands constitute a "self-governing territory in free association with New Zealand," are admitted as New Zealand citizens, and islanders from Western Samoa, a former New Zealand trust territory and a nation on its own since 1962, are admitted at the rate of fifteen hundred a year by a special arrangement. These islanders' form a colony of nearly forty thousand, mostly in the older section of Auckland, where they crowd densely into housing vacated by Maoris moving to better areas. Freshly arrived islanders have a reputation for drinking, violence, and trouble with the police, but their children are said to settle in well. A problem of this migration is that it tends to drain the islands of their richest and most talented people, who prefer Auckland's bright lights to small-island boredom. New Zealand requires that Samoan girls, before immigrating, must prove by a medical test that they are not pregnant. The girls object to this as "social stigmatization," which indeed it would seem to be, since scores of Australian girls go untested to New Zealand every year to have illegitimate babies inconspicuously in the free government hospitals.

One final note on this complex race and migration problem, where some people are decidedly more equal than others. Australia admits Maoris freely, but demands a visa for Cook Islanders, who are equally New Zealand citizens. New Zealand, conversely, raises no barriers against black Papuans, who are Australian citizens, but these same Papuans are not allowed into Australia, despite their citizenship, without a permit.

The Economy

Grass! English grass, growing in ten thousand meadows. Or natural tussock, covering the hillsides. To this country, grass, turned into wool, lamb, and mutton by that four-footed factory, the sheep, or into milk, butter, cheese, and powdered milk by that equally industrious animal, the cow, is like oil

to Kuwait. Beautifully, simply, and endlessly, the soil has poured out the wealth that let New Zealand live its easy, cushioned life. Every New Zealander—man, woman, child, Pakeha, Maori—is backed up by two dozen tireless animals busily getting fat or producing milk or extruding a valuable fiber. The land is, to be sure, extraordinarily well tended. Cobalt and superphosphates are added in the tiny proportions that energize the soil like the discharge of an electric eel. The country leads the world in the use of airplane-dropped top dressing on pasture land. Planes operating off of *eight thousand* airstrips drop fertilizer on five million acres of land every year. The soil, in fact, has been completely rebuilt from the condition in which the first English settlers found it. Considering that the two islands are both mountainous, the New Zealanders have managed to use an amazingly high proportion of it: two-thirds.

The most typical case of a product raised to the English taste is the fat lamb ("fat" because they look blocky, not because the meat is fatty). Americans will accept meat from a lamb that weighs as much as forty-five pounds, dressed, but the English pay enough more per pound for baby lambs, only three months old, that New Zealand aims to ship lamb carcasses at just thirty-two pounds. At Invercargill, which with fifty thousand people is the most southerly city in the world,* five *million* fat lambs are slaughtered every season, between early September and the middle of May. The men in the meat works are a doughty crew, slitting the throats of the lambs and hanging them on moving chains. These independent men will quit in a flash if they don't like the "tucker" (food) in the "tucker box," the slaughterhouse lunchroom.

Farmers breed sheep by putting a ram in a paddock with the ewes for almost six weeks, which is certain to include two menstrual cycles for each ewe. The gestation period is 148 days, give or take a few; thus a sheep raiser may breed the ewes in April so that they will drop their lambs in August and reach the slaughtering age of three months by November to hit the English market for Christmas. A lamb is a lamb until it is eight months old; then it becomes a hogget, and at sixteen months a mature sheep. Good sheep land in New Zealand supports ten sheep to an acre, poor land one sheep to ten acres. A farmer needs at least two thousand sheep to make a profit, and a big farm may have eight thousand. The huge sheep stations of the past, broken up, as we have seen, in the 1890's, occupied as much as fifty thousand acres. Now stations average two or three thousand acres.

About half of the country's forty-four million acres of farmland is leased

* Provided that Chile's more southerly Punta Arenas is defined as a town, not a city.

from the Crown, and about half owned outright. The price of good sheep land is $150 an acre, and of poorer land $60 or $70 an acre. In general, the South Island raises sheep and grain; the North Island breeds sheep more intensively, and yet specializes in dairying. By law milk processing companies are cooperatives. At Te Rapa, south of Auckland, one of the fifty plants of the New Zealand Cooperative Dairy Company uses 400,000 gallons of milk a day, an amount so big, the manager says, that "we mine our own coal, we need so much for the steam processing." Eight hundred farmers supply the raw milk, getting thirty-two cents a pound for the butterfat content, and the product, powdered milk, is sent to Southeast Asia—some also to, of all places, Denmark. The workers wear white caps, white overalls, and white boots; they get $36 for a forty-hour week, and the factory manager makes $5,000 a year. The machinery, fully automated, came from West Germany.

*

Lying ahead for this economy of wool and meat and cheese is something that could be akin to disaster, and its initials are EEC. After Britain joins the European Economic Community, a wall of tariffs will eventually rise to shut out New Zealand's products. The thought causes anguish. One New Zealand politician, visiting London early in 1970, called the EEC a "gigantic swindle of the British public." He passionately asked, "What has happened to our ancient dignity and pride of race? Our Queen and all she stands for as a focal point of Commonwealth loyalty are being sacrificed for a few pieces of silver. Will her robes be divided among the directors of the great international moneymakers of Wall and Threadneedle streets? The cause of the British housewife and the New Zealand producer are the same cause. We must unite and march together so that we can continue as brothers to toast as our sovereign lady, the Queen. God bless her."

Farm products form nearly 90 percent of New Zealand's exports; it is the world's largest exporter of butter, and for many years the largest exporter of meat. All told, farm exports bring in $1 billion a year. Twenty percent of the Gross National Product is exported, as compared to 4 percent in the U.S. Almost half of these exports go to Britain, including four-fifths of the butter,* cheese, and lamb. The loss of these markets would be a dreadful blow. During negotiations in mid-1971, the continental members

---

* The British are mad about butter. They eat almost 20 pounds per capita per year, along with almost 12 pounds of margarine. Americans eat almost 11 pounds of margarine, but only 5½ pounds of butter.

of the European Economic Community agreed to let Britain continue buying these New Zealand products during a five-year phaseout after Britain's entrance into the EEC. Finding new markets for butter, cheese, and lamb may prove difficult. To most of the world, particularly New Zealand's Asian neighbors, these are luxury foods, to be consumed only in small quantities. In other markets, where she could compete powerfully, there are import restrictions. New Zealand farmers could put a pound of butter on U.S. tables for 40 percent less than domestic prices. New Zealand's other agricultural exports, happily, are in less peril. Three-fourths of its beef goes to the United States and Canada, at high prices. Mutton goes to Japan for sausages. Big buyers of wool are Britain and other European countries, the United States, and Japan, but of course wool suffers ever-growing competition from synthetics.

Luckily, ten years of De Gaulle excluding Britain from the European Economic Community gave New Zealand a breathing spell in which to plot new incomes. The Comalco aluminum smelter at Bluff is, in effect, an exporter of electricity from the numerous hydro plants in the Southern Alps. Here electric power, added to alumina (from Australia), makes aluminum, to be sent to Australia, Japan, and elsewhere. Radiata pine is exported as lumber and pulp. Iron sands, found along hundreds of miles of the New Zealand coastline, have succumbed to a technology that permits smelting it into iron for the local steel industry, and for export to Japan. Sand also yields titanium and ilmenite. Oil and gas have been found off the coast of Taranaki, on the North Island, in big enough quantities to give New Zealand hope for self-sufficiency. Presently New Zealand imports oil worth $100 million a year from the Middle East and Indonesia; now, with these discoveries by British Petroleum, Shell, and Todd Consortium (a company backed by one of the few millionaires in New Zealand), it may be able to start a petrochemical industry. A similar economy might come from a proposed shipping line; until now all the country's export shipment has been left to the British. Tourism, too, is a developing possibility; Americans departing from the West Coast can get Hawaii, Samoa, Fiji, Tahiti, Australia, and New Zealand all on the same airline ticket—taking the Tahiti leg aboard Air New Zealand.

New Zealand might also become a workshop country, along the lines of Sweden; already it manufactures goods of a variety all out of proportion to its small population. Wire, nylon cloth, bathtubs, hydraulic trench diggers, acrylic paint, TV tubes, gin, thermostats, bicycles, road rollers, ships, adhesives—these are some of the products of New Zealand factories. Exports of manufactured goods (newsprint, farm machinery, woodworking equip-

ment) reached $100 million in 1969—not bad. Most of it went to Australia, which has a limited free-trade agreement with New Zealand. But Europeans now buy New Zealand lawn mowers, and Singapore buys New Zealand carpets.

The spread of socialism in New Zealand is perhaps best measured by how much business the government runs, and it runs a lot; telephones, radio and TV, mines, railways, natural-gas distribution, hotels, domestic and international airlines, insurance, electricity. The power goes from the cheap hydroelectric sources in the South Island to the big consumers on the North Island through a unique, 500,000-volt submarine cable, twenty-five miles long, under Cook Strait. The Bank of New Zealand, the country's biggest, which competes directly against commercial banks, is government-owned. "I hold all the shares," says Minister of Finance Robert Muldoon. Sometimes, though, the state pioneers in some industry, and then turns it over to private enterprise. The government discovered how to make steel by pelletizing the iron sands of the West Coast of the North Island, and set up a private company to carry on from there.

On the other hand, the government stays completely out of farming. The whole dairy industry, for example, is based on one-man farms, working through cooperative marketing organizations. Two minor industries of note are brewing (the DB Export beer is strong and hearty) and wine making (the whites are superb, and the New Zealanders are slowly being weaned away from their dreadful fortified reds).

Pottery, Painting, and the Enzedbeecee

We are a young nation with enough leisure, education and material security to start looking for a culture of our own, and we want more human values than usually found in suburbia.

—HELEN MASON

New Zealand frets a good deal over its cultural development, lamenting, for example, that such a writer as Katherine Mansfield and such a painter as Frances Hodgkins felt, in the early part of this century, that they must work out their careers in London and Paris. Ernest Rutherford, later Lord Rutherford, Nobel prizewinner and first man to split the atom, was another New Zealander who had to go abroad, to Cambridge, before he could do the work that won him fame. Perhaps the same might be said of Sir Edmund Hillary, but of course he had to go to where the highest mountain is before he could climb it.

But of late the country seems to be a quite adequate home for artists,

poets, dancers, writers, and musicians. Strangely, though, the pre-eminent art is pottery. In fifteen years, New Zealand has come to the fore as the workshop of dozens of highly talented potters turning out jugs and bowls and jars in huge variety and quantity. Kilns and studios abound. Great events shake the pottery world from time to time, such as the arrival of expert Japanese potters to teach and demonstrate. In turn the local potters go abroad, most frequently to Japan but also to England, the United States, and Mexico. Their work shows Japanese and Maori influences.

The Professional Potters Association has about a hundred members, a leading light being Helen Mason, who also edited the *New Zealand Potter* for many years. Mrs. Mason went to Fiji a few years ago and lived in a village there for a month, working with Fijian potters. In her house-studio in the bush fifteen miles west of Auckland, she still uses techniques and motifs that she learned in Fiji. She can produce $350 worth of pottery from one firing of her kiln, even though she consciously holds prices low because "we want people to use our things every day. If they break a dish, they come back and buy another." She also teaches the potting craft: "A young-ster will arrive on my doorstep and stay on for a year to work with me."

Painting in New Zealand inspires much lyricism about the quality of the light. The poet A. R. D. Fairburn calls it a "hard, clear light" that "reveals the bones, the sheer form, of hills, trees, stones and scrub. We must draw rather than paint, even if we are using a brush, or we shall not be perfectly truthful." Landscape, rather than the figure, has been the customary theme of the New Zealand painter, from Charles Heaphy, official artist and drafts-man of the New Zealand Company in 1840, to Colin McCahon, a present-day South Islander. Painting has a respectable history in New Zealand. In the Auckland City Gallery, the National Art Gallery in Wellington, and the Hocken Library in Dunedin, many of the paintings are unusual, spe-cifically because they are not derivative. Distance kept artists from firsthand knowledge of French painting. The Auckland gallery has a large collection of European Old Masters. Peter Bromhead, the bearded, long-haired curator, says that his eighteenth- and nineteenth-century Italian collection is "better and larger" than Melbourne's. This gallery also has sculpture by Rodin, Maillol, Epstein, Henry Moore, and others.

A lively private gallery in Auckland is the Mill, a century-old building that once housed a flour mill. While a folk singer performs, the people of Auckland buy art and craftwork. One Auckland gallery owner, Barry Lett, contracted with twelve leading artists to do prints, which he sells in forty-dollar sets (one from each artist) to "get away from the idea that art is for people with money." Paintings here are much cheaper than in Australia.

*

The press in New Zealand (about twenty daily newspapers) must by law be New Zealand owned—no Lord Thomsons or Rupert Murdocks allowed here. The result is a constricted ownership that, among other defects, leaves the Labor Party without daily-newspaper support. A narrow and conservative press. NZBC (pronounced "enzedbeecee"), the government-subsidized New Zealand Broadcasting Corporation, is scarcely more venturesome. The rows that it gets into are usually over what it refuses to air, rather than what it does air on its four TV and forty-four radio stations. New Zealand viewers, who have had television only since 1960, now get: one channel only, no color, a surfeit of American programs, and the obligation of paying a license fee of thirteen dollars a year. They even have to put up with commercials, but only on Tuesday, Wednesday, Thursday, and Saturday. Broadcasting hours are 2 P.M. to 11 P.M.—midnight Fridays and Saturdays.

Enzedbeecee has a much-admired ninety-piece symphony orchestra, which broadcasts and also plays in concert halls all over the country. The New Zealand Ballet Company and the New Zealand Opera Company, both government subsidized, also play to the whole country and pull enthusiastic full houses. The theater is mainly amateur, but well attended.

New Zealand has had any number of competent male writers, but the internationally notable aspect of the country's literature is its domination by women. Katherine Mansfield was a master of the short story; Ngaio Marsh is a ranking writer of detective stories; Janet Frame plants broad themes of tragedy in a New Zealand setting; and Sylvia Ashton-Warner has won over the world with her poetic, moving, and touchingly funny novels, such as *Spinster*.

New Zealanders take their sports obsessively, and with approximately equal dedication to watching and participating. Rugby football is virtually the private property of New Zealanders, whose revered All Blacks won fifteen straight annual "Tests" with the few other nations that play this occult game (Australia, France, South Africa, and Britain). The other big spectator draw is horse racing, not only galloping but also trotting and pacing. This little country has five thousand brood mares, producing stock so good that one horse was recently sold to some Americans for a reported one million dollars. But when New Zealanders aren't watching, they're doing. What they do best is fish for trout, which are plenteous; climb mountains, which are also plenteous; and sail in the beautiful bays. They also have a special thing about camping.

*

There is abroad in the world today, particularly in the United States, a dark frustration, a feeling that the quality of life is deteriorating and progress is illusory. In most U.S. cities, safety, civility, standards of all kinds, decline in spite of the speeches and promises of the politicians. The citizenry despairs, frustrated by the vast immobility of an overgrown system. Perhaps a summary judgment of New Zealand is that it is happily free of this curse. Probably because of its smallness, government does respond to the people's needs. The individual is protected. People are not afraid. Land tenure is well handled. Taxes are high, but used in ways that are visibly beneficial, or at least not destructive.

New Zealand: you would not think you'd want to live there, down at the bottom of the world, twenty tiresome hours by jet from Kennedy Airport. But if you did live there, you might enjoy it hugely.

CHAPTER 24

How Australasia Fits with Asia

*Australia is anchored off Asia and
cannot be towed away, no matter how
awkward its Asian neighbors may become.*
—PROFESSOR J. D. B. MILLER

WITHIN RECENT MEMORY, Australia and New Zealand were essentially detached parts of Europe, filially linked to England by blood and loyalty and the teak-decked vessels of the P. & O. Line and the New Zealand Shipping Company. Asia might be physically on the doorstep, but in the Australasian public consciousness it was as remote as Venus. Then Japan, hungry for resources and markets, decided in 1941 to incorporate Australasia by war into the Greater East Asia Co-Prosperity Sphere (as Japan euphemistically called its plan of conquest). The might of England, based in Singapore, crumbled in a few moments. The might of the United States, assembled with panicky speed in the Coral Sea, rescued Australia and New Zealand from certain invasion. After the war India, Pakistan, and Burma opted to become republics. China, for decades dismissed as a famine surrounded by Western exploiters, emerged as a heavy, hostile, Communist power, with atomic bombs and a population on the order of fifty times that of Australasia—all not much farther away than the distance between Sydney and Perth. Dienbienphu fell. Then the revived Japanese economy proceeded to buy Australasian produce in gulps so voracious that Australians nowadays ruefully refer to their country as "the Japanese quarry." Nice irony here— the defeated Japanese gained after defeat a good deal of what they set out to achieve by war. Finally, the United States failed to win the war in Indochina, raising Australasian fears that American power in 1971 was no

more reliable than English power in 1941. Is an attack on Australasia still equivalent, in the U.S. view, to an attack on Ohio, as Walter Lippmann once put it?

So now Australia and New Zealand look to England from a position less daughterly than cousinly, and to the United States with gratitude for past favors tainted by doubts about whether the United States can be, or even wants to be, a stout ally. And they view Asia, large and close and omni-present, with fascination, apprehension, and a good deal of incomprehension. These two countries share geography with Asia and history with Europe, and they sense an urgent problem of deciding how the twain shall be made to meet.

*

Australia's navy still sails "Her Majesty's Australian Ships," and New Zealand has a "Royal New Zealand Air Force," but the process of separation from Britain—begun with federation in 1901 for Australia, and with Do-minion status for New Zealand in 1907—is now virtually complete. Way back in 1919, Australian Prime Minister William M. Hughes asserted his country's independence from Britain, demanding, "What is there that we cannot do?"; the same year New Zealand independently signed the Charter of the League of Nations.

The hard ties that remain are in trade and investment. The United King-dom buys $400 million worth of Australian wool, meat, grain, fruit, sugar, and ores per year, and sells Australia $700 million-plus worth of machinery, electrical equipment, vehicles, and chemicals. These figures amount to about one-seventh of Australia's exports and one-fifth of its imports—which is, of course, a steep decline from decades past when the U.K. was Australia's major supplier and customer. From New Zealand the U.K. buys $380 million worth of lambs, wool, and butter, and to New Zealand it sells $208 million worth of cars and assorted manufactures, thus holding its traditional position as most important trader both ways. As we have seen, Britain, upon en-tering the European Common Market, will eventually cut its purchases of New Zealand lamb and dairy products. Yet the British still control almost half of overseas investment in Australia, and a great deal in New Zealand.

The relationship between Australia–New Zealand and the United States has always been a pretty easy one. World War II, drawing hundreds of thousands of American G.I.s to Australian staging bases, produced a gen-eration of Americans half in love with Australia. American history and geography are well known in both countries. The Brisbane *Courier Mail*

runs a quote from Jefferson on its masthead; ordinary Australians and New Zealanders speak of U.S. states and cities as familiarly as they do of their own. There is a closeness based on the three countries' similarities as former British colonies built largely by people of British stock.

When Australia and New Zealand initiated and signed the 1951 ANZUS (Australia–New Zealand–United States) treaty for mutual defense in the Pacific, they pointedly shunned British participation. It was, in fact, the first pact ever negotiated between members of the British Commonwealth and a foreign power. Americans may not think about ANZUS very often, but it has been a sort of religion down-under. "ANZUS is critically important to us because it contains a formal recognition by the U.S. that an armed attack upon Australia or its territories or its defense forces would be dangerous to the United States' own peace and safety," says Australia's former Foreign Affairs Minister Leslie Bury. A concept often evoked in these two nations is that the U.S. is a "Pacific country," as, of course, they themselves are. New Zealanders are quite capable of quipping that their country is "the fifty-first state," while realistically recognizing that New Zealand is not indispensable to the U.S. An Australian official says, "There is very little feeling that the U.S. is Big Brother, breathing down our necks."

Trade with the United States is a strong link, too. U.S. purchases from Australia, chiefly beef, run to more than $500 million a year, and its exports to Australia (transport equipment, machinery, chemicals) are more than $1 billion. Trading with New Zealand, the U.S. imports wool, meat, and other products worth something under $200 million, and exports airplanes, tobacco, chemicals, computers, and so on, worth about $75 million. American investment there is relatively low.

The Vietnam war, by the time it wound down, had become a severe irritant in relations between the U.S. and Australia and New Zealand. The troops they had sent to fight there represented, in bald terms, an IOU cashed by Washington for past and future protection—or, as one radio commentator put it, the two countries "sacrificed twenty-year-olds as an insurance premium to the American alliance." The Prime Ministers of Australia and New Zealand, paying the premium, had to follow the U.S. lead at every twist and turn of the war, while, for domestic consumption, they tried to make it appear that the initiatives were theirs, or at least that they were amply "consulted" by Washington. When the U.S. escalated, so did they. Former Prime Minister Harold Holt allied himself strongly with President Johnson, and when Johnson renounced his job (and, implicitly, his faith in his policy), the Australian government was left with egg on its face. President Nixon's decision to begin pulling U.S. troops out

of Indochina forced Australian and New Zealand leaders to mumble that they just happened to be planning to do the same thing.

Beyond that, the two governments try to weigh the extent to which the U.S. pullout leaves them vulnerable to Communist China—or even, should history repeat itself, to Japan. In January 1970, Vice President Spiro Agnew went to Australia and New Zealand, and tried to reassure both countries that "the United States will keep its treaty commitments. Our policy is neither one of gradual withdrawal from Asia nor of unwarranted intervention in Asian affairs."

Relations with Washington are the No. 1 concern of the foreign affairs departments of both countries. And both have been run for many years by parties that represent rightist, pro-American positions. They suffer assorted slings and arrows. Senator Fulbright says: "Both Australia and New Zealand have been rewarded with lavish praise and gratitude from the U.S. Gratitude for what? For making a tiny contribution to what they themselves consider their own defense."

The Brown-Skinned Neighbors

For a bunch of Anglo-Saxons like the Australians and New Zealanders, though, diplomacy with Britain and the United States is easy compared to relating to the neighbors. The near neighbors are friendly. Australia (fearing to do otherwise, lest it create a next-door enemy) helped Sukarno grab West Irian in 1962, and felt its judgment justified when the agreeable government of General Suharto replaced Sukarno. The Australians managed to remain quite unperturbed over the fact that Sukarno's ouster required the massacre of 300,000 Indonesian Communists and sympathizers.

The future nation of Niugini will doubtless be an Australian client state, and the new countries of the near South Pacific are amicable. Melanesian Fiji (population: 519,000—many of them, to be sure, Chinese and Indians) gets along passably well with partly Polynesian New Zealand, 1,149 miles away. So does Western Samoa. The stout little South Pacific nation of Nauru (area: 8 square miles; population: 6,500) seems favorably disposed to its large southern neighbors, at least until its mines of phosphate, exported to New Zealand for an average income of $6,000 per capita per annum, run out in 1990.

Malaysia and Singapore, sister members of the dissolving Commonwealth, are power bases land-linked to tumultuous Indochina. Britain, confessing its helplessness to sustain Singapore, virtually withdrew in 1971, and the United States, singed in Vietnam, would like to settle for monetary support

only—namely, commitments to repair naval vessels in the Singapore dry-docks. It seems to be up to Australia and New Zealand to establish a credible medium-power presence there, both in a military way and as a Eurasian force for collaboration and sanity.

Then there is Japan, once mortal enemy, now foremost customer, third-largest industrial power of the world, harborer of vaulting ambitions, land of inscrutable yellow men who may want to emigrate to White Australia but on the other hand may be perfectly content to use it as a quarry. Just as enigmatic is the People's Republic of China, which may aspire to occupy Australia's empty spaces, or, conversely, may be a lucrative market for Australian goods. Many Australians visit China as tourists, and Australian ships call at Chinese ports. India, Pakistan, and Ceylon relate to Australia and New Zealand as fellow members of the Commonwealth, but they are more strongly inclined toward neutralism and have less critical attitudes toward Communism. There is envy, too, based on the unignorable fact that the Australians have been able to make so much out of their barren continent. As Australians helped India with the Pakistani refugee problem in mid-1971, one Indian official turned to an Australian aid administrator in Calcutta and blazed: "You bloody big white Australian. You bloody bastard. Get out of here. Damn your bloody big rich country for its great help." This was an irrational explosion, but it demonstrates one Indian attitude toward Australia.

Thus, since World War II, Australia has been drawn into Asia, partly by the opportunity to secure a measure of dominance, partly to be a helpful neighbor, partly as a half-thought-out "forward defense" strategy. The Far East, as Australians illogically refer to it, has become what it really is, the Near North. The Australian population build-up is a conscious response to the dangers of being an empty continent in crowded Asia. Government leaders and journalists increasingly visit Asian countries. The weekly *Bulletin* scrupulously covers Southeast Asia, Fiji, and the Philippines, in a department called "The Neighborhood," and prints a column called "This Week in Japan." The Australian diplomatic service is said to be doing a highly skilled job of reporting and representation in its Asian embassies, particularly Tokyo. The Sydney and Melbourne symphonies play in Asian capitals, and the seven short-wave stations of Radio Australia broadcast in Japanese, Malay, Mandarin, Thai, Indonesian, and Vietnamese.

Australia and New Zealand make such contributions as geodetic surveys for the Philippines, water systems for Vietnam, military training for Asian officers. Both also founded and take part in the Colombo Plan for economic aid: Australia spent $160 million up to June 1968, chiefly in Indonesia and

Pakistan, but also in such remote spots as Brunei, Nepal, and Bhutan. In fact, according to Prime Minister McMahon, Australia's foreign aid is proportionately higher than the United States'—.56 percent of the GNP as compared to .38 percent for the U.S. New Zealand, typically, ran dairy projects in Bombay and Dhulia in India. Both countries provide higher education for Asians; more than twelve thousand Asian students are regularly enrolled in Australian universities, and up to mid-1969 New Zealand had brought in 2,662 Asians for college educations under the Colombo Plan.* A criticism of this effort is that the Asians are often pushed into courses that are underenrolled rather than courses they want, and go home to unemployment because they are miseducated.

*

We Australians have been for long an unbearable collection of Pommies, but may now become a rather likable collection of second-rate Japanese.

—PROFESSOR GEOFFREY SAWER

Japan is a word that has popped up again and again in this book on Australia and New Zealand, chiefly as the engine of the Australian boom. Japan buys over one-fourth of what Australia exports, taking wool, iron ore, nickel, copper, bauxite, coal, grain, and meat. Australia, therefore, is greatly dependent on Japan—a thought that causes much wry comment. "It was bad enough being run by the Brits and the Americans," says an Australian journalist, "but at least we could talk to our bosses in English." Australia still has a higher standard of living than Japan, but the Japanese have better technology, and are thus in the position of being able to decide, for example, whether they will continue to import and then smelt Australian

---

* The regional organizations in this part of the world are confusing and overlapping. The Colombo Plan (so called because it was set up at a meeting in Colombo, the capital of Ceylon, in 1950) tries to do for non-Communist Asia roughly what the Alliance for Progress tries to do for Latin America. Members, besides eighteen Asian nations, are Australia, New Zealand, Canada, the United Kingdom, and the U.S. The South East Asian Treaty Organization (SEATO), which includes the United States, is the primary defense mechanism. The United Nations Economic Commission for Asia and the Far East (ECAFE), backed by the Asian Development Bank, undertakes such projects as the utilization of the waters of the Indus and Mekong rivers. The Asian and Pacific Council (ASPAC) promotes solidarity among Korea, Japan, Taiwan, the Philippines, Malaysia, Thailand, Australia and New Zealand. The Association of South East Asian Nations (ASEAN) is a consultative group which includes Thailand, Indonesia, Malaysia, Singapore, and the Philippines.

iron ore, or whether they will let the Australians handle part of this profitable business of making their own ore into steel.

Australian euphoria at every new billion-dollar contract with Japan is not quite enough to make the Australians *accept* the Japanese. The Australian National University in Canberra teaches the Japanese language, and so do some high schools, but only about eight hundred Australians are learning the language, nothing like as many as commerce demands. Back of all this is that old White Australia policy. High-ranking Japanese businessmen have to go through an elaborate procedure to get a visa to visit Australia, and as a rule those who settle there for business reasons are allowed to stay only three years. The bars are high enough so that at any given time only about seventeen hundred Japanese have residence permits. As of early 1971, only 717 Japanese had acquired Australian citizenship; by and large the Japanese may not emigrate to either Australia or New Zealand. These are powerful and poisonous ironies in what has been called "the first European country to be largely dependent on an Asian nation." The Japanese Ambassador in Canberra, Shizuo Saito, in a book published in Tokyo in mid-1971, criticized Australian immigration policy, and charged that the Australians are unwilling to have warm and friendly relations with the Japanese.

The volume of Japanese imports from Australia is huge and growing— from $470 million in 1966 to $1.1 billion in 1970. What Japan sells to Australia—cars, cameras, electronic goods, textiles, and machinery—amounts to only half of this sum, but the imbalance does not bother Japan. It has a big (if growingly reluctant) customer, the U.S., which buys more from Japan than it sells to Japan. To complete this triangular relationship, U.S. sales to Australia are double U.S. purchases from Australia. In short, the Americans give dollars to Japan for Toyotas and Sonys; then the Japanese give the dollars to Australia for wool and iron ore; then the Australians give the dollars back to the Americans for airplanes and computers—it works something like that. Japan is also important to New Zealand, as its third-biggest customer.

Besides buying Australian raw materials, the Japanese are increasingly investing in Australian ranching, mining, and manufacturing—zippers, TV sets, microwave equipment, ball bearings. They plan to push their investment stake from $120 million to $420 million, much of it in fifty-fifty deals with Australian companies—an arrangement more amenable to Australians than the branch-office operations that the English and Americans prefer. Partnership might ultimately reach diplomatic levels. With complementary economies, Australia and Japan have common interests in keeping Japanese exports high, and preserving favorable markets—which may indicate a need for a formal alliance.

Australasia in the South Pacific

Some characteristics of the smaller islands of Melanesia and Polynesia are these: their economies are mostly feeble; they are a drain on the treasuries of their parent nations, chiefly Britain, the U.S., New Zealand, and France; and from these indifferent parents they can all get (or have already got) autonomy (except for France's jealously held possessions). A widely accepted hypothesis is that Australia and New Zealand should take responsibility for the social and economic advancement of all these islands, except France's. As a matter of fact, the two countries already exercise an important stewardship. The currency of Nauru, for example, is Australian, and Australia controls 70 percent of the foreign investment in Fiji. New Zealand takes charge of the defense and international relations of the Cook Islands, and New Zealanders still staff many government departments there. New Zealand sends Niueans and Tokelau Islanders to medical, dental, nursing, and agriculture schools of the University of the South Pacific in Fiji. Various free-trade pacts are in force. Both Australia and New Zealand resist the immigration of islanders, for essentially racist reasons, but this at least spares the islands from being stripped of their most talented and ambitious people. The machinery for organizing the islands is the South Pacific Commission, created in 1947 by Australia, New Zealand, Britain, France, and the United States.

The Trans-Tasman Relationship

For more than a century, Australia and New Zealand had an amazing diplomatic relationship with one another: they talked only through their representatives in London. Not until mid-war, in 1943, did they exchange high commissioners. Now, according to former Prime Minister Holyoake, they are probably more intimately related than any two other governments in the world.

The two nations cannot trade in great volume, because many of their products—wool, meat, butter—are the same. Still, Australia manages to sell New Zealand around $160 million worth of cars, steel, medicine, gasoline, and chemicals every year, and New Zealand sells Australia around $75 million worth of newsprint, lumber, fish, and peas—all under a free-trade agreement. New Zealand was originally administered from the colony of New South Wales, and was once considered to be a possible member of the Australian federation—it even took a tentative part in talks that led

to federation. As late as 1969 a cabinet minister in New Zealand had to rebut proponents of union with Australia. To people in other parts of the world (many of whom are so uninformed as to think that Australia and New Zealand *are* politically united), the advantages of union in facing the world might seem imperative. But most New Zealanders take tremendous pride in their country's independence, and simply won't hear of union. Besides, they think it would upset their delicately balanced system of subsidies and welfare. Australians don't seem to give the idea much thought at all. When Sir Keith Holyoake visited Canberra in 1970, he was rather conspicuously ignored by the Australian press and public.

Still, New Zealand often takes Australia's lead, and the two countries have been united since 1944 by the Anzac Agreement.* This rather vaguely pledges each to keep in close touch with the other, through a secretariat, on matters of regional interest.

Militarily Speaking

Anzac military policy is to assert the two countries' determination to defend themselves while realizing that the U.S. must provide the bulk of their protection. They know that they do not have the population to fend off a big attack, and hope that the U.S. will forever regard Australasia as a vital bastion of mutual defense.

The main strategic issue is "forward defense" versus "armed neutrality." "Forward" means Singapore and Malaysia. Anzac forces have been there ever since the British war against Communist guerrillas in Malaya in the 1950's, and for a long time they took the comfortable role of backing up Britain. In other words, forward defense *depended* on Britain—on British command. Then, in 1967 and 1968, Britain, overstrained, announced that it would pull out. Australia and New Zealand boldly declared that they would take over the military responsibility in that area, and they did. The boldness began to fade in 1969 when President Nixon proclaimed in his Guam Doctrine that the United States would try to coexist with Asian Communists after the Vietnam war ended. This seemed to leave the Anzacs alone in Singapore-Malaysia, not so much as a considered decision, but to plug the hole the British left. To put a good face on the matter, Australia, Britain, Malaysia, New Zealand, and Singapore invented yet another acronym, ANZUK, signifying a joint military force in Singapore-Malaysia that includes a submarine on rotation from Britain's Royal Navy.

* An extension of the earlier collaboration in the Australia–New Zealand Army Corps in World War I.

"Armed neutrality" amounts to withdrawal from forward bases in favor of a flexible, mobile force based, probably, in northern Australia. The Labor Party favors this approach. The essence of this strategy is that the home force could stop any small conventional attack; therefore an aggressor would have to attack in such force as to constitute a sizable war, and the U.S. would have no choice but to come to Australasia's aid. This is a posture of weakness, which probably cannot be helped given the two countries' small populations; it creates a sense of insecurity. Nowadays the alarming possibility is that in some major dispute Australia and New Zealand could become principals, instead of loyal little allies in Great Power conflicts. A variation of armed neutrality is the proposal for an Israeli-like semimilitary society, in which every citizen is a soldier; such a nation, though small in population, can be fearfully dangerous to attack. The difficulty with this approach is that it sacrifices the delights of the consumer society, and requires persuading people that they are living in a fool's paradise. Prime Minister Gorton toyed with this idea for a year, but dropped it in favor of "forward defense."

Toward these somewhat ambivalent defense goals, Australia, at least, has made a persistent effort in the 1960's to strengthen itself. The size of the services grew from 112,000 men to 133,000. The defense budget doubled in five years to $1.2 billion, 4.3 percent of the GNP. The first peacetime draft, for two-year terms of service, was begun in 1966. Russian interest in the Indian Ocean, evidenced by fleet maneuvers there, brought a strong Australian response. The navy spent $40 million for berths for submarines (it has three, with six on order) and escort vessels at Cockburn Sound, near the Northwest Cape. Here there is also an air force base, as well as the U.S. Navy communications station mentioned earlier.

The Australian Navy's largest vessel is the carrier *Melbourne*,* equipped with Skyhawks; there are also three guided-missile destroyers and a couple of dozen smaller ships. The Australian Air Force has Canberra bombers, 110 Mirage fighters, and numerous transports and trainers. In 1963 the air force ordered twenty-four American F-111s, but then backed away after eighteen such planes crashed in the U.S. Only late in 1971 did the government finally decide to accept the planes, and in the meantime the cost had risen from $229 million to $344 million. New Zealand's armed forces have four frigates, a squadron of Canberras, and a total of 13,000 men; $85

---

* This is the ship that sliced through an American destroyer, the *Frank Evans*, during night maneuvers in the South China Sea in June 1969, killing seventy-four American sailors. Two officers from the *Evans* were later court-martialed and convicted for negligence.

million a year, or 2 percent of the GNP (versus 2.3 in 1960), goes into defense. Men are subject to conscription, but draftees need serve only fourteen weeks.

Australia provides the real teeth for ANZUK: forty-two Mirages at Butterworth Air Force Base in Malaysia—nearly half of the troops in the command; New Zealand has some transport planes and a battalion there; and between them the two countries keep one naval ship in or around Singapore. When Australia committed troops to Vietnam, New Zealand felt obliged to commit a token force of 500, later raised to 750. Australia had three battalions of men in Vietnam. Both countries have now withdrawn all of their combat troops; Australian losses in the war were 487 dead and 3,000 wounded. By fighting in Vietnam, incidentally, both countries were involved for the first time in a war not entered on Britain's behalf.

*

To pull the threads together, Australia and New Zealand seem to be increasingly responsible in an area of decreasing danger. The struggle in South Vietnam taught the U.S. the perils of getting mired in distant guerrilla wars, and neither Australia nor New Zealand, as American allies, would repeat the experience. But the domino theory of Communist encroachment on Australasia has been somewhat discredited, and the danger from China, which worried Australasians for so long, now seems more clearly to be a factor of the Russia–China–United States equation. The United States pullback will leave Australasia more exposed but freer to aid and protect Southeast Asia. Yet the United States will keep its base in Guam and its attachments to Korea, Japan, the Pacific Islands Terrritory, and Australia and New Zealand themselves—in short, it will not disappear from Asia. Japan may develop a foreign policy aimed at Asian stability, and Australia and New Zealand will benefit both from the lead that Japan may give and from the possibility of allying themselves in some way to it. Two factors will give Australia and New Zealand new responsibilities and opportunities in the Pacific area: the end of their adolescent dependence on Britain and the United States, and the inevitable termination of colonial relationships throughout the Pacific islands. The two nations will remain relatively helpless, simply from lack of population, but they should be able to work more effectively in their Asian environment, and worry less about the possible perils it contains.

CHAPTER 25

Some Thoughts in Conclusion

If you are looking for the Utopia of the 20th Century, it is there Down Under.

—ASIA MAGAZINE

AUSTRALIANS are knockers and scoffers, and when the poet Robert Heanley speaks of them as a people who will prove to be "a living bridge between east and west," by and large they find such talk a pretty piece of nonsensical idealism. Yet—to repeat and emphasize a point touched upon often in this book—the major fact of present-day Australasian life is that two historically British nations have been forced by geography to become political and sociological parts of the vast and varied Asian and Pacific region. This represents not only a belated abandonment of pro-British snobbery, and an equally belated recognition of the possibilities that the neighborhood affords Australia and New Zealand; it is also an acknowledgment of the brute fact that Europe, specifically the European Economic Community, is casting Australasia out of its trade arrangements. Australasia is being both pushed and pulled into Asia. The consequences will transform both countries.

1. Economically, the link to Japan can only grow closer. What is not clearly realized in much of the world is that Australian prosperity is now a component part of Japanese prosperity; the two nations have common interests—such as opposing U.S. barriers to the importation of Japanese goods—that could lead to a formal alliance. As a consequence of this tight link Japan would react to Australia in three ways: by selling Australia more goods, such as the heavy machinery that now comes from England, by investing more in Australia, and by sharing its great technological resources with Australia. This sharing would mean more intensive processing of Aus-

tralia's raw materials before they are shipped to Japan. If the European Economic Community takes away from Australasia its English market for butter and meat, the certain result will be termination of the trade preference that both nations have always extended to Great Britain.

2. Politically, closer ties to Asian neighbors should have the effect of making Australasia less defensive—literally. The emphasis in foreign policy will shift from the military to trade and aid. If China becomes more of a market, it should be less of a threat. Perhaps in collaboration with Japan, Australia and New Zealand will feel obliged to reduce the stunning contrast between their wealth and the poverty of Indonesia and some of the Pacific islands. The three nations have the means and technology to help in development and education on a scale much bigger than at present. Indonesia and Australia, in particular, have an existing friendship not necessarily to be expected from nations of their sharply contrasting cultures.

3. Sociologically, the White Australia policy must eventually go by the boards. Beyond a doubt the Japanese will politely but insistently press for treatment equal to that accorded to Englishmen. Japan's pollution alone may drive the Japanese to emigrate, and it is also quite possible that Australia may find itself pleading for skilled Japanese immigrants. Australia, if not New Zealand, may also become desirous of its share of the stream of skilled and educated Filipinos who presently emigrate in droves to the United States and Canada. What should become clear to Australians is that the greatly increased immigration of Japanese, Hong Kong Chinese, Filipinos, Indians, Pakistanis, American Negroes, and so on will not trouble, but rather enrich, a nation where currently only one person in one hundred is not white. Closer relations to the various kinds of Asians will also require changes in Australasian education, specifically much more teaching of Japanese and other languages.

*

Are these prospective developments a little too visionary? Is Australia (plus, in a lesser role, New Zealand) ready to be a catalyst as well as a junior partner to Japan and a senior partner to her poverty-ridden neighbors?

To judge from their politicians, the answer is "not yet." Australia's cautious Prime Minister McMahon seems to want to steer clear of possibly dangerous adventures. The opposition Labor Party, partly inspired by the Ping-Pong initiative of the Chinese in April 1971, but also reflecting a traditional Labor position, proposed that Australia recognize Peking and help it get a seat in the United Nations. Mr. McMahon replied that Australia's

"major objective" was to keep Taiwan in the United Nations, and that recognition of China was not a paramount concern. McMahon thus shared with Nixon the rebuff delivered by most of the world's nations when they cast Taiwan out of the United Nations as part of the process of admitting the People's Republic. He was still holding basically to something resembling the old domino theory, contending that China's ideology is the promotion of people's revolutions, and that if relations with China became relaxed or friendly, Australia might ultimately be left surrounded by Communist nations.

The bulk of Australians would seem to be with the Prime Minister, fearful of China and Communism generally. Here, however, we get back to the Australian character, and particularly to a strange aspect of it. One civil servant who has profoundly studied his fellow citizens feels that Australians have a deep-seated fear that the European occupation of Australia will be only momentary, that tomorrow they could be pushed into the sea, that their domain cannot be expected to last, that they are essentially alien to this continent—it was never in the cards for them to stay. This philosophy, if indeed widespread, might make Australia an easy conquest. On the other hand, instead of reinforcing the White Australia policy, it might encourage Australians to collaborate with their neighbors. They would be good at it if they really tried because white Australians relate easily, even enthusiastically, to dark-skinned immigrants on a one-to-one basis. The trouble has not been racial, but the fear of being engulfed.

Joining Asia will have to involve, most importantly, recognition of the People's Republic of China. The grounds for fearing China are certainly not unreasonable. But the reality of the Chinese threat has to be weighed off against the benefits of getting along better. It is also to be argued that even if Communist China is the most apparent threat, there is little point in relatively weak nations like Australia and New Zealand making a show of belligerence at a moment when Britain has withdrawn protection and the United States is lowering its level of commitment in Asia. Over and over in Australia one hears the questions, "How far would the U.S. go to protect us? Would it get into a major war for us?" Many Australians think not.

Moreover, neutralist and nonaligned nations in Southeast Asia might concede Australia and New Zealand a little more scope for leadership if the two took a less suspicious stance. Australia has been criticized by a newspaper in Thailand for creating "the impression that the basis of Australia's relations with the Asian region is apprehension and unease"; critics of present policy argue that this should not be the case. Some Australasians find it difficult to square the tough anti-Communist pacts, ANZUS and SEATO, with the Australian–New Zealand need to be a closer neighbor to Asia.

In New Zealand, criticism of the big-power alliance policy takes another twist. This most remote nation, it is argued, should assume that it is of very little interest to possible aggressors, and adopt a Swiss-style policy of nonalignment. This would mean weaker ties to Australia and recognition of China, and might clear the way to sales of meat, butter, and wool to the Communist nations of Eastern Europe, and to Russia, when England stops buying these commodities.

Despite political caution and the ordinary man's reluctance, the overriding fact is that the forces of economics in Japan and Europe have pushed Australia and New Zealand into an unpredictable multiple marriage with the countries on and near the Asian continent. They are, by virtue of possessing economies several times stronger than those of their near neighbors put together, the natural fount of leadership in this sea of brown and yellow faces.

*

Australians and New Zealanders, we are repeatedly told, do not hold their own cultures in high regard; things are done better "overseas," they say, and what is well done at home was really not worth doing in the first place. Many Australians believe, with the English writer James Morris, that their country's origins are unsavory, its history disagreeable, its suburbs hideous, its politics crooked, its buildings plain, its people's voices rasping, and its art movement spurious. The world originates, and Australia and New Zealand follow. As a general observation, they do lag behind the United States in attitudes and concerns by somewhere between two and twelve years; an American senses something of the Eisenhower era in present-day Australia. This lag implies a nation not in a rage, not frustrated, not angry (at least yet), a nation that still sets a slower pace and avoids widespread dissension, a nation that can still afford complacency and reserves ample time for play— to this extent the lag is a blessing; may the future never arrive!

But in point of fact, Australians and New Zealanders should be happy for reasons better than merely not being in the plight of the U.S. The more positive reasons for national satisfaction are the genuine accomplishments of both countries. What an achievement, for example, the full-employment economy is. This is a system that values what human beings do, and assures them of a chance to do it. So much needs to be done that labor must be imported, and the newly arrived labor creates needs for still more to be done. Obviously, these economies, particularly New Zealand's, may stumble, but they will always be better for being guided by the national determination that a man or a woman can train for a job and then get a job. Moreover,

the availability of work is probably related to the low rate of crime. Fewer homicides are committed in all of Australia in a typical year (the rate runs around two hundred or less) than in Atlanta alone, or Baltimore, or Dallas— and much fewer than New York City, which has nearly two thousand.

Another achievement (and how the social critics of Australia and New Zealand will dispute this one!) is suburbia. Granted that the suburbs of Melbourne and Sydney are too extensive and monotonous, granted that new cities should be created before allowing these monsters to get bigger, granted that the "mod-cons" (modern-construction houses) are banal, granted that suburban life may be bound by television, gossip, and high-interest mortgages—it is nevertheless a respectable accomplishment to have two nations housed this well. Not much can be said against a garden unless it becomes a total obsession. A house is, by and large, likely to be a better place to live in than an apartment. And unlike the United States, Australia and New Zealand do not have painful contrasts between suburbs for the rich and center-city slums for the poor. The two countries have few rich, few poor, and few slums. In the 1960's, when Australia attempted to imitate the United States in a war on poverty, only 8 percent of the population met standards of poorness that applied to 20 percent of the people of the United States.

More broadly, the world must respect Australia, and perhaps New Zealand even more so, as places where one of humanity's most ancient ideals, democracy, manages to function better than in most of the rest of the globe. The leaders do not get assassinated; the streets are not full of tumult. Small population, no doubt, has a lot to do with this; a man can reach the politician who represents him. The average Australian or New Zealander does not delude himself that he can affect the course of the world, and the world may eventually snuff out his dreams. But in the operation of his own country, he still feels that he has at least a slender grip on the handle of his own fate.

Acknowledgments and Bibliography

If John Gunther had settled down to write these acknowledgments, he would, I do not doubt, have thumbed through the record he kept of persons interviewed, and after that he would have run an eye along his shelves of books about Australia and New Zealand. Then he would have been forced to conclude that this book is in debt to more people than it is possible to list. For example, the number of persons he talked with, each of whom contributed a greater or lesser part of the book, is five hundred and five. This does not include hundreds of other informed people whom he met briefly on social occasions or in the course of travel. The books and booklets that Gunther and I collected add up to almost two hundred. So I will start this, as Gunther might have, with gratitude to the contributors of the facts and insights of which the book is made—and despair over how to do justice to them all.

Sir Patrick Shaw, who headed the Australian delegation at the United Nations when the Gunthers started their trip, and John V. Scott, the New Zealand Ambassador to the United Nations, sent them off with full and lucid briefings, introductions to figures high in government, and excellent advice about other people whom they might consult. Thanks must be extended to the members of the government in both Australia and New Zealand who gave interviews to the Gunthers, the Governors of the states of Australia, the United States Ambassadors to both countries, and the various United States Consuls who were hospitable and informative. Without the invaluable help of all these people this book would never have been written.

To greet the Gunthers on arrival was Anthony Paul, then editor-in-chief in Australia for the *Reader's Digest*, a magazine with which Gunther had enjoyed a close association for many years and which published an article by him about Australia. Mr. Paul put the resources of the *Digest* in Sydney at their disposal, introduced them to his colleagues all over Australia, and gave them welcome hospitality. Mrs. Bryce Fraser, a member of Mr. Paul's staff in Sydney, was exceedingly helpful, and Alex Freeleagus in Brisbane, John Bulbeck in Western Australia, John Bennetts in Canberra, Lindsay Ellis in Darwin and Alice Springs, and Peter Cole-Adams in Melbourne were of invaluable assistance.

Sir Colin Syme, chairman of the board of Broken Hill Proprietary, opened doors for the Gunthers in Melbourne and in the whole mining world. John Dorrance, then second secretary of the American Embassy in Canberra, shared the insights he had.

352 ACKNOWLEDGMENTS AND BIBLIOGRAPHY

gained from his experience and study of New Guinea. In New Zealand, Paul Tipping of the Department of External Affairs accompanied the Gunthers on all their travels, smoothing the way and providing information in the process. In New Guinea, the man who helped Gunther most was his namesake (but no relation), John Gunther, Vice Chancellor of the University of Papua and New Guinea (a "most glorious human being," in Jane Gunther's words), who briefed them, arranged interviews, laid out an itinerary, and provided many contacts there and in Australia. A chance meeting at an airport put the Gunthers in touch with Captain Percival Trezise, a pilot and, more relevantly to the author's purpose, a man interested in and knowledgeable about the Aborigines.

A few others must be singled out: some because their discourse with Gunther (or with me) ranged helpfully beyond the set scope of an interview; others because they provided the kind of hospitality that allows the outsider to see a little distance inside the society he is studying; still others because they supplied information for substantial parts of this book. These benefactors include Donald Horne, John Pringle, Geoffrey Dutton, Lady Lloyd-Jones, Mrs. Edna Ayres, Miss Sheila Scotter, Mrs. Ronald Stewart, Sir Keith Park, Sir James Plimpsoll, W. C. Wentworth, Thomas Larkin, George T. Walker, Luke Hazlett, Sir Frank and Lady Packer, and Clem Jones.

Quite a few of those mentioned in the preceding paragraphs, having been interviewed by Gunther, were kind enough to see me too when, later, my wife and I toured Australia, and I talked to at least another fifty people. I want to express a special indebtedness to my wife Deborah for sensitive reporting and observation while we were in Australia and New Zealand, for criticism of the manuscript that forfended some egregious errors, and for creating an atmosphere propitious for writing. Another Forbis, a distant cousin unknown to me until I met him in Auckland, also helped. He turned out to be a professional journalist, and much of the material in the chapters on New Zealand is the result of a happy collaboration with John Forbis, who supplied many of the facts and verified others. In Australia, assistance of this kind was ably given by Geraldine Kirshenbaum, a skilled researcher formerly with *Time* magazine. Margaret Butterfield, a Harper editor, helped tighten and shape the book, and the meticulous eye of Richard Passmore, another Harper editor, caught many redundancies, contradictions, solecisms, and other mistakes.

For certain verses used in this book, printed with permission of the copyright owners, I would like to thank: (1) Angus and Robertson Pty. Ltd. and author A. B. Paterson for lines from "Waltzing Matilda" (on page 6), taken from *The Collected Verse of A. B. Paterson*. (2) Ure Smith Pty. Ltd. and Ronald M. Berndt, Professor of Anthropology at the University of Western Australia, for his translation of the verse beginning "Up and up soars the Evening Star" (on page 77), taken from *The World of the First Australians*, by Professor Berndt and Catherine H. Berndt. (3) Hill of Content Publishing Co. for a couple of lines from "Jim Jones" (on page 18) and a verse of "Herbert Hoover's Love Song" (on page 204), taken from *Great Australian Folk Songs*, collected by John Lahey. (4) The Viking Press for two lines from "The Arctic Ox (or Goat)" (on page 119), from *O to Be a Dragon*, by Marianne Moore. Copyright © 1958 by Marianne Moore. First appeared in *The New Yorker*, 1958. (5) The Department of External Territories for the lines beginning "Hark and behold" by A. P. Allan Natachee (on page 264). (6) The Estate of A. R. D. Fairburn for a quotation from "Conversation in the Bush" (on page 290), published by Penguin Books.

By way of bibliography, a number of books were significantly helpful in providing background knowledge about physical characteristics, history, politics, sociology, and culture. The principal over-all reference works used were the *Australian Encyclopaedia* (ten volumes, Angus and Robertson, Sydney, and Michigan State University Press, Lansing, 1958), Osmar White's *Guide to Australia* (Heinemann, Melbourne, 1968), *Yearbook Australia 1969* (Commonwealth Bureau of Census and Statistics, Canberra), *New Zealand Official Yearbook 1970* (Department of Statistics, Wellington), *Australian Handbook 1971* and *Australian Facts and Figures* (Australian News and Information Bureau, Canberra), the *Encyclopaedia Britannica*, and the *World Almanac 1971* (Newspaper Enterprise Association, New York).

A number of periodicals and newspapers were essential sources of information. Among them the *Bulletin* and the *Sunday Australian*, as well as Robert Trumbull's dispatches in the *New York Times*, provided up-to-date news, as did the *Australian Weekly News*, published in San Francisco. Equally indispensable were the many articles and news stories which had appeared in recent years in literally dozens of newspapers and periodicals, both English and American. Much useful information was also obtained from pamphlets, guidebooks, speeches, and brochures.

Listed below are most of the books which John Gunther and I enjoyed and found useful:

AUSTRALIA

Abbie, A. A. *The Original Australians.* Reed, Sydney, 1970.

Australian Economy, The. The Treasury, Canberra, 1970.

Baker, Sidney J. *The Australian Language.* Sun Books, Melbourne, 1970.

Bergamini, David. *The Land and Wildlife of Australia.* Life Nature Library, New York, 1964.

Berndt, Ronald M. and Catherine H. *The World of the First Australians.* Ure Smith, Sydney, 1968.

Blainey, Geoffrey. *The Rise of Broken Hill.* Macmillan of Australia, South Melbourne, 1968.

———. *The Tyranny of Distance.* Sun Books, Melbourne, 1966.

Boyd, Robin. *The New Architecture.* Longmans, Croyden, Victoria, 1963.

Brash, Donald T. *American Investment in Australian Industry.* Australian National University Press, Canberra.

Clemens, Samuel Langhorne. *Following the Equator.* Harper & Bros., 1899.

Clifford, Beverley; and Barbara Richards. *More Than a Harbor.* Murray, Sydney and Melbourne, 1969.

Cobley, John. *The Crimes of the First Fleet Convicts.* Australian National University Press, Canberra, 1970.

Commonwealth Electoral Law. Chief Electoral Office, Canberra, 1966.

Commonwealth Scientific and Industrial Research Organization, *Annual Report,* 1968–69.

Coulls, A. "Charles Rasp, Founder of Broken Hill." Unpublished paper, 1952.

Covell, Roger. *Australia's Music.* Sun Books, Melbourne, 1967.

Crawford, R. M. *Australia.* Hutchinson University Library, London, 1952.

Darwin, Charles. *The Voyage of the Beagle.* P. F. Collier & Son, New York, 1909.

Davies, A. F., and S. Encel (eds.). *Australian Society: A Sociological Introduction.* Atherton Press, New York, 1964.

Durrell, Gerald. *Two in the Bush.* Collins, London, 1966.

Elkin, A. P. *The Australian Aborigines.* Anchor Books, Garden City, New York, 1964.

Establishing a Business in Australia. Bank of New South Wales, Sydney, 1968.

Farwell, George. *Around Australia on Highway One.* Nelson, Melbourne and Sydney, 1966.

Finemore, Brian. *Painting.* Longmans, Croyden, Victoria, 1961.

Gordon, Richard (ed.). *The Australian New Left.* Heinemann, Melbourne, 1970.

Gritscher, Helmut, and Craig McGregor. *To Sydney with Love.* Nelson, Melbourne, 1968.

Hallows, John. *The Dreamtime Society.* Collins, Sydney, 1970.

Hancock, Sir W. Keith. *Australia.* Tri-Ocean, 2nd ed., 1930.

Harvey, Frank R. *Theatre.* Longmans, Croyden, Victoria, 1965.

Herman, Morton. *Early Colonial Architecture.* Longmans, Croyden, Victoria, 1963.

Hood, Kenneth. *Pottery.* Longmans, Croyden, Victoria, 1961.

Hope, A. D. *Selected Poems.* Angus and Robertson, Sydney, 1963.

Horne, Donald. *The Lucky Country.* Penguin, 2nd rev. ed., Ringwood, Victoria, 1968.

———. *The Next Australia.* Angus and Robertson, Sydney, 1970.

Howitt, Alfred William. *The Native Tribes of Southeast Australia.* Macmillan, London, 1904.

Hughes, Robert. *The Art of Australia.* Penguin, Harmondsworth, England, 1970.

Huxley, Elspeth. *Their Shining Eldorado.* William Morrow, New York, 1967.

Idriess, Ion L. *The Cattle King.* Angus and Robertson, Pacific Books, Sydney, reprinted 1967.

Kaberry, Phyllis. *Aboriginal Women Sacred and Profane.* Blakiston, Philadelphia, 1939.

Keesing, Nancy (ed.). *Transition.* Angus and Robertson, Sydney, 1970.

King, Philip Parker. *Survey of Australia 1818–1822.* Reproduced by the Libraries Board of South Australia, Murray, London, 1969.

Knorr, Hilde and Hans. *Religious Art in Australia.* Longmans, Croyden, Victoria, 1967.

Lauder, Afferbeck. *Let Stalk Strine.* Ure Smith, Sydney.

———. *Nose Tone Unturned.* Ure Smith, Sydney, 1966.

Lawrence, D. H. *Kangaroo.* Penguin ed., Harmondsworth, England, 1968.

William Light. Libraries Board of South Australia, 1966.

Luck, R. K. *Modern Australian Painting.* Sun Books, Melbourne, 1969.

McGregor, Craig. *Profile of Australia.* Henry Regnery, Chicago, 1967.

Melville, Robert. *The Legend of Ned Kelly,* with paintings by Sidney Nolan. Viking Press, New York, 1964.

Miller, J. D. B. *Australian Government and Politics.* Duckworth, London, 1959.

Moorehead, Alan. *Cooper's Creek.* Harper & Row, New York, 1963.

———. *The Fatal Impact.* Harper & Row, New York, 1966.

Mountford, Charles P. *Aboriginal Art.* Longmans, Croyden, Victoria, 1961.

Mulvaney, Derek John. *The Prehistory of Australia.* Praeger, New York, 1969.

Murphy, D. F. *The Australian Wine Guide.* Sun Books, Melbourne, 1966.

O'Grady, John. *Aussie English.* Ure Smith, Sydney, 1969.

Parr, Lenton. *Sculpture.* Longmans, Croyden, Victoria, 1961.

Pearl, Cyril. *Beer, Glorious Beer!* Nelson, Melbourne and Sydney, 1969.
———. *So, You Want to Be an Australian.* Ure Smith, London, 1968.
Perkins, Kevin. *Menzies, Last of the Queen's Men.* Rigby Limited, Adelaide, 1968.
Pike, Douglas, *Australia: The Quiet Continent.* Cambridge University Press, Cambridge, 1970.
Preston, Richard (ed.). *Contemporary Australia: Studies in History, Politics and Economics.* Duke University Press, Durham, 1969.
Pringle, John Douglas. *Australian Accent.* Chatto & Windus, London, 1965.
Reed, John. *Australian Landscape Painting.* Longmans, Croyden, Victoria, 1965.
Rivett, Rohan. *Australia.* Oxford University Press, Melbourne, 1968.
Scott, Ernest. *Australian Discovery by Sea.* Vol. I, J. M. Dent & Sons, London, 1929.
———. *Australian Discovery by Land.* Vol. II, J. M. Dent & Sons, London, 1929.
Seidel, Brian. *Printmaking.* Longmans, Croyden, Victoria, 1965.
Shaw, A. G. L. *The Story of Australia.* Faber & Faber, London, 1966.
Snowy Mountains of Australia, The. Nucolorvue Productions, Mentone, Victoria, nd.
Splatt, W. J. *Architecture.* Longmans, Croyden, Victoria, 1962.
Trollope, Anthony. *Australia.* Republished by the University of Queensland Press, Brisbane, 1965.
Turner, Ian. *The Australian Dream.* Sun Books, Melbourne, 1968.
Van Sommers, Tess, and Unk White. *Sydney Sketchbook.* Rigby Limited, Adelaide, 1965.
Ward, Russell. *The Australian Legend.* Oxford University Press, Melbourne, 1966.
———. *Australia.* Walkabout Pocketbooks, London, 1969.
White, Patrick. *The Vivisector.* Jonathan Cape, London, 1970.
———. *Voss.* Longmans, London, 1965.
Whittington, Don, and Bob Chalmer. *Inside Canberra.* Rigby Limited, Kent Town, S.A., 1971.
Wright, Judith. *Selected Poems.* Angus and Robertson, Sydney, 1963.
Younger, R. M. *Australia and the Australians.* Rigby Limited, Adelaide, 1970.

NEW GUINEA

Bjerre, Jens. *The Last Cannibals.* Michael Joseph, London, 1956.
Blackwood, Beatrice. *The Technology of Modern Stone Age People in New Guinea.* Oxford University Press, Oxford, 1950.
Champion, Ivan F. *Across New Guinea from the Fly to the Sepik.* Landsoune, Melbourne, revised 1966.
Clune, Frank. *Prowling Through Papua.* Angus and Robertson, Sydney, 1948.
Davies, Donald M. *Journey into the Stone Age.* Robert Hale, London, 1969.
Durack, Elizabeth. *Face Value.* Ure Smith, Sydney, 1970.
Essai, Brian. *Papua and New Guinea: A Contemporary Survey.* Oxford University Press, Melbourne, 1961.
Fiske, E. K. (ed.). *New Guinea on the Threshold.* Australian National University Press, Canberra, 1966.
Handbook of Papua and New Guinea. Jacaranda Press, Brisbane, 1969.
Hastings, Peter. *New Guinea, Problems and Prospects.* Cheshire, Melbourne, 1969.
Lea, D. A. M., and P. G. Irwin. *New Guinea: The Territory and Its People,* Oxford University Press, Melbourne, 1967.

Mallord, Ann. *A Traveller's Guide to Papua New Guinea*. Jacaranda Press, Milton, Queensland, 1969.

Murray, Sir Hubert. *Papua of Today*. P. S. King & Son, Westminster, 1925.

Price, Sir A. Grenfell. *The Challenge of New Guinea*. Angus and Robertson, London, 1965.

Ruhen, Olaf. *Mountains in the Clouds*. Rigby Limited, Adelaide, 1963.

Ryan, John. *The Hot Land*. Macmillan of Australia, South Melbourne, 1969.

Salisbury, R. F. *From Stone to Steel*. Cambridge University Press, Cambridge, 1962.

Simpson, Colin. *Adam with Arrows*. Praeger, New York, 1968.

Souter, Gavin. *New Guinea, the Last Unknown*. Angus and Robertson, London, 1964.

White, Osmar. *Parliament of a Thousand Tribes*. Heinemann, London, 1965.

Williams, Maslyn. *In One Lifetime*. Cheshire, Melbourne, 1970.

NEW ZEALAND

Armstrong, Ala. *Say It in Maori*. Seven Seas, Wellington, 1968.

Ashton-Warner, Sylvia. *Bell Call*. Simon and Schuster, New York, 1964.

――――. *Spinster*. Simon and Schuster, New York, 1959.

Bigwood, Kenneth and Jean. *New Zealand in Color*, Vols. I and II. A. H. and A. W. Reed, Wellington, 1961 and 1962.

Brewer, Ian H. *Te Rauparaha*. R. E. Owen, Government Printer, Wellington, 1966.

Brown, Gordon, and Hamish Keith. *New Zealand Painting: An Introduction*. Collins, London and Auckland, 1969.

Buck, Sir Peter Henry. *Vikings of the Pacific*. University of Chicago Press, Chicago, 1959.

Cape, Peter. *Artists and Craftsmen in New Zealand*. Collins, Auckland and London, 1969.

Condliffe, J. B., and W. T. G. Airey. *A Short History of New Zealand*. Whitcomb and Tombs Limited, Christchurch, reprinted 1968.

Duff, Roger (ed.). *No Sort of Iron*. Art Galleries and Museums Association of New Zealand, Christchurch, 1968.

Grey, George. *Ko nga Whakapepeha me nga Whakaahuareka a nga Tupuna*.

Growth and Development of Social Security in New Zealand, The. Social Security Department, Wellington, 1950.

Jackson, Keith, and John Harré. *New Zealand*. Thames and Hudson, London, 1969.

Keam, R. F. *Volcanic Wonderland*. A. H. and A. W. Reed, Wellington, 1965.

Maori Place Names. A. H. and A. W. Reed, Wellington, 1962.

Maori Proverbs. A. H. and A. W. Reed, Wellington, 1963.

Mason, Helen. *10 Years of Pottery in New Zealand*. The New Zealand Potter, Wellington, 1968.

National Parks of New Zealand. R. E. Owen, Government Printer, Wellington, 1965.

Oliver, W. H. *The Story of New Zealand*. Faber & Faber, London, 1968.

Poole, A. L. *Forestry in New Zealand*. Hodder and Stoughton, Auckland, 1969.

――――. *New Zealand Forestry*. R. E. Owen, Government Printer, Wellington, 1964.

Powles, Sir Guy. "The W. Clifford Clark Memorial Lectures." Reprinted in *Canadian Public Administration*, June, 1966.

――――. *Report of the Ombudsman*. Government Printer, Wellington, 1969.

Royal Commission of Inquiry. *Background Paper*. Social Security Department, Wellington, 1969.

Shadbolt, Maurice. *The Shell Guide to New Zealand*. Whitcomb and Tombs Limited, Christchurch, reprinted 1969.

Sinclair, Keith. *A History of New Zealand*. Penguin, Harmondsworth, England, rev. ed., 1969.

Index